Conceptual Foundations of Genetics

Selected Readings

Conceptual Foundations of Genetics

Selected Readings

Harry O. Corwin

University of Pittsburgh

John B. Jenkins

Swarthmore College

Houghton Mifflin Company · Boston

Atlanta   Dallas   Geneva, Illinois   Hopewell, New Jersey   Palo Alto   London

This book is dedicated to Elof Axel Carlson whose teaching and friendship have been an inspiration.

COVER PICTURE CREDITS

Top, Courtesy of E.B. Lewis, California Institute of Technology
Right, Carolina Biological Supply Company
Bottom, Courtesy of A.M. Lawn, Lister Institute
Left, Stock/Boston

Printed in the U.S.A.
Library of Congress Catalog Card Number: 75-26092
ISBN: 0-395-24064-6

Contents

Part 4

Function of the genetic material 303

Preface

This book is the product of two geneticists who have been extensively involved in the presentation of genetics to college students. Although the specifics of our course content and our manners of presentation differ, we agree that the student of modern genetics should have access to a reference book that provides: a collection of classical experiments to illustrate the historical evolution of major concepts; a sufficient variety of contemporary papers to inform the student of the field's current level of sophistication; and, where possible, a selection of papers presenting alternative viewpoints to illustrate the gradual emergence of concepts from tentative hypotheses by means of rigorous experimentation.

In the field of genetics there are two main reasons why a collection of this sort is possible. First, each step toward our current understanding in genetics has been challenged experimentally; each confrontation has been well documented; and, each resolution is still open for further experimentation. Second, most of the major conceptual developments have taken place since 1900 (excluding, for example, Gregor Mendel's unique insight in 1865). The pioneers in the field of genetics are still fresh in the minds of today's investigators—their works often forming the basis of our current research.

Rarely has there been a field in which it is possible to so easily document the historical background of its emerging conceptual framework. There are remarkably clear examples which illustrate the evolution of hypotheses leading to today's experimentally sound concepts, i.e., the path leading to the discovery of the chemical basis of heredity. In addition, these examples emphasize the tenuous nature of hypotheses. They show how an apparently sound concept can be modified or left by the wayside when it is subjected to more rigorous levels of experimentation. Certainly the continual changes in the 1940 "one gene-one enzyme" hypothesis show the vulnerability of concepts to experimentally based modification. Finally, this collection of readings shows how, in a rapidly emerging field of study, there must be people able to step back from the details; to carefully assimilate and pull together the wide spectrum of observations available; to formulate clear statements about the status of the field; and, to unify its concepts. In so doing, these people provide fresh guidelines for new experimentation. Certainly the Sutton paper of 1903, the Muller paper of 1922, and the Watson and Crick paper of 1953 are documentation of this phenomenon.

Obviously, it would not be possible to demonstrate adequately the features that led to the current understanding of heredity if one presented only a classical or contemporary collection of papers. The authors feel that the student of genetics must have available both the current resolutions in the field and their historical bases.

This book is directed toward a one-term course in genetics, which necessitates an abbreviated examination of the field. Only a sampling of the different areas will be considered. Omissions do not reflect the authors' attempt to estimate the importance of different areas in the study of inheritance. Instead, the topics we have included reflect a sequence we have used successfully to illustrate the general features of inheritance. Although the sequence of parts in the book may not coincide with the approach of all genetics courses, we fell that it does provide one logical approach to the presentation of the subject material. The organization within the parts is as follows.

Part 1 focuses first on a selection of experiments which were instrumental in defining the chemical and structural nature of the genetic information. A central issue raised in these early papers concerns the manner which the genetic material is replicated. A hypothetical model of replication emerged—in conjunction with the proposed structure of the genetic material—but the details of the replication mechanism were obscure. Hence a sequence of papers is included that documents the gradual emergence of the molecular events involved in DNA replication. In addition to its ability to replicate, the genetic information had to possess the capacity to change (*mutate*). The concluding series of papers examines the phenomenon of mutation and reveals how the studies of mutation helped to expose the structural organization of the gene.

Part 2 includes papers that define the manner in which the genetic material is transmitted. Vital to that definition was an experimental approach that could analyze the mechanisms of transmission. Mendel's work was fundamental to this area; and his contributions are presented here through the work of Hugo de Vries. Mendel's work not only provided a technique to analyze the mechanics of transmission, but also revealed pertinent information concerning the key mechanical attributes of genetic transmission in eucaryote organisms. Papers that questioned the validity and universality of the Mendelian theories are also presented. The answers to two of these questions resulted in the development of two major fields of genetic investigation: population and quantitative genetics. A third problem—the universality of Mendel's "law" of independent assortment—had to be qualified to account for conflicting observations. This qualification was ultimately explained by the chromosome theory of inheritance (genes are associated with chromosomes). The last papers in this part document the confirmation of the chromosome theory. With this experimental confirmation, the distribution and transmission of the genetic material can now be explained in terms of the orderly segregation, assortment, and distribution of chromosomes during meiosis and mitosis.

Part 3 examines the early experiments fundamental to understanding the mechanics of transmission of hereditary material in bacteria and viruses. An analysis of these papers shows that prokaryotic genetic systems have evolved distinct and unique mechanisms for generating recombinants. Recombination in prokaryotes is then contrasted with recombination in eukaryotes. And finally, a comparison of Parts 2 and 3 shows that although the actual mechanisms for generating genetic recombinants may vary, the end results are fundamentally the same.

Part 4 explores the connection between all the hereditary units in an organism (*genome*) and the final expression of the genome's activity—the organism's phenotype. Papers are presented that investigate gene function and the regulation of that function. The first papers depict the historical emergence of knowledge, establishing the manner in which the genetic information controls the structure and function of the organism. The next series of papers examines one specific aspect of this relationship: the form and content of the information used to control the heritable characteristics of the organism—the genetic code. The phenotypic expression, although controlled by a specific genotype, is not based solely on gene function but also on the specific timing for the activity of particular genes throughout the genome. The concluding papers in this part discuss evidence that supports the existence of differential gene activity and particular molecular models for gene regulation.

The authors would like to acknowledge those persons who have been helpful in the preparation of this book. First, an acknowledgement is due the students of genetics who have passed through our courses. They have greatly influenced the organization of the book. Their comments and suggestions as to the instructive value of many papers used in this reader were a major consideration in the establishment of the final sequence of papers. Second, we wish to acknowledge the valuable assistance of our colleagues who reviewed the manuscript and to the editorial staff of Houghton Mifflin who provided invaluable assistance during the preparation of this book. Third, we wish to express our deep appreciation to the authors and publishers who so graciously gave permission to have their articles reprinted in this book. And finally, to our wives, Judy and Fereshteh, and to our children, Laurie, Jennifer, John, and Soraya, we wish to express our love and thanks for their encouragement and understanding during the many hours that this book was taking shape.

On the other hand, if these d'Herelle bodies were really genes, fundamentally like our chromosome genes, they would give us an utterly new angle from which to attack the gene problem. They are filterable, to some extent insoluble, can be handled in test tubes, and their properties, as shown by their effects on the bacteria, can then be studied after treatment. It would be very rash to call these bodies genes, and yet at present we must confess that there is no distinction known between the genes and them. Hence we cannot categorically deny that perhaps we may be able to grind genes in a mortar and cook them in a beaker after all. Must we geneticists become bacteriologists, physiological chemists and physicists, simultaneously with being zoologists and botanists? Let us hope so.

H. J. Muller 1921

Part 1. The identification of the genetic material and its implications

In 1921, H. J. Muller addressed the thirty-ninth annual meeting of the American Society of Naturalists on the topic of organismic variation due to changes in individual genes.[21] Muller stressed that the genetic material had to possess two unique properties. First, the gene had to be able to direct the formation of exact copies of itself. Second, the gene had to undergo infrequent structural changes, and the changes could not interfere with the capacity of the gene to make copies of the new structural variant. Part 1 begins with an examination of some of the key steps taken toward elucidating the chemical basis of inheritance. This is followed by the presentation of papers devoted to work instrumental in revealing how the genetic information could direct the formation of copies or structural variants of the original gene.

The research that led to the identification of DNA as the genetic material began in 1869 when Miescher described the chemical properties of a heretofore unknown biological macromolecule.[18] This material, called *nuclein* by Miescher, is now called *deoxyribonucleic acid,* or *DNA* for short. The connection between this nuclein and inheritance was not made until well after the discovery of Mendel's paper of 1865. However, there was speculation on the biological significance of nuclein during the late nineteenth century. For example, during the early 1880s, there were two separate references to the possible role of nuclein in heredity and fertilization. In both instances, the two investigators, E. Zacharias[32] and O. Hertwig,[14] failed to follow through with their inferences; and their inferences were not picked up by others in the field.

It is especially interesting to note that while Miescher was the first to actually isolate DNA, the course of research that he spawned was not, in fact, the definitive research that showed DNA to be the genetic material. The work that unequivocally demonstrated DNA to be the chemical basis of inheritance was born out of mutation studies with bacteria during the 1920s. And, in all likelihood, these studies operated without the knowledge or, at least, the influence of Miescher's work.

The lines of research that did develop out of Miescher's pioneering efforts led primarily to studies of the chemical composition of nuclei and chromosomes. Representative of these investigations are the papers of Mirsky,[19] of Pollister and Ris,[25] of Boivin, Vendrely, and Vendrely,[3] and of Swift.[28] In these papers, the nucleic acid–protein complex was examined cytologically and biochemically, and the cellular DNA content was studied as a function of meiosis and mitosis. All the studies were consistent with the theory that DNA was the genetic material, but they in no way showed that DNA specified an inherited phenotype. The research showing this developed separately, from studies of mutation in pathogenic bacteria.

The first, though not definitive, study that led to the identification of DNA as the genetic material was that of Griffith[12] in 1928. He showed that a nonpathogenic strain of *Diplococcus pneumoniae* (better known as *Pneumococcus*) could change to a pathogenic strain if it were grown in the presence of heat-killed pathogenic cells. Pathogenicity itself was an inherited phenotype. Griffith suggested that a "transforming principle" was released from the heat-killed pathogenic cells and was picked up by the nonpathogenic cells, thus converting or transforming them into genetically stable pathogenic cells. The work immediately following Griffith's suggested that such transformation could occur *in vitro*,[6] and, indeed, could occur using an extract of the heat-killed pathogenic cells.

In 1944, the chemical nature of the transforming principle was determined. Avery, MacLeod, and McCarty showed that it was the DNA from the pathogenic cells that could transform the nonpathogenic cells[1] — an inherited trait was shown to be the result of transmitted DNA. The researchers who might be considered Miescher's intellectual descendants did not give much attention to Griffith's work or that of Avery, MacLeod, and McCarty. Perhaps this was so because canonical knowledge had relegated nucleic acids to basically structural functions having no pertinence to the informational content of the gene. This stigma was removed in 1950 when Chargaff[4] clearly demonstrated the species-specific nature of the chemical composition of different DNA molecules. His work played a key role in dispelling the naive notion that nucleic acids were simply structural in their function.

If, after Chargaff's findings, there remained serious reservations about the specific role of DNA in an organism's heredity, they disappeared in 1952. In that year, Hershey and Chase[13] published their work showing that for the T2 bacteriophage, composed only of protein and DNA, it is the DNA that enters the host cell and that it alone directs the assembly of new progeny T2 phage.

Thus, by 1952, experimentation had clearly demonstrated that nucleic acid molecules (DNA in all cases except for RNA viruses) provided the chemical basis of heredity. However, knowledge was lacking as to the exact structure of DNA and its ability to self-replicate and change. In 1953, Watson and Crick[29,30] published two papers that (1) presented a model for the structure of DNA, and (2) based on this hypothetical structure, discussed how DNA revealed both a molecular mechanism for replication and the origin of spontaneous structural variations. Their ideas were born of the research of many of their colleagues, and it is this research on which we briefly dwell in this introduction.

What was known about DNA in the two or three years preceding the discovery of its structure? Above, it was noted that Chargaff dispelled the notion that DNA was a simple, repetitive sequence of nucleotides. In essence, he showed that the nucleotide content of DNA varied from species to species — making possible the idea that DNA differed qualitatively from one species to the next on the basis of unique nucleotide sequences. Of course, the chemical composition of DNA subunits (*nucleotides*) was also known — a subunit consisting of one of the four nitrogenous bases: adenine and guanine (*purines*); cytosine and thymine (*pyrimidines*); a pentose sugar (*deoxyribose*); and inorganic phosphate.

A second source of information about DNA emerged from x-ray crystallography studies.[31] To begin with, the fact that DNA could crystallize suggested that the molecule possessed a regular symmetry. Furthermore, analysis of the x-ray diffraction patterns suggested a double-stranded molecule with a helical conformation.

A third body of knowledge that contributed in a major way to Watson and Crick's efforts was Pauling's work on hydrogen bonding and helical configurations of biological macromolecules. Pauling worked on proteins primarily, but his ideas about hydrogen bonding and helices were easily extended to DNA. Pauling and Corey[24] suggested their own model of DNA which, to the surprise of many, was quite far off base.

Watson and Crick, utilizing the resources available to them, proposed a structure for DNA in 1953,[29] for which they drew heavily on the work performed by Chargaff, Wilkins, Franklin, and Pauling. Their model basically showed a double-stranded helical molecule, with one strand complementary to the other. The complementarity was determined by requisite base pairing as specified by hydrogen-bonding potentials: adenine on one strand would pair with thymine on the other; and cytosine would pair with guanine.

Watson and Crick's ideas and data are now known to be an actual indication of DNA's structure. However, it is important to bear in mind that in 1953, although there existed data to construct a model, there was no experimental data to support it. Experimental support came later. For example, to show that DNA was double-stranded, the molecule was heated to a point where the hydrogen bonds holding the two strands together were disrupted, producing single strands. Each single strand was one-half of DNA's molecular weight. To show that DNA was composed of complementary strands, the strands were separated and then reannealed, a process called *hybridization.* DNA strands from the same source, i.e., the same species, would hybridize, but DNA strands from different sources would not.

A final note on DNA structure before we discuss replication. DNA molecules can be circular or linear. Also, DNA need not always be double-stranded. The bacterial virus ϕX174 contains a single-stranded, circular DNA molecule. These facts emphasize the variety of forms the genetic information can assume.

As was mentioned earlier, a model for the replication of DNA emerged from Watson and Crick's proposed structure. Each DNA strand served as the template for the synthesis of a complementary strand. Base-pairing restrictions maximized the accuracy of the replication process. The suggested model of DNA replication was called the *semiconservative* model because it predicted that after one round of replication, daughter molecules would each contain one strand of parental DNA and one strand of newly synthesized DNA.

Other models also existed. In one, the *conservative* model, the parental molecule remained intact during the replication process — the result being one daughter molecule composed of all parental DNA and one composed of all newly synthesized DNA. In another, the *dispersive* model, the parental DNA broke up in such a way as to generate daughter molecules composed of parental and newly synthesized DNA, but with both types distributed throughout all four daughter strands.

All three models had predictive value if parental DNA could be distinguished from newly synthesized DNA. In 1957, Meselson, Stahl, and Vinograd[17] showed that DNA synthesized using the heavy isotope of nitrogen (^{15}N) could be separated by density-gradient centrifugation from DNA molecules synthesized using ordinary ^{14}N. This enabled Meselson and Stahl in 1958[16] to test the three models and show that DNA replicated semiconservatively.

Unfortunately, our understanding of the events leading to the synthesis of new DNA molecules is not nearly so complete as is our understanding of the general replication mode. In 1956, Kornberg and his colleagues[15] reported the discovery of an enzyme that could polymerize nucleotides. Following this announcement, Kornberg worked on the isolation, purification, and characterization of properties of this enzyme — called *DNA polymerase.* In 1967, Goulian, Kornberg and Sinsheimer[11] succeeded in demonstrating that DNA polymerase could synthesize a fully infectious bacteriophage ϕX174 chromosome *in vitro.* But what the enzyme did *in vitro* may have no bearing on what it could do inside a living cell. In 1969, Cairns[7] isolated a mutant of *Escherichia coli* that lacked Kornberg's enzyme yet was able to replicate its DNA in a normal fashion. Our current interpretation of Kornberg's enzyme, now called *DNA polymerase I,* is that it is not the enzyme normally used to replicate the cell's DNA. It appears that DNA polymerase I serves a repair function, filling in naturally occurring gaps in DNA molecules. This may well be tied in with the replication process, but so far this is not clear.

Since the discovery of DNA polymerase I, two other polymerase enzymes have been discovered: DNA polymerase II and DNA polymerase III.[10] We know the latter is essential to DNA replication, but, to date, do not understand the role of DNA polymerase II.

In addition to the polymerizing enzymes, there are two other groups of proteins which might be involved in the replication process. One, the unwinding proteins, is a group of proteins that seems to relax supercoiled, double-stranded DNA. The second group, the ligases, heals breaks in single strands of DNA. They appear to be essential in DNA replication because the process is discontinuous, with short segments of DNA being synthesized first, followed by the joining of segments to each other vis-à-vis the ligase.

Finally, while discussing mechanisms by which DNA is produced, we must mention the manner in which certain RNA viruses replicate through a DNA intermediate. For years, the ability of certain RNA viruses to induce malignant tumors was well known. However, the manner in which the viral RNA was stabilized, replicated, and passed on was unknown. Until 1970, all identified mechanisms of DNA replication involved the synthesis of new DNA off template DNA, involving enzymes classified as DNA-dependent DNA polymerases. In 1970, Temin, Baltimore, Mizutani, and others showed that these viruses first produced complementary DNA in the presence of a viral-specific, RNA-dependent DNA polymerase, which then generated more viral RNA with the host's DNA-dependent RNA polymerase.

The last several papers in this section deal with experiments analyzing the process of gene variability and the use of gene variability to analyze the structure of the individual gene. The origin and impact of mutations on biological systems first attracted attention because of the role mutants play in the course of evolutionary change. For example: Lamarck visualized the process of evolution taking place by means of changes in the structure and function of organisms — these changes directed by the action of environmental stresses. According to Lamarck, these changes were transmissible, but he failed to speculate how the changes occurred and how they passed to the next generation.

Charles Darwin[5] attempted to clarify the origin of transmissible change and how changes could be passed on to progeny. Darwin observed the existence of a large amount of genetic variation in natural populations. The environment screened these genetic variants in terms of their ability to survive and reproduce. Since the mutations were heritable, those individuals favorably screened would pass on the optimal genetic constitution to their progeny. In turn, their progeny would be favored in their reproductive potential. In this way, new, genetically based changes would be established in the population. Given sufficient time and change, speciation would ensue.

Darwin's concept of a mechanism of inheritance assumed that all portions of the organism contributed formative elements (*gemmules*) to the gametes so, at the time of conception, maternal and paternal gemmules would come together and control the emergence of the new individual. Since they were contributed by the parents, these gemmules would lead to the production of features similar to the parental expressions. Darwin suggested that genetic variation arose by means of quantitative changes in the number of gemmules controlling a specific character — or by qualitative changes in the gemmule itself. He viewed mutation as a modification of discrete units of inheritance. Initially, he felt that genetic change was independent of the environment but he gradually altered his thinking until, eventually, he accepted a Lamarckian interpretation.

Between 1865 and 1900, cytological examination of cellular and organismal reproduction supported the hypothesis that the genetic information resided within the nucleus.[14] Mutations would have to take place within the nucleus of the cell. These theories and the failure to environmentally channel transmissible change cast doubt upon Lamarck's hypothesis; and, in conjunction with the reappearance of Mendel's work in 1900, paved the way for the next major advance in mutation theory. This took the form of Hugo de Vries's two-volume work entitled *Mutation Theory*.[8]

De Vries, like Darwin, felt that mutations arose as changes in separate units of inheritance (*pangenes*). He attributed the distribution of these pangenes throughout a natural population to a Mendelian mode of inheritance. He believed that mutations arose from two different types of pangene modification. First, he felt that heritable variations could arise by means of a redistribution of pre-existing pangenes, through assortment and genetic exchange, during sexual reproduction. Mendelian laws of inheritance would control the appearance of this form of variation. However, since this type of variation could never exceed the existing natural variation within the population, de Vries felt that it was not instrumental in evolutionary change.

The second form of genetic variation, the appearance of which was not predictable on the basis of Mendelian laws, was based upon the creation of new pangenes. This produced expressions distinctly different from the continuous spectrum of expressions in the natural population. DeVries called these *discontinuous variations*. On the basis of his experiments with *Oenothera,* he felt a single discontinuous variation could give rise to a new species. Thus, he visualized Darwin's environmental screening working at the level of species selection and not intraspecies selection. Because de Vries proposed that evolutionary change rested on the occurrence of easily observable, discontinuous changes, his ideas were readily amenable to experimental scrutiny. He placed both evolution and mutation in the realm of experimental research and many laboratories rapidly undertook the detection and study of mutations.

Thomas H. Morgan was one of the investigators who took up the search for de Vriesian discontinuous mutations. Morgan introduced the fruit fly, *D. melanogaster,* as a system in which the occurrence of such mutations could be screened. He and his students did not discover the existence of species-forming mutations but they did accumulate a significant number of spontaneous mutations that allowed them to make some generalizations about the phenomenon. By 1920, a view of mutation had emerged that pictured the event as exceptionally rare. And, once a mutation had taken place, the modified form of the gene was again exceptionally resistant to additional change. Also, mutations appeared to be restricted to single genes, and the size of the mutable gene seemed to equal the size of the transmissible gene.

In 1927, a new dimension was added to the study of mutation. Up to this time, mutation research dealt exclusively with spontaneous mutations. Even with Muller's techniques for more efficiently sifting out spontaneous mutations,[20] their low frequency made any type of mutation rate or recombination analysis impossible. In 1927, Muller[22] and Stadler[27] demonstrated that the frequency of mutation could be enhanced above the spontaneous rate by the application of ionizing radiation. This greatly increased the number of mutants one could collect. Their work clearly demonstrated that major chromosomal rearrangements could mimic single gene mutations in phenotype and transmission pattern; however, it shed little light on other mechanisms of mutation. Even the discovery, in the early 1940s, that chemicals could increase the frequency of mutation helped little in revealing its mechanisms. Obviously the reason for the lack of progress was the void in information regarding the chemical and structural nature of the gene.

As set forth earlier in this introduction, the chemical nature of the genetic material gradually emerged between 1944 and 1950, and culminated in 1953 with Watson and Crick's presentation of a structural model. With this model, the major question of how mutated genes could still preserve their capacity for self-replication was resolved. By altering the pairing capacities of the nitrogenous bases, nucleotide changes could take place without destroying the ability of the bases to direct the synthesis of complementary strands. Using this idea, models were advanced to show how spontaneous and induced mutations might arise. Following the establishment of the genetic code, it became possible to identify codon changes based upon amino acid substitutions in mutant proteins. This provided a means for experimentally examining many of the proposed mechanisms of mutation.

One last point should be made before leaving this topic. Recently, gene mutations have been identified that lead to the subsequent enhancement of mutations throughout the genome due to alterations in physiological reactions involving DNA. Experiments have suggested that a wholesale elevation of the frequency of mutation[9] could be effected by those mutations arising initially in genes coding for the proteins involved in replication, recombination, or repair. When one considers the possibility that natural selection can screen spontaneous mutation rates for optimal frequencies, the importance to evolution is clear.

Interest in the use of mutation as a tool to study the structure and function of the gene did not arise until the 1920s. In 1921, Muller, in a prophetic talk, advocated the use of mutation to study the gene. He acknowledged the paucity of mutants available for such studies but he projected how special genetic techniques could enhance the detection of mutations. Also, he predicted the possibility of artificially enhancing the frequency of mutation. Beginning in 1927, the application of Muller's screening techniques to detect induced mutations provided a means of accumulating large numbers of mutations. Through recombination mapping, many gene loci were found to possess numerous and separable mutant sites — a mutant site often being the equivalent of a single base pair. These mutants could then be placed in a heterozygous condition and analyzed for their capacity to functionally complement (display a normal phenotype) and genetically recombine. Initially, such analyses failed to show complementation or recombination. The failure to observe complementation or recombination firmly established the idea that only the entire gene, due to its structural organization, could mutate and recombine. Thus, prior to 1940, the gene took on an almost mystical quality about its structural integrity. It was viewed as the genetic unit that could not be subdivided further by mutation or recombination. This interpretation received added support from Muller's work, in Russia, in which he identified the existence of preferential break points on the X chromosome following X irradiation.[23] These break points fell *between* functionally discrete genes and not within them.

The concept of exceptional structural integrity began to come under attack in the 1940s, when investigators found recombination between previously established allelic mutants. These findings suggested that the portion of the gene that could mutate and recombine was smaller than the functionally defined gene. There was strong resistance to the idea of intragenic-recombination. Certainly, examples of intragenic-recombination were extremely rare in *Drosophila*. Hypothetical models were suggested that explained the inconsistent results without invoking intragenic-recombination. However, investigators working with *Neurospora* and other Ascomycetes[26] — biological systems that made it possible to screen large numbers of individuals for rare recombinational events — demonstrated that recombinations between allelic mutants was the rule rather than the exception, if one worked with large numbers of organisms.

The resolving power of recombination studies continued to increase in sophistication with the advent of viral genetics. Screening techniques made it possible to identify exceptional phenotypes in a hundred million or more progeny counts. Using these techniques, Seymour Benzer conclusively demonstrated that intragenic-recombination could take place within a single functional unit of a virus particle.[2] Using Watson and Crick's model for the structure of DNA, Benzer calculated recombination frequencies that suggested mutations could be as small as a single base pair, and that recombination was possible between adjacent nucleotide pairs. Thus, the use of mutations was instrumental in defining the structural organization of the gene in lower organisms.

In higher organisms the picture of the gene's structure is much more obscure. Recombination studies have revealed a low level of genetic exchange that can be interpreted as being intragenic in nature. However, even though large-scale recombination studies reveal the existence of intragenic-recombination, it may be exceptional. Much of what appears to be intragenic-recombination in higher organisms may actually be exchange between operational and structural gene mutants in regulatory complexes. Intergenic-recombination may be the more common type of exchange in higher organisms. The last paper of Part 1 deals with the complexity of this situation.

REFERENCES

1. Avery, O. T., C. M. MacLeod, and M. McCarty. 1944. Studies on the chemical nature of the substance inducing transformation of pneumococcal types. *J. Exp. Med.* 79:137–158.

2. Benzer, S. 1956. The elementary units of heredity. In *Symposium on the chemical basis of heredity,* ed. W. D. McElroy and B. Glass. Baltimore: Johns Hopkins Press.

3. Boivin, A., R. Vendrely, and C. Vendrely. 1948. *L'acide desoxyribonucleique du noyau cellulaire, depositaire des caracteres hereditaires; arguments d'ordre analytique. C. R. Acad. Sci. Paris.* 226:1061–1063.

4. Chargaff, E. 1950. Chemical specificity of nucleic acids and mechanism of their enzymatic degradation. *Experientia* 6:201–209.

5. Darwin, C. 1868. Provisional hypothesis of pangenesis. In *Animals and plants under domestication.* 2:428–483.

6. Dawson, M. H., and R. H. P. Sia. 1931. *In vitro* transformation of pneumococcal types: I. a technique for inducing transformation of pneumococcal types *in vitro. J. Exp. Med.* 54:681–699.

7. De Lucia, P., and J. Cairns. 1969. Isolation of an *E. coli* strain with a mutation affecting DNA polymerase. *Nature* 224:1164–1166.

8. de Vries, H. 1901–1903. *Die Mutationstheorie.* Leipzig: Veit and Co.

9. Drake, J. W. 1973. The genetic control of spontaneous and induced mutation rates in bacteriophage T4. *Gen. Supp.* 73:45–64.

10. Gefter, M. L., Y. Hirota, T. Kornberg, J. A. Wechsler, and C. Barnoux. 1971. Analysis of DNA polymerase II and III in mutants of *Escherichia coli* thermo-sensitive for DNA synthesis. *Proc. Nat. Acad. Sci. U.S.,* 68:3150–3153.

11. Goulian, M., A. Kornberg, and R. Sinsheimer. 1967. Enzymatic synthesis of DNA: XXIV. Synthesis of infectious phage ϕX174 DNA. *Proc. Nat. Acad. Sci. U.S.* 58:2321–2328.

12. Griffith, F. 1928. The significance of pneumococcal types. *J. Hyg.* 27:113–159.

13. Hershey, A., and M. Chase, 1952. Independent functions of viral protein and nucleic acid in growth of bacteriophage. *J. Gen. Physiol.* 36:39–56.

14. Hertwig, O. 1885. The problem of fertilization and isotropy of the egg, a theory of inheritance. *Jenaische Zeitschrift* 18:276–318. Partially translated and included in Voeller's, *The Chromosome Theory of Inheritance,* 28:33. New York: Appleton Century Crofts.

15. Kornberg, A., I. R. Lehman, M. J. Bessman, and E. S. Simms. 1956. Enzymic synthesis of deoxyribonucleic acid. *Biochem. et Biophys. Acta* 21:197.

16. Meselson, M., and F. W. Stahl. 1958. The replication of DNA in *Escherichia coli. Proc. Nat. Acad. Sci. U.S.* 44:671–682.

17. Meselson, M., F. W. Stahl, and J. Vinograd. 1957. Equilibrium sedimentation of macromolecules in density gradients. *Proc. Nat. Acad. Sci. U.S.* 43:581–588.

18. Miescher, F. 1871. On the chemical composition of pus cells. From *Hoppe-Seyler's medizinische-chemische untersuchungen* 4:441–460.

19. Mirksy, A. E. 1947. Chemical properties of isolated chromosomes. *Cold Springs Harbor Symp. Quant. Biol.* 12:143–146.

20. Muller, H. J. 1919. The rate of change of hereditary factors in *Drosophila. Proc. of the Soc. for Exp. Biol. and Med.* 17:10–14.

21. Muller, H. J. 1922. Variation due to change in the individual gene. *Amer. Nat.* 61:156.

22. Muller, H. J. 1927. Artificial transmutation of the gene. *Science* 66:84–87.

23. Muller, H. J. and A. Prokofjeva. 1934. Continuity and discontinuity of the heredity material. *Comptes Rendus de l'academie des Sciences de l'URRS* 4:1–2.

24. Pauling, L., and R. B. Corey. 1953. Structure of the nucleic acids. *Nature:* 171:346.

25. Pollister, A. W., and H. Ris. 1947. Nucleoprotein determinations in cytological preparations. *Cold Spring Harbor Symp. Quant. Biol.* 12:147–157.

26. Pontecorvo, G., and J. A. Roper. 1956. The resolving power of genetic analysis. *Nature* 178:83–84.

27. Stadler, L. J. 1928. Mutations in barley induced by x-ray and radium. *Science* 68:186–187.

28. Swift, H. 1950. The constancy of DNA in plant nuclei. *Proc. Nat. Acad. Sci. U.S.* 36:643–654.

29. Watson, J. D., and F. H. C. Crick. 1953. Molecular structure of nucleic acids: a structure for deoxyribose nucleic acid. *Nature* 171:737–738.

30. Watson, J. D., and F. H. C. Crick. 1953. Genetical implications of the structure of deoxyribonucleic acid. *Nature* 171:964–967.

31. Wilkins, M. H. F., A. R. Stokes, and H. R. Wilson. 1953. Molecular structure of deoxypentose nucleic acids. *Nature* 171:738–740.

32. Zacharias, E. 1881. *Botanische Zeitung* 39:169–176.

Studies on the chemical nature of the substance inducing transformation of pneumococcal types

Induction of transformation by a desoxyribonucleic acid fraction isolated from Pneumococcus Type III

OSWALD T. AVERY, COLIN M. MACLEOD, and MACLYN MCCARTY

Reprinted by authors' and publisher's permission from Journal of Experimental Medicine, *vol. 79, 1944, pp. 137–158.*

This paper is a true classic. The experiments described here are logical extensions of Griffith's observations and techniques and, later, those of Dawson and Sia. It is clearly shown here that DNA is the transforming principle. Contrary to what many textbooks imply, this paper is a stronger argument for DNA being the genetic material than Hershey and Chase's paper of 1952. It is stronger because it is less ambiguous about any possible role proteins may play in inheritance. Protein was considered by many to be the major candidate for the genetic material, and DNA was viewed as a rather monotonous molecule playing a structural role.

What is also interesting about this paper is the reception it received. It was widely discussed, unlike Mendel's paper of 1865, but there was certainly a general lack of appreciation for its main conclusion, i.e., that DNA was the transforming principle hence the basis of an inherited phenotype. So, in that sense, the two papers are alike: both, in retrospect, were premature because their implications were not obviously pertinent to accepted ideas on the biochemical nature of the gene. Between 1944 and 1950, the "tetranucleotide theory," in which DNA was seen as a regularly repeating macromolecule, best embraced the concept of DNA. Chargaff, in 1950, convincingly dispelled this interpretation.

Biologists have long attempted by chemical means to induce in higher organisms predictable and specific changes which thereafter could be transmitted in series as hereditary characters. Among microorganisms the most striking example of inheritable and specific alterations in cell structure and function that can be experimentally induced and are reproducible under well defined and adequately controlled conditions is the transformation of specific types of Pneumococcus. This phenomenon was first described by Griffith[1] who succeeded in transforming an attenuated and nonencapsulated (R) variant derived from one specific type into fully encapsulated and virulent (S) cells of a heterologous specific type. A typical instance will suffice to illustrate the techniques originally used and serve to indicate the wide variety of transformations that are possible within the limits of this bacterial species.

Griffith found that mice injected subcutaneously with a small amount of living R culture derived from Pneumococcus Type II together with a large inoculum of heat-killed Type III (S) cells frequently succumbed to infection, and that the heart's blood of these animals yielded Type III pneumococci in pure culture. The fact that the R strain was avirulent and incapable by itself of causing fatal bacteremia and the additional fact that the heated suspension of Type III cells contained no viable organisms brought convincing evidence that the R forms growing under these conditions had newly acquired the capsular structure and biological specificity of Type III pneumococci.

The original observations of Griffith were later confirmed by Neufeld and Levinthal,[2] and by Baurhenn[3] abroad, and by Dawson[4] in this laboratory. Subsequently Dawson and Sia[5] succeeded in inducing transformation *in vitro*. This they accomplished by growing R cells in a fluid medium containing anti-R serum and heat-killed encapsulated S cells. They showed that in the test tube as in the animal body transformation can be selectively induced, depending on the type specificity of the S cells used in the reaction system. Later, Alloway[6] was able to cause specific transformation *in vitro* using sterile extracts of S cells from which all formed elements and cellular debris had been removed by Berkefeld filtration. He thus showed that crude extracts containing active transforming material in soluble form are as effective in inducing specific transformation as are the intact cells from which the extracts were prepared.

Another example of transformation which is analogous to the interconvertibility of pneumococcal types lies in the field of viruses. Berry and Dedrick[7] succeeded in changing the virus of rabbit fibroma (Shope) into that of infectious myxoma (Sanarelli). These investigators innoculated rabbits with a mixtue of active fibroma virus together with a suspension of heat-inactivated myxoma virus and produced in the animals the symptoms and pathological lesions characteristic of infectious myxomatosis. On subsequent animal passage the transformed virus was transmissible and induced myxomatous infection typical of the naturally occurring disease. Later Berry[8] was successful in inducing the same transformation using a heat-inactivated suspension of washed elementary bodies of myxoma virus. In the case of these viruses the methods employed were similar in principle to those used by Griffith in the transformation of pneumococcal types. These observations have subsequently been confirmed by other investigators.[9]

The present paper is concerned with a more detailed analysis of the phenomenon of transformation of specific types of Pneumococcus. The major interest has centered in attempts to isolate the active principle from crude bacterial extracts and to identify if possible its chemical nature or at least to characterize it sufficiently to place it in a general group of known chemical substances. For purposes of study, the typical example of transformation chosen as a working model was the one with which we have had most experience and which consequently seemed best suited for analysis. This particular example represents the transformation of a nonencapsulated R variant of Pneumococcus Type II to Pneumococcus Type III.

EXPERIMENTAL

Transformation of pneumococcal types *in vitro* requires that certain cultural conditions be fulfilled before it is possible to demonstrate the reaction even in the presence of a potent extract. Not only must the broth medium be optimal for growth but it must be supplemented by the addition of serum or serous fluid known to possess certain "special properties." Moreover, the R variant, as will be shown later, must be in the reactive phase in which it has the capacity to respond to the transforming stimulus. For purposes of convenience these several components as combined in the transforming test will be referred to as the *reaction system*. Each constituent of this system presented problems which required clarification before it was possible to obtain consistent and reproducible results. The various components of the system will be described in the following order: (1) nutrient broth, (2) serum or serous fluid, (3) strain of R Pneumococcus, and (4) extraction, purification, and chemical nature of the transforming principle.

1. Nutrient Broth Beef heart infusion broth containing 1 percent neopeptone with no added dextrose and adjusted to an initial pH of 7.6–7.8 is used as the basic medium. Individual lots of broth show marked and unpredictable variations in the property of supporting transformation. It has been found, however, that charcoal adsorption according to the method described by MacLeod and Mirick[10] for removal of sulfonamide inhibitors, eliminates to a large extent these variations; consequently this procedure is used as routine in the preparation of consistently effective broth for titrating the transforming activity of extracts.

2. Serum or Serous Fluid In the first successful experiments on the induction of transformation *in vitro*, Dawson and Sia[5] found that it was essential to add serum to the medium. Anti-R pneumococcal rabbit serum was used because of the observation that reversion of an R pneumococcus to the homologous S form can be induced by growth in a medium containing anti-R serum. Alloway[6] later found that ascitic or chest fluid and normal swine serum, all of which contain R antibodies, are capable of replacing antipneumococcal rabbit serum in the reaction system. Some form of serum is essential, and to our knowledge transformation *in vitro* has never been effected in the absence of serum or serous fluid.

In the present study human pleural or ascitic fluid has been used almost exclusively. It became apparent, however, that the effectiveness of different lots of serum varied and that the differences observed were not necessarily dependent upon the content of R antibodies, since many sera of high titer were found to be incapable of supporting transformation. This fact suggested that factors other than R antibodies are involved.

It has been found that sera from various animal species, irrespective of their immune properties, contain an enzyme capable of destroying the transforming principle in potent extracts. The nature of this enzyme and the specific substrate on which it acts will be referred to later in this paper. This enzyme is inactivated by heating the serum at 60°–65°C, and sera heated at temperatures known to destroy the enzyme are often rendered effective in the transforming system. Further analysis has shown that certain sera in which R antibodies are present and in which the enzyme has been inactivated may nevertheless fail to support transformation. This fact suggests that still another factor in the serum is essential. The content of this factor varies in different sera, and at present its identity is unknown.

There are at present no criteria which can be used as a guide in the selection of suitable sera or serous fluids except that of actually testing their capacity to support transformation. Fortunately, the requisite properties are stable and remain unimpaired over long periods of time; and sera that have been stored in the refrigerator for many months have been found on retesting to have lost little or none of their original effectiveness in supporting transformation.

The recognition of these various factors in serum and their role in the reaction system has greatly facilitated the standardization of the cultural conditions required for obtaining consistent and reproducible results.

3. The R Strain (R36A)   The unencapsulated R strain used in the present study was derived from a virulent "S" culture of Pneumococcus Type II. It will be recalled that irrespective of type derivation all "R" variants of Pneumococcus are characterized by the lack of capsule formation and the consequent loss of both type specificity and the capacity to produce infection in the animal body. The designation of these variants as R forms has been used to refer merely to the fact that on artificial media the colony surface is "rough" in contrast to the smooth, glistening surface of colonies of encapsulated S cells.

The R strain referred to above as R36A was derived by growing the parent S culture of Pneumococcus Type II in broth containing Type II antipneumococcus rabbit serum for 36 serial passages and isolating the variant thus induced. The strain R36A has lost all the specific and distinguishing characteristics of the parent S organisms and consists only of attenuated and nonencapsulated R variants. The change S → R is often a reversible one provided the R cells are not too far "degraded." The reversion of the R form to its original specific type can frequently be accomplished by successive animal passages or by repeated serial subculture in anti-R serum. When reversion occurs under these conditions, however, the R culture invariably reverts to the encapsulated form of the same specific type as that from which it was derived.[11] Strain R36A has become relatively fixed in the R phase and has never spontaneously reverted to the Type II S form. Moreover, repeated attempts to cause it to revert under the conditions just mentioned have in all instances been unsuccessful.

The reversible conversion of S ⇌ R within the limits of a single type is quite different from the transformation of one specific type of Pneumococcus into another specific type through the R form. Transformation of types has never been observed to occur spontaneously and has been induced experimentally only by the special techniques outlined earlier in this paper. Under these conditions, the enzymatic synthesis of a chemically and immunologically different capsular polysaccharide is specifically oriented and selectively determined by the specific type of S cells used as source of the transforming agent.

In the course of the present study it was noted that the stock culture of R36 on serial transfers in blood broth undergoes spontaneous dissociation giving rise to a number of other R variants which can be distinguished one from another by colony form. The significance of this in the present instance lies in the fact that of four different variants isolated from the parent R culture only one (R36A) is susceptible to the transforming action of potent extracts, while the others fail to respond and are wholly inactive in this regard. The fact that differences exist in the responsiveness of different R variants to the same specific stimulus emphasizes the care that must be exercised in the selection of a suitable R variant for use in experiments on transformation. The capacity of this R strain (R36A) to respond to a variety of different transforming agents is shown by the readiness with which it can be transformed to Types I, III, VI, or XIV, as well as to its original type (Type II), to which, as pointed out, it has never spontaneously reverted.

Although the significance of the following fact will become apparent later on, it must be mentioned here that pneumococcal cells possess an enzyme capable of destroying the activity of the transforming principle. Indeed, this enzyme has been found to be present and highly active in the autolysates of a number of different strains. The fact that this intracellular enzyme is released during autolysis may explain, in part at least, the observation of Dawson and Sia[5] that it is essential in bringing about transformation in the test tube to use a small inoculum of young and actively growing R cells. The irregularity of the results and often the failure to induce transformation when large inocula are used may be attributable to the release from autolyzing cells of an amount of this enzyme sufficient to destroy the transforming principle in the reaction system.

In order to obtain consistent and reproducible results, two facts must be borne in mind: first, that an R culture can undergo spontaneous dissociation and give rise to other variants which have lost the capacity to respond to the transforming stimulus; and secondly, that pneumococcal cells contain an intracellular enzyme which when released destroys the activity of the transforming principle. Consequently, it is important to select a responsive strain and to prevent as far as possible the destructive changes associated with autolysis.

Method of Titration of Transforming Activity In the isolation and purification of the active principle from crude extracts of pneumococcal cells it is desirable to have a method for determining quantitatively the transforming activity of various fractions.

The experimental procedure used is as follows: Sterilization of the material to be tested for activity is accomplished by the use of alcohol since it has been found that this reagent has no effect on activity. A measured volume of extract is precipitated in a sterile centrifuge tube by the addition of 4 to 5 volumes of absolute ethyl alcohol, and the mixture is allowed to stand 8 or more hours in the refrigerator in order to effect sterilization. The alcohol precipitated material is centrifuged, the supernatant discarded, and the tube containing the precipitate is allowed to drain for a few minutes in the inverted position to remove excess

alcohol. The mouth of the tube is then carefully flamed and a dry, sterile cotton plug is inserted. The precipitate is redissolved in the original volume of saline. Sterilization of active material by this technique has invariably proved effective. This procedure avoids the loss of active substance which may occur when the solution is passed through a Berkefeld filter or is heated at the high temperatures required for sterilization.

To the charcoal-adsorbed broth described above is added 10 percent of the sterile ascitic or pleural fluid which has previously been heated at 60°C for 30 minutes, in order to destroy the enzyme known to inactivate the transforming principle. The enriched medium is distributed under aseptic conditions in 2.0 cc amounts in sterile tubes measuring 15 × 100 mm. The sterilized extract is diluted serially in saline neutralized to pH 7.2–7.6 by addition of 0.1 N NaOH, or it may be similarly diluted in $M/40$ phosphate buffer, pH 7.4. 0.2 cc of each dilution is added to at least 3 or 4 tubes of the serum medium. The tubes are then seeded with a 5- to 8-hour blood broth culture of R36A. 0.05 cc of a 10^{-4} dilution of this culture is added to each tube, and the cultures are incubated at 37°C for 18 to 24 hours.

The anti-R properties of the serum in the medium cause the R cells to agglutinate during growth, and clumps of the agglutinated cells settle to the bottom of the tube leaving a clear supernatant. When transformation occurs, the encapsulated S cells, not being affected by these antibodies, grow diffusely throughout the medium. On the other hand, in the absence of transformation the supernatant remains clear, and only sedimented growth of R organisms occurs. This difference in the character of growth makes it possible by inspection alone to distinguish tentatively between positive and negative results. As routine all the cultures are plated on blood agar for confirmation and further bacteriological identification. Since the extracts used in the present study were derived from Pneumococcus Type III, the differentiation between the colonies of the original R organism and those of the transformed S cells is especially striking, the latter being large, glistening, mucoid colonies typical of Pneumococcus Type III. Figures 1.1 & 1.2 illustrate these differences in colony form.

A typical protocol of a titration of the transforming activity of a highly purified preparation is given in Table 4.

Preparative Methods

Source Material In the present investigation a stock laboratory strain of Pneumococcus Type III (A66) has been used as source material for obtaining the

Figs. 1 and 2.   (1) Colonies of the R variant (R36A) derived from Pneumococcus Type II. Plated on blood agar from a culture grown in serum broth in the absence of the transforming substance. X3.5.   (2) Colonies on blood agar of the same cells after induction of transformation during growth in the same medium with the addition of active glistening, mucoid colonies shown are characteristic of Pneumococcus Type III and readily distinguishable from the small, rough colonies of the parent R strain illustrated in Fig. 1. X3.5. *(The photograph was made by Mr. Joseph B. Haulenbeek.)*

active principle. Mass cultures of these organisms are grown in 50- to 75-liter lots of plain beef heart infusion broth. After 16 to 18 hours' incubation at 37°C the bacterial cells are collected in a steam-driven sterilizable Sharples centrifuge. The centrifuge is equipped with cooling coils immersed in ice water so that the culture fluid is thoroughly chilled before flowing into the machine. This procedure retards autolysis during the course of centrifugation. The sedimented bacteria are removed from the collecting cylinder and resuspended in approximately 150 cc of chilled saline (0.85 percent NaCl), and care is taken that all clumps are thoroughly emulsified. The glass vessel containing the thick, creamy suspension of cells is immersed in a water bath, and the temperature of the suspension rapidly raised to 65°C. During the heating process the material is constantly stirred, and the temperature maintained at 65°C for 30 minutes. Heating at this temperature inactivates the intracellular enzyme known to destroy the transforming principle.

Extraction of Heat-Killed Cells   Although various procedures have been used, only that which has been found most satisfactory will be described here. The heat-killed cells are washed with saline 3 times. The chief value of the washing process is to remove a large excess of capsular polysaccharide together with much of the protein, ribonucleic acid, and somatic "C" polysaccharide. Quantitative titrations of transforming activity have shown that not more than 10 to 15 percent of the active material is lost in the washing, a loss which is small in comparison to the amount of inert substances which are removed by this procedure.

After the final washing, the cells are extracted in 150 cc of saline containing sodium desoxycholate in final concentration of 0.5 percent by shaking the mixture mechanically 30 to 60 minutes. The cells are separated by centrifugation and the extraction process is repeated 2 or 3 times. The desoxycholate extracts prepared in this manner are clear and colorless. These extracts are combined and precipitated by the addition of 3 to 4 volumes of absolute ethyl alcohol. The sodium desoxycholate being soluble in alcohol remains in the supernatant and is thus removed at this step. The precipitate forms a fibrous mass which floats to the surface of the alcohol and can be removed directly by lifting it out with a spatula. The excess alcohol is drained from the precipitate which is then redissolved in about 50 cc of saline. The solution obtained is usually viscous, opalescent, and somewhat cloudy.

Deproteinization and Removal of Capsular Polysaccharide The solution is then deproteinized by the chloroform method described by Sevag.[12] The procedure is repeated 2 or 3 times until the solution becomes clear. After this preliminary treatment the material is reprecipitated in 3 to 4 volumes of alcohol. The precipitate obtained is dissolved in a larger volume of saline (150 cc) to which is added 3 to 5 mg of a purified preparation of the bacterial enzyme capable of hydrolyzing the Type III capsular polysaccharide.[13] The mixture is incubated at 37°C, and the destruction of the capsular polysaccharide is determined by serological tests with Type III antibody solution prepared by dissociation of immune precipitate according to the method described by Liu and Wu.[14] The advantages of using the antibody solution for this purpose are that it does not react with other serologically active substances in the extract and that it selectively detects the presence of the capsular polysaccharide in dilutions as high as 1:6,000,000. The enzymatic breakdown of the polysaccharide is usually complete within 4 to 6 hours, as evidenced by the loss of serological reactivity. The digest is then precipitated in 3 to 4 volumes of ethyl alcohol, and the precipitate is redissolved in 50 cc of saline. Deproteinization by the chloroform process is again used to remove the added enzyme protein and remaining traces of pneumococcal protein. The procedure is repeated until no further film of protein-chloroform gel is visible at the interface.

Alcohol Fractionation Following deproteinization and enzymatic digestion of the capsular polysaccharide, the material is repeatedly fractionated in ethyl alcohol as follows. Absolute ethyl alcohol is added dropwise to the solution with constant stirring. At a critical concentration varying from 0.8 to 1.0 volume of alcohol the active material separates out in the form of fibrous strands that wind themselves around the stirring rod. This precipitate is removed on the rod and washed in a 50 percent mixture of alcohol and saline. Although the bulk of active material is removed by fractionation at the critical concentration, a small but appreciable amount remains in solution. However, upon increasing the concentration of alcohol to 3 volumes, the residual fraction is thrown down together with inert material in the form of a flocculent precipitate. This flocculent precipitate is taken up in a small volume of saline (5 to 10 cc) and the solution again fractionated by the addition of 0.8 to 1.0 volume of alcohol. Additional fibrous material is obtained which is combined with that recovered from the original solution. Alcoholic fractionation is repeated 4 to 5 times. The yield of fibrous material obtained by this method varies from 10 to 25 mg per 75 liters of culture and represents the major portion of active material present in the original crude extract.

Effect of Temperature As a routine procedure all steps in purification were carried out at room temperature unless specifically stated otherwise. Because of the theoretical advantage of working at low temperature in the preparation of biologically active material, the purification of one lot (preparation 44) was carried out in the cold. In this instance all the above procedures with the exception of desoxycholate extraction and enzyme treatment were conducted in a cold room maintained at 0–4°C. This preparation proved to have significantly higher activity than material similarly prepared at room temperature.

Desoxycholate extraction of the heat-killed cells at low temperature is less efficient and yields smaller amounts of the active fraction. It has been demonstrated that higher temperatures facilitate extraction of the active principle, although activity is best preserved at low temperatures.

Analysis of Purified Transforming Material

General Properties Saline solutions containing 0.5 to 1.0 mg per cc of the purified substance are colorless and clear in diffuse light. However, in strong transmitted light the solution is not entirely clear and when stirred exhibits a silky sheen. Solutions at these concentrations are highly viscous.

Purified material dissolved in physiological salt solution and stored at 2–4°C retains its activity in

Table 1 Elementary chemical analysis of purified preparations of the transforming substance

Preparation no.	Carbon percent	Hydrogen percent	Nitrogen percent	Phosphorus percent	N/P ratio
37	34.27	3.89	14.21	8.57	1.66
38B	—	—	15.93	9.09	1.75
42	35.50	3.76	15.36	9.04	1.69
44	—	—	13.40	8.45	1.58
Theory for sodium desoxyribonucleate	34.20	3.21	15.32	9.05	1.69

undiminished titer for at least 3 months. However, when dissolved in distilled water, it rapidly decreases in activity and becomes completely inert within a few days. Saline solutions stored in the frozen state in a CO_2 ice box ($-70°C$) retain full potency for several months. Similarly, material precipitated from saline solution by alcohol and stored under the supernatant remains active over a long period of time. Partially purified material can be preserved by drying from the frozen state in the lyophile apparatus. However, when the same procedure is used for the preservation of the highly purified substance, it is found that the material undergoes changes resulting in decrease in solubility and loss of activity.

The activity of the transforming principle in crude extracts withstands heating for 30 to 60 minutes at 65°C. Highly purified preparations of active material are less stable, and some loss of activity occurs at this temperature. A quantitative study of the effect of heating purified material at higher temperatures has not as yet been made. Alloway,[6] using crude extracts prepared from Type III pneumococcal cells, found that occasionally activity could still be demonstrated after 10 minutes' exposure in the water bath to temperatures as high as 90°C.

The procedures mentioned above were carried out with solutions adjusted to neutral reaction, since it has been shown that hydrogen ion concentrations in the acid range result in progressive loss of activity. Inactivation occurs rapidly at pH 5 and below.

Qualitative Chemical Tests The purified material in concentrated solution gives negative biuret and Millon tests. These tests have been done directly on dry material with negative results. The Dische di-

phenylamine reaction for desoxyribonucleic acid is strongly positive. The orcinol test (Bial) for ribonucleic acid is weakly positive. However, it has been found that in similar concentrations pure preparations of desoxyribonucleic acid of animal origin prepared by different methods give a Bial reaction of corresponding intensity.

Although no specific tests for the presence of lipid in the purified material have been made, it has been found that crude material can be repeatedly extracted with alcohol and ether at $-12°C$ without loss of activity. In addition, as will be noted in the preparative procedures, repeated alcohol precipitation and treatment with chloroform result in no decrease in biological activity.

Elementary Chemical Analysis * Four purified preparations were analyzed for content of nitrogen, phosphorus, carbon, and hydrogen. The results are presented in Table 1. The nitrogen-phosphorus ratios vary from 1.58 to 1.75 with an average value of 1.67 which is in close agreement with that calculated on the basis of the theoretical structure of sodium desoxyribonucleate (tetranucleotide). The analytical figures by themselves do not establish that the substance isolated is a pure chemical entity. However, on the basis of the nitrogen-phosphorus ratio, it would appear that little protein or other substances containing nitrogen or phosphorus are present as impurities since if they were this ratio would be considerably altered.

---

*The elementary chemical analyses were made by Dr. A. Elek of The Rockefeller Institute.

Table 2 The inactivation of transforming principle by crude enzyme preparations

Crude enzyme preparations	Enzymatic activity			
	Phosphatase	Tributyrin esterase	Depolymerase for desoxyribonucleate	Inactivation of transforming principle
Dog intestinal mucosa	+	+	+	+
Rabbit bone phosphatase	+	+	–	–
Swine kidney phosphatase	+	–	–	–
Pneumococcus autolysates	–	+	+	+
Normal dog and rabbit serum	+	+	+	+

Enzymatic Analysis Various crude and crystalline enzymes* have been tested for their capacity to destroy the biological activity of potent bacterial extracts. Extracts buffered at the optimal pH, to which were added crystalline trypsin and chymotrypsin or combinations of both, suffered no loss in activity following treatment with these enzymes. Pepsin could not be tested because extracts are rapidly inactivated at the low pH required for its use. Prolonged treatment with crystalline ribonuclease under optimal conditions caused no demonstrable decrease in transforming activity. The fact that trypsin, chymotrypsin, and ribonuclease had no effect on the transforming principle is further evidence that this substance is not ribonucleic acid or a protein susceptible to the action of tryptic enzymes.

In addition to the crystalline enzymes, sera and preparations of enzymes obtained from the organs of various animals were tested to determine their effect on transforming activity. Certain of these were found to be capable of completely destroying biological activity. The various enzyme preparations tested included highly active phosphates obtained from rabbit bone by the method of Martland and Robison[15] and from swine kidney as described by H. and E. Albers.[16] In addition, a preparation made from the intestinal mucosa of dogs by Levene and Dillon[17] and containing a polynucleotidase for thymus nucleic acid was used. Pneumococcal autolysates and a commercial preparation of pancreatin

were also tested. The alkaline phosphatase activity of these preparations was determined by their action on β-glycerophosphate and phenyl phosphate, and the esterase activity by their capacity to split tributyrin. Since the highly purified transforming material isolated from pneumococcal extracts was found to contain desoxyribonucleic acid, these same enzymes were tested for depolymerase activity on known samples of desoxyribonucleic acid isolated by Mirsky† from fish sperm and mammalian tissues. The results are summarized in Table 2 in which the phosphatase, esterase, and nucleodepolymerase activity of these enzymes is compared with their capacity to destroy the transforming principle. Analysis of these results shows that irrespective of the presence of phosphatase or esterase only those preparations shown to contain an enzyme capable of depolymerizing authentic samples of desoxyribonucleic acid were found to inactivate the transforming principle.

Greenstein and Jenrette[18] have shown that tissue extracts, as well as the milk and serum of several mammalian species, contain an enzyme system which causes depolymerization of desoxyribonucleic acid. To this enzyme system Greenstein has later given the name desoxyribonucleodepolymerase.[19] These investigators determined depolymerase activity by following the reduction in viscosity of solutions of sodium desoxyribonucleate. The nucleate and enzyme were mixed in the viscosimeter and viscosity measurements made at intervals during

*The authors are indebted to Dr. John H. Northrop and Dr. M. Kunitz of The Rockefeller Institute for Medical Research, Princeton, N. J., for the samples of crystalline trypsin, chymotrypsin, and ribonuclease used in this work.

†The authors express their thanks to Dr. A. E. Mirsky of the Hospital of The Rockefeller Institute for these preparations of desoxyribonucleic acid.

incubation at 30°C. In the present study this method was used in the measurement of depolymerase activity except that incubation was carried out at 37°C and, in addition to the reduction of viscosity, the action of the enzyme was further tested by the progressive decrease in acid precipitability of the nucleate during enzymatic breakdown.

The effect of fresh normal dog and rabbit serum on the activity of the transforming substance is shown in the following experiment.

Sera obtained from a normal dog and normal rabbit were diluted with an equal volume of physiological saline. The diluted serum was divided into three equal portions. One part was heated at 65°C for 30 minutes, another at 60°C for 30 minutes, and the third was used unheated as control.

A partially purified preparation of transforming material which had previously been dried in the lyophile apparatus was dissolved in saline in a concentration of 3.7 mg per cc. 1.0 cc of this solution was mixed with 0.5 cc of the various samples of heated and unheated diluted sera, and the mixtures at pH 7.4 were incubated at 37°C for 2 hours. After the serum had been allowed to act on the transforming material for this period, all tubes were heated at 65°C for 30 minutes to stop enzymatic action. Serial dilutions were then made in saline and tested in triplicate for transforming activity according to the procedure described under method of titration. The results given in Table 3 illustrate the differential heat inactivation of the enzymes in dog and rabbit serum which destroy the transforming principle.

From the data presented in Table 3 it is evident that both dog and rabbit serum in the unheated

Table 3 Differential heat inactivation of enzymes in dog and rabbit serum which destroy the transforming substance

	Heat treatment of serum	Dilution*	Triplicate tests					
			1		2		3	
			Diffuse growth	*Colony form*	*Diffuse growth*	*Colony form*	*Diffuse growth*	*Colony form*
Dog serum	Unheated	Undiluted	–	R only	–	R only	–	R only
		1:5	–	R only	–	R only	–	R only
		1:25	–	R only	–	R only	–	R only
	60°C for 30 min.	Undiluted	+	SIII	+	SIII	+	SIII
		1:5	+	SIII	+	SIII	+	SIII
		1:25	+	SIII	+	SIII	+	SIII
	65°C for 30 min.	Undiluted	+	SIII	+	SIII	+	SIII
		1:5	+	SIII	+	SIII	+	SIII
		1:25	+	SIII	+	SIII	+	SIII
Rabbit serum	Unheated	Undiluted	–	R only	–	R only	–	R only
		1:5	–	R only	–	R only	–	R only
		1:25	–	R only	–	R only	–	R only
	60°C for 30 min.	Undiluted	–	R only	–	R only	–	R only
		1:5	–	R only	–	R only	–	R only
		1:25	–	R only	–	R only	–	R only
	65°C for 30 min.	Undiluted	+	SIII	+	SIII	+	SIII
		1:5	+	SIII	+	SIII	+	SIII
		1:25	+	SIII	+	SIII	+	SIII
Control (no serum)	None	Undiluted	+	SIII	+	SIII	+	SIII
		1:5	+	SIII	+	SIII	+	SIII
		1:25	+	SIII	+	SIII	+	SIII

*Dilution of the digest mixture of serum and transforming substance.

state are capable of completely destroying transforming activity. On the other hand, when samples of dog serum which have been heated either at 60°C or at 65°C for 30 minutes are used, there is no loss of transforming activity. Thus, in this species the serum enzyme responsible for destruction of the transforming principle is completely inactivated at 60°C. In contrast to these results, exposure to 65°C for 30 minutes was required for complete destruction of the corresponding enzyme in rabbit serum.

The same samples of dog and rabbit serum used in the preceding experiment were also tested for their depolymerase activity on a preparation of sodium desoxyribonucleate isolated by Mirsky from shad sperm.

A highly viscous solution of the nucleate in distilled water in a concentration of 1 mg per cc was used. 1.0 cc amounts of heated and unheated sera diluted in saline as shown in the preceding protocol were mixed in Ostwald viscosimeters with 4.0 cc of the aqueous solution of the nucleate. Determinations of viscosity were made immediately and at intervals over a period of 24 hours during incubation at 37°C.

The results of this experiment are graphically presented in Chart 1. In the case of unheated serum of both dog and rabbit, the viscosity fell to that of water in 5 to 7 hours. Dog serum heated at 60°C for 30 minutes brought about no significant reduction in viscosity after 22 hours. On the other hand, heating rabbit serum at 60°C merely reduced the rate of depolymerase action, and after 24 hours the viscosity was brought to the same level as with the unheated serum. Heating at 65°C, however, completely destroyed the rabbit serum depolymerase.

Thus, in the case of dog and rabbit sera there is a striking parallelism between the temperature of inactivation of the depolymerase and that of the enzyme which destroys the activity of the transforming principle. The fact that this difference in temperature of inactivation is not merely a general property of all enzymes in the sera is evident from experiments on the heat inactivation of tributyrin esterase in the same samples of serum. In the latter instance, the results are the reverse of those observed with depolymerase since the esterase of rabbit serum is almost completely inactivated at 60°C while that in dog serum is only slightly affected by exposure to this temperature.

Of a number of substances tested for their capacity

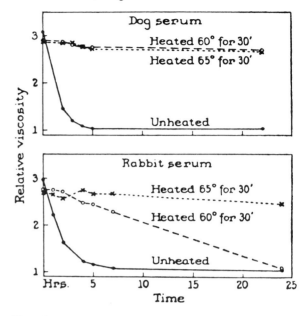

Differential Heat Inactivation of Desoxyribonucleodepolymerase of Dog and Rabbit Serum

Chart 1.

to inhibit the action of the enzyme known to destroy the transforming principle, only sodium fluoride has been found to have a significant inhibitory effect. Regardless of whether this enzyme is derived from pneumococcal cells, dog intestinal mucosa, pancreatin, or normal sera its activity is inhibited by fluoride. Similarly it has been found that fluoride in the same concentration also inhibits the enzymatic depolymerization of desoxyribonucleic acid.

The fact that transforming activity is destroyed only by those preparations containing depolymerase for desoxyribonucleic acid and the further fact that in both instances the enzymes concerned are inactivated at the same temperature and inhibited by fluoride provide additional evidence for the belief that the active principle is a nucleic acid of the desoxyribose type.

Serological Analysis In the course of chemical isolation of the active material it was found that as crude extracts were purified, their serological activity in Type III antiserum progressively decreased without corresponding loss in biological activity. Solutions of the highly purified substance itself gave only faint trace reactions in precipitin tests with high titer Type III antipneumococcus rabbit serum.* It is well known that pneumococcal protein can be detected by serological methods in dilutions as high as 1:50,000 and the capsular as well as the somatic polysaccharide in dilutions of at least 1:5,000,000. In view of these facts, the loss of serological reactivity indicates that these cell constituents have been almost completely removed from the final preparations. The fact that the transforming substance in purified state exhibits little or no serological reactivity is in striking contrast to its biological specificity in inducing pneumococcal transformation.

Physiochemical Studies[†] A purified and active preparation of the transforming substance (preparation 44) was examined in the analytical ultracentrifuge. The material gave a single and unusually sharp boundary indicating that the substance was homogeneous and that the molecules were uniform in size and very asymmetric. Biological activity was found to be sedimented at the same rate as the optically observed boundary, showing that activity could not be due to the presence of an entity much different in size. The molecular weight cannot be accurately determined until measurements of the diffusion constant and partial specific volume have been made. However, Tennent and Vilbrandt[20] have determined the diffusion constant of several preparations of thymus nucleic acid the sedimentation rate of which is in close agreement with the values observed in the present study. Assuming that the asymmetry of the molecules is the same in both instances, it is estimated that the molecular weight of the pneumococcal preparation is of the order of 500,000.

Examination of the same active preparation was carried out by electrophoresis in the Tiselius apparatus and revealed only a single electrophoretic component of relatively high mobility comparable to that of a nucleic acid. Transforming activity was associated with the fast-moving component giving the optically visible boundary. Thus in both the electrical and centrifugal fields, the behavior of the purified substance is consistent with the concept that biological activity is a property of the highly polymerized nucleic acid.

Ultraviolet absorption curves showed maxima in the region of 2600 Å and minima in the region of 2350 A. These findings are characteristic of nucleic acids.

Quantitative Determination of Biological Activity
In its highly purified state the material as isolated has been found to be capable of inducing transformation in amounts ranging from 0.02 to 0.003 μg. Preparation 44, the purification of which was carried out at low temperature and which had a nitrogen-phosphorus ratio of 1.58, exhibited high transforming activity. Titration of the activity of this preparation is given in Table 4.

A solution containing 0.5 mg per cc was serially diluted as shown in the protocol. 0.2 cc of each of these dilutions was added to quadruplicate tubes containing 2.0 cc of standard serum broth. All tubes were then inoculated with 0.05 cc of a 10^{-4} dilution of a 5- to 8-hour blood broth culture of R36A. Transforming activity was determined by the procedure described under method of titration.

The data presented in Table 4 show that on the basis of dry weight 0.003 μg of the active material brought about transformation. Since the reaction system containing the 0.003 μg has a volume of 2.25 cc, this represents a final concentration of the purified substance of 1 part in 600,000,000.

DISCUSSION

The present study deals with the results of an attempt to determine the chemical nature of the substance inducing specific transformation of pneumococcal types. A desoxyribonucleic acid fraction has been isolated from Type III pneumococci which is capable of transforming unencapsulated R variants derived from Pneumcoccus Type II into fully encapsulated Type III cells. Thompson and Dubos[21] have

---

*The Type III antipneumococcus rabbit serum employed in this study was furnished through the courtesy of Dr. Jules T. Freund, Bureau of Laboratories, Department of Health, City of New York.
†Studies on sedimentation in the ultracentrifuge were carried out by Dr. A. Rothen; the electrophoretic analyses were made by Dr. T. Shedlovsky, and the ultraviolet absorption curves by Dr. G. I. Lanvin. The authors gratefully acknowledge their indebtedness to these members of the staff of The Rockefeller Institute.

Table 4 Titration of transforming activity of preparation 44

| Transforming principle preparation 44* | | Quadruplicate tests | | | | | | | |
| | | 1 | | 2 | | 3 | | 4 | |
Dilution	Amount added (μg)	Diffuse growth	Colony form	Diffuse growth	Colony form	Diffuse growth	Colony form	Diffuse growth	Colony form
10^{-2}	1.0	+	SIII	+	SIII	+	SIII	+	SIII
$10^{-2.5}$	0.3	+	SIII	+	SIII	+	SIII	+	SIII
10^{-3}	0.1	+	SIII	+	SIII	+	SIII	+	SIII
$10^{-3.5}$	0.03	+	SIII	+	SIII	+	SIII	+	SIII
10^{-4}	0.01	+	SIII	+	SIII	+	SIII	+	SIII
$10^{-4.5}$	0.003	−	R only	+	SIII	−	R only	+	SIII
10^{-5}	0.001	−	R only	−	R only	−	R only	−	R only
Control	None	−	R only	−	R only	−	R only	−	R only

*Solution from which dilutions were made contained 0.5 mg per cc of purified material. 0.2 cc of each dilution added to quadruplicate tubes containing 2.0 cc of standard serum broth. 0.05 cc of a 10^{-4} dilution of a blood broth culture of R36A is added to each tube.

isolated from pneumococci a nucleic acid of the ribose type. So far as the writers are aware, however, a nucleic acid of the desoxyribose type has not heretofore been recovered from pneumococci nor has specific transformation been experimentally induced *in vitro* by a chemically defined substance.

Although the observations are limited to a single example, they acquire broader significance from the work of earlier investigators who demonstrated the interconvertibility of various pneumococcal types and showed that the specificity of the changes induced is in each instance determined by the particular type of encapsulated cells used to evoke the reaction. From the point of view of the phenomenon in general, therefore, it is of special interest that in the example studied, highly purified and protein-free material consisting largely, if not exclusively, of desoxyribonucleic acid is capable of stimulating unencapsulated R variants of Pneumococcus Type II to produce a capsular polysaccharide identical in type specificity with that of the cells from which the inducing substance was isolated. Equally striking is the fact that the substance evoking the reaction and the capsular substance produced in response to it are chemically distinct, each belonging to a wholly different class of chemical compounds.

The inducing substance, on the basis of its chemical and physical properties, appears to be a highly polymerized and viscous form of sodium desoxyribonucleate. On the other hand, the Type III capsular substance, the synthesis of which is evoked by this transforming agent, consists chiefly of a nonnitrogenous polysaccharide constituted of glucoseglucuronic acid units linked in glycosidic union.[22] The presence of the newly formed capsule containing this type-specific polysaccharide confers on the transformed cells all the distinguishing characteristics of Pneumococcus Type III. Thus, it is evident that the inducing substance and the substance produced in turn are chemically distinct and biologically specific in their action and that both are requisite in determining the type specificity of the cell of which they form a part.

The experimental data presented in this paper strongly suggest that nucleic acids, at least those of the desoxyribose type, possess different specificities as evidenced by the selective action of the transforming principle. Indeed, the possibility of the existence of specific differences in biological behavior of nucleic acids has previously been suggested[23,24] but has never been experimentally demonstrated owing in part at least to the lack of suitable biological methods. The techniques used in the study of transformation appear to afford a sensitive means of testing the validity of this hypothesis, and the results thus far obtained add supporting evidence in favor of this point of view.

If it is ultimately proved beyond reasonable doubt that the transforming activity of the material described is actually an inherent property of the nucleic acid, one must still account on a chemical basis for the biological specificity of its action. At first glance, immunological methods would appear to offer the ideal means of determining the differential specificity of this group of biologically important substances. Although the constituent units and general pattern of the nucleic acid molecule have been defined, there is as yet relatively little known of the possible effect that subtle differences in molecular configuration may exert on the biological specificity of these substances. However, since nucleic acids free or combined with histones or protamines are not known to function antigenically, one would not anticipate that such differences would be revealed by immunological techniques. Consequently, it is perhaps not surprising that highly purified and protein-free preparations of dexosyribonucleic acid, although extremely active in inducing transformation, showed only faint trace reactions in precipitin tests with potent Type III antipneumococcus rabbit sera.

From these limited observations it would be unwise to draw any conclusion concerning the immunological significance of the nucleic acids until further knowledge on this phase of the problem is available. Recent observations by Lackman and his collaborators[25] have shown that nucleic acids of both the yeast and thymus type derived from hemolytic streptococci and from animal and plant sources precipitate with certain antipneumococcal sera. The reactions varied with different lots of immune serum and occurred more frequently in antipneumococcal horse serum than in corresponding sera of immune rabbits. The irregularity and broad cross reactions encountered led these investigators to express some doubt as to the immunological significance of the results. Unless special immunochemical methods can be devised similar to those so successfully used in demonstrating the serological specificity of simple nonantigenic substances, it appears that the techniques employed in the study of transformation are the only ones available at the present for testing possible differences in the biological behavior of nucleic acids.

Admittedly there are many phases of the problem of transformation that require further study and many questions that remain unanswered largely because of technical difficulties. For example, it would be of interest to know the relation between rate of reaction and concentration of the transforming substance; the proportion of cells transformed to those that remain unaffected in the reaction system. However, from a bacteriological point of view, numerical estimations based on colony counts might prove more misleading than enlightening because of the aggregation and sedimentation of the R cells agglutinated by the antiserum in the medium. Attempts to induce transformation in suspensions of resting cells held under conditions inhibiting growth and multiplication have thus far proved unsuccessful, and it seems probable that transformation occurs only during active reproduction of the cells. Important in this connection is the fact that the R cells, as well as those that have undergone transformation, presumably also all other variants and types of pneumococci, contain an intracellular enzyme which is released during autolysis and in the free state is capable of rapidly and completely destroying the activity of the transforming agent. It would appear, therefore, that during the logarithmic phase of growth when cell division is most active and autolysis least apparent, the cultural conditions are optimal for the maintenance of the balance between maximal reactivity of the R cell and minimal destruction of the transforming agent through the release of autolytic ferments.

In the present state of knowledge any interpretation of the mechanism involved in transformation must of necessity be purely theoretical. The biochemical events underlying the phenomenon suggest that the transforming principle interacts with the R cell giving rise to a coordinated series of enzymatic reactions that culminate in the synthesis of the Type III capsular antigen. The experimental findings have clearly demonstrated that the induced alterations are not random changes but are predictable, always corresponding in type specificity to that of the encapsulated cells from which the transforming substance was isolated. Once transformation has occurred, the newly acquired characteristics are thereafter transmitted in series through innumerable transfers in artificial media without any further addition of the transforming agent. Moreover from the transformed cells themselves, a substance of identical activity can again be recovered in amounts far in excess of that originally added to induce the change. It is evident, therefore, that not only is the capsular material reproduced in successive generations but that the primary factor, which controls the occurrence and specificity of capsular development is also reduplicated in the daughter cells. The induced changes are not temporary modifications but are permanent alterations which persist provided the cultural conditions are favorable for the maintenance of capsule formation. The transformed cells can be readily distinguished from the parent R forms not alone by serological reactions but by the presence of a newly formed and visible capsule which is the immunological unit of type specificity and the accessory structure essential in determining the infective capacity of the microorganism in the animal body.

It is particularly significant in the case of pneumococci that the experimentally induced alterations are definitely correlated with the development of a new morphological structure and the consequent acquisition of new antigenic and invasive properties. Equally if not more significant is the fact that these changes are predictable, type-specific, and heritable.

Various hypotheses have been advanced in explanation of the nature of the changes induced. In his original description of the phenomenon Griffith[1] suggested that the dead bacteria in the inoculum might furnish some specific protein that serves as a "pabulum" and enables the R form to manufacture a capsular carbohydrate.

More recently the phenomenon has been interpreted from a genetic point of view.[26,27] The inducing substance has been likened to a gene, and the capsular antigen which is produced in response to it has been regarded as a gene product. In discussing the phenomenon of transformation Dobzhansky[27] has stated that "If this transformation is described as a genetic mutation — and it is difficult to avoid so describing it — we are dealing with authentic cases of induction of specific mutations by specific treatments. . . ."

Another interpretation of the phenomenon has been suggested by Stanley[28] who has drawn the analogy between the activity of the transforming agent and that of a virus. On the other hand, Murphy[29] has compared the causative agents of fowl tumors with the transforming principle of Pneumococcus. He has suggested that both these groups of agents be termed "transmissible mutagens" in order to differentiate them from the virus group. Whatever may prove to be the correct interpretation, these differences in viewpoint indicate the implica-

tions of the phenomenon of transformation in relation to similar problems in the fields of genetics, virology, and cancer research.

It is, of course, possible that the biological activity of the substance described is not an inherent property of the nucleic acid but is due to minute amounts of some other substance adsorbed to it or so intimately associated with it as to escape detection. If, however, the biologically active substance isolated in highly purified form as the sodium salt of desoxyribonucleic acid actually proves to be the transforming principle, as the available evidence strongly suggests, then nucleic acids of this type must be regarded not merely as structurally important but as functionally active in determining the biochemical activities and specific characteristics of pneumococcal cells. Assuming that the sodium desoxyribonucleate and the active principle are one and the same substance, then the transformation described represents a change that is chemically induced and specifically directed by a known chemical compound. If the results of the present study on the chemical nature of the transforming principle are confirmed, then nucleic acids must be regarded as possessing biological specificity the chemical basis of which is as yet undetermined.

SUMMARY

1. From Type III pneumococci a biologically active fraction has been isolated in highly purified form which in exceedingly minute amounts is capable under appropriate cultural conditions of inducing the transformation of unencapsulated R variants of Pneumococcus Type II into fully encapsulated cells of the same specific type as that of the heat-killed microorganisms from which the inducing material was recovered.

2. Methods for the isolation and purification of the active transforming material are described.

3. The data obtained by chemical, enzymatic, and serological analyses together with the results of preliminary studies by electrophoresis, ultracentrifugation, and ultraviolet spectroscopy indicate that, within the limits of the methods, the active fraction contains no demonstrable protein, unbound lipid, or serologically reactive polysaccharide and consists

principally, if not solely, of a highly polymerized, viscous form of desoxyribonucleic acid.

4. Evidence is presented that the chemically induced alterations in cellular structure and function are predictable, type-specific, and transmissible in series. The various hypotheses that have been advanced concerning the nature of these changes are reviewed.

CONCLUSION

The evidence presented supports the belief that a nucleic acid of the desoxyribose type is the fundamental unit of the transforming principle of Pneumococcus Type III.

BIBLIOGRAPHY

1. Griffith, F., *J. Hyg.,* Cambridge, Eng., 1928, 27:113.
2. Neufeld, F., and Levinthal, W., *Z. Immunitätsforsch.,* 1928, 55:324.
3. Baurhenn, W., *Centr. Bakt., 1. Abt., Orig.,* 1932, 126:68.
4. Dawson, M. H., *J. Exp. Med.,* 1930, 51:123.
5. Dawson, M. H., and Sia, R. H. P., *J. Exp. Med.,* 1931, 54:681.
6. Alloway, J. L., *J. Exp. Med.,* 1932, 55:91; 1933, 57:265.
7. Berry, G. P., and Dedrick, H. M., *J. Bact.,* 1936, 31:50.
8. Berry, G. P., *Arch. Path.,* 1937, 24:533.
9. Hurst, E. W., *Brit. J. Exp. Path.*, 1937, 18:23. Hoffstadt, R. E., and Pilcher, K. S., *J. Infect. Dis.,* 1941, 68:67. Gardner, R. E., and Hyde, R. R., *J. Infect. Dis.,* 1942, 71:47. Houlihan, R. B., *Proc. Soc. Exp. Biol. and Med.,* 1942, 51:259.
10. MacLeod, C. M., and Mirick, G. S., *J. Bact.,* 1942, 44:277.
11. Dawson, M. H., *J. Exp. Med.,* 1928, 47:577; 1930, 51:99.
12. Sevag, M. G., *Biochem. Z.,* 1934, 273:419.

Sevag, M. G., Lackman, D. B., and Smolens, J., *J. Biol. Chem.,* 1938, 124:425.
13. Dubos, R. J., and Avery, O. T., *J. Exp. Med.,* 1931, 54:51. Dubos, R. J., and Bauer, J. H., *J. Exp. Med.,* 1935, 62:271.
14. Liu, S., and Wu, H., *Chinese J. Physiol.,* 1938, 13:449.
15. Martland, M., and Robison, R., *Biochem. J.,* 1929, 23:237.
16. Albers, H., and Albers, E., *Z. physiol. Chem.,* 1935, 232:189.
17. Levene, P. A., and Dillon, R. T., *J. Biol. Chem.,* 1933, 96:461.
18. Greenstein, J. P., and Jenrette, W. Y., *J. Nat. Cancer Inst.,* 1940, 1:845.
19. Greenstein, J. P., *J. Nat. Cancer Inst.,* 1943, 4:55.
20. Tennent, H. G., and Vilbrandt, C. F., *J. Am. Chem. Soc.,* 1943, 65:424.
21. Thompson, R. H. S., and Dubos, R. J., *J. Biol. Chem.,* 1938, 125:65.
22. Reeves, R. E., and Goebel, W. F., *Biol. Chem.,* 1941, 139:511.
23. Schultz, J., in Genes and chromosomes. Structure and organization, *Cold Spring Harbor symposia on quantitative biology,* Cold Spring Harbor, Long Island Biological Association, 1941, 9:55.
24. Mirsky, A. E., in *Advances in enzymology and related subjects of biochemistry,* (F. F. Nord and C. H. Werkman, editors), New York: Interscience Publishers, Inc., 1943, 3:1.
25. Lackman, D., Mudd, S., Sevag, M. G., Smolens, J., and Wiener, M., *J. Immunol.,* 1941, 40:1.
26. Gortner, R. A., *Outlines of biochemistry,* New York, Wiley, 2nd edition, 1938, 547.
27. Dobzhansky, T., *Genetics and the origin of the species,* New York: Columbia University Press, 1941, 47.
28. Stanley, W. M., in Doerr, R., and Hallauer, C., *Handbuch der Virusforschung,* Vienna: Julius Springer, 1938, 1:491.
29. Murphy, J. B., *Tr. Assn. Am. Physn.,* 1931, 46:182; *Bull. Johns Hopkins Hosp.,* 1935, 56:1.

Chemical specificity of nucleic acids and mechanism of their enzymatic degradation*

ERWIN CHARGAFF†

Reprinted by author's and publisher's permission from Experimentia, *vol. 6, 1950, pp. 201–209.*

This is a crucial paper because it dramatically changed canonical thinking in respect to the amount of structural variability possible within DNA molecules. This paper paved the way for a more general acceptance of the implication of Hershey and Chase's experiments of 1952 — that DNA alone carried the genetic information. Chargaff showed that DNA from different sources had different base compositions, making any simple repetitive structure of DNA, like the tetranucleotide theory, impossible. Thus, we see this paper sitting at a pivotal point in the thinking about DNA. It is a good possibility that if this work had preceded that of Avery and his colleagues in 1944, researchers would have more carefully reviewed that work and subsequently realized its implications.

I. INTRODUCTION

The last few years have witnessed an enormous revival in interest for the chemical and biological properties of nucleic acids, which are components essential for the life of all cells. This is not particularly surprising, as the chemistry of nucleic acids represents one of the remaining major unsolved problems in biochemistry. It is not easy to say what provided the impulse for this rather sudden rebirth. Was it the fundamental work of E. Hammarsten[1] on the highly polymerized desoxyribonucleic acid of calf thymus? Or did it come from the biological side, for instance the experiments of Brachet[2] and Caspersson?[3] Or was it the very important research of Avery[4] and his collaborators on the transformation of pneumococcal types that started the avalanche?

It is, of course, completely senseless to formulate a hierarchy of cellular constituents and to single out certain compounds as more important than others. The economy of the living cell probably knows no conspicuous waste; proteins and nucleic acids, lipids and polysaccharides, all have the same importance. But one observation may be offered. It is impossible to write the history of the cell without considering its geography; and we cannot do this without attention to what may be called the chronology of the cell, i.e., the sequence in which the cellular constituents are laid down and in which they develop from each other. If this is done, nucleic acids will be found pretty much at the beginning. An attempt to say more leads directly into empty speculations in which almost no field abounds more than the chemistry of the cell. Since an ounce of proof still weighs more than a pound of prediction, the important genetic functions, ascribed — probably quite rightly — into the nucleic acids by many workers, will not be discussed here. Terms such as "template" or "matrix" or "reduplication" will not be found in this lecture.

II. IDENTITY AND DIVERSITY IN HIGH MOLECULAR CELL CONSTITUENTS

The determination of the constitution of a complicated compound, composed of many molecules of a number of organic substances, evidently requires the

*This article is based on a series of lectures given before the Chemical Societies of Zurich and Basle (June 29th and 30th, 1949), the Société de chimie biologique at Paris, and the Universities of Uppsala, Stockholm, and Milan.
†Department of Biochemistry, College of Physicians and Surgeons, Columbia University, New York. The author wishes to thank the *John Simon Guggenheim Memorial Foundation* for making possible his stay in Europe. The experimental work has been supported by a research grant from the *United States Public Health Service.*

exact knowledge of the nature and proportion of all constituents. This is true for nucleic acids as much as for proteins or polysaccharides. It is, furthermore, clear that the value of such constitutional determinations will depend upon the development of suitable methods of hydrolysis. Otherwise, substances representing an association of many chemical individuals can be described in a qualitative fashion only; precise decisions as to structure remain impossible. When our laboratory, more than four years ago, embarked upon the study of nucleic acids, we became aware of this difficulty immediately.

The state of the nucleic acid problem at that time found its classical expression in Levene's monograph.[5] (A number of shorter reviews, indicative of the development of our conceptions concerning the chemistry of nucleic acids, should also be mentioned.)[6] The old tetranucleotide hypothesis — it should never have been called a theory — was still dominant; and this was characteristic of the enormous sway that the organic chemistry of small molecules held over biochemistry. I should like to illustrate what I mean by one example. If in the investigation of a disaccharide consisting of two different hexoses we isolate 0.8 mole of one sugar and 0.7 mole of the other, this will be sufficient for the recognition of the composition of the substance, provided its molecular weight is known. The deviation of the analytical results from simple, integral proportions is without importance in that case. But this will not hold for high-molecular compounds in which variations in the proportions of their several components often will provide the sole indication of the occurrence of different compounds.

In attempting to formulate the problem with some exaggeration one could say: The validity of the identification of a substance by the methods of classical organic chemistry ends with the mixed melting point. When we deal with the extremely complex compounds of cellular origin, such as nucleic acids, proteins, or polysaccharides, a chemical comparison aiming at the determination of identity or difference must be based on the nature and the proportions of their constituents, on the sequence in which these constituents are arranged in the molecule, and on the type and the position of the linkages that hold them together. The smaller the number of components of such a high-molecular compound is, the greater is the difficulty of a decision. The occurrence of a very large number of different proteins was recognized early; no one to my

knowledge ever attempted to postulate a protein as a compound composed of equimolar proportions of 18 or 20 different amino acids. In addition, immunological investigations contributed very much to the recognition of the multiplicity of proteins. A decision between identity and difference becomes much more difficult when, as is the case with the nucleic acids, only few primary components are encountered. And when we finally come to high polymers, consisting of one component only, e.g., glycogen or starch, the characterization of the chemical specificity of such a compound becomes a very complicated and laborious task.

While, therefore, the formulation of the tetranucleotide conception appeared explainable on historical grounds, it lacked an adequate experimental basis, especially as regards "thymonucleic acid." Although only two nucleic acids, the desoxyribose nucleic acid of calf thymus and the ribose nucleic acid of yeast, had been examined analytically in some detail, all conclusions derived from the study of these substances were immediately extended to the entire realm of nature; a jump of a boldness that should astound a circus acrobat. This went so far that in some publications the starting material for the so-called thymonucleic acid was not even mentioned or that it was not thymus at all, as may sometimes be gathered from the context, but, for instance, fish sperm or spleen. The animal species that had furnished the starting material often remained unspecified.

Now the question arises: How different must complicated substances be, before we can recognize their difference? In the multiformity of its appearances nature can be primitive and it can be subtle. It is primitive in creating in a cell, such as the tubercle bacillus, a host of novel compounds, new fatty acids, alcohols, etc., that are nowhere else encountered. There, the recognition of chemical peculiarities is relatively easy. But in the case of the proteins and nucleic acids, I believe, nature has acted most subtly; and the task facing us is much more difficult. There is nothing more dangerous in the natural sciences than to look for harmony, order, regularity, before the proper level is reached. The harmony of cellular life may well appear chaotic to us. The disgust for the amorphous, the ostensibly anomalous — an interesting problem in the psychology of science — has produced many theories that shrank gradually to hypotheses and then vanished.

We must realize that minute changes in the nucleic

acid, e.g., the disappearance of one guanine molecule out of a hundred, could produce far-reaching changes in the geometry of the conjugated nucleoprotein; and it is not impossible that rearrangements of this type are among the causes of the occurrence of mutations.[7]

The molecular weight of the pentose nucleic acids, especially of those from animal tissue cells, is not yet known; and the problem of their preparation and homogeneity still is in a very sad state. But that the desoxypentose nucleic acids, prepared under as mild conditions as possible and with the avoidance of enzymatic degradation, represent fibrous structures of high molecular weight, has often been demonstrated. No agreement has as yet been achieved on the order of magnitude of the molecular weight, since the interpretation of physical measurements of largely asymmetric molecules still presents very great difficulties. But regardless of whether the desoxyribonucleic acid of calf thymus is considered as consisting of elementary units of about 35,000 which tend to associate to larger structures[8] or whether it is regarded as a true macromolecule of a molecular weight around 820,000,[9] the fact remains that the desoxypentose nucleic acids are high-molecular substances which in size resemble, or even surpass, the proteins. It is quite possible that there exists a critical range of molecular weights above which two different cells will prove unable to synthesize completely identical substances. The enormous number of diverse proteins may be cited as an example. *Duo non faciunt idem* is, with respect to cellular chemistry, perhaps an improved version of the old proverb.

III. PURPOSE

We started in our work from the assumption that the nucleic acids were complicated and intricate high polymers, comparable in this respect to the proteins, and that the determination of their structures and their structural differences would require the development of methods suitable for the precise analysis of all constituents of nucleic acids prepared from a large number of different cell types. These methods had to permit the study of minute amounts, since it was clear that much of the material would not be readily available. The procedures developed in our laboratory make it indeed possible to perform a complete constituent analysis on 2 to 3 mg of nucleic acid, and this in six parallel determinations.

The basis of the procedure is the partition chromatography on filter paper. When we started our experiments, only the qualitative application to amino acids was known.[10] But it was obvious that the high and specific absorption in the ultraviolet of the purines and pyrimidines could form the basis of a quantitative ultramicro method, if proper procedures for the hydrolysis of the nucleic acids and for the sharp separation of the hydrolysis products could be found.

IV. PREPARATION OF THE ANALYTICAL MATERIAL

If preparations of desoxypentose nucleic acids are to be subjected to a structural analysis, the extent of their contamination with pentose nucleic acid must not exceed 2 to 3%. The reason will later be made clearer; but I should like to mention here that all desoxypentose nucleic acids of animal origin studied by us so far were invariably found to contain much more adenine than guanine. The reverse appears to be true for the animal pentose nucleic acids: in them guanine preponderates. A mixture of approximately equal parts of both nucleic acids from the same tissue, therefore, would yield analytical figures that would correspond, at least as regards the p·rines, to roughly equimolar proportions. Should the complete purification — sometimes an extremely difficult task — prove impossible in certain cases, one could think of subjecting preparations of both types of nucleic acid from the same tissue specimen to analysis and of correcting the respective results in this manner. This, however, is an undesirable device and was employed only in some of the preparations from liver which will be mentioned later.

It is, furthermore, essential that the isolation of the nucleic acids be conducted in such a manner as to exclude their degradation by enzymes, acid, or alkali. In order to inhibit the desoxyribonucleases which require magnesium,[11] the preparation of the desoxypentose nucleic acids was carried out in the presence of citrate ions.[12] It would take us here too far to describe in detail the methods employed in our laboratory for the preparation of the desoxypentose nucleic acids from animal tissues. They represent in general a combination of many procedures, as described recently for the isolation of yeast desoxyribonucleic acid.[13] In this manner, the desoxypentose nucleic acids of thymus, spleen, liver, and also yeast were prepared. The corresponding compound from tubercle bacilli was isolated *via* the nucleoprotein.[14] The procedures leading to the preparation of desoxypentose nucleic acid from human sperm will soon be published.[15] All desoxypentose nucleic acids used in the analytical studies were prepared as the sodium salts (in one case the potassium salt was used); they were free of protein, highly polymerized, and formed extremely viscous solutions in water. They were homogeneous electrophoretically and showed a high degree of monodispersity in the ultracentrifuge.

The procedure for the preparation of pentose nucleic acids from animal tissues resembled, in its first stages, the method of Clarke and Schryver.[16] The details of the isolation procedures and related experiments on yeast ribonucleic acid are as yet unpublished. Commercial preparations of yeast ribonucleic acid also were examined following purification. As has been mentioned before, the entire problem of the preparation and homogeneity of the pentose nucleic acids, and even of the occurrence of only one type of pentose nucleic acid in the cell, urgently requires reexamination.

V. SEPARATION AND ESTIMATION OF PURINES AND PYRIMIDINES

Owing to the very unpleasant solubility and polar characteristics of the purines, the discovery of suitable solvent systems and the development of methods for their quantitative separation and estimation[17] presented a rather difficult problem in the solution of which Dr. Ernst Vischer had an outstanding part. The pyrimidines proved somewhat easier to handle. The choice of the solvent system for the chromatographic separation of purines and pyrimidines will, of course, vary with the particular problem. The efficiency of different solvent systems in effecting separation is illustrated schematically in Fig. 1. Two of the solvent systems listed there are suitable for the separation of the purines found in nucleic acids, i.e., adenine and quanine, namely (1) n-butanol, morpholine, diethylene glycol, water (column 5 in Fig. 1); and (2) n-butanol, diethylene glycol, water in a NH₃ atmosphere (column 11). The second system listed proved particularly convenient. The separation of the pyrimidines is carried out in aqueous butanol (column 1).

Following the separation, the location of the var-

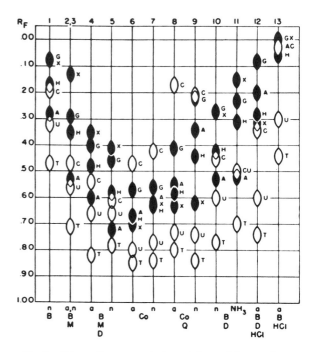

Fig. 1. Schematic representation of the position on the paper chromatogram of the purines and pyrimidines following the separation of a mixture. *A* adenine, *G* guanine, *H* hypoxanthine, *X* xanthine, *U* uracil, *C* cytosine, *T* thymine. The conditions under which the separations were performed are indicated at the bottom, *a* acidic, *n* neutral, *B* n-butanol, *M* morpholine, *D* diethylene glycol, *Co* collidine, *Q* quinoline. *(Taken from E. Vischer and E. Chargaff, J. Biol. Chem. 176: 704 [1948].)*

ious adsorption zones on the paper must be demonstrated. Our first attempts to bring this about in ultraviolet light were unsuccessful, probably because of inadequate filtration of the light emitted by the lamp then at our disposal. For this reason, the expedient was used in fixing the separated purines or pyrimidines on the paper as mercury complexes which then were made visible by their conversion to mercuric sulfide. The papers thus developed served as guide strips for the removal of the corresponding zones from untreated chromatograms that were then extracted and analyzed in the ultraviolet spectrophotometer. The development of the separated bases as mercury derivatives has, however, now become unnecessary, except for the preservation of permanent records, since there has for some time been available commercially an ultraviolet lamp emitting short wave ultraviolet ("Mineralight", Ultraviolet Products Corp., Los Angeles, California). With the help of this lamp it is now easy to demonstrate directly the position of the separated purines and pyrimidines (and also of nucleosides and nucleotides,[18] which appear as dark adsorption shadows on the background of the fluorescing filter paper and can be cut apart accordingly. (We are greatly indebted to Dr. C. E. Carter, Oak Ridge National Laboratory, who drew our attention to this instrument.)[19]

The extracts of the separated compounds are then studied in the ultraviolet spectrophotometer. The measurement of complete adsorption spectra permits the determination of the purity of the solutions and at the same time the quantitative estimation of their contents. The details of the procedures employed have been published.[20] In this manner, adenine, guanine, uracil, cytosine, and thymine (and also hypoxanthine, xanthine, and 5-methylcytosine)[21] can be determined quantitatively in amounts of 2–40 γ. The precision of the method is ±4% for the purines and even better for the pyrimidines, if the average of a large series of determinations are considered. In individual estimations the accuracy is about ±6%.

Procedures very similar in principle served in our laboratory for the separation and estimation of the ribonucleosides uridine and cytidine and for the separation of desoxyribothymidine from thymine. Methods for the separation and quantitative determination of the ribonucleotides in an aqueous ammonium isobutyrate-isobutyric acid system have likewise been developed.[22]

VI. METHODS OF HYDROLYSIS

It has long been known that the purines can be split off completely by a relatively mild acid hydrolysis of the nucleic acids. This could be confirmed in our laboratory in a more rigorous manner by the demonstration that heating at 100° for 1 hour in N sulfuric acid effects the quantitative liberation of adenine and quanine from adenylic and guanylic acids respectively.[23] The liberation of the pyrimidines, however, requires much more energetic methods of cleavage. Heating at high temperatures with strong mineral acid under pressure is usually resorted to. To what extent these procedures brought about the destruction of the pyrimidines, could not be ascer-

Table 1 Resistance of pyrimidines to treatment with strong acid. A mixture of pyrimidines of known concentration was dissolved in the acids indicated below and heated at 175° in a bomb tube. The concentration shifts of the individual pyrimidines were determined through a comparison of the recoveries of separated pyrimidines before and after the heating of the mixture.

Experiment no.	Acid	Heating time min.	Concentration shift, percent of starting concentration		
			Uracil	Cytosine	Thymine
1	HC1 (10%)	90	+62	−63	+3
2	10 N HCOOH {	60	+ 3	− 5	0
3	+N HCl	120	+24	−19	0
	(1:1)				
4	HCOOH (98 {	60	0	− 1	−2
5	to 100%)	120	0	+ 2	+1

tained previously owing to the lack of suitable analytical procedures. The experiments summarized in Table 1, which are quoted from a recent paper,[23] show that the extremely robust cleavage methods with mineral acids usually employed must have led to a very considerable degradation of cytosine to uracil. Uracil and also thymine are much more resistant. For this reason, we turned to the hydrolysis of the pyrimidine nucleotides by means of concentrated formic acid. For the liberation of the purines N sulfuric acid (100°, 1 hour) is employed; for the liberation of the pyrimidines, the purines are first precipitated as the hydrochlorides by treatment with dry HCl gas in methanol and the remaining pyrimidine nucleotides cleaved under pressure with concentrated formic acid (175°, 2 hours). This procedure proved particularly suitable for the investigation of the desoxypentose nucleic acids. For the study of the composition of pentose nucleic acids a different procedure, making use of the separation of the ribonucleotides, was developed more recently, which will be mentioned later.

VII. COMPOSITION OF DESOXYPENTOSE NUCLEIC ACIDS

It should be stated at the beginning of this discussion that the studies conducted thus far have yielded no indication of the occurrence in the nucleic acids examined in our laboratory of unusual nitrogenous constituents. In all desoxypentose nucleic acids investigated by us the purines were adenine and guanine, the pyrimidines cytosine and thymine. The occurrence in minute amounts of other bases, e.g., 5-methylcytosine, can, however, not yet be excluded. In the pentose nucleic acids uracil occurred instead of thymine.

A survey of the composition of desoxyribose nucleic acid extracted from several organs of the ox is provided in Table 2. The molar proportions reported in each case represent averages of several hydrolysis experiments. The composition of desoxypentose nucleic acids from human tissues is similarly illustrated in Table 3. The preparations from human liver were obtained from a pathological specimen in which it was possible, thanks to the kind cooperation of M. Faber, to separate portions of unaffected hepatic tissue from carcinomatous tissue consisting of metastases from the sigmoid colon, previous to the isolation of the nucleic acids.[24]

In order to show examples far removed from mammalian organs, the composition of two desoxyribonucleic acids of microbial origin, namely from yeast[25] and from avian tubercle bacilli,[26] is summarized in Table 4.

The very far-reaching differences in the composition of desoxypentose nucleic acids of different species are best illustrated by a comparison of the ratios of adenine to guanine and of thymine to cytosine as given in Table 5. It will be seen that in all cases where enough material for statistical analysis was available highly significant differences were found. The analytical figures on which Table 5 is based were derived by comparing the ratios found for individual nucleic acid hydrolysates of one species regardless of the organ from which the preparation was isolated. This procedure assumes that there is no organ specificity with respect to the composition of desoxypentose nucleic acids of the same species. That this appears indeed to be the case may be gathered from Tables 2 and 3 and even better from Table 6 where the average purine and pyrimidine ratios in individual tissues of the same species are compared. That the isolation of nucleic acids did not entail an appreciable fractionation is shown by the finding that when whole defatted human spermatozoa, after being washed with cold 10% trichloroacetic acid, were analyzed, the same ratios of adenine to guanine and of thymine to cytosine were found as are reported in Tables 5 and 6. It

Table 2 Composition of desoxyribonucleic acid of ox (in moles of nitrogenous constituent per mole of P).

Constituent	Thymus			Spleen		Liver
	Prep. 1	Prep. 2	Prep. 3	Prep. 1	Prep. 2	
Adenine	0.26	0.28	0.30	0.25	0.26	0.26
Guanine	0.21	0.24	0.22	0.20	0.21	0.20
Cytosine	0.16	0.18	0.17	0.15	0.17	
Thymine	0.25	0.24	0.25	0.24	0.24	
Recovery	0.88	0.94	0.94	0.84	0.88	

From E. Chargaff, E. Vischer, R. Doniger, C. Green, and F. Misani, J. Biol. Chem. 179: 405 (1949); and unpublished results.

Table 3 Composition of desoxypentose nucleic acid of man (in moles of nitrogenous constituent per mole of P).

Constituent	Sperm		Thymus	Liver	
	Prep. 1	Prep. 2		Normal	Carcinoma
Adenine	0.29	0.27	0.28	0.27	0.27
Guanine	0.18	0.17	0.19	0.19	0.18
Cytosine	0.18	0.18	0.16		0.15
Thymine	0.31	0.30	0.28		0.27
Recovery	0.96	0.92	0.91		0.87

From E. Chargaff, S. Zamenhof, and C. Green, Nature (in press); and unpublished results.

Table 4 Composition of two microbial desoxyribonucleic acids.

Constituent	Yeast		Avian tubercle bacilli
	Prep. 1	Prep. 2	
Adenine	0.24	0.30	0.12
Guanine	0.14	0.18	0.28
Cytosine	0.13	0.15	0.26
Thymine	0.25	0.29	0.11
Recovery	0.76	0.92	0.77

From E. Vischer, S. Zamenhof, and E. Chargaff, J. Biol. Chem. 177, 429 (1949); and unpublished results.

Table 5 Molar proportions of purines and pyrimidines in desoxypentose nucleic acids from different species

Species	Number of different organs	Number of different preparations	Adenine/Guanine			Thymine/Cytosine		
			Number of hydrolyses‡	Mean ratio	Standard error	Number of hydrolyses‡	Mean ratio	Standard error
Ox*	3	7	20	1.29	0.013	6	1.43	0.03
Man†	2	3	6	1.56	0.008	5	1.75	0.03
Yeast	1	2	3	1.72	0.02	2	1.9	
Avian tubercles bacillus	1	1	2	0.4		1	0.4	

*Preparations from thymus, spleen, and liver served for the purine determinations, the first two organs for the estimation of pyrimidines.

†Preparations from spermatozoa and thymus were analyzed.

‡In each hydrolysis between 12 and 24 determinations of individual purines and pyrimidines were performed.

Table 6 Molar proportions of purines and pyrimidines in desoxypentose nucleic acids from different organs of one species.

Species	Organ	Adenine/ Guanine	Thymine/ Cytosine
Ox	Thymus	1.3	1.4
	Spleen	1.2	1.5
	Liver	1.3	
Man	Thymus	1.5	1.8
	Sperm	1.6	1.7
	Liver (normal)	1.5	1.8
	Liver (carcinoma)	1.5	1.8

should also be mentioned that all preparations, with the exception of those from human liver, were derived from pooled starting material representing a number, and in the case of human spermatozoa a very large number, of individuals.

The desoxypentose nucleic acids extracted from different species thus appear to be different substances or mixtures of closely related substances of a composition constant for different organs of the same species and characteristic of the species.

The results serve to disprove the tetranucleotide hypothesis. It is, however, noteworthy — whether this is more than accidental, cannot yet be said — that in all desoxypentose nucleic acids examined thus far the molar ratios of total purines to total pyrimidines, and also of adenine to thymine and of guanine to cytosine, were not far from 1.

VIII. COMPOSITION OF PENTOSE NUCLEIC ACIDS

Here a sharp distinction must be drawn between the prototype of all pentose nucleic acid investigations — the ribonucleic acid of yeast — and the pentose nucleic acids of animal cells. Nothing is known as yet about bacterial pentose nucleic acids. In view of the incompleteness of our information on the homogeneity of pentose nucleic acids, which I have stressed before, I feel that the analytical results on these preparations do not command the same degree of confidence as do those obtained for the desoxypentose nucleic acids.

Three procedures, to which reference is made in Tables 7 and 8, were employed in our laboratory for the analysis of pentose nucleic acids. In *Procedure 1,* the pentose nucleic acid was hydrolysed to the nucleotide stage with alkali, at pH 13.5 and 30°, and the nucleotides, following adjustment to about pH 5, separated by chromatography with aqueous ammonium isobutyrate-isobutyric acid as the solvent. Under these conditions, guanylic acid shares its position on the chromatogram with uridylic acid; but it is possible to determine the concentrations of the two components in the eluates by simultaneous equations based on the ultraviolet absorption of the pure nucleotides.[27] The very good recoveries of nucleotides obtained in terms of both nucleic acid phosphorus and nitrogen show the cleavage by mild alkali treatment of pentose nucleic acids to be practically quantitative. In *Procedure 2*, the purines are first liberated by gaseous HCl in dry methanol and the evaporation residue of the reaction mixture is adjusted to pH 13.5 and then treated as in Procedure 1. In this manner, uridylic and cytidylic acids, adenine and guanine are separated and determined on one chromatogram. The determinations of free purines and pyrimidines in acid hydrolysates of pentose nucleic acids, following the methods outlined before for the desoxypentose nucleic acids, are listed as *Procedure 3*. It will be seen that it is mainly uracil which in this procedure escapes quantitative determination. This is due to the extreme refractoriness of uridylic acid to complete hydrolysis by acids, a large portion remaining partially unsplit as the nucleoside uridine. As matters stand now, I consider the values for purines yielded by Procedures 1 and 3 and those for pyrimidines found by Procedures 1 and 2 as quite reliable.

A survey of the composition of yeast ribonucleic acid is provided in Table 7. Preparations 1 and 2, listed in this table, were commercial preparations that had been purified in our laboratory and had been subjected to dialysis; Preparation 3 was isolated from baker's yeast by B. Magasanik in this laboratory by procedures similar to those used for the preparation of pentose nucleic acids from animal tissues and had not been dialyzed. It will be seen that the results are quite constant and not very far from the proportions required by the presence of equimolar quantities of all four nitrogenous constituents.

Table 8* Composition of pentose nucleic acids from animal tissues.

Constituent	Calf liver	Ox liver	Sheep liver	Pig liver	Pig pancreas
Guanylic acid	16.3	14.7	16.7	16.2	22.5
Adenylic acid	10	10	10	10	10
Cytidylic acid	11.1	10.9	13.4	16.1	9.8
Uridylic acid	5.3	6.6	5.6	7.7	4.6
Purines: pyrimidines	1.6	1.4	1.4	1.1	2.5

*Unpublished results.

An entirely different picture, however, was encountered when the composition of pentose nucleic acids from animal cells was investigated. A preliminary summary of the results, in all cases obtained by *Procedure 1,* is given in Table 8. Here guanylic acid was the preponderating nucleotide followed, in this order, by cytidylic and adenylic acids; uridylic acid definitely was a minor constituent. This was true not only of the ribonucleic acid of pancreas which has been known to be rich in guanine,[28] but also of all pentose nucleic acids isolated by us from the livers of three different species (Table 8).

In the absence of a truly reliable standard method for the isolation of pentose nucleic acid from animal tissue, generalizations are not yet permitted; but it would appear that pentose nucleic acids from the same organ of different species are more similar to each other at least in certain respects (e.g., the ratio of guanine to adenine), than are those from different organs of the same species. (Compare the pentose nucleic acids from the liver and the pancreas of pig in Table 8.)

Table 7 Composition of yeast ribonucleic acid (in moles of nitrogenous constituent per mole of P).

Constituent	Preparation 1			Preparation 2			Preparation 3		
	Procedure 1	Procedure 2	Procedure 3	Procedure 1	Procedure 2	Procedure 3	Procedure 1	Procedure 2	Procedure 3
Adenylic acid	0.29	0.26	0.26	0.27		0.24	0.25	0.23	0.24
Guanylic acid	0.28	0.29	0.26	0.25		0.25	0.26	0.28	0.26
Cytidylic acid	0.18	0.17	0.24	0.20			0.21	0.21	
Uridylic acid	0.20	0.20	0.08	0.18	0.19		0.20	0.25	
Recovery	0.95	0.92	0.84	0.90			0.92	0.97	

From E. Vischer and E. Chargaff, J. Biol. Chem. *176*, 715 (1948). E. Chargaff, B. Magasanik, R. Doniger, and E. Vischer, J. Amer. Chem. Soc. *71*, 1513 (1949); and unpublished results.

IX. SUGAR COMPONENTS

It is deplorable that such designations as desoxyribose and ribose nucleic acids continue to be used as if they were generic terms. Even the "thymus nucleic acid of fish sperm" is encountered in the literature. As a matter of fact, only in a few cases have the sugars been identified, namely, d-2-desoxyribose as a constituent of the guanine and thymine nucleosides of the desoxypentose nucleic acid from calf thymus, D-ribose as a constituent of the pentose nucleic acids from yeast, pancreas, and sheep liver.

Since the quantities of novel nucleic acids usually will be insufficient for the direct isolation of their sugar components, we attempted to employ the very sensitive procedure of the filter paper chromatography of sugars[29] for the study of the sugars isolated from minute quantities of nucleic acids. It goes without saying that identifications based on behavior in adsorption or partition are by no means as convincing as the actual isolation, but they will at least permit a tentative classification of new nucleic acids. Thus far the pentose nucleic acids of pig pancreas[30] and of the avian tubercle bacillus[31] have been shown to contain ribose, the desoxypentose nucleic acids of ox spleen,[32] yeast and avian tubercle bacilli[33] desoxyribose. It would seem that the free play with respect to the variability of components that nature permits itself is extremely restricted, where nucleic acids are concerned.

X. DEPOLYMERIZING ENZYMES

Enzymes capable of bringing about the depolymerization of both types of nucleic acids have long been known; but it is only during the last decade that crystalline ribonuclease[34] and desoxyribonuclease[35] from pancreas have become available thanks to the work of Kunitz. Important work on the latter enzyme was also done by McCarty.[36]

We were, of course, interested in applying the chromatographic micromethods for the determination of nucleic acid constituents to studies of enzymatic reaction mechanisms for which they are particularly suited. The action of crystalline desoxyribonuclease on calf thymus desoxyribonucleic acid resulted in the production of a large proportion of dialyzable fragments (53 percent of the total after 6 hours digestion), without liberation of ammonia or inorganic phosphate. But even after extended digestion there remained a nondialyzable core whose composition showed a significant divergence from both the original nucleic acid and the bulk of the dialyzate.[37] The preliminary findings summarized in Table 9 indicate a considerable increase in the molar proportions of adenine to guanine and especially to cytosine, of thymine to cytosine, and of purines to pyrimidines. This shows that the dissymmetry in the distribution of constituents found in the original nucleic acid (Table 2), is intensified in the core. The most plausible explanations of this interesting phenomenon, the study of which is being continued, are that the preparations consisted of more than one desoxypentose nucleic acid or that the nucleic contained in its chain clusters of nucleotides (relatively richer in adenine and thymine) that were distinguished from the bulk of the molecule by greater resistance to enzymatic disintegration.

In this connection another study, carried out in collaboration with S. Zamenhof, should be mentioned briefly that dealt with the desoxypentose nuclease of yeast cells.[38] This investigation afforded a possibility of exploring the mechanisms by which

Table 9 Enzymatic degradation of calf thymus desoxyribonucleic acid.

	Digestion hours	Dialysis hours	Distribution of fractions % of original	Composition of fractions (molar proportions)			
				Adenine Guanine	Thymine Cytosine	Adenine Cytosine	Pyrimidines Purines
Original	0	0	100	1.2	1.3	1.6	1.2
Dialysate	6	6	53	1.2	1.2	1.2	1.0
Dialysis residue	24	72	7	1.6	2.2	3.8	2.0

From S. Zamenhof and E. Chargaff, J. Biol. Chem. *178:* 531 (1949).

an enzyme concerned with the disintegration of desoxypentose nucleic acid is controlled in the cell. Our starting point again was the question of the specificity of desoxypentose nucleic acids; but the results were entirely unexpected. Since we had available a number of nucleic acids from different sources, we wanted to study a pair of desoxypentose nucleic acids as distant from each other as possible, namely that of the ox and that of yeast, and to investigate the action on them of the two desoxypentose nucleases from the same cellular sources. The desoxyribonuclease of ox pancreas has been thoroughly investigated, as was mentioned before. Nothing was known, however, regarding the existence of a yeast desoxypentose nuclease.

It was found that fresh salt extracts of crushed cells contained such an enzyme in a largely inhibited state, due to the presence of a specific inhibitor protein. This inhibitor specifically inhibited the desoxypentose nuclease from yeast, but not that from other sources, such as pancreas. The yeast enzyme depolymerized the desoxyribose nucleic acids of yeast and of calf thymus, which differ chemically, as I have emphasized before, at about the same rate. In other words, the enzyme apparently exhibited inhibitor specificity, but not substrate specificity. It is very inviting to assume that such relations between specific inhibitor and enzyme, in some ways reminiscent of immunological reactions, are of more general biological significance. In any event, a better understanding of such systems will permit an insight into the delicate mechanisms through which the cell manages the economy of its life, through which it maintains its own continuity and protects itself against agents striving to transform it.

XI. CONCLUDING REMARKS

Generalizations in science are both necessary and hazardous; they carry a semblance of finality which conceals their essentially provisional character; they drive forward, as they retard; they add, but they also take away. Keeping in mind all these reservations, we arrive at the following conclusions. The desoxypentose nucleic acids from animal and microbial cells contain varying proportions of the same four nitrogenous constituents, namely, adenine, guanine, cytosine, thymine. Their composition appears to be characteristic of the species, but not of the tissue, from which they are derived. The presumption, therefore, is that there exists an enormous number of structurally different nucleic acids; a number, certainly much larger than the analytical methods available to us at present can reveal.

It cannot yet be decided, whether what we call the desoxypentose nucleic acid of a given species is one chemical individual, representative of the species as a whole, or whether it consists of a mixture of closely related substances, in which case the constancy of its composition merely is a statistical expression of the unchanged state of the cell. The latter may be the case if, as appears probable, the highly polymerized desoxypentose nucleic acids form an essential part of the hereditary processes; but it will be understood from what I said at the beginning that a decision as to the identity of natural high polymers often still is beyond the means at our disposal. This will be particularly true of substances that differ from each other only in the sequence, not in the proportion, of their constituents. The number of possible nucleic acids having the same analytical composition is truly enormous. For example, the number of combinations exhibiting the same molar proportions of individual purines and pyrimidines as the desoxyribonucleic acid of the ox is more than 10^{56}, if the nucleic acid is assumed to consist of only 100 nucleotides; if it consists of 2,500 nucleotides, which probably is much nearer the truth, then the number of possible "isomers" is not far from 10^{1500}.

Moreover, desoxypentose nucleic acids from different species differ in their chemical composition, as I have shown before; and I think there will be no objection to the statement that, as far as chemical possibilities go, they could very well serve as one of the agents, or possibly as the agent, concerned with the transmission of inherited properties. It would be gratifying if one could say — but this is for the moment no more than an unfounded speculation — that just as the desoxypentose nucleic acids of the nucleus are species–specific and concerned with the maintenance of the species, the pentose nucleic acids of the cytoplasm are organ–specific and involved in the important task of differentiation.

I should not want to close without thanking my colleagues who have taken part in the work discussed here; they are, in alphabetical order, Miss R.

Doniger, Mrs. C. Green, Dr. B. Magasanik, Dr. E. Vischer, and Dr. S. Zamenhof.

REFERENCES

1. E. Hammarsten, Biochem Z. *144,* 383 (1924).
2. J. Brachet in *Nucleic Acid,* Symposia Soc. Exp. Biol. No. 1 (Cambridge University Press, 1947), p. 207. Cp. J. Brachet, in *Nucleic Acids and Nucleoproteins,* Cold Spring Harbor Symp. Quant. Biol. *12,* 18 (Cold Spring Harbor, N.Y., 1947).
3. T. Caspersson, in *Nucleic Acid,* Symp. Soc. Exp. Biol., No. 1 (Cambridge University Press, 1947), p. 127.
4. O. T. Avery, C. M. MacLeod, and M. McCarty, J. Exp. Med. *12,* 137 (1944).
5. P. A. Levene and L. W. Bass, Nucleic Acids (Chemical Catalog Co., New York, 1931).
6. H. Bredereck, Fortschritte der Chemie organischer Naturstoffe *1*:121 (1938). — F. G. Fischer, Naturwissensch. 30:377 (1942). — R. S. Tipson, Adv. Carbohydrate Chem. *1*:193 (1945). — J. M. Gulland, G. R. Barker, and D. O. Jordan, Ann. Rev. Biochem. *14*:175 (1945). — E. Chargaff and E. Vischer, Ann. Rev. Biochem. *17*: 201 (1948). — F. Schlenk, Adv. Enzymol. *9*:455 (1949).
7. For additional remarks on this problem, compare E. Chargaff, in *Nucleic Acids and Nucleoproteins,* Cold Spring Harbor Symp. Quant. Biol. *12*:28 (Cold Spring Harbor, N.Y., 1947).
8. E. Hammarsten, Acta med. Scand. Suppl. *196*: 634 (1947). — G. Jungner, I. Jungner, and L. G. Allgén, Nature *163*:849 (1949).
9. R. Cecil and A. G. Ogston, J. Chem. Soc. 1382 (1948).
10. R. Consden, A. H. Gordon, and A. J. P. Martin, Biochem J. *38*: 224 (1944).
11. F. G. Fischer, I. Böttger, and H. Lehmann–Echternacht, Z. physiol. Chem. *271*:246 (1941).
12. M. McCarty, J. Gen. Physiol. *29*:123 (1946).
13. E. Chargaff and S. Zamenhof, J. Biol. Chem. *173*:327 (1948).
14. E. Chargaff and H. F. Saidel, J. Biol. Chem. *177*:417 (1949).
15. S. Zamenhof, L. B. Shettles, and E. Chargaff, Nature (in press).
16. G. Clarke and S. B. Schryver, Biochem. J. *11*: 319 (1917).
17. E. Vischer and E. Chargaff, J. Biol. Chem. *168*: 781 (1947); *176*:703 (1948).
18. E. Chargaff, B. Magasanik, R. Doniger, and E. Vischer. J. Amer. Chem. Soc. *71*:1513 (1949).
19. A similar arrangement was recently described by E. R. Holiday and E. A. Johnson, Nature *163*:216 (1949).
20. E. Vischer and E. Chargaff, J. Biol. Chem. *176*: 703 (1948).
21. J. Kream and E. Chargaff, unpublished experiments.
22. E. Vischer, B. Magasanik, and E. Chargaff, Federation Proc. *8*:263 (1949). — E. Chargaff, B. Magasanik, R. Doniger, and E. Vischer, J. Amer. Chem. Soc. *71*:1513 (1949).
23. E. Vischer and E. Chargaff, J. Biol. Chem. *176*: 715 (1948).
24. Unpublished experiments.
25. E. Chargaff and S. Zamenhof, J. Biol. Chem. *173*:327 (1948).
26. E. Chargaff and H. F. Saidel, J. Biol. Chem. *177*:417 (1949).
27. E. Vischer, B. Magasanik, and E. Chargaff, Federation Proc. *8*:263 (1949). — E. Chargaff, B. Magasanik, R. Doniger, and E. Vischer, J. Amer. Chem. Soc. *71*:1513 (1949).
28. E. Hammarsten, Z. physiol. Ch. *109*:141 (1920). P. A. Levene and E. Jorpes, J. Biol. Chem. *86*:389 (1930). E. Jorpes, Biochem. J. *28*:2102 (1934).
29. S. M. Partridge, Nature *158*:270 (1946). -- S. M. Partridge and R. G. Westall, Biochem. J. *42*:238 (1948). — E. Chargaff, C. Levine, and C. Green, J. Biol. Chem. *175*:67 (1948).
30. E. Vischer and E. Chargaff, J. Biol. Chem. *176*: 715 (1948).
31. E. Vischer, S. Zamenhof, and E. Chargaff, J. Biol. Chem. *177*:429 (1949).
32. E. Chargaff, E. Vischer, R. Doniger, C. Green, and F. Misani, J. Biol. Chem. *177*:405 (1949).
33. E. Vischer, S. Zamenhof, and E. Chargaff, J. Biol. Chem. *177*:429 (1949).
34. M. Kunitz, J. Gen. Physiol. *24*:15 (1940).
35. M. Kunitz, Science *108*:19 (1948).
36. M. McCarty, J. Gen. Physiol. *29*:123 (1946).
37. S. Zamenhof and E. Chargaff, J. Biol Chem. *178*:531 (1949).
38. S. Zamenhof and E. Chargaff, Science *108*:628 (1948); J. Biol. Chem. *180*:727 (1949).

Independent functions of viral protein and nucleic acid in growth of bacteriophage*

A. D. HERSHEY and MARTHA CHASE†

Reprinted by authors' and publisher's permission from Journal of General Physiology, *vol. 36(1), 1952, pp. 39–56.*

This paper, like Avery's, has also become a classic. It can be said to have launched the mushrooming study of molecular genetics. The paper's main strength lies in its demonstration that it is the DNA, and not the protein, that directs the assembly of progeny virus particles during the life cycle of the bacterial virus. There are also other interesting features to note here.

First, a desire to more completely understand the role of proteins and nucleic acids in the intracellular phase of the T2 life cycle provided the main impetus for Hershey and Chase's research. Thus, the authors' intellectual antecedents do not directly include those investigators whose work we have already discussed.

A second interesting feature is the data this paper presents. (For example, look carefully at Figure 1, page 45.) The work is not as equivocal as that done by Avery and his coworkers. It shows only that about 80 percent of the T2 DNA went inside the host cell while 20 percent remained outside. In addition, it shows that 90 percent of the T2 protein was recovered outside the host cell while 10 percent was unaccounted for.

While these and other issues of the research reported here require careful thinking, they in no way detract from the basic findings. We may ask, however, if some protein *is* injected into a host cell by T2, what function might it serve?

The work of Doermann (1948), Doermann and Dissoway (1949), and Anderson and Doermann (1952) has shown that bacteriophages T2, T3, and T4 multiply in the bacterial cell in a noninfective form. The same is true of the phage carried by certain lysogenic bacteria (Lwoff and Gutmann 1950). Little else is known about the vegetative phase of these viruses. The experiments reported in this paper show that one of the first steps in the growth of T2 is the release from its protein coat of the nucleic acid of the virus particle, after which the bulk of the sulfur-containing protein has no further function.

Materials and Methods Phage T2 means in this paper the variety called T2H (Hershey 1946); T2h means one of the host range mutants of T2; UV-phage means phage irradiated with ultraviolet light from a germicidal lamp (General Electric Co.) to a fractional survival of 10^{-5}.

Sensitive bacteria means a strain (H) of *Escherichia coli* sensitive to T2 and its *h* mutant; resistant bacteria B/2 means a strain resistant to T2 but sensitive to its *h* mutant; resistant bacteria B/2h means a strain resistant to both. These bacteria do not adsorb the phages to which they are resistant.

"Salt-poor" broth contains per liter 10 gm bacto-peptone, 1 gm glucose, and 1 gm NaCl. "Broth" contains, in addition, 3 gm bacto-beef extract and 4 gm NaCl.

Glycerol-lactate medium contains per liter 70 mM sodium lactate, 4 gm glycerol, 5 gm NaCl, 2 gm KCl, 1 gm NH$_4$Cl, 1 mM MgCl$_2$, 0.1 mM CaCl$_2$, 0.01 gm gelatin, 10 mg P (as orthophosphate), and 10 mg S (as MgSO$_4$), at pH 7.0.

*This investigation was supported in part by a research grant from the National Microbiological Institute of the National Institutes of Health, Public Health Service. Radioactive isotopes were supplied by the Oak Ridge National Laboratory on allocation from the Isotopes Division, United States Atomic Energy Commission.
†From the Department of Genetics, Carnegie Institution of Washington, Cold Spring Harbor, Long Island. Received for publication, April 9, 1952.

Adsorption medium contains per liter 4 gm NaCl, 5 gm K_2SO_4, 1.5 gm KH_2PO_4, 3.0 gm Na_2HPO_4, 1 mM $MgSO_4$, 0.1 mM $CaCl_2$, and 0.01 gm gelatin, at pH 7.0.

Veronal buffer contains per liter 1 gm sodium diethyl-barbiturate, 3 mM $MgSO_4$, and 1 gm gelatin, at pH 8.0.

The HCN referred to in this paper consists of molar sodium cyanide solution neutralized when needed with phosphoric acid.

Adsorption of isotope to bacteria was usually measured by mixing the sample in adsorption medium with bacteria from 18-hour broth cultures previously heated to 70°C for 10 minutes and washed with adsorption medium. The mixtures were warmed for 5 minutes at 37°C, diluted with water, and centrifuged. Assays were made of both sediment and supernatant fractions.

Precipitation of isotope with antiserum was measured by mixing the sample in 0.5 percent saline with about 10^{11} per ml of nonradioactive phage and slightly more than the least quantity of antiphage serum (final dilution 1:160) that would cause visible precipitation. The mixture was centrifuged after 2 hours at 37°C.

Tests with DNase (desoxyribonuclease) were performed by warming samples diluted in veronal buffer for 15 minutes at 37°C with 0.1 mg per ml of crystalline enzyme (Worthington Biochemical Laboratory).

Acid-soluble isotope was measured after the chilled sample had been precipitated with 5 percent trichloracetic acid in the presence of 1 mg/ml of serum albumin, and centrifuged.

In all fractionations involving centrifugation, the sediments were not washed, and contained about 5 percent of the supernatant. Both fractions were assayed.

Radioactivity was measured by means of an end-window Geiger counter, using dried samples sufficiently small to avoid losses by self-absorption. For absolute measurements, reference solutions of P^{32} obtained from the National Bureau of Standards, as well as a permanent simulated standard, were used. For absolute measurements of S^{35} we relied on the assays (±20 percent) furnished by the supplier of the isotope (Oak Ridge National Laboratory).

Glycerol-lactate medium was chosen to permit growth of bacteria without undesirable pH changes at low concentrations of phosphorus and sulfur, and proved useful also for certain experiments described in this paper. 18-hour cultures of sensitive bacteria grown in this medium contain about 2×10^9 cells per ml, which grow exponentially without lag or change in light-scattering per cell when subcultured in the same medium from either large or small seedings. The generation time is 1.5 hours at 37°C. The cells are smaller than those grown in broth. T2 shows a latent period of 22 to 25 minutes in this medium. The phage yield obtained by lysis with cyanide and UV-phage (described in context) is one per bacterium at 15 minutes and 16 per bacterium at 25 minutes. The final burst size in diluted cultures is 30 to 40 per bacterium, reached at 50

minutes. At 2×10^8 cells per ml, the culture lyses slowly, and yields 140 phage per bacterium. The growth of both bacteria and phage in this medium is as reproducible as that in broth.

For the preparation of radioactive phage, P^{32} of specific activity 0.5 mc/mg or S^{35} of specific activity 8.0 mc/mg was incorporated into glycerol-lactate medium, in which bacteria were allowed to grow at least 4 hours before seeding with phage. After infection with phage, the culture was aerated overnight, and the radioactive phage was isolated by three cycles of alternate slow (2000 G) and fast (12,000 G) centrifugation in adsorption medium. The suspensions were stored at a concentration not exceeding 4μc/ml.

Preparations of this kind contain 1.0 to 3.0×10^{-12} μg S and 2.5 to 3.5×10^{-11} μg P per viable phage particle. Occasional preparations containing excessive amounts of sulfur can be improved by absorption with heat-killed bacteria that do not adsorb the phage. The radiochemical purity of the preparations is somewhat uncertain, owing to the possible presence of inactive phage particles and empty phage membranes. The presence in our preparations of sulfur (about 20 percent) that is precipitated by antiphage serum (Table 1) and either adsorbed by bacteria resistant to phage, or not adsorbed by bacteria sensitive to phage (Table 7) indicates contamination by membrane material. Contaminants of bacterial origin are probably negligible for present purposes as indicated by the data given in Table 1. For proof that our principal findings reflect genuine properties of viable phage particles, we rely on some experiments with inactivated phage cited at the conclusion of this paper.

The Chemical Morphology of Resting Phage Particles Anderson (1949) found that bacteriophage T2 could be inactivated by suspending the particles in high concentrations of sodium chloride, and rapidly diluting the suspension with water. The inactivated phage was visible in electron micrographs as tadpole-shaped "ghosts." Since no inactivation occurred if the dilution was slow he attributed the inactivation to osmotic shock, and inferred that the particles possessed an osmotic membrane. Herriott (1951) found that osmotic shock released into solution the DNA (desoxypentose nucleic acid) of the phage particle, and that the ghosts could adsorb to bacteria and lyse them. He pointed out that this was a beginning toward the identification of viral functions with viral substances.

We have plasmolyzed isotopically labeled T2 by suspending the phage (10^{11} per ml) in 3M sodium chloride for 5 minutes at room temperature, and rapidly pouring into the suspension 40 volumes of distilled water. The plasmolyzed phage, containing not more than 2 percent survivors, was then analy-

Table 1 Composition of ghosts and solution of plasmolyzed phage

Percent of isotope	Whole phage labeled with		Plasmolyzed phage labeled with	
	P^{32}	S^{35}	P^{32}	S^{35}
Acid-soluble	—	—	1	—
Acid-soluble after treatment with DNase	1	1	80	1
Adsorbed to sensitive bacteria	85	90	2	90
Precipitated by antiphage	90	99	5	97

zed for phosphorus and sulfur in the several ways shown in Table 1. The results confirm and extend previous findings as follows:

1. Plasmolysis separates phage T2 into ghosts containing nearly all the sulfur and a solution containing nearly all the DNA of the intact particles.

2. The ghosts contain the principal antigens of the phage particle detectable by our antiserum. The DNA is released as the free acid, or possibly linked to sulfur-free, apparently nonantigenic substances.

3. The ghosts are specifically adsorbed to phage-susceptible bacteria; the DNA is not.

4. The ghosts represent protein coats that surround the DNA of the intact particles, react with antiserum, protect the DNA from DNase (desoxyribonuclease), and carry the organ of attachment to bacteria.

5. The effects noted are due to osmotic shock, because phage suspended in salt and diluted slowly is not inactivated, and its DNA is not exposed to DNase.

Sensitization of Phage DNA to DNase by Adsorption to Bacteria The structure of the resting phage particle described above suggests at once the possibility that multiplication of virus is preceded by the alteration or removal of the protective coats of the particles. This change might be expected to show

Table 2 Sensitization of phage DNA to DNase by adsorption to bacteria

Phage adsorbed to		Phage labeled with	Nonsedimentable isotope, *percent*	
			After DNase	*No DNase*
Live bacteria		S^{35}	2	1
Live bacteria		p^{32}	8	7
Bacteria heated before infection		S^{35}	15	11
Bacteria heated before infection		p^{32}	76	13
Bacteria heated after infection		S^{35}	12	14
Bacteria heated after infection		p^{32}	66	23
Heated unadsorbed phage: acid-soluble P^{32}	$70°$	p^{32}	5	
	$80°$	p^{32}	13	
	$90°$	p^{32}	81	
	$100°$	p^{32}	88	

Phage adsorbed to bacteria for 5 minutes at $37°$C in adsorption medium, followed by washing.
Bacteria heated for 10 minutes at $80°$C in adsorption medium (before infection) or in veronal buffer (after infection).
Unadsorbed phage heated in veronal buffer, treated with DNase, and precipitated with trichloroacetic acid.
All samples fractionated by centrifuging 10 minutes at 1300 G.

itself as a sensitization of the phage DNA to DNase. The experiments described in Table 2 show that this happens. The results may be summarized as follows:

1. Phage DNA becomes largely sensitive to DNase after adsorption to heat-killed bacteria.

2. The same is true of the DNA of phage adsorbed to live bacteria, and then heated to $80°C$ for 10 minutes, at which temperature unadsorbed phage is not sensitized to DNase.

3. The DNA of phage adsorbed to unheated bacteria is resistant to DNase, presumably because it is protected by cell structures impervious to the enzyme.

Graham and collaborators (personal communication) were the first to discover the sensitization of phage DNA to DNase by adsorption to heat-killed bacteria.

The DNA in infected cells is also made accessible to DNase by alternate freezing and thawing (followed by formaldehyde fixation to inactivate cellular enzymes), and to some extent by formaldehyde fixation alone, as illustrated by the following experiment.

Bacteria were grown in broth to 5×10^7 cells per ml, centrifuged, resuspended in adsorption medium, and infected with about two P^{32}-labeled phage per bacterium. After 5 minutes for adsorption, the suspension was diluted with water containing per liter 1.0 mM MgSO$_4$, 0.1 mM CaCl^2, and 10 mg gelatin, and recentrifuged. The cells were resuspended in the fluid last mentioned at a concentration of 5×10^8 per ml. This suspension was frozen at $-15°C$ and thawed with a minimum of warming, three times in succession. Immediately after the third thawing, the cells were fixed by the addition of 0.5 percent (v/v) of formalin (35 percent HCHO). After 30 minutes at room temperature, the suspension was dialyzed free from formaldehyde and centrifuged at 2200 G for 15 minutes. Samples of P^{32}-labeled phage, frozen-thawed, fixed, and dialyzed, and of infected cells fixed only and dialyzed, were carried along as controls.

The analysis of these materials, given in Table 3, shows that the effect of freezing and thawing is to make the intracellular DNA liable to DNase, without, however, causing much of it to leach out of the cells. Freezing and thawing and formaldehyde fixation have a negligible effect on unadsorbed phage, and formaldehyde fixation alone has only a mild effect on infected cells.

Both sensitization of the intracellular P^{32} to DNase, and its failure to leach out of the cells, are constant features of experiments of this type, independently of visible lysis. In the experiment just described, the frozen suspension cleared during the period of dialysis. Phase-contrast microscopy showed that the cells consisted largely of empty membranes, many apparently broken. In another experiment, samples of infected bacteria from a culture in salt-poor broth were repeatedly frozen and thawed at

Table 3   Sensitization of intracellular phage to DNase by freezing, thawing, and fixation with formaldehyde

	Unadsorbed phage frozen, thawed, fixed	Infected cells frozen, thawed, fixed	Infected cells fixed only
Low speed sediment fraction			
Total P^{32}	—	71	86
Acid-soluble	—	0	0.5
Acid-soluble after DNase	—	59	28
Low speed supernatant fraction			
Total P^{32}	—	29	14
Acid-soluble	1	0.8	0.4
Acid-soluble after DNase	11	21	5.5

The figures express percent of total P^{32} in the original phage, or its adsorbed fraction.

various times during the latent period of phage growth, fixed with formaldehyde, and then washed in the centrifuge. Clearing and microscopic lysis occurred only in suspensions frozen during the second half of the latent period, and occurred during the first or second thawing. In this case the lysed cells consisted wholly of intact cell membranes, appearing empty except for a few small, rather characteristic refractile bodies apparently attached to the cell walls. The behavior of intracellular P^{32} toward DNase, in either the lysed or unlysed cells, was not significantly different from that shown in Table 3, and the content of P^{32} was only slightly less after lysis. The phage liberated during freezing and thawing was also titrated in this experiment. The lysis occurred without appreciable liberation of phage in suspensions frozen up to and including the 16th minute, and the 20 minute sample yielded only five per bacterium. Another sample of the culture formalinized at 30 minutes, and centrifuged without freezing, contained 66 percent of the P^{32} in nonsedimentable form. The yield of extracellular phage at 30 minutes was 108 per bacterium, and the sedimented material consisted largely of formless debris but contained also many apparently intact cell membranes.

We draw the following conclusions from the experiments in which cells infected with P^{32}-labeled phage are subjected to freezing and thawing.

1. Phage DNA becomes sensitive to DNase after adsorption to bacteria in buffer under conditions in which no known growth process occurs. (Benzer 1952; Dulbecco 1952).

2. The cell membrane can be made permeable to DNase under conditions that do not permit the escape of either the intracellular P^{32} or the bulk of the cell contents.

3. Even if the cells lyse as a result of freezing and thawing, permitting escape of other cell constituents, most of the P^{32} derived from phage remains inside the cell membranes, as do the mature phage progeny.

4. The intracellular P^{32} derived from phage is largely freed during spontaneous lysis accompanied by phage liberation.

We interpret these facts to mean that intracellular DNA derived from phage is not merely DNA in solution, but is part of an organized structure at all times during the latent period.

Liberation of DNA from Phage Particles by Adsorption to Bacterial Fragments The sensitization of phage DNA to specific depolymerase by adsorption to bacteria might mean that adsorption is followed by the ejection of the phage DNA from its protective coat. The following experiment shows that this is in fact what happens when phage attaches to fragmented bacterial cells.

Bacterial debris was prepared by infecting cells in adsorption medium with four particles of T2 per bacterium, and transferring the cells to salt-poor broth at 37°C. The culture was aerated for 60 minutes, $M/50$ HCN was added, and incubation continued for 30 minutes longer. At this time the yield of extracellular phage was 400 particles per bacterium, which remained unadsorbed because of the low concentration of electrolytes. The debris from the lysed cells was washed by centrifugation at 1700 G, and resuspended in adsorption medium at a concentration equivalent to 3×10^9 lysed cells per ml. It consisted largely of collapsed and fragmented cell membranes. The adsorption of radioactive phage to this material is described in Table 4. The following facts should be noted.

1. The unadsorbed fraction contained only 5 percent of the original phage particles in infective form, and only 13 percent of the total sulfur. (Much of this sulfur must be the material that is not adsorbable to whole bacteria.)

2. About 80 percent of the phage was inactivated. Most of the sulfur of this phage, as well as most of the surviving phage, was found in the sediment fraction.

3. The supernatant fraction contained 40 percent of the total phage DNA (in a form labile to DNase) in addition to the DNA of the unadsorbed surviving phage. The labile DNA amounted to about half of the DNA of the inactivated phage particles, whose sulfur sedimented with the bacterial debris.

4. Most of the sedimentable DNA could be accounted for either as surviving phage, or as DNA

Table 4 Release of DNA from phage adsorbed to bacterial debris

	Phage labeled with	
	S^{35}	P^{32}
Sediment fraction		
Surviving phage	16	22
Total isotope	87	55
Acid-soluble isotope	0	2
Acid-soluble after DNase	2	29
Supernatant fraction		
Surviving phage	5	5
Total isotope	13	45
Acid-soluble isotope	0.8	0.5
Acid-soluble after DNase	0.8	39

S^{35} and P^{32} labeled T2 were mixed with identical samples of bacterial debris in adsorption medium and warmed for 30 minutes at 37°C. The mixtures were then centrifuged for 15 minutes at 2200 G, and the sediment and supernatant fractions were analyzed separately. The results are expressed as percent of input phage or isotope.

labile to DNase, the latter amounting to about half the DNA of the inactivated particles.

Experiments of this kind are unsatisfactory in one respect: one cannot tell whether the liberated DNA represents all the DNA of some of the inactivated particles, or only part of it.

Similar results were obtained when bacteria (strain B) were lysed by large amounts of UV-killed phage T2 or T4 and then tested with P^{32}-labeled T2 and T4. The chief point of interest in this experiment is that bacterial debris saturated with UV-killed T2 adsorbs T4 better than T2, and debris saturated with T4 adsorbs T2 better than T4. As in the preceding experiment, some of the adsorbed phage was not inactivated and some of the DNA of the inactivated phage was not released from the debris.

These experiments show that some of the cell receptors for T2 are different from some of the cell receptors for T4, and that phage attaching to these specific receptors is inactivated by the same mechanism as phage attaching to unselected receptors. This mechanism is evidently an active one, and not merely the blocking of sites of attachment to bacteria.

Removal of Phage Coats from Infected Bacteria
Anderson (1951) has obtained electron micrographs

indicating that phage T2 attaches to bacteria by its tail. If this precarious attachment is preserved during the progress of the infection, and if the conclusions reached above are correct, it ought to be a simple matter to break the empty phage membranes off the infected bacteria, leaving the phage DNA inside the cells.

The following experiments show that this is readily accomplished by strong shearing forces applied to suspensions of infected cells, and further that infected cells from which 80 percent of the sulfur of the parent virus has been removed remain capable of yielding phage progeny.

Broth-grown bacteria were infected with S^{35}- or P^{32}-labeled phage in adsorption medium, the unadsorbed material was removed by centrifugation, and the cells were resuspended in water containing per liter 1 mM MgSO$_4$, 0.1 mM CaCl$_2$, and 0.1 gm gelatin. This suspension was spun in a Waring blender (semimicro size) at 10,000 rpm. The suspension was cooled briefly in ice water at the end of each 60 second running period. Samples were removed at intervals, titrated (through antiphage serum) to measure the number of bacteria capable of yielding phage, and centrifuged to measure the proportion of isotope released from the cells.

The results of one experiment with each isotope are shown in Fig. 1. The data for S^{35} and survival of

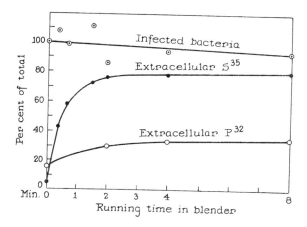

Fig. 1. Removal of S^{35} and P^{32} from bacteria infected with radioactive phage, and survival of the infected bacteria, during agitation in a Waring blender.

infected bacteria come from the same experiment, in which the ratio of added phage to bacteria was 0.28, and the concentrations of bacteria were 2.5×10^8 per ml infected, and 9.7×10^8 per ml total, by direct titration. The experiment with P^{32}-labeled phage was very similar. In connection with these results, it should be recalled that Anderson (1949) found that adsorption of phage to bacteria could be prevented by rapid stirring of the suspension.

At higher ratios of infection, considerable amounts of phage sulfur elute from the cells spontaneously under the conditions of these experiments, though the elution of P^{32} and the survival of infected cells are not affected by multiplicity of infection (Table

5). This shows that there is a cooperative action among phage particles in producing alterations of the bacterial membrane which weaken the attachment of the phage. The cellular changes detected in this way may be related to those responsible for the release of bacterial components from infected bacteria (Prater 1951; Price 1952).

A variant of the preceding experiments was designed to test bacteria at a later stage in the growth of phage. For this purpose infected cells were aerated in broth for 5 or 15 minutes, fixed by the addition of 0.5 percent (v/v) commercial formalin, centrifuged, resuspended in 0.1 percent formalin in water, and subsequently handled as described above. The results were very similar to those already presented, except that the release of P^{32} from the cells was slightly less, and titrations of infected cells could not be made.

The S^{35}-labeled material detached from infected cells in the manner described possesses the following properties. It is sedimented at 12,000 G, though less completely than intact phage particles. It is completely precipitated by antiphage serum in the presence of whole phage carrier. 40 to 50 percent of it readsorbs to sensitive bacteria, almost independently of bacterial concentration between 2×10^8 and 10^9 cells per ml, in 5 minutes at 37°C. The adsorption is not very specific: 10 to 25 percent adsorbs to phage-resistant bacteria under the same conditions. The adsorption requires salt, and for this reason the efficient removal of S^{35} from infected bacteria can be accomplished only in a fluid poor in electrolytes.

The results of these experiments may be summarized as follows:

Table 5 Effect of multiplicity of infection on elution of phage membranes from infected bacteria

Running time in blender (min.)	Multiplicity of infection	P^{32}-labeled phage		S^{35}-labeled phage	
		Isotope eluted percent	Infected bacteria surviving percent	Isotope eluted percent	Infected bacteria surviving percent
0	0.6	10	120	16	101
2.5	0.6	21	82	81	78
0	6.0	13	89	46	90
2.5	6.0	24	86	82	85

The infected bacteria were suspended at 10^9 cells per ml in water containing per liter $1mM$ $MgSO_4$, 0.1 mM $CaCl_2$, and 0.1 gm gelatin. Samples were withdrawn for assay of extracellular isotope and infected bacteria before and after agitating the suspension. In either case the cells spent about 15 minutes at room temperature in the eluting fluid.

1. 75 to 80 percent of the phage sulfur can be stripped from infected cells by violent agitation of the suspension. At high multiplicity of infection, nearly 50 percent elutes spontaneously. The properties of the S^{35}-labeled material show that it consists of more or less intact phage membranes, most of which have lost the ability to attach specifically to bacteria.

2. The release of sulfur is accompanied by the release of only 21 to 35 percent of the phage phosphorus, half of which is given up without any mechanical agitation.

3. The treatment does not cause any appreciable inactivation of intracellular phage.

4. These facts show that the bulk of the phage sulfur remains at the cell surface during infection, and takes no part in the multiplication of intracellular phage. The bulk of the phage DNA, on the other hand, enters the cell soon after adsorption of phage to bacteria.

Transfer of Sulfur and Phosphorus from Parental Phage to Progeny We have concluded above that the bulk of the sulfur-containing protein of the resting phage particle takes no part in the multiplication of phage, and in fact does not enter the cell. It follows that little or no sulfur should be transferred from parental phage to progeny. The experiments described below show that this expectation is correct, and that the maximal transfer is of the order 1 percent.

Bacteria were grown in glycerol-lactate medium overnight and subcultured in the same medium for 2 hours at 37°C with aeration, the size of seeding being adjusted nephelometrically to yield 2×10^8 cells per ml in the subculture. These bacteria were sedimented, resuspended in adsorption medium at a concentration of 10^9 cells per ml, and infected with S^{35}-labeled phage T2. After 5 minutes at 37°C, the suspension was diluted with 2 volumes of water and resedimented to remove unadsorbed phage (5 to 10 percent by titer), and S^{35} (about 15 percent). The cells were next suspended in glycerol-lactate medium at a concentration of 2×10^8 per ml and aerated at 37°C. Growth of phage was terminated at the desired time by adding in rapid succession 0.02 mM HCN and 2×10^{11} UV-killed phage per ml of cul-

ture. The cyanide stops the maturation of intracellular phage (Doermann 1948), and the UV-killed phage minimizes losses of phage progeny by adsorption to bacterial debris, and promotes the lysis of bacteria (Maaløe and Watson 1951). As mentioned in another connection, and also noted in these experiments, the lysing phage must be closely related to the phage undergoing multiplication (*e.g.,* T2H, its *h* mutant, or T2L, but not T4 or T6, in this instance) in order to prevent inactivation of progeny by adsorption to bacterial debris.

To obtain what we shall call the maximal yield of phage, the lysing phage was added 25 minutes after placing the infected cells in the culture medium, and the cyanide was added at the end of the 2nd hour. Under these conditions, lysis of infected cells occurs rather slowly.

Aeration was interrupted when the cyanide was added, and the cultures were left overnight at 37°C. The lysates were then fractionated by centrifugation into an initial low speed sediment (2500 G for 20 minutes), a high speed supernatant (12,000 G for 30 minutes), a second low speed sediment obtained by recentrifuging in adsorption medium the resuspended high speed sediment, and the clarified high speed sediment.

The distribution of S^{35} and phage among fractions obtained from three cultures of this kind is shown in Table 6. The results are typical (except for the excessively good recoveries of phage and S^{35}) of lysates in broth as well as lysates in glycerol-lactate medium. The striking result of this experiment is that the distribution of S^{35} among the fractions is the same for early lysates that do not contain phage progeny, and later ones that do. This suggests that little or no S^{35} is contained in the mature phage progeny. Further fractionation by adsorption to bacteria confirms this suggestion.

Adsorption mixtures prepared for this purpose contained about 5×10^9 heat-killed bacteria (70°C for 10 minutes) from 18-hour broth cultures, and about 10^{11} phage (UV-killed lysing phage plus test phage), per ml of adsorption medium. After warming to 37°C for 5 minutes, the mixtures were diluted with 2 volumes of water, and centrifuged. Assays were made from supernatants and from unwashed resuspended sediments.

The results of tests of adsorption of S^{35} and phage to bacteria (H) adsorbing both T2 progeny and *h*-mutant lysing phage, to bacteria (B/2) adsorbing

Table 6 Percent distributions of phage and S^{35} among centrifugally separated fractions of lysates after infection with S^{35}-labeled T2

Fraction	Lysis at $t = 0$ S^{35}	Lysis at $t = 10$ S^{35}	Maximal yield	
			S^{35}	Phage
1st low speed sediment	79	81	82	19
2nd low speed sediment	2.4	2.1	2.8	14
High speed sediment	8.6	6.9	7.1	61
High speed supernatant	10	10	7.5	7.0
Recovery	100	100	96	100

Infection with S^{35}-labeled T2, 0.8 particles per bacterium. Lysing phage UV-killed h mutant of T2. Phage yields per infected bacterium: < 0.1 after lysis at $t = 0$; 0.12 at $t = 10$; maximal yield 29. Recovery of S^{35} means percent of adsorbed input recovered in the four fractions; recovery of phage means percent of total phage yield (by plaque count before fractionation) recovered by titration of fractions.

lysing phage only, and to bacteria (B/2h) adsorbing neither, are shown in Table 7, together with parallel tests of authentic S^{35}-labeled phage.

The adsorption tests show that the S^{35} present in the seed phage is adsorbed with the specificity of the phage, but that S^{35} present in lysates of bacteria infected with this phage shows a more complicated behavior. It is strongly adsorbed to bacteria adsorbing both progeny and lysing phage. It is weakly adsorbed to bacteria adsorbing neither. It is moderately well adsorbed to bacteria adsorbing lysing phage but not phage progeny. The latter test shows that the S^{35} is not contained in the phage progeny, and explains the fact that the S^{35} in early lysates not containing progeny behaves in the same way.

The specificity of the adsorption of S^{35}-labeled material contaminating the phage progeny is evidently due to the lysing phage, which is also adsorbed much more strongly to strain H than to B/2, as shown both by the visible reduction in Tyndall

Table 7 Adsorption tests with uniformly S^{35}-labeled phage and with products of their growth in nonradioactive medium

Adsorbing bacteria	Percent adsorbed				
	Uniformly labeled S^{35} phage		Products of lysis at $t = 10$	Phage progeny (maximal yield)	
	+ UV-h	No UV-h			
	S^{35}	S^{35}	S^{35}	S^{35}	Phage
Sensitive (H)	84	86	79	78	96
Resistant (B/2)	15	11	46	49	10
Resistant (B/2h)	13	12	29	28	8

The uniformly labeled phage and the products of their growth are respectively the seed phage and the high speed sediment fractions from the experiment shown in Table 6.
The uniformly labeled phage is tested at a low ratio of phage to bacteria: + UV-h means with added UV-killed h mutant in equal concentration to that present in the other test materials.
The adsorption of phage is measured by plaque counts of supernatants, and also sediments in the case of the resistant bacteria, in the usual way.

scattering (due to the lysing phage) in the supernatants of the test mixtures, and by independent measurements. This conclusion is further confirmed by the following facts.

1. If bacteria are infected with S^{35} phage, and then lysed near the midpoint of the latent period with cyanide alone (in salt-poor broth, to prevent readsorption of S^{35} to bacterial debris), the high speed sediment fraction contains S^{35} that is adsorbed weakly and non-specifically to bacteria.

2. If the lysing phage and the S^{35}-labeled infecting phage are the same (T2), or if the culture in salt-poor broth is allowed to lyse spontaneously (so that the yield of progeny is large), the S^{35} in the high speed sediment fraction is adsorbed with the specificity of the phage progeny (except for a weak nonspecific adsorption). This is illustrated in Table 7 by the adsorption to H and B/2h.

It should be noted that a phage progeny grown from S^{35}-labeled phage and containing a larger or smaller amount of contaminating radioactivity could not be distinguished by any known method from authentic S^{35}-labeled phage, except that a small amount of the contaminant could be removed by adsorption to bacteria resistant to the phage. In addition to the properties already mentioned, the contaminating S^{35} is completely precipitated with the phage by antiserum, and cannot be appreciably separated from the phage by further fractional sedimentation, at either high or low concentrations of electrolyte. On the other hand, the chemical contamination from this source would be very small in favorable circumstances, because the progeny of a single phage particle are numerous and the contaminant is evidently derived from the parents.

The properties of the S^{35}-labeled contaminant show that it consists of the remains of the coats of the parental phage particles, presumably identical with the material that can be removed from unlysed cells in the Waring blender. The fact that it undergoes little chemical change is not surprising since it probably never enters the infected cell.

The properties described explain a mistaken preliminary report (Hershey *et al.* 1951) of the transfer of S^{35} from parental to progeny phage.

It should be added that experiments identical to those shown in Tables 6 and 7, but starting from

phage labeled with P^{32}, show that phosphorus is transferred from parental to progeny phage to the extent of 30 percent at yields of about 30 phage per infected bacterium, and that the P^{32} in prematurely lysed cultures is almost entirely nonsedimentable, becoming, in fact, acid-soluble on aging.

Similar measures of the transfer of P^{32} have been published by Putnam and Kozloff (1950) and others. Watson and Maaløe (1952) summarize this work, and report equal transfer (nearly 50 percent) of phosphorus and adenine.

A Progeny of S^{35}-Labeled Phage Nearly Free from the Parental Label The following experiment shows clearly that the obligatory transfer of parental sulfur to offspring phage is less than 1 percent, and probably considerably less. In this experiment, the phage yield from infected bacteria from which the S^{35}-labeled phage coats had been stripped in the Waring blender was assayed directly for S^{35}.

Sensitive bacteria grown in broth were infected with five particles of S^{35}-labeled phage per bacterium, the high ratio of infection being necessary for purposes of assay. The infected bacteria were freed from unadsorbed phage and suspended in water containing per liter 1 mM MgSO$_4$, 0.1 mM CaCl$_2$, and 0.1 gm gelatin. A sample of this suspension was agitated for 2.5 minutes in the Waring blender, and centrifuged to remove the extracellular S^{35}. A second sample not run in the blender was centrifuged at the same time. The cells from both samples were resuspended in warm salt-poor broth at a concentration of 10^8 bacteria per ml, and aerated for 80 minutes. The cultures were then lysed by the addition of 0.02 mM HCN, 2×10^{11} UV-killed T2, and 6 mg NaCl per ml of culture. The addition of salt at this point causes S^{35} that would otherwise be eluted (Hershey *et al.* 1951) to remain attached to the bacterial debris. The lysates were fractionated and assayed as described previously, with the results shown in Table 8.

The data show that stripping reduces more or less proportionately the S^{35}-content of all fractions. In particular, the S^{35}-content of the fraction containing most of the phage progeny is reduced from nearly 10 percent to less than 1 percent of the initially adsorbed isotope. This experiment shows that the bulk of the S^{35} appearing in all lysate fractions is derived from the remains of the coats of the parental phage particles.

Properties of Phage Inactivated by Formaldehyde
Phage T2 warmed for 1 hour at 37°C in adsorption

Table 8 Lysates of bacteria infected with S^{35}-labeled T2 and stripped in the Waring blender

Percent of adsorbed S^{35} or of phage yield:	Cells stripped		Cells not stripped	
	S^{35}	Phage	S^{35}	Phage
Eluted in blender fluid	86	—	39	—
1st low speed sediment	3.8	9.3	31	13
2nd low speed sediment	(0.2)	11	2.7	11
High speed sediment	(0.7)	58	9.4	89
High speed supernatant	(2.0)	1.1	(1.7)	1.6
Recovery	93	79	84	115

All the input bacteria were recovered in assays of infected cells made during the latent period of both cultures. The phage yields were 270 (stripped cells) and 200 per bacterium, assayed before fractionation. Figures in parentheses were obtained from counting rates close to background.

medium containing 0.1 percent (*v*/*v*) commercial formalin (35 percent HCHO), and then dialyzed free from formaldehyde, shows a reduction in plaque titer by a factor 1000 or more. Inactivated phage of this kind possesses the following properties.

1. It is adsorbed to sensitive bacteria (as measured by either S^{35} or P^{32} labels), to the extent of about 70 percent.

2. The adsorbed phage kills bacteria with an efficiency of about 35 percent compared with the original phage stock.

3. The DNA of the inactive particles is resistant to DNase, but is made sensitive by osmotic shock.

4. The DNA of the inactive particles is not sensitized to DNase by adsorption to heat-killed bacteria, nor is it released into solution by adsorption to bacterial debris.

5. 70 percent of the adsorbed phage DNA can be detached from infected cells spun in the Waring blender. The detached DNA is almost entirely resistant to DNase.

These properties show that T2 inactivated by formaldehyde is largely incapable of injecting its DNA into the cells to which it attaches. Its behavior in the experiments outlined gives strong support to our interpretation of the corresponding experiments with active phage.

DISCUSSION

We have shown that when a particle of bacteriophage T2 attaches to a bacterial cell, most of the phage DNA enters the cell, and a residue containing at least 80 percent of the sulfur-containing protein of the phage remains at the cell surface. This residue consists of the material forming the protective membrane of the resting phage particle, and it plays no further role in infection after the attachment of phage to bacterium.

These facts leave in question the possible function of the 20 percent of sulfur-containing protein that may or may not enter the cell. We find that little or none of it is incorporated into the progeny of the infecting particle, and that at least part of it consists of additional material resembling the residue that can be shown to remain extracellular. Phosphorus and adenine (Watson and Maaløe 1952) derived from the DNA of the infecting particle, on the other hand, are transferred to the phage progeny to a considerable and equal extent. We infer that sulfur-containing protein has no function in phage multiplication, and that DNA has some function.

It must be recalled that the following questions remain unanswered. (1) Does any sulfur-free phage material other than DNA enter the cell? (2) If so, is it transferred to the phage progeny? (3) Is the transfer of phosphorus (or hypothetical other substance) to progeny direct — that is, does it remain at all times in a form specifically identifiable as phage substance — or indirect?

Our experiments show clearly that a physical separation of the phage T2 into genetic and nongenetic parts is possible. A corresponding functional separation is seen in the partial independence of phenotype and genotype in the same phage (Novick and Szilard 1951; Hershey *et al.* 1951). The chemical identification of the genetic part must wait, however, until some of the questions asked above have been answered.

Two facts of significance for the immunologic method of attack on problems of viral growth should be emphasized here. First, the principal antigen of the infecting particles of phage T2 persists unchanged in infected cells. Second, it remains attached to the bacterial debris resulting from lysis of the cells. These possibilities seem to have been overlooked in a study by Rountree (1951) of viral antigens during the growth of phage T5.

SUMMARY

1. Osmotic shock disrupts particles of phage T2 into material containing nearly all the phage sulfur in a form precipitable by antiphage serum, and capable of specific adsorption to bacteria. It releases into solution nearly all the phage DNA in a form not precipitable by antiserum and not adsorbable to bacteria. The sulfur-containing protein of the phage particle evidently makes up a membrane that protects the phage DNA from DNase, comprises the sole or principal antigenic material, and is responsible for attachment of the virus to bacteria.

2. Adsorption of T2 to heat-killed bacteria, and heating or alternate freezing and thawing of infected cells, sensitize the DNA of the adsorbed phage to DNase. These treatments have little or no sensitizing effect on unadsorbed phage. Neither heating nor freezing and thawing releases the phage DNA from infected cells, although other cell constituents can be extracted by these methods. These facts suggest that the phage DNA forms part of an organized intracellular structure thoughout the period of phage growth.

3. Adsorption of phage T2 to bacterial debris causes part of the phage DNA to appear in solution, leaving the phage sulfur attached to the debris. Another part of the phage DNA, corresponding roughly to the remaining half of the DNA of the inactivated phage, remains attached to the debris but can be

separated from it by DNase. Phage T4 behaves similarly, although the two phages can be shown to attach to different combining sites. The inactivation of phage by bacterial debris is evidently accompanied by the rupture of the viral membrane.

4. Suspensions of infected cells agitated in a Waring blender release 75 percent of the phage sulfur and only 15 percent of the phage phosphorus to the solution as a result of the applied shearing force. The cells remain capable of yielding phage progeny.

5. The facts stated show that most of the phage sulfur remains at the cell surface and most of the phage DNA enters the cell on infection. Whether sulfur-free material other than DNA enters the cell has not been determined. The properties of the sulfur-containing residue identify it as essentially unchanged membranes of the phage particles. All types of evidence show that the passage of phage DNA into the cell occurs in nonnutrient medium under conditions in which other known steps in viral growth do not occur.

6. The phage progeny yielded by bacteria infected with phage labeled with radioactive sulfur contain less than 1 percent of the parental radioactivity. The progeny of phage particles labeled with radioactive phosphorus contain 30 percent or more of the parental phosphorus.

7. Phage inactivated by dilute formaldehyde is capable of adsorbing to bacteria, but does not release its DNA to the cell. This shows that the interaction between phage and bacterium resulting in release of the phage DNA from its protective membrane depends on labile components of the phage particle. By contrast, the components of the bacterium essential to this interaction are remarkably stable. The nature of the interaction is otherwise unknown.

8. The sulfur-containing protein of resting phage particles is confined to a protective coat that is responsible for the adsorption to bacteria, and functions as an instrument for the injection of the phage DNA into the cell. This protein probably has no function in the growth of intracellular phage. The DNA has some function. Further chemical inferences should not be drawn from the experiments presented.

REFERENCES

Anderson, T. F., 1949, The reactions of bacterial viruses with their host cells, *Bot. Rev.,* 15, 464.

Anderson, T. F., 1951, *Tr. New York Acad. Sc.,* 13, 130.

Anderson, T. F., and Doermann, A. H., 1952., *J. Gen. Physiol.,* 35, 657.

Benzer, S., 1952, *J. Bact.,* 63, 59.

Doermann, A. H., 1948, *Carnegie Institution of Washington Yearbook, No. 47,* 176.

Doermann, A. H., and Dissoway, C., 1949, *Carnegie Institution of Washington Yearbook, No. 48,* 170.

Dulbecco, R., 1952, *J. Bact.,* 63, 209.

Herriott, R. M., 1951, *J. Bact.,* 61, 752.

Hershey, A. D., 1946, *Genetics,* 31, 620.

Hershey, A. D., Roesel, C., Chase, M., and Forman, S., 1951, *Carnegie Institution of Washington Yearbook, No. 50,* 195.

Lwoff, A., and Gutmann, A., 1950, *Ann. Inst. Pasteur,* 78, 711.

Maaløe, O., and Watson, J. D., 1951, *Proc. Nat. Acad. Sc.,* 37, 507.

Novick, A., and Szilard, L., 1951, *Science,* 113, 34.

Prater, C. D., 1951, Thesis, University of Pennsylvania.

Price, W. H., 1952, *J. Gen. Physiol.,* 35, 409.

Putnam, F. W., and Kozloff, L., 1950, *J. Biol. Chem.,* 182, 243.

Rountree, P. M., 1951, *Brit. J. Exp. Path.,* 32, 341.

Watson, J. D., and Maaløe, O., 1952, *Acta path. et microbiol. scand.,* in press.

Genetical implications of the structure of deoxyribonucleic acid

*J. D. WATSON and F. H. C. CRICK**

Reprinted by authors' and publisher's permission from Nature, *vol. 171, 1953,*
pp. 964–967.

This paper is, in many ways, the more important of the two published by Watson and Crick in 1953. It set the tone for the application of DNA's structure to problems concerning (1) the replication of the genetic material; (2) the origin of mutations; and (3) the storage of genetic information — sometimes called the *genetic code*. It is little wonder that the study of molecular genetics mushroomed after 1953. It is also not surprising that Watson, Crick, and Wilkins[†] received the Nobel Prize for their work on DNA.

The importance of deoxyribonucleic acid (DNA) within living cells is undisputed. It is found in all dividing cells, largely if not entirely in the nucleus, where it is an essential constituent of the chromosomes. Many lines of evidence indicate that it is the carrier of a part of (if not all) the genetic specificity of the chromosomes and thus of the gene itself. Until now, however, no evidence has been presented to show how it might carry out the essential operation required of a genetic material, that of exact self-duplication.

We have recently proposed a structure[1] for the salt of deoxyribonucleic acid which, if correct, immediately suggests a mechanism for its self-duplication. X-ray evidence obtained by the workers at King's College, London,[2] and presented at the same time, gives qualitative support to our structure and is incompatible with all previously proposed structures.[3] Though the structure will not be completely proved until a more extensive comparison has been made with the x-ray data, we now feel sufficient confidence in its general correctness to discuss its genetical implications. In doing so we are assuming that fibers of the salt of deoxyribonucleic acid are not artifacts arising in the method of preparation, since it has been shown by Wilkins and his coworkers that similar x-ray patterns are obtained from both the isolated fibers and certain intact biological materials such as sperm head and bacteriophage particles.[3,4]

The chemical formula of deoxyribonucleic acid is now well established. The molecule is a very long chain, the backbone of which consists of a regular alternation of sugar and phosphate groups, as shown in Fig. 1. To each sugar is attached a nitrogenous base, which can be of four different types. (We have considered 5-methyl cytosine to be equivalent to cytosine, since either can fit equally well into our structure.) Two of the possible bases — adenine and guanine — are purines, and the other two — thymine and cytosine — are pyrimidines. So far as is known, the sequence of bases along the chain is irregular. The monomer unit, consisting of phosphate, sugar and base, is known as a nucleotide.

The first feature of our structure which is of biological interest is that it consists not of one chain, but of two. These two chains are both coiled around a common fiber axis, as is shown diagrammatically in Fig. 2. It has often been assumed that since there was only one chain in the chemical formula there would only be one in the structural unit. However, the density, taken with the x-ray evidence,[2] suggests very strongly that there are two.

The other biologically important feature is the manner in which the two chains are held together. This is done by hydrogen bonds between the bases,

*Medical Research Council Unit for the Study of the Molecular Structure of Biological Systems, Cavendish Laboratory, Cambridge.

†Franklin died before the 1962 Nobel Prize was awarded; she probably would have shared it.

D.N.A.

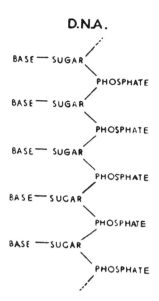

Fig. 1. Chemical formula of a single chain of deoxyribonucleic acid.

as shown schematically in Fig. 3. The bases are joined together in pairs, a single base from one chain being hydrogen-bonded to a single base from the other. The important point is that only certain pairs of bases will fit into the structure. One member of a pair must be a purine and the other a pyrimidine in order to bridge between the two chains. If a pair consisted of two purines, for example, there would not be room for it.

We believe that the bases will be present almost entirely in their most probable tautomeric forms. If this is true, the conditions for forming hydrogen bonds are more restrictive, and the only pairs of bases possible are:

adenine with thymine
guanine with cytosine

The way in which these are joined together is shown in Figs. 4 and 5. A given pair can be either way round. Adenine, for example, can occur on either chain; but when it does, its partner on the other chain must always be thymine.

This pairing is strongly supported by the recent analytical results,[5] which show that for all sources of deoxyribonucleic acid examined the amount of adenine is close to the amount of thymine, and the amount of guanine close to the amount of cytosine, although the cross-ratio (the ratio of adenine to guanine) can vary from one source to another. Indeed, if the sequence of bases on one chain is irreg-

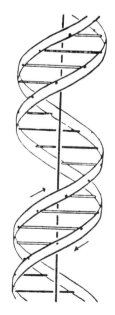

Fig. 2. This figure is purely diagrammatic. The two ribbons symbolize the two phosphate-sugar chains, and the horizontal rods the pairs of bases holding the chains together. The vertical line marks the fiber axis.

Fig. 3. Chemical formula of a pair of deoxyribonucleic acid chains. The hydrogen bonding is symbolized by dotted lines.

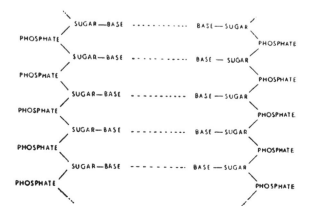

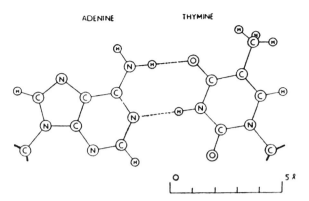

Fig. 4. Pairing of adenine and thymine. Hydrogen bonds are shown dotted. One carbon atom of each sugar is shown.

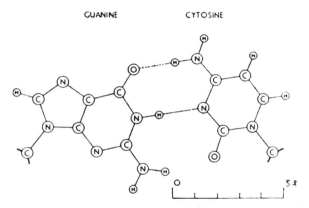

Fig. 5. Pairing of guanine and cytosine. Hydrogen bonds are shown dotted. One carbon atom of each sugar is shown.

ular, it is difficult to explain these analytical results except by the sort of pairing we have suggested.

The phosphate-sugar backbone of our model is completely regular, but any sequence of the pairs of bases can fit into the structure. It follows that in a long molecule many different permutations are possible, and it therefore seems likely that the precise sequence of the bases is the code which carries the genetical information. If the actual order of the bases on one of the pair of chains were given, one could write down the exact order of the bases on the other one, because of the specific pairing. Thus one chain is, as it were, the complement of the other, and it is this feature which suggests how the

deoxyribonucleic acid molecule might duplicate itself.

Previous discussions of self-duplication have usually involved the concept of a template, or mold. Either the template was supposed to copy itself directly or it was to produce a "negative," which in its turn was to act as a template and produce the original "positive" once again. In no case has it been explained in detail how it would do this in terms of atoms and molecules.

Now our model for deoxyribonucleic acid is, in effect, a *pair* of templates, each of which is complementary to the other. We imagine that prior to duplication the hydrogen bonds are broken, and the two chains unwind and separate. Each chain then acts as a template for the formation on to itself of a new companion chain, so that eventually we shall have *two* pairs of chains, where we only had one before. Moreover, the sequence of the pairs of bases will have been duplicated exactly.

A study of our model suggests that this duplication could be done most simply if the single chain (or the relevant portion of it) takes up the helical configuration. We imagine that at this stage in the life of the cell, free nucleotides, strictly polynucleotide precursors, are available in quantity. From time to time the base of a free nucleotide will join up by hydrogen bonds to one of the bases on the chain already formed. We now postulate that the polymerization of these monomers to form a new chain is only possible if the resulting chain can form the proposed structure. This is plausible, because steric reasons would not allow nucleotides "crystallized" on to the first chain to approach one another in such a way that they could be joined together into a new chain, unless they were those nucleotides which were necessary to form our structure. Whether a special enzyme is required to carry out the polymerization, or whether the single helical chain already formed acts effectively as an enzyme, remains to be seen.

Since the two chains in our model are intertwined, it is essential for them to untwist if they are to separate. As they make one complete turn around each other in 34 Å, there will be about 150 turns per million molecular weight, so that whatever the precise structure of the chromosome a considerable amount of uncoiling would be necessary. It is well known from microscopic observation that much coiling and uncoiling occurs during mitosis, and though this is on a much larger scale it probably reflects similar processes on a molecular level. Although it is difficult at the moment to see how these processes occur without everything getting tangled, we do not feel that this objection will be insuperable.

Our structure, as described,[1] is an open one. There is room between the pair of polynucleotide chains (see Fig. 2) for a polypeptide chain to wind around the same helical axis. It may be significant that the distance between adjacent phosphorus atoms, 7.1 A, is close to the repeat of a fully extended polypeptide chain. We think it probable that in the sperm head and in artificial nucleoproteins, the polypeptide chain occupies this position. The relative weakness of the second layer line in the published x-ray pictures[3a,4] is crudely compatible with such an idea. The function of the protein might well be to control the coiling and uncoiling, to assist in holding a single polynucleotide chain in a helical configuration, or some other nonspecific function.

Our model suggests possible explanations for a number of other phenomena. For example, spontaneous mutation may be due to a base occasionally occurring in one of its less likely tautomeric forms. Again, the pairing between homologous chromosomes at meiosis may depend on pairing between specific bases. We shall discuss these ideas in detail elsewhere.

For the moment, the general scheme we have proposed for the reproduction of deoxyribonucleic acid must be regarded as speculative. Even if it is correct, it is clear from what we have said that much remains to be discovered before the picture of genetic duplication can be described in detail. What are the polynucleotide precursors? What makes the pair of chains unwind and separate? What is the precise role of the protein? Is the chromosome one long pair of deoxyribonucleic acid chains, or does it consist of patches of the acid joined together by protein?

Despite these uncertainties we feel that our proposed structure for deoxyribonucleic acid may help to solve one of the fundamental biological problems — the molecular basis of the template needed for genetic replication. The hypothesis we are suggesting is that the template is the pattern of bases formed by one chain of the deoxyribonucleic acid and that the gene contains a complementary pair of such templates.

One of us (J.D.W.) has been aided by a fellowship from the National Foundation for Infantile Paralysis (U.S.A.).

REFERENCES

1. Watson, J. D., and Crick, F. H. C., *Nature,* 171: 737 (1953).
2. Wilkins, M. H. F., Stokes, A. R., and Wilson, H. R., *Nature,* 171:738 (1953). Franklin, R. E., and Gosling, R. G., *Nature,* 171:740 (1953).
3. (*a*) Astbury, W. T., Symp. No. 1 Soc. Exp. Biol., 66 (1947). (*b*) Furberg, S., *Acta Chem. Scand.,* 6, 634 (1952). (*c*) Pauling, L., and Corey, R. B., *Nature,* 171:346 (1953); *Proc. U.S. Nat. Acad. Sci.,* 39:84 (1953). (*d*) Fraser, R. D. B. (In preparation).
4. Wilkins, M. H. F., and Randall, J. T., *Biochim. et Biophys. Acta,* 10:192 (1953).
5. Chargaff, E., for references see Zamenhof, S., Brawerman, G., and Chargaff, E., *Biochim. et Biophys. Acta,* 9:402 (1952). Wyatt, G. R., *J. Gen. Physiol.,* 36:201 (1952).

The replication of DNA in *Escherichia coli**

MATTHEW MESELSON and FRANKLIN W. STAHL†

Reprinted by authors' and publisher's permission from the Proceedings of the
National Academy of Sciences, *U. S., vol. 44(7), 1958, pp. 671–682.*

This paper was published five years after the papers presented by Watson and Crick. It is a logical extension of their work since it is the first, unambiguous confirmation of their prediction: that double-stranded DNA molecules would replicate in a semiconservative manner. As was noted in the introduction, the delay in confirmation was due primarily to the slow development of an experimental protocol that could clearly differentiate between parental and progeny DNA molecules. Ten years after the publication of this work, Meselson in a meeting at Cold Spring Harbor, projected a slide showing DNA replication as semiconservative — but with a fig leaf covering the replicating fork. We are still unsure about the structure of DNA at the replicating fork.

Introduction Studies of bacterial transformation and bacteriophage infection[1-5] strongly indicate that deoxyribonucleic acid (DNA) can carry and transmit hereditary information and can direct its own replication. Hypotheses for the mechanism of DNA replication differ in the predictions they make concerning the distribution among progeny molecules of atoms derived from parental molecules.[6]

Radiosotopic labels have been employed in experiments bearing on the distribution of parental atoms among progeny molecules in several organisms.[6-9] We anticipated that a label which imparts to the DNA molecule an increased density might permit an analysis of this distribution by sedimentation techniques. To this end, a method was developed for the detection of small density differences among macromolecules.[10] By use of this method, we have observed the distribution of the heavy nitrogen isotope N^{15} among molecules of DNA following the transfer of a uniformly N^{15}-labeled, exponentially growing bacterial population to a growth medium containing the ordinary nitrogen isotope N^{14}.

Density–Gradient Centrifugation A small amount of DNA in a concentrated solution of cesium chloride is centrifuged until equilibrium is closely approached. The opposing processes of sedimentation and diffusion have then produced a stable concentration gradient of the cesium chloride. The concentration and pressure gradients result in a continuous increase of density along the direction of centrifugal force. The macromolecules of DNA present in this density gradient are driven by the centrifugal field into the region where the solution density is equal to their own buoyant density.[‡] This concentrating tendency is opposed by diffusion, with the result that at equilibrium a single species of DNA is distributed over a band whose width is inversely related to the molecular weight of that species (Fig. 1).

If several density species of DNA are present, each will form a band at the position where the density of the CsCl solution is equal to the buoyant density of that species. In this way DNA labeled with heavy nitrogen (N^{15}) may be resolved from unlabeled DNA. Figure 2 shows the two bands formed as a result of centrifuging a mixture of ap-

*Aided by grants from the National Foundation for Infantile Paralysis and the National Institutes of Health.
†Gates and Crellin Laboratories of Chemistry (Contribution No. 2344), and Norman W. Church Laboratory of Chemical Biology, California Institute of Technology, Pasadena, California. [Communicated by Max Delbrück, May 14, 1958.]

‡The buoyant density of a molecule is the density of the solution at the position in the centrifuge cell where the sum of the forces acting on the molecule is zero.

HOURS

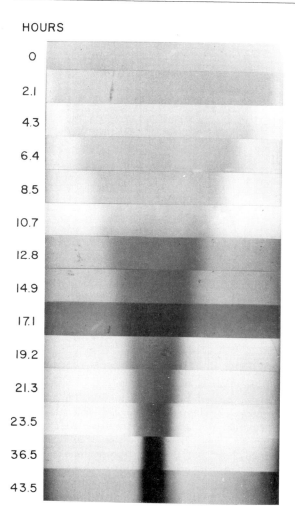

0

2.1

4.3

6.4

8.5

10.7

12.8

14.9

17.1

19.2

21.3

23.5

36.5

43.5

Fig. 1. Ultraviolet absorption photographs showing successive stages in the banding of DNA from *E. coli*. An aliquot of bacterial lysate containing approximately 10^8 lysed cells was centrifuged at 31,410 rpm in a CsCl solution as described in the text. Distance from the axis of rotation increases toward the right. The number beside each photograph gives the time elapsed after reaching 31,410 rpm.

proximately equal amounts of N^{14} and N^{15} *Escherichia coli* DNA.

In this paper reference will be made to the apparent molecular weight of DNA samples determined by means of density-gradient centrifugation. A discussion has been given[10] of the considerations

upon which such determinations are based, as well as of several possible sources of error.*

Experimental Escherichia coli B was grown at 36°C with aeration in a glucose salts medium containing ammonium chloride as the sole nitrogen source.[†] The growth of the bacterial population was followed by microscopic cell counts and by colony assays (Fig. 3).

Bacteria uniformly labeled with N^{15} were prepared by growing washed cells for 14 generations (to a titer of 2×10^8/ml) in medium containing 100 μg/ml of $N^{15}H_4Cl$ of 96.5 percent isotopic purity. An abrupt change to N^{14} medium was then accomplished by adding to the growing culture a tenfold excess of $N^{14}H_4Cl$, along with ribosides of adenine and uracil in experiment 1 and ribosides of adenine, guanine, uracil, and cytosine in experiment 2, to give a concentration of 10 μg/ml of each riboside. During subsequent growth the bacterial titer was kept between 1 and 2×10^8/ml by appropriate additions of fresh N^{14} medium containing ribosides.

Samples containing about 4×10^9 bacteria were withdrawn from the culture just before the addition of N^{14} and afterward at intervals of several generations. Each sample was immediately chilled and centrifuged in the cold for 5 minutes at 1,800× *g*. After resuspension in 0.40 ml of a cold solution 0.01 *M* in NaCl and 0.01 *M* in ethylenediaminetetra-acetate (EDTA) at pH 6, the cells were lysed by the addition of 0.10 ml of 15 percent sodium dodecyl sulfate and stored in the cold.

For density-gradient centrifugation, 0.010 ml of the dodecyl sulfate lysate was added to 0.70 ml of CsCl solution buffered at pH 8.5 with 0.01 *M* tris(hydroxymethyl) aminomethane. The density of

---

*Our attention has been called by Professor H. K. Schachman to a source of error in apparent molecular weights determined by density-gradient centrifugation which was not discussed by Meselson, Stahl, and Vinograd. In evaluating the dependence of the free energy of the DNA component upon the concentration of CsCl, the effect of solvation was neglected. It can be shown that solvation may introduce an error into the apparent molecular weight if either CsCl or water is bound preferentially. A method for estimating the error due to such selective solvation will be presented elsewhere.

[†]In addition to NH_4Cl, this medium consists of 0.049 *M* Na_2HPO_4, 0.022 *M* KH_2PO_4, 0.05 *M* NaCl, 0.01 *M* glucose, 10^{-3} *M* $MgSO_4$, and 3×10^{-6} *M* $FeCl_3$.

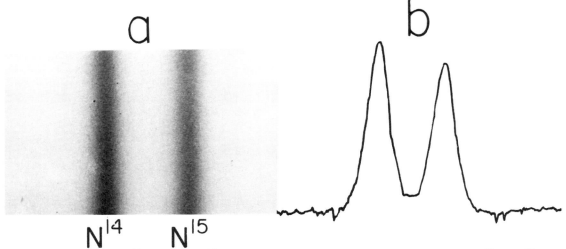

Fig. 2. *a:* The resolution of N^{14} DNA from N^{15} DNA by density-gradient centrifugation. A mixture of N^{14} and N^{15} bacterial lysates, each containing about 10^8 lysed cells, was centrifuged in CsCl solution as described in the text. The photograph was taken after 24 hours of centrifugation at 44,770 rpm. *b:* A microdensitometer tracing showing the DNA distribution in the region of the two bands of Fig. 2a. The separation between the peaks corresponds to a difference in buoyant density of 0.014 gm cm^{-3}.

the resulting solution was 1.71 gm cm^{-3}. This was centrifuged at 140,000X *g* (44,770 rpm) in a Spinco model E ultracentrifuge at 25° for 20 hours, at which time the DNA had essentially attained sedimentation equilibrium. Bands of DNA were then found in the region of density 1.71 gm cm^{-3}, well isolated from all other macromolecular components of the bacterial lysate. Ultraviolet absorption photographs taken during the course of each centrifugation were scanned with a recording microdensitometer (Fig. 4).

The buoyant density of a DNA molecule may be expected to vary directly with the fraction of N^{15} label it contains. The density gradient is constant in the region between fully labeled and unlabeled DNA bands. Therefore, the degree of labeling of a partially labeled species of DNA may be determined directly from the relative position of its band between the band of fully labeled DNA and the band of unlabeled DNA. The error in this procedure for the determination of the degree of labeling is estimated to be about 2 percent.

Results   Figure 4 shows the results of density-gradient centrifugation of lysates of bacteria sampled at various times after the addition of an excess of

N^{14}-containing substrates to a growing N^{15}-labeled culture.

It may be seen in Figure 4 that, until one generation time has elapsed, half-labeled molecules accumulate, while fully labeled DNA is depleted. One generation time after the addition of N^{14}, these half-labeled or "hybrid" molecules alone are observed. Subsequently, only half-labeled DNA and completely unlabeled DNA are found. When two generation times have elapsed after the addition of N^{14}, half-labeled and unlabeled DNA are present in equal amounts.

Discussion   These results permit the following conclusions to be drawn regarding DNA replication under the conditions of the present experiment.

1. *The nitrogen of a DNA molecule is divided equally between two subunits which remain intact through many generations.*

The observation that parental nitrogen is found only in half-labeled molecules at all times after the passage of one generation time demonstrates the existence in each DNA molecule of two subunits containing equal amounts of nitrogen. The finding that at the second generation half-labeled and

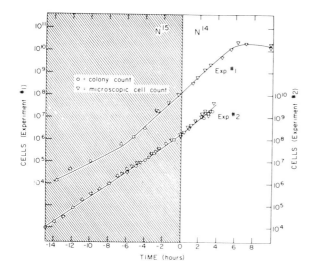

Fig. 3. Growth of bacterial populations first in N^{15} and then in N^{14} medium. The values on the ordinates give the actual titers of the cultures up to the time of addition of N^{14}. Thereafter, during the period when samples were being withdrawn for density-gradient centrifugation, the actual titer was kept between 1 and 2×10^8 by additions of fresh medium. The values on the ordinates during this later period have been corrected for the withdrawals and additions. During the period of sampling for density-gradient centrifugation, the generation time was 0.81 hours in Experiment 1 and 0.85 hours in Experiment 2.

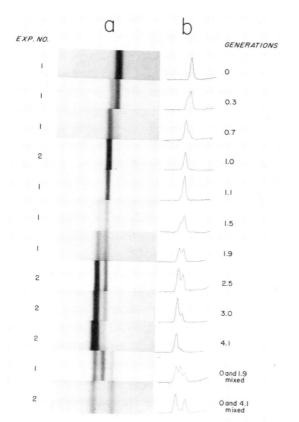

Fig. 4. *a:* Ultraviolet absorption photographs showing DNA bands resulting from density-gradient centrifugation of lysates of bacteria sampled at various times after the addition of an excess of N^{14} substrates to a growing N^{15}-labeled culture. Each photograph was taken after 20 hours of centrifugation at 44,770 rpm under the conditions described in the text. The density of the CsCl solution increases to the right. Regions of equal density occupy the same horizontal position on each photograph. The time of sampling is measured from the time of the addition of N^{14} in units of the generation time. The generation times for Experiments 1 and 2 were estimated from the measurements of bacterial growth presented in Fig. 3. *b:* Microdensitometer tracings of the DNA bands shown in the adjacent photographs. The microdensitometer pen displacement above the base line is directly proportional to the concentration of DNA. The degree of labeling of a species of DNA corresponds to the relative position of its band between the bands of fully labeled and unlabeled DNA shown in the lowermost frame, which serves as a density reference. A test of the conclusion that the DNA in the band of intermediate density is just half-labeled is provided by the frame showing the mixture of generations 0 and 1.9. When allowance is made for the relative amounts of DNA in the three peaks, the peak of intermediate density is found to be centered at 50 ± 2 percent of the distance between the N^{14} and N^{15} peaks.

unlabeled molecules are found in equal amounts shows that the number of surviving parental subunits is twice the number of parent molecules initially present. That is, the subunits are conserved.

2. *Following replication, each daughter molecule has received one parental subunit.*

The finding that all DNA molecules are half-labeled one generation time after the addition of N^{14} shows that each daughter molecule receives one parental subunit.* If the parental subunits had segregated in any other way among the daughter molecules, there would have been found at the first generation some fully labeled and some unlabeled DNA molecules, representing those daughters which received two or no parental subunits, respectively.

3. *The replicative act results in a molecular doubling.*

This statement is a corollary of conclusions 1 and 2 above, according to which each parent molecule passes on two subunits to progeny molecules and each progeny molecule receives just one parental subunit. It follows that each single molecular reproductive act results in a doubling of the number of molecules entering into that act.

The above conclusions are represented schematically in Figure 5.

The Watson-Crick Model A molecular structure for DNA has been proposed by Watson and Crick.[11] It has undergone preliminary refinement[12] without alteration of its main features and is supported by physical and chemical studies.[13] The structure consists of two polynucleotide chains wound helically about a common axis. The nitrogen base (adenine, guanine, thymine, or cytosine) at each level on one chain is hydrogen-bonded to the base at the same level on the other chain. Structural requirements allow the occurrence of only the hydrogen-bonded base pairs adenine-thymine and guanine-cytosine, resulting in a detailed complementariness between the two chains. This suggested to Watson and Crick[14] a definite and structurally plausible hypothesis for the duplication of the

DNA molecule. According to this idea, the two chains separate, exposing the hydrogen-bonding sites of the bases. Then, in accord with the base-pairing restrictions, each chain serves as a template for the synthesis of its complement. Accordingly, each daughter molecule contains one of the parental chains paired with a newly synthesized chain (Fig. 6).

The results of the present experiment are in exact accord with the expectations of the Watson-Crick model for DNA duplication. However, it must be emphasized that it has not been shown that the molecular subunits found in the present experiment are single polynucleotide chains or even that the DNA molecules studied here correspond to single DNA molecules possessing the structure proposed by Watson and Crick. However, some information has been obtained about the molecules and their subunits; it is summarized below.

The DNA molecules derived from *E. coli* by detergent-induced lysis have a buoyant density in CsCl of 1.71 gm cm^{-3}, in the region of densities found for T2 and T4 bacteriophage DNA, and for purified calf-thymus and salmon-sperm DNA. A highly viscous and elastic solution of N^{14} DNA was

Fig. 5. Schematic representation of the conclusions drawn in the text from the data presented in Fig. 4. The nitrogen of each DNA molecule is divided equally between two subunits. Following duplication, each daughter molecule receives one of these. The subunits are conserved through successive duplications.

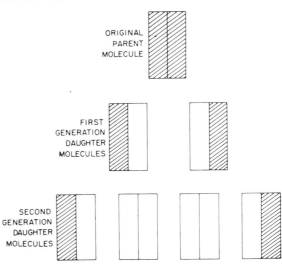

ORIGINAL PARENT MOLECULE

FIRST GENERATION DAUGHTER MOLECULES

SECOND GENERATION DAUGHTER MOLECULES

---

*This result also shows that the generation time is very nearly the same for all DNA molecules in the population. This raises the questions of whether in any one nucleus all DNA molecules are controlled by the same clock and, if so, whether this clock regulates nuclear and cellular division as well.

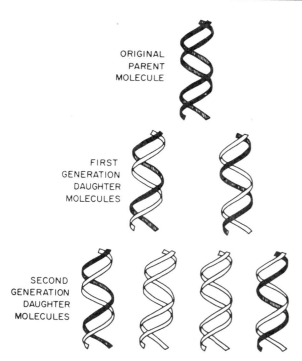

ORIGINAL
PARENT
MOLECULE

FIRST
GENERATION
DAUGHTER
MOLECULES

SECOND
GENERATION
DAUGHTER
MOLECULES

Fig. 6. Illustration of the mechanism of DNA duplication proposed by Watson and Crick. Each daughter molecule contains one of the parental chains (*black*) paired with one new chain (*white*). Upon continued duplication, the two original parent chains remain intact, so that there will always be found two molecules each with one parental chain.

prepared from a dodecyl sulfate lysate of *E. coli* by the method of Simmons[15] followed by deproteinization with chloroform. Further purification was accomplished by two cycles of preparative density-gradient centrifugation in CsCl solution. This purified bacterial DNA was found to have the same buoyant density and apparent molecular weight, 7×10^6, as the DNA of the whole bacterial lysates (Figs. 7, 8).

Heat Denaturation It has been found that DNA· from *E. coli* differs importantly from purified salmon-sperm DNA in its behavior upon heat denaturation.

Exposure to elevated temperatures is known to bring about an abrupt collapse of the relatively rigid and extended native DNA molecule and to make

available for acid base titration a large fraction of the functional groups presumed to be blocked by hydrogen-bond formation in the native structure.[15, 16, 17, 18] Rice and Doty[18] have reported that this collapse is not accompanied by a reduction in molecular weight as determined from light scattering. These findings are corroborated by density-gradient centrifugation of salmon-sperm DNA.[19] When this material is kept at 100° for 30 minutes either under the conditions employed by Rice and Doty or in the CsCl centrifuging medium, there results a density increase of 0.014 gm cm^{-3} with no change in apparent molecular weight. The same results are obtained if the salmon-sperm DNA is pretreated at pH 6 with EDTA and sodium dodecyl sulfate. Along with the density increase, heating brings about a sharp reduction in the time required for band formation in the CsCl gradient. In the absence of an increase in molecular weight, the decrease in banding time must be ascribed[10] to an increase in the diffusion coefficient, indicating an extensive collapse of the native structure.

Fig. 7. Microdensitometer tracing of an ultraviolet absorption photograph showing the optical density in the region of a band of N^{14} *E. coli* DNA at equilibrium. About 2 μg of DNA purified as described in the text was centrifuged at 31,410 rpm at 25° in 7.75 molal CsCl at pH 8.4. The density gradient is essentially constant over the region of the band and is 0.057 gm/cm^4. The position of the maximum indicates a buoyant density of 1.71 gm cm^{-3}. In this tracing the optical density above the base line is directly proportional to the concentration of DNA in the rotating centrifuge cell. The concentration of DNA at the maximum is about 50 μg/ml.

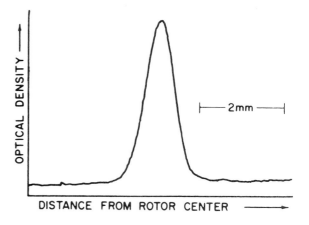

OPTICAL DENSITY ⟶

⊢—— 2mm ——⊣

DISTANCE FROM ROTOR CENTER ⟶

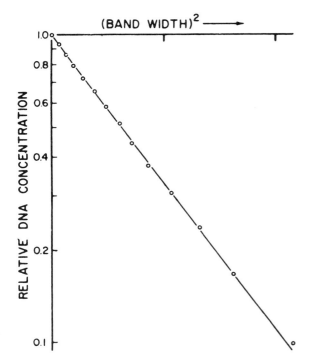

Fig. 8. The square of the width of the band of Fig. 7 plotted against the logarithm of the relative concentration of DNA. The divisions along the abscissa set off intervals of 1 mm². In the absence of density heterogeneity, the slope at any point of such a plot is directly proportional to the weight average molecular weight of the DNA located at the corresponding position in the band. Linearity of this plot indicates monodispersity of the banded DNA. The value of the slope corresponds to an apparent molecular weight for the Cs–DNA salt of 9.4×10^6, corresponding to a molecular weight of 7.1×10^8 for the sodium salt.

The decrease in banding time and a density increase close to that found upon heating salmon-sperm DNA are observed (Fig. 9, A) when a bacterial lysate containing uniformly labeled N^{15} or N^{14} $E.\ coli$ DNA is kept at 100°C for 30 minutes in the CsCl centrifuging medium; but the apparent molecular weight of the heated bacterial DNA is reduced to approximately half that of the unheated material.

Half-labeled DNA contained in a detergent lysate of N^{15} $E.\ coli$ cells grown for one generation in N^{14} medium was heated at 100°C for 30 minutes in the CsCl centrifuging medium. This treatment results in the loss of the original half-labeled material and

in the appearance in equal amounts of two new density species, each with approximately half the initial apparent molecular weight (Fig. 9, B). The density difference between the two species is 0.015 gm cm⁻³, close to the increment produced by the N^{15} labeling of the unheated DNA.

This behavior suggests that heating the hybrid molecule brings about the dissociation of the N^{15}-containing subunit from the N^{14} subunit. This possibility was tested by a density-gradient examination of a mixture of heated N^{15} DNA and heated N^{14} DNA (Fig. 9, C). The close resemblance between the products of heating hybrid DNA (Fig. 9 B) and the mixture of products obtained from heating N^{14} and N^{15} DNA separately (Fig. 9, C) leads to the conclusion that the two molecular subunits have indeed dissociated upon heating. Since the apparent molecular weight of the subunits so obtained is found to be close to half that of the intact molecule, it may be further concluded that the subunits of the DNA molecule which are conserved at duplication are single, continuous structures. The scheme for DNA duplication proposed by Delbrück[20] is thereby ruled out.

To recapitulate, both salmon-sperm and $E.\ coli$ DNA heated under similar conditions collapse and undergo a similar density increase, but the salmon DNA retains its initial molecular weight, while the bacterial DNA dissociates into the two subunits which are conserved during duplication. These findings allow two different interpretations. On the one hand, if we assume that salmon DNA contains subunits analogous to those found in $E.\ coli$ DNA, then we must suppose that the subunits of salmon DNA are bound together more tightly than those of the bacterial DNA. On the other hand, if we assume that the molecules of salmon DNA do not contain these subunits, then we must concede that the bacterial DNA molecule is a more complex structure than is the molecule of salmon DNA. The latter interpretation challenges the sufficiency of the Watson-Crick DNA model to explain the observed distribution of parental nitrogen atoms among progeny molecules.

Conclusion The structure for DNA proposed by Watson and Crick brought forth a number of proposals as to how such a molecule might replicate. These proposals[6] make specific predictions concerning the distribution of parental atoms among

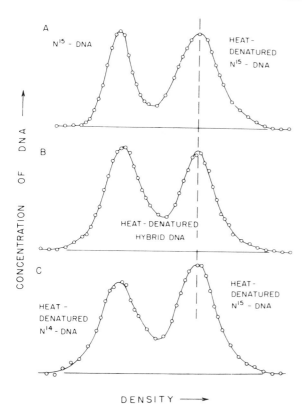

CONCENTRATION OF DNA →

DENSITY ⟶

Fig. 9. The dissociation of the subunits of *E. coli* DNA upon heat denaturation. Each smooth curve connects points obtained by microdensitometry of an ultraviolet absorption photograph taken after 20 hours of centrifugation in CsCl solution at 44,770 rpm. The baseline density has been removed by subtraction. *A:* a mixture of heated and unheated N[15] bacterial lysates. Heated lysate alone gives one band in the position indicated. Unheated lysate was added to this experiment for comparison. Heating has brought about a density increase of 0.016 gm cm^{-3} and a reduction of about half in the apparent molecular weight of the DNA. *B:* heated lysate of N[15] bacteria grown for one generation in N[14] growth medium. Before heat denaturation, the hybrid DNA contained in this lysate forms only one band, as may be seen in Fig. 4. *C:* a mixture of heated N[14] and heated N[15] bacterial lysates. The density difference is 0.015 gm cm^{-3}

progeny molecules. The results presented here give a detailed answer to the question of this distribution and simultaneously direct our attention to other problems whose solution must be the next step in

progress toward a complete understanding of the molecular basis of DNA duplication. What are the molecular structures of the subunits of *E. coli* DNA which are passed on intact to each daughter molecule? What is the relationship of these subunits to each other in a DNA molecule? What is the mechanism of the synthesis and dissociation of the subunits *in vivo*?

Summary By means of density-gradient centrifugation, we have observed the distribution of N[15] among molecules of bacterial DNA following the transfer of a uniformly N[15]-substituted exponentially growing *E. coli* population to N[14] medium. We find that the nitrogen of a DNA molecule is divided equally between two physically continuous subunits; that, following duplication, each daughter molecule receives one of these; and that the subunits are conserved through many duplications.

REFERENCES

1. R. D. Hotchkiss, in *The Nucleic Acids*, ed. E. Chargaff and J. N. Davidson (New York: Academic Press, 1955), p. 435; and in *Enzymes: Units of Biological Structure and Function*, ed. O. H. Gaebler (New York: Academic Press, 1956), p. 119.
2. S. H. Goodgal and R. M. Herriott, in *The Chemical Basis of Heredity*, ed. W. D. McElroy and B. Glass (Baltimore: Johns Hopkins Press, 1957), p. 336.
3. S. Zamenhof, in *The Chemical Basis of Heredity*, ed. W. D. McElroy and B. Glass (Baltimore: Johns Hopkins Press, 1957), p. 351.
4. A. D. Hershey and M. Chase, *J. Gen. Physiol.*, 36: 39, 1952.
5. A. D. Hershey, *Virology*, 1: 108, 1955; 4: 237, 1957.
6. M. Delbrück and G. S. Stent, in *The Chemical Basis of Heredity,* ed. W. D. McElroy and B. Glass (Baltimore: Johns Hopkins Press, 1957), p. 699.
7. C. Levinthal, these Proceedings, 42: 394, 1956.
8. J. H. Taylor, P. S. Woods, and W. L. Huges, these Proceedings, 43: 122, 1957.
9. R. B. Painter, F. Forro, Jr., and W. L. Hughes, *Nature*, 181: 328, 1958.
10. M. S. Meselson, F. W. Stahl, and J. Vinograd, these Proceedings, 43: 581, 1957.

11. F. H. C. Crick and J. D. Watson, *Proc. Roy. Soc. London, A*, 223: 80, 1954.

12. R. Langridge, W. E. Seeds, H. R. Wilson, C. W. Hooper, M. H. F. Wilkins, and L. D. Hamilton, *J. Biophys. and Biochem. Cytol.*, 3: 767, 1957.

13. For reviews see D. O. Jordan, in *The Nucleic Acids*, ed. E. Chargaff and J. D. Davidson (New York: Academic Press, 1955), 1: 447; and F. H. C. Crick, in *The Chemical Basis of Heredity*, ed. W. D. McElroy and B. Glass (Baltimore: Johns Hopkins Press, 1957), p. 532.

14. J. D. Watson and F. H. C. Crick, *Nature*, 171: 964, 1953.

15. C. E. Hall and M. Litt, *J. Biophys. and Biochem. Cytol.*, 4: 1, 1958.

16. R. Thomas, *Biochim. et Biophys. Acta*, 14: 231, 1954.

17. P. D. Lawley, *Biochim. et Biophys. Acta*, 21: 481, 1956.

18. S. A. Rice and P. Doty, *J. Am. Chem. Soc.*, 79: 3937, 1957.

19. Kindly supplied by Dr. Michael Litt. The preparation of this DNA is described by Hall and Litt (*J. Biophys. and Biochem. Cytol.*, 4: 1, 1958).

20. M. Delbrück, these Proceedings, 40: 783, 1955.

The next two papers illustrate the path which began to reveal the biochemical basis of replication. In 1956, Kornberg and his associates discovered an enzyme system that could polymerize nucleotide triphosphates into chains of DNA. The polymerizing enzyme was termed DNA polymerase. [This discovery won Kornberg the Nobel Prize in 1959 — three years before Watson, Crick, and Wilkins were awarded the prize for their discovery of DNA's structure.]

Although the enzyme could synthesize DNA *in vitro*, it seemed incapable of synthesizing biologically active DNA. The DNA synthesized *in vitro* bore little resemblance to native DNA: it was highly branched; it had abnormal denaturation properties; and, it added on nucleotides in only one direction (i.e., $5' \rightarrow 3'$). The latter presented a major problem. The polymerase work had already confirmed that the DNA strands were oppositely polar.

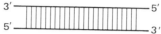

This meant that new DNA molecules would be generated by a synthetic process beginning at opposite ends.

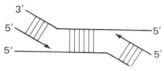

However, Cairns had already demonstrated by *in vivo* studies that DNA synthesis proceeds along both strands in the same direction.

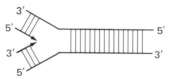

Okazaki worked out this discrepancy by suggesting a discontinuous scheme for the unidirectional synthesis of DNA. And, in 1967, Goulian, Kornberg, and Sinsheimer demonstrated that Kornberg's polymerase could synthesize biologically active DNA in the presence of a polydeoxyribonucleotide-joining enzyme (ligase). Temporarily, it seemed that Kornberg's polymerase could be the enzyme responsible for DNA synthesis *in vivo*.

By 1969, research appeared supporting the idea that Kornberg's DNA polymerase was not the replicating enzyme, but rather, a repair enzyme. In the opening paragraph of the first paper, we see that *E. coli* mutants were found that had normal levels of DNA polymerase, but could not replicate their DNA. Also, cells unusually sensitive to ultraviolet light were shown to be defective in Kornberg's polymerase (DNA polymerase I), yet they had normal replication. Since ultraviolet-induced genetic damage can be repaired by many bacteria, this and other observations strengthened the view that DNA polymerase I was primarily a repair enzyme.

De Lucia and Cairns isolated an *E. coli* mutant that had less than one percent of the normal DNA polymerase I activity, yet replicated its DNA in a normal fashion. Their discovery again raised the question: What enzyme was responsible for the replication of the *E. coli* genome? In 1970 and 1971, two more DNA polymerases were discovered (DNA polymerase II and DNA polymerase III). Could one of these function as the main polymerase? To a significant degree, the second paper answers this question. It summarizes much of the polymerase work up to the present and presents many new ideas on how the enzymes actually build new strands of DNA. The paper reveals that, today, most scientists recognize Kornberg's enzyme (DNA polymerase I) as having a twofold function: the excision of nucleotides followed by repair. This activity could be instrumental during the removal of the RNA primer responsible for the initiation of DNA synthesis and adding back deoxyribonucleotides; or, it could be involved with the excision and repair of errors of replication. In addition, the paper discusses the possible participation of seven nonpolymerase proteins during the replication process. Thus, the paper begins to illustrate the complexity of the DNA replication process.

Isolation of an *E. coli* strain with a mutation affecting DNA polymerase

PAULA DE LUCIA and JOHN CAIRNS*

Reprinted by authors' and publisher's permission from Nature, *vol. 224, 1969, pp. 1164–1166.*

By testing indiscriminately several thousand colonies of mutagenized *E. coli*, a mutant has been isolated that on extraction proves to have less than 1 percent of the normal level of DNA polymerase. The mutant multiplies normally but has acquired an increased sensitivity to ultraviolet light.

Kornberg's discovery of an enzyme that could faithfully copy DNA *in vitro*[1] was a crucial step in the history of molecular biology because it firmly established the fact that only a small part of a cell's DNA is needed to code for a mechanism that can duplicate the whole. Whether this is the enzyme responsible for DNA duplication *in vivo* was rightly thought, at that time, to be of secondary importance. Since then, however, circumstantial evidence has accumulated suggesting that, at least in bacteria, this particular enzyme is used for the repair of DNA rather than for its duplication. The various mutants of *Escherichia coli* and *Bacillus subtilis* that are unable to duplicate their DNA at high temperature have all been shown to contain normal polymerase and normal deoxyribonucleoside triphosphate pools at the nonpermissive temperature,[2-6] and at least one of them has been shown to carry out repair synthesis at high temperature.[7] Repair replication and the process of DNA duplication apparently differ in the extent to which they discriminate against 5-bromo-uracil as an acceptable substitute for thymine, suggesting that the two reactions involve different polymerases.[8] Finally, the 5'-exonucleolytic activity, recently shown to be an intrinsic property of the *E. coli* polymerase,[9] is clearly a desirable attribute for an enzyme responsible for excision and repair but is of no obvious advantage for an enzyme carrying out semiconservative replication.

These and other less persuasive arguments prompted us to look for mutants of the polymerase, in the hope that they would either establish a role for the polymerase in DNA duplication or exclude it and, at the same time, provide convenient strains in which to search for the right enzyme. Although we have not succeeded in these more distant objectives, we have isolated such a mutant and here describe the method of isolation and some of its properties. The accompanying article describes a genetic study of the mutation.

THE SELECTIVE PROCEDURE

The successful isolation of mutants of *E. coli* lacking ribonuclease I[10] demonstrated that it is possible to find the mutant one wants simply by testing individually several hundred colonies grown from a heavily mutagenized stock. Because we wished to avoid having to guess what symptoms, if any, would result from a lack of DNA polymerase, we decided to follow that example and assay the polymerase in clones of a mutagenized stock until we found what we were looking for. We had to allow for the possibility that the mutation we sought might be a conditional lethal, so we began by assaying at 45° extracts made from clones grown at 25° or 30°; later we tested clones grown at 37°, thinking that temperature-sensitive mutants of the polymerase might be more readily detectable if the enzyme had been assembled at a higher temperature. As it turned out, the mutant we eventually isolated would have been found whatever approach had been adopted, and we shall therefore simply give the history of the mutant when we describe its isolation and properties.

EXTRACTION OF POLYMERASE

Because we expected to have to test many hundred colonies, we required a very simple method for preparing extracts. In addition, we needed a pro-

*Cold Spring Harbor Laboratory, Cold Spring Harbor, New York 11724

cedure which made the bacteria incapable of incorporating deoxyribonucleosides, to ensure that labeled triphosphates could not enter DNA by way of breakdown to nucleosides and incorporation by those few cells that might have survived the extraction procedure. These two requirements were satisfied by the slight modification of a method devised for extracting polysomes, using the nonionic detergent Brij-58.[11] E. coli is suspended at a concentration of about 3×10^9/ml in ice cold 10 percent sucrose 0.1 M Tris (pH 8.5): lysozyme and EDTA are added to final concentrations of 50 μg/ml and 0.005 M, respectively, and the mixture is kept on ice for 30 min; addition of a mixture of Brij and $MgSO_4$ (at room temperature) to give final concentrations of 5 percent and 0.05 M, respectively, results in partial clearing; following centrifugation (1,500g for 30 min), the deposit contains 99.9 percent of the DNA and the supernatant contains the polymerase, which may then be assayed simply by adding sonicated calf thymus DNA (to 50 μg/ml) and the four deoxyribonucleoside triphosphates (to a final concentration of 4 nmoles/ml dATP, dGTP, dCTP and 2 nmoles/ml ^{3}H-TTP).

This extraction procedure demonstrates one point of interest: any method of lysis that liberates fragmented DNA will automatically create sites for the attachment of polymerase and therefore cannot give a true picture of the location of the polymerase *in vivo*.[12] Extraction with Brij yields cells which still contain their DNA but, on resuspension, have little if any ability to incorporate deoxyribonucleoside triphosphates. Because Brij apparently does not dissociate polymerase from its template (the polymerase being assayable in the presence of Brij), we can conclude that most of the polymerase in E. coli is normally not attached to DNA but lies free within the cell — as might befit an enzyme awaiting the summons to repair synthesis. This conclusion is supported by the observation that when E. coli segregates daughter cells which lack DNA these cells nevertheless retain their full quota of DNA polymerase.[13,14]

ISOLATION OF THE MUTANT

E. coli W3110 *thy*[-], growing in minimal medium, was washed and suspended in 0.15 M acetate (pH 5.5), treated with N-methyl-N'-nitro-N-nitrosoguanidine (1 mg/ml) for 30 min, and then centrifuged

and suspended in Penassay broth.[15] Following growth at 25°C for 18 h, the culture was plated; after incubation overnight at 37°C, the colonies were picked into 1 ml lots of Penassay broth which were incubated overnight at 37°C and then centrifuged and extracted with lysozyme and Brij.

After testing a few thousand colonies in this way we found a clone, p3478, that appeared to lack polymerase activity. It was therefore tested again using a more conventional method for extracting the enzyme. According to this test (Fig. 1), extracts of the mutant have 0.5–1.0 percent of the normal activity. This decrease in activity does not seem to arise from the presence of an inhibitor.

SOME PROPERTIES OF THE MUTANT

As far as we can determine, the mutant multiplies at the same rate as the parent strain, in minimal and complete media, and at temperatures from 25° to 42°C. On plating, it forms slightly smaller colonies than those of the parent strain, and occasionally it seems to have difficulty in getting out of stationary phase, but we have not investigated further either of these phenomena.

Parent and mutant are equally susceptible to infection with T4, T5, T7 and λ bacteriophages. When converted to spheroplasts, they are equally susceptible to ϕX174 DNA and produce equal yields of phage (personal communication from David Dressler). This finding was somewhat surprising, but it should be remembered that all stages in the replication of ϕX174 DNA are temperature sensitive in a mutant that is temperature sensitive for normal DNA replication[17] but not for repair synthesis.[7]

With regard to host cell reactivation, there is no detectable increase in the rate of inactivation of T7 by ultraviolet light, when the survivors are assayed on the mutant rather than the parent. Thus the mutant is *her*[+].

The mutant has a marked increase in sensitivity to ultraviolet light. For convenience, this effect will be documented in the following article,[18] where the sensitivities of various derivative strains are compared.

The parent strain will form colonies normally in the presence of 0.04 percent methylmethanesulphonate, whereas the mutant plates with an efficiency of about 10^{-7}. We assume that these rare methylmethanesulphonate-resistant cells are revertants that

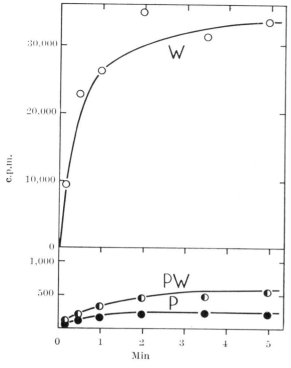

Fig. 1. Triphosphate incorporation by extracts of the parent strain (W), the mutant strain (P) and a mixture of 90 percent mutant and 1 percent parent (PW). *E. coli* W3110 *thy-* and the mutant derivative, P3478, were grown with aeration in Penassay broth at 37° to about 5×10^8/ml. Each culture was then chilled, centrifuged and suspended in 0.1 M Tris–0.01 M $MgSO_4$ (*p*H 7.4) at a concentration of 6×10^9/ml. A mixture of 1 percent parent strain, 99 percent mutant strain was prepared. This and the two unmixed suspensions were centrifuged and suspended in Tris-Mg^{2+} at a concentration of 1×10^{10}/ml. The three suspensions were disrupted by sonic vibration and mildly centrifuged (1,000*g* for 10 min.). To 0.9 ml of each supernatant at 25° 0.1 ml sonicated calf thymus DNA was added (final concentration 20 μg/ml) and, 5 min later, 0.3 ml triphosphate solution (final concentrations 100 nmoles/ml dGTP, dATP, dCTP, and 0.6 nmoles/ml, 2.5 μCi/ml, ^{3}H-TTP). Samples of 0.2 ml were taken from each reaction mixture into 5 ml 5 percent trichloroacetic acid–1 percent sodium pyrophosphate.[16] These samples were then washed on Whatman GFA filters with 5 percent trichloroacetic acid and with 5 percent acetic acid, dried and counted in a scintillation counter.

have either arisen spontaneously or been created by the methylmethanesulphonate. Because every one of twenty such independently arising revertants

exhibited normal sensitivity to ultraviolet light and had normal or near-normal levels of polymerase, it is clear that the three basic properties of the mutant (UVs, MMSs and lack of polymerase) are the result of a single mutational step.

REPAIR OR REPLICATION

The accompanying article[18] demonstrates that we are dealing with an amber mutation which is recessive in partial diploids. We assume that it is in the gene coding for DNA polymerase, although proof will require the demonstration that it — or other similar mutations — results in changes in the polymerase protein. Because the mutation produces an increased sensitivity to ultraviolet light, it seems likely that recovery from the effects of ultraviolet light is partly the responsibility of this polymerase.

Unfortunately, it is not going to be easy, by a study of this or other such mutants, to show that this polymerase plays no part in normal DNA duplication. Because *E. coli* contains several hundred polymerase molecules per bacterium,[19] the residual activity found in extracts of our mutant could represent perhaps 5–10 molecules per cell — a number that could well be sufficient for normal duplication. Even if we could somehow prove that the residual activity were entirely that of another enzyme (in other words, that this amber mutation is not measurably leaky), we should still not have proved that duplication is carried out by some other enzyme, for it could readily be argued that those few polymerase molecules concerned with duplication are necessarily incorporated into some larger enzyme complex the activity of which is not assayable *in vitro*. It could even be argued that more of the polymerase molecule must be intact for it to serve as a repair enzyme (and, incidentally, to survive extraction) than for it to act when part of the replicating machinery. We therefore believe that the question will be resolved either by engineering a total deletion of the polymerase gene or by determining, in some direct manner, which enzymes and what precursors are used for normal DNA duplication. It is our hope that each of these exercises will have been made easier now that the polymerase gene has probably been located[18] and a mutant is generally available.

We thank Dr. Raymond Gesteland (who pioneered this kind of mutant hunt) for encouragement; Dr. David Dressler for testing our mutant with ϕX174 and for permission to cite his results; and Drs. Julian and Marilyn Gross for arranging to stay on at Cold Spring Harbor to conduct most of the experiments reported in the next article. The work was supported by a grant from the U.S. National Science Foundation.*

*Received November 26, 1969.

REFERENCES

1. Lehman, I. R., Bessman, M. J., Simms, E. S., and Kornberg, A., *J. Biol. Chem.*, 233: 163 (1958).
2. Bonhoeffer, F., *Z. Vererbungslehre*, 98: 141 (1966).
3. Buttin, G., and Wright, M., *Cold Spring Harbor Symp. Quant. Biol.*, 33: 259 (1968).
4. Fangman, W. L., and Novick, A., *Genetics*, 60: 1 (1968).
5. Gross, J. D., Karamata, D., and Hempstead, P. G., *Cold Spring Harbor Symp. Quant. Biol.*, 33: 307 (1968).
6. Hirota, Y., Ryter, A., and Jacob, F., *Cold Spring Harbor Symp. Quant. Biol.*, 33: 677 (1968).
7. Couch, J., and Hanawalt, P. C., *Biochem. Biophys. Res. Commun.*, 29: 779 (1967).
8. Kanner, L., and Hanawalt, P. C., *Biochim. Biophys. Acta*, 157: 532 (1968).
9. Kornberg, A., *Science*, 163: 1410 (1969).
10. Gesteland, R. F., *J. Mol. Biol.*, 16: 67 (1966).
11. Godson, G. N., and Sinsheimer, R. L., *Biochim. Biophys. Acta*, 149: 476 (1967).
12. Billen, D., *Biochim. Biophys. Acta*, 68: 342 (1963).
13. Cohen, A., Fisher, W. D., Curtiss, R., and Adler, H. I., *Cold Spring Harbor Symp. Quant. Biol.*, 33: 635 (1968).
14. Hirota, Y., Jacob, F., Ryter, A., Buttin, G., and Nakai, T., *J. Mol. Biol.*, 35: 175 (1968).
15. Adelberg, E. A., Mandel, M., and Chien Ching Chen, G., *Biochem. Biophys. Res. Commun.*, 18: 788 (1965).
16. Hurwitz, J., Gold, M., and Anders, M., *J. Biol. Chem.*, 239: 3462 (1964).
17. Steinberg, R. A., and Denhardt, D. T., *J. Mol. Biol.*, 37: 525 (1968).
18. Gross, J. D., and Gross, M., *Nature*, 224: 1166 (1969).
19. Richardson, C. C., Schildkraut, C. L., Aposhian, H. V., and Kornberg, A., *J. Biol. Chem.*, 239: 222 (1964).

Multienzyme systems of DNA replication

Proteins required for chromosome replication are resolved with the aid of a simple viral DNA template.

RANDY SCHEKMAN, ALAN WEINER, and ARTHUR KORNBERG*

Reprinted by authors' and publisher's permission from Science, *vol. 186, 1974, pp. 987–993. Copyright © 1974 by the American Association for the Advancement of Science.*

Template direction of nucleic acid synthesis was seen first in DNA synthesis catalyzed by DNA polymerase and only later was observed in RNA synthesis by RNA polymerase. Nevertheless, work on RNA polymerase during the last 10 years has explained key physiological features of transcription,[1] whereas the reconstruction *in vitro* of replication by DNA polymerase action alone has remained inadequate. The basis for this discrepancy in progress is rooted in a biochemical fortuity. By means of enzyme fractionation procedures, RNA polymerase was isolated from cell extracts as a large multisubunit transcriptase. Isolated DNA polymerase, on the other hand, is only one component of a multienzyme DNA replicase. The purpose of this article is to describe the first stage in our efforts to identify and reassemble the pieces of this multienzyme system.

Polymerase Actions The basic elements in synthesis of a nucleic acid are the same whether the chain produced is DNA or RNA, and whether template directions are taken from DNA or RNA.[2] Elongation of a $3',5'$-phosphodiester-linked polynucleotide invariably has these features: (i) Substrates are a $3'$-hydroxyl terminated chain (primer terminus) and a $5'$-nucleoside triphosphate. Nucleophilic attack by the primer terminus adds a $5'$-nucleoside monophosphate to the chain. (ii) Selection of the specific triphosphate depends on its forming a Watson-Crick base pair with the template. (iii) Growth of the chain is necessarily in a $5' \rightarrow 3'$ direction, antiparallel to the template.

Among the more than 20 DNA polymerases of viral, bacterial, and animal origin isolated to date,

none can start a chain *in vitro*. This feature distinguishes DNA polymerases most clearly from RNA polymerases. The essence of transcription is the highly selective copying of passages from the chromosome record, and the capacity of RNA polymerases to start chains at defined "promoter" sequences is a prominent part of its function. Thus, DNA polymerases, remarkable for their error-free copying of the entire chromosome, are apparently blind to initiation signals, including the one promoting the origin of a replication cycle. Until recently the enzymatic mechanism of starting DNA chains remained an enigma.

The Chain Initiation Enigma Two kinds of DNA chain starts need to be considered: (i) initiation of chromosome replication at its unique origin, and (ii) initiation of the short replication fragments that are synthesized discontinuously (Okazaki pieces) at the growing fork or nascent region of the chromosome. This distinction is indicated by genetic and biochemical experiments. Several *Escherichia coli* mutants, thermosensitive in DNA replication [*dna*A and *dna*C,[3] and, more recently, *dna*H and *dna*I],[4] are defective in initiating replication at the chromosome origin. They are unlike other mutants [*dna*B, *dna*E, and *dna*G][3] whose DNA replication stops abruptly when the temperature is raised to a restrictive level; the origin-defective mutants continue DNA synthesis until the chromosome duplication under way is completed. Among the abrupt-stop mutants, *dna*G has been implicated in nascent chain starts[5] and *dna*E shows a defect in DNA polymerase III.[6] Fruitful as these genetic studies are, elucidating the mechanism of chain initiation requires a biochemical approach.

Discovery of these numerous thermosensitive, *E. coli* replication mutants illustrates the multiplicity of gene products needed for chromosome replication.

---

*Mr. Schekman is a graduate student, Dr. Weiner is a Helen Hay Whitney postdoctoral fellow, and Dr. Kornberg is a professor in the Department of Biochemistry, Stanford University School of Medicine, Stanford, California 94305.

This was anticipated since the phage T4 chromosome, only a twentieth the size of *E. coli*, still induces formation of at least six proteins essential for its replication. We now find from our studies reported below that much of the complexity of the replicative machinery resides in the events of chain initiation. This recent progress is based on two things: the use of small phage chromosomes as probes and the development of a cell extract consisting of soluble enzymes and capable of phage chromosome replication.

Phage Probes and Soluble Enzymes Attempts to understand how DNA chains are started and elongated had been frustrated by the use of large chromosomes, such as those of bacteria and medium-sized phages, and, in addition, by the fragmentation of the multienzyme replication system. Enzymological studies have been greatly aided by the use of the single-stranded (SS) circular chromosome of the small DNA phages.[2]

The filamentous phage M13 (also fl and fd) and polyhedral phage ϕX174 (also S13 and G4) offer several crucial advantages. (i) The tiny viral chromosome contains only eight to ten genes, devoted mostly to coat proteins and assembly and so must rely on the replicative systems of the host cell. Thus, in appropriating host enzymes for its replication, the phage chromosome may illuminate mechanisms by which more complex chromosomes are handled. (ii) The viral chromosome is a relatively simple template molecule preparable in large quantity, is easily purified and characterized, and becomes part of a readily analyzed product. (iii) The initial event upon infection is conversion of a SS circle of DNA to a double-stranded replicative form (RF). Therefore, initiation of a new chain on the viral template *in vitro* should be less complicated by artifact than with a duplex template. In the latter, endonucleolytic scission of one strand can provide a primer terminus for copying the other strand. (iv) The template may be so small as to contain only one initiation signal.

In a series of investigations in which the bacterial chromosome served as template, DNA replication could be observed only with permeable cells or with membranous, immobilized lysates.[7] Many valuable insights into the role of external substrates and factors emerged from the studies with these complex subcellular preparations, but resolution of the replicative system into its molecular components was

not attained. The choice of a tiny phage chromosome offered not only the practical advantages just enumerated, but, equally important, the psychological impetus needed to undertake the resolution of a complete replication system. It seemed reasonable that the apparently simple copying of a small DNA circle could be attained with purified enzymes and that, without such success, attempts with large chromosomes would be futile. A soluble enzyme system which could be resolved into its components was clearly necessary.

Preparing such a soluble enzyme extract from *E. coli*, which efficiently converts the M13 or ϕX174 to RF, depended on several exacting conditions[8, 9] (i) the harvesting of young growing cells without chilling, (ii) the use of gentle lysozyme lysis, with minimal disruption of DNA, and (iii) the removal of virtually all the host DNA with particulate material by sedimentation at high speed. Extracts prepared in this way have since been applied successfully to replicate the larger phage chromosomes of T7 and λ[10] and, most recently, the intact, folded *E. coli* chromosome itself.[11]

RNA Priming of DNA Synthesis The ability of RNA polymerases to start new chains, and the capacity of DNA polymerase to extend a polyribonucleotide primer, led us to explore the possibility that a brief RNA transcript might prime DNA synthesis. The effect of rifampicin (a specific inhibitor of RNA polymerase) on the conversion of the M13 viral circle to its duplex form was chosen as a test of this hypothesis. The drug blocked the conversion of M13 DNA both *in vivo* and in cell extracts.[8, 12] Rifampicin-resistant RNA polymerase mutants did not show this effect. Additional evidence then indicated that the action of RNA polymerase was needed to provide an RNA primer. A stage of transcription was separable from a subsequent stage of DNA synthesis. A dramatic verification of RNA priming by RNA polymerase action has come with the isolation of an intermediate form in rifampicin-sensitive replication of the DNA plasmid colicin E1, with the RNA piece still in place.[13]

The initial choice of M13 replication (SS to RF conversion) as an object for examining rifampicin inhibition was fortunate, inasmuch as a test with ϕX174 proved negative.[8] The ϕX174 result, because it came later, did not discourage our belief in a generalized status for the RNA-priming hypothesis.

Instead, the lack of dependence of φX174 DNA synthesis on initiation by RNA polymerase spurred experiments which disclosed RNA priming by a novel RNA synthetic system. Although rifampicin-resistant, the φX174 reaction nevertheless required all four ribonucleoside triphosphates leading to a covalent attachment of RNA to the DNA product.[14, 15]

With the use of M13, φX174, and, more recently, of G4 – which is a φX174-related phage[16] – as probes, we have obtained evidence, presented below, that at least three distinctive replicative systems exist in E. coli, each with a characteristic RNA-priming mechanism. Beyond these examples illustrated by the small coliphages, RNA priming of DNA synthesis has been observed in many instances, such as in the nascent replication fragments of the E. coli chromosome[17] and in animal cells utilizing as template either tumor virus DNA[18] or their own DNA.[19]

Resolution of a Multienzyme System   There are two approaches to fractionation of a multienzyme system responsible for a complex reaction. (In this case it is the incorporation of deoxynucleotides into DNA, dependent on M13, φX174, or G4 DNA templates.) One is by complementation of an extract prepared from mutant cells. Assay at an elevated temperature, at which the thermosensitive enzyme is inactive, determines the amount of added wild-type enzyme available. In this way each of the components, for which a mutant deficiency has been identified, can be isolated from wild-type cells. The second approach is direct resolution and reconstitution of fractionated wild-type enzymes, where protein resolving reagents are used to sort out and eventually to isolate each of the components. There are serious limitations to each approach.

Complementation assays have suffered from the following drawbacks. (i) The thermosensitive protein is always present and may be stabilized by a variety of nonspecific compounds. As an example, complementation assays of the *dna*A gene product were confused by the capacity of polyethylene glycol (introduced during a phase partition purification step) to specifically stabilize the mutant extract at a nonpermissive temperature. (ii) If the wild-type protein is normally part of a multienzyme package, the mutant component must be physically replaced. (iii) Multiple deficiencies are often manifested in a

mutant extract. Defective enzyme Y may interfere with enzyme Z activity. If Z proves rate-limiting in the assay, Z rather than Y will be purified from wild-type cells. In one instance a *dna*G mutant extract was more deficient in another required activity, DNA polymerase III*, than in the mutant *dna*G protein. Complementation assays with this extract resulted in the mistaken purification of DNA polymerase III*. (iv) When an enzyme has finally been purified, it is unlikely that its mechanism of action can be deduced from complementation of a crude extract. (v) When all the enzymes for which mutants are available have been purified, essential enzymes may still be lacking for which mutants are still unknown. Furthermore, enzymes purified for their capacity to complement a crude extract may depend on activating factors in the extract lacking in a collection of pure enzymes.

There are also serious drawbacks to resolution and reconstitution of a wild-type system. First of all, adequate criteria are necessary to establish an authentic reconstitution. Drug resistance or sensitivity, substrate specificity, or specific enzyme requirements may be used. Once the criteria are available, techniques that give sharp enough resolution are required to provide a reliable assay for each component. Assays are complicated by the need for supplies of each of the numerous purified components, some of which are rather unstable. Order of addition of the individual components to the incubation mixture and their relative amounts may be crucial. Even when all the essential purified components appear to be in hand, the list may lack some that function in the system *in vivo* or may include some that do not.

Resolution of the φX174 System   Earlier studies showed that the SS to RF reaction with φX174 DNA required many proteins also necessary for E. coli chromosome replication. Soluble extracts prepared from temperature-sensitive DNA synthesis mutants *dna*A, B, C, E, and G were defective compared to those from temperature-resistant revertants.[14, 20] The *dna*B, C, E, and G gene products have been partially purified by means of a complementation assay;[21-23] however, these proteins are not sufficient for the overall reaction.

We have used a combination of both complementation and total fractionation to resolve and identify the components of the φX174 system. The flow diagram in Fig. 1 indicates the major steps used to segregate the essential components. The purification procedure for a component was reexamined, starting in each case with an independent procedure from the first step to optimize yield and purity.

The *dna*C protein was separated from other essential components by ammonium sulfate fractionation and assayed by complementation of a mutant (*dna*C) extract.[22] After further purification by phosphocellulose chromatography, this fraction serves in rifampicin-resistant synthesis of ϕX174 DNA by the mutant extract but has no influence on M13 DNA synthesis (Fig. 2). Requirement for *dna*C protein was the touchstone for authentic reconstitution in subsequent fractionation. The other components were divided into two groups by passage through a DNA-cellulose column (Fig. 1). At this stage three fractions are needed for ϕX174 DNA synthesis: the fraction unadsorbed to DNA-cellulose, the fraction bound to and eluted from it, and the *dna*C protein fraction.

The fraction unadsorbed to DNA-cellulose was further divided by chromatography on Sephadex G-150 into an excluded and an included fraction. Each of these was then purified by further chromatography to yield, respectively, *dna*B protein (identified by complementation with a *dna*B mutant extract) and protein i (not identified with any of the *dna*A, B, C, E, or G gene products). At this

stage, the components required for activity were DNA-cellulose binding fraction, *dna*B protein, protein i, and *dna*C protein.

The fraction adsorbed to, and eluted from, DNA-cellulose was dissected by sensitivity to the sulfhydryl-blocking agent N-ethylmaleimide (NEM) and further resolved by chromatography on Sephadex G-150. Excluded from the gel was the NEM-sensitive, DNA polymerase III holoenzyme (a complex containing DNA polymerase III* and copolymerase III* activities, as is discussed below). Included in the gel were three essential proteins. Two were NEM-resistant and separable into *dna*G protein — identified by complementation[23] — and the DNA-unwinding protein;[24] the other was an NEM-sensitive protein (protein n), corresponding to none of the *dna*A, B, C, E, or G gene products.

Properties of the Reconstituted ϕX174 System
The seven protein fractions incubated with the four ribo- and four deoxyribonucleoside triphosphates, spermidine, and Mg^{2+} replicated the ϕX174 viral template (Table 1). Each of the purified proteins was required in the reconstituted reaction; omission of any one reduced synthesis by 5- to 50-fold. The extent of dependence on each protein has varied with degrees of cross contamination in different preparations. In no instance could any pair of proteins be omitted, as might be seen in the action of

Fig. 1. Scheme for resolution of the ϕX174 enzymes. Soluble extracts were prepared from *E. coli* H560 as described before.[8,9] The outlined procedure[22] of subdividing the essential components provided assays for purifying each of the proteins.

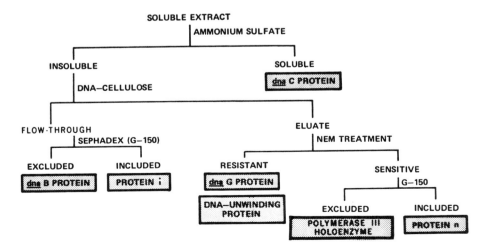

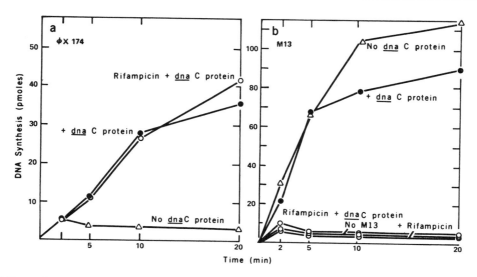

Fig. 2. Partially purified *dna*C protein required for φX174 (a) but not M13 (b) replication. Partially purified *dna*C protein[22] was added to a soluble enzyme fraction prepared from *E. coli* PC79 (*dna*C mutant, formerly *dna*D) as described before for wild-type *E. coli*.[8,9] Assays (25-μl volume) included 0.3 mg of mutant-extract protein, 0.3 nmole of φX174 or M13 DNA, 20 nmole of ATP, 2.5 nmole each of GTP, UTP, and CTP, 1.25 nmole each of dATP, dGTP, and TTP, 0.45 nmole of [^{32}P] dCTP (400 count/min pmole), 125 nmole of MgCl$_2$, 100 nmole of spermidine hydrochloride; 100 ng of rifampicin and 2.5 μg of *dna*C protein were present where indicated. Incubation was at 30°C.

an anti-inhibitor overcoming the effect of an inhibitor. The reaction is resistant to rifampicin. At 30°C, DNA synthesis progressed at a constant rate for about 40 minutes (Fig. 3a) with as much as 50 percent of the DNA copied. The product was RFII, in which a nearly full-length synthetic strand complements the circular viral strand. A small gap between the 3′ and 5′ ends of the synthetic strand remains. Whether this gap is at a unique location relative to the template, as is true for G4 and M13 (see below), is yet to be determined. Conversion of the RFII to the covalently closed duplex RFI requires excision of the RNA-priming fragment at the 5′ end by the 5′ →3′ exonuclease function of DNA polymerase I coupled with its replicative gap-filling activity, and DNA ligase action to seal the circle.[15]

Division of the standard reaction into two stages was obtained by dilution. Incubation with the ribonucleoside triphosphates but without deoxynucleoside triphosphates was followed by a tenfold dilution with a buffer containing the latter. Formation of a primed single-strand was measured by its ability to support subsequent DNA synthesis. The rate of the

first-stage, priming reaction was constant for 10 minutes then decreased (Fig. 3b), with dilution at any intermediate time reducing the subsequent extent of DNA synthesis. In the absence of new initiations, the second stage was complete within 2 minutes. Optimal reaction in the first stage required all four ribonucleoside triphosphates (Table 2); addition of omitted ribonucleoside triphosphates in the DNA synthesis stage did not suffice. Two-stage reactions were only 20 to 30 percent as efficient as standard reactions. This may be due to degradation of the primer by contaminating ribonucleases in the absence of immediate extension by DNA.

DNA polymerase III holoenzyme[25, 26] was absolutely required in the reconstituted system (Fig. 4). This multisubunit enzyme contains two *dna*E polypeptides (90,000-dalton chains) and two molecules of copolymerase III* (77,000-dalton chain). The holoenzyme is active on long, SS templates, such as RNA-primed φX174 viral DNA, as well as on templates with short (10 to 20 nucleotides) exposed regions. By contrast, DNA polymerase III can utilize only the latter.[27]

Table 1 Enzyme requirements for the reconstituted ϕX174 and G4 reactions. All the purified proteins were assembled in the complete reaction or omitted individually, as indicated. Assays were performed in 25-μl volumes at 30°C for 20 minutes. Components were mixed in the following order: 10 μl of dilution buffer (10 percent sucrose, 50 mM tris-HCl, pH 7.5, 20 mM dithiothreitol, 0.2 mg of bovine serum albumin per milliliter) 0.13 μg of protein n, 0.73 μg of DNA-unwinding protein, a mixture of protein i (40 ng), dnaB protein (20 ng), dnaG protein (40 ng) and DNA polymerase III holoenzyme (0.33 μg), and then 2.5 μg of dnaC protein. A mixture of other components—1.25 nmole each of dATP, dGTP and dCTP,[*] 0.45 nmole of [³H]TTP (240 count/min pmole), 125 nmole of MgCl$_2$, 20 nmole of ATP, 2.5 nmole each of GTP, UTP, and CTP, 40 nmole of spermidine HCl, 0.33 nmole of ϕX174 or G4 DNA and 100 ng of rifampicin—was added to start the reaction. The DNA synthesis was measured by incorporation of the labeled deoxynucleotide into an acid-insoluble form.

Omitted item	Reconstitution (% relative to complete reaction)[†]	
	ϕX174	G4
None (complete[†])	100	100
dnaB protein	1	150
dnaC protein	7	150
DNA polymerase III holoenzyme	2	12
dnaG protein	7	13
Protein i	12	130
Protein n	4	150
Spermidine	2	75
Unwinding protein,	15	140
spermidine	1	12

[*]Abbreviations: ATP, GTP, UTP, CTP, adenosine, guanosine, uridine, and cytidine triphosphates; dATP, dGTP, TTP and dCTP, the corresponding deoxynucleoside triphosphates; dNTP, deoxynucleoside triphosphate; rNTP, ribonucleoside triphosphate; SS, single-stranded DNA; RF, replicative-form DNA.

[†]In the complete reactions 120 pmole of deoxynucleotide was incorporated with ϕX174 DNA as template, 42 pmole with G4 DNA, but only 2.5 pmole with M13 DNA; the M13 DNA was used under conditions identical to those used for ϕX174 and G4 DNA's.

Either spermidine or the DNA-unwinding protein serve the holoenzyme in its replication (elongation) of a primed, SS template.[28] However, replication of the ϕX174 template (unprimed) required both

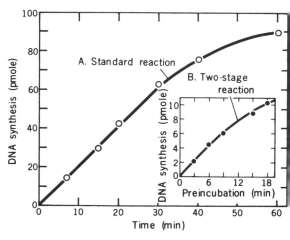

Fig. 3. Time course of standard and two-stage reconstituted ϕX174 reactions. (a) Standard reaction. Components were the same as those described in Table 1. (b) Two-stage reaction. Formation of primed single-strands took place in a reaction mixture of 25 μl lacking only the deoxytriphosphates. After incubation for the indicated times the first stage was stopped by dilution, and the product was measured by its capacity to support DNA synthesis. This was performed by addition of 250 μl of buffer containing the deoxynucleoside triphosphates (4.25 nmole of [³H] TTP, 12.5 nmole each of dATP, dGTP, and dCTP), 1.25 μmole of MgCl$_2$, and 1.7 μg of DNA polymerase III holoenzyme. Diluted reactions were further incubated for 10 minutes.

(Fig. 5). Thus both agents may be needed in the initiation stage. Optimal levels of unwinding protein (4 to 7 μg per microgram of DNA template) corresponded to 1 monomer per 8 to 14 nucleotide residues. Binding studies indicate that DNA is saturated at a level of 1 monomer per 8 nucleotides.[29] The known antagonism of spermidine to unwinding of the DNA duplex needs to be reconciled with its complementary action with unwinding protein in this replication system. These agents behave differently, depending on the secondary structure of the DNA, and may act at different substages of the initiation reaction.

With the components of the ϕX174 replicative system largely resolved, we have come closer to a direct examination of what each component contributes to the mechanism of the replicative reaction. Its dissection into partial reactions and intermediates should provide important clues. However, it would

Table 2 Ribonucleoside triphosphates required for initiation. Reactions were largely as described in Table 1. Reactions 2 to 5 contained 10 nmole of ATP (purified on Dowex 1X-8) in addition to 2.5 nmole of the other rNTP's* as indicated. Dilutions were performed as in Fig. 3b, except for reaction 6 which contained a tenfold complement of rNTP's in the dilution mixture. Reactions 7 and 8 were not diluted; a single complement of rNTP's or dNTP's or both was added to start the second-stage reaction. First-stage reactions were incubated for 10 minutes; second-stage reactions (1 to 5) were incubated for 10 minutes; reactions 6 to 8 were incubated for 15 minutes.

Exp. no.	rNTP requirements			DNA synthesis (pmole)
	First-stage†	Tenfold dilution	Second-stage addition	
1	None	+	dNTP's	1.3
2	A	+	dNTP's	2.1
3	A, G	+	dNTP's	6.2
4	A, G, U	+	dNTP's	7.0
5	A, G, U, C	+	dNTP's	16.0
6	None	+	rNTP's + dNTP's	1.1
7	None	None	rNTP's + dNTP's	84.4
8	A, G, U, C	None	dNTP's	110

*Abbreviations: ATP, GTP, UTP, CTP, adenosine, guanosine, uridine, and cytidine triphosphates; dATP, dGTP, TTP, and dCTP, the corresponding deoxynucleoside triphosphates; dNTP, deoxynucleoside triphosphate; rNTP, ribonucleoside triphosphate; SS, single-stranded DNA; RF, replicative-form DNA.
†rNTP added.

be a delusion to assume that replicative systems will be free of complexities like those which have bedeviled analysis of the multienzyme systems of transcription and translation for so many years. For example, it is already perplexing that the need for the *dna*A gene product demonstrated in crude extracts[14] is no longer manifest in the reconstituted system. Neither protein i nor protein n, both candidates for the *dna*A protein role, were temperature-sensitive when purified from a *dna*A mutant, nor did they complement the mutant extract. Possibly the *dna*A protein serves by counteracting an inhibitor that is present in the crude extract but absent from the reconstituted system. Or perhaps the mutant *dna*A protein itself inhibits the normal reaction in a crude extract.

Resolution and Reconstitution of the G4 System
When the foregoing studies (conversion of SS to RF) were extended to a ϕX174-like phage, called G4,[16] the results were rewarding in an unexpected way. Whereas the template from the closely related phage S13 was indistinguishable from ϕX174 in its replicative requirements, DNA from the less similar phage G4 was strikingly and gratifyingly simpler (Table 1).

Fig. 4. DNA polymerase III holoenzyme is not replaced by DNA polymerase III in the ϕX174 reaction. The DNA polymerase III (purified by the method of Kornberg and Gefter)[27] and DNA polymerase III holoenzyme[26] were compared in reactions on activated calf thymus DNA[27] and in the reconstituted ϕX174 reaction. Reactions (as in Table 1) contained 0.04, 0.08, or 0.16 μg of DNA polymerase III holoenzyme (fraction 4) or 0.04, 0.09, 0.17, 0.34, or 0.68 μg of DNA polymerase III (fraction 4).

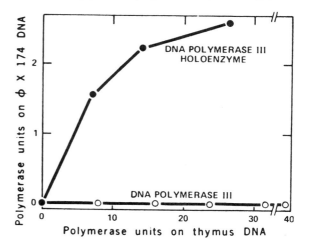

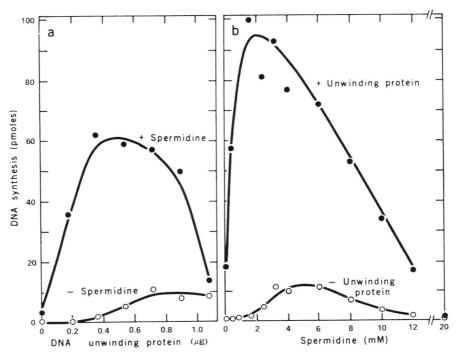

Fig. 5. Requirements for DNA-unwinding protein and spermidine. Reactions were performed as described in Table 1 with DNA-unwinding protein or spermidine concentrations as indicated. (a) Reactions contained 40 nmole of spermidine hydrochloride where indicated. (b) Reactions contained 0.73 μg of DNA-unwinding protein where indicated.

Rifampicin-resistant DNA synthesis on a G4 template was achieved merely with *dna*G protein, DNA-unwinding protein (or spermidine), and the DNA polymerase III holoenzyme.

The reaction required ribonucleoside triphosphates and was divisible into two stages, much as the φX174 reaction. The RFII produced in the presence of DNA-unwinding protein contained a nearly full-length complementary strand with a uniquely located gap, as was indicated by analysis of cleavage products generated by the *Eco*RI restriction endonuclease. We have evidence[23] suggesting that with this template, *dna*G may be serving as an RNA polymerase to generate the priming fragment for DNA synthesis. The difference in enzyme requirements between φX174 and G4 encourages us to search for other templates which may elicit initiation requirements intermediate in complexity between G4 and φX174.

Resolution and Reconstitution of the M13 System
Replication of M13 viral DNA to RF depends on

RNA polymerase, and consequently is completely inhibited by rifampicin. The unique location of the small gap in the synthetic complementary strand[30] has been taken to indicate the promoter region in the template for the start of RNA polymerase action (Fig. 6). Unique initiation requires that sufficient DNA-unwinding protein be present to cover the template. Elongation of the chain from the RNA primer depends specifically on DNA polymerase III holoenzyme; and closure of RFII depends on DNA polymerase I and DNA ligase.[18, 28] This reconstituted M13 system, unlike the crude extract, fails to distinguish M13 from φX174 DNA.

Fig. 6. Scheme for conversion of M13 single-stranded DNA to RFI.

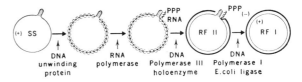

This set of purified enzymes supports rifampicin-sensitive, and therefore RNA polymerase-mediated, replication of M13 and ϕX174 DNA.[28] Evidently a discriminatory factor present in the crude system has been lost upon resolution and purification of the enzymes.[31]

Template Selection by Replicative Systems The three distinctive systems illustrated by the phage template probes are alike in their utilization of the same DNA polymerase III holoenzyme for chain elongation, and DNA polymerase I and DNA ligase for completion and closure of the chain (Table 3). *They are distinguished by their requirements for the initiation event* which primes DNA polymerase action by RNA synthesis. The templates all rely upon the DNA-unwinding protein to mask all but the significant promoter region. *In vivo*, decapsidation is coupled to replication, and the SS template never enters the cell as such. Therefore, phage coat proteins may function as the DNA-unwinding protein does *in vitro*.

The G4 template is least demanding; action by the 65,000-dalton *dna*G protein suffices. In this regard, the promoter signals for starting RNA synthesis may resemble those that direct the start of

nascent fragments which have been judged to depend only on a few of the *dna* gene products. The ϕX174 template has far more complex needs for initiation, and the additional proteins may serve to modify the promoter or its recognition for *dna*G protein action. The M13 template is inert with any combination of these proteins (legend to Table 1) and is utilized only by RNA polymerase.

The ϕX174 template promoter requires a battery of replication proteins including those needed for starting a new cycle of replication of the *E. coli* chromosome. The artifactual priming of ϕX174 DNA by RNA polymerase *in vitro* posed a serious question of specificity until the recent discovery of a new form of RNA polymerase, named RNA polymerase III (and so distinguished from the classic form, RNA polymerase I) which primes DNA synthesis on M13 DNA but not ϕX174 DNA.[32] This discrimination is achieved by the presence of a distinctive, small subunit in RNA polymerase III. This small subunit released from RNA polymerase III (by rifampicin or certain procedures used conventionally to isolate RNA polymerase I) can be separated and used to endow RNA polymerase I with the discriminatory capacity of RNA polymerase III. Masking the ϕX174 template with DNA-unwinding protein is essential for display of specificity by RNA polymerase III. The mechanisms for these effects remain to be elucidated.

The M13 template which appropriates RNA polymerase (either I or III) is clearly distinct from rifampicin-resistant host replicative systems. In this sense the M13 system resembles the replicative systems for plasmids, such as the one for colicinogenic factor E1, or the fertility factor, F. Possibly other extrachromosomal DNA's will be found to fall in this group as well.

Table 3 Summary of ϕX174, M13, and G4 replication properties.

Item	Template		
	ϕX174	G4	M13
Initiation			
Rifampicin-sensitive *dna*B, *dna*C	−	−	+
protein i, protein n	+	−	−
*dna*G	+	+	−
RNA polymerase III	−	−	+
Initiation and elongation			
DNA-unwinding protein	+	+	+
Spermidine	+	−	−
Elongation			
DNA polymerase III holoenzyme	+	+	+
Termination			
DNA polymerase I + DNA ligase	+	+	+
Related host replication	Origin	Nascent	Plasmid

CONCLUSIONS

Replication is accomplished by multienzyme systems whose operations are usefully considered in respect to three stages of the process: initiation, elongation, and termination.

1. Initiation entails synthesis of a short RNA fragment that serves as primer for the elongation step of DNA synthesis. This stage, probed by SS phage DNA templates, reveals three distinctive and highly specific systems in *E. coli*. The M13 DNA utilizes

RNA polymerase in a manner that may reflect how plasmid elements are replicated in the cell. The ϕX174 DNA does not rely on RNA polymerase, but requires instead five distinctive proteins which may belong to an apparatus for initiating a host chromosome replication cycle at the origin. The G4 DNA, also independent of RNA polymerase, needs simply the *dna*G protein for its distinctive initiation and may thus resemble the system that initiates the replication fragments at the nascent growing fork. In each case it is essential that *in vitro* the DNA-unwinding protein coat the viral DNA and influence its structure.

2. Elongation is achieved in every case by the multisubunit, holoenzyme form of DNA polymerase III. Copolymerase III*, which is an enzyme subunit, and adenosine triphosphate are required to form a proper complex with the primer template but appear dispensable for the ensuing chain growth by DNA polymerase.[33]

3. Termination requires excision of the RNA priming fragment, filling of gaps and sealing of interruptions to produce a covalently intact phosphodiester backbone. DNA polymerase I has the capacity for excision and gap-filling and DNA ligase is required for sealing.

What once appeared to be a simple DNA polymerase-mediated conversion of a single-strand to a duplex circle[34] is now seen as a complex series of events in which diverse multienzyme systems function. Annoyance with the difficulties in resolving and reconstituting these systems is tempered by the conviction that these are the very systems used by the cell in replicating its chromosome and extrachromosomal elements. Thus, understanding of the regulation of replication events in the cell, their localization at membrane surfaces and integration with cell division, and their coordination with phage DNA maturation and particle assembly will all be advanced by knowledge of the components of the replicative machinery.

REFERENCES

1. M. J. Chamberlin, *Annu. Rev. Biochem.* 43: 721 (1974).

2. A. Kornberg, *DNA Synthesis* (Freeman, San Francisco, 1974).

3. J. D. Gross, *Curr. Top. Microbiol. Immunol.* 57: 39 (1972).

4. T. Komano and H. Sakai, in *Molecular Biology Meeting of Japan* (Kyoritsu Shuppan, Tokyo, 1972); D. Beyersmann, W. Messer, M. Schlicht, *J. Bacteriol.* 118: 783 (1974).

5. K. G. Lark, *Nat. New Biol.* 240: 237 (1972).

6. M. L. Gefter, Y. Hirota, T. Kornberg, J. A. Wechsler, C. Barnoux, *Proc. Natl. Acad. Sci. U.S.A.* 68:3150 (1971).

7. A. Klein and F. Bonhoeffer, *Annu. Rev. Biochem.* 41:301 (1972).

8. W. Wickner, D. Brutlag, R. Schekman, A. Kornberg, *Proc. Natl. Acad. Sci. U.S.A.* 69:965 (1972).

9. J. H. Weiner, D. Brutlag, K. Geider, R. Schekman, W. Wickner, A. Kornberg, in *Methods in Molecular Biology*, R. B. Wickner, ed. (Dekker, New York, 1974), vol. 7.

10. W. Strätling, F. J. Ferdinand, E. Krause, R. Knippers, *Eur. J. Biochem.* 38:160 (1973); D. C. Hinkle and C. C. Richardson, *J. Biol. Chem.* 249:2974 (1974); H. Shizuya and C. C. Richardson, *Proc. Natl. Acad. Sci. U.S.A.* 71: 1758 (1974).

11. T. Kornberg and A. Worcel, *Proc. Natl. Acad. Sci. U.S.A.* 71:3189 (1974).

12. D. Brutlag, R. Schekman, A. Kornberg, *ibid.* 68:2826 (1971).

13. P. A. Williams, H. W. Boyer, D. R. Helinski, *ibid.* 70:3744 (1973).

14. R. Schekman, W. Wickner, O. Westergaard, D. Brutlag, K. Geider, L. Bertsch, A. Kornberg, *ibid.* 69:2691 (1972).

15. O. Westergaard, D. Brutlag, A. Kornberg, *J. Biol. Chem.* 248:1361 (1973).

16. G. N. Godson, *Virology* 58:272 (1974).

17. A. Sugino and R. Okazaki, *Proc. Natl. Acad. Sci. U.S.A.* 70:88 (1973).

18. G. Magnusson, V. Pigiet, E.-L. Winnacker, R. Abrams, P. Reichard, *ibid.*, p. 412.

19. J. A. Huberman, H. Horwitz, R. Minkoff, M. A. Wagar, *Cold Spring Harbor Symp. Quant. Biol.* 38:323 (1973).

20. R. B. Wickner, M. Wright, S. Wickner, J. Hurwitz, *Proc. Natl. Acad. Sci. U.S.A.* 69:3233 (1972).

21. S. Wickner, M. Wright, J. Hurwitz, *ibid.* 70:1613

(1973); S. Wickner, I. Berkower, M. Wright, J. Hurwitz, *ibid.*, p. 2369; M. Wright, S. Wickner, J. Hurwitz, *ibid.*, p. 3120.

22. R. Schekman, A. M. Weiner, J. H. Weiner, A. Kornberg, in preparation.
23. K. Zechel, J.-P. Bouche, A. Kornberg, in preparation.
24. N. Sigal, H. Delius, T. Kornberg, M. L. Gefter, B. Alberts, *Proc. Natl. Acad. Sci. U.S.A.* 69: 3537 (1972).
25. W. Wickner, R. Schekman, K. Geider, A. Kornberg, *ibid.* 70:1764 (1973).
26. W. Wickner and A. Kornberg, *J. Biol. Chem.* 249:6244 (1974).
27. T. Kornberg and M. L. Gefter, *ibid.* 247:5369 (1972).
28. K. Geider and A. Kornberg, *ibid.* 249:3999 (1974).
29. J. H. Weiner and A. Kornberg, in preparation.
30. H. F. Tabak, J. Griffith, K. Geider, H. Schaller, A. Kornberg, *J. Biol. Chem.* 249:3049 (1974).
31. J. Hurwitz, S. Wickner, M. Wright, *Biochem. Biophys. Res. Commun.* 51:257 (1973).
32. W. Wickner and A. Kornberg, *Proc. Natl. Acad. Sci. U.S.A.*, in press.
33. —, *ibid.* 70:3679 (1973).
34. M. Goulian and A. Kornberg, *ibid.* 58:1723 (1967).

An RNA-dependent DNA polymerase in virions of Rous Sarcoma Virus

SATOSHI MIZUTANI and HOWARD M. TEMIN *

Reprinted by authors' and publisher's permission from Cold Spring Harbor
Symposia on Quantitative Biology, *vol. 35, 1970, pp. 847–849.*

We include this paper because it provided a new direction for our understanding of nucleic acid replication and its application to cancer research. The paper demonstrates that RNA tumor viruses replicate by generating a DNA intermediate which then serves as the template for the production of RNA progeny molecules. The DNA intermediate helps clear up some of the mystery of how an RNA virus could (1) take over a cell's function and make it cancerous; and (2) replicate and pass on to all progeny cancer cells.

Many spontaneous tumors in animals are associated with viruses of a particular class — the RNA tumor viruses. These viruses have virions of about 100 mμ diameter, a lipid-containing envelope, and an internal ribonucleoprotein core or nucleoid containing single-stranded RNA. The RNA tumor viruses are non-cytopathic; they do not kill infected cells, but replicate along with the infected cells. The replication may or may not include production of infectious virus, and may or may not be associated with the conversion of the infected cell to a tumor cell. The replication of the nucleic acid of RNA tumor viruses has appeared to be different from that of other RNA viruses since early experiments with inhibitors. First, it was found that virus production was sensitive to actinomycin D (Temin 1963), then that inhibition of DNA synthesis soon after infection blocked virus production (Bader 1964; Temin 1964a), and third, that there appeared to be increased hybridization of viral RNA with DNA from infected cells (Temin 1964b). These observations led to the "DNA provirus hypothesis": RNA tumor viruses replicate through a DNA intermediate (Temin 1964c). Later experiments confirmed the actinomycin D sensitivity of virus RNA synthesis (Baluda and Nayak 1969; Bader 1970); they showed that the reported effect of inhibition of DNA synthesis was probably an effect on cell division, but also showed that there was a new DNA synthesis different from

S-phase DNA synthesis (Temin 1967, 1968, 1970a; Murray and Temin 1970; Boettiger and Temin 1970); and, finally, they confirmed the hybridization experiments with much more convincing data (Baluda and Nayak 1970).

The DNA provirus hypothesis requires transfer of information from RNA to DNA. Early experiments looking for such an enzyme in extracts of uninfected cells were negative (Helgeland and Temin, unpubl. observ., 1965). More recently, it was shown that formation of the provirus in stationary cells could take place normally in the presence of concentrations of cycloheximide that inhibit 90% of protein synthesis (Mizutani and Temin 1970). Since polymerases had been found in virions of vaccinia and reoviruses (Kates and McAuslan 1967; Munyon et al. 1967; Borsa and Graham 1968; Shatkin and Sipe 1968), we decided to look for an RNA-directed DNA polymerase in virions of Rous sarcoma virus (RSV). Such a polymerase was found (Temin and Mizutani 1970). At the same time, David Baltimore independently found such a polymerase in virions of Rauscher leukemia virus (Baltimore 1970).

Schmidt-Ruppin (SRV), B77, and avian myeloblastosis virus were prepared from the supernatant medium of cultures of infected chicken fibroblasts. Most preparations were purified by equilibrium sucrose density-gradient centrifugation.

The polymerase assay consisted of 0.125 μmoles each of dATP, dCTP, and dGTP (in 0.02 *M* Tris-HCl buffer at pH 8.0, containing 0.33 m*M* ethylene diamine-tetracetate and 1.7 m*M* 2-mercaptoethanol),

*McArdle Laboratory for Cancer Research, University of Wisconsin, Madison, Wisconsin

1.25 μmoles of MgCl$_2$ and 2.5 μmoles of KCl, 2.5 μg phosphoenol pyruvate, 10 μg pyruvate kinase, 2.5 μc of ^{3}H-TTP (Schwarz) (12 c/mM), and 0.025 ml of enzyme (usually 10^8 focus forming units of disrupted RSV, OD$_{280}$ = 0.30; about 5 μg protein) in a total volume of 0.125 ml. Incubation was usually at 40°C for 1 hr. Using Furlong's method (1967) 0.025 ml of the reaction mixture was withdrawn and assayed for acid-insoluble counts. Reduction of the levels of the cold deoxyribonucleotides by 10-fold or increasing the level of the labeled dTTP 20-fold had no effect on the reaction rate. Addition of 5 or more mμmoles of ATP to the reaction mixture caused a 30% increase in rate.

For full activity the virions had to be pretreated with detergent (Table 1). After it was found that lower concentrations did not inactivate virus infectivity, 0.25% Nonidet P-40 was used. The presence of dithiothreitol was necessary if the Nonidet treatment was carried out at 40°.

When disrupted virions of SRV were incubated in a polymerase assay, ^{3}H-TTP was incorporated as shown in Fig. 1. The reaction was biphasic and continued for about 6 hr. By 6 hr about 10^{13} nucleotides had been incorporated. This would make about

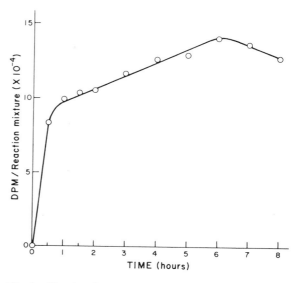

Fig. 1. Kinetics of incorporation. Disrupted virus was assayed in a standard polymerase assay and at different times portions were analyzed for incorporation.

Table 1 Requirements for enzyme activity

	dpm ^{3}H-TTP incorporated
Complete system without virions	0
Complete system with nondisrupted virions	1410
Complete system with disrupted virions	5675
Complete system without MgCl$_2$	186
Complete system without MgCl$_2$, with MnCl$_2$	5570
Complete system without dATP	897
Complete system without dGTP	2190
Complete system without dCTP	1780
Complete system	10,700
Complete system without dXTPs, with XTPs*	0

A standard polymerase assay was performed with the substitutions listed.

*A standard polymerase assay was performed with ATP, GTP and CTP substituted for the deoxyribonucleoside triphosphates, and 2.5 μc of ^{3}H-UTP (17.1 c/mM; Schwarz) instead of ^{3}H-TTP.

10^9 virus-sized nucleic acids. There were about 10^{10} virions in the reaction mixture.

The data in Table 1 demonstrate some other requirements for activity. A divalent cation, either Mg^{2+} or Mn^{2+}, is required. For full activity all deoxyribonucleoside triphosphates must be present. When deoxyribonucleoside triphosphates were removed and ribonucleoside triphosphates added with labeled UTP, no incorporation was found. When deoxyribonucleoside triphosphates were removed and ribonucleoside triphosphates added with labeled TTP, no incorporation was found. The pH optimum was 8 to 9.5. The temperature optimum was 40° to 50°.

The data in Table 2 demonstrate that pretreatment with heated RNase destroys the activity. Low concentrations destroy 70% of activity. Much higher concentrations are necessary for complete destruction.

The data in Table 3 demonstrate that the product is resistant to KOH and RNase, and sensitive to DNase. Higher concentrations of DNase cause 95% of the incorporated counts to become acid-soluble. We have found (Mizutani et al. 1970) that the product of the virion polymerase system is double-stranded DNA, and that there is a double-stranded

Table 2 RNA-dependency of polymerase activity

RNase concentration	Incorporation of ^{3}H-TTP (% of control)
0	100
1 μg/ml	94
3 μg/ml	27
5 μg/ml	26
50 μg/ml	29
1 mg/ml	<1

Disrupted virions were incubated at 20° for 30 min with the indicated amount of preheated (80° for 10 min) RNase A (Worthington) and standard polymerase assays were performed.

Table 3 Nature of product

Treatment	Residual acid-insoluble ^{3}H-TMP (dpm)
Buffer	7580
RNase A, 1 mg/ml	6150
RNase T$_1$, 2500 units/ml	7370
DNase 1, 50 μg/ml	1520
Buffer	8350
KOH	8250

A standard polymerase assay was performed with disrupted virions. The product was incubated in buffer or 0.3 M KOH at 37° for 20 hr; or with 1 mg/ml of RNase A (Worthington), or 2500 units/ml RNase T$_1$ (Calbiochem), or 50 μg/ml DNase I (Worthington) for 1 hr at 37°. Portions were removed and tested for acid-insoluble counts.

nucleic acid-dependent polymerase, an endonuclease for DNA, and possibly, a ligase in the virions of SRV.

These results suggest that virions of RSV contain an RNA-directed DNA polymerase. Further work is underway in several laboratories to characterize this enzyme further.

We did not find the enzyme in supernatant of normal cells, but have found it in virions of avian myeloblastosis virus. It has now also been found in virions of murine and feline RNA tumor viruses (Baltimore 1970; Spiegelman, pers. commun.; Green, pers. commun.). It appears to be a general component of virions of RNA tumor viruses. Biological

experiments suggest that the new DNA has an informational role (Boettiger and Temin 1970). Therefore, the DNA provirus hypothesis appears likely to be correct.

In terms of this hypothesis, it is useful to look at the uninfected chickens positive for the avian tumor virus group-specific (gs) antigen (Dougherty and DiStefano 1966). Work by Payne and Chubb (1968) showed that the genetic determinant for this antigen was inherited in the chickens as a single Mendelian dominant. Work by Weiss (1969) and Hanafusa (1970) showed that complementation occurred between certain types of Rous sarcoma virus and some factor in these gs-positive cells. Hanafusa (1970) further showed that this factor could be transferred to gs-negative cells by avian tumor viruses.

The chromosomal determinant for the gs antigen in these chickens is probably, therefore, a DNA related to a DNA provirus of Rous sarcoma virus — either as precursor or descendant.

The possible existence in uninfected cells of such DNA and the widespread occurrence of C-type RNA viruses led to the suggestion that uninfected cells contain genetic determinants which normally replicate DNA to RNA to DNA and serve to carry information from cell to cell in a multicellular animal; the protovirus hypothesis (Temin 1970b). These protoviruses could control normal cell multiplication and their variation could lead to cancer.

ACKNOWLEDGMENTS

This investigation was supported by Public Health Service Research Grant CA 07175 from the National Cancer Institute. H. M. T. holds Research Career Development Award 10K 3-CA8182 from the National Cancer Institute.

REFERENCES

Bader, J. P. 1964. The role of deoxyribonucleic acid in the synthesis of Rous sarcoma virus. Virology 22:462.

——. 1970. Synthesis of the RNA of RNA-containing tumor viruses. I. The interval between synthesis and envelopment. Virology 40:494.

Baltimore, D. 1970. RNA-dependent DNA polymerase in virions of RNA tumor viruses. Nature 226:1209.

Baluda, M. B. and D. P. Nayak. 1969. Incorporation of precursors into ribonucleic acid, protein, glycoprotein, and lipoprotein of avian myeloblastosis virions. J. Virol. 5:554.

—. —. 1970. Homology between RNA from avian myeloblastosis virus and DNA from leukemic and normal chicken cells. In (R. Barry and B. Mahy, ed.) Biology of Large RNA Viruses, Academic Press, London.

Boettiger, D. and H. M. Temin. 1970. The inactivation by visible light of focus formation by chicken embryo fibroblasts infected with avian sarcoma virus in the presence of 5-bromodeoxyuridine. Nature. 228:424

Borsa, J. and A. F. Graham. 1968. Reovirus: RNA polymerase activity in purified virions. Biochem. Biophys. Res. Commun. 33:895.

Dougherty, R. M. and H. S. DiStefano. 1966. Lack of relationship between infection with avian leukosis virus and the presence of COFAL antigen in chick embryos. Virology 29:586.

Furlong, N. B. 1967. DNA polymerase. Methods in Cancer Res. 3:27.

Hanafusa, H., T. Miyamoto, and T. Hanafusa. 1970. A cell-associated factor essential for formation of an infectious form of Rous sarcoma virus. Proc. Nat. Acad. Sci. 66:314.

Kates, J. R. and B. R. McAuslan, 1967. Messenger RNA synthesis by a "coated" viral genome. Proc. Nat. Acad. Sci. 57:314.

Mizutani, S., D. Boettiger, and H. M. Temin. 1970. A DNA-dependent DNA polymerase and a DNA endonuclease in virions of Rous sarcoma virus. Nature. In press.

Munyon, W., E. Paoletti, and J. T. Grace. 1967. RNA polymerase activity in purified infectious vaccinia virus. Proc. Nat. Acad. Sci. 58:2280.

Murray, R. K. and H. M. Temin. 1970. Carcinogenesis by RNA sarcoma viruses. XIV. Infection of stationary cultures with murine sarcoma virus (Harvey). Int. J. Cancer 5:320.

Payne, L. N. and R. C. Chubb. 1968. Studies on the nature and genetic control of an antigen in normal chick embryos which reacts in the COFAL test. J. Gen. Virol. 3:379.

Shatkin, A. J. and J. D. Sipe. 1968. RNA polymerase activity in purified reoviruses. Proc. Nat. Acad. Sci. 61: 1462.

Temin, H. M. 1963. The effects of actinomycin D on growth of Rous sarcoma virus in vitro. Virology 20:577.

—. 1964a. The participation of DNA in Rous sarcoma virus production. Virology 23:486.

—. 1964b. Homology between RNA from Rous sarcoma virus and DNA from Rous sarcoma virus-infected cells. Proc. Nat. Acad. Sci. 52:323.

—. 1964c. The nature of the provirus of Rous sarcoma. Nat. Cancer Inst. Monogr. 17:557.

—. 1967. Studies on carcinogenesis by avian sarcoma viruses. V. Requirement for new DNA synthesis and for cell division. J. Cell. Physiol. 69:53.

—. 1968. Carcinogenesis by avian sarcoma viruses. Cancer Res. 28:1835.

—. 1970a. Formation and activation of the provirus of RNA sarcoma viruses. In (R. Barry and B. Mahy, ed.) Biology of Large RNA Viruses, p. 37. Academic Press, London.

—. 1970b. Malignant transformation of cells by viruses. Pers. in Biol. and Med. In press.

Temin, H. M. and S. Mizutani. 1970. RNA-dependent DNA polymerase in virions of Rous sarcoma virus. Nature 226:1211.

Weiss, R. A. 1969. The host range of Bryan strain Rous sarcoma virus synthesized in the absence of helper virus. J. Gen. Virol. 5:511.

Variation due to change in the individual gene*

DR. H. J. MULLER

Reprinted with permission from American Naturalist, *vol. 56, 1922, pp. 32–50. Copyright © 1922 by the University of Chicago.*

Certain investigators have the unique quality of being able to assess the status of a field of study based upon a comprehensive survey of the available literature. By doing this, they call attention to areas of ambiguity that would be fruitful topics for future study. In rare instances, the investigators suggest techniques and experimental protocol that might be employed to resolve the questions. These suggestions indicate that the investigators are aware of both the theoretical aspects of the particular field and the details of the various experiments employed in the establishment of the theories.

In the following paper, Muller proposes that the study of the gene's internal structure would be a fruitful area of investigation. He enumerates newly developed methods for more efficiently detecting mutations with the hope of gaining knowledge about the structure of the gene by analyzing its manner of change. He questions the feasibility of basing such studies on the meager supply of spontaneous mutations; and, he predicts the need for inducing higher frequencies of mutation in order to provide a more extensive catalogue of mutants with which to work. Five years after he presented this paper, Muller demonstrated that he could enhance the frequency of mutation in *Drosophila,* above the spontaneous level, by exposing the flies to ionizing radiation. At the same time, Lewis J. Stadler reported similar findings using barley. Time has revealed that Muller's and Stadler's works were of major importance to the ultimate fulfillment of Muller's prophesy that mutation analysis would reveal the internal structure of the gene.

I. THE RELATION BETWEEN THE GENES AND THE CHARACTERS OF THE ORGANISM

The present paper will be concerned rather with problems, and the possible means of attacking them, than with the details of cases and data. The opening up of these new problems is due to the fundamental contribution which genetics has made to cell physiology within the last decade. This contribution, which has so far scarcely been assimilated by the general physiologists themselves, consists in the demonstration that, besides the ordinary proteins, carbohydrates, lipoids, and extractives, of their several types, there are present within the cell *thousands* of distinct substances — the "genes"; these genes exist as ultramicroscopic particles; their influences nevertheless permeate the entire cell, and they play a fundamental role in determining the nature of all cell substances, cell structures, and cell activities. Through these cell effects, in turn, the genes affect the entire organism.

It is not mere guesswork to say that the genes are ultramicroscopic bodies. For the work on *Drosophila* has not only proved that the genes are in the chromosomes, in definite positions, but it has shown that there must be hundreds of such genes within each of the larger chromosomes, although the length of these chromosomes is not over a few microns. If, then, we divide the size of the chromosome by the minimum number of its genes, we find that the latter are particles too small to give a visible image.

---

*In symposium on "The Origin of Variations" at the thirty-ninth annual meeting of the American Society of Naturalists, Toronto, December 29, 1921.

The chemical composition of the genes, and the formulae of their reactions, remain as yet quite unknown. We do know, for example, that in certain cases a given pair of genes will determine the existence of a particular enzyme (concerned in pigment production), that another pair of genes will determine whether or not a certain agglutinin shall exist in the blood, a third pair will determine whether homogentisic acid is secreted into the urine ("alkaptonuria"), and so forth. But it would be absurd, in the third case, to conclude that on this account the gene itself consists of homogentisic acid, or any related substance, and it would be similarly absurd, therefore, to regard cases of the former kind as giving any evidence that the gene is an enzyme, or an agglutininlike body. The reactions whereby the genes produce their ultimate effects are too complex for such inferences. Each of these effects, which we call a "character" of the organism, is the product of a highly complex, intricate, and delicately balanced system of reactions, caused by the interaction of countless genes, and every organic structure and activity is therefore liable to become increased, diminished, abolished, or altered in some other way, when the balance of the reaction system is disturbed by the alteration in the nature or the relative quantities of any of the component genes of the system. To return now to these genes themselves.

II. THE PROBLEM OF GENE MUTABILITY

The most distinctive characteristic of each of these ultramicroscopic particles — that characteristic whereby we identify it as a gene — is its property of self-propagation: the fact that, within the complicated environment of the cell protoplasm, it reacts in such a way as to convert some of the common surrounding material into an end product identical in kind with the original gene itself. This action fulfills the chemist's definition of "autocatalysis"; it is what the physiologist would call "growth"; and when it passes through more than one generation it becomes "heredity." It may be observed that this reaction is in each instance a rather highly localized one, since the new material is laid down by the side of the original gene.

The fact that the genes have this autocatalytic power is in itself sufficiently striking, for they are undoubtedly complex substances, and it is difficult to understand by what strange coincidence of chemistry a gene can happen to have just that very special series of physico-chemical effects upon its surroundings which produces — of all possible end products — just this particular one, which is identical with its own complex structure. But the most remarkable feature of the situation is not this oft-noted autocatalytic action in itself — it is the fact that, when the structure of the gene becomes changed, through some "chance variation," the catalytic property of the gene may* become correspondingly changed, in such a way as to leave it still *auto*catalytic. In other words, the change in gene structure — accidental though it was — has somehow resulted in a change of exactly *appropriate* nature in the catalytic reactions, so that the new reactions are now accurately adapted to produce more material just like that in the new changed gene itself. It is this paradoxical phenomenon which is implied in the expression "variation due to change in the individual gene," or, as it is often called, "mutation."

What sort of structure must the gene possess to permit it to mutate in this way? Since, through change after change in the gene, this same phenomenon persists, it is evident that it must depend upon some general feature of gene construction — common to all genes — which gives each one a *general* autocatalytic power — a "carte blanche" — to build material of whatever specific sort it itself happens to be composed of. This general principle of gene structure might, on the one hand, mean nothing more than the possession by each gene of some very simple character, such as a particular radicle or "side chain" — alike in them all — which enables each gene to enter into combination with certain highly organized materials in the outer protoplasm, in such a way as to result in the formation, "by" the protoplasm, of more material like this gene which is in combination with it. In that case the gene itself would only initiate and guide the direction of the reaction. On the other hand, the extreme alternative to such a conception has been generally assumed, perhaps gratuitously, in nearly all previous theories concerning hereditary units; this postulates that the chief feature of the autocatalytic mechanism resides in the structure of the genes themselves, and that

---

*It is of course conceivable, and even unavoidable, that *some* types of changes do destroy the gene's autocatalytic power, and thus result in its eventual loss.

the outer protoplasm does little more than provide the building material. In either case, the question as to what the general principle of gene construction is, that permits this phenomenon of mutable autocatalysis, is the most fundamental question of genetics.

The subject of gene variation is an important one, however, not only on account of the apparent problem that is thus inherent in it, but also because this same peculiar phenomenon that it involves lies at the root of organic evolution, and hence of all the vital phenomena which have resulted from evolution. It is commonly said that evolution rests upon two foundations — inheritance and variation; but there is a subtle and important error here. Inheritance by itself leads to no change, and variation leads to no permanent change, unless the variations themselves are heritable. Thus it is not inheritance *and* variation which bring about evolution, but the inheritance of variation, and this in turn is due to the general principle of gene construction which causes the persistence of autocatalysis despite the alteration in structure of the gene itself. Given, now, any material or collection of materials having this one unusual characteristic, and evolution would automatically follow, for this material would, after a time, through the accumulation, competition and selective spreading of the self-propagated variations, come to differ from ordinary inorganic matter in innumerable respects, in addition to the original difference in its mode of catalysis. There would thus result a wide gap between this matter and other matter, which would keep growing wider, with the increasing complexity, diversity and so-called adaptation of the selected mutable material.

III. A POSSIBLE ATTACK THROUGH CHROMOSOME BEHAVIOR

In thus recognizing the nature and the importance of the problem involved in gene mutability have we now entered into a *cul de sac,* or is there some way of proceeding further so as to get at the physical basis of this peculiar property of the gene? The problems of growth, variation and related processes seemed difficult enough to attack even when we thought of them as inherent in the organism as a whole or the cell as a whole — how now can we get at them when they have been driven back, to some extent at least, within the limits of an invisible par-

ticle? A gene cannot effectively be ground in a mortar, or distilled in a retort, and although the physicochemical investigation of other biological substances may conceivably help us, by analogy, to understand its structure, there seems at present no method of approach along this line.

There is, however, another possible method of approach available: that is, to study the behavior of the chromosomes, as influenced by their contained genes, in their various physical reactions of segregation, crossing over, division, synapsis, etc. This may at first sight seem very remote from the problem of getting at the structural principle that allows mutability in the gene, but I am inclined to think that such studies of synaptic attraction between chromosomes may be especially enlightening in this connection, because the most remarkable thing we know about genes — besides their mutable autocatalytic power — is the highly specific attraction which like genes (or local products formed by them) show for each other. As in the case of the autocatalytic forces, so here the attractive forces of the gene are somehow exactly adjusted so as to react in relation to more material of the same complicated kind. Moreover, when the gene mutates, the forces become readjusted, so that they may now attract material of the new kind; this shows that the attractive or synaptic property of the gene, as well as its catalytic property, is not primarily dependent on its specific structure, but on some general principle of its make-up, that causes whatever specific structure it has to be autoattractive (and autocatalytic).

This autoattraction is evidently a strong force, exerting an appreciable effect against the nonspecific mutual repulsions of the chromosomes, over measurable microscopic distances much larger than in the case of the ordinary forces of so-called cohesion, adhesion and adsorption known to physical science. In this sense, then, the physicist has no parallel for this force. There seems, however, to be no way of escaping the conclusion that in the last analysis it must be of the same nature as these other forces which cause inorganic substances to have specific attractions for each other, according to their chemical composition. These inorganic forces, according to the newer physics, depend upon the arrangement and mode of motion of the electrons constituting the molecules, which set up electromagnetic fields of force of specific patterns. To find the principle peculiar to the construction of the force-field pattern of genes would accordingly be requisite for solving the problem of their tremendous auto-attraction.

Now, according to Troland (1917), the growth of crystals from a solution is due to an attraction between the solid crystal and the molecules in solution caused by the similarity of their force field patterns, somewhat as similarly shaped magnets might attract each other — north to south poles — and Troland maintains that essentially the same mechanism must operate in the autocatalysis of the hereditary particles. If he is right, each different portion of the gene structure must — like a crystal — attract to itself from the protoplasm materials of a similar kind, thus molding next to the original gene another structure with similar parts, identically arranged, which then become bound together to form another gene, a replica of the first. This does not solve the question of what the general principle of gene construction is, which permits it to retain, like a crystal, these properties of autoattraction,* but if the main point is correct, that the autocatalysis is an expression of specific attractions between portions of the gene and similar protoplasmic building blocks (dependent on their force-field patterns), it is evident that the very same forces which cause the genes to grow should also cause the genes to attract each other, but much more strongly, since here all the individual attractive forces of the different parts of the gene are summated. If the two phenomena are thus really dependent on a common principle in the make-up of the gene, progress made in the study of one of them should help in the solution of the other.

Great opportunities are now open for the study of the nature of the synaptic attraction, especially through the discovery of various races having abnormal numbers of chromosomes. Here we have already the finding by Belling, that where three like chromosomes are present, the close union of any two tends to exclude their close union with the third. This is very suggestive, because the same thing is found in the cases of specific attractions between inorganic particles, that are due to their force-field patterns. And through Bridges' finding of triploid *Drosophila,* the attraction phenomena can now be brought down to a definitely genic basis, by the introduction of specific genes — especially those known to influence chromosome behavior — into one of the chromosomes of a triad. The amount of influence of this gene on attraction may then be tested quantitatively, by genetic determination of the frequencies of the various possible types of segregation. By extending such studies to include the effect of various conditions of the environment — such as temperature, electrostatic stresses, etc. — in the presence of the different genetic situations, a considerable field is opened up.

This suggested connection between chromosome behavior and gene structure is as yet, however, only a possibility. It must not be forgotten that at present we cannot be sure that the synaptic attraction is exerted by the genes themselves rather than by local products of them, and it is also problematical whether the chief part of the mechanism of autocatalysis resides within the genes rather than in the "protoplasm." Meanwhile, the method is worth following up, simply because it is one of our few conceivable modes of approach to an all-important problem.

It may also be recalled in this connection that besides the genes in the chromosomes there is at least one similarly autocatalytic material in the chloroplastids, which likewise may become permanently changed, or else lost, as has been shown by various studies on chlorophyll inheritance. Whether this plastic substance is similar to the genes in the chromosomes we cannot say, but of course it cannot be seen to show synaptic attraction, and could not be studied by the method suggested above.*

IV. THE ATTACK THROUGH STUDIES OF MUTATION

There is, however, another method of attack, in a sense more direct, and not open to the above criticisms. That is the method of investigating the individual gene, and the structure that permits it to

---

*It can hardly be true, as Troland intimates, that all similar fields attract each other more than they do dissimilar fields, otherwise all substances would be autocatalytic, and, in fact, no substances would be soluble. Moreover, if the parts of a molecule are in any kind of "solid," three dimensional formation, it would seem that those in the middle would scarcely have opportunity to exert the moulding effect above mentioned. It therefore appears that a special manner of construction must be necessary, in order that a complicated structure like a gene may exert such an effect.

---

*It may be that there are still other elements in the cell which have the nature of genes, but as no critical evidence has ever been adduced for their existence, it would be highly hazardous to postulate them.

change, through a study of the changes themselves that occur in it, as observed by the test of breeding and development. It was through the investigation of the *changes* in the chromosomes — caused by crossing over — that the structure of the chromosomes was analyzed into their constituent genes in line formation; it was through study of molecular changes that molecules were analyzed into atoms tied together in definite ways, and it has been finally the rather recent finding of changes in atoms and investigation of the resulting pieces, that has led us to the present analysis of atomic structure into positive and negative electrons having characteristic arrangements. Similarly, to understand the properties and possibilities of the individual gene, we must study the mutations as directly as possible, and bring the results to bear upon our problem.

(a) The Quality and Quantity of the Change

In spite of the fact that the drawing of inferences concerning the gene is very much hindered, in this method, on account of the remoteness of the gene-cause from its character-effect, one salient point stands out already. It is that the change is not always a mere loss of material, because clear-cut reverse mutations have been obtained in corn, *Drosophila, Portulaca,* and probably elsewhere. If the original mutation was a loss, the reverse must be a gain. Secondly, the mutations in many cases seem not to be quantitative at all, since the different allelomorphs formed by mutations of one original gene often fail to form a single linear series. One case, in fact, is known in which the allelomorphs even affect totally different characters: this is the case of the truncate series, in which I have found that different mutant genes at the same locus may cause either a shortening of the wing, an eruption on the thorax, a lethal effect, or any combination of two or three of these characters. In such a case we may be dealing either with changes of different types occurring in the same material or with changes (possibly quantitative changes, similar in type) occurring in different component parts of one gene. Owing to the universal applicability of the latter interpretation, even where allelomorphs do not form a linear series, it cannot be categorically denied, in any individual case, that the changes may be merely quantitative changes of some *part* of the gene. If all changes were

thus quantitative, even in this limited sense of a loss or gain of part of the gene, our problem of why the changed gene still seems to be autocatalytic would in the main disappear, but such a situation is excluded *a priori* since in that case the thousands of genes now existing could never have evolved.

Although a given gene may thus change in various ways, it is important to note that there is a strong tendency for any given gene to have its changes of a particular kind, and to mutate in one direction rather than in another. And although mutation certainly does not always consist of loss, it often gives effects that might be termed losses. In the case of the mutant genes for bent and eyeless in the fourth chromosome of *Drosophila* it has even been proved, by Bridges, that the effects are of exactly the same kind, although of lesser intensity, than those produced by the entire loss of the chromosome in which they lie, for flies having bent or eyeless in one chromosome and lacking the homologous chromosome are even more bent, or more eyeless, than those having a homologous chromosome that also contains the gene in question. The fact that mutations are usually recessive might be taken as pointing in the same direction, since it has been found in several cases that the loss of genes — as evidenced by the absence of an entire chromosome of one pair — tends to be much more nearly recessive than dominant in its effect.

The effect of mutations in causing a loss in the characters of the organism should, however, be sharply distinguished from the question of whether the gene has undergone any loss. It is generally true that mutations are much more apt to cause an apparent loss in character than a gain, but the obvious explanation for that is, not because the gene tends to lose something, but because most characters require for proper development a nicely adjusted train of processes, and so any change in the genes — no matter whether loss, gain, substitution or rearrangement — is more likely to throw the developmental mechanism out of gear, and give a "weaker" result, than to intensify it. For this reason, too, the most frequent kind of mutation of all is the lethal, which leads to the loss of the entire organism, but we do not conclude from this that all the genes had been lost at the time of the mutation. The explanation for this tendency for most changes to be degenerative, and also for the fact that certain other kinds of changes — like that from red to pink eye in *Dro-*

sophila — are more frequent than others — such as red to brown or green eye — lies rather in developmental mechanics than in genetics. It is because the developmental processes are more unstable in one direction than another, and easier to push "downhill" than up, and so any mutations that occur — no matter what the gene change is like — are more apt to have these *effects* than the other *effects*. If now selection is removed in regard to any particular character, these character changes which occur more readily must accumulate, giving apparent orthogenesis, disappearance of unused organs, of unused physiological capabilities, and so forth. As we shall see later, however, the changes are not so frequent or numerous that they could ordinarily push evolution in such a direction against selection and against the immediate interests of the organism.

In regard to the magnitude of the somatic effect produced by the gene variation, the *Drosophila* results show that there the smaller character changes occur oftener than large ones. The reason for this is again probably to be found in developmental mechanics, owing to the fact that there are usually more genes slightly affecting a given character than those playing an essential role in its formation. The evidence proves that there are still more genes whose change does not affect the given character at all — no matter what this character may be, unless it is life itself — and this raises the question as to how many mutations are absolutely unnoticed, affecting no character, or no detectable character, to any appreciable extent at all. Certainly there must be many such mutations, judging by the frequency with which "modifying factors" arise, which produce an effect only in the presence of a special genetic complex not ordinarily present.

(b) The Localization of the Change

Certain evidence concerning the causation of mutations has also been obtained by studying the relations of their occurrence to one another. Hitherto it has nearly always been found that only one mutation has occurred at a time, restricted to a single gene in the cell. I must omit from consideration here the two interesting cases of deficiency, found by Bridges and by Mohr, in each of which it seems certain that an entire region of a chromosome, with its whole cargo of genes, changed or was lost, and also a certain peculiar case, not yet cleared up, which has recently been reported by Nilsson-Ehle; these important cases stand alone. Aside from them, there are only two instances in which two (or more) new mutant genes have been proved to have been present in the same gamete. Both of these are cases in *Drosophila* — reported by Muller and Altenburg (1921) — in which a gamete contained two new sex-linked lethals; two cases are not a greater number than was to have been expected from a random distribution of mutations, judging by the frequency with which single mutant lethals were found in the same experiments. Ordinarily, then, the event that causes the mutation is specific, affecting just one particular kind of gene of all the thousands present in the cell. That this specificity is due to a spatial limitation rather than a chemical one is shown by the fact that when the single gene changes the other one, of identical composition, located nearby in the homologous chromosome of the same cell, remains unaffected. This has been proved by Emerson in corn, by Blakeslee in *Portulaca* and I have shown there is strong evidence for it in *Drosophila*. Hence these mutations are not caused by some general pervasive influence, but are due to "accidents" occurring on a molecular scale. When the molecular or atomic motions chance to take a particular form, to which the gene is vulnerable, then the mutation occurs.

It will even be possible to determine whether the entire gene changes at once, or whether the gene consists of several molecules or particles, one of which may change at a time. This point can be settled in organisms having determinate cleavage, by studies of the distribution of the mutant character in somatically mosaic mutants. If there is a group of particles in the gene, then when one particle changes it will be distributed irregularly among the descendant cells, owing to the random orientation of the two halves of the chromosome on the mitotic spindles of succeeding divisions,* but if there is only one

---

*This depends on the assumption that if the gene does consist of several particles, the halves of the chromosomes, at each division, receive a random sample of these particles. That is almost a necessary assumption, since a gene formed of particles each one of which was separately partitioned at division would tend not to persist as such, for the occurrence of mutation in one particle after the other would in time differentiate the gene into a number of different genes consisting of one particle each.

particle to change, its mutation must affect all of the cells in a *bloc* that are descended from the mutant cell.

(c) The Conditions under which the Change Occurs

But the method that appears to have most scope and promise is the experimental one of investigating the conditions under which mutations occur. This requires studies of mutation frequency under various methods of handling the organisms. As yet, extremely little has been done along this line. That is because, in the past, a mutation was considered a windfall, and the expression "mutation frequency" would have seemed a contradiction in terms. To attempt to study it would have seemed as absurd as to study the conditions affecting the distribution of dollar bills on the sidewalk. You were simply fortunate if you found one. Not even controls, giving the "normal" rate of mutation — if indeed there is such a thing — were attempted.* Of late, however, we may say that certain very exceptional banking houses have been found, in front of which the dollars fall more frequently — in other words, especially mutable genes have been discovered, that are beginning to yield abundant data at the hands of Nilsson-Ehle, Zeleny, Emerson, Anderson and others. For some of these mutable genes the rate of change is found to be so rapid that at the end of a few decades half of the genes descended from those originally present would have become changed. After these genes have once mutated, however, their previous mutability no longer holds. In addition to this "banking house method" there are also methods, employed by Altenburg and myself, for — as it were — automatically sweeping up wide areas of the streets and sifting the collections for the valuables. By these special genetic methods of reaping mutations we have recently shown that the ordinary genes of *Drosophila* — unlike the mutable genes above — would usually require at least a thousand years — probably very much more — before half of them became changed. This puts their stability about on a par with, if not much higher than, that of atoms of radium — to use a fairly familiar analogy. Since,

even in these latter experiments, many of the mutations probably occurred within a relatively few rather highly mutable genes, it is likely that most of the genes have a stability far higher than this result suggests.

The above mutation rates are mere first gleanings — we have yet to find how different conditions affect the occurrence of mutations. There had so far been only the negative findings that mutation is not confined to one sex (Muller and Altenburg 1919; Zeleny 1921), or to any one stage in the life cycle (Bridges 1919; Muller 1920; Zeleny 1921), Zeleny's finding that bar-mutation is not influenced by recency of origin of the gene (1921), and the as yet inconclusive differences found by Altenburg and myself for mutation rate at different temperatures (1919), until at this year's meeting of the botanists Emerson announced the definite discovery of the influence of a genetic factor in corn upon the mutation rate in its allelomorph, and Anderson the finding of an influence upon mutation in this same gene, caused by developmental conditions — the mutations from white to red of the mutable gene studied occurring far more frequently in the cells of the more mature ear than in those of the younger ear. These two results at least tell us decisively that mutation is not a sacred, inviolable, unapproachable process: it may be altered. These are the first steps; the way now lies open broad for exploration.

It is true that I have left out of account here the reported findings by several investigators, of genetic variations caused by treatments with various toxic substances and with certain other unusual conditions. In most of these cases, however, the claim has not been made that actual gene changes have been caused: the results have usually not been analyzed genetically and were in fact not analyzable genetically; they could just as well be interpreted to be due to abnormalities in the distribution of genes — for instance, chromosome abnormalities like those which Mayor has recently produced with x-rays — as to be due to actual gene mutations. But even if they were due to real genic differences, the possibility has in most cases by no means been excluded (1) that these genic differences were present in the stock to begin with, and merely became sorted out unequally, through random segregation; or (2) that other, invisible genic differences were present which, after random sorting out, themselves caused differences in mutation rate between the different lines. Certain recent results by Altenburg and myself suggest that

---

*Studies of "mutation frequency" had of course been made in the OEnotheras, but as we now know that these were not studies of the rate of gene change but of the frequencies of crossing over and of chromosome aberrations they may be neglected for our present purposes.

genic differences, affecting mutation rate, may be not uncommon. To guard against either of these possibilities it would have been necessary to test the stocks out by a thorough course of inbreeding beforehand, or else to have run at least half a dozen different pairs of parallel lines of the control and treated series, and to have obtained a definite difference in the same direction between the two lines of *each* pair; otherwise it can be proved by the theory of "probable error" that the differences observed may have been a mere matter of random sampling among genic differences originally present. Accumulating large numbers of abnormal or inferior individuals by selective propagation of one or two of the treated lines — as has been done in some cases — adds nothing to the significance of the results.

At best, however, these genetically unrefined methods would be quite insensitive to mutations occurring at anything like ordinary frequency, or to such differences in mutation rate as have already been found in the analytical experiments on mutation frequency. And it seems quite possible that larger differences than these will not easily be hit upon, at least not in the early stages of our investigations, in view of the evidence that mutation is ordinarily due to an accident on an ultramicroscopic scale, rather than directly caused by influences pervading the organism. For the present, then, it appears most promising to employ organisms in which the genetic composition can be controlled and analyzed, and to use genetic methods that are sensitive enough to disclose mutations occurring in the control as well as in the treated individuals. In this way relatively slight variations in mutation frequency, caused by the special treatments, can be determined, and from the conditions found to alter the mutation rate slightly we might finally work up to those which affect it most markedly. The only methods now meeting this requirement are those in which a particular mutable gene is followed, and those in which many homozygous or else genetically controlled lines can be run in parallel, either by parthenogenesis, self-fertilization, balanced lethals or other special genetic means, and later analyzed, through sexual reproduction, segregation and crossing over.

V. OTHER POSSIBILITIES

We cannot, however, set fixed limits to the possibilities of research. We should not wish to deny that some new and unusual method may at any time be found of directly producing mutations. For example, the phenomena now being worked out by Guyer may be a case in point. There is a curious analogy between the reactions of immunity and the phenomena of heredity, in apparently fundamental respects,* and any results that seem to connect the two are worth following to the limit.

Finally, there is a phenomenon related to immunity, of still more striking nature, which must not be neglected by geneticists. This is the d'Hérelle phenomenon. D'Hérelle found in 1917 that the presence of dysentery bacilli in the body caused the production there of a filterable substance, emitted in the stools, which had a lethal and in fact dissolving action on the corresponding type of bacteria, if a drop of it were applied to a colony of the bacteria that were under cultivation. So far, there would be nothing to distinguish this phenomenon from immunity. But he further found that when a drop of the affected colony was applied to a second living colony, the second colony would be killed; a drop from the second would kill a third colony, and so on indefinitely. In other words, the substance, when applied to colonies of bacteria, became multiplied or increased, and could be so increased indefinitely; it was self-propagable. It fulfills, then, the definition of an autocatalytic substance, and although it may really be of very different composition and work by a totally different mechanism from the genes in the chromosomes, it also fulfills our definition of a gene.† But the resemblance goes further — it has been found by Gratia that the substance may, through appropriate treatments on other bacteria, become changed (so as to produce a somewhat different effect than before, and attack different

---

*I refer here to the remarkable specificity with which a particular complex antigen calls forth processes that construct for it an antibody that is attracted to it and fits it "like lock and key," followed by further processes that cause more and more of the antibody to be reproduced. *If the antigen were a gene, which could be slightly altered by the cell to form the antibody that neutralized it — as some enzymes can be slightly changed by heating so that they counteract the previous active enzyme — and if this antibody-gene then became implanted in the cell so as to keep on growing, all the phenomena of immunity would be produced.*

†D'Hérelle himself thought that the substance was a filterable virus parasitic on the bacterium, called forth by the host body. It has since been found that various bacteria each cause the production of d'Hérelle substances which are to some extent specific for the respective bacteria.

bacteria) and still retain its self-propagable nature.

That two distinct kinds of substances — the d'Hérelle substances and the genes — should both possess this most remarkable property of heritable variation or "mutability," each working by a totally different mechanism, is quite conceivable, considering the complexity of protoplasm, yet it would seem a curious coincidence indeed. It would open up the possibility of two totally different kinds of life, working by different mechanisms. On the other hand, if these d'Hérelle bodies were really genes, fundamentally like our chromosome genes, they would give us an utterly new angle from which to attack the gene problem. They are filterable, to some extent isoluble, can be handled in test tubes, and their properties, as shown by their effects on the bacteria, can then be studied after treatment. It would be very rash to call these bodies genes, and yet at present we must confess that there is no distinction known between the genes and them. Hence we cannot categorically deny that perhaps we may be able to grind genes in a mortar and cook them in a beaker after all. Must we geneticists become bacteriologists, physiological chemists and physicists, simultaneously with being zoologists and botanists? Let us hope so.

I have purposely tried to paint things in the rosiest possible colors. Actually, the work on the individual gene, and its mutation, is beset with tremendous difficulty. Such progress in it as has been made has been by minute steps and at the cost of infinite labor. Where results are thus meager, all thinking becomes almost equivalent to speculation. But we cannot give up thinking on that account, and thereby give up the intellectual incentive to our work. In fact, a wide, unhampered treatment of all possibilities is, in such cases, all the more imperative, in order that we may direct these labors of ours where they have most chance to count. We must provide eyes for action.

The real trouble comes when speculation masquerades as empirical fact. For those who cry out most loudly against "theories" and "hypotheses" — whether these latter be the chromosome theory, the factorial "hypothesis," the theory of crossing over, or any other — are often the very ones most guilty of stating their results in terms that make illegitimate *implicit* assumptions, which they themselves are scarcely aware of simply because they are opposed to dragging "speculation" into the open. Thus they may be finally led into the worst blunders of all. Let us, then, frankly admit the uncertainty of many of the possibilities we have dealt with, using them as a spur to the real work.

The theory of mutagenesis

S. BRENNER,* L. BARNETT,* F. H. C. CRICK* and A. ORGEL†

Reprinted by authors' and publisher's permission from Journal of Molecular Biology, vol. 3, 1961, pp. 121–124. [Received December 16, 1960.]

Excluding Watson and Crick's suggestion in 1953, that tautomeric shifts might lead to the origin of spontaneous mutations, Freese set forth the first attempt at a comprehensive theory of mutation in 1959. Scientific theories are only so good as the number of experimental challenges they have faced and survived. Some, when scrutinized experimentally are found wanting and must be discarded. Others survive the rigors of extensive probing and become platforms on which the construction of more complex theories can be initiated. Still others are shown to be partially in error, but when modified to account for the experimentally detected inconsistencies eventually emerge as sound theories. Freese's theory was of the last type. In this paper Brenner and his colleagues point to the experimental inconsistencies in Freese's theory. However, by expanding the number of classes of mutational events to include a third type, the authors successfully modified the initial theory to one that has remained essentially intact up to the present time.

In this preliminary note we wish to express our doubts about the detailed theory of mutagenesis put forward by Freese (1959b), and to suggest an alternative.

Freese (1959b) has produced evidence that shows that for the r_{II} locus of phage T4 there are two mutually exclusive classes of mutation and we have confirmed and extended his work (Orgel & Brenner, in manuscript). The technique used is to start with a standard wild type and make a series of mutants from it with a particular mutagen. Each mutant is then tested with various mutagens to see which of them will back-mutate it to wild type.

It is found that the mutations fall into two classes. The first, which we shall call the base analogue class, is typically produced by 5-bromodeoxyuridine (BD) and the second, which we shall call the acridine class, is typically produced by proflavin (PF). In general a mutant made with BD can be reverted by BD, and a mutant made with PF can be reverted by PF. A few of the PF mutants do not appear to revert with either mutagen, but the strong result is that no mutant has been found which reverts identically with both classes of mutagens, and that (with a few possible exceptions) mutants produced by one class cannot be reverted by the other.

Freese also showed that 2-aminopurine falls into the base analogue class, and that most (85%) spontaneous mutants at the r_{II} locus were not of the base analogue type. We have confirmed this and shown that they are in fact revertible by acridines. We have also shown that a number of other acridines, and in particular 5-aminoacridine, act like proflavin (Orgel & Brenner, in manuscript).

Freese has produced an ingenious explanation of these results, which should be consulted in the original for fuller details. In brief he postulated that the base analogue class of mutagens act by altering an A — T base pair on the DNA (A = adenine, T = thymine) into a G — C pair, or *vice versa* (G = guanine, C = cytosine, or, in the T even phages, hydroxymethylcytosine). The fact that BD, which replaces thymine, could act both ways (from A — T to G — C or from G — C to A — T) was accounted for (Freese 1959a) by assuming that in the latter case there was an error in pairing of the BD (such that it accidentally paired with guanine) while *entering* the DNA, and in the former case after it was already in the DNA.

*Medical Research Council Unit for Molecular Biology, Cavendish Laboratory, Cambridge University, England.
†Pathology Laboratory, Cambridge University, England.

Such alterations only change a purine into another purine, or a pyrimidine into another pyrimidine. Freese (1959*b*) has called these "transitions." He suggested that other conceivable changes, which he called "transversions" (such as, for example, from $A - T$ to $C - G$) which change a purine into a pyrimidine and *vice versa*, occurred during mutagenesis by proflavin. This would neatly account for the two mutually exclusive classes of mutagens, since it is easy to see that a transition cannot be reversed by a transversion, and *vice versa*.

We have been led to doubt this explanation for the following reasons.

Our suspicions were first aroused by the curious fact that a comparison between the *sites* of mutation for one set of mutants made with BD and another set made with PF (Brenner, Benzer & Barnett 1958) showed there were no sites in the r_{II} gene, among the samples studied, common to both groups.

Now this result alone need not be incompatible with Freese's theory of mutagenesis, since we have no good explanation for "hot spots" and this confuses quantitative argument. However it led us to the following hypothesis:

that acridines act as mutagens because they cause the insertion or the deletion of a base pair.

This idea springs rather naturally from the views of Lerman (1960) and Luzzati (in preparation) that acridines are bound to DNA by sliding *between* adjacent base pairs, thus forcing them 6.8 Å apart, rather than 3.4 Å. If this occasionally happened between the bases on *one* chain of the DNA, but not the other, during replication, it might easily lead to the addition or subtraction of a base.

Such a possible mechanism leads to a prediction. We know practically nothing about coding (Crick 1959) but on most theories (except overlapping codes which are discredited because of criticism by Brenner (1957)) the deletion or the addition of a base pair is likely to cause not the substitution of just one amino acid for another, but a much more substantial alteration, such as a break in the polypeptide chain, a considerable alteration of the amino acid sequence, or the production of no protein at all.

Thus one would not be surprised to find on these ideas that mutants produced by acridines were not capable of producing a slightly modified protein, but usually produced either no protein at all or a grossly altered one.

Somewhat to our surprise we find we already have data from two separate genes supporting this hypothesis.

1. The *o* locus of phage T4 (resistance to osmotic shock) is believed to control a protein of the finished phage, possibly the head protein, because it shows phenotypic mixing (Brenner, unpublished). Using various base analogues we have produced mutants of this gene, though these map at only a small number of sites. We have failed on several occasions to produce any *o* mutants with proflavin. On another occasion two mutants were produced; one never reverted to wild type, while the other corresponded in position and spontaneous reversion rate to a base analogue site. We suspect therefore that these two mutants were not really produced by proflavin, but were the rarer sort of spontaneous mutant (Brenner & Barnett, unpublished).

2. We have also studied mutation at the *h* locus in T2L, which controls a protein of the finished phage concerned with attachment to the host (Streisinger & Franklin 1956).

Of the six different spontaneous h^+ mutants tested, all were easily induced to revert to *h* with 5-bromouracil (BU).* This is especially significant when it is recalled that 85% of the spontaneous r_{II} mutants could not be reverted with base analogues (Freese 1959*b*).

We have also shown (Brenner & Barnett, unpublished) that it is difficult to produce h^+ mutants from *h* by proflavin, though relatively easy with BU. The production of *r* mutants was used as a control.

It can be seen from Table 1 that if the production of h^+ mutants by BU and proflavin were similar to the production of *r* mutants we would expect to have obtained

$$\frac{57 \times 26}{108} = 13h^+ \text{ mutants with proflavin}$$

whereas in fact we only found 1, and this may be spontaneous background.

Let us underline the difference between the *r* loci and the *o* and *h* loci. The former appear to produce proteins which are probably *not* part of the finished phage. For both the *o* and the *h* locus, however, the protein concerned forms part of the

---

*(Added in proof.) Five of these have now been tested and have been shown not to revert with proflavin.

Table 1

	r	h+
BU	108	57
Proflavin	26	1

finished phage, which presumably would not be viable without it, so that a mutant can be picked up only if it forms an *altered* protein. A mutant which deleted the protein could not be studied.

It is clear that further work must be done before our generalization — that acridine mutants usually give no protein, rather than a slightly modified one — can be accepted. But if it turns out to be true it would support our hypothesis of the mutagenic action of the acridines, and this may have serious consequences for the naïve theory of mutagenesis, for the following reason.

It has always been a theoretical possibility that the reversions to wild type were not true reversions but were due to the action of "suppressors" (within the gene), possibly very closely linked suppressors. The most telling evidence against this was the existence of the two mutually exclusive classes of mutagens, together with Freese's explanation.

For clearly if the forward mutation could be made at one base pair and the reverse one at a different base pair, we should expect, on Freese's hypothesis, exceptions to the rule about the two classes of mutagens. Since these were not found it was concluded that even close suppressors were very rare.

Unfortunately our new hypothesis for the action of acridines destroys this argument. Under this new theory an alteration of a base pair at one place *could* be reversed by an alteration at a different base pair, and indeed from what we know (or guess) of the structure of proteins and the dependence of structure on amino acid sequence, we should be surprised if this did not occur.

It is all too easy to conceive, for example, that at a certain point on the polypeptide chain at which there is a glutamic residue in the wild type, and at which the mutation substituted a proline, a further mutation might alter the proline to aspartic acid

and that this might appear to restore the wild phenotype, at least as far as could be judged by the rather crude biological tests available. If several base pairs are needed to code for one amino acid the reverse mutation might occur at a base pair close to but not identical with the one originally changed.

On our hypothesis this could happen, and yet one would still obtain the two classes of mutagens. The one, typified by base analogues, would produce the substitution of one base for another, and the other, typically produced by acridines, would lead to the addition or subtraction of a base pair. Consequently the mutants produced by one class could not be easily reversed by the mutagens of the other class.

Thus our new hypothesis reopens in an acute form the question: which back-mutations to wild type are truly to the original wild type, and which only appear to be so? And on the answers to this question depends our interpretation of all experiments on back-mutation.

We suspect that this problem can most easily be approached by work on systems for which the amino acid sequence of the protein can be studied, such as the phage lysozyme of Dreyer, Anfinsen & Streisinger (personal communications) or the phosphatase from *E. coli* of Levinthal, Garen & Rothman (Garen 1960). Meanwhile we are continuing our genetic studies to fill out and extend the preliminary results reported here.

REFERENCES

Brenner, S. (1957). *Proc. Nat. Acad. Sci., Wash.* 43, 687.

Brenner, S., Benzer, S. & Barnett, L. (1958). *Nature*, 182, 983.

Crick, F. H. C. (1959). In *Brookhaven Symposia in Biology*, 12, 35.

Freese, E. (1959a). *J. Mol. Biol.* 1, 87.

Freese, E. (1959b). *Proc. Nat. Acad. Sci., Wash.* 45, 622.

Garen, A. (1960). 10th Symposium *Soc. Gen. Microbiol.*, London, 239.

Lerman, L. (1961). *J. Mol. Biol.* 3, 18.

Streisinger, G. & Franklin, N. C. (1956). In *Cold Spr. Harb. Sym. Quant. Biol.* 21, 103.

The genetic control of spontaneous and induced mutation rates in bacteriophage T4*,†

JOHN W. DRAKE‡

Reprinted by author's and publisher's permission from Genetics Supplement, *vol. 73, 1973, pp. 45–64.*

Radiation experiments in the thirties pointed to the existence of gross chromosomal rearrangements that could structurally alter more than one gene and, in so doing, alter the function of several genes. However, these multiple genic modifications arose as a consequence of the extensive damage introduced by the initial experimental treatment. Recently, interest has centered on single gene mutations that possess the ability to influence the rate of mutation throughout the entire genome in *subsequent* replications. Mutations can lead to the enhancement of the frequency of mutation (*mutator gene mutations*) or to the reduction in the frequency of mutation (*antimutator gene mutations*). The following paper discusses the experimental support for the existence of such mutational events and considers the possible impact that these types of mutations might have on determining the spontaneous rate of mutation.

ABSTRACT

A number of genes of bacteriophage T4 whose functions in DNA replication and repair have been at least partially identified have been implicated in the genetic determination of both spontaneous and induced mutation rates. Temperature-sensitive alleles of these genes markedly increase or decrease mutation rates, the strongest effects being observed with mutations in genes *px* (generalized repair and recombination), *32* (Alberts protein), *42* (dCMP hydroxymethylase), and *43* (DNA polymerase). Smaller but significant effects have also been observed with mutations in genes *30* (DNA ligase), *td* (thymidylate synthetase), *v* (pyrimidine dimer exonuclease), and *hm* (function unknown, but probably involved in generalized repair), and with mutations in several other genes of unknown function. Both increased and decreased spontaneous and induced mutation rates have been observed in these systems, indicating that bacteriophage T4 maintains an "optimal" mutation rate. Studies of mutator and antimutator alleles have important implications for mechanisms of DNA replication and repair and for basic mutational mechanisms. A clear need has developed for sophisticated chemical studies of the mutation process as influenced by local base pair sequences and by replication and repair.

Recent years have witnessed a sharply increased number of studies of mutator and antimutator mutations in a variety of organisms, the most penetrating analyses being conducted in prokaryotes and fungi. These studies have proceeded in parallel with analyses of mechanisms of DNA replication and repair, at the same time as genes responsible for the latter functions have been identified as conditional lethal mutations and a few of the corresponding enzymes have been subjected to close chemical study. The properties of mutator and antimutator mutations are relevant to three central questions:

---

*Abbreviations: A:T = adenine-thymine base pair; G:C = guanine-cytosine base pair (or guanine-hydroxymethylated base pair when reference is made to bacteriophage T4); py:pu = pyrimidine-purine base pair (in the indicated position); FUdR = 5-fluorodeoxyuridine; UV = ultraviolet; MMS = methyl methanesulfonate; EMS = ethyl methanesulfonate; EES = ethyl ethanesulfonate.

†Many undocumented statements which appear here are discussed, with references, in Drake (1970).

‡Department of Microbiology, University of Illinois, Urbana, Illinois 61801

What are the basic molecular mechanisms of mutation? What are the basic enzymological mechanisms of DNA synthesis during both replication and repair? And what is the evolutionary significance of the evident ability of organisms to determine their own mutation rates?

Mutator mutations are now easily obtained, either by direct selection procedures or by searches among conditional lethal mutants of DNA synthesis and among mutants exhibiting increased sensitivities to a variety of inactivating agents. While elegant studies have been performed with mutants identified only by their mutator characteristics, these mutants have generally lacked additional characteristics (such as conditional lethality or radiation sensitivity) which could facilitate their manipulation as genetic markers or their identification with a specific enzymatic function. Mapping studies, for instance, are decidedly difficult when the only characteristic which can be scored is a mutation rate, and identification of the altered gene product can only proceed by inspired guesses as to the affected function. Furthermore, direct selection for antimutator mutations has not yet proved possible. On the other hand, conditional lethal mutations are already available for many of the genes of bacteriophage T4 which have been implicated in DNA synthesis. Many of these mutants display mutator and antimutator activities. Studies in this laboratory have therefore mainly concentrated upon *ts* alleles of T4 genes which function in DNA replication or repair, and for which at least some information is available concerning the nature of this function.

All of the viral studies to be described here utilize bacteriophage T4. There appear to be no comparable studies in other viruses, and for very good reasons: the dense genetic mapping of T4, the availability of several good systems for scoring both forward and reverse mutation, and the many elegant enzymological studies on T4 DNA replication, make this organism highly suitable for detailed studies of mutation. *Escherichia coli* cannot yet boast such a wealth of background information about DNA replication (although many repair systems are well described), and bacteriophage lambda, for which extensive genetic and molecular information is already available, unfortunately lacks a suitable system for the study of forward and reverse mutation. It would be of the greatest significance to discover and develop such a system in phage lambda, since it

apparently employs many of the host enzymes for replication and repair, and since it can be studied during both prophage and lytic replication (Dove 1968).

SITE-SPECIFIC MUTATION RATES

The unique suitability of bacteriophage T4 for studies of the genetic determination of mutation rates was first demonstrated by Benzer's (1961) description of strongly and weakly mutable sites in the *rII* cistrons. It seemed likely that two quite different types of phenomena were involved in the determination of mutational "hot spots" and "cold spots."

The number of base pairs in the T4*rII* region far exceeds the number of mutational sites which can be identified from exhaustive mapping experiments, and it appeared that many base pair substitution mutations escape detection, either because they produce no amino acid substitution, or else because they produce an innocuous substitution. That this was indeed the case was first demonstrated by Koch and Drake (1970), who "sensitized" particular regions of the *rIIA* cistrons by inserting *ts* mutations and then collecting and mapping numerous nitrous-acid-induced mutants. Many of these mutants mapped at hitherto undiscovered sites within the sensitized regions, and when separated from the sensitizing mutation, were observed to be extremely leaky. Hot-spotting was exhibited by these "cryptic" mutants themselves, which in some cases even fell within the adjacent *rIIB* cistron, even though the sensitizing mutation resided in the *rIIA* cistron.

True hot-spotting, that is, high versus low mutability at sites where mutations could easily be recovered, has also been investigated in bacteriophage T4. Marked differences in the rates of hydroxylamine induction (G:C → A:T) of amber and ochre *rII* mutants at different sites have been reported (Brenner, Stretton, and Kaplan 1965), and marked differences in the rates of interconversion of nonsense codons (A:T → G:C) in the *rII* region have also been observed (Salts and Ronen 1971). The first experimental modification of highly defined base pair substitution mutation rates by alterations of neighboring base pairs was reported by Koch (1971), who showed that the rate of an A:T → G:C transition induced by 2-aminopurine (and probably also the spontaneous rate) was increased about twentyfold by changing the adjacent base pair from A:T to G:C, while the rate of another A:T → G:C transition was increased about threefold by changing the base pair two positions away from T:A to C:G. For reasons which remain obscure at present, a run of A:T base pairs, as observed, for instance, within the ochre codon, appears to be much less mutable than are A:T base pairs closer to G:C base pairs.

Okada *et al.* (1972) have also presented an analysis of the role of base pair sequence in determining the rate of frameshift mutagenesis in the T4*e* (lysozyme) cistron. A very highly revertible site, extending in this case over several base pairs, was shown to consist of a run of six A:T base pairs, and high frameshift mutation rates at other sites in the *e* cistron seem likely to arise at similar A:T runs. This result is in excellent accord with the theory of frameshift mutagenesis (Streisinger *et al.* 1966), which assigns a critical role to localized repeating base pair sequences. In this regard, it is also significant that the two giant hot spots in the *rII* region, *r117* and *r131,* which together compose 50% of all spontaneous *rII* mutations, are also composed of frameshift mutations (J. W. Drake, unpublished results). Since bacteriophage T4 is an A:T-rich organism, it will be of great interest to investigate frameshift mutation rates in G:C-rich organisms, and to compare the results with determinations of local base pair sequences.

ROLE OF DNA POLYMERASE

The next exciting discovery concerning the genetic determination of mutation rates in bacteriophage T4 arose out of the observation that two *ts* mutations of gene *43,* the structural gene for T4 DNA polymerase, exhibited powerful mutator activities (Speyer 1965). This discovery gave renewed credence to speculations that the accuracy of DNA synthesis depends upon enzymatic function(s) as well as upon base-pairing and stacking interactions. Although the mechanism by which the DNA polymerase achieves its control over mutation rates still remains obscure (see Drake *et al.* 1969; Kornberg 1969; Hershfield and Nossal 1972; and the reports by Nossal and by Bessman in this volume), gene *43* is clearly a major determinant of the accuracy of DNA synthesis in bacteriophage T4, as may be gene

dnaE in *E. coli* (R. Hall, personal communication). In an extensive survey of *ts* mutations of gene *43,* most were observed to produce easily recognized effects upon mutation rates (Drake *et al.* 1969; Allen, Albrecht, and Drake, 1970), and it seems likely that any destabilizing mutation in this gene affects mutation rates along one or another pathway.

Shortly after the discovery of mutagenic DNA polymerases in T4, Drake and Allen (1968) reported that several gene *43 ts* mutations exhibited powerful *antimutator* activities, particularly along the transition pathways A:T ↔ G:C. (Recent unpublished results, however, suggest that the rate of *transversion* mutagenesis is sometimes *increased* by these *ts* mutations.) This was the first evidence to suggest that the spontaneous mutation rate of an organism may be optimized by natural selection: the antimutator activities of these *ts* mutations could be retained among some of the *ts*[+] revertants. Various genotypes equivalent to these revertants, which appeared perfectly healthy under laboratory conditions, can be estimated to arise in nature in very substantial numbers (at frequencies on the order of 10^{-14} or possibly much higher, compared with total natural T-even populations greatly in excess of 10^{-14}). The balancing forces which determine the optimal mutation rate for a prokaryote are at best poorly understood, but restrictions on the amino acid sequence of the enzymes of DNA replication appear to be no hindrance to setting mutation rates.

An additional puzzling observation is also of significance in this respect: an approximately constant mutation rate per genome replication is observed among several common microbes, including bacteriophages, bacteria and fungi (Drake 1969). This mutation rate is on the order of 1%, and since their genomes vary in size by a factor of more than 1000, average mutation rates per base pair or per gene also vary (inversely) by a factor of more than 1000 in these organisms. The common 1% mutation rate clearly requires both chemical and theoretical explanations, which are not yet forthcoming. (The much greater mutation rates recorded for both *Drosophila* and man, which may approach or exceed 100% per genome per *sexual* generation, also require explanation.)

In addition to their effects upon spontaneous mutation rates, gene *43 ts* mutations also affect the mutational responses of bacteriophage T4 to a variety of mutagens. Both increased and decreased induced mutation rates have been observed (Drake and Greening 1970), but while considerably more is known about the action of some mutagens than is known about spontaneous mutation, we still lack sufficient understanding of the mechanism of action of DNA polymerases to specify how these effects are brought about.

Finally, in order to emphasize how general is the role of gene 43 in the determination of mutation rates, it should be noted that all types of point mutation rates can be affected by ts mutations in gene 43, including transitions, transversions, and even frameshift mutations (Bernstein 1971; deVries, Swart-Idenburg, and deWaard 1972; Drake and Allen 1968; Drake et al. 1969; Speyer, Karam, and Lenny 1966). This result is particularly notable in light of the current state of the theory of mutational mechanisms, which holds that a variety of different mechanisms operate to produce the different types of point mutations. Since DNA synthesis is a priori an obligatory step in mutagenesis, it is not surprising to find that DNA polymerases are important determinants of mutation rates; but it is somewhat surprising that the polymerases seem to be important in the determination of each kind of mutation rate.

ROLE OF OTHER GENES OF DNA REPLICATION AND REPAIR

Critique of Methods Most methods are detailed in the cited publications, and only a few additional points will be stressed here. The most common technique involves constructing ts-rII (or ts-e) double mutants by recombination, and then comparing the reversion rates of the rII or e mutation in the wild type and in the ts background. (In a few cases, however, the $r^+ \rightarrow r$ mutation rate has been measured, and the induced rII mutations have then been characterized by reversion analysis.) Mutational pathways are determined from the characteristics of the rII or e tester mutations, specifically their revertibility by base analogues, hydroxylamine, and proflavin. (The characteristics of the mutants employed in experiments described in this paper are described in Drake [1963] and in Drake and McGuire [1967a]). Additional information is sometimes obtained from analyses of the genetic or amino acid compositions of revertants. The later methods are particularly useful for determining

transversion mutation rates (Ripley and Drake 1972); amino acid composition measurements have been performed almost exclusively with the e cistron, since the rII polypeptide is relatively inaccessible.

The accuracy of measurements of this type depends strongly upon the kind of measurement performed. Spontaneous mutation rates are estimated from mutant frequencies in several stocks, since frequencies in individual stocks vary greatly as a result of the clonal origin of mutants during the growth of the population (Luria 1951; see also Chapter 5 of Drake 1970). Mutation rates estimated from the mean, median, or minimum mutant frequency among several stocks are somewhat variable, and only changes greater than two- or threefold are significant, depending in part upon the number of replicate stocks which are grown. Furthermore, at least in this laboratory, day-to-day variations of the same order of magnitude are routinely observed, so that it is important to compare only those tests which are performed strictly in parallel. Induced mutation rates, on the other hand, are usually relatively free of such variations, and changes as small as 20–30% may be significant, particularly if kinetic or mutagen concentration dependence data are collected.

It is also important to conduct a number of control experiments when studying the reversion of the double mutants. The composition of the double mutants should be confirmed by appropriate recombination tests, and it is also advisable to split the double mutant, recovering and identifying the rII component and retesting its reversion rate. It is also informative to select ts^+-rII revertants from the ts-rII double mutants, to determine that the ts mutation itself is responsible for the observed effect upon mutation rates. (Note, however, that even the ts^+ revertant sometimes retains mutator or antimutator activity, and that it may be necessary to collect and test several independent ts^+ revertants.) This control is particularly important in view of the fact that most of the ts mutants in common use arose from heavily mutagenized stocks, and undoubtedly contain additional mutations which may not even be removed by extensive backcrosses; see, for instance, the section below on the px and hm mutations, and also Allen, Albrecht, and Drake (1970). Finally, reconstruction selection controls should be performed, to show that the r^+

revertant grows at the same rate in the *ts* and in the *ts*+ background: such is certainly not the case, for instance, in studies on gene *30* (DNA ligase) mutants (Koch and Drake 1972).

Gene 30 (*DNA Ligase*) The theory of frameshift mutagenesis (Streisinger *et al.,* 1966) stipulates that the final step in the fixation of a mutational heterozygote is brought about by polynucleotide ligase, and that the site of ligase action is likely to be very close to an irregularity in the double helix. It therefore seemed likely that aberrant DNA ligases might accept such substrates with altered efficiencies compared to the wild-type enzyme. Credence for this notion was gained by a report (Sarabhai and Lamfrom 1969) that the rate of proflavin-induced reversion of a T4rII frameshift mutation was increased in a gene *30 ts* background. Bernstein (1971), however, observed no effect of ligase defects upon frameshift mutation rates. We have performed an extensive search for such effects (Koch and Drake 1972), but we observed only limited stimulation of either spontaneous or proflavin-induced frameshift mutagenesis in either the *rII* or the *ac* cistrons. Small but significant increases were sometimes observed, usually on the order of two- to threefold, under conditions where the suppression of gene *30* defects by *rII* mutations was carefully taken into account, but the results of Sarabhai and Lamfrom (1969) could not be duplicated even when competing host ligase activity was minimized. Evidence that gene *30* plays an important role in the rate of frameshift mutagenesis is therefore still lacking, but might, of course, appear in the future.

Since frameshift mutagenesis in bacteriophage T4 probably proceeds somewhat differently from frameshift mutagenesis in organisms which do not employ circularly permuted chromosomes, the mutations in T4 arising specifically at chromosome tips (Lindstrom and Drake 1970), it is still possible that the DNA ligase genes of other organisms will turn out to be important determinants of frameshift mutation rates, and tests of this possibility should be performed.

Gene 32 (*Alberts Protein*) Gene *32* of bacteriophage T4 encodes a protein which functions in DNA replication, recombination, and repair (Alberts 1970). The gene *32* protein binds cooperatively to singlestranded DNA, and maintains it in the extended configuration which is required for optimal base pairing with a complementary strand and, presumably, entering progeny bases. Furthermore, the gene *32* protein binds weakly but specifically to T4 DNA polymerase (Huberman, Kornberg, and Alberts 1971), and thereby inhibits its 3' exonuclease activity (Huang and Lehman 1972), suggesting a very intimate role for this protein in DNA replication. We (M. McGaw and J. W. Drake) therefore tested the effects of gene *32 ts* mutations upon the reversion of well-characterized *rII* mutants. Some typical results appear in Table 1. (Similar results have also been obtained by Bernstein 1971, and by Bernstein *et al.* 1972). An unanticipated general trend is emerging from these studies: mutation rates at A:T base pairs are frequently increased, while mutation rates at G:C base pairs are frequently decreased. Reversion rates of frameshift mutations show both

Table 1 Gene *32* mutator and antimutator activities

Reverting *rII* mutant	Probable path of reversion	*r*+ revertants per 10^8 particles		
		ts+	*tsG26*	*tsP7*
rUV183	A:T → G:C	23	850	290
rUV199	A:T → G:C	63	130	100
rUV13	G:C → A:T	24	9	8
rUV363	G:C → A:T	38	3.3	≤2
rUV27	frameshift	1070	≤340	550
rUV353	frameshift	510	37	200
rUV58	frameshift	13	40	46
rUV113	frameshift	2.0	3.3	25

Three to twelve independent stocks were grown in parallel from single particles using BB cells and L broth at 32°C. Revertant frequencies were determined by differential platings on KB and BB cells; median values are listed.

increases and decreases, but we do not yet know whether this dichotomy reflects the underlying process of base addition and base deletion. It should also be noted that although we have found the effect to disappear in ts^+ revertants of the ts-rII double mutants, we have not yet performed the appropriate reconstruction selection controls.

We have no information as yet which would help to decide whether the gene *32* effects are due to alterations in the configuration of the template (or primer) strands at the replication fork, or whether the effects are due to the interaction of the gene *32* protein with the DNA polymerase itself. The results do, however, raise the interesting possibility that the exact configuration of the template or primer strands during DNA replication is just as important to the accuracy of DNA replication as is the specificity of base pairing and the functional state of the DNA polymerase.

Genes 42 and td (*Pyrimidine Metabolism*)   The T4 genome contains at least nine genes concerned with deoxyribonucleotide metabolism, several of which are involved in the synthesis of dTTP and dHMCTP. Since thymine deprivation is mutagenic in *E. coli,* it seemed likely that pyrimidine deprivation would also be mutagenic in T4, and that this mutagenicity would display interesting specificities. The supply of dHMCTP for T4 DNA synthesis requires the function of gene *42,* which encodes dCMP hydroxymethylase. We (W. E. Williams and J. W. Drake) have therefore tested the effects of gene *42 ts* mutations upon the spontaneous reversion rates of *rII* mutations. Typical results appear in Table 2. It is clear

that gene *42 ts* mutations can act as mutator alleles (as has also been observed in unpublished experiments of G. R. Greenberg, personal communication), and that they do so with marked specificity: the reversion of A:T and frameshift mutations is relatively unaffected, but the reversion of G:C mutations is enhanced. The reversion of one of these mutants, *rSM94,* is promoted neither by base analogues nor by proflavin, and we strongly suspect that this mutant contains a G:C base pair which reverts by transversion but not by transition; the other G:C mutants are reverted by hydroxylamine, and are therefore capable of reverting by transition. We hope that the extension of this method will provide a system capable of scoring for transversions at G:C base pairs, which is not yet possible in bacteriophage T4 by any other method short of amino acid analyses of lysozyme mutants.

The most straightforward interpretation of these results is that dHMCTP deprivation increases all base-pair substitution rates at G:C base pairs by mass action, but two considerations suggest caution until more refined tests are performed. First, the increased mutation rates we observed were determined at 32°C, at which temperature the burst size approaches that of the wild type and the synthesis of dHMCTP must therefore not be severely limiting for DNA synthesis. Thymineless mutagenesis both in bacteria and in T4, on the other hand, is only observed at levels of thymine deprivation which sharply inhibit DNA replication. Increasing the growth temperature of the gene *42 ts* mutants did not result in marked increases in mutator effect, even though burst sizes were eventually sharply reduced. Second, Chiu and Greenberg (1968) have observed that the wild-type dCMP hydroxymethylase activity in crude extracts sediments very rapidly, much more so than is observed with the purified enzyme; and in addition, certain gene *42 ts* mutants produce an enzyme which displays high activity at elevated temperatures, but which no longer appears in the rapidly sedimenting complex. It is therefore possible that the dCMP hydroxymethylase interacts rather directly with the DNA polymerase, perhaps even receiving specific instructions when to provide residues to dHMCMP kinase, and that this interaction is the basis of the observed mutagenicity, rather than simple mass action. It is difficult at present to decide between these two alternatives. We have observed little or no increase in mutator

Table 2   Gene *42* mutator activities

Reverting *rII* mutant	Probable path of reversion	*r+* revertants per 10^8 particles	
		ts+	*tsLB3*
rUV13	G:C → A:T	24	280
rUV363	G:C → A:T	14	280
rSM94	G:C → py:pu	30	61
rUV183	A:T → G:C	45	39
rUV373	A:T → G:C	600	340
rUV27	frameshift	550	750
rUV28	frameshift	44	60

Three to 24 independent stocks were grown in parallel from single particles as in table 1.

activity over the temperature range 32–37° C, despite a corresponding decrease in burst sizes, which argues against the mass action interpretation. Preliminary experiments involving coinfection with *ts* and *ts*[+] alleles of gene *42,* however, failed to reveal any dominance in the mutator activity of the *ts* allele, which argues against the polymerase interaction interpretation.

No single genetic block is able to induce total deprivation of dTTP for DNA synthesis, since this substrate is available both from the continued functioning of preexisting cellular enzymes and also from the breakdown of host DNA. Thymine deprivation can, however, be achieved by either of two methods. FUdR efficiently inhibits both the viral and the cellular thymidylate synthetase activities, and fortuitously also inhibits the breakdown and reutilization of host DNA. Alternatively, one can employ a host strain carrying a *thy* mutation rendering its own thymidylate synthetase inactive, together with viral strains carrying both *td* and *den* mutations, which inactivate the viral thymidylate synthetase and a viral nuclease required for host DNA breakdown, respectively. We (D. Smith and J. W. Drake) have therefore employed both of these methods, usually with similar results (Table 3). The allele specificity of thymineless mutagenesis is strictly complementary to the allele specificity of cytosineless mutagenesis: the reversion of G:C and frameshift mutations is unaffected, but the reversion

of A:T mutations is enhanced. Furthermore, the reversion of one mutant, *rUV74,* is also promoted neither by base analogues nor by proflavin, but is promoted by thymine deprivation; and we strongly suspect that this mutant contains an A:T base pair which reverts by transversion but not by transition. Together with studies on the reversion of putative G:C transversions, and with applications of the transversion assay of Ripley and Drake (1972) which is specific for A:T transversions, these results offer considerable hope for the analysis of the mechanism of transversion mutation, which remains obscure at present.

Thymineless mutagenesis in bacteriophage T4 is again most simply interpreted as an increase in all base-pair substitution rates at A:T base pairs by mass action. (As in bacteria, thymineless mutagenesis is effective only under conditions which severely inhibit replication, as illustrated in Table 4.) Under this interpretation, both types of pyrimidineless mutagenesis would enhance whatever mechanisms already operate to produce base-pair substitutions, and synergism with chemical mutagenesis is also to be expected, although it has not yet been sought in T4 (but see Pauling 1968). This interpretation does not, however, speak to the question of whether pyrimidineless mutagenesis occurs during replication or repair DNA synthesis. Bridges, Law, and Munson (1968) have suggested that thymineless mutagenesis in *E. coli* occurs during repair synthesis, and since

Table 3 Thymineless mutagenesis

Reverting *rII* mutant	Probable path of reversion	Method	$r+$ revertants per 10^8 particles	
			$+Td$	$-Td$
rUV183	A:T → G:C	genetic	430	7400
		chemical	73	360
rUV191	A:T → G:C	genetic	21	104
		chemical	5.8	41
rUV74	A:T → py:pu	genetic	29	85
		chemical	52	280
rUV48	G:C → A:T	either	80	66
rUV28	frameshift	either	105	117

The "genetic" method consisted of infecting 011 *'thy* cells at 37°C in M9CA medium plus 10 μg/ml thymidine (Td) with an average of five particles of *rII-td8-denAS112* phage; the properties of the *td* and *denA* mutants are described by Simon and Tessman (1963) and by Hercules *et al.* (1971). At four minutes the complexes were washed with cold M9CA and resuspended in 37°C M9CA with or without Td at 10 μg/ml. Lysis was completed at 70 minutes with chloroform. The "chemical" method was the same, except that the infecting phages carried only the *rII* marker, and 10 μg/ml FUdR plus 20 μg/ml uridine were included throughout.

Table 4 Relationship between thymine deprivation and thymineless mutagenesis

Thymidine μg/ml	Burst size	r+ revertants per 10^8 progeny
40	270	25
10	150	42
1	150	59
0.05	2.9	150
0.00	0.44	330

The "chemical" method was applied to rUV183; see Table 3.

we have some rather indirect evidence that base analogue mutagenesis in bacteriophage T4 may also occur preferentially during repair synthesis (see below), it is perfectly possible that the enzymes of DNA repair are somewhat more error-prone than are the enzymes of DNA replication. It is significant in this regard that, although the T4 DNA polymerase probably acts primarily during DNA replication and has not yet been implicated as a component of DNA repair, the antimutagenic DNA polymerases of T4 suppress thymineless mutagenesis very efficiently indeed (Drake and Greening 1970).

Genes hm, px, *and* v *(Repair)*   There is considerable evidence from studies with *E. coli* that many mutagens act by triggering an error-prone repair system (Watkin 1969; Kondo *et al.* 1970), and there are now two lines of evidence that a similar process operates in bacteriophage T4. The first of these arises from a comparison of UV and acetophenone-photosensitized mutagenesis (Meistrich and Drake 1972). One of the main mutational pathways promoted by UV irradiation in T4 is the G:C → A:T transition (Drake 1963, 1966), and the same pathway is also efficiently promoted by acetophenone-sensitized irradiation, which produces nearly exclusively T-T cyclobutane dimers. The induction of mutations at G:C sites by photochemical reactions which affect only A:T base pairs is therefore evidence for an indirect route of mutagenesis. The second line of evidence, to be described below, derives from the properties of the mutants px and hm.

Harm (1963) isolated a mutant of T4D which he called x, and which displays a 1.7-fold increased UV sensitivity and a 3-fold decreased rate of recombination (Harm 1964). We have backcrossed this mutant extensively into T4B, and have observed that it segregates two different mutations affecting mutagenesis, px and hm. The px mutant exhibits the increased UV and MMS sensitivities, the reduced rate of recombination, and the slightly decreased growth rate characteristic of the original strain. Preliminary mapping studies place px somewhere in the region containing genes 41 and 56, that is, between the rII region and gene 43 (DNA polymerase). The hm mutant exhibits the wild-type UV and MMS sensitivities, recombination rate, and growth rate, and maps in the neighborhood of gene 1. Further studies have indicated that both px and hm affect spontaneous, UV- and MMS-induced mutation rates. (MMS has hitherto been considered not to be a mutagen for T4 (Loveless 1959), but recent studies indicate that it is in fact a fairly strong mutagen for T4, particularly when the scoring procedures developed for UV mutagenesis in T4 (Drake 1966) are employed.)

Neither px nor hm is an easy marker to manipulate in recombination experiments. The px allele is routinely scored by picking plaques and testing their UV sensitivities. The hm marker is usually scored as hypermutable to UV irradiation (see below), individual plaques being picked, grown into stocks, irradiated to survivals of about 0.1%, and plated to obtain about 3000 plaques; the wild type sports about 4–15 r mutants among these survivors, while hm sports about 20–45 r mutants. The hm character is also sometimes scored in an rII background by growing low titer stocks and scoring reversion by spot tests, or where necessary, by quantitative tests. These procedures have considerably slowed the investigation of these two mutants.

Measurements of spontaneous mutation rates indicate that hm exhibits mild mutator activity (Table 5). Preliminary evidence suggests that px, on the other hand, slightly reduces spontaneous mutation rates.

The rates of UV-induced mutagenesis are shown in Table 6, and are about 6×10^{-4} r mutants per phage lethal hit for the wild type, about 0.9×10^{-4} for px, and about 19×10^{-4} for hm. UV mutagenesis is therefore decreased about 7-fold (per lethal hit) or about 4-fold (per erg per mm^2) by px, and increased about 3-fold by hm. For the wild type and for both hm and px, the rate of mutation induction is linear with dose except at the highest doses applied, where

Table 5 Effect of *hm* on spontaneous mutation rates

	Mutation frequencies	
Mutational pathway	*Wild type*	*hm*
$r+ \rightarrow r$	6.1×10^4	20.0×10^4
$\rightarrow rI$	1.7×10^4	12.5×10^4
$\rightarrow rII$	4.4×10^4	7.5×10^4
$rUV183 \rightarrow r^+$	3.1×10^7	10.8×10^7
$rUV199 \rightarrow r+$	5.7×10^7	51.3×10^7

Mutation frequencies were determined from three or more independent stocks. The two *rII* mutants are reverted by base analogues but not by hydroxylamine (A:T → G:C).

Table 6 Effects of *px* and *hm* on ultraviolet mutagenesis

Seconds of	r per 10^8 survivors		
irradiation	*Wild type*	*px*	*hm*
0	0.54	0.24	1.45
20	1.7	. . .	7.3
30	. . .	0.82	. . .
40	4.1	. . .	14.3
60	~3.6	. . .	~11

Irradiation was performed with a 15-watt low-pressure mercury lamp delivering 0.155 phage lethal hits per second. Scoring procedures are described in Drake (1966).

scoring becomes poorly reproducible because of the great irregularity of plaque sizes produced by the survivors; there is no reason at present to believe that the mutation induction curves really deviate downwards at the higher doses, and in fact they sometimes appear to deviate upwards, probably because of multiplicity reactivation on the plate. (A report by Azizbekyan and Pogosov (1971) claiming that the reversion of *r131* by UV irradiation displays markedly curvilinear kinetics, and depends upon the mutations *x* and *v*, could not be reproduced in this laboratory. Indeed, no induced reversion of *r131* could be achieved in any genetic background by UV irradiation, whereas artifactual reversion due to reactivation phenomena on the plate could be demonstrated by reconstruction experiments. The

data of Azizbekyan and Pogosov is also rendered suspect by their use of stationary phase plating cells, and their value for the spontaneous reversion rate of *r131*, which is 100-fold lower than that recorded by Champe and Benzer (1962) and also observed in this laboratory.)

The spectra of spontaneous and UV-induced *rII* mutants is shown in Table 7. The results are generally similar to those of previous reports (Drake 1963, 1966) in that few or no deletions and G:C mutants are generated, while A:T mutants and probably also frameshift mutants are generated in roughly equal amounts (when account is taken of the contribution of the spontaneous mutant background to the UV-induced mutant population). The spontaneous mutants arising in both the wild type

Table 7 Characteristics of spontaneous and ultraviolet-induced *rII* mutants

Genotype irradiated	UV	Number tested	Percent deletions	Percent G:C	Percent A:T	Percent other
Wild type	−	30	17	0	7	77
	+	43	19	2	21	58
hm	−	56	2	2	25	71
	+	122	9	11	43	37
px	−	12	8	0	8	83
	+	27	4	0	30	67

Data were pooled from several experiments; the weighted average contributions of the spontaneous mutant backgrounds, considering only the *rII* mutants, were 43% for the wild type, 28% for *hm*, and 52% for *px*, and the percentage values are not corrected for these contributions. The mutants were characterized by base analogue spot tests (Drake 1966): mutants which never exhibited revertants are classified as deletions, mutants which exhibited stronger 5-bromouracil than 2-aminopurine responses are classified as G:C, mutants which exhibited 2-aminopurine responses equal to or greater than their 5-bromouracil responses are classified as A:T, and mutants which reverted spontaneously but did not respond to either base analogue are classified as "other," and as judged from extensive tests with many other *rII* mutants, contain mainly frameshift mutations together with a few transversions.

and in *px* are predominantly nonrevertible by base analogues and probably mainly contain frameshift mutations), but *hm* produces a somewhat higher frequency of A:T mutants (most of which probably arose from G:C → A:T transitions). UV mutagenesis also seems to induce relatively more base pair substitution mutants in *hm* than it does in the wild type or in *px*, and an appreciable proportion of these consist of G:C mutants. The specificity, as well as the rate of UV mutagenesis, is therefore altered by *hm*.

The *px* and *hm* mutations affect MMS mutagenesis in directions parallel to those by which they affect UV mutagenesis (Table 8). The averate rates of *r* mutation induction per phage lethal hit are about 8×10^{-4} for the wild type, about 2×10^{-4} for *px*, and about 13×10^{-4} for *hm*, values remarkably similar to those observed following UV irradiation.

The *hm* mutation therefore increases MMS mutagenesis about 1.7-fold. The *px* mutation decreases MMS mutagenesis about 3-fold per lethal hit, or about 1.3-fold per minute of treatment.

The spectrum of *rII* mutations induced by MMS is shown in Table 9, and is virtually identical to the spectrum of *rII* mutants induced by UV irradiation, despite the very different types of lesions produced in DNA by these two agents. This is a particularly remarkable result when one considers that the closely related monofunctional alkylating agents EMS and EES produce almost exclusively transition mutations, a considerable proportion being of the type A:T → G:C (Freese 1961; Krieg 1963; U. Ray and J. W. Drake, unpublished results).

A few tests have also been conducted to explore the effects of *hm* upon the action of other mutagens (Table 10). While the rate of proflavin (frameshift) mutagenesis was unaffected, the rate of 2-aminopurine-induced reversion of an A:T mutant was substantially increased.

Our interpretation of the mechanism of action of *px* and *hm* is inspired by studies of the *rec* mutants of *E. coli* (Witkin 1969; Kondo *et al.* 1970). Like *recA, recB, recC* and *exr*, *px* affects both recombination and mutagenesis, and also displays increased sensitivity to a variety of noxious agents, including UV irradiation, MMS (Baldy, Strom, and Bernstein 1971), EMS (Ray, Bartenstein, and Drake 1972), and photodynamic action (Geissler 1970). It therefore appears that *px* is a mutation which at least partially inactivates a generalized error-prone repair process which is also involved in genetic recombination. (Note that recombination appears to be an indispensable function for the replication of bacteriophage T4, presumably because of the circularly

Table 8 Effects of *px* and *hm* on MMS mutagenesis

Minutes of treatment	*r* per 10^8 survivors		
	Wild type	*px*	*hm*
0	1.1	0.72	1.0
6	2.0	...	3.8
9	...	2.3	...
12	4.4	...	6.2
18	5.7	...	9.2

Phages were treated with 0.1 M MMS in 0.2 M phosphate buffer, pH 7.0, at 43°C. The reaction was stopped by a 5-fold dilution into 1 M sodium thiosulfate in the same buffer, and the survivors were assayed within minutes. Scoring procedures were the same as those employed for UV mutagenesis (Drake 1966).

Table 9 Characteristics of spontaneous and MMS-induced *rII* mutants

Genotype treated	MMS	Number tested	Percent deletions	Percent G:C	Percent A:T	Percent other
Wild type	–	16	6	0	13	81
	+	51	14	4	25	57
hm	–	6	17	0	17	67
	+	103	2	9	43	47
px	–	9	0	0	0	100
	+	14	0	21	36	43

Data were pooled from several experiments: the weighted average contributions of the spontaneous *rII* mutant background to the induced *rII* mutant populations were 22% for the wild type, 7% for *hm*, and 46% for *px*. The mutants were characterized as described in Table 7.

Table 10 Effect of *hm* on proflavin and 2 aminopurine mutagenesis

Mutational pathway	Mutagen	r^- revertants per 10^7 particles	
		Wild type	hm
$rUV58 \rightarrow r+$	none	1.3	2.5
	proflavin	1010	1090
$rUV183 \rightarrow r+$	none	18	140
	2-aminopurine	260	1060

Proflavin mutagenesis was performed by infecting BB cells at 37°C in broth at pH 7.8 with an average of 5 particles per cell of the frameshift mutant *rUV58*. At 5 minutes the complexes were diluted into broth with or without proflavin hemisulfate at 2 μg/ml, and were then superinfected with an additional five particles per cell. At 34 minutes the complexes were diluted 50-fold into broth, and lysis was completed with chloroform at 50 minutes. 2-aminopurine mutagenesis was performed by infecting BB cells at 37°C in M9CA medium with an average of 4 particles per cell of the A:T mutant *rUV183*. At 5 minutes the complexes were diluted into M9CA with or without 2-aminopurine at 800 μg/ml. Lysis was completed with chloroform at 40 minutes.

redundant nature of the T4 chromosome. Fully recombination-defective mutants of T4 are therefore unlikely to be recovered, as has been observed by many workers; indeed, the only other published report of a recombination-defective mutant of T4 (van der Ende and Symonds 1972) also describes a weakly defective strain.) Whether *px* is analogous to one of the known *E. coli rec* functions, none of which affect either UV sensitivity or recombination rates in T4, remains to be seen.

It would be expected that the mutagenic specificities of gene *px*-dependent mutagenesis primarily reflect the particular error tendencies of the repair system itself, rather than the nature of the premutational lesion. As a result, a variety of different mutagens would be expected to produce an approximately identical spectrum of mutational types. Studies of this type have still to be performed in *E. coli,* but the present results (Tables 7 and 9) indicate no significant differences in the types of mutants produced by UV irradiation and by MMS. The great difference between MMS on the one hand and EMS and EES on the other probably arises from the ability of the later agents to carry out 0^6 alkylation of guanine (Loveless 1969), and probably also of

thymine; these reactions would be expected to produce the observed transitions. It may, however, be significant that photodynamic mutagenesis in bacteriophage T4 (Drake and McGuire 1967b), although not yet shown to depend upon gene *px,* produces a somewhat different mutational spectrum compared to UV irradiation.

The role of *hm* in error-prone repair remains obscure. It could, for instance, encode one of the components of the *px* system, which might be an undiscovered DNA polymerase or a structural protein analogous to the product of gene *32*. It is significant that *hm* also promotes 2-aminopurine mutagenesis: this result suggests that base analogue mutagenesis may normally proceed preferentially during repair synthesis, which no doubt normally accompanies T4 replication both during the (obligatory) process of genetic recombination and probably also during the repair of spontaneously arising lesions in the DNA. While this interpretation casts considerable doubt on the meaning of experiments which assume that base analogue mutagenesis occurs specifically at the replication fork, it does not affect the basic theory of base analogue-induced mispairing (Drake 1970).

Gene *v* of bacteriophage T4 encodes an endonuclease which is specific for pyrimidine dimers (Friedberg and King 1971). Spontaneous mutation rates in *v* have never, in numerous incidental tests in this laboratory, been observed to be different from those in the wild type. The UV-induced mutation rate in *v* is the same per *lethal hit* as in the wild type (Meistrich and Drake 1972). Since T4*v* exhibits a UV sensitivity about twice as great as the wild type, it follows that the UV-induced mutation rate in T4*v* is increased about twofold *per erg per mm² of irradiation.* This situation is entirely analogous to what is observed in *E. coli* (Hill 1965), and whether or not *v* is to be considered a mutator mutation is a simple matter of definition. It appears that *v* merely acts, in the present context, to reduce the effective UV dose, those lesions not excised by the *v* system being subject to *px*-dependent, error-prone repair.

PROSPECTS FOR PROGRESS

It is possible to recognize a number of problems related to the genetic determination of mutation rates which are ripe for analysis. The first of these is the basic chemistry of action of important mutagens (chemicals, radiations, heat) upon DNA, particularly as it bears upon the question of whether the reacted bases directly mispair during DNA replication, or whether they merely stimulate error-prone repair processes. The theory of action of numerous chemical mutagens has remained in a rather stagnant and dogmatic state for the better part of a decade now (excepting at least the exciting work on the monofunctional alkylating agents), and much too little is known, for instance, about the specific kinds of mispairings which are induced by 2-aminopurine, by hypoxanthine (produced by the action of nitrous acid upon adenine), by 0^6-alkylated bases, and by 4:5-saturated cytosine derivatives. Even the mechanism of action of 5-bromouracil remains unestablished: it is unclear whether the enolized or the ionized form of the base mispairs, just as it is for the normal bases themselves. Any detailed analysis of how the effects of these altered residues are reduced or promoted by the enzymes of DNA synthesis must clearly await a greater understanding of the basic chemical configurations of primary mutagenic lesions themselves.

Just as we lack basic information about the chemical lesions produced by mutagens, we also lack basic information about the enzymology of DNA replication and repair. The numerous recent advances which have defined some of the enzymes of DNA synthesis in prokaryotes, and to a lesser extent partitioned them between DNA replication and DNA repair, have raised many more questions than they have answered. It remains unclear, for example, whether the DNA polymerase actively selects bases for incorporation based upon instructions from the template strand (Drake *et al.* 1969), whether the polymerase merely checks the fit of stacked but unpolymerized residues (Kornberg 1969), whether the 3' exonuclease activity associated with all DNA polymerases isolated to date acts in an editorial function to remove misincorporated bases (see for instance the articles by Nossal and by Bessman in this volume), or whether some combination of these mechanisms operates. It also remains unclear whether gene *32* protein merely maintains template (and primer?) strands in an extended form, or whether it also mediates the alignment of numerous progeny bases prior to polymerase action.

Our understanding of the *rec-* and *px*-mediated repair processes is not sufficiently advanced as yet

to ask questions such as those just elaborated. Are relatively intact single strands transferred between daughter chromosomes in a manner somewhat akin to genetic recombination, as suggested for the *rec* function (Rupp *et al.* 1971)? If so, do the repair-dependent mutations then arise in the donor or in the recipient chromosome, and how? Which specific enzymatic steps are error-prone? Are they DNA polymerases such as polymerase II of *E. coli* or an undiscovered polymerase of bacteriophage T4, or does the already well characterized gene *43* DNA polymerase of T4 also act during repair, but under conditions which are less optimal for accuracy than those encountered during DNA replication? Furthermore, many genes whose activities are required for DNA synthesis both in *E. coli* and in bacteriophage T4, are still unassociated with specific enzymatic or structural functions. As these functions are discovered, it will become even more worthwhile to examine the roles of their genes in the determination of mutation rates.

There is continued need for *in vitro* studies on the accuracy of DNA synthesis. Hall and Lehman (1968), for example, attempted to compare the error rates of wild type and mutator DNA polymerase from bacteriophage T4, with somewhat disappointing results: most of the mutational pathways strongly promoted by the mutator polymerase *in vivo* were not enhanced *in vitro*. It is unclear whether this was because it is still difficult to measure background error rates sufficiently accurately, or whether the purification of the mutator polymerase altered its properties. Improved radiochemical methods for measuring error rates during *in vitro* DNA synthesis are therefore required, as is the development of systems incorporating other components of DNA synthesis, such as the gene *32* protein (see Huberman, Kornberg, and Alberts 1971), RNA polymerase (Sugino, Hirose, and Okazaki 1972), and others. Furthermore, since these studies may profitably employ defined polydeoxyribonucleotides as templates, the range of defined substrates also needs to be increased.

In vitro studies to date have concentrated upon the process of base pair substitution. It should now be possible to extend these studies to frameshift mutagenesis, using both defined polynucleotides and DNA transformation systems. The main advantage of transformation systems is that the induced mutations can then be characterized genetically. A T4

DNA transformation system suitably modified for the study of mutation has already been devised (Baltz and Drake 1972), and it has proved a simple matter to induce mutations *in vitro* with hydroxylamine and other agents.

Finally, it should be possible to gather much more information about base-pair sequences in order to characterize base-pair substitution hot spots. The T4rII proteins are sufficiently intractable so that the required information is more likely to come from the T4e (lysozyme), the *E. coli* tryptophan synthetase, or the yeast cytochrome *c* systems, in which detailed amino acid sequence information is available both for the wild type and for numerous mutants. The extension of preliminary *in vitro* experiments should also provide information about nearest neighbor effects, particularly when coupled with basic chemical studies describing base pairing and base stacking interactions.

ACKNOWLEDGEMENT

Previously unreported work described here was supported by grants E59 and VC-5L from the American Cancer Society, grants GB15139 and GB30604X from the National Science Foundation, and grant AI04886 from the Public Health Service.

LITERATURE CITED

Alberts, B. M., 1970. Function of gene 32-protein, a new protein essential for the genetic recombination and replication of T4 bacteriophage DNA. Fed. Proc. 29:1154–1163.

Allen, E. R., I. Albrecht and J. W. Drake, 1970. Properties of bacteriophage T4 mutants defective in DNA polymerase. Genetics 65:187–200.

Azizbekyan, R. R. and V. Z. Pogosov, 1971. Repair of UV-induced mutations in phage T4. Biochem. Biophys. Res. Commun. 43:1402–1407.

Baldy, M. W., B. Strom and H. Bernstein, 1971. Repair of alkylated bacteriophage T4 deoxyribonucleic acid by a mechanism involving polynucleotide ligase. J. Virol. 7:407–408.

Baltz, R. H. and J. W. Drake, 1972. Bacteriophage T4 transformation: an assay for mutations induced *in vitro*. Virology 49:462–474.

Benzer, S., 1961. On the topography of the genetic fine structure. Proc. Nat. Acad. Sci. U.S. 47:403–415.

Bernstein, C., H. Bernstein, S. Mufti and B. Strom, 1972. Stimulation of mutation in phage T4 by lesions in gene 32 and by thymidine imbalance. Mutation Res. 16:113–119.

Bernstein, H., 1971. Reversion of frameshift mutations stimulated by lesions in early function genes of bacteriophage T4. J. Virol. 7:460–466.

Brenner, S., A. O. W. Stretton and S. Kaplan, 1965. Genetic code: the 'nonsense' triplets for chain termination and their suppression. Nature 206:994–998.

Bridges, B. A., J. Law and R. J. Munson, 1968. Mutagenesis in *Escherichia coli*. II. Evidence for a common pathway for mutagenesis by ultraviolet light, ionizing radiation and thymine deprivation. Molec. Gen. Genet. 103:266–273.

Champe, S. P. and S. Benzer, 1952. Reversal of mutant phenotypes by 5-fluorouracil: an approach to nucleotide sequences in messenger-RNA. Proc. Nat. Acad. Sci. U.S. 48:532–546.

Chiu, C. S. and G. R. Greenberg, 1968. Evidence for a possible direct role of dCMP hydroxymethylase in T4 phage DNA synthesis. Cold Spring Harbor Symp. Quant. Biol. 33:351–359.

de Vries, F. A. J., C. J. H. Swart-Idenburg and A. de Waard, 1972. An analysis of replication errors made by a defective T4 DNA polymerase. Molec. Gen. Genet. 117:60–71.

Dove, W. F., 1968. The genetics of the lambdoid phages. Ann. Rev. Genet. 2:305–340.

Drake, J. W., 1963. Properties of ultraviolet-induced rII mutants of bacteriophage T4. J. Mol. Biol. 6:268–283. ——, 1966. Ultraviolet mutagenesis in bacteriophage T4. I. Irradiation of extracellular phage particles. J. Bacteriol. 91:1775–1780. ——, 1969. Comparative rates of spontaneous mutation. Nature 221:1132. ——, 1970. *The Molecular Basis of Mutation*. Holden-Day, Inc., San Francisco.

Drake, J. W. and E. F. Allen, 1968. Antimutagenic DNA polymerase of bacteriophage T4. Cold Spring Harbor Symp. Quant. Biol. 33:339–344.

Drake, J. W., E. F. Allen, S. A. Forsberg, R. M. Preparata and E. O. Greening, 1969. Genetic control of mutation rates in bacteriophage T4. Nature 221:1128–1132.

Drake, J. W. and E. O. Greening, 1970. Suppression

of chemical mutagenesis in bacteriophage T4 by genetically modified DNA polymerases. Proc. Nat. Acad. Sci. U.S. 66:823–829.

Drake, J. W. and J. McGuire, 1967a. Characteristics of mutations appearing spontaneously in extracellular particles of bacteriophage T4. Genetics 55:387–398. ———, 1967b. Properties of r mutants of bacteriophage T4 photodynamically induced in the presence of thiopyronin and psoralen. J. Virol. 1:260–267.

Freese, E. B., 1961. Transitions and tranversions induced by depurinating agents. Proc. Nat. Acad. Sci. U.S. 47:540–545.

Friedberg, E. C. and J. L. King, 1971. Dark repair of ultraviolet-irradiated deoxyribonucleic acid by bacteriophage T4: Purification and characterization of a dimer-specific phage-induced endonuclease. J. Bacteriol. 106:500–507.

Geissler, E., 1970. Different sensitivities of T4D and lambda mutants to photodynamic action. Molec. Gen. Genet. 109:264–268.

Hall, Z. W. and I. R. Lehman, 1968. An *in vitro* transversion by a mutationally altered T4-induced RNA polymerase. J. Mol. Biol. 36:321–333.

Harm, W., 1963. Mutants of phage T4 with increased sensitivity to ultraviolet. Virology 19:66–71. ———. 1964. On the control of UV-sensitivity of phage T4 by the gene x. Mutation Res. I:344–354.

Hercules, K., J. L. Munro, S. Mendelsohn and J. S. Wiberg, 1971. Mutants in a nonessential gene of bacteriophage T4 which are defective in the degradation of *Escherichia coli* deoxyribonucleic acid. J. Virol. 7:95–105.

Hershfield, M. S. and N. G. Nossal, 1972. Hydrolysis of template and newly synthesized deoxyribonucleic acid by the 3′ to 5′ exonuclease activity of the T4 deoxyribonucleic acid polymerase. J. Biol. Chem. 247:3393–3404.

Hill, R. F., 1965. Ultraviolet-induced lethality and reversion to prototrophy in *Escherichia coli* strains with normal and reduced dark repair ability. Photochem. Photobiol. 4:563–568.

Huang, W. M. and I. R. Lehman, 1972. On the exonuclease activity of phage T4 deoxyribonucleic acid polymerase. J. Biol. Chem. 247:3439–3446.

Huberman, J. A., A. Kornberg and B. M. Alberts, 1971. Stimulation of T4 bacteriophage DNA polymerase by the protein product of T4 gene 32. J. Mol. Biol. 62:39–52.

Koch, R. E., 1971. The influence of neighboring base pairs upon base-pair substitution rates. Proc. Nat. Acad. Sci. U.S. 68:773–776.

Koch, R. E., and J. W. Drake, 1970. Cryptic mutants of bacteriophage T4. Genetics 65:379–390. ———, 1972. Ligase-defective bacteriophage T4. I. Effects on mutation rates. J. Virol. 11:35–40.

Kondo, S., H. Ichikawa, K. Iwo and T. Kato, 1970. Base-change mutagenesis and prophage induction in strains of *Escherichia coli* with different DNA repair capacities. Genetics 66:187–217.

Kornberg, A., 1969. Active center of DNA polymerase. Science 163:1410–1418.

Krieg, D. R., 1963. Ethyl methanesulfonate-induced reversion of bacteriophage T4rII mutants. Genetics 48:561–580.

Lindstrom, D. M. and J. W. Drake, 1970. Mechanics of frameshift mutagenesis in bacteriophage T4: role of chromosome tips. Proc. Nat. Acad. Sci. U.S. 65:617–624.

Loveless, A., 1959. The influence of radiomimetic substances on deoxyribonucleic acid synthesis and function studied in *Escherichia coli*/phage systems. III. Mutation of T2 bacteriophage as a consequence of alkylation *in vitro*: the uniqueness of ethylation. Proc. Roy. Soc. Lond. Ser. B 150:497–508. ———, 1969. Possible relevance of 0-6 alkylation of deoxyguanosine to the mutagenicity and carcinogenicity of nitrosamines and nitrosamides. Nature 223:206–207.

Luria, S. E., 1951. The frequency distribution of spontaneous bacteriophage mutants as evidence for the exponential rate of phage reproduction. Cold Spring Harbor Symp. Quant. Biol. 16:463–470.

Meistrich, M. L. and J. W. Drake, 1972. Mutagenic effects of thymine dimers in bacteriophage T4. J. Mol. Biol. 66:107–114.

Okada, Y., G. Streisinger, J. Owen, J. Newton, A. Tsugita and M. Inouye, 1972. Molecular basis of mutational hot spot in the lysozyme gene of bacteriophage T4. Nature 236:338–341.

Pauling, C., 1968. The specificity of thymineless mutagenesis, pp. 383–398. In: *Structural Chemistry and Molecular Biology*. Edited by A. Rich and N. Davidson. W. H. Freeman and Co., San Francisco.

Ray, U. L., Bartenstein and J. W. Drake, 1972. Inactivation of bacteriophage T4 by ethyl methanesulfonate: influence of host and viral genotypes. J. Virol. 9:440–447.

Ripley, L. S. and J. W. Drake, 1972. A genetic assay for transversion mutations in bacteriophage T4. Molec. Gen. Genet. 118:1–10.

Rupp, W. D., C. E. Wilde, D. L. Reno and P. Howard-Flanders, 1971. Exchange between DNA strands in ultraviolet-irradiated *Escherichia coli.* J. Mol. Biol. 61:25–44.

Salts, Y. and A. Ronen, 1971. Neighbor effects in the mutation of *ochre* triplets in the T4rII gene. Mutation Res. 13:109–113.

Sarabhai, A. and H. Lamfrom, 1969. Mechanism of proflavin mutagenesis. Proc. Nat. Acad. Sci. U.S. 63:1196–1197.

Simon, E. H. and I. Tessman, 1963. Thymidine-requiring mutants of phage T4. Proc. Nat. Acad. Sci. U.S. 50:526–532.

Speyer, J. F., 1965. Mutagenic DNA polymerase. Biochem. Biophys. Res. Commun. 21:6–8.

Speyer, J. F., J. D. Karam and A. B. Lenny, 1966. On the role of DNA polymerase in base selection. Cold Spring Harbor Symp. Quant. Biol. 31:693–697.

Streisinger, G., Y. Okada, J. Emrich, J. Newton, A. Tsugita, E. Terzaght and M. Inouye, 1966. Frame-shift mutations and the genetic code. Cold Spring Harbor Symp. Quant. Biol. 31:77–84.

Sugino, A., S. Hirose and R. Okazaki, 1972. RNA-linked nascent DNA fragments in *Escherichia coli.* Proc. Nat. Acad. Sci. U.S. 69:1863–1867.

van den Ende, P. and N. Symonds, 1972. The isolation and characterization of a T4 mutant partially defective in recombination. Molec. Gen. Genet. 115:239–247.

Witkin, E. M., 1969. Ultraviolet-induced mutation and RNA repair. Ann. Rev. Genet. 3:525–552.

DISCUSSION

B. Strauss: The proposals made by Dr. Drake and others about the mechanisms for mutation induction by alkylating agents make some observations by Dr. Scudiero in my laboratory particularly pertinent. DNA synthesized by HEp. 2 cells treated with methyl methanesulfonate contains regions of single strandedness. DNA with such single-stranded regions adsorbs to columns of benzoylated naphthoylated DEAE cellulose and is only eluted when the columns are washed with caffeine. DNA fragments with such single-stranded regions are converted to purely double-stranded fragments by treatment with exonuclease. The fragments produced by alkylated cells are qualitatively similar to those produced by control cells given a pulse of radioactivity, but differ in both their relative number and in the extent of the regions of single strandedness. The single-stranded regions from both control and alkylated cells are gradually converted to purely double-stranded DNA, although the conversion occurs more slowly in alkylated cells. We concluded that the fragments arise in the course of normal DNA synthesis, but have a longer life time when DNA synthesis is delayed by a methyl methanesulfonate-induced lesion. The disappearance of the single-stranded regions from the newly synthesized DNA formed by alkylated cells is probably a reflection of the processes of post replication repair.

A reversion to wild-type associated with crossing over in *Drosophila melanogaster*

C. P. OLIVER *

Reprinted by author's and publisher's permission from the Proceedings of the National Academy of Sciences, *U.S., vol. 26, 1940, pp. 452–454. [Communicated May 31, 1940.]*

Starting in the late 1920s, in conjunction with the ability to artificially enhance the frequency of mutations, mutants were used as tools to study gene structure. Prior to 1940, crossover analyses of mutants, established as being allelic by means of a complementation analysis, failed to show recombination that would relocate the two allelic mutants on the same chromosome. Likewise, the reciprocal wild-type recombinant was never found. These failures firmly entrenched the viewpoint that the functional gene could not be subdivided by means of mutation and recombination. Allelism came to be defined in terms of the inability of a pair of mutant alleles to complement or show recombination. The gene took on an almost mystical degree of structural integrity, which shielded it from subdivision. Thus, C. P. Oliver's paper was received with great surprise and skepticism. It set forth the first case in which two noncomplementing mutant alleles, displaying a mutant phenotype in the heterozygous condition, yielded wild-type progeny. This suggested that recombination falling between the two mutant alleles was generating a normal gene. If his interpretation were correct, then the gene could be subdivided by means of mutation and intragenic-recombination.

In a study of two alleles of lozenge eye in *Drosophila melanogaster*, a low frequency of reversion to the wild-type has been observed. The reversion involves the color and structure of the eye, and also the genital tracts which are abnormal in the mutant females.[1] In each case in which the reversion has occurred, crossing over between the X chromosomes has also occurred.

Glossy and spectacle are sex-linked, recessive mutants, alleles of lozenge, which were induced by irradiation. Glossy (lz^g) individuals have eyes which are blood red in color, with fused facets making a glossy surface. Spectacle (lz^s), as reported by Dr. J. T. Patterson,[2] is characterized by a light brown color of the eye, and the facets are run together to cause a smooth surface of the eye. In the compound, heterozygous glossy-spectacle females, glossy is more dominant in its expression; and the character of the eye is nearly that of homozygous glossy. Spectacle is associated with the *dl*-49 inversion, and the mutant gene is located within that inversion. Glossy is also located within an inversion which is very similar to, probably the same as, the *dl*-49 inversion.

From the mating of lz^g Bx/lz^s f females to lz^g Bx males, a total of 5584 offspring have been inspected. Most of the offspring were of the expected types. Males were glossy or spectacle. Females were phenotypically alike, although genotypically they were either homozygous glossy or heterozygous glossy-spectacle. Eleven of the 5584 offspring, 2 males and 9 females, were wild-type for the mutant eye and genital traits. The wild-type traits persist whether the individuals are heterozygous for spectacle or glossy. Two of the exceptional flies were tested for crossing over with a combination of genes in a non-inverted X chromosome. The results indicated that the inversion had not been lost.

The appearance of each exceptional, wild-type offspring was associated with crossing over. Ten of the exceptional offspring were Beadex (Bx), and carried, therefore, the right end of the glossy-bearing X chromosome. The other one, forked (f), probably was the result of a double crossover. In a test-cross

*Department of Zoology, University of Minnesota.

using yellow and Hairy wing as markers for the left end of the chromosome ($y\ IIw\ lz^s\ f/lz^g\ Bx$), 2 out of 305 offspring were of the reversed type. One, a male, was $y\ IIw\ Bx$; and the other, a female, proved to be heterozygous for that combination. The flies showing the reversion to wild-type have an X chromosome which is composed of the right end of the glossy-bearing chromosome and the left end of the spectacle-bearing chromosome. Moreover, the associated crossing over occurs within the inverted regions. The loci for miniature (m) and vermilion (v) are located in that order within the inverted regions. From the mating of $v\ lz^s\ f/m\ lz^g\ Bx$ females to glossy males, 2 exceptional males occurred among 1285 offspring, and both were $v\ Bx$. Apparently the crossover occurs between v and one of the alleles of lozenge.

Although crossing over seems to be associated with the reversion, and one crossover type appears, the complementary type has not been recovered. It is not known whether the failure to recover the complementary type is due to the inability of that type to live, or to the inability of the observer to recognize the combination. Glossy is almost completely dominant over spectacle. The presence of glossy and spectacle in one chromosome, with either glossy or spectacle in the other chromosome, may produce a phenotype so similar to homozygous glossy and the compound glossy-spectacle that it will be difficult, if not impossible, to differentiate the three genotypes.

It does not seem possible to explain the reversions on the basis of mutations in the genic sense. The exceptional type has not appeared among the offspring of either homozygous glossy or spectacle; nor has it been observed among the offspring from the females heterozygous for either glossy or spectacle and lozenge. Lozenge is present in a chromosome which has no inversion. The only mating which has given the exceptional, wild-type offspring has been the compound glossy-spectacle. Each of the mutants is present in an inversion, and the inversions are at least almost identical. Under such conditions, crossing over is expected to occur throughout the chromosome. The crossing over associated with the reversion occurs within the inverted region, and as shown by the appearance of the $v\ Bx$ crossover type involves the region near the alleles glossy and spectacle.

Although crossing over is an active factor in the reversion of the alleles to the wild-type, it is not possible as yet to determine the exact nature of the phenomenon. The condition can be a case of unequal crossing over;[3] but it can as likely be a case which involves the "repeat" hypothesis developed by Bridges,[4] in which different primary loci of the chromosome are involved in the expression of the two mutants.

SUMMARY

1. A reversion to wild-type occurs with a frequency of 0.2% among offspring from a female which had glossy on one and spectacle on the other X chromosome. Both traits are recessive and are allelic to lozenge.

2. The reversion is always associated with crossing over.

3. The associated crossing over occurs within the inverted regions of the chromosome, and at or near the lozenge locus.

4. The suggestion is made that the results probably involve either unequal crossing over or crossing over between "repeats."

REFERENCES

1. Oliver and Green, *Anat. Rec.*, 75 (Supp.):100–101 (1939).
2. Patterson and Muller, *Genetics*, 15:495–578 (1930).
3. Sturtevant, *Ibid.*, 10:117–147 (1925).
4. Bridges, *Jour. Heredity*, 29:11–13 (1938).

The elementary units of heredity*

SEYMOUR BENZER†

Reprinted with permission from A Symposium on the Chemical Basis of Heredity, *1956, The Johns Hopkins Press, pp. 70–93.*

C. P. Oliver's initial observations were soon followed by similar, independently made findings. All refuted the universality of the definition of allelism and questioned the belief in the nonsubdivisible nature of the gene. Attempts were made to explain the inconsistent findings as representative cases of infrequent recombination between tightly linked genes. This interpretation allowed the gene to retain its exceptional stability against intragenic-recombination and the unit of mutation to remain the same size as the functional gene. However, an explanation was required: How do two nonallelic mutant genes fail to functionally complement in the heterozygous condition? Creative models advanced to explain the appearance of mutant phenotypes in the presence of two tightly linked heterozygous loci were forthcoming. Then, in order to acknowledge the belief that the exceptions were based on erroneously established allelisms, the "false" allelic combinations that displayed recombination were named *pseudoalleles.*

By 1950, pseudoalleles had been found in other organisms besides *Drosophila*. Several different Ascomycetes, employed in recombination studies, were important because their life cycle facilitated the screening of a large number of progeny. While examples of pseudoallelism were rare in *Drosophila*, almost any set of functionally defined allelic mutants in Ascomycetes displayed wild-type recombinants if large progeny samples were observed. In the early 1950s, several papers reviewed the extant literature on pseudoalleles in an attempt to evaluate, from a biochemical standpoint, whether pseudoalleles could best be interpreted in terms of intergenic- or intragenic-recombination. Although it appeared that many cases of pseudoalleles were based upon intragenic-recombination, definitive evidence was not available to conclusively support that interpretation until Seymour Benzer presented this paper in 1956.

Benzer's work combined a masterful knowledge of phage genetics with a biochemical view of gene structure based upon the double-helix model of DNA. He pushed the resolution of mutant recombination analyses to the point where he concluded that he was detecting recombinational events within the smallest segment of the gene possible, in respect to the molecular model of DNA. His theory was based upon the acceptance of the probable exchange between two adjacent mutant nucleotide pairs. With the completion of this work, the unit of mutation (*muton*) and the unit of recombination (*recon*) were reduced to small subdivisions of the intact functional unit (*cistron*). All mutants falling within the same functional unit (i.e., two mutants that show a mutant phenotype when present in the same host bacterium but on different DNA segments) were said to belong to the same "cistron." While reading the following paper, keep in mind Muller's prediction of 1922 that the internal structure of the gene would be resolved through an analysis of its potential for change.

*This research has been supported by grants from the American Cancer Society, upon recommendation of the Committee on Growth of the National Research Council, and from the National Science Foundation.

†Biophysical Laboratory, Purdue University, Lafayette, Indiana.

INTRODUCTION

The techniques of genetic experiments have developed to a point where a highly detailed view of the hereditary material is attainable. By the use of selective procedures in recombination studies with certain organisms, notably fungi,[14] bacteria,[21] and viruses,[1] it is now feasible to "resolve" detail on the molecular level. In fact, the amount of observable detail is so enormous as to make an exhaustive study a real challenge.

A remarkable feature of genetic fine structure studies has been the ability to construct (by recombination experiments) genetic maps which remain one-dimensional down to the smallest levels. The molecular substance (DNA) constituting the hereditary material in bacteria and bacterial viruses is also one-dimensional in character. It is therefore tempting to seek a relation between the linear genetic map and its molecular counterpart which would make it possible to convert "genetic length" (measured in terms of recombination frequencies) to molecular length (measured in terms of nucleotide units).

The classical "gene," which served at once as the unit of genetic recombination, of mutation, and of function, is no longer adequate. These units require separate definition. A lucid discussion of this problem has been given by Pontecorvo.[13]

The unit of recombination will be defined as the smallest element in the one-dimensional array that is interchangeable (but not divisible) by genetic recombination. One such element will be referred to as a "recon." The unit of mutation, the "muton," will be defined as the smallest element that, when altered, can give rise to a mutant form of the organism. A unit of function is more difficult to define. It depends upon what level of function is meant. For example, in speaking of a single function, one may be referring to an ensemble of enzymatic steps leading to *one* particular physiological end effect, or of the synthesis of *one* of the enzymes involved, or of the specification of *one* peptide chain in one of the enzymes, or even of the specification of *one* critical amino acid.

A functional unit can be defined genetically, independent of biochemical information, by means of the elegant *cis-trans* comparison devised by Lewis.[12] This test is used to tell whether two mutants, having apparently similar defects, are indeed defective in the same way. For the *trans* test, both mutant genomes are inserted in the same cell (e.g., in heterocaryon form, or, in the case of a bacterial virus, the equivalent obtained by infecting a bacterium with virus particles of both mutant types). If the resultant phenotype is defective, the mutants are said to be noncomplementary, i.e., defective in the same "function." As a control, the same genetic material is inserted in the *cis* configuration, i.e., as the genomes from one double mutant and one nonmutant. The *cis* configuration usually produces a nondefective phenotype (or a close approximation to it). It turns out that a group of noncomplementary mutants falls within a limited segment of the genetic map. Such a map segment, corresponding to a function which is unitary as defined by the *cis-trans* test applied to the heterocaryon, will be referred to as a "cistron."

The experiments to be described in this paper represent an attempt to place limits on the sizes of these three genetic units in the case of a specific region of the hereditary material of the bacterial virus T4. A group of *"rII"* mutants of T4 has particularly favorable properties for this kind of analysis. Mutants are easily isolated. Recombinants can be detected, even in extremely low frequency, by a selective technique. The system is sufficiently sensitive to permit extension of genetic mapping down to the molecular (nucleotide) level, so that the recon and muton become accessible to measurement. The *rII* mutants are defective in the sense of being unable to multiply in cells of a certain host bacterium (although they do infect and kill the cell). The *cis-trans* test can therefore be readily applied.

GENETIC MAPS

Method of Construction

The construction of a genetic map of an organism starts with the selection of a standard ("wild") type. From the progeny of the wild type, mutant forms can be isolated on the basis of some heritable difference. When two mutants are crossed, there is a possibility that a wild-type organism will be formed as a result of recombination of genetic material. The reciprocal recombinant, containing both mutational alterations, also occurs. The proportion of progeny constituting such recombinant types is characteristic of the particular mutants used. The results of crosses involving a group of mutants can be plotted on a one-dimensional diagram where each mutant is represented by a point. The interval between two points signifies the proportion of recombinants occurring in a cross between the two corresponding mutants. Usually, it is not possible to construct a single map for all the mutants of an organism; instead the mutants must be broken up into "linkage groups." A linear map may be constructed within each linkage group, but the mutant characters assigned to different linkage groups assort randomly among the progeny. The number of linkage groups, in some cases, has been shown to correspond to the number of visible chromosomes.

The procedure for constructing a genetic map for a bacterial virus is much the same.[9] A genetically uniform population of a mutant can readily be grown from a single individual. Two mutants are crossed by infecting a susceptible bacterium with both types and examining the resulting virus progeny for recombinant types. Virus T4 has been mapped in some detail,[3, 1], and behaves as a haploid organism with a single linkage group.[18]

Relativity of Genetic Maps

A genetic map is an image composed of individual points. Each point represents a mutation which has been localized with respect to other mutations by recombination experiments. The image thus obtained is a highly colored representation of the hereditary material. Alterations in the hereditary material will lead to noticeable mutations only if they affect some phenotypic characteristic to a visible degree. Innocuous changes may pass unnoticed, leaving their corresponding regions on the map blank. At the other extreme, alterations having a lethal effect will also be missed (in a haploid organism). The map represents, therefore, only cases which fall between these extremes under the conditions of observation. By varying these conditions, a given mutational event may be shifted from one of these categories (innocuous, noticeable, or lethal) to another, thereby appearing on, or disappearing from the map.

This effect may be illustrated by the "r" mutants of bacterial virus T4. Wild-type T4 produces small, fuzzy plaques on *Escherichia coli* B (Fig. 1). From plaques of wild-type T4, r-type mutants can be isolated which produce a different sort of plaque. Fig.

1 shows the plaques of nine r mutants, each isolated from a different plaque of the wild type in order to assure independent origin. The similarity of plaque type of these r's on B disappears when they are plated on another host strain, *E. coli* K (a lysogenic K12 strain[10] carrying phage lambda), as shown in Fig. 2. Here, they split into three groups: two mutants form r-type plaques, one forms wild-type plaques, and the remaining six do not register. Thus, with B as host, all three types of mutation lead to visible effects, while with K as host, the effects may be visible, innocuous, or lethal.

When the same set of mutants is plated on a third strain, *E. coli* S (K12S, a nonlysogenic derivative[10] of K12) or BB (a "Berkeley" derivative[17] of B) the pattern of plaque morphology is different from that on either B or K (Table 1).

If a genetic map is constructed for these mutants, using B as host, the three groups fall into different map regions, as indicated in Fig. 3. On Strain S, the rII and rIII types of mutation are innocuous. Thus, if S had been used as host in the isolation of r mutants, only the rI region would have appeared on the map. On K as host, the rIII mutation is innocuous, and the rII mutation is (usually) lethal, so that only the rI region would appear. Actually, a few rII mutants are able to multiply somewhat on K, producing visible tiny plaques. If K were used as host in the isolation and testing of mutants from wild-type T4, these mutants could be noticed and would probably be designated by some other name, perhaps "minute." The map would then appear as in the bottom row of Fig. 3. The distribution of points on the map within this "minute" region would be very different from those for the rII region using B as the host.

The appearance of a genetic map also depends on the choice of the standard type, which is, after all, arbitrary. For example, suppose an r form were taken as the standard type and non-r mutants were isolated from it. Then a completely different map would result. An example of this is to be found in the work of Franklin and Streisinger[5] on the $h \rightarrow h^+$ mutation in T2, as compared with that of Hershey and Davidson[7] on the $h^+ \rightarrow h$ mutation.

Another way in which the picture is weighted is by local variations in the stability of the genetic material. Certain types of structural alterations may occur more frequently than others. Thus, a perfectly stable genetic element (i.e., one which never

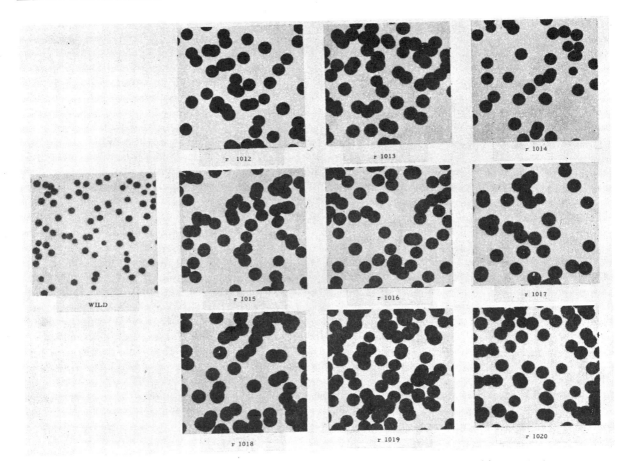

Fig. 1. Photographs of plaques formed on *E. coli* B by T4 "wild type" and nine independently arising *r* mutants.

errs during replication) would not be represented by any point on the map.

Determination of the Sizes of the Hereditary Units by Mapping

Determination of the recon requires "running the map into the ground" (Delbrück's expression), that is, isolation and mapping of so large a linear density of mutants that their distances apart diminish to the point of being comparable to the indivisible unit. With a finite set of mutants, only an upper limit can be set upon the recon, which must be smaller than (or equal to) the smallest nonzero interval observed between pairs of mutants.

To determine the length of map involved in a mutational alteration, a group of three closely linked mutants is needed. Since map distances are (approximately) additive, a calculation of the "length" of th

Table 1 Plaque morphology of T4 strains (isolated in B) plated on various hosts

Phage strain	Bacterial host strain		
	B	*S*	*K*
wild	wild	wild	wild
r I	r	r	r
r II	r	wild	—
r III	r	wild	wild

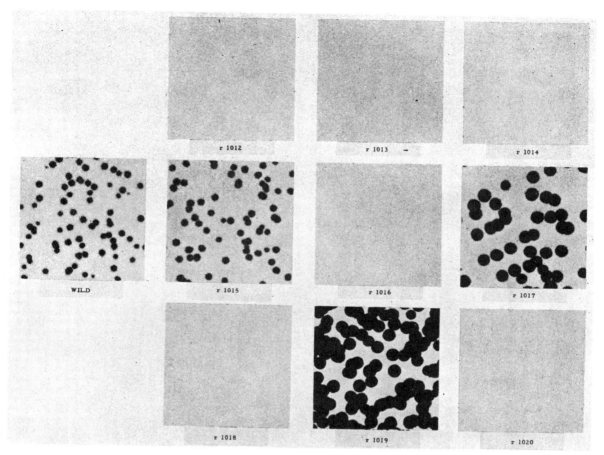

Fig. 2. The same mutants as used in Fig. 1, plated on *E. coli* K.

central mutation can be attempted[15] from the discrepancy observed between the longest distance and the sum of the two shorter ones, as shown in Fig. 4. The upper limit to the size of the muton would be the smallest discrepancy observed by this method, which can be determined accurately only if the three mutants are very closely linked. It should be noted that since the degree to which the genetic structure can be sliced by recombination experiments is limited by the size of recon, the size of the muton will register as zero by this method if it is equal to or smaller than one recon. A second method for determining the muton size is by the maximum number

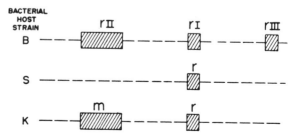

Fig. 3. Dependence of the genetic map of T4 upon the choice of host. Three regions of the map are shown as they probably would appear if *E. coli* strains B, S, or K were used as the host.

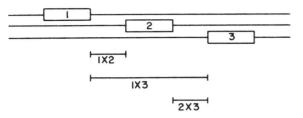

Fig. 4. Method for determining the "length" of a mutation. The discrepancy between the long distance and the sum of the two short distances measures the length of the central mutation.

of mutations, separable by recombination, that can be packed into a definite length of the map.

For the cistron size, only a *lower* limit can be set with a finite group of mutants. The cistron must be at least as large as the distance between the most distant pair within it. Its boundaries become more sharply defined the larger the number of points which are shown to lie inside them.

Thus, the determination of the sizes of all three units requires the isolation and crossing of large numbers of mutants. The magnitude of this undertaking increases with the square of the number of mutants, since to cross n mutants in all possible pairs requires $n(n-1)/2$ crosses, or approximately $n^2/2$. Fortunately, however, the project can be shortened considerably by means of a trick.

Fig. 5. Illustration of the behavior of an "anomalous" mutant. Mutant no. 6 is anomalous with respect to the segment of the map indicated by the bar; it fails to give wild recombinants with mutants (nos. 2, 3, and 4) located within that segment. A + signifies production, and 0 lack of production, of wild recombinants in a cross.

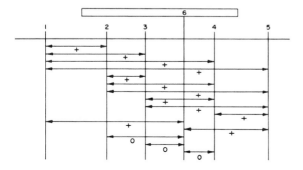

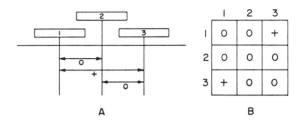

Fig. 6. The method of overlapping deletions. Three mutants are shown, each differing from wild type in the deletion of a portion of the genetic material. Mutants no. 1 and no. 3 can recombine with each other to produce wild type, but neither of them can produce wild recombinants when crossed to mutant no. 2. The matrix B represents the results obtained by crossing three such mutants in pairs and testing for wild recombinants; the results uniquely determine the order of the mutations on the map A.

The Method of Overlapping "Deletions"

Certain *rII* mutants are anomalous in the sense that they cannot be represented as *points* on the map. The anomalous mutants give no detectable wild recombinants with any of several other mutants which *do* give wild recombinants with each other. An anomalous mutant can be represented (Fig. 5) as covering a segment of the map. Reversion of such an *rII* mutant has never been observed; also, no mutant which does revert has been found to have this anomalous character. The properties of an anomalous mutant can be explained as owing to the deletion (i.e., loss) of a segment of hereditary material corresponding to the map span covered. However, anomalous behavior and stability against reversion are not sufficient to establish that a deletion has occurred. Similar properties could be expected of a double mutant when crossed with either of two different single mutants located at the same points. An inversion also would show the same behavior. However, the occurrence of a deletion seems to be the only reasonable explanation in the cases of several of the *rII* mutants, since they fail to give recombinants with any of three or more (in one case as many as 20) well-separated mutants.

Whether a given mutation belongs in the region covered by a deletion can be determined by the appropriate cross. If wild recombinants are produced, the mutant must have a map position *outside* the region of the deletion. This eliminates the need to cross that mutant with any of the mutants whose map positions lie *within* the region of the deletion. The problem of mapping a large number of mutants is greatly simplified by this system of "divide and conquer." The mutants can first be classified into groups that fall into different regions on the basis of crosses with mutants of the deletion-type. Further crossing in all possible pairs is then necessary only within each group.

Suppose that three deletions occur in overlapping configuration, as shown in Fig. 6A. Fig. 6B represents the results that would be obtained in crosses of pairs of these three mutants. A diagonal element (representing a cross of a mutant with itself) is, of course, zero, since no wild recombinants can be produced. An overlap is reflected by the pattern of nondiagonal zeros. These results would establish a unique *order* of the deletions (without resort to the three-factor crosses that would ordinarily be neces-

sary). With a sufficient number and appropriate distribution of deletions, one could hope to order a large length of map. The reader will note an analogy (not altogether without significance!) to the technique used by Sanger[16] to order the amino acids in a polypeptide chain by means of overlapping peptide segments.

MAPPING THE *rII* REGION OF T4

Taxonomy of *r* Mutants

In classifying the mutants of T4, classical terminology may be conveniently used for the taxonomic scheme shown in Fig. 7. Mutants of the *r* "kingdom," isolated on B, can be separated into three "phyla" by testing on K. For the present purposes,

Fig. 7. Classification scheme for *r* mutants of T4.

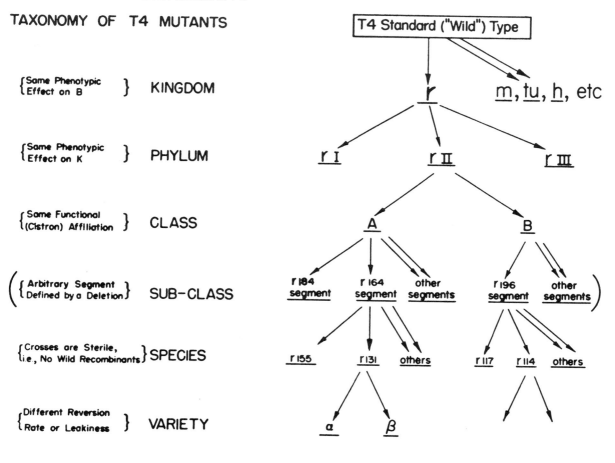

TAXONOMY OF T4 MUTANTS

our attention will be limited to mutants of the *rII* phylum, which are inactive on K.

A pair of *rII* mutants may be subjected to the *cis-trans* test. The *cis* configuration (mixed infection of K with double-mutant and wild-type particles) is active, since the presence of a wild particle in the cell enables both types to multiply. The *trans* configuration (mixed infection of K with the two single mutants) may be active or inactive. If inactive, the two mutants are placed in the same "class." Since the members of a class fail to complement each other, they can be considered as belonging to a single functional group. On the basis of this test, the *rII* mutants divide into two clear-cut classes. The map positions of the mutants in each class have been found to be restricted to separate map segments, the A and B "cistrons."

Arbitrary subclasses can be chosen (from among the available deletions) for convenience in mapping: mutants falling within the map region encompassed by a particular deletion form a subclass.

Reverting mutants are considered as of different "species" if crosses between them yield wild recombinants. Among a group of mutants which have not yielded to resolution by recombination tests, "varieties" can in some cases be distinguished by other criteria (reversion rate or degree of ability to grow on K).

Procedures in the Classification of *r* Mutants of T4

1. Isolation of Mutants Each mutant is isolated from a separate plaque of wild-type T4 (plated on B) and freed from contaminating wild-type particles by replating. Stocks of mutants are prepared by growth on S (to avoid the selective advantage which wild-type revertants would have on B). Mutants are numbered in the order of isolation, starting with 101 to avoid confusion with mutants previously isolated by others.

2. Spot Test on K In this first test of a new *r* mutant, 10^8 particles are plated on K and then the plate is spotted with one drop (10^6 particles) of *r164* (a mutant having a "deletion" in an A cistron) and one drop of *r196* (a mutant having a deletion located in the B cistron). Typical examples of the results of this test are shown in Fig. 8. If the new mutant belongs either to the *rI* or the *rII* phylum, the plating bacteria will be completely lysed (except for a background of colonies formed by mutants of K which are resistant to T4), as typified by mutant X in Fig. 8.

Mutant Y in Fig. 8 is typical of a stable *rII* mutant. The background shows no plaques, indicating that the proportion of revertants in the stock is less than

Fig. 8. Spot test used in classification of *r* mutants. To test a mutant, 10^8 mutant particles are plated on bacterial strain K and the plate is spotted with one drop of *r164* (left) and one drop of *r196* (right). Mutants *X*, *Y*, and *Z* illustrate typical results. Mutant *X* is of the *rI* phylum. Mutant *Y* is a stable mutant of the *rII* phylum and the A class, but is not in the *r164* subclass. Mutant *Z* is a reverting mutant (wild-type plaques in background) of the *rII* phylum, B class, and *r196* subclass.

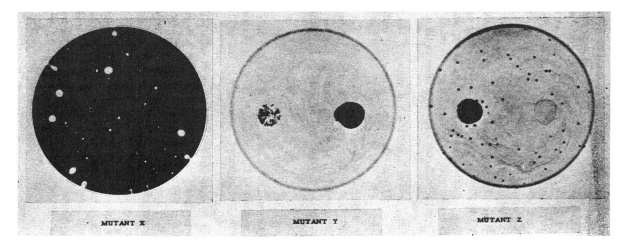

MUTANT X   MUTANT Y   MUTANT Z

10^{-8}. The spot of *r196* is completely clear, in contrast to the *r164* spot. This massive lysis is caused by the ability of mutant *Y* and *r196* to complement each other for growth on K. From this result, it may be concluded that mutant *Y* belongs to the A class. Within the *r164* spot, however, some plaques may be seen. These are due to wild recombinants, arising from *r164* and mutant *Z* (by virtue of very feeble growth of *rII* mutants on K). Therefore, mutant *Y* is not in the subclass defined by *r164*.

The third test plate is typical of a reverting *rII* mutant. The stock contains a fraction 10^{-6} of wild-type particles which produce the plaques seen in the background. It is evident from the spot tests that the mutant *Z* belongs in the B class, and appears to lie within the *r196* subclass.

3. Spot Test on a Mixture of K and B Cells   Once the class of a new mutant is known, it can be tested on a single plate against several mutants of the same class. For this purpose, the sensitivity of the test may be increased enormously by the addition of some B cells (about one part in a hundred) to the K used for plating. The additional growth possible for the mutants on B cells enhances their opportunity to produce wild recombinants. This test gives a positive response down to the level of around 0.01 percent recombination. A negative result does not, of course, eliminate the possibility that recombination occurs with a lower frequency.

4. Preliminary Crosses   A semiquantitative measure of recombination frequency may be obtained by mixedly infecting B with two mutants and plating the infected cells on K. B cells which liberate one or more wild-type particles can produce plaques. This method is convenient for preliminary testing for recombination in the range from 0.0001 to 0.1 percent. With higher frequencies of recombination, approaching the point where a large fraction of the mixedly infected cells liberate recombinants, saturation sets in.

5. Standard Crosses   Standard measurements of recombination frequency are made in conventional crosses. B cells are infected with an average of three particles per cell of each phage. The infected cells are allowed to burst in a liquid medium, and the progeny are plated on K and on B to determine the proportion of wild-type particles. The reciprocal

recombinant (double *rII* mutant) does not, in general, produce plaques on K, but since the two recombinant types are produced in statistically equal numbers,[9] the proportion of recombinants in the progeny can be taken as twice the ratio of plaques on K to plaques on B (corrected for the relative efficiency of plating of wild type on these two strains, which is close to unity).

6. Reversion Rates   The reversion rate of a mutant is reflected in the proportion of wild-type particles present in a stock. This value is an important characteristic of each mutant, varying over an enormous range for different mutants. It may be less than 10^{-8} for "stable" (i.e., nonreverting) mutants or as high as several percent. (For one exceedingly unstable mutant the proportion of revertants averages 70 percent, even in stocks derived from individual mutant particles.) The precision with which a mutant can be localized on the map is inversely related to its reversion rate; only relatively stable mutants are useful for mapping. In the experiments here reported, it has been assumed that the reversion rate of a mutant is not altered during a cross; the reversion contribution is subtracted from the observed percentage of wild particles in the progeny. In most cases, this correction is negligible.

7. "Leakiness" of rII *Mutants*   rII mutants differ greatly in their ability to grow on K cells. A sensitive measure of this ability can be obtained by infecting K cells and plating them on B. Any K cell that liberates one or more virus particles can give rise to a plaque. The fraction of infected cells yielding virus progeny, which is a characteristic property of each mutant (when measured under fixed conditions), may vary from almost 100 percent down to less than 1 percent for different mutants.

Leakiness has the effect of limiting the sensitivity of K as a tool for selection of wild recombinants, thereby hampering the mapping of very leaky mutants.

Classification of a Set of 241 *r* Mutants

A set of *r* mutants was isolated, using B as host, and given numbers from *r101* to *r338*; the mutants *r47*, *r48*, and *r51*, isolated by Doermann,[3] were added to this set, making a total of 241 *r* mutants. These were analyzed according to methods already described.

The results are shown in Table 2. Of these mutants, 134 fell into the *rII* phylum. Each of these (with the exception of one very leaky mutant) could be assigned unambiguously to either of two classes on the basis of the test for complementary action of pairs of mutants for growth on K. Mutants within each class were crossed with stable mutants of the same class; those giving no detectable wild recombinants with a particular stable mutant were assigned to the same subclass. Mutants of each subclass were crossed in all pairs. When two or more mutants were found to be of the same species (i.e., showed, in a "preliminary" type cross, recombination of less than about 0.001 percent, or less than the uncertainty level set by the reversion rate, whichever was greater), one was used to represent the species in

Table 2 Classification of an unselected group of 241 *r* mutants of T4

Kingdom	Phylum	Class	Subclass	Species	Variety
r mutants (241)	*rI* (96)				
	rII (134)	IIA (73)	*r164** (27)	1 sp. (20) 1 sp. (2) 4 sp. (1 ea.)	1 var. (11) 1 var. (9)
			r184* (5)	1 sp. (2) 1 sp. (2)*	1 var. (1) 1 var. (1)*
			r221* (4)	1 sp. (3)	1 var. (1) 1 var. (1) 1 var. (1)
			r47* (4)	1 sp. (2)* 1 sp. (1)*	
			r197* (2)	1 sp. (1)	
			others (29)	1 sp. (3)* 1 sp. (2) 1 sp. (2) 1 sp. (2) 1 sp. (2) 1 sp. (2) 1 sp. (2) 14 sp. (1 ea.)	1 var. (1) 1 var. (1) 1 var. (1) 1 var. (1)
			not determined (1)		
		IIB (60)	*r196** (34)	1 sp. (21) 1 sp. (9) 3 sp. (1 ea.)	1 var. (19) 1 var. (2) 1 var. (5) 1 var. (4)
			r187* (6)	1 sp. (3)* 1 sp. (2)	1 var. (1) 1 var. (1)
			others (16)	1 sp. (3) 1 sp. (3) 1 sp. (2) 8 sp. (1 ea.)	1 var. (2) 1 var. (1)
			not determined (4)		
	rIII (11)	not determined (1)			

The number of mutants in each classification is given in parentheses. An asterisk indicates that reversion of the mutant to wild-type has not been detected. A few mutants (indicated as "not determined") could not be further classified due to excessively high reversion rate or leakiness.

further crosses. Those mutants not falling into any of the subclasses defined by the available stable mutants were crossed with each other in pairs. By these procedures, the classification was carried to the species level for the entire set of mutants, except for six highly revertible or leaky mutants whose subclass was not established.

Several of the species showed evidence of splitting into varieties distinguishable by reversion rate or degree of leakiness. Some mutant varieties recurred frequently (e.g., 19, 11, 9 times). These recurrences were far outside the expectation for a Poisson distribution, and are indicative of local variations of mutability. The fact that many species were represented by only one occurrence suggests that many other species remain to be found.

The 33 species found in the A cistron and the 18 species found in the B cistron are sufficient to define reasonably well the limits of each cistron. The minimum size of a cistron in recombination units is determined by the maximum amount of recombination observed in standard crosses between pairs of

mutants within it. On the basis of the standard crosses performed so far, this value is about 4 percent recombination for the A cistron, and 2 percent for the B cistron.

Study of 923 r Mutants

While the study of the foregoing 241 mutants yielded a good idea of the sizes and complexity of the A and B cistrons, it fell short of "saturating" the map sufficiently to provide the close clusters of mutants required for the determination of the sizes of the recon and muton. To this end, it was decided to isolate many more mutants. By confining attention to those falling into the *r164* subclass, a more exhaustive study could be made of a selected portion of the map.

In a group of 923 r mutants (*r101* through *r1020*, plus Doermann's three), 149 were found to belong to the *r164* subclass. Four of those were stable. The remaining 145 mutants separated into the 11 species shown in Fig. 9. One of the species accounted for

Fig. 9. Grouping into species of mutants within the segment defined by the "deletion" *r164*. The arrow indicates 103 additional mutants of the *r131* species. The different-sized boxes of the *r131* species indicate two distinct varieties. The number assigned to each location is that of the first mutant to indicate it.

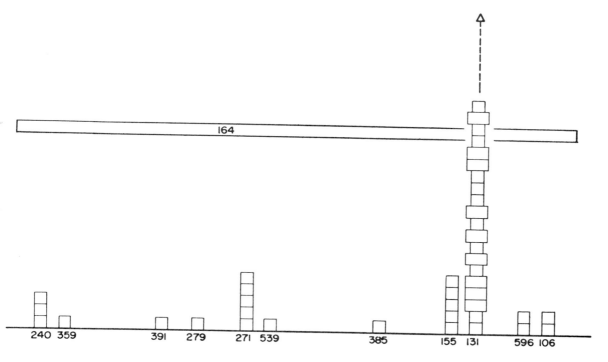

123 of the mutants! As shown in Table 3, this species included three varieties as distinguished by their reversion rates; two were of roughly equal abundance, while the third occurred only once.

Results of standard crosses between the mutants of the *r164* subclass are presented in Fig. 10. The smallest recombination distance, setting an upper limit to the size of the recon, is around 0.02 percent (between *r240* and *r359*). Since only one interval has this value, the possibility of smaller values is not ruled out.

One procedure for measuring the size of the muton requires a group of three closely linked mutants in order to compare the long distance with the sum of the two shorter ones. If the central mutation has an appreciable size, there should be a discrepancy between these values. There are eight cases in Fig. 10 for which the distances have been measured for three adjacent mutants: 240-359-391, 359-391-279, 391-279-271, 279-271-539, 271-539-385, 385-155-131, 155-131-596, 131-596-106. The discrepancies for these groups are, reading from left to right, +0.14, –0.03, –0.02, –0.03, –0.02, –0.03, +0.03, and +0.05 percent recombination. The average of

these values is +0.01, with an average deviation of ±0.05. Since each measurement of recombination frequency is subject to experimental error of the order of 20 percent of its magnitude, these determinations of mutation size (each derived from three measurements) are uncertain to plus or minus about 0.05 percent recombination. Therefore, the latter is the smallest upper limit that can be set upon the size of the muton by these data.

Another measure of muton size can be attempted by finding the number of species that can exist within a given length of the map. As shown in Fig. 10, a map length of 0.8 percent recombination includes nine separable mutant species, or no more than 0.09 percent per species. Both this determination and the previous one suffer uncertainty due to imperfect additivity of map distances ("negative interference" — see Discussion).

Stable Mutants

Since the problem of mapping large numbers of mutants is greatly facilitated by the use of "dele-

Table 3 Reversion data for mutants of the *r131* species

	Mutant	Reversion index (units of 10^{-6})			
variety α	*r200*	0.47	0.91	0.17	0.25
	r220	0.55	0.41	0.25	0.21
	r274	0.24	0.29	0.61	(10.)
	r930	0.17	0.66	0.19	0.21
	r1012	0.58	0.27	0.22	0.15
variety β	*r245*	420.	490.	490.	1120.
	r353	540.	530.	3900.	(500,000.)
	r376	520.	360.	1240.	450.
	r510	2000.	640.	460.	860.
	r888	610.	530.	250.	570.
variety γ	*r978*	0.01*	0.005*	0.005*	0.01*

The "reversion index" is the proportion of wild-type particles in a lysate prepared from a few mutant particles (to avoid introduction of any revertants present in the original stock) using S as host. The measurement is subject to large fluctuations due to the clonal growth of the revertants formed. Therefore, four separate lysates are made for each mutant. Parentheses indicate extreme fluctuations. An asterisk indicates a background of tiny plaques (smaller than those produced by wild-type) when the lysate is plated on K.

In addition to the examples listed in the table, 67 other mutants of this species are also of variety α, as judged by the proportion of revertants (from 0.2×10^{-6} to 4.0×10^{-6}) in single lysates; 45 additional mutants apparently are of variety β (having values from 300×10^{-6} to $4,000 \times 10^{-6}$).

The mutants of variety α give less than 0.001 percent recombination with *r274*. For the more highly revertible mutants of variety β, this limit can only be set at less than 0.02 percent recombination with *r274*. The mutant *r973*, of variety γ, gives less than 0.005 percent (the limit set by background on K) recombination with *r274*.

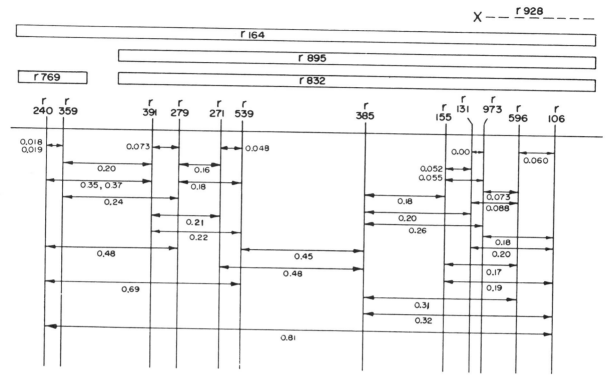

Fig. 10. Map of the mutants in the *r164* segment. The numbers give the percentage of recombination observed in standard crosses between pairs of mutants. The arrangement on this map is that suggested by these recombination values; it has not yet been verified by three-point tests. Stable mutants are represented as bars above the axis; the span of the bar covers those mutants with which the stable mutant produces no detectable wild recombinants. The stable mutant *r928* appears to be a double mutant having one mutation at the highly mutable *r131* location and a second mutation at a point in the B cistron. Mutants *r131* and *r973* are separated on the map so that the data for each can be indicated. Some of the data here given differ from (and supersede) previously published data based upon unconventional crosses which turned out to be incorrect.

tions," particular attention has been paid to non-reverting mutants, in the hope of obtaining a complete set of overlapping deletions. Among the series of 923 r mutants, 72 stable rII mutants were found, 47 in the A cistron and 24 in the B cistron. One mutant (*r928*) was exceptional: it failed to complement mutants of either the A or B cistrons and therefore belongs to *both* classes.

The stable mutants of the B cistron have been crossed (by spot tests on K plus B) in all possible pairs. The results are shown in Fig. 11. Overlapping relationships are indicated by nondiagonal zeros. Fig. 12 shows the genetic map representation of these results together with the results derived from an analysis (as yet incomplete) of the stable mutants

of the A cistron. Unfortunately, gaps still remain in the map.

The mutant *r638* is of particular note. No B class mutant has been found that gives wild recombinants with it, so that it appears to be due to deletion of the entire B cistron. In spite of this gross defect, it is capable of normal reproduction on *E. coli* strains B and S.

In order to characterize a stable mutant of a "deletion" type, it is necessary to show that it gives no wild recombinants with at least three other mutants that do give recombination with each other (to exclude the possibilities that it is a double mutant or has an inversion). This criterion cannot be applied unless a suitable set of three mutants is available.

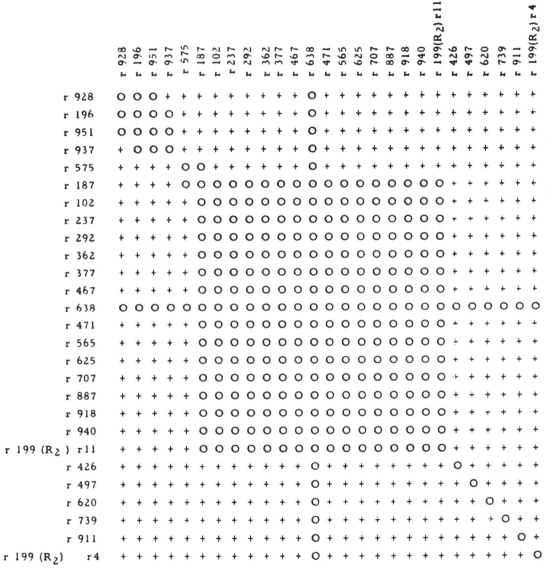

Fig. 11. Recombination matrix for stable mutants of the B class. A + indicates production of wild recombinants (around 0.01 percent recombination or more would be detected) in the cross between the indicated pair of mutants (by spot test on K plus B). All diagonal elements (self-crosses) are zero; nondiagonal zeros indicate overlaps. Two of the mutants are derived, not from the original wild-type, but from a revertant of *r199*.

Only some of the stable mutants (*164, 184, 221, 196, 782, 638, 832, 895, 951*) have as yet been shown to satisfy this criterion.

Stable mutations tend to occur "all over the map." However, as in the case of reverting mutants, certain localities show a strikingly high recurrence tendency, as illustrated by *r102*, et al., and by *r145*, et al.

DISCUSSION

Relation of Genetic Length to Molecular Length

We would like to relate the genetic map, an abstract construction representing the results of recombination experiments, to a material structure. The most promising candidate in a T4 particle is its DNA component, which appears to carry the hereditary information.[6] DNA also has a linear geometry.[20] The problem, then, is to derive a relation between genetic map distance (a probability measurement) and molecular distance. In order to have a unit of molecular distance which is invariant to changes in molecular configuration, the interval between two points along the (paired) DNA structure will be expressed in nucleotide (pair) units, which are more meaningful for our purposes than, say, ängstrom units.

Unfortunately, present information is inadequate to permit a very accurate calculation to be made of map distance in terms of nucleotide units. First, it is not known whether the probability of recombination is constant (per unit of molecular length) along the entire genetic structure. Second, there is the question of what portion of the total DNA of a T4 particle constitutes hereditary material (for a discussion of this problem, see the paper by Delbrück and Stent in this volume). The result of Levinthal's elegant experiment[11] suggests a value of 40 percent. Since the total DNA content of a T4 particle is 4×10^5 nucleotides,[8] it would seem that the hereditary information of T4 is carried in 1.6×10^5 nucleotides. We do not know, however, whether the information exists in one or in many copies. If there is just one copy, and if it has the paired structure of the model of Watson and Crick,[20] the total length of

hereditary material should be 8×10^4 nucleotide pairs.

There are difficulties on the genetic side as well. The total length of the genetic map is not well established. The determination of this length requires a number of genetic markers sufficient to define the ends of the map. It also requires a favorable distribution of markers in order that the intervals between them can be summated; if the distance between two markers is sufficiently large, the frequency of recombination between them approaches that for unlinked markers and therefore loses its value as a measure of the linkage distance. Unfortunately, the linkage data presently available for T4 leave much to be desired. The experiments of Streisinger[18] indicate that the map of T4 consists of a single-linkage group. Adding up the intervals between markers (corrected for successive rounds of mating according to the theory of Visconti and Delbrück[19]) leads to a total value of the order of 200 percent recombination units. This estimate is very rough, since the number of available markers upon which it is based is small.

A further difficulty arises from the fact that the map distances measured in standard crosses are not quite additive: a large distance tends to be less than the sum of its component smaller distances. For distances of the order of 10 percent recombination units and more, the deviations from additivity, referred to as "negative interference," can be

Fig. 12. Preliminary locations of stable *rII* mutants. Mutants producing no wild recombinants with each other are drawn in overlapping configuration. Pairs which produce small amounts are placed near each other. Since there remain some gaps, the order shown depends upon that established by Doermann[4] for the mutants shown on the axis. The scale is somewhat distorted in order to show the overlap relationships clearly. Brackets indicate groups, the internal order of which is not established. Ten stable mutants of the A class and six of the B class were not sufficiently close to any others to permit them to be placed on the map. (A class equals A cistron; B class equals B cistron.)

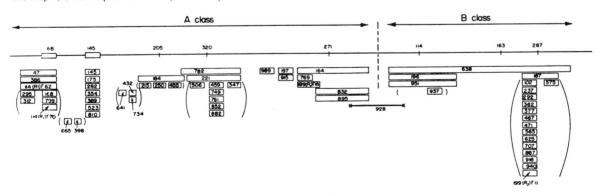

accounted for by the Visconti-Delbrück considerations. However, a "negative interference" effect, not accountable for by their theory, persists, and apparently gets worse, at very small distances.[4] This presents a serious obstacle for our purposes, since we are interested in knowing what fraction of the total map is represented by a small distance. According to preliminary data on this point, summation of the smallest available distances between *rII* mutants yields a total length for the *rII* region which is severalfold greater than that found for crosses involving distant *rII* markers. If the total T4 map length could be obtained by a similar summation of small distances, the indications are that it might be of the order of 800 percent recombination units in length.

Thus, there are plenty of uncertainties involved in relating the genetic map quantitatively to the DNA structure. The best we can do at present is to make a rough estimate based upon the following assumptions: (1) the genetic information of T4 is carried in 1 copy consisting of a DNA thread 80,000 nucleotide pairs long; (2) the genetic map has a total length of about 800 percent recombination units; (3) the probability of recombination per unit molecular length is uniform. According to these assumptions, the ratio of recombination probability (at small distances) to molecular distance would be 800 percent recombination divided by 80,000 nucleotide pairs, or 0.01 percent recombination per nucleotide pair. That is to say, if two mutants, having mutations one nucleotide pair apart, are crossed, the proportion of recombinants in the progeny should be 0.01 percent. This estimate is greater, by a factor of ten, than one made a year ago, in which it was assumed that all the DNA was genetic material and that the effect of negative interference was negligible. It should become possible to improve this calculation as more information becomes available.

The estimate indicates that the level of genetic fine structure which has been reached in these experiments is not far removed from that of the individual nucleotides. Furthermore, the estimate is useful in that it defines an "absolute zero" for recombination probabilities: if a cross between two (single) T4 mutants does not give at least 0.01 percent recombination, the locations of the two mutations probably are not separated by even one nucleotide pair.

Molecular Sizes of the Genetic Units

Recon The smallest nonzero recombination value so far observed among the *rII* mutants of T4 is around 0.02 percent recombination. If the estimate of 0.01 percent recombination per nucleotide pair should prove to be correct, the size of the recon would be limited to no more than two nucleotide pairs.

Muton Evidently, among the stable mutants, mutations may involve varied lengths of the map. The muton is defined as the *smallest* element, alteration of which can be effective in causing a mutation. In the case of reverting mutants, it has not been possible, so far, to demonstrate any appreciable mutation size greater than around 0.05 percent recombination. This would indicate that alteration of very few nucleotides (no more than five, according to the present estimate) is capable of causing a visible mutation.

Cistron A cistron turns out to be a very sophisticated structure. The function to which it corresponds can be impaired by mutation at many different locations. In the study of 241 *r* mutants, 33 species were found to be located in the A cistron, of which 6 were in the *r164* subclass. In extending the survey to 923 *r* mutants, the number of known species in the *r164* subclass was doubled. Consequently, it may be expected that about 60 A cistron species will be found — when the analysis of the 923 mutants is completed. Since many species are represented by only one occurrence, implying that many more are yet to be found, it seems safe to conclude that in the A cistron alone there are over 100 "sensitive" points, i.e., locations at which a mutational event leads to an observable phenotypic effect. Just as in the case of the entire genetic map of an organism, the portrait of a cistron is weighted by considerations of which alterations are effectual. It should be fascinating to try to translate the "topography" within a cistron into that of a physiologically active structure, such as a polypeptide chain folded to form an enzyme.

REFERENCES

1. Benzer, S., *Proc. Natl. Acad. Sci. U.S.,* 41:344 (1955).

2. Demerec, M., Blomstrand, I., and Demerec, Z. E., *Proc. Natl. Acad. Sci. U.S.,* 41:359 (1955).
3. Doermann, A. H., and Hill, M. B., *Genetics,* 33:79 (1953).
4. Doermann, A. H., and collaborators, pers. commun.
5. Franklin, N., and Streisinger, G., pers. commun.
6. Hershey, A. D., and Chase, M., *J. Gen. Physiol.,* 36:39 (1952).
7. Hershey, A. D., and Davidson, H., *Genetics,* 36:667 (1951).
8. Hershey, A. D., Dixon, J., and Chase, M., *J. Gen. Physiol.,* 36:777 (1953).
9. Hershey, A. D., and Rotman, R., *Genetics,* 34:44 (1949).
10. Lederberg, E. M., and Lederberg, J., *Genetics,* 38:51 (1953).
11. Levinthal, C., *Proc. Natl. Acad. Sci. U.S.,* 42:394 (1956).
12. Lewis, E. B., *Cold Spring Harbor Symposia Quant. Biol.,* 16:159 (1951).
13. Pontecorvo, G., *Advances in Enzymol.,* 13:121 (1952).
14. Pritchard, R. H., *Heredity,* 9:343 (1955).
15. Roper, J. A., *Nature,* 166:956 (1950).
16. Sanger, F., *Advances in Protein Chem.,* 7:1 (1952).
17. Stent, G. S., pers. commun.
18. Streisinger, G., pers. commun.
19. Visconti, N., and Delbrück, M., *Genetics,* 38:5 (1953).
20. Watson, J. D., and Crick, F. H. C., *Cold Spring Harbor Symposia Quant. Biol.,* 18:123 (1953).

Analysis of a gene in *Drosophila*

With variations, the genes of microorganisms and those of *Drosophila* are much the same

W. J. WELSHONS

Reprinted by author's and publisher's permission from Science, *vol. 150, 1965, pp. 1122–1129. Copyright © 1965 by the American Association for the Advancement of Science.*

Finally, the question remains: What have mutation studies revealed about the internal structure of the gene in higher organisms? You have seen that the insight into the structure of the prokaryote gene was obtained through intragenic-recombination studies. However, in 1940, it appeared that the eucaryote gene could not be subdivided by recombination. Has the eucaryote gene maintained its structural uniqueness or has it fallen to fine-structure recombination experiments similar to Benzer's analysis? Is there evidence to suggest the size relationship between the mutationally, recombinationally, and functionally defined gene in eucaryotes? The final paper in this part addresses itself to these questions. Like Benzer's paper, there is an attempt to extrapolate to the molecular level of genetic organization by means of monitoring phenotypic phenomena. It is important to keep in mind the hazards involved in making such a conceptual leap.

Since the discovery of the helical structure of DNA,[1] microorganisms have played an increasingly important role in investigations of gene structure and function. Discoveries have come forth so rapidly that one hardly grasps the significance of one before being confronted by another. Because of the flood of exciting disclosures it has been necessary to focus attention upon the gene as it exists in the microbial world. A divergence will be made here to consider how well the concept of the gene derived from studies with microorganisms applies to higher organisms. To do this, the fine structure of the sex-linked Notch locus in *Drosophila* will be examined. Comparisons will then be made with complex loci of microorganisms, from which it is concluded that the functional gene or cistron can be most easily identified in higher organisms if mutants classified as amorphs and lacking all or virtually all of their genetic activity are used for the analysis. When Notch is compared with other complex loci in *Drosophila*, one sees that amorphic mutants may be identified as recessive lethals at one locus and as recessive visibles at another. Cytogenetic considerations suggest that if the Notch locus is correctly characterized as a single functional gene, it is represented cytologically as a single salivary band.

Since the Notch locus had been the object of intensive study many years ago, a body of sound genetic and cytogenetic information was already available to serve as a springboard for investigation.[2] To summarize briefly, Notch, symbolized as *N*, is located near the distal end of the X chromosome. Cytological analyses showed that the locus of Notch corresponded to salivary band 3C7, and that a change to the mutant *N* condition was accompanied either by a visible deletion of the band or, in other instances, by an inactivation of the genetic material within it, perhaps as a result of the presence of a subvisible deficiency. *N* mutants are generally classified as dominants since females heterozygous for *N* and a wild allele, that is *N*/+ ♀ ♀, have an abnormal phenotype consisting of variably notched wings, thickened wing veins, and minor bristle abnormalities. Alternatively, *N*'s can be classified as recessive lethals because females that carry *N* on both X chromosomes and males that carry *N* on their single X do not survive. It is the recessive lethality of *N* mutants that is to be stressed in this discussion.

Fortunately, duplications for 3C7 as well as deficiencies and inactivations of that salivary band are available. Whereas a deficiency causes the Notch phenotype, the dominant phenotype Confluens

Table 1 Genotypes and phenotypes of animals with various doses of *N* mutants and duplications.

Female genotype	Phenotypes	Male genotype
+/+; +/+	Wild ♀, ♂	+; +/+
+/+; Dp/+	Confluens ♀, ♂	+; Dp/+
+/N; +/+	Notch ♀	
+/N; Dp/+	Wild ♀, ♂	N; Dp/+
N/N; +/+	Lethal ♀, ♂	N; +/+
N/N; Dp/+	Notch ♀	
N/N; Dp/Dp	Wild ♀*	N; Dp/Dp
	Confluens ♂*	

*Predicted but not tested.

(*Co*) appears when 3C7 is present as an extra dose.[3] A particularly serviceable form of the duplication is one in which a short segment of the X including 3C7 has been inserted into an autosome.[4] To demonstrate its usefulness, let us adopt the symbolism +/+; *Dp*/+, in which +/+ indicates that two normal X chromosomes are present, and *Dp*/+, following the semicolon, represents the autosomal constitution. In this case, one of the autosomes carries the additional short segment of the X chromosome (*Dp*), including a normal 3C7 band. By simply remembering (i) that a normal female has two X chromosomes, hence two 3C7 bands, whereas a male has only one, and (ii) that the Notch phenotype stands for a deficiency or inactivation of 3C7, whereas Confluens is due to the presence of an extra band, one can derive the phenotypes of animals with various combinations of X chromosomes and duplication as listed in Table 1.[5] The N^x/N^y; *Dp*/+ condition, Notch in phenotype, is the important one for our purposes, since the recessive lethality can be circumvented while the possibility is tested that two *N* mutants of the inactivated type, not

visibly deficient for 3C7, occupy separable mutant sites within the gene.

MUTANT INTERACTIONS

The Notch phenotype is associated only with females (Table 1). This is expected since one obtains a Notch female by deleting or inactivating one of two 3C7 bands. If one 3C7 is deleted from a male, nothing remains, and the lethal condition is attained. However, a viable Notch-like male could result from a partial rather than complete inactivation of the gene. If a hypomorphic mutant had genetic activity approximating one-half that of the normal allele, it could resemble *N* phenotypically. A recessive visible called facet-notchoid (*fa^no*), viable and fertile in homozygotes and hemizygotes, seems to approach this condition;[6] males and females have notched wings, although the phenotype is not identical to that of *N*/+ ♀. Furthermore, *N*/*fa^no* heterozygotes are lethal, consistent with the idea that this constitution should be more hypomorphic than *N*/+.

The mutant *fa^no* is not the only recessive visible that shows interactions with *N*. At present there are three eye mutants and three wing mutants (Table 2). Of the eye mutants, *fa* is very mild, causing a slight roughening of the eye. The *fa^g* mutant is characterized by a more extreme roughening, and, in addition, the eye has a glossy appearance. The heterozygous combination *fa/fa^g* is noncomplementary, that is, mutant in phenotype; the eye is rough but not glossy. This condition should be contrasted with the complementary condition of either facet mutant with split; *fa/spl* is virtually wild-type. Although there might be a slight roughness in the eye of this heterozygote, the expression is so unreliable that for operational purposes the combination is considered nonmutant. All combinations of the subtly

Table 2 The recessive visible mutants at the Notch locus. Braces indicate noncomplementary heterozygotes.

Eye mutants		Wing mutants	
Name	*Symbol*	*Name*	*Symbol*
Facet	*fa* ⎫	Facet-notchoid	fa^{no} ⎫
Facet-glossy	fa^g ⎬	Notchoid	*nd* ⎬
Split	*spl* ⎭	Notchoid-2	nd^2 ⎭

different wing abnormalities are noncomplementary and show an intermediate mutant expression. All wing mutants are complementary with eye mutants. The breakdown of mutants into the two classes eye and wing, as well as the classification of some heterozygotes as complementary and others as noncomplementary, is not as clear-cut as it sounds, but it is an operational necessity. Probably, most or all of the mutants are interacting functionally but usually at a level below the visual perception of the investigator.[6]

Except for fa^{no}, little evidence has been presented that there is a functional relationship between the recessive lethal Notches and the recessive visible mutants. That a relationship exists is shown most simply by making heterozygotes. The combination N/fa is phenotypically Notch and facet. If N and fa were unrelated, one would not expect to see facet since it is a recessive visible present in one dose. This pseudodominant expression of fa can be understood by considering N as a deficiency for salivary band 3C7, wherein lies the wild-type allele of fa; the facet phenotype is expressed because the wild-type allele is missing. Inactivation of the genetic material in band 3C7 would similarly result in the expression of fa.

An interesting aspect of the pseudodominant expression of N/fa heterozygotes is that the facet phenotype is exaggerated in comparison to homozygotes. This is understandable if one thinks of fa as a hypomorphic mutant,[7] as was done previously for the mutant fa^{no}. The mutation causes a reduction in genetic activity; hence, as far as total activity is concerned, the combination N/fa produces less product than fa/fa and, accordingly, is more mutant in phenotype. The mutations spl and fa^g are expressed pseudodominantly when heterozygous with Notch, but spl shows no phenotypic exaggeration;* the mutant fa^g is relatively new, and I cannot say whether or not its expression is more extreme in N/fa^g heterozygotes.

As for wing mutants, we already know that N/fa^{no} is lethal. In the case of N/nd and N/nd^2, pseudo-

dominance is seen again, and an exaggeration effect may be present also, but these combinations are difficult to evaluate in this respect. The heterozygotes of Notch with either notchoid have reduced viability and fertility; evidently, nd and nd^2 are much like fa^{no} but represent less hypomorphic mutant states.

The recessive visible eye mutants show little evidence of functional relationship to wing mutants, but, paradoxically, both mutant types interact with N. Do the recessive visibles and lethals occupy mutant sites separable by crossing over, and, if so, how are they distributed on the genetic map?

THE GENETIC MAP

Finding the linear order of a mixture of recessive visible and recessive lethal mutants has required three distinct types of crosses. The simplest of the three is a cross between two of the recessive visible mutants. An example is shown in Table 3. Only the genotype of the parental female and the expected recombinant types of chromosomes are shown. The outside or peripheral markers w^a (white-apricot, 1.5 units to the left of fa^{no}) and rb (ruby, 4.5 to the right of spl) are used to give the order of the internal mutants in the event an exchange occurs between them. If the order indicated by the genotype is correct, one will obtain only the recombinant chromosomes shown in Table 3; if the internal order were reversed, the peripheral markers would alter their association. Multiple exchanges which could

*The mutant spl does not appear to be a hypomorph, but I will not try to classify it. The discussion of mutant types will be limited throughout to hypomorphs and amorphs, because they are relatively frequent and one is simply the extreme condition of the other. It should be understood that at Notch a relatively extreme hypomorph may be indistinguishable from an amorph.

Table 3 General format for the three types of crosses from which the genetic map is derived.

Female genotypes						Recombinal chromosomes				
w^a	fa^{no}	+	+			w^a	fa^{no}	spl	rb	
+	+	spl	rb			+	+	+	+	
+	spl	+	rb			w^a	+	+	rb	*
w^a	+	N	+			+	spl	N	+	
y	w^a	N	+	+	Dp	+	w^a	+	+	+	*
+	w^a	+	N	rb	+	y	w^a	N	N	rb

*Undetectable recombinants.

confuse the issue are rare. By varying the visible mutants in various crosses, one obtains the linear order shown in Fig. 1, but for the moment a discussion of these results will be postponed.

Crosses between N (the genetically inactivated type of N mutant that has salivary band 3C7) and a recessive visible follow the format shown on the second line of Table 3. In this case only one of the two reciprocal recombinant products can be detected, because the double mutant recombinant, that is the $+ spl N +$ chromosome, is indistinguishable from a chromosome possessing only the recessive lethal N. This is in accord with the previous interpretation that the pseudodominant expression of spl seen in N/spl females results from an inactivation of the genetic material carried on the N chromosome. This being the case, the spl mutant as well as its wild-type allele should be inactivated when coupled to N; hence $+ + N +$ should be indistinguishable from $+ spl N +$.

To circumvent the recessive lethality of N's (inactivated type) in Table 3, it is necessary to introduce the autosomally carried Dp into females that are heterozygous for the X-linked lethals.[8] The use of Dp requires that certain other changes be made in the genotype of the parental female. The principle is the same as in the previous crosses, the details of which have been described.[9] As in the second cross, only one of two recombinants can be detected. Since one mutant site already produces an amorphic condition, the coupled N's can hardly be expected to differ from the single N.

Fig. 1. Map of the Notch locus. *Top line:* Distribution of recessive lethals (N's) and recessive visibles. Total map distance = 0.14 units. The N mutants from left to right are N^{55e11}, N^{264-40}, N^{Nic}, $N^{264-103}$, N^{j24}, N^{Co}, N^{60g11}. *Bottom two lines:* Approximate localizations of mutants not critically positioned. Proper symbolism for non-Notched lethals: $l^N = l(1)N$, $l^{NB} = l(1)N^B$, and so on. With the exceptions of $l(1)N$ and nd^3 all mutants have been examined cytologically and found to be normal. The recessive visible nd^2 has not been examined, and $l(1)N$ has failed to give satisfactory preparations.

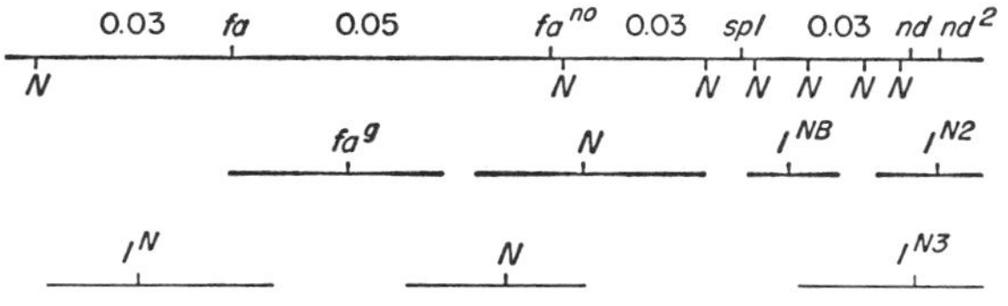

Let us examine the genetic map and confine our interest momentarily to the distribution of the recessive visibles. It can be seen from the top line of Fig. 1 that the localization of these mutants relative to one another does not make much sense in terms of phenotypes. For example, fa and spl, eye mutants, are separated by the mutant site of fa^{no}, a wing mutant. To further complicate the situation, fa complements spl, operationally at least, and both eye mutants complement the wing mutant. The noncomplementary wing mutants fa^{no}, nd, and nd^2 are in turn separated by the eye mutant spl.

Excluding the recessive visibles and confining attention to the recessive lethals, one finds a much more satisfactory situation. Although there is a clustering of mutant sites at the right end,* N's are distributed over almost the entire length of the map. All N's are similar in that they show pseudodominance with the recessive visibles. With some minor variations,[9] they are phenotypically similar in the $N/+$ condition, and, while not all combinations of N have been tested, most have and are lethal. Results with recessive lethal N mutants suggest that Notches represent mutant sites within the same functional gene or cistron. Should one hesitate to draw this conclusion because the recessive visibles confuse the picture? I think not, but one should find an interpretation for them, and to do this it is necessary to examine the cistron concept and consider the genetic criteria for the identification of the functional gene.

THE CISTRON

A *cis-trans* effect and its use for discerning gene relationships is not new to *Drosophila*.[10] I will demonstrate it only with reference to the Notch locus. For example, $N/+$ is phenotypically Notch, and $+/fa$ is wild-type. If N and fa were unrelated, the heterozygote $+ N/fa+$ should simply be Notch in phenotype, whereas this trans condition of the mutants is actually Notch and facet; exactly the same situation exists for N and spl. A comparison

*This concentration of N's is probably a consequence of using only radiation-induced and spontaneous mutants. The use of chemical mutagens might yield a different array of mutant sites.

should be made with the *cis* condition of the mutants, and, in this case, the *fa N*/+ + and *N spl*/+ + conditions are simply Notch in phenotype. The expression of *fa* and *spl* in the *trans* condition with *N* and lack of expression in the *cis* compound implies a functional relationship between *N, fa,* and *spl.* Presumably *cis-trans* comparisons between *N*'s and any of the other recessive visibles will yield similar results, although the coupled mutants required for the *cis* test are not available.

If two *N* mutants separable by crossing over are functionally related, *trans* should differ from *cis.* The *trans* condition, N^x +/+ N^y, is lethal. Presuming that $N^x N^y$/+ + is viable, the difference between *trans* and *cis* indicates a functional relationship. It was presumed that $N^x N^y$/+ + was not a lethal condition, but the presumption is supported by the fact that even a visible cytological deficiency is viable as a heterozygote, hence the coupled mutant condition $N^x N^y$/+ + can hardly be otherwise.[9] Thus when *cis* is viable and the *trans* configuration is lethal, it is assumed that the mutant *N* sites occur within the same *cistron* or *functional gene.*[11]

Heterozygotes of some of the recessive visibles in the *trans* condition are complementary, that is, wild-type, and they do not differ significantly from the *cis* condition. This seems to indicate that they belong to different functional genes or cistrons, yet all these visibles are related to the *N* mutants and are presumably related to each other, although the *cis-trans* test sometimes fails when only recessive visibles are used. It is suggested that the use of amorphic mutants (*N*'s) will clearly identify a cistron, whereas the use of hypomorphic mutants can lead one astray.[9] To support this contention, it will be necessary to look at the cistron and cistron-protein relationship as pictured in microorganisms, and particularly in *Neurospora,* since it tends to bridge the gap between phage, bacteria, and so forth, and an organism like *Drosophila.*

In Fig. 2, the intermediate steps between DNA and protein are shown. In a highly diagrammatic way, Fig. 3 depicts the same basic situation in a form that is more suitable for this discussion of functional genes. Only a single strand of the DNA double helix is shown in Fig. 3, since coding for a protein seems to occur on only one strand.[12] The short vertical lines may represent any one of the four bases: thymine, adenine, guanine, or cytosine. Starting at one end, the messenger RNA makes a comple-

mentary copy of the DNA series of bases except that uracil replaces thymine. This message is carried to ribosomes of the cytoplasm, where the protein is constructed through the action of transfer RNA (shown in Fig. 2 but not in Fig. 3). Commencing at the left and progressing to the right, a sequence of three bases specifies a particular amino acid.[13] With this in mind, one can imagine the mutation process in various ways: Part or all of the locus could be deleted. One base of a series of three could be substituted for another, and consequently a single different amino acid might be inserted into the protein.[14] The deletion or addition of a base could be responsible for extensive damage by causing a whole series of erroneous amino acid substitutions from the point of change onward (a reading frame mutant) or by stopping the message at the point of alteration.[15]

Rather frequently in an organism like *Neurospora* one finds two or more mutants that will not survive in the homokaryotic condition (to be compared to the homozygous condition in *Drosophila* with certain reservations), but are partially complementary, that is, viable in heterokaryons (the *trans* condition in *Drosophila*). Strictly speaking, this is unexpected if the two mutants are in the same cistron — they should be noncomplementary, that is, inviable. However, without jeopardy to the cistron concept, the cause can be imagined as follows. Each of the two mutants forms an inactive protein, and, in heterokaryons, each is represented. The requirement for activity is that two (or more) proteins be associated in a dimer (or polymer) condition as shown in Fig. 3. In mixed dimers, the error in one protein partially corrects the other, and some function is restored.[16] In *Neurospora,* confusion is avoided by using completely noncomplementary mutants to define by recombination the genetic limits of the cistron, for then the partially complementary mutants will fall within them.[17]

Now the curious situation at the Notch locus can be explained. The genetic criterion for the functional gene or cistron as developed in microorganisms utilizes the amorphic mutants; the methodology virtually requires it. Using the recessive lethal (amorphic) *N* mutants, one recognizes a single cistron at the Notch locus, something that cannot be done with the recessive visibles. To use them at Notch for this purpose would be similar to defining the cistron in *Neurospora* with the leaky hypomorphic mutants

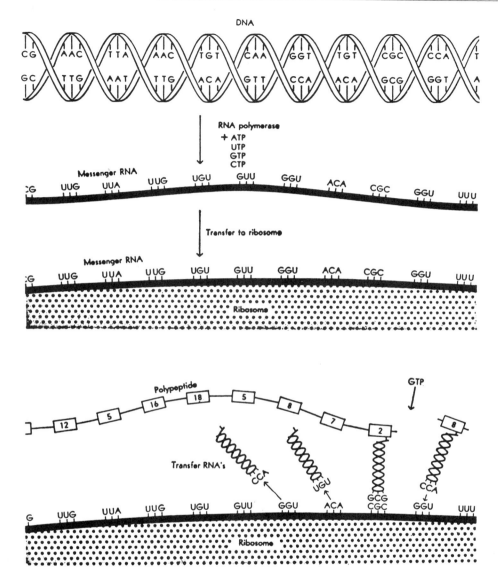

Fig. 2. Conceptualization of the chain of events in the synthesis of protein. The messenger RNA is a complementary copy of one strand of DNA, commencing at the lower left end of the double helix. *A-T*, adenine-thymine; *G-C*, guanine-cytosine; *U* (uracil) is substituted for thymine in mRNA. Linked rectangles represent amino acid residues of the growing polypeptide chain. For additional information see Nirenberg.[13] (After McElroy)[34]

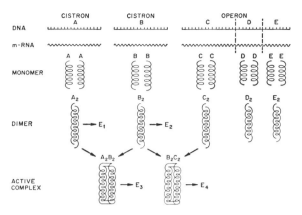

Fig. 3. Hypothetical relationships between cistrons and proteins. In the polycistronic operon the mRNA unit specifies three different proteins. Proteins might have activity as monomers, dimers, and so on, or as tetramerlike complexes (A_2B_2). Dimerlike complexes (AB) may also exist, but are not shown.

that are less amenable to study. In a strict sense one cannot equate the complementary *trans* condition of recessive visibles at Notch to the intracistronic complementing mutants of *Neurospora*, since the homokaryons will not survive on unsupplemented medium, whereas the homozygous visibles in *Drosophila* are viable. At Notch, a viable heterozygote composed of two recessive lethals would be a more equivalent comparison.[9]

COMPARISONS WITHIN *DROSOPHILA*

In a previous paper we made comparisons with other loci in *Drosophila* and concluded that, if only recessive visible mutant sites were used to the exclusion of lethal sites, the issue would be as confused as it is at Notch.[9] It was assumed that at most pseudoallelic loci in *Drosophila*, known only by their recessive visible mutants, a mutation to an amorphic lethal condition was possible. However, recent developments have indicated that there are more loci than believed earlier at which an amorphic condition is represented by a viable recessive visible. It must mean that in such cases the DNA specifies a nonessential protein, and unless some special conditions are invoked, it follows that the production of a recessive lethal at the locus would be an improbable

event. At rosy (*ry*), a series of separable noncomplementary mutant sites defining the cistron have been demonstrated,[18] but the investigators were unable to produce a lethal.[19] Evidently the amorphic mutants represented as recessive lethals at Notch are recessive visibles at rosy. Complementing and noncomplementing visibles comparable to what one might find in *Neurospora*[20] are known to exist at rudimentary (*r*). At least, the noncomplementing *r* mutants could represent amorphic viable mutants like *ry*. At the white locus a deficiency for the single salivary band 3C2 is phenotypically white (*w*) but not lethal.[21] Apparently *w* mutants can represent viable amorphs.

There are complex loci in *Drosophila* that contain a mixture of recessive visible and recessive lethal loci. The locus of dumpy (*dp*) makes an interesting comparison. The symbolism developed for dumpy is intimately related to the phenotypes: The symbol *o* represents a recessive visible wing abnormality (oblique wing); *v* (vortex), an abnormality of the thorax (a small pit surrounded by bristles); *cm* (comma), a bilateral groove of the anterior thorax; *l*, a recessive lethal condition. The linear array of mutants is shown in Fig. 4.[22, 23] The locus contains a mixture of phenotypically different recessive visibles and recessive lethals like Notch, although unlike the *N* mutants the recessive lethals at dumpy would not be alternatively classified as dominants. The symbol *olv* indicates that these particular lethals show pseudodominance when heterozygous with *o*, *v*, or *ov*; thus the *olv*'s, like the *N*'s, show pseudodominance with all the recessive visibles. At dumpy, the lethals *ol*, *lv*, and *l* showing pseudodominance with one recessive visible but not with another (*ol* and *lv*), or not displaying it at all (*l*), differ from the condition at the Notch locus.

I assumed that lethals comparable to *l*, *ol*, or *lv* could occur at Notch but had escaped detection because mutants had been selected on the basis of the dominant Notch phenotype. If so, one might find them by selecting for recessive lethals rather than for dominant Notches. Subsequently, three atypical types were found in a sample of x-ray-induced spermatogonial lethals.* Another was found preexisting on a chromosome that already carried a typical *N* mutant. Complete information is not yet available, but what there is can be briefly discussed.

*Courtesy of Dr. S. Abrahamson, University of Wisconsin.

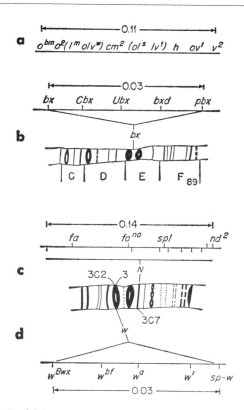

Fig. 4. (a) Genetic map of the dumpy (*dp*) locus, modified from Southin and Carlson.[23] The order of mutants within parentheses is not critically demonstrated. The *h* mutant is *ov*-like. (b) The genetics and cytology of the bithorax (*bx*) locus modified from Lewis.[10] The *Ubx* mutant is associated with the leftmost double-banded structure in 89E; mutant *bxd* with the rightmost doublet. (c) The Notch (*N*) locus associated with salivary band 3C7. (d) The white (*w*) locus. The genetic map is modified from Judd.[33] The association with 3C2 is from LeFevre and Wilkins.[21]

The lethals are partially mapped as shown in Fig. 1. The mutant *l(1)N* (preexisting with an *N*) is similar to the one at dumpy that fails to show pseudodominance with any of the visibles. It is lethal in combination with *N*'s and with the remaining atypical lethals at this locus. The mutants *l(1)N²* and *l(1)N³* allow pseudodominant expression of the recessive visibles, but unlike *N* they are viable and fertile with *fa^{no}*. With the rough-eyed glossy mutant (*fa^g*) they are facet-like but not glossy. The mutant *l(1)N^B* is similar, except that with *fa^g* it is facet-

glossy.* It has one other interesting aspect: it could be classified as a dominant like *N* but not as *N*. The lethal has associated with it an abnormal condition of the thoracic microchaetae in which the small "hairs" are frequently missing, sometimes leaving bald areas on the thorax. The abnormality seems to be enhanced in males carrying the duplication.

The discovery of non-Notch recessive lethals increases the similarity to dumpy, but the parallelism is not exact. At both loci the existence of strikingly different phenotypes confuses the picture. Later an explanation will be advanced to explain the phenomenon, but at the moment it is more important to notice that at both loci recessive lethals occur, and that the *olv*'s resemble the amorphic *N*'s. If both loci are single cistrons, one would expect *olv*'s to be spread over the entire map, but at present they are not. I believe that Notch is a cistron. The dumpy locus could be the same, but there is a plausible alternative for dumpy, and that is an operon. We will have to look at it as pictured in microorganisms to understand why.

THE OPERON

An operon appears to be a complex of two or more cistrons controlled by an operator so that they function coordinately rather than independently.[24] The operator and the more recently described promoter[25] will not be discussed, since it is the polycistronic condition of the operon that is to be emphasized. From an investigation of the histidine-3 region of *Neurospora*,[26] it is inferred that at the level of messenger RNA (mRNA) the cistron and operon have some similarity; there is in a sense a unitary mRNA in a cistron or an operon, although in the latter case the mRNA carries a message specifying more than one protein. Because of the situation at the mRNA level, a *cis-trans* test in *Drosophila* could recognize a cistron or operon without distinguishing between the two. Imagine one recessive lethal mutant in cistron C that interferes with the proper coding of C, D, and E (Fig. 3). Let another

---

*These lethals, with the exception of *l(1)N*, could represent hypomorphic states intermediate between the recessive visibles and the amorphic or near amorphic *N*'s. They demonstrate the diversity of mutant types that can occur at this locus.

recessive lethal occur in E so that only E is affected. In the *trans* condition the heterozygote could lack a functional E protein, and the lethality would suggest a functional relationship between the mutants although they are in different cistrons.

In *Neurospora* and *Salmonella*, mutations within an operon display a kind of polarization effect in that the damage spreads in one direction through the cistrons of an operon.[26, 27] In terms of Fig. 3, mutant sites in C may cause the loss of functions C, D, and E; D mutants lose D and E functions but maintain C; and E mutants may lose only the E function. The operon concept has been applied in *Drosophila*; the case will be examined and compared to the cistronic condition at Notch.

CYTOGENETIC CONSIDERATIONS

Recently it has been shown that at the bithorax locus an older sequential reaction scheme and the concept of an operon fit the known facts about equally well.[10] For the purposes of this discussion the condition of an operon will be assumed. To contrast it with Notch, I will take only a small portion of the genetic information so as to make a point that is primarily cytogenetic.

The genetic and cytological situation at the bithorax locus is shown in Fig. 4. One need only consider the two mutants *Ubx* (ultrabithorax) and *bxd* (bithoraxoid). The former is a dominant with a recessive lethal effect, and *bxd* is a recessive visible. It has been possible to assign *Ubx* to the left doublet in section 89E of the salivary chromosome map and *bxd* to the right. The *Ubx* mutant acts like an inactivation of *bxd*; that is, *Ubx* +/+ *bxd* is phenotypically ultrabithorax and bithoraxoid (a pseudodominant effect) just as *N* +/+ *spl* is Notch and split. In the *cis* condition *Ubx bxd*/+ + is only *Ubx* phenotypically, just as the *N spl*/+ + is only Notch. An interesting series of observations is the following: The *Ubx* +/*Ubx*+ homozygote is lethal, but a duplication carrying the wild allele of *Ubx* may be added; in which case the lethality is circumvented and the animals are *Ubx* and *bxd* in phenotype. The addition of a second wild allele of *Ubx* completely eliminates the *Ubx* phenotype, but the animals remain *bxd* even though the wild alleles are present on the *Ubx* chromosomes of the homozygote.[28] An interpretation can be made by referring to the operon of Fig. 3 but assuming the existence of only two cistrons C and D. The *Ubx* mutant is in cistron C, and it prevents the proper transcription of both C and D. Hence the *Ubx* +/*Ubx* + homozygote lacks C and D proteins. The addition of one or two wild alleles of *Ubx* allows the formation of C, but the D protein is still missing and the animals remain phenotypically *bxd*. If *Ubx* and *bxd* had belonged to the same cistron, the addition of duplications should have eliminated both mutant expressions.

If a polarization effect existed at Notch, one might expect to make an observation like the following: A lethal at the right end of the map (Fig. 1) would show pseudodominance with all the recessive visibles *nd²* to *fa* (reading from right to left on the genetic map). Another lethal site between *spl* and *faⁿᵒ* would show pseudodominance with visibles to the left but not with *spl, nd,* and *nd²* localized to the right, and so on. Polarized mutants appropriate to an operon have been demonstrated at bithorax, whereas at Notch the scattering and interspersion of recessive lethal and visible sites over the genetic map best fit a cistron concept. Bithorax, interpretable as an operon of three cistrons,[10] is cytologically associated with four salivary bands. Notch is associated with a single band. The inference is that at the cytological level a cistron may be distinguished from an operon when genetic evidence fails to discriminate. It also suggests that a single band is a single functional gene or cistron.

It should be clearly understood that the equivalence of salivary bands and cistrons derives entirely from comparisons between the single-banded Notch condition, genetically interpreted as a cistron, and the multiple-banded condition at bithorax, interpreted as an operon. At some future time, the genetic explication of one or both systems may alter as additional information becomes available. However, the cytological state of *N* (and white) compared to bithorax will likely remain as a meaningful difference whatever codicils are finally appended to the current interpretation of these complex loci.

It is noteworthy that Notch and white, each conforming to a single salivary band, have quite different genetic lengths. Even more striking is the contrast between Notch and the multiple-banded condition of bithorax (Fig. 4). If genetic lengths are correlated with gene sizes, Notch is relatively large. Alternatively, recombination within genes (or between them) may be subject to a control ultimately referable to the sophisticated structure of the chromosome.

The implication of the cytogenetic aspects is that cistrons can be represented as single salivary bands at least on some occasions. Does this mean that the functional units are delimited by the band to the exclusion of the interband spaces? The existence of Notches with an apparently intact salivary band and a chromosomal breakpoint adjacent to 3C7 suggests that the gene might not be so strictly confined. Upon noting that a cytologically visible deficiency could recombine with closely adjacent loci at the miniature-dusky complex,[29] we made an attempt to recombine a Notch deficient for 3C7 (N^{264-39}) with other mutant sites in the cistron. The results were positive and implied that the interband regions bordering on 3C7 were part of the functional gene.[30] Unfortunately, subsequent cytological examination revealed that the mutant was no longer cytologically deficient, and a valid test must wait until an appropriate deficiency becomes available.

NONAMORPHIC MUTANTS

Referring back to the dumpy locus, one can see why it might conform to a cistron or operon. If amorphic mutants are scattered over the map, dumpy will look like a cistron. If a polarization effect is found, the situation may be that of an operon. The recessive visibles at dumpy, as at Notch, may initially confuse the issue if they are not amorphic.

One still has to explain why at Notch the intra-cistronic visibles have such different phenotypes while presumably affecting the same protein or polypeptide. Some known conditions in a microorganism and in man suggest an answer. In *Escherichia coli*, two contiguous cistrons code for two separate proteins A and B. Each single protein has some catalytic activity in one of two different reactions, but the enzyme tryptophan synthetase, formed by the union of A and B, functions in a third reaction as well as in the first two.[14] The formation of tryptophan synthetase in *E. coli* is similar to the condition found in human hemoglobin. The production of α and β proteins is controlled by two separate loci, but α and β dimers are joined to form hemoglobin.[31] With these examples in mind, one can

suggest a way in which two different changes in a protein controlled by a single cistron could result in two quite dissimilar phenotypes. For example, consider a mutant change in cistron B (Fig. 3). One to three different systems (E_2 to E_4) could be affected by each individual change, and two different intra-cistronic mutants need not affect the same reaction. That is, one mutant is detectable only in E_3, the other in E_4. Some similar scheme would be required to explain the situation at the white (w) locus (Fig. 4), which functions in the formation of two different eye pigments and shows some interesting interactions with zest (z).[32, 33]

SUMMARY

In following the fine-structure analysis of the Notch locus in *Drosophila* and utilizing the recessive lethal *N* mutants, one gets a picture of the functional gene or cistron that does not differ materially from our picture of the gene of phage, *Neurospora*, and other microorganisms. The reason for this seems clear. The concept of the functional gene is a product of the genetics of microorganisms, and the methodology requires the use of mutants that in *Drosophila* are classified as amorphs. When the same mutant type is used in *Drosophila*, whether it be a recessive lethal like *N* or a recessive visible amorph like *ry*, the genes of macro- and microorganisms are similar.

When hypomorphic mutants are used to the exclusion of the recessive lethal amorphs, the cistron cannot be clearly identified. It is suggested that the gene product of the Notch locus ultimately complexes with other products to form two or more different structural proteins or enzymes. Since two different mutant varieties of the same Notch protein (derived from two different hypomorphic mutants) need not affect the same enzyme system of the two or more systems to which the protein contributes, the *trans* heterozygotes may have a normal phenotype.

Cytogenetic comparisons based upon Notch as a cistron and bithorax as an operon suggest that a single salivary band corresponds to a functional gene. However, it is the association of the genetic complexity at bithorax with multiple bands and the affinity of Notch with a single band that is of central importance, for, even if genetic interpretations change in time, the cytological difference must retain some meaning in its new context.

REFERENCES AND ACKNOWLEDGMENT

1. J. D. Watson and F. H. C. Crick, *Nature* 171, 737 (1953); *ibid.*, p. 964.
2. O. L. Mohr, *Genetics* 4, 275 (1919); —, *Z. Verebungslehre* 33, 108 (1923); M. Demerec, *Proc. Intern. Congr. Genet. 7th, Edinburgh* (1939), p. 99; H. Slizynska, *Genetics* 23, 291 (1938).
3. T. H. Morgan, J. Schultz, V. Curry, *Carnegie Inst. Wash. Year Book* 40, 283 (1941).
4. G. Lefevre, Jr., *Drosophila Inform. Serv.* 26, 66 (1952).
5. This treatment evades a discussion of dosage compensation — Why does a male deficient for an entire X chromosome not suffer from the loss? For a discussion of dosage compensation see C. Stern, *Can. J. Genet. Cytol.* 2, 105 (1960).
6. H. Bauer, *Z. Vererbungslehre* 81, 374 (1943).
7. H. J. Muller, *Proc. Intern. Congr. Genet., 6th, New York* 1, 213 (1932). According to Muller's classification there are *amorphic, hypomorphic, hypermorphic, antimorphic,* and *neomorphic* mutants.
8. W. J. Welshons, *Cold Spring Harbor Symp. Quant. Biol.* 23, 171 (1958).
9. W. J. Welshons and E. S. Von Halle, *Genetics* 47, 743 (1962).
10. E. B. Lewis, *Am. Zool.* 3, 33 (1963).
11. S. Benzer, in *The Chemical Basis of Heredity*, W. D. McElroy and H. B. Glass, Eds. (Johns Hopkins Press, Baltimore, 1957), pp. 70–93.
12. M. Hayashi, M. N. Hayashi, S. Spiegelman, *Proc. Natl. Acad. Sci. U.S.* 50, 664 (1963); M. H. Green, *ibid.* 52, 1388 (1964).
13. M. W. Nirenberg, *Sci. Am.* 208, 80 (March 1963).
14. C. Yanofsky, U. Henning, D. Helinski, B. Carlton, *Federation Proc.* 22, 75 (1963).
15. F. H. C. Crick, *Science* 139, 461 (1963).
16. M. E. Case and N. H. Giles, *Proc. Natl. Acad. Sci. U.S.* 46, 659 (1960).
17. N. H. Giles, *Proc. Intern. Congr. Genet. 11th, The Hague* 2, 17 (1963).
18. A. Chovnick, A. Schalet, R. P. Kernaghan, M. Krauss, *Genetics* 50, 1245 (1964).
19. A. Schalet, R. P. Kernaghan, A. Chovnick, *ibid.*, p. 1261.
20. O. G. Fahmy and M. J. Fahmy, *Nature* 184, 1927 (1959); M. M. Green, *Genetics* 34, 242 (1963).
21. G. Lefevre, Jr., and M. D. Wilkins, *Genetics* 50, 264 (1964).
22. E. A. Carlson, *ibid.* 44, 347 (1958).
23. J. L. Southin and E. A. Carlson, *ibid.* 47, 1017 (1962).
24. F. Jacob and J. Monod, *J. Mol. Biol.* 3, 318 (1961).
25. F. Jacob, A. Ullman, J. Monod, *Compt. Rend.* 258, 3125 (1964).
26. A. Ahmed, M. E. Case, N. H. Giles, *Brookhaven. Symp. Biol.* 17, 53 (1964).
27. J. C. Loper, M. Grabnar, R. C. Stahl, Z. Hartman, P. E. Hartman, *ibid.*, p. 15.
28. E. B. Lewis, *Cold Spring Harbor Symp. Quant. Biol.* 16, 159 (1961).
29. G. L. Dorn and A. B. Burdick, *Genetics* 47, 503 (1962).
30. W. J. Welshons, E. S. Von Halle, B. J. Scandlyn, *Proc. Intern. Congr. Genet. 11th, The Hague* 1, 1 (1963).
31. H. S. Rhinesmith, W. A. Schroeder, N. Martin, *J. Am. Chem. Soc.* 80, 3358 (1958); V. M. Ingram, *Nature* 189, 704 (1961); C. Baglioni, in *Molecular Genetics*, J. H. Taylor, Ed. (Academic Press, New York, 1963), vol. 1, p. 405.
32. M. Gans, *Bull. Biol. France Belg.* (Suppl.) 38, 1 (1953); M. M. Green, *Heredity* 13, 302 (1962).
33. B. H. Judd, *Genetics* 49, 253 (1963).
34. W. D. McElroy, in *Cell Physiology and Biochemistry* (Prentice-Hall, Englewood Cliffs, N.J., ed. 2, 1964), p. 120.

Work performed in the Biology Division, Oak Ridge National Laboratory, Oak Ridge, Tennessee, operated by Union Carbide Corp. for the AEC.

Part 2. Transmission of genetic material in eucaryote organisms

In 1866, Mendel published his paper on the inheritance of pea plant characteristics, which essentially formulated the theoretical foundation of inheritance as we conceive it today.[21] However, his was not the first interest in heredity. Since the transition of humanity from a nomadic to an agrarian mode of existence, about 10,000 B.C., humans had consciously and unconsciously dealt with the phenomenon of inheritance. From the earliest attempts to maintain selected characteristics — initially in crops, and later in domesticated animals — humans had been involved with the outcome of inheritance, although, during most of this period the implications of selective breeding programs were not fully understood. The first documented inquiries into the mechanics of inheritance originated in the Socratic and Aristotelian schools of philosophy, around the fourth century B.C.[28] — approximately 23 centuries before Mendel's presentation.

In this period, the ideas concerning the mechanics of transmission of characteristics passed through many different formulations. However, by 1850, most plant and animal hybridizers agreed that the following attributes were characteristic of the genetic information. First, the interaction of the genetic information contributed by both parents produced the expression of a hybrid individual. Second, the genetic material lacked stability — evidenced by the rapid change of the overall appearance of organisms when followed from generation to generation. Third, since many hybridizations yielded offspring that displayed intermediate expressions to those of the parents, it was assumed that the parental genetic information blended when combined within the offspring. The appearance of the organism was viewed as a whole and was taken to be the result of the unified activity of the entire genetic complement of both parents. It was, therefore, impossible for the investigators of the 1860s to conceptually visualize the genetic information in terms of discrete units whose mode of transmission could be observed and mathematically predicted.

The latter is exactly what Mendel's experiments suggested. His work advanced the idea that an organism could be viewed as a composite of separate, distinct traits and that these traits were controlled by stable units of inheritance. He based his theory on carefully controlled hybridization data which, when analyzed, produced mathematical formulas describing patterns of transmission contingent upon the existence of stable units of heredity.

With perspective we can understand why the investigators of the 1860s found it difficult to accept Mendel's proposals. We know his work was overlooked for 35 years. Unquestionably, one of the major reasons for this oversight was the scientific community's inability to grasp the importance of the implications of Mendel's data. It cannot be denied that a few of the top hybridizers of the nineteenth century were exposed to Mendel's results by means of reprints and/or letters. Certainly, if such radical proposals were published today, one would immediately expect to encounter cynicism, followed by attempts to reproduce the reported results. We have no reports of investigators attempting to verify or disprove Mendel's pea plant observations. Because his observations required a radical departure in thinking the results were simply ignored.

It was not until 1900 that three geneticists, Hugo de Vries,[12] Carl Correns,[10] and Erich Tchermak,[32] independently observed experimental results that supported the existence of a system of inheritance based upon separate units controlling distinct features. By this time, the scientific community was prepared for the acceptance of such a mechanism of inheritance. Between 1865 and 1900, the importance of the nucleus and its contents, in respect to inheritance, had been documented independently by several different investigators.[18,27,34] The existence of structurally discrete bodies (*chromosomes*) and the elaborate movements of these chromosomes during cell division[15] and gametogenesis[4] were reported. The mechanics of meiosis had illustrated the existence of a mechanism whereby an orderly reduction in the number of chromosomes could be explained — a process predicted on the basis of Mendel's observations. In addition, the emergence of the biometric school, a group of investigators that attempted to study heredity through mathematics, had established an atmosphere in which investigators could more easily accept the mathematical examination of the patterns of transmission. For these and additional reasons, all of which are discussed in several recent historical surveys of genetics,[7,13,30] the works of de Vries, Corren, and Tchermak were immediately subjected to close scrutiny. This intensive experimental investigation was destined to clarify and to expand the application of a Mendelian system of inheritance to several different areas of heredity.

One of the first objections raised to the universal applicability of a mechanism of heredity based upon stable, discrete units of control was advanced by William Castle.[8,9] Castle was investigating the pattern of transmission of coat colors in mice. He found what he interpreted to be experimental evidence supporting the existence of the physical mixing of materials between two units (*genes*) controlling different coat colors in a hybrid mouse. The coat color of the F_2 hybrid mouse had an intermediate appearance between the two parental expressions. The F_2 progeny displayed the expected 3:1 Mendelian ratios; however, the recessive parental phenotype (albino) had patches of black. This suggested the possibility of gene contamination in the F_1 hybrid, and, if true, would have required modifying or discarding the concept of structurally stable, discrete genes. The discrepancy was eventually explained by Nilsson-Ehle[24] and East,[14] each demonstrating that traits could be controlled by more than one set of genes. Instead of only two genes controlling the expression of a single phenotypic character, they established the existence of polygenic traits — traits controlled by more than one pair of genes which possessed a number of possible alternative expressions and traits dependent on the number and type of genes present. They explained, then, how phenotypes could display continuous patterns of expression, i.e., height in the human population; and with their demonstrations that the polygenic systems could be analyzed mathematically, the field of quantitative inheritance was born.

Udny Yule voiced a second objection to the exclusive application of the Mendelian type of inheritance to explain the transmission of dominant traits in a population of organisms. He asserted that a population of organisms, heterozygous for a particular pair of genes that exhibited a dominant relationship in the control of their specific phenotype, would rapidly lose the recessive expression after several generations of unselected breeding. Yule questioned how a Mendelian type of inheritance could maintain phenotypic variability in a population. His experimentally unsupported speculation was refuted by two investigators, independently: G. Hardy[17] and W. Weinberg.[33] They mathematically demonstrated that in an ideal situation of unselected mating in a large population, the frequency of genotypes (neither Hardy nor Weinberg differentiated between individuals and alleles) would remain constant, thus ensuring the maintainance of genetic variability. Once again, the resolution of a criticism of the Mendelian system of inheritance stimulated the appearance of a second major area of genetics: population genetics.

A third objection was raised to the universal applicability of a unitary system of inheritance. Many investigators reported F_2 phenotypic ratios that differed significantly from the 9:3:3:1 ratio expected on the basis of independent assortment of discrete units of inheritance. The first report was presented by Bateson and Punnett in a paper for the Evolutionary Committee of the Royal Society.[2,3] While crossing sweet peas, established to be heterozygous for two sets of alleles controlling two different characteristics, the distribution of the phenotypic combinations for the dominant and recessive forms of the two characters did not resemble the 9:3:3:1 ratio expected if assortment was independent. Bateson, a strong advocate of Mendelism, attempted to maintain his belief in the discrete nature of the controlling units by invoking hypothetical explanations other than a simple structural association between the units. First, he proposed a hierarchy of attracting and repelling forces between sets of alleles. When this proposal became experimentally unacceptable, he invoked a selective postmeiotic increase in certain gametes (*reduplication*).[1] However, to create a compatible relationship between experimental observations and Mendel's hypotheses, a portion of Mendel's proposal had to be modified. Ultimately, this led to the formulation of the second major conceptual advancement in the newly emerging field of genetics: the chromosome theory of inheritance.

In 1902 and 1903, Walter Sutton and Theodor Boveri each advanced a hypothesis that ultimately would explain the exceptional ratios reported by Bateson and Punnett.[31] In their proposals, they drew information from both cytological analyses of chromosome movements in gametogenesis and hybridization studies. They were able to see the correlation between the stable nature and pattern of distribution of the chromosomes, from parent to offspring, and the stable nature and pattern of distribution of the units of inheritance, from parent to offspring. They were the first to clearly state that the units of inheritance were associated with the chromosomes — the chromosome theory of inheritance. At the time of the theory's presentation, there was no experimental support for its validity. Such experimental support would not appear for another eight years.

The initial experimental support originated in an unlikely spot: Thomas Hunt Morgan's Columbia University laboratory. Morgan, strongly influenced by Castle's "blending" experiments, at first questioned the discrete nature of the hereditary units. However, because of the multifactoral explanation of coat-color control, which had dispelled Castle's objection, he gradually accepted Mendelism. Still, as late as 1910, he found no reason for linking the genes to the chromosomes, as he stated in the *American Naturalist.*[22] Despite his bias, the first solid support for the chromosome theory of inheritance was to appear in Morgan's laboratory.

Morgan was interested in accumulating de Vriesian types of discontinuous mutants. To this end, he introduced the fruit fly, *D. melanogaster,* into his laboratory as an experimental tool. These organisms were ideal to use in the detection of rare discontinuous variations because (1) they were prolific; (2) they were small enough to store in large numbers in restricted space but large enough to easily scan for phenotypic changes; and (3) they had a short generation time.

The first experimental support of the chromosome theory of inheritance resulted from Morgan's detection of a white-eyed mutant which displayed a unique transmission pattern.[23] Morgan verified experimentally that the white-eyed condition was recessive to the normal red-eyed color. However, when mating red- and white-eyed flies, the white eye color did not always disappear in the first generation as one would expect on the basis of Mendel's pea plant experiments. The transmission patterns for the eye color were dependent upon the sex of the parents carrying the two eye colors. Thus, it seemed that eye color was transmitted in conjunction with the determinants for sex. Morgan was well aware of the work being done by McClung,[20] Stevens,[26] Wilson,[35] and others that attributed sex determination to an individual's chromosomal constitution. So, it would seem a logical step to connect the Mendelizing units controlling eye color and the sex-determining chromosomes as being one and the same; but Morgan remained opposed to this idea. However, during the same year (1910), he detected the mutants yellow-body and miniature–wing, both of which also showed a sex-lined pattern of inheritance. After this, there was no question in Morgan's mind — the correlation of these factors with sex demanded their presence in the single X chromosome of male *Drosophila.*

Morgan assumed the validity of (1) the chromosome theory of inheritance; (2) the corollary concept that the genes were arrayed in a linear order along the chromosome; and (3) the hypothesis that crossing over took place between homologous chromosomes (*genetic recombination*). He suggested, therefore, that genetic exchange would be more likely to recombine sets of genes widely spaced on the chromosome than genes residing close together. A prediction based on his hypothesis would be that progeny, displaying recombinant phenotypes which reflect recombinant gametes, should be more frequent in crosses dealing with distantly separated genes, than in crosses dealing with closely linked genes. One of Morgan's students, A. H. Sturtevant, took this hypothesis and proposed that one should be able to mathematically derive an indirect measure of the distance between sets of genes on the same chromosome based upon the number of recombinant progeny in a population of flies. He extended this proposal to include the prediction that distances between sets of genes located on the same chromosome should be additive. He said that the distance between two sets of linked genes should equal the sum of the map distances calculated for the two outer sets in respect to a set of genes located between them. His data, with some qualifications, supported his predictions and, thus, supported the validity of the chromosome theory of inheritance and the idea that genes were arranged in a linear sequence along the chromosomes.[28]

Calvin Bridges, another of Morgan's students, published observations that also supplied indirect proof for the validity of the chromosome theory of inheritance.[5,6] Bridges, assuming that genes were associated with chromosomes, explained the occurrence of flies with rare phenotypes in sex-linked inheritance studies on the basis of infrequent, abnormal disjunction (*lack of segregation*) between the sex chromosomes. The abnormal phenotypes were due to the presence of unusual numbers of chromosomes. Bridges advanced the obvious, testable prediction based on the above hypothesis — that those individuals who displayed the rare, unexpected phenotypes would also possess abnormal numbers of chromosomes. Cytological examinations revealed the existence of chromosome numbers other than the expected diploid number. These examinations confirmed Bridge's hypothesis and supported the chromosome theory of inheritance.

Evidence suggesting the existence of structural linkage was not limited to *Drosophila.* Bateson's early data in sweet peas, now seen in light of the chromosome theory, became supportive evidence for linkage in plants.[2,3] By 1935, some 400 genes had been identified in maize and most of them had been placed in 10 linkage groups corresponding to the 10 chromosomes of the haploid complement. The first report of linkage in humans, made by Haldane[16] in 1937, was for the genes causing color blindness and hemophilia. Also during the 1930s, linkage was demonstrated in the fungus, *Neurospora crassa*.[19] Recombination studies with *Neurospora* and other fungi have been fundamental in defining the major features characterizing the physical exchange of genes between homologous chromosomes.

Finally, direct cytological proof for the chromosome theory of inheritance, in which specific regions of chromosomes could be associated with genetically established alleles, did not appear until the early 1930s. At that time, Stern using *Drosophila,*[25] and Creighton and McClintock using maize,[11] successfully demonstrated that the presence or absence of specific traits in those two organisms directly corresponded to the presence or absence of certain chromosomal areas. These chromosomal regions were structurally unique, enabling one to follow their distribution. By the middle of the 1930s, the chromosome theory of inheritance was firmly established as an experimentally supported concept of major importance to the field of genetics. The genes were no longer conceptual entities completely lacking physical dimension. Instead, genes became a region of the chromosome, and gene transmission became a function of the distribution of chromosomes during mitosis and meiosis. Therefore, the deviations from a 9:3:3:1 ratio, observed by Bateson and Punnett, could be explained in terms of the chromosomal distance between two sets of genes linked to the same chromosome.

REFERENCES

1. Bateson, W., and R. C. Punnett. 1911. On gametic series involving reduplication of certain terms. *J. of Gen.* 1:239–302.

2. Bateson, W., and E. R. Saunders. 1902. Experimental studies in the physiology of heredity. From *Reports to the Evolution Committee of the Royal Society,* 1:1–160.

3. Bateson, W., E. R. Saunders, and R. C. Punnett. 1904. Experimental studies in the physiology of heredity. From *Reports to the Evolution Committee of the Royal Society* 2:1–53.

4. Beneden, E. 1883. Researches on the maturation of the egg and fertilization. *Archives de Biologie* 4:265–640.

5. Bridges, C. 1914. Direct proof through nondisjunction that the sex-linked genes of *Drosophila* are borne by the X chromosome. *Science* 40:107–109.

6. Bridges, C. 1916. Nondisjunction as proof of the chromosome theory of heredity. *Genetics* 1:1–52, 107–163.

7. Carlson, E. A. 1966. *The gene: a critical history.* Philadelphia: Saunders.

8. Castle, W. 1905. Recent discoveries in heredity and their bearing on animal breeding. *Pop. Sci. Mon.* 66:193–208.

9. Castle, W., and G. M. Allen. 1903. Mendel's law and the heredity of albinism. *Mark's Anniversary Volume* 379–398. Cambridge: Harvard University Press.

10. Correns, C. 1900. *G. Mendels Regel über das Verhalten der Nachkommenschaft der Rassenbastarde. Ber. Dtsch. Bot. Ges.* 18:158–168.

11. Creighton, H. B., and B. McClintock. 1931. A correlation of cytological and genetical crossing over in *Zea mays. Science* 17:492–97.

12. de Vries, H. 1900. *Sur le loi de disjonction des hybrides. C. R. Acad. Sci., Par.* 130:845–47.

13. Dunn, L. C. 1965. *A short history of genetics.* New York: McGraw-Hill.

14. East, E. M. 1910. A Mendelian interpretation of variation that is apparently continuous. *Am. Nat.* 44:633–95.

15. Flemming, W. 1879. Contributions to the knowledge of the cell and its life appearance. *Ark. für Mik. Anat.* 16:302–406.

16. Haldane, J. B. S. 1954. *The biochemistry of genetics.* London: Allen & Unwin.

17. Hardy, G. 1908. Mendelian proportions in a mixed population. *Science* 28:49–50.

18. Hertwig, O. 1885. The problem of fertilization and isotropy of the egg, a theory of inheritance. *Jen. Zeit.* 18:276–318.

19. Lindegren, C. C. 1936. A six-point map of the sex-chromosome of *Neurospora crassa. J. of Gen.* 52:243–56.

20. McClung, C. E. 1902. Notes on the accessory chromosome. *Anat. Anz.* 20:220–226.

21. Mendel, G. J. 1865. *Versuche über pflanzen-hybriden* (Experiments on plant hybrids). *Verh. naturf. Vers. in Brünn* 4:3–47.

22. Morgan, T. H. 1910. Chromosomes and heredity. *Am. Nat.* 44:449–96.

23. Morgan, T. H. 1910. Sex-limited inheritance in *Drosophila. Science* 32:120–22.

24. Nilsson-Ehle, H. 1909. *Kreuzungsuntersuchungen an Hafer und Weizen. Lunds Universit. Arsskr., N. F. Afd.* 2:1–122.

25. Stern, C. 1931. *Zytolotisch-genetische Untersuchungen ab Beweise für die Morgansche Theorie des Factorenaustauchs. Biol. Zent.* 51:547–87.

26. Stevens, N. M. 1905. Studies in spermatogenesis with especial reference to the "accessory chromosome". *Carn. Inst. Wash., publ.* 36:1–33.

27. Strasburger, E. 1884. New investigations on the course of fertilization in the phanerogams as basis for a theory of inheritance. From *Neue Untersuchungen über den Befruchtungsvorgang bei den Phanerogamen als Grundlage für eine Theorie der Zeugung.* Jena: G. Fischer.

28. Stubbe, H. 1972. *History of genetics.* Cambridge, London: The MIT Press.

29. Sturtevant, A. H. 1913. The linear arrangement of six sex-linked factors in *Drosophila,* as shown by their mode of association. *J. Exp. Zool.* 14:43–59.

30. Sturtevant, A. H. 1965. *A history of genetics.* New York: Harper and Row.

31. Sutton, W. S. 1903. The chromosomes in heredity. *Biol. Bull.* 4:231–51.

32. Tchermak, E. 1900. *Über Kunstliche Kreuzung bei Pisum sativum. Ber. Dtsch. Bot. Ges.* 18:232–39.

33. Weinberg, W. 1908. On the demonstration of heredity in man. *Natur. in Würt., Stut.* 64:368–82.

34. Weismann, A. 1891. The continuity of the germ plasm as the foundation of a theory of heredity. From *Essays Upon Heredity and Kindred Biological Problems,* vol. 1.

35. Wilson, E. B. 1909. Recent researches on the determination and heredity of sex. *Science, N.S.* 29:53–70.

The law of segregation of hybrids

Das Spaltungsgesetz der Bastarde (Preliminary Communication)*

HUGO DE VRIES (Translated by Evelyn Stern)

In 1865, in the face of 23 centuries of spontaneous and controlled hybridization data, none of which was well interpreted, Mendel presented his pea plant hybridization data. From his data, Mendel extracted experimental support for the existence of stable units of inheritance. He argued that these units maintained their structural integrity when two units controlling alternative forms of the same expression were present in the same plant. Even though associated closely in the plant, each unit would segregate from the other during gametogenesis. Why did Mendel place more faith in his interpretations than in other interpretations which were based upon the extensive catalogue of previous hybridization data — all of which seemed to contradict his interpretations? He clearly explained the reasons for his success and his predecessors' failures. First, his experimentation necessitated the use of pure breeding parents. These parents had to differ significantly in respect to one or more characteristics. The differences had to be distinct and easily traceable from one generation to the next. Second, careful records had to be kept for each generation in respect to the presence or absence of different expressions and the number of plants displaying these expressions. Careful adherence to the above protocol yielded results that supported the existence of a system of inheritance in which separate, stable units control individually analyzable characteristics.

As pointed out in the introduction to this part, the general scientific community did not share Mendel's faith in his work. Thirty-five years passed before his work was discovered and appreciated for its importance to the study of inheritance. In 1900 de Vries, one of the three researchers who reobtained Mendel's results, was prepared to grasp the significance of the data. Guided by his interest in Darwin's proposal of the existence of gemmules that circulated in the body and controlled the structure and function of the organism, de Vries was convinced of the need for a genetic system founded upon separate units of control. Darwin had proposed that the gemmules could be altered by environmental effects, and that these gemmules accumulated in the germinal tissue via the circulatory system and were passed on to the zygote at conception. In the zygote they directed the gradual emergence of the new individual. The progeny could display different characteristics than those of their parents — possibly because of environmental modification of the gemmules. Strongly influenced by Darwin's hypothesis, de Vries presented a modified model incorporating the idea of separate units of inheritance that were not free to circulate throughout the body but were restricted to the cell's nucleus. He designated his hypothetical model *intracellular pangenesis*. He postulated that the units of inheritance (*pangenes*) were shielded from environmental change. The modifi-

[The original paper was published in Berichte der deutschen botanischen Gesellschaft 18 (1900): 83–90. *Submitted March 14, 1900.*]

*Detailed description of my experiments and the theoretical analysis I intend to publish in a rather extensive work on the empirical units of species traits and their origin: *Mutation Theory*.

cations of the original Darwinian model were introduced in light of cytological evidence which had localized the hereditary substance to the nucleus and of experimental data which had rejected the concept of acquired characteristics.

By the turn of the century, in addition to those in de Vries' hypothesis, additional physiological processes were linked to hypothetical regulatory systems based upon separate units of control: Weismann's "biophores," Hertwig's "idioblasts," Nageli's "micella groups." These speculations and cytological work that suggested the existence of a constant number of structurally discrete units in the nuclei (*chromosomes*) provided the ideal climate for the reintroduction of Mendel's experimental results. De Vries presented in this paper — one of the three works published in 1900 that experimentally supported the discrete nature of the hereditary elements controlling specific characteristics — the same proposal advanced 35 years earlier by Gregor Mendel.

According to pangenesis the total character of a plant is built up of distinct units. These so-called elements of the species, or its elementary characters, are conceived of as tied to bearers of matter, a special form of material bearer corresponding to each individual character.[1] Like chemical molecules, these elements have no transitional stages between them.

For many years this principle has represented the starting point for my investigations. Many important consequences can be deduced from it and may be tested experimentally. My experiments lie in part in the realm of variability[2] and mutability and in part in that of hybridization.

In this latter area, however, a complete change in the viewpoint from which the investigation proceeds is necessary. The altered viewpoint requires that *"the concept of species recede into the background in favor of the consideration of a species as a composite of independent factors."*[3]

Current doctrine regarding hybrids considers the species, the subspecies, and the varieties as the units whose combinations create hybrids and which should be studied. One distinguishes between mixtures of varieties and the true species hybrids. Depending on the number of parental types one speaks of diphyletic or polyphyletic hybrids, of triple or quadruple hybrids, and so on.

In my opinion this way of looking at the problem should be abandoned in physiological investigation, for although it suffices for systematic and horticultural purposes, it is inadequate for the purpose of obtaining more basic knowledge of species.

The *principle of the crossing of species-specific traits* should replace it. The units of species-specific traits are to be seen in this connection as sharply separate entities and should be studied as such. They should be treated as independent of each other everywhere, as long as there is no basis for doing otherwise. In every crossing experiment only a single character or a definite number of them is to be taken into consideration: the others can be disregarded temporarily. Or rather, it is a matter of indifference whether the parents are distinguishable from each other in still further ways. However, for experimental purposes the simplest conditions are presented by hybrids whose parents differ from each other in one trait only (*monohybrids,* in contrast to the *di-* or *polyhybrids*).

If the parents of a hybrid differ from each other in one point only, or if only one or a few of their points of difference are selected for consideration, in these characteristics they are *antagonistic,* while in all other respects they are alike or indifferent for the analysis. The crossing experiment is thereby limited to the antagonistic characteristics.

My experiments have led me to make the two following statements:*

1. *Of the two antagonistic characteristics, the hybrid carries only one,* and that in complete development. Thus in this respect the hybrid is indistinguishable from one of the two parents. There are no transitional forms.

2. *In the formation of pollen and ovules the two antagonistic characteristics separate,* following for the most part simple laws of probability.

---

*The "false hybrids" of Millardet are temporarily disregarded altogether in what follows.

These two statements, in their most essential points, were drawn up long ago by Mendel for a special case (peas).[4] These formulations have been forgotten and their significance misunderstood.[5] As my experiments show, they possess generalized validity for true hybrids.

The lack of transitional forms between any two simple antagonistic characters in the hybrid is perhaps the best proof that such characters are well-delimited units.[6]

And to support the correctness of this statement innumerable cases can be advanced, partly from my own experience, and partly from the literature. The fact that polyhybrids so frequently represent intermediate forms obviously depends on the fact that they inherited one part of their traits from the father and the other part from the mother. In monohybrids such is not possible.

Of the two antagonistic characters, Mendel calls the one visible in the hybrid the *dominating,* the latent one *recessive.*

Ordinarily the character higher in the systematic order is the dominating one, or, in cases of known ancestry, it is the older one. For example,

dominating	recessive
Papaver somniferum (tall form)	P. s. nanum
Antirrhinum majus, red	A. m. album.
Polemonium coeruleum, blue	P. c. album.

And where the ancestral background is known, for example:

dominating	recessive	known since
Chelidonium majus	C. laciniatum	±1590
Oenothera Lamarckiana	O. brevistylis	±1880
Lychnis vespertina (hairy)	L. v. glabra	±1880

Using this rule analogously in other cases, at times one arrives at contradictions of the prevailing systematic interpretation, for example:

dominating	recessive
Datura Tatula	D. Stramonium
Zea Mays (naked seed)	Z. cryptosperma

In species hybrids (polyhybrids), where the relative ages of the parental forms are usually unknown, possibly conclusions may be drawn from crossing experiments, for example, with reference to flower color:

dominating	recessive
Lychnis diurna (red)	L. vespertina (white)

THE LAW OF SEGREGATION OF HYBRIDS

In the hybrid the two antagonistic characters lie next to each other as anlagen. In vegetative life only the dominating one is usually visible. Exceptions occur seldom; an example is presented by some sectional segregations. Thus *Veronica longifolia* (blue) × *V. longifolia alba* in my experiments not infrequently forms racemes whose flowers are white on one side and blue on the other.

In the formation of pollen grains and ovules these characters separate. The individual pairs of antagonistic characters behave independently during this process. From this separation the law can be deduced:

The pollen grains and ovules of monohybrids are not hybrids but belong exclusively to one or the other of the two parental types. The same holds for di- to polyhybrids with reference to each character taken by itself.*

The composition of the progeny can be calculated from this statement, and by means of this calculation the validity of the statement can be proven experimentally. In the simplest case segregation will take place obviously into two equal halves and so one obtains:

50% dom. + 50% rec. pollen grains, and
50% dom. + 50% rec. ovules

If dominating is designated by d and recessive by $r,$ fertilization yields:

$$(d + r)(d + r) = d^2 + 2dr + r^2$$

or

$$25\% \, d + 50\% \, dr + 25\% \, r$$

The individual d and d^2 have only the dominating character, those of r and r^2 constitution possess only the recessive character, while the dr plants are obviously hybrids.

---

*The combinations take place according to probability calculus.

In self-fertilization, whether this takes place in isolation or in groups, the hybrids of the first generation yield, with reference to each single trait,

25% of individuals with the paternal trait
25% of individuals with the maternal trait
50% of individuals that are again hybrids

According to our first main statement the hybrids possess the dominating trait, so that one obtains

75% of individuals with the dominating trait
25% of individuals with the recessive trait

I found this composition confirmed in many experiments. For example:

in 1896 with *Papaver somniferum* Mephisto × Danebrog and obtained accordingly for the composition of the first generation of 1897 the following:

24% Dominating (Mephisto)
51% Hybrids (with ±25% Danebrog)
25% Recessive (Danebrog)

This result is concordant with the formula cited above, or more correctly expressed, it was from these numerical relations that I first deduced the formula.

The dominating and the recessive traits are shown to be constant in the progeny, so far as they were isolated by segregation. However, the hybrids segre-

A. Following artificial crossing:

Dominating	Recessive	Rec.	Year of crossing
Agrostemma Githago	nicaeenis	24%	1898
Chelidonium majus	laciniatum	26%	1898
Hyoscyamus niger	pallidus	26%	1898
Lychnis diurna	L. vespert. (white)	27%	1892
Lychnis vespertina (hairy)	glabra	28%	1892
Oenothera Lamarckiana	brevistylis	22%	1898
Papaver somnif. Mephisto	Danebrog	28%	1893
Papaver somnif. nanum (simple)	filled	24%	1894
Zea Mays (starchy)	saccharata	25%	1898

B. Following free crossing, for example:

Dominating	Recessive	Rec.	Year of crossing
Aster Tripolium	album	27%	1897
Chrysanthemum Roxburghi (yellow)	album	23%	1896
Coreopsis tinctoria	brunnea	25%	1896
Solanum nigrum	chlorocarpum	24%	1894
Veronica longifolia	alba	22%	1894
Viola cornuta	alba	23%	1899

The mean of all these experiments is 24.93%.

The experiments usually included several hundred plants and at times about a thousand. I obtained corresponding results with many other species.

Distinguishing the remaining 75% into the two groups listed is much more troublesome. This requires that a number of plants bearing the dominating trait be fertilized with their own pollen and that in the succeeding year the progeny be cultivated and counted for each plant. I carried out this experiment

gated again according to the same law. In this experiment they yielded an average of 77% with the dominating and 23% with the recessive trait.

Such behavior remains unchanged during the course of years. I extended this experiment through two further generations. The 50% of hybrids segregated out while the 25% with dominating trait remained constant.

From the main statement of the law of segregation

various other inferences may be drawn, by means of which experimental testing becomes possible.

For example, by fertilizing a hybrid with the pollen of one of the two parents, or, in reverse, fertilizing one of the parental types with the hybrid, one obtains:

$$(d + r)d = d^2 + dr$$

and

$$(d + r)r = dr + r^2$$

Thus in the first case some plants that develop are hybrids, some pure forms, but all of them exhibit the dominating trait. In the second case there are equal numbers of hybrids with the dominating trait and of pure specimens, so that one sees:

50% dominating (hybrids)
50% recessive (pure)

I found, for instance:

	Recessive	Year
Clarkia pulchella **XX** white	50%	1896
Oenothera Lamarckiana **XX** brevistylis	55%	1895
Silene Armeria (red) **XX** white	50%	1895

The same law holds also, as already mentioned, if one investigates dihybrids or studies two pairs of antagonistic traits in polyhybrids. I choose as an example a cross of the thorny *Datura Tatula* with *Datura Stramonium inermis* which I made in 1897. In accord with a well-known rule the hybrids are all alike, regardless of which form contributed the ovules and which the pollen. They bloom blue and bear prickly fruit. Some flowers were fertilized by their own pollen and the resulting seed sown in 1899. By the time of germination the blue-blooming plants were distinguishable from the white-blooming ones by the color of the stems. I found:

Blue (domin. + hybr.) 72%
White (recessive) 28%

as was confirmed by the flower. With reference to the fruit there were produced:

Thornless, among the blues 26.8%
Thornless, among the whites 28.0%
Mean 27.4%

From this the composition of the progeny can be calculated for nearly all cases. If, for instance, one pair of antagonistic characters is called A and the other pair B, one obtains for dihybrids:

A. 25% Dom.

B. 6.25 *d*, 12.5 *dr*, 6.25 *r*

A. 50% D × R 25% Rec.

B. 12.5 *d*, 25 *dr*, 12.5 *r* 6.25 *d*, 12.5 *dr*, 6.25 *r*

Thus 6.25% of the cases are pure dominating in both respects, and an equal number pure recessive in both, etc.

If one applies the rule that hybrids exhibit the dominating trait, one finds for the visible characteristics of the progeny:

1. A. dom. + B. rec. 18.75%
2. A. rec. + B. dom. 18.75%
3. A. dom. + B. dom. 56.25%
4. A. rec. + B. rec. 6.25%

As evidence I cite the following experiment. *Trifolium pratense album* was crossed with *Trifolium pratense quinquefolium;* the white flowers and the ternate leaves are recessive to the antagonistic species traits. I found for the progeny of the hybrids in approximately 220 plants:

		Calculated
1. Red and ternate	13%	19%
2. White and pentad	21%	19%
3. Red and pentad	61%	56%
4. White and ternate	5%	6%

In similar fashion calculations and experiments are to be applied to tri- to polyhybrids.

Success is frequently had in separating simple characters into a number of factors by means of segregation. For example, the color of flowers is often composite, and after crossing one obtains the individual factors, partly separate and partly in various mixtures. I have carried out such analyses with *Antirrhinum majus, Silene Armeria,* and *Brunella vulgaris* and found in them the above numerical relationships confirmed. *Antirrhinum majus* red, for example, may be split by crossing with white into both these colors and into yellow with red (Brilliant) and white with red (Delila); *Silene Armeria* may be split into red, pink, and white. *Brunella vulgaris* forms a constant intermediate form with a white flower and a brown calyx.

From these and numerous other experiments I draw the conclusion that the law of segregation of hybrids as discovered by Mendel for peas finds very general application in the plant kingdom and that it has a basic significance for the study of the units of which the species character is composed.

REFERENCES

1. [Hugo de Vries] Intracelluläre Pangenesis [1889] pp. 60–75. For the opposite point of view, that every bearer of matter represents the total species character, compare pp. 47–60 of *Mutation Theory*.

2. Berichte der deutschen botanischen Gesellschaft 12 (1894): 197.

3. Intracelluläre Pangenesis, p. 25.

4. Gregor Mendel, Versuche über Pflanzen-Hybriden, in Verhandlungen des naturforschenden Vereines in Brünn 4 (1865): 3. This important treatise is so seldom cited that I first learned of its existence after I had completed the majority of my experiments and had deduced from them the statements communicated in the text.

5. See G. and A. Focke, Die Pflanzenmischlinge, p. 110. [Reference is to W. O. Focke, Die Pflanzen-Mischlinge, Eds.]

6. Intracelluläre Pangenesis.

A Mendelian interpretation of variation that is apparently continuous

EDWARD M. EAST[*][†]

Reprinted with permission from American Naturalist, *vol. 44, 1910, pp. 65–82. Copyright © 1910 by the University of Chicago.*

In 1866 there was no experimentation to defend or refute Mendel's observations; in 1900, there were immediate efforts by large numbers of investigators, utilizing different organisms, to carefully examine the validity and universality of Mendel's two hypotheses. Within three years, information began to appear which questioned the universality of the hypothesis that predicted the segregation of hereditary units during gametogenesis in hybrids. In England, in 1905, William Castle based his disagreement on his and other investigators' inability to obtain pure coat colors in the F_2 generation from crosses involving F_1 hybrid guinea pigs. Instead, F_2 offspring with modified parental coat colors were consistently bred from the cross of F_1 hybrid guinea pigs. Castle suggested that at least for some characteristics there was a lack of complete segregation of the hereditary units housed within the hybrid. He proposed that those characters expressing distinct alternatives would undergo a Mendelian type of segregation (*Mendelize*), while those characters displaying a spectrum of intermediate expressions would undergo a different, more complex system of dispersal. He called for the reinstitution of a contamination mechanism of inheritance in conjunction with the concept of segregation of discrete units of inheritance at the time of gametogenesis.

Nilsson-Ehle, working in Sweden, and East, working in the United States, rejected Castle's interpretation. From their tests they hypothesized that particular characters were controlled by more than one set of "factors" (*units of inheritance*). This multiple-factor hypothesis stipulated that when different numbers of pairs of genetic factors, controlling the same characteristic, were present in different individuals, quantitatively different expressions would result. However, during the formation of gametes, each pair of factors would segregate in a Mendelian fashion. So, the intermediate expression in the coat colors of guinea pigs did not negate the universality of segregation; in fact, it opened the door to the realization of the existence of polygenic systems and the study of quantitative inheritance.

There are two objects in writing this paper. One is to present some new facts of inheritance obtained from pedigree cultures of maize; the other is to discuss the hypothesis to which an extension of this class of facts naturally leads. This discussion is to be regarded simply as a suggestion toward a working hypothesis, for the facts are not sufficient to support a theory. They do, however, impose certain limitations upon speculation which should receive careful consideration.

The facts which are submitted have to do with independent allelomorphic pairs which cause the formation of like or similar characters in the zygote. Nilsson-Ehle[1] has just published facts of the same character obtained from cultures of oats and of wheat. My own work is largely supplementary to his, but it had been given these interpretations previous to the publication of his paper.

Contributions from the Laboratory of Genetics, Bussey Institution, Harvard University, No. 4. Read before the annual meeting of the American Society of Naturalists, Boston, December 29, 1909.
[*]Harvard University, Cambridge, Massachusetts.
[†]Deceased.

In brief, Nilsson-Ehle's results are as follows: He found that while in most varieties of oats with black glumes blackness behaved as a simple Mendelian monohybrid, yet in one case there were two definite independent Mendelian unit characters, each of which was allelomorphic to its absence. Furthermore, in most varieties of oats having a ligule, the character behaved as a monohybrid dominant to absence of ligule, but in one case no less than four independent characters for presence of ligule, each being dominant to its absence, were found. In wheat a similar phenomenon occurred. Many crosses were made between varieties having red seeds and those having white seeds. In every case but one the F_2 generation gave the ordinary ratio of three red to one white. In the one exception — a very old red variety from the north of Sweden — the ratio in the F_2 generation was 63 red to 1 white. The reds of the F_2 generation gave in the F_3 generation a very close approximation to the theoretical expectation, which is 37 constant red, 8 red and white separating in the ratio of 63:1, 12 red and white separating in the ratio of 15:1, 6 red and white separating in the ratio of 3:1, and one constant white. He did not happen to obtain the expected constant white, but in the total progeny of 78 F_2 plants his other results are so close to the theoretical calculation that they quite convince one that he was really dealing with three indistinguishable but independent red characters, each allelomorphic to its absence. Nor can the experimental proof of the two colors of the oat glumes be doubted. The evidence of four characters for presence of ligule in the oat is not so conclusive.

In my own work there is sufficient proof to show that in certain cases the endosperm of maize contains two indistinguishable, independent yellow colors, although in most yellow races only one color is present. There is also some evidence that there are three and possibly four independent red colors in the pericarp, and two colors in the aleurone cells. The colors in the aleurone cells when pure are easily distinguished, but when they are together they grade into each other very gradually.

Fully fifteen different yellow varieties of maize have been crossed with various white varieties, in which the crosses have all given a simple monohybrid ratio. In the other cases that follow it is seen that there is a dihybrid ratio.

No. 5–20, a pure white eight-rowed flint, was pollinated by No. 6, a dent pure for yellow endosperm. An eight-rowed ear was obtained containing 159 medium yellow kernels and 145 light yellow kernels. The pollen parent was evidently a hybrid homozygous, for one yellow which we will call Y_1 and heterozygous for another yellow Y_2. The gametes $Y_1 Y_2$ and Y_1 fertilized the white in equal quantities, giving a ratio of approximately one medium yellow to one light yellow. The F_2 kernels from the dark yellow were as shown in Table 1. The ratios of light yellows to dark yellows is very arbitrary, for there was a fine gradation of shades. The ratio of total yellows to white, however, is unmistakably 15:1.

In the next table (Table 2) are given the results of F_2 kernels from the light yellows of F_1. Only ear No. 8, which was really planted with the dark yellows, showed yellows dark enough to be mistaken for kernels containing both Y_1 and Y_2. The remaining ears are clearly monohybrids with reference to yellow endosperm.

In a second case the female parent possessed the yellow endosperm. No. 11, a twelve-rowed yellow flint, was crossed with No. 8, a white dent. The F_2 kernels in part showed clearly a monohybrid ratio, and in part blended gradually into white. Two of these indefinite ears proved in the F_3 generation to have had the 15:1 ratio in the F_2 generation. Ear 7 of the F_2 generation calculated from the results of the entire F_3 crop must have had about 547 yellow to 52 white kernels, the theoretical number being

Table 1* F_2 seeds from cross of no. 5–20, white flint × no. 6 yellow dent, homozygous for Y_1 and heterozygous for Y_2

Ear No.	Dark Y	Light Y	Total Y	No Y
		Dark seeds heterozygous for both yellows planted		
1	270	56	326	29
2	101	215	316	27
3	261	52	313	28
5	273	284	557	35
10	358	117	475	25
12	296	72	368	19
13	207	156	363	35
14	387	102	489	29
Total	2153	1054	3207	227
Ratio			14.1	1

*In these tables only hand pollinated ears are given.

Table 2 F_2 seeds from same cross as shown in Table 1

	Light yellow seeds heterozygous for Y_1 planted		
Ear No.	Dark Y	Light Y	No Y
1		359	117
2		144	54
3		173	63
4		433	136
6		316	120
8	331		109
8a		229	86
9		325	115
10		227	87
11*		4	434
12		318	118
13		256	93
Total		3111	1098
Ratio		2.8	1

*Discarded from average. This ear evidently grew from one kernel of the original white mother that was accidentally self-pollinated. The four yellow kernels all show zenia from accidental pollination in the next generation.

561 to 31. The hand-pollinated ears of the F_3 generation (yellow seeds) gave the results shown in Table 3.

The F_3 generation grown from the other ear, Ear No. 8, showed that the ratio of yellows to whites in the F_2 generation was about 227 to 47. As the theoretical ratio is 257 to 17, the ratio obtained is somewhat inconclusive. A classification of the open field crop could not be made accurately on account of the light color of the yellows and the presence of many kernels showing zenia. Table 4, however, showing the hand-pollinated kernels of the interbred yellows of the F_2 generation, settles beyond a doubt the fact that the two yellows were present.

In a third case an eight-rowed yellow flint, No. 22, was crossed with a white dent, No. 8. Only four selfed ears were obtained in the F_2 generation. Ear 1 had 72 yellow to 37 white kernels. This ear was poorly developed and undoubtedly had some yellow kernels which were classed as whites. Ear 4 had 158 yellow and 42 white kernels. It is very likely that both of these ears were monohybrids, but the F_3 generation was not grown. Ear 5 had 148 yellow and 15 white kernels. Ear 7 had 78 yellow and 5 white kernels. It seems probable that both of these

ears were dihybrids, but only Ear 5 was grown another generation. The kernels classed as white proved to be pure; the open field crop from the yellow kernels gave 14 pure yellow ears and 14 hybrid yellow. Theoretically the ratio should be 7 pure yellows (that is, pure for either one or both yellows) and 8 hybrid yellows (4 giving 15 yellows to 1 white and 4 giving 3 yellows to 1 white). Five hand-pollinated selfed ears were obtained. Three of these gave monohybrid ratios, with a total of 607 yellows to 185 white kernels. One ear was a pure dark yellow (probably $Y_1 Y_1 Y_2 Y_2$). The other ear was poorly filled, but had 27 dark yellows (probably $Y_1 Y_2$) and 7 light yellow kernels (Y_1 or Y_2). Unfortunately no 15:1 ratio was obtained in this generation, but this is quite likely to happen when only five selfed ears are counted. The gradation of colors and the general appearance of the open field crop, however, led me to believe that we were again dealing with a dihybrid.

Two yellows appeared in still another case, that of white sweet No. 40♀ × yellow dent No. 3♂. Only one selfed ear was obtained in the F_2 generation giving 599 yellow to 43 white kernels. Of these kernels 486 were starchy and 156 sweet, which complicated matters in the F_3 generation because it was very difficult to separate the light yellow sweet from the white sweet kernels. Among the selfed ears were three pure to the starchy character, and in these ears the dark yellows, the light yellows and whites stood out very distinctly. Ear 12 had 156 dark yellow; 47 light yellow; 14 white kernels. Ear 13 had 347 dark yellow; 93 light yellow; 25 white kernels. The third

Table 3 No. 11 yellow × no. 8 white

	F_3 generation from yellow seeds of F_2 generation				
Ear No.	Dark Y	Light Y	Total Y	No Y	Ratio they approximate
1	116	95	211	19	$15Y$:1 no Y
14			88	5	$15Y$:1 no Y
5	181	122			$3Y_1 Y_2$:1 $Y_{1 \text{ or } 2}$
4		253		68	$3Y$:1 no Y
6		193		73	$3Y$:1 no Y
8		163		79	$3Y$:1 no Y
11		108		35	$3Y$:1 no Y
9		456			Constant $Y_{1 \text{ or } 2}$

Table 4 Progeny of ear no. 8 of the same cross as shown in Table 3

					F₃ generation from yellow seeds of F₂ generation				

Ear No.	Dark Y	Light Y	Total Y	No Y	Ratio they approximate
10	101	188	289	25	15Y:1 no Y
11	89	219	308	23	15Y:1 no Y
3		233			constant light Y
9	dark and light		331		3 dark: 1 light Y
13	dark and light		350		3 dark: 1 light Y
8		294		108	3 light: 1 no Y
15		221		87	3 light: 1 no Y
1*		197		203	

*Kernel from which this ear grew was evidently pollinated by no Y.

starchy ear, No. 6, had 320 light yellow; 97 white kernels. Two ears, therefore, were dihybrids, and one ear a monohybrid.

The ears which were heterozygous for starch and no starch and those homozygous for no starch, could not all be classified accurately, but it is certain that some pure dark yellows, some pure light yellows, some showing segregation of yellows and whites at the ratio 15:1, and some showing segregation of yellows and whites at the ratio of 3:1, were obtained.

One other case should be mentioned. One ear of a dent variety of unknown parentage obtained for another purpose was found to have some apparently heterozygous yellow kernels. Seven selfed ears were obtained from them, of which two were pure yellow. The other five ears each gave the dihybrid ratio. There was a total of 1906 yellow seeds to 181 white seeds, which is reasonably close to the expected ratio, 1956 yellow to 131 white.

It is to be regretted that I can present no other case of this class that has been fully worked out, although several other characters which I have under observation in both maize and tobacco seem likely to be included ultimately. Nevertheless, the fact that we have to deal with conditions of this kind in studying inheritance is established; granting only that they will be somewhat numerous, it opens up an entirely new outlook in the field of genetics.

In certain cases it would appear that we may have several allelomorphic pairs each of which is inherited independently of the others, and each of which is separately capable of forming the same character. When present in different numbers in different individuals, these units simply form quantitative differences. It may be objected that we do not know that two colors that appear the same physically are exactly the same chemically. That is true; but Nilsson-Ehle's case of several unit characters for presence of ligule in oats is certainly one where each of several Mendelian units forms exactly the same character. It may be that there is a kind of biological isomerism, in which, instead of molecules of the same formula having different physical properties, there are isomers capable of forming the same character, although, through difference in construction, they are not allelomorphic to each other. At least it is quite a probable supposition that through imperfections in the mechanism of heredity an individual possessing a certain character should give rise to different lines of descent so that in the F_n generation when individuals of these different lines are crossed, the character behaves instead of as a monohybrid. In other words, it is more probable that these units arise through variation in different individuals and are combined by hybridization, than that actually different structures for forming the same character arise in the same individual.

On the other hand, there is a possibility of an action just the opposite of this. Several of these quantitative units which produce the same character may become attached like a chemical radical and again behave as a single pair. Nilsson-Ehle gives one case which he does not attempt to explain, where the same cross gave a 4:1 ratio in one instance and 8:4:1 ratio in another instance. In his other work characters always behaved the same way; that is, either as one pair, two pairs, three pairs, etc. In my work, the yellow endosperm of maize has behaved differently in the same strain, but it is probably because the yellow parent is homozygous for one yellow and heterozygous for the other. They were known to be pure for one yellow, but it would take a long series of crosses to prove purity in two yellows.

Let us now consider what is the concrete result of the interaction of several cumulative units affecting the same character. Where there is simple presence dominant to absence of a number n of such factors, in a cross where all are present in one parent and all absent in the other parent, there must be 4^n individuals to run an even chance of obtaining a single F_2 individual in which the character is absent. When four such units, $A_1A_2A_3A_4$ are crossed with $a_1a_2a_3a_4$, their absence, only one pure recessive is expected in 256 individuals. And 256 individuals is a larger number than is usually reported in genetic publications. When a smaller population is considered, it will appear to be a blend of the two parents with a fluctuating variability on each side of its mode. Of course if there is absolute dominance and each unit appears to affect the zygote in the same manner that they do when combined, the F_2 generation will appear like the dominant parent unless a very large number of progeny are under observation and pure recessives are obtained. This may be an explanation of the results obtained by Millardet; it is certainly as probable as the hypothesis of the nonformation of homozygotes. Ordinarily, however, there is not perfect dominance, and variation due to heterozygosis combined with fluctuating variation makes it almost impossible to classify the individuals except by breeding. The two yellows in the endosperm of maize is an example of how few characters are necessary to make classification difficult. First, there is a small amount of fluctuation in different ears due to varying light conditions owing to differences in thickness of the husk; second, all the classes having different genetic formulae differ in the intensity of their yellow in the following order, $Y_1Y_1Y_2Y_2$, $Y_1y_1Y_2Y_2$ or $Y_1Y_1Y_2y_2$, Y_1Y_1, Y_2Y_2, Y_1y_1, Y_2y_2, y_1y_2. As dominance becomes less and less evident, the Mendelian classes vary more and more from the formula $(3 + 1)^n$, and approach the normal curve, with a regular gradation of individuals on each side of the mode. When there is no dominance and open fertilization, a state is reached in which the curve of variation simulates the fluctuation curve, with the difference that the gradations are heritable.

One other important feature of this class of genetic facts must be considered. If units $A_1A_2A_3a_4$ meet units $a_1a_2a_3A_4$, in the F_2 generation there will be one pure recessive, $a_1a_2a_3a_4$, in every 256 individuals. This explains an apparent paradox. Two individuals are crossed, both seemingly pure for presence of the same character, yet one individual out of 256 is a pure recessive. When we consider the rarity with which pure dominants or pure recessives (for all characters) are obtained when there

are more than three factors, we can hardly avoid the suspicion that here is a perfectly logical way of accounting for many cases of so-called atavism. Furthermore, many apparently new characters may be formed by the gradual dropping of these cumulative factors without any additional hypothesis. For example, in *Nicotiana tabacum* varieties there is every gradation* of loss of leaf surface near the base of the sessile leaf, until in *N. tabacum fruticosa* the leaf is only one step removed from a petioled condition. If this step should occur the new plant would almost certainly be called a new species; yet it is only one degree further in a definite series of loss gradations that have already taken place. If it should be assumed that in other instances slight qualitative as well as quantitative changes take place as units are added, then it becomes very easy, theoretically, to account for quite different characters in the individual homozygous for presence of all dominant units, and in the individual in which they are all absent.

Unfortunately for these conceptions, although I feel it extremely probable that variations in *some* characters that seem to be continuous will prove to be combinations of segregating characters, it is exceedingly difficult to demonstrate the matter beyond a reasonable doubt. As an illustration of the difficulties involved in the analysis of pedigree cultures embracing such characters, I wish to discuss some data regarding the inheritance of the number of rows of kernels on the maize cob.

The maize ear may be regarded as a fusion of four or more spikes, each joint of the rachis bearing two spikelets. The rows are, therefore, distinctly paired, and no case is known where one of the pair has been aborted. This is a peculiar fact when we consider the great number of odd kinds of variations that occur in nature. The number of rows per cob has been considered to belong to continuous variations by de Vries, and a glance at the progeny from the seeds of a single selfed ear as shown in Table 5 seems to confirm this view.

There is considerable evidence, however, that this character is made up of a series of cumulative units, independent in their inheritance. There is no reason why it should not be considered to be of the same nature as various other size characters in which

---

*It is not known at present how this character behaves in inheritance.

Table 5 Progeny of a selfed ear of leaming maize having 20 rows

Classes of rows	12	14	16	18	20	22	24	26	28	30
No. of ears	1	0	5	4	53	35	19	5	2	1

variation seems to be continuous, but in which relatively constant gradations may be isolated, each fluctuating around a particular mode. But this particular case possesses an advantage not held by most phenomena of its class, in that there is a definite discontinuous series of numbers by which each individual may be classified.

Previous to analyzing the data from pedigree cultures, however, it is necessary to take into consideration several facts. In the first place, what limits are to be placed on fluctuations?* From the variability of the progeny of single ears of dent varieties that have been inbred for several generations, it might be concluded that the deviations are very large. But this is not necessarily the case; these deviations may be due largely to gametic structure in spite of the inbreeding, since no conscious selection of homozygotes has been made. There is no such variation in eight-rowed varieties, which may be considered as the last subtraction form in which maize appears and therefore an extreme homozygous recessive. In a count of the population of an isolated maize field where Longfellow, an eight-rowed flint, had been grown for many years, 4 four-rowed, 993 eight-rowed, 2 ten-rowed and 1 twelve-rowed ears were found. Only seven aberrant ears out of a thousand had been produced, and some of these may have been due to vicinism.

On the other hand, a large number of counts of the number of rows of both ears on stalks that bore two ears has shown that it is very rare that there is a change greater than ±2 rows. If conditions are more favorable at the time when the upper ear is laid down it will have two more rows than the second ear; if conditions are favorable all through the season, the ears generally have the same number of rows; while if conditions are unfavorable when the upper ear is laid down, the lower ear may have two more

rows than the upper ear. Furthermore, seeds from the same ear have several times been grown on different soils and in different seasons, and in each case the frequency distribution has been the same. Hence it may be concluded that in the great majority of cases fluctuation is not greater than in ± 2 rows, although fluctuations of ± 4 rows have been found.

A second question worthy of consideration is: Do somatic variations due to varying conditions during development take place with equal frequency in individuals with a large number of rows and in individuals with a small number of rows? From the fact that several of my inbred strains that have been selected for three generations for a constant number of rows, increase directly in variability as the number of rows increases, the question should probably be answered in the negative. This answer is reasonable upon other grounds. The eight-rowed ear may vary in any one of four spikes, the sixteen-rowed ear may vary in any one of eight spikes; therefore the sixteen-rowed ear may vary twice as often as the eight-rowed ear. By the same reasoning, the sixteen-rowed ear may sometimes throw fluctuations twice as wide as the eight-rowed ear.

A third consideration is the possibility of increased fluctuation due to hybridization. Shull[2] and East[3] have shown that there is an increased stimulus to cell division when maize biotypes are crossed — a phenomenon apart from inheritance. There is no evidence, however, that increased gametic variability results. Johannsen[4] has shown that there is no such increase in fluctuation when close-pollinated plants are crossed. I have crossed several distinct varieties of maize where the modal number of rows of each parent was twelve, and in every instance the F_1 progeny had the same mode and about the same variability.

Finally, a possibility of gametic coupling should be considered. Our common races of flint maize all have a low number of rows, usually eight but sometimes twelve; dent races have various modes running from twelve to twenty-four rows. When crosses between the two subspecies are made, the tendency is to separate in the same manner.

Attention is not called to these obscuring factors with the idea that they are universally applicable in the study of supposed continuous variation. But there are similar conditions always present that make analysis of these variations difficult, and the facts given here should serve to prevent premature

---

*The word fluctuation is used to designate the somatic changes due to immediate environment, and which *are not inherited.*

decision that they do not show segregation in their inheritance.

Table 6 shows the results from several crosses between maize races with different modal values for number of rows. Several interesting points are noticeable. The modal number is always divisible by four. This is also the case with some twenty-five other races that I have examined but which are not shown in the table. I suspect that through the presence of pure units zygotes having a multiple of four rows are formed, while heterozygous units cause the dropping of two rows. The eight-rowed races are pure for that character, the twelve-rowed races vary but little, but the races having a higher number of rows are exceedingly variable.

When twelve-rowed races are crossed with those having eight rows, the resulting F_1 generation always — or nearly always — has the mode at twelve rows. In one case cited in Table 6, No. 24 × No. 53, nearly all the F_1 progeny were eight-rowed. It might appear from this, either that the low number of rows was in this case dominant, or that the female parent has more influence on the resulting progeny than the male parent. I prefer to believe, however,

that the individual of No. 53 which furnished the pollen was due to produce eight-rowed progeny. Unfortunately no record was kept of the ear borne by this plant, but No. 53 sometimes does produce eight-rowed ears.

When a race with a mode higher than twelve is crossed with an eight-rowed race, the F_1 generation is always intermediate, although it tends to be nearer the high-rowed parent. Only one example is given in the table, but it is indicative of the class. These results are rather confusing, for there seems to be a tendency to dominance in the twelve-rowed form that is not found in the forms with a higher number of rows. I have seen cultures of other investigators where 12-row × 8-row resulted in a ten-rowed F_1 generation, so the complication need not worry us at present.

The results of the F_2 generation show a definite tendency toward segregation and reproduction of the parent types. I might add that in at least two cases I have planted extracted eight-rowed ears and have immediately obtained an eight-rowed race which showed only slight departures from the type. Selection from those ears having a high number of

Table 6 Crosses between maize strains with different numbers of rows

Parents (female given first)	Gen.	Row classes						
		8	10	12	14	16	18	20
Flint No. 5		100						
Flint No. 11		1	4	387	7	1		
Flint No. 24		100						
Flint No. 15		100						
Dent No. 6				6	31	51	18	4
Dent No. 8			3	54	36	12	2	
Sweet No. 53*		1	5	25	4			
Sweet No. 54*		25	2	1				
No. 5 × No. 53	F_1	1	7	13				
No. 5 × No. 6	F_1	11	18	27	3			
No. 11 × No. 5	F_1	2	4	18				
No. 11 × No. 53	F_1	2	5	17				
No. 24 × No. 53	F_1	57	8	3				
No. 15 × No. 8	F_1	1	14	26	3	1		
No. 15 × No. 8 (from 10-row ear)	F_2	14	15	28	9	1		
No. 15 × No. 8 (from 12-row ear)	F_2	4	13	25	6	3		
No. 8 × No. 54	F_1	1	6	14				
No. 8 × No. 54 (from 12-row ear)	F_2	11	25	38	2	1		

*Approximately.

ows has also given races like the high-rowed parent without recrossing with it. It is regretted that commercial problems were on hand at the time and no exact data were recorded. It can be stated with confidence, however, that ears like each parent are obtained in the F_2 generation, from which with care races like each parent may be produced. *Segregation seems to be the best interpretation of the matter.*

These various items may seem disconnected and uninteresting, but they have been given to show the tangible basis for the following theoretical interpretation. No hard and fast conclusion is attempted, but I feel that this interpretation with possibly slight modifications will be found to aid the explanation of many cases where variation is apparently continuous.

Suppose a basal unit to be present in the gametes of all maize races, this unit to account for the production of eight rows. Let additional independent interchangeable units, each allelomorphic to its own absence, account for each additional four rows; and let the heterozygous condition of any unit represent only half of the homozygous condition, or two rows. Then the gametic condition of a homozygous twenty-rowed race would be 8 + $AABBCC$, each letter actually representing two rows. When crossed with an eight-rowed race, the F_2 generation will show ears of from eight to twenty rows, each class being represented by the number of units in the coefficients in the binomial expansion where the exponent is twice the number of characters, or in this case $(a + b)^6$.

The result appears to be a blend between the characters of the two parents with a normal frequency distribution of the deviants. Only one twenty-rowed individual occurs in 64 instead of the 27 expected by the interaction of three dominant factors in the usual Mendelian ratios. The remainder of the 27 will have different numbers of rows, and, by their gametic formulae, different expectations in future breeding as follows:

$$1\ AABBCC = 20 \text{ rows}$$
$$2\ AaBBCC = 18 \text{ rows}$$
$$2\ AABbCC = 18 \text{ rows}$$
$$2\ AABBCc = 18 \text{ rows}$$
$$4\ AaBbCC = 16 \text{ rows}$$
$$4\ AaBBCc = 16 \text{ rows}$$
$$4\ AABbCc = 16 \text{ rows}$$
$$8\ AaBbCc = 14 \text{ rows}$$

There are four visibly different classes and eight gametically different classes. It must also be remembered that the probability that the original twenty-rowed ear in actual practise may have had more than three units in its gametes has not been considered. This point is illustrated clearly if we work out the complete ratio for the three characters, and note the number of gametically different classes which compose the modal class of fourteen rows in Table 7. It actually contains seven gametically different classes and not a single homozygote. If this conception of independent allelomorphic pairs affecting the same character proves true, it will sadly upset the biometric belief that the modal class is *the type* around which the variants converge, for there is actually less chance of these individuals breeding true than those from *any other* class.

The conception is simple and is capable theoretically of bringing in order many complicated facts, although the presence of fluctuating variation will be a great factor in preventing analysis of data. I have thought of only one fact that is difficult to bring into line. If $8AA$, $8BB$ and $8CC$ all represent homozygous twelve-rowed ears — to continue the maize illustration — and none of these factors are allelomorphic to each other, sixteen-rowed ears should sometimes be obtained when crossing two twelve-rowed ears. I am not sure but that this would happen if we were to extract all the homozygous twelve-rowed strains after a cross between sixteen-row and eight-row, and after proving their purity cross them. In some cases the additional four-row units would probably be allelomorphic to each other and in other cases independent of each other. On the other hand, this is only an hypothesis, and while I have faith in its foundation facts, the details may need change.

Castle has raised the point that greater variation should be expected in the F_1 generation than in the P_1 generations when crossing widely deviating individuals showing variation apparently continuous. If

Table 7 Theoretical expectation in F_2 when a homozygous twenty-rowed maize ear is crossed with an eight-rowed ear

Classes	8	10	12	14	16	18	20
No. ears	1	6	15	20	15	6	1

the parents are strictly pure for a definite number of units, say for size, a greater variation should certainly be expected in the F_1 generation after crossing. But considering the difficulties that arise when even five independent units are considered, can it be said that anything has heretofore been known concerning the actual gametic status of parents which it is known do vary in the character in question and in which the variations are inherited, for the race can be changed by selection within it. It may be, too, that the correct criterion has not been used in size measurements, for, as others have suggested, solids vary as the cube root of their mass, whereas the sum of the weights of the body cells has usually been measured and compared directly with similar sums.

Attention should be called to one further point. Many characters in all probability are truly blending in their inheritance, but there is another interpretation which may apply in certain cases. I have repeatedly tried to cross Giant Missouri Cob Pipe maize (14 feet high) and Tom Thumb pop maize (2 feet high), but have always failed. They both cross readily with varieties intermediate in size, but are sterile between themselves. We may imagine that the gametes of each race, though varying in structure, are all so dissimilar that none of them can unite to form zygotes. Other races may be found where only part of the gametes of varying structure are so unlike that they will not develop after fusion. The zygotes that do develop will be from those more alike in construction. An apparent blend results, and although segregation may take place, no progeny as extreme as either of the parents will ever occur.

I may say in conclusion that the effect of the truth of this hypothesis would be to add another link to the increasing chain of evidence that the word mutation may properly be applied to any inherited variation, however small; and the word fluctuation should be restricted to those variations due to immediate environment which do not affect the germ cells, and which — it has been shown — are not inherited. In addition it gives a rational basis for the origin of *new* characters, which has hitherto been somewhat of a Mendelian stumbling-block; and also gives the term unit-character less of an irrevocably fixed-entity conception, which is more in accord with other biological beliefs.

REFERENCES

1. Nilsson-Ehle, H., Kreuzungsuntersuchungen an Hafer und Weizen. Lunds Universitets Årsskrift, N. F. Afd. 2., Bd. 5, No. 2, 1909.
2. Shull, G. H., "A Pure-line Method in Corn Breeding," Rept. Amer. Breeders' Assn. 5:51–59, 1909.
3. East, E. M., "The Distinction between Development and Heredity in Inbreeding," Amer. Nat. 43:173–181, 1909.
4. Johannsen, W., "Does Hybridization Increase Fluctuating Variability?" Rept. Third Inter. Con. on Genetics, 98–113, London, Spottiswoode, 1907.

On the demonstration of heredity in man
(Über den Nachweis der Vererbung beim Menschen)

W. WEINBERG

Reprinted from Jahreshefte des Vereins für Vaterländische *Naturkunde in Württemberg, Stuttgart.* 1908. 64:368–382. *Lecture at the scientific evening at Stuttgart, January 13, 1908.*

A second objection was voiced by Yule to the validity of a system of inheritance based upon discrete units of control. Unlike Castle, whose argument was founded on empirical results, Yule argued on the basis of speculation. He suggested that if Mendel's segregation hypothesis were valid, the brachydactyl condition, a human disorder caused by a dominant gene, would spread to approximately three-fourths of the human population over a period of time. His speculation was based on the projected 3:1 second generation phenotypic ratio observed for characters possessing distinct alternative expressions having a dominant-recessive relationship. Since it was apparent that brachydactyly was found in an extremely small segment of the population, Yule felt that Mendelian segregation could not apply, in all instances, to man.

Although Castle, in 1903, and Pearson, in 1904, hinted at the error in Yule's logic, the clearest and most comprehensive rebuttals appeared in 1908. In that year, Hardy, a British mathematician, and Weinberg, a German physician, each demonstrated the fallacy in Yule's reasoning. Hardy presented a purely theoretical argument. He demonstrated mathematically that the genotypic frequencies for the distribution of a pair of genes present in a large Mendelian population that reproduced panmictically would remain constant in subsequent generations. Weinberg used data on the frequency of the occurrence of the genetic propensity for human twinning to suggest that a recessive trait is transmitted throughout the population by means of a Mendelian mechanism of transmission. In conjunction with this analysis, he also predicted mathematically that the genotypic frequencies for sets of genes controlling expressions having other than the dominant condition for the abnormal trait (like brachydactyly) would remain constant in subsequent generations. Weinberg's paper not only refuted Yule's argument but expanded the idea of genetic equilibrium in populations to include sets of genes which did not necessarily show a dominant mutant expression.

Both Hardy's and Weinberg's papers refuted Yule's objection to the concept of segregation and provided an introduction into the field of population genetics. In so doing, they helped to establish a theoretical means for predicting gene frequencies in a randomly mating Mendelian population. With this prediction, one could study those factors present in the environment responsible for instituting changes in gene frequency within a population — the raw material of evolutionary change.

Under the term heredity we understand the fact that on fertilization of the egg by the semen, species and individual qualities of the parents are received by the forming individual. In this procedure the most substantial part is attributed to the nucleus and, specifically, to the chromosomes of the germ cells, a view which, to be sure, is again hotly disputed. The mature sex cell undergoes a double division just before fertilization. According to a broad view, a part of the genetic composition originating in the two parents is liberated by this process. This process is of the utmost importance in regulating the relationship of the individual to his ancestors and in establishing genetic hypotheses. If we construct the genealogical table of an individual, that is, a schematic synopsis of his ancestors, we have a synopsis of those individuals who may have influenced, through the germ plasm, certain qualities of the individual under consideration. However, among all theoretical possibilities only a few are actually to be taken into consideration. There is no continuity of the germ plasm for all ancestors with reference to all qualities, and in the concurrence most of the ancestors are eliminated for the determination of the individual with regard to each particular quality. We do not know how many ancestors actually determine the individual with regard to a certain quality; we can only say that it must be at least two, one on the paternal and one on the maternal side. The more ancestors that are actually involved, the larger is the number of gradations or variations produced. From the significance of reduction division it further follows that the connection with a certain ancestor becomes more improbable the further removed the degree of kinship. Since the number of ancestors is doubled through each step of relationship, the probability that any one ancestor will influence a certain quality is halved in each degree of kinship. According to the Mendelian rule it appears that each quality in the final analysis is determined by only two ancestors and thereby the most rigorous selection of ancestors is expressed.

These are the most essential viewpoints we must take if we wish to explore and evaluate the facts of human genetics. It may be said at once that the limits of genetic research on man are considerably narrower than can be set in general biology. Essentially we can only be concerned with establishing in what cases heredity is present, what influence heredity has as compared to other factors which influence a certain manifestation, and what specific genetic rules enter into consideration of a particular quality. Only general biology can offer us an understanding of the nature of heredity from the point of view of cellular history; man, in particular, is not a profitable subject for its study since it is not possible to expose him to numerous well thought out breeding experiments as can be done with lower plants and animals and which have led to such beautiful results as the discovery of Mendel's rule of inheritance. The study of human heredity can only be a matter of

subsequent evaluation of experiments which life has provided without deliberation. In the case of man, the statistical evaluation of mass phenomena must substitute, in a poor way, for experiments. The question of whether a selection has occurred in a positive or a negative direction, whether there has been partial inbreeding or random mixing — panmixis — and to what degree, brings an element of uncertainty into the analysis of the findings. The consequences of strictest inbreeding, that is, sib matings, with which the classical experiment of Mendel dealt, cannot be determined for man and the relatively rare matings of more distant relatives can offer no complete substitute.

But this is only a part of the difference between research on plants and animals and that on man.

The essential difference is due more to the manner of obtaining the material to be investigated and the method of analyzing it. In the case of plants and animals one can personally observe the results of breeding experiments through several generations. In the case of man, one observer can know only fractional portions of the history of two generations of one family, except possibly for those qualities such as color which are established at birth. Even pathological manifestations, as well as some normal ones, can first come to observation at an age when the individual lives far distant from his parents. Many familial traits can therefore be known only by way of tradition, which is frequently incomplete and even with the best intentions (which may not always be assumed), deceptive. How many people are unable to give correctly and completely even the number of children their mother gave birth to, or the causes of death or even the names of their grandparents?

Many questions can therefore be solved conclusively only with the aid of documentary material, the exploration of which will essentially be a thing of the future. Up to the present two directions have become important in this regard. One, as representatives of which I would like to mention Goehlert and Ottokar Lorenz, seeks to analyze the history of prominent families, namely, princely houses and families of the nobility. But these kinds of investigations offer neither a sufficiently large material nor assure a uniform reliability and completeness of data. Moreover, such families represent the product of selection, the analysis of which can never give a picture of average situations.

The other direction, in which Ammon and Riffel

have been active, seeks to discover the anthropological and pathological relationships of the total population of entire regions and communities over a rather long period of time. This is also the method I have adopted in my investigations. I was fortunate in being spared the great effort of compiling the composition of the families from church and local registers, since the Württemberg family registers permitted me to learn the demographic history of a family household and its connection with ancestors and descendants. I had only to enter the causes of death on the abstracts placed at my disposal.

Although it is true that the unreliability of the material obtained from studies in man can be considerably corrected or sometimes even removed by the choice of a suitable method, there remain considerable difficulties. For the most part it is impossible to follow the descendants of the personally observed persons for a rather long time or over several generations. In contrast to investigations which follow the descendants of crosses in plants and animals, one is very substantially and often predominantly dependent on establishing the ratios in the ascendant and collateral kinships. Insofar as numerical determinations are concerned, the results must suffer not unimportant numerical displacements, as I will show you later.

The numerous errors which explain the decade-long standstill of genetics are situated in the domain of method.

One of these errors was due to the fact that the relatives of an individual were followed only insofar as pathological conditions could be ascertained. The pedigrees thus obtained were mostly very incomplete and offered only one aspect. Ottokar Lorenz was correct in drawing attention to the fact that such pedigrees are worthless and in pointing out the difference between pedigrees and genealogical tables. However, instead of recommending the correct analysis, i.e., complete pedigrees, he believed that the genealogical table should, on the whole, be given preference and expected from it especially a reduction of exaggerated views on the importance of pathological inheritance. A comparison with the biological research on plants and animals should have been able to teach him that successful investigation examines the results of certain crossings in the descendants. His view that very extensive genealogical tables give a more correct picture of the influence of heredity is based on two mistakes. First of all, he misinterprets the influence of reduction

division and the importance of the different degrees of relationship. Secondly, he disregards the favorable selection due to marriage which, especially in the mentally ill, leads to the exclusion in rather high degree of the opportunity for, and possibility of, reproduction. Therefore one seldom finds idiots among the parents of idiots. For this reason a correctly constructed pedigree is preferable to a genealogical table since it offers greater security against a biased selection.

The failure of research in pathological genetics was, in fact, founded on biased casuistics and on the initially erroneous statistical method of inspection which gradually took its place, without completely suppressing it. It was not enough to recognize that the negative case had the same value as the positive one. For example, instead of counting up the typical cases there was a tendency, by extensive investigation of the kinship, to seek the trait among other family members and thereafter to calculate an absolute high percentage of the trait. This was the reason why the studies of Riffel on the inheritance of tuberculosis were harshly criticized by bacteriologists. The same error, i.e., the misinterpretation of the different importance of the different degrees of kinship to the genetic composition of an individual, also led Lorenz and Riffel to entirely opposite and equally false views on the meaning of transmission. Only by the comparative method was it possible to obtain a measure of the influence of heredity. The first attempts in this direction, by Koller on mental diseases and by Kuthry on tuberculosis, however, still gave too little consideration of the influence of age and external living conditions.

The demonstration of a familially increased trait is not directly identical with inheritance in the genetic sense. Familial occurrence of a trait can also be based on common factors of environmental conditions and habits. Thus the family history of persons with gout or diabetes do not readily prove that these illnesses have a genetic foundation. In mental diseases the influence of environmental conditions has been studied too little, whereas in tuberculosis there probably has been overemphasis. In all these diseases the influence of heredity can be recognized only insofar as a trait, whose nature we do not yet know in tuberculosis, remains as a familial trait even after elimination of the influence of age and external environment. In order to study this in tuberculosis, I compared the incidence of familial tuberculosis in affected people and in their spouses. I found familial evidence of tuberculosis among affected subjects to be 50% higher than in their spouses. In well off persons familial evidence was 100% higher among the patients.

A certain inbreeding among tuberculous persons is the necessary result of a definite, if not strong, effort on the part of healthy persons to avoid marriages with susceptible individuals. Therefore my figures representing the relative familial burden of tuberculous persons may be too low, and for the same reason, the mortality from phthisis among spouses of tuberculous persons, which I likewise established, perhaps appears somewhat too high. With this I believe I have given you a picture of the difficulties encountered in determining and evaluating the influence of pathological inheritance in man.

Of particular interest for the theory of heredity are those qualities whose inheritance is, or appears to be, linked more or less to a certain sex. Color blindness and hemophilia belong here. According to the literature to date, both diseases occur many times more frequently in men than in women. Women play a role as intermediaries in transmission from grandfather to grandson. However, recently some doubts have arisen, at least with regard to color blindness, whether color blindness does not occur considerably more often in women than has hitherto been believed, and it has been especially recommended that school physicians should look for it in their examinations. It is not excluded that color blindness has been found more often in men only because it is troublesome in their occupations. Such demonstration would throw some light on the inheritance of color blindness and in this event the question of the carrier would become more complicated. Hemophilia is also not entirely limited to the male sex and here too the question arises whether the male, as a consequence of his occupation and different mode of life from childhood on, more often gives the disease an opportunity to become manifest.

In contrast to these diseases, the ability to bear twins from two ova is a quality in which the male plays only the role of agent of transmission. According to my investigations, begun over 7 years ago, it is quite certain that this quality is doubtless inheritable, as Darwin assumed only on the basis of the casuistic literature. This quality is due to a difference in the ovaries at least in a portion of mothers of nonidentical twins in such a way that the abundance

of eggs, existing in childhood in all females, persists in some adult women and thereby allows a more frequent loss of eggs from the adult ovaries. In contrast, the majority of adult women have ovaries which are relatively poor in eggs. The ovaries of mothers of twins therefore resemble those of multiparous animals and in this respect a twin birth represents an atavism not only physiologically but also from the viewpoint of comparative anatomy. From this we see, as I stated at that time, that there is no fundamental difference between heredity and atavism. The fact that children of unlike sex occur only among binovular twins and the justified assumption that the frequency of unlike sex twins is almost exactly 50% of the total binovular twins have made it possible for me to determine a series of qualities of binovular twins and their mothers in a large material of population statistics and in specially collected family registers. From this the fact emerges that unlike sex pairs, and consequently binovular twins, occur in very different frequencies among the different peoples of Europe, when compared with the total births. In particular, nations of German origin are distinguished by a high frequency of binovular twins, whereas these are relatively rare among the Latin peoples. For this reason it is probably not justified to consider the occurrence of twins in a family as a sign of degeneration, as Rosenfeld in Vienna tried to do. The fact that the frequency of nonidentical twins presents racial variation supports a Mendelian form of inheritance for twinning. At the suggestion of Professor Hacker I studied my previously collected material to find out whether evidence could be found which would support a Mendelian mode of inheritance in twins. However, it would have been very difficult to collect a sufficient number of cases in which children of mothers of twins had married each other and then determine the frequency of twin births among their children. I have therefore tried to construct a formula for the frequency of dominant and recessive traits among the mothers, daughters, and sibs of persons affected with such traits of the same character, under the assumption that absolute panmixis is present.

Before I pursue the problem of Mendelian behavior, I must explain how I demonstrated the fact of inheritance among twins. The proof was a double one.

First, I established that not every woman has the

ability to bear twins in the same measure. I could demonstrate this by studying how frequently twins occur again among the later or earlier births of mothers of twins. Among the binovular I found a frequency of 1/30, that is, one would have to examine on the average 30 further births to mothers of twins before finding another monovular or binovular twin birth. This apparently low ratio appears significant, however, if one considers that in Württemberg a twin birth occurs, on the average, once in 75 births and in Stuttgart only once in 90 births. Among the mothers of triplets, quadruplets and quintuplets, of which I collected a total of about 400 cases in Württemberg, the figure for repeated multiple births rose to 1/19, 1/13, and 1/3, respectively.

Second, I was able to establish that there is a direct correlation between the frequency with which multiple births are repeated by a mother, and the frequency of multiple births among mother, sisters and daughters. In these relatives the frequency of multiple births rises to about double, and among triplets the value of 1/9 is reached.

From this relatively low figure for repeatability, to the cause of which I may perhaps still have the opportunity to return, I concluded that since the average number of births to a mother of twins is only about 4–5, numerous women with a tendency to twin births do not manifest this trait because they have not performed the experiment often enough.

Moreover, I have proved that among mothers, sisters, and daughters of mothers of twins, triplets, etc., multiple births occur considerably more frequently than among the total number of births. Here it was not that a daughter of a woman who had had twins once in 5 births would have had, on the average, twins among 10 children (because the intensity of the trait among the children represented the mean of that found in the parents); rather the ratio was such that it could be assumed that the mothers of twins had transmitted their ability with that frequency with which, on the average, they repeated having twins, that is, the frequency of 1/30, while the father inherited, on the average, the frequency of 1/90, in the Stuttgart cases. Thus by means of mixing, the actual frequency with which twins were repeated among mothers, sisters, and daughters of twins was 1/30 – 1/90 = 1/45. Seven years ago I, like many others, knew nothing about Mendelian genetics, and I was inclined to consider

these figures as proof that we were dealing with a simple mixture of characters. In so doing I had overlooked that the ratio among the relatives of mothers of triplets, quadruplets and quintuplets did not agree with this; such discrepancy I attributed to the smallness of the numbers. For these one does not obtain among mothers and sisters (I have not been able to investigate the daughters) the simple average of the repetition figure and the general figure, but rather considerably smaller values, which, moreover, are somewhat higher among the sisters than among the mothers.

I have now asked myself whether these striking phenomena among the triplets could not perhaps be related to the action of Mendel's rule and arrived at the conclusion that this is actually the case. However, in the presentation of the train of thought which led me to this I must refer to what was said at the beginning of my lecture.

As far as my investigations up to the present could show — I hope soon to be able to extend them considerably — neither a conscious nor unconscious selection takes place with reference to the inherited factors for twinning. The existence of a rather extensive panmixis with regard to this character is also supported on theoretical grounds.

I was therefore confronted with the question: How does the numerical influence of Mendelian inheritance behave under the influence of panmixis? The typical Mendelian rule represents only the effect of the splitting of the inherited factors in the germ cells under the influence of the most absolute inbreeding. Such inbreeding does not occur in man.

If one continues the exclusive crossing of pure types and hybrids for several generations and further counts the hybrids AB among the dominant type AA (whereas the recessive is designated BB), then the relative frequency in the nth generation after the first cross is*

$$A = 2^{n-1} + 1$$
$$B = 2^{n-1} - 1.$$

The difference then shows the excess of hybrids, which is 2 each time, and becomes relatively less frequent with each generation, since for each generation increasing numbers of A and B are obtained.

The situation is entirely different when Mendelian

---

*Editor's note: A and B represent phenotype frequencies. The first cross represents the hybrid generation where $n = 1$.

inheritance is examined under the influence of panmixis. I start with the general hypothesis that there are initially m males and females who are pure representatives of type A, and likewise n individuals who are pure representatives of type B present. If these are crossed at random the composition of the daughter generation is obtained by using the symbolism of the binomial theorem:

$$(m\,AA + n\,BB)^2$$

$$= \frac{m^2}{(m+n)^2}AA + \frac{2mn}{(m+n)^2}AB + \frac{n^2}{(m+n)^2}BB$$

or if $m + n = 1$

$$m^2\,AA + 2mn\,AB + n^2\,BB$$

If the male and female members of the first generation are crossed at random the following frequency of the different combinations of crosses are obtained:*

$$m^2 \cdot m^2\ (AA \times AA) = m^4\,AA$$
$$2 \cdot m^2 \cdot 2mn\ (AA \times AB) = 2m^3n\,AA + 2m^3n\,AB$$
$$2 \cdot m^2 \cdot n^2\ (AA \times BB) = 2m^2n^2\,AB$$
$$2mn \cdot 2mn\ (AB \times AB) = m^2n^2\,AA + 2m^2n^2\,AB + m^2n^2\,BB$$
$$2 \cdot 2mn \cdot n^2\ (AB \times BB) = 2mn^3\,AB + 2mn^3\,BB$$
$$n^2 \cdot n^2\ (BB \times BB) = n^4\,BB$$

or the relative frequencies are for

$$AA:m^2\ (m+n)^2$$
$$AB:2mn\ (m+n)^2$$
$$BB:n^2\ (m+n)^2$$

and the composition of the second or daughter generation is again

$$m^2\,AA + 2mn\,AB + n^2\,BB$$

We thus obtain the same distribution of pure types and hybrids for each generation under the influence of panmixis and therewith the possibility of calculating for each generation how the representation of these types is arranged in panmixis and Mendelian

inheritance among the parents, sibs and children of the various types and hybrids.

If the original distribution of the two types and hybrids is

$$m^2\,AA + 2mn\,AB + n^2\,BB$$

and among the relatives the representatives of the dominant type (AA) and hybrids (AB) are considered together in that they are designated by the same single letter (A), then when A is dominant the frequency of A and B is

among the parents of A: $(1 + mn)A{:}n^2B$
among the children of A: $(1 + mn)A{:}n^2B$
among the sibs of A: $[4(1 + mn) + mn^2]\ A{:}n^2(3 + n)B$

(Editor's note: The derivation of these ratios may not be immediately obvious and the following elaboration is provided.

An individual with the dominant phenotype A may have genotypes AA or AB. The relative frequencies of phenotypes A and B, i.e., the A:B ratio, among the parents of A individuals may be determined by a series of steps which combine the Hardy-Weinberg expectations with those of Mendelism.

The first step is to set out the mating types which are potentially productive of an A child. This is done in the first column of the table below. Only the $BB \times BB$ mating type is incapable of producing an A phenotype child. This mating is therefore omitted. The second step, shown in the second column below, is to indicate the frequency with which these mating types occur in the population. The next step is firstly to indicate, in each mating type, the probability that one parent will have phenotype A and secondly the probability that the other parent will have phenotype B. These probabilities are shown in columns three and four. The fourth step is to provide, as done in the fifth column, the Mendelian expectation that a specific mating will produce an A child.

The probability that a parent will be phenotype A when given that a child is phenotype A may be symbolized Pr (A parent | A child) and may be calculated by compound probabilities. Thus by cross multiplication

Pr(A parent | A child) $= m^4(1)(1) + 4m^3n(1)(1) + 2m^2n^2$ $(\frac{1}{2})(1) + 4m^2n^2(1)(\frac{3}{4}) + 4mn^3(\frac{1}{2})(\frac{1}{2})$

Summed this expression becomes

$$m^4 + 4m^3n + 4m^2n^2 + mn^3$$

Simplification depends on the use of identities such as $(m + n) = 1$. Thus $1 + n = m + n + n = m + 2n$. In turn,

---

*Editor's note: The expression to the left of the equality sign represents the frequency of the cross (mating type) while the figure to the right represents the relative proportions of types among children of these matings determined for Mendelian expectation. The operation thus combines the facts of binomial expectation with those of Mendelism.

Genotype mating	Frequency genotype mating	Probability one parent is phenotype A	Probability one parent is phenotype B	Probability child is phenotype A
$AA \times AA$	m^4	1	0	1
$AA \times AB$	$4m^3n$	1	0	1
$AA \times BB$	$2m^2n^2$	1/2	1/2	1
$AB \times AB$	$4m^2n^2$	1	0	3/4
$AB \times BB$	$4mn^3$	1/2	1/2	1/2

$m^4 + 4m^3n + 4m^2n^2 = m^2(m^2 + 4mn + 4n^2) = m^2(m + 2n)^2 = m^2(1 + n)^2$.

Thereby Pr (A parent | A child) $= m^2(1 + n)^2 + mn^3$.

The Pr (B parent | A child), obtained in a similar fashion, $= 2m^2n^2(1/2)(1) + 4mn^3(1/2)(1/2) = m^2n^2 + mn^3$.

The ratio of the Pr (A parent) : Pr (B parent) thus becomes

$$\frac{m^2(1 + n)^2 + mn^3}{m^2n^2 + mn^3} = \frac{m(1 + n)^2 + n^3}{mn^2 + n^3}$$

$$= \frac{m(1 + n)^2 + n^3}{n^2(m + n)} = \frac{m(1 + n)^2 + n^3}{n^2}$$

$$= \frac{m + 2mn + mn^2 + n^3}{n^2} = m + 2mn + n^2(m + n)$$

$$= \frac{m + 2mn + n^2}{n^2} = \frac{m + mn + n(m + n)}{n^2}$$

$$= \frac{m + mn + n}{n^2} = \frac{1 + mn}{n^2}.$$

The balance of the expressions for the relative proportions of A and B phenotypes among other classes of relatives and in the case of other forms of inheritance may be derived in a similar manner. Such derivations are still occasionally employed today for the demonstration of simple inheritance in those instances where data, for example, is limited to parent-child combinations.]

But if A is recessive, one obtains:

for the parents of A: mA:nB
for the children of A: mA:nB
for the sibs of A: $(1 + m)^2$ A:$n(3 + m)$B

Thus it appears that in Mendelian inheritance the representation of types obtained for sibs differs from the proportion among parents and children. In qualities that are measurable, as in our case, this must lead to different average values among parents and sibs. In a non-Mendelian character, where the hybrids represent several intermediary stages, the average representation of measurable qualities would be the same in parents and sibs. There is, therefore, a real difference between Mendelian and non-Mendelian characters. I have found with respect to the tendency for triplet birth that the sibs and parents exhibit essentially different numbers, which speaks for a Mendelian trait. There are similar small differences in the tendency for twinning.

It is now apparent that where A is dominant, type A is always represented in at least half of the parents.

Where A is recessive one obtains limiting values for frequencies of A, of 1:0 and 0:1 for the parents and 1:0 and 1:3 for sibs. At the same time we learn that rare recessive characters can be detected easier in sibs than in parents. The possibility of calculating the expected figures for Mendelian inheritance has enabled me to utilize not only the absolutely few cases in which children of mothers of twins married each other but also my entire earlier material.

It is now a question of determining the value of m. For this the following consideration is needed. A binovular twin birth occurs once in 35 births among mothers with a tendency to twinning, but only once in 140 among all mothers, as in Stuttgart, then the former represent only one fourth of all mothers. If their frequency is placed at $m^2 = 1/4$, then that of the other women is $2mn + n^2 = 3/4$, and we obtain $m = n = 1/2$ where A is recessive. In the case of dominance of twinning we would obtain the ratio $m:n = 1:6.5$, if $m^2 + 2mn:n^2 = 1:3$.

Likewise, if a triplet birth occurs in ca. 6000 births, and in mothers with a tendency to triplet births, a triplet birth occurs once to every 200 single births, the value

$m = 1/5$ in the case of recessivity
$m = 1/60$ in the case of dominance

In each of these cases, however, a twin birth would occur in 84 births in Württemberg women without a tendency to triplet births.

If the value thus found is inserted into the above formula for the hereditary trait through parents, sibs, and children, one obtains, from the comparison of the calculated probable number with the actual ratios, the assumption which comes closest to the mode of inheritance.

According to whether the trait for multiple births as opposed to single birth is (I) recessive, (II) dominant, or (III) equivalent, the following expected numbers are obtained as the expected frequency of multiple births in the kinships of mothers of twins and triplets in Stuttgart and Württemberg:

	I	*II*	*III*
(a) among the mothers of mothers of twins	1/52	1/46	1/45
(b) among the daughters of mothers of twins	1/52	1/46.8	1/45
(c) among the sisters of mothers of twins	1/49	1/46.6	1/45
(d) among the mothers of mothers of triplets	1/52	1/29	1/29
(e) among the sisters of mothers of triplets	1/37	1/29	1/29

The observed births in the individual groups were

a) 1365   d) 2638
b) 1464   e) 1666
c) 1022

Accordingly the absolute number of expected multiple births is:

In group	According to the assumption			Observed
	I	*II*	*III*	
(a)	26	27	30	27
(b)	28	31	33	24
(c)	21	22	23	27
(d)	51	91	91	45
(e)	45	57	57	36
Totals	171	228	234	159

The assumption that the trait for twin births is recessive is therefore obtained from the expected numbers which are closer to the observed than under any other assumption. The difference in 12 cases lies within the mean error, which is close to

$$\sqrt{171} = 13$$

The situation found in the inheritance of twinning best finds its explanation in the assumption that the trait for twinning is inherited according to the Mendelian rule and is recessive.

This investigation (a more thorough presentation based on a new collection of material that I have in the meantime obtained from another source will follow) may show that one can gain an insight into the nature of human inheritance by suitable changes in the investigative methods.

The chromosomes in heredity

WALTER S. SUTTON

Reprinted with permission from Biological Bulletin, *vol. 4, 1903, pp. 231–251.*

Probably the first published report to cast doubt on the universality of independent assortment between two sets of alleles controlling different characters was presented by Bateson and his colleagues to the Evolutionary Committee of the Royal Society. Bateson reported that the phenotypic ratios involving the color of sweet peas and the shape of their pollen grains suggested that the distribution of the hereditary units controlling these two traits was not truly independent. In an attempt to preserve the structural individuality of the controlling units, Bateson proposed the existence of attracting forces that existed between the structurally distinct units; and that these forces led certain hereditary units to pass into the same gametes (*coupling*). Bateson could not explain the mechanistic basis of this coupling force. His speculation, however, explained observations that had contradicted the hypothesis that all sets of hereditary units were structurally separate.

An alternative hypothesis was available to Bateson. He found it unacceptable because it meant discarding the idea of structurally separate units of inheritance. That alternative hypothesis was based upon a paper published several years earlier by Walter Sutton. In his paper, Sutton presented a statement proposing that the units of heredity resided in or on the chromosomes. He based his statement on two observations: (1) the parallel between the distribution of the units of inheritance when analyzed by means of plant and animal hybridizations, and (2) the cytological behavior of chromosomes in meiosis and fertilization. Sutton illustrated the close relationship between the fields of cytology and genetics by bringing data from both disciplines to bear on the same question. In light of Bateson's struggle to explain his phenotypic ratios, which deviated from the expected 9:3:3:1, it is interesting that Sutton predicted certain sets of hereditary units, located on the same chromosome, should show preferential linked assortment.

In a recent announcement of some results of a critical study of the chromosomes in the various cell generations of *Brachystola*[1] the author briefly called attention to a possible relation between the phenomena there described and certain conclusions first drawn from observations on plant hybrids by Gregor Mendel[2] in 1865, and recently confirmed by a number of able investigators. Further attention has already been called to the theoretical aspects of the subject in a brief communication by Professor E. B. Wilson.[3] The present paper is devoted to a more detailed discussion of these aspects, the speculative character of which may be justified by the attempt to indicate certain lines of work calculated to test the validity of the conclusions drawn. The general conceptions here advanced were evolved purely from cytological data, before the author had knowledge of the Mendelian principles, and are now presented as the contribution of a cytologist who can make no pretensions to complete familiarity with the results of experimental studies on heredity. As will appear hereafter, they completely satisfy the conditions in typical Mendelian cases, and it seems that many of the known deviations from the Mendelian type may be explained by easily conceivable variations from the normal chromosomic processes.

It has long been admitted that we must look to the organization of the germ cells for the ultimate determination of hereditary phenomena. Mendel fully appreciated this fact and even instituted special experiments to determine the nature of that organization. From them he drew the brilliant conclusion that, while, in the organism, maternal and paternal potentialities are present in the field of each character, *the germ cells in respect to each character are pure*. Little was then known of the nature of cell division, and Mendel attempted no comparisons in that direction; but to those who in recent years have revived and extended his results the probability of a relation between cell organization and cell division has repeatedly occurred. Bateson[4] clearly states his impression in this regard in the following words: "It is impossible to be presented with the fact that in Mendelian cases the crossbred produces on an average *equal* numbers of gametes of each kind, that is to say, a symmetrical result, without suspecting that this fact must correspond with some symmetrical figure of distribution of the gametes in the cell divisions by which they are produced."

Nearly a year ago it became apparent to the author that the high degree of organization in the chromosome-group of the germ cells as shown in *Brachystola* could scarcely be without definite significance in inheritance, for, as shown in the paper[5] already referred to, it had appeared that:

1. The chromosome group of the presynaptic germ cells is made up of two equivalent chromosome series, and that strong ground exists for the conclusion that one of these is paternal and the other maternal.

2. The process of synapsis (pseudoreduction) consists in the union in pairs of the homologous members (i.e., those that correspond in size) of the two series.[6]

3. The first postsynaptic or maturation mitosis is equational and hence results in no chromosomic differentiation.

4. The second postsynaptic division is a reducing division, resulting in the separation of the chromosomes which have conjugated in synapsis, and their relegation to different germ cells.

5. The chromosomes retain a morphological individuality throughout the various cell divisions.

It is well known that in the eggs of many forms the maternal and paternal chromosome groups remain distinctly independent of each other for a considerable number of cleavage mitoses, and with this fact in mind the author was at first inclined to conclude that in the reducing divisions all the maternal chromosomes must pass to one pole and all the paternal ones to the other, and that the germ cells are thus divided into two categories which might be described as maternal and paternal respectively. But this conception, which is identical with that recently brought forward by Cannon,[7] was soon seen to be at variance with many well known facts of breeding; thus:

1. If the germ cells of hybrids are of pure descent, no amount of crossbreeding could accomplish more than the condition of a first cross.

2. If any animal or plant has but two categories of germ cells, there can be only four different combinations in the offspring of a single pair.

3. If either maternal or paternal chromosomes are entirely excluded from every ripe germ cell, an individual cannot receive chromosomes (qualities) from more than one ancestor in each generation of each of the parental lines of descent, e.g., could not inherit chromosomes (qualities) from both paternal or both maternal grandparents.

Moved by these considerations a more careful study was made of the whole division process, including the positions of the chromosomes in the nucleus before division, the origin and formation of the spindle, the relative positions of the chromosomes and the diverging centrosomes, and the point of attachment of the spindle fibers to the chromosomes. The results gave no evidence in favor of parental purity of the gametic chromatin as a whole. On the contrary, many points were discovered which strongly indicate[8] that the position of the bivalent chromosomes in the equatorial plate of the reducing division is purely a matter of chance — that is, that any chromosome pair may lie with maternal or paternal chromatid indifferently toward either pole irrespective of the positions of other pairs — and hence that a large number of different combinations of maternal and paternal chromosomes are possible in the mature germ products of an individual. To illustrate this, we may consider a form having eight chromosomes in the somatic and presynaptic germ cells and consequently four in the ripe germ products. The germ cell series of the species in general may be designated by the letters A, B, C, D, and any cleavage nucleus may be considered as containing chromosomes A, B, C, D from the father and a, b, c, d, from the mother. Synapsis being the union of homologues would result in the formation of the bivalent chromosomes Aa, Bb, Cc, Dd, which would again be resolved into their components by the reducing division. Each of the ripe germ cells arising from the reduction divisions must receive one member from each of the synaptic pairs, but there are sixteen possible combinations of maternal and paternal chromosomes that will form a complete series, to wit: a, B, C, D; A, b, C, D; A, B, c, D; A, B, C, d; a, b, C, D; a, B, c, D; a, B, C, d; a, b, c, d; and their conjugates A, b, c, d; a, B, c, d; a, b, C, d; a, b, c, D; A, B, c, d; A, b, C, d; A, b, c, D; A, B, C, D. Hence instead of two kinds of gametes an organism with four chromosomes in its reduced series may give rise to 16 different kinds; and the offspring of two unrelated individuals may present

16 × 16 or 256 combinations, instead of the four to which it would be limited by a hypothesis of parental purity of gametes. Few organisms, moreover, have so few as 8 chromosomes, and since each additional pair doubles the number of possible combinations in the germ products* and quadruples that of the zygotes it is plain that in the ordinary form having from 24 to 36 chromosomes, the possibilities are immense. Table 1 below shows the number of possible combinations in forms having from 2 to 36 chromosomes in the presynaptic cells.

Thus if Bardeleben's estimate of sixteen chromosomes for man (the lowest estimate that has been made) be correct, each individual is capable of producing 256 different kinds of germ products with reference to their chromosome combinations, and the numbers of combinations possible in the offspring of a single pair is 256 × 256 or 65,536; while *Toxopneustes*, with 36 chromosomes, has a possibility of 262,144 and 68,719,476,736 different

---

*The number of possible combinations in the germ products of a single individual of any species is represented by the simple formula 2^n in which n represents the number of chromosomes in the reduced series.

Table 1

Chromosomes		Combinations in gametes	Combinations in zygotes
Somatic Series	*Reduced Series*		
2	1	2	4
4	2	4	16
6	3	8	64
8	4	16	256
10	5	32	1,024
12	6	64	4,096
14	7	128	16,384
16	8	256	65,536
18	9	512	262,144
20	10	1,024	1,048,576
22	11	2,048	4,194,304
24	12	4,096	16,777,216
26	13	8,192	67,108,864
28	14	16,384	268,435,456
30	15	32,768	1,073,741,824
32	16	65,536	4,294,967,296
34	17	131,072	17,179,869,184
36	18	262,144	68,719,476,736

combinations in the gametes of a single individual and the zygotes of a pair respectively. It is this possibility of so great a number of combinations of maternal and paternal chromosomes in the gametes which serves to bring the chromosome theory into final relation with the known facts of heredity; for Mendel himself followed out the actual combinations of two and three distinctive characters and found them to be inherited independently of one another and to present a great variety of combinations in the second generation.

The constant size differences observed in the chromosomes of *Brachystola* early led me to the suspicion, which, however, a study of spermatogenesis alone could not confirm, that the individual chromosomes of the reduced series play different *roles* in development. The confirmation of this surmise appeared later in the results obtained by Boveri[9] in a study of larvae actually lacking in certain chromosomes of the normal series, which seem to leave no alternative to the conclusion that the chromosomes differ qualitatively and as individuals represent distinct potentialities. Accepting this conclusion we should be able to find an exact correspondence between the behavior in inheritance of any chromosome and that of the characters associated with it in the organism.

In regard to the characters, Mendel found that, if a hybrid produced by crossing two individuals differing in a particular character be self-fertilized, the offspring, in most cases, conform to a perfectly definite rule as regards the differential character. Representing the character as seen in one of the original parents by the letter A and that of the other by a, then all the offspring arising by self-fertilization of the hybrid are represented from the standpoint of the given character by the formula AA : 2Aa : aa — that is, one fourth receive only the character of one of the original purebred parents, one fourth only that of the other; while one half the number receive the characters of both original parents and hence present the condition of the hybrid from which they sprang.

We have not heretofore possessed graphic formulae to express the combinations of chromosomes in similar breeding experiments, but it is clear from the data already given that such formulae may now be constructed. The reduced chromosome series in *Brachystola* is made up of eleven members, no two of which are exactly of the same size. These I dis-

tinguished in my previous paper by the letters A, B, C, . . . K. In the unreduced series there are twenty-two elements* which can be seen to make up two series like that of the mature germ cells, and hence may be designated as A, B, C . . . K + A, B, C . . . K. Synapsis results in the union of homologues and the production of a single series of double elements thus: AA, BB, CC . . . KK, and the reducing division affects the separation of these pairs so that one member of each passes to each of the resulting germ products.

There is reason to believe that the division products of a given chromosome in *Brachystola* maintain in their respective series the same size relation as did the parent element; and this, taken together with the evidence that the various chromosomes of the series represent distinctive potentialities, makes it probable that a given size-relation is characteristic of the physical basis of a definite set of characters. But each chromosome of any reduced series in the species has a homologue in any other series, and from the above consideration it should follow that these homologues cover the same field in development. If this be the case chromosome *A* from the father and its homologue, chromosome *a*, from the mother in the presynaptic cells of the offspring may be regarded as the physical bases of the antagonistic unit characters A and a of father and mother respectively. In synapsis. copulation of the homologues gives rise to the bivalent chromosome *Aa*, which as is indicated above would, in the reducing division, be separated into the components *A* and *a*. These would in all cases pass to different germ products and hence in a monœcious form we should have four sorts of gametes,

$$A \, \male \qquad a \, \male$$
$$A \, \female \qquad a \, \female$$

which would yield four combinations,

$$A \, \male + A \, \female = AA$$
$$A \, \male + a \, \female = Aa$$
$$a \, \male + A \, \female = aA$$
$$a \, \male + a \, \female = aa$$

Since the second and third of these are alike the result would be expressed by the formula $AA : 2Aa : aa$ which is the same as that given for any character

---

*Disregarding the accessory chromosome which takes no part in synapsis.

in a Mendelian case. *Thus the phenomena of germ-cell division and of heredity are seen to have the same essential features, viz., purity of units (chromosomes, characters) and the independent transmission of the same;* while as a corollary, it follows in each case that each of the two antagonistic units (chromosomes, characters) is contained by exactly half the gametes produced.

The observations which deal with characters have been made chiefly upon hybrids, while the cytological data are the result of study of a purebred form; but the correlation of the two is justified by the observation of Cannon[10] that the maturation mitoses of fertile hybrids are normal. This being the case it is necessary to conclude, as Cannon has already pointed out, that the course of variations in hybrids either is a result of normal maturation processes or is entirely independent of the nature of those divisions. If we conclude from the evidence already given that the double basis of hybrid characters is to be found in the pairs of homologous chromosomes of the presynaptic germ-cells, then we must also conclude that in purebred forms likewise, the paired arrangement of the chromosomes indicates a dual basis for each character. In a hypothetical species breeding absolutely true, therefore, all the chromosomes or subdivisions of chromosomes representing any given character would have to be exactly alike, since the combination of any two of them would produce a uniform result. As a matter of fact, however, specific characters are not found to be constant quantities but vary within certain limits; and many of the variations are known to be inheritable. Hence it seems highly probable that homologous chromatin entities are not usually of strictly uniform constitution, but present minor variations corresponding to the various expressions of the character they represent. In other words, it is probable that specific differences and individual variations are alike traceable to a common source, which is a difference in the constitution of homologous chromatin entities. Slight differences in homologues would mean corresponding, slight variations in the character concerned — a correspondence which is actually seen in cases of inbreeding, where variation is well known to be minimized and where obviously in the case of many of the chromosome pairs both members must be derived from the same chromosome of a recent common ancestor and hence be practically identical.

In the various forms of parthenogenesis we meet the closest kind of inbreeding and a brief consideration of the variability to be expected in each, from the standpoint of the chromosome theory, may serve as a guide to such research as will test the validity of the latter. The simplest form, of which chemical parthenogenesis in sea urchins is an example, is that in which the organism has only a single chromosome series, to be represented by $A, B, C, D \ldots N$. Thus far no recognized cases of this type have been reared to sexual maturity, but it is to be expected that no reducing division will be found in the maturation of such forms, and that their parthenogenetic offspring will exactly resemble the immediate parent.

In cases of natural parthenogenesis which are accompanied by the reentrance of the second polar body and its fusion with the egg nucleus (or its failure to form) there must be a double chromosome series; but we may distinguish two classes according as the reducing process is accomplished in the first or the second maturation division.[*] If reduction is accomplished in the first division, one half the chromosomes of the oögonia are thrown out and lost in the first polar body. The second division, being equational, would result in a polar body which would be the exact duplicate of the egg nucleus as far as chromosomes are concerned and which accordingly, by its reentrance would add nothing new to the egg series. The series after fusion would, therefore, be represented by the letters $A, B, C, D \ldots N + A, B, C, D \ldots N$. If such a type of parthenogenesis were to follow sexual reproduction, the first generation of offspring might be expected to differ materially from the parent by reason of the casting out, in the first polar body, of chromosomes representing certain dominant characters, and the consequent appearance in the offspring of the corresponding recessives. Subsequent parthenogenetic generations, however, would in each case be endowed with a chromosome series exactly similar to that of the immediate parent and accordingly might be expected to show the same characters.

In case the second division of a parthenogenetic

---

[*]Either must be regarded as possible in cases where we have no definite knowledge since it is regularly described as the second in the Orthoptera (McClung, Sutton) and Copepoda (Rückert, Häcker) while in the Hemiptera-Heteroptera it is believed to be the first (Paulmier, Montgomery).

egg were the reducing division, the reentrance or suppression of the second polar body would accomplish the restoration of the oögonial chromosome series. In this case the first parthenogenetic generation might be expected to duplicate the characters of the parent (if environmental conditions remained unchanged) and little or no variability would be expected as long as parthenogenesis persisted.

In relation to these problems there is great need of a simultaneous study of the germ-cell divisions and the variation of periodically parthenogenetic forms.

We have seen reason, in the foregoing considerations, to believe that there is a definite relation between chromosomes and allelomorphs* or unit characters but we have not before inquired whether an entire chromosome or only a part of one is to be regarded as the basis of a single allelomorph. The answer must unquestionably be in favor of the latter possibility, for otherwise the number of distinct characters possessed by an individual could not exceed the number of chromosomes in the germ products; which is undoubtedly contrary to fact. We must, therefore, assume that some chromosomes at least are related to a number of different allelomorphs. If then, the chromosomes permanently retain their individuality, it follows that all the allelomorphs represented by any one chromosome must be inherited together. On the other hand, it is not necessary to assume that all must be apparent in the organism, for here the question of dominance enters and it is not yet known that dominance is a function of an entire chromosome. It is conceivable that the chromosome may be divisible into smaller entities (somewhat as Weismann assumes), which represent the allelomorphs and may be dominant or recessive independently. In this way the same chromosome might at one time represent both dominant and recessive allelomorphs.

Such a conception infinitely increases the number of possible combinations of characters *as actually seen* in the individuals and unfortunately at the same time increases the difficulty of determining what characters are inherited together, since usually recessive chromatin entities (allelomorphs?) constantly associated in the same chromosome with usually dominant ones would evade detection for generations and then becoming dominant might appear as reversions in a very confusing manner.

In their experiments on *Matthiola*, Bateson and Saunders[11] mention two cases of correlated qualities which may be explained by the association of their physical bases in the same chromosome. "In certain combinations there was close correlation between (a) green color of seed and hoariness, (b) brown color of seed and grabrousness. In other combinations such correlation was entirely wanting." Such results may be due to the association in the same chromosomes of the physical bases of the two characters. When close correlation was observed, both may be supposed to have dominated their homologues; when correlation was wanting, one may have been dominant and the other recessive. In the next paragraph to that quoted is the statement: "The rule that plants with flowers either purple or claret arose from green seeds was universal." Here may be a case of constant dominance of two associated chromatin entities.

Dominance is not a conception which grows out of purely cytological consideration. Cytology merely shows us the presence in a cell of two chromosomes, either of which is capable of producing some expression of a given character, and it is left to experiment in each case to show what the effect of this combined action will be. The experiment[12] has shown that any one of the three theoretical possibilities may be realized, viz: (1) One or the other may dominate and obscure its homologue. (2) The result may be a compromise in which the effect of each chromosome is to be recognized. (3) The combined action of the two may result in an entirely new cast of character. In cases belonging to the first category, the visible quality (allelomorph, chromatin entity) was described by Mendel as dominant and the other as recessive, and the experiments of Bateson and Saunders and others, as well as those of Mendel, have shown that in many cases a dominant character tends to remain dominant during successive generations if the environment is not materially changed. Nevertheless, some experiments cited by Bateson[13] go to show that dominance may be variable or defective. Furthermore, it is not only conceivable, but highly probable that in most, if not all cases, there are many different expressions of each character (i.e., many different allelomorphs as suggested by Bateson in regard to human stature), which on various combinations would necessarily exhibit relative dominance. The experiments with peas show an almost constant

---

*Bateson's term.

dominance of certain allelomorphs, such as round over wrinkled in seeds, and of yellow over green in cotyledons; but it is worthy of note that here, as in most Mendelian experiments, only two antagonistic characters have been used. Investigations on varieties, in general similar, but exhibiting different expressions of some particular character, will certainly yield instructive results. Bateson's observations on crosses between single-, rose- and pea-combed fowls, represent a simple form of such a case and may be expected on completion to add much to our knowledge of the nature of dominance.

In addition to the many examples brought forward by Bateson in support of the Mendelian principle he cites three types of cases which are to be regarded as non-Mendelian. These are:

1. The ordinary blended inheritance of continuous variation.

2. Cases in which the form resulting from the first cross breeds true.

3. The "false hybrids" of Millardet.

1. *Blended Inheritance*   In treating of this class Bateson clearly states the possibility that the case may be one entirely "apart from those to which Mendel's principles apply," but goes on to show how it may possibly be brought into relation with true Mendelian cases. He says in part: "It must be recognized that in, for example, the stature of a civilized race of man, a typically continuous character, there must certainly be on any hypothesis more than one pair of possible allelomorphs. There may be many such pairs, but we have no certainty that the number of such pairs and consequently of the different kinds of gametes are altogether *unlimited*, even in regard to stature. If there were even so few as, say, four or five pairs of possible allelomorphs, the various homo- and heterozygous combinations might, on seriation, give so near an approach to a continuous curve that the purity of the elements would be unsuspected, and their detection practically impossible." This hypothesis, which presents no difficulties from the point of view of the chromosome theory, is sufficient in the present state of our knowledge to bring many cases of apparently continuous variation into definite relation with strictly Mendelian cases; but, on the other hand, it

seems probable, as already noted, that the individual variation in many characters now thought to be strictly Mendelian may prove to be due to the existence in the species of many variations of what may be regarded as the type allelomorphs, accompanying similar variations of the homologous chromatin entities representing those types.

2. *First Crosses that Breed True*   It is obvious that in the germ cells of true breeding hybrids[14] there can be no qualitative reduction. In the normal process synapsis must be accounted for by the assumption of an affinity existing between maternal and paternal homologues, and conversely reduction is the disappearance of that affinity or its neutralization by some greater force. Now in *Hieracium* the characters of the hybrid are frequently intermediate between those of the two parents, showing that both allelomorphs (or chromatin entities) are at work, but on self-fertilization there is no resolution of allelomorphs (reduction division). On the contrary, all the germ cells are equivalent, as shown by the fact that all combinations produce similar offspring which in turn are similar to the parent. The suggestion made by Bateson in another connection, that "if one allelomorph were alone produced by the male and the other by the female we should have a species consisting *only* of heterozygotes," which would come true as long as bred together, at first sight seems logically applicable to these cases. For such an idea, however, we can find no cytological justification, since if any reduction occurs both chromosomes occur in both male and female germ cells in equal numbers; and further, the evidence is in favor of a great variety of combinations of maternal and paternal chromosomes in the germ cells so that the exact chromosome group of a hybrid parent could hardly be duplicated except by fusion of the very pair of cells separated by the reducing division. A more plausible explanation from the cytological standpoint is that the union of the chromosomes in synapsis is so firm that no reduction can take place, i.e., that in each case, a paternal and a maternal chromosome fuse permanently to form a new chromosome which subsequently divides only equationally. The result must be germ cells which are identical with one another and with those of the parents, and hence self-fertilization would produce offspring practically without variation. If this explanation be the correct one the

process is distinctly pathological and hence it is not surprising that such cases, as noted by Bateson, should often present "a considerable degree of sterility."

3. *The "False Hybrids" of Millardet* Millardet, de Vries and Bateson have all described experiments in which the offspring resulting from a cross between dissimilar individuals showed the character of one parent only, those of the other parent being shown by further experiment to be lost permanently. The obvious cytological explanation of such a phenomenon is hinted at by Bateson in the words "Such phenomena may perhaps be regarded as fulfilling the conception of Strasburger and Boveri, that fertilization may consist of two distinct operations, the stimulus to development and the union of characters in the zygote."[15] Division of the egg without fusion of the pronuclei is a well-known phenomenon having been observed in eggs treated with chloral (Hertwig brothers) or ether (Wilson) and may be supposed to occur under certain unusual conditions in nature. In the experiments mentioned, however, both pronuclei continue to divide separately, while for a cytological explanation of the occurrence of "false hybrids" it is necessary to conceive not only the failure of the nuclei to copulate but the entire disappearance of one of them. Such a case would be comparable to that of chemically induced parthenogenesis or to the fertilization of enucleate egg fragments, according as the nucleus remaining was maternal or paternal. Speculation in this connection, however, is unprofitable excepting so far as it may serve as a guide to research. A careful study of the fertilization of such cases as Millardet's strawberries, de Vries's *Oenothera* and Bateson's *Matthiola* crosses will no doubt be productive of immediate and positive results.

4. *Mosaics* A fourth class of non-Mendelian cases, the "mosaics" or "piebalds" constitute a group in relation to which, as I believe, only negative evidence is to be expected from direct cytological study. A good example of the class is the "mosaic" fruit of *Datura* obtained by Bateson and Saunders, which, although in general exhibiting the thornless recessive condition, showed in exceptional cases a thorny patch. Of this case Bateson says: "Unless this is an original sport on the part of the individual, such a phenomenon may be taken as indicating that the

germ cells may also have been mosaic." I must confess my failure to comprehend just what is here meant by mosaic germ cells. I have attempted to show that in all probability the germ cells are normally a mosaic of maternal and paternal chromosomes, but very evidently this is not Bateson's meaning.

From the standpoint of the chromosome theory I would suggest a possible explanation of the conditions as follows: We have already assumed that the somatic chromosome group, having a similar number of members to that of the cleavage nucleus and derived from it by equation divisions, is made up in the same way of pairs of homologous chromosomes. Every somatic cell, by this conception, must contain a double basis in the field of each character it is capable of expressing. In strictly Mendelian cases one of the homologues is uniformly dominant throughout the parts of the organism in which the character is exhibited. As already noted, however, it is unlikely that all the descendants of a dominant chromatin entity will be dominant. This is shown by the experiment of de Vries with sugar beets, which are normally biennial but always produce a small percentage of annual plants or "runners," which latter are regarded as recessives. The percentage of these runners may be increased by rearing the plants under unfavorable conditions and this is taken as evidence that the recessive allelomorphs may become dominant under such conditions.[16]

If each cell contains maternal and paternal potentialities in regard to each character, and if dominance is not a common function of one of these, there is nothing to show why as a result of some disturbing factor one body of chromatin may not be called into activity in one group of cells and its homologue in another. This would produce just the sort of a mosaic which Bateson and Saunders found in *Datura* or as Tchermak's pied yellow and green peas obtained by crossing the *Telephone* pea with yellow varieties. Correns describes the condition as *pœcilodynamous* and his conception of the causes of the phenomenon as I understand it is parallel with that which I have outlined above. The logical possibility suggested by Bateson[17] that the recessive islands in such cases as the mosaic pea may be due to recessive allelomorphs in the paired state does not accord with the theory of a chromosomic basis for those allelomorphs, since the chromosome groups, both of cells showing the recessive character and of neighboring cells showing the dominant one, are derived, so far as we know, by longitudinal or equational division from the chromosomes of the same original cleavage nucleus and hence must be alike.

The application of the theory here suggested may be put to test by an experiment in which hybrids of dissimilar truebreeding parentage are crossed and a third generation of "quarter-bloods" produced. Mosaics occurring in such an organism, if this theory be correct, would show one character resembling that of one of the maternal grandparents and one resembling that of one of the original purebreds of the paternal side. If both characters of the mosaic should be clearly paternal or maternal the theory as outlined is proven inadequate, since one of each pair of chromosomes, and hence the corresponding character-group, is thrown out by the reduction-division in each generation.

In considering the behavior of the two chromosomes forming the basis of any given character, it was noted that in some cases the heterozygote character resulting from the combinations of dissimilar allelomorphs is sometimes totally unlike either of the latter. Thus Mendel found that in crosses between peas respectively 1 and 6 feet in height the offspring ranged from 6 to 7½ feet. In discussing similar cases, Bateson calls attention to the light which would be thrown on the phenomenon if we ventured to assume that the bases of the two allelomorphs concerned are chemical compounds; and he compares the behavior of the allelomorphs to the reaction of sodium and chlorine in the formation of salt. The results of chemical analysis show that one of the most characteristic features of chromatin is a large percentage content of highly complex and variable chemical compounds, the nucleo-proteids, and therefore if, as assumed in the theory here advanced, the chromosomes are the bases of definite hereditary characters, the suggestion of Bateson becomes more than a merely interesting comparison.

We have seen reason in the case of the truebreeding hybrids to suspect that the transmission by the hybrid of heterozygote characters may be due to permanent union of the homologous chromosomes. From this it is but a short step to the conclusion that even if, as is normally the case, the chromosomes do not fuse permanently, the very fact of their association in the same liquid medium may allow a possibility of a certain degree of chemical interaction. This must normally be slight, since its effects do not appear to be visible in a single generation; but the slightest of variation as a result of repeated new association, even though it tend in diverse directions, must in time, guided by natural selection, result in an appreciable difference in a definite direction between a chromosome and its direct descendant and hence between the characters associated with them. In this we have a suggestion of a possible cause of individual variation in homologous chromosomes which we have already seen reason to suspect.

Finally, we may briefly consider certain observations which seem at first sight to preclude the general applicability of the conclusions here brought out. If it be admitted that the phenomenon of character reduction discovered by Mendel is the expression of chromosome reduction, it follows that forms which vary according to Mendel's law must present a reducing division. But the vertebrates and flowering plants — the very forms from which most of the Mendelian results have been obtained — have been repeatedly described as not exhibiting a reducing division. Here, therefore, is a discrepancy of which I venture to indicate a possible explanation in the suggestion first made by Fick[18] and more recently by Montgomery.[19] This is to the effect that in synapsis as it occurs in vertebrates and other forms possessing loop-shaped chromosomes, the union is side by side instead of end-to-end as in Arthropods. In vertebrates, two parallel longitudinal splits, the forerunners of the two following divisions, appear in the chromosomes of the primary spermatocyte prophases. Both being longitudinal, they have been described as equational divisions, but if it shall be found possible to trace one to the original line of union of the two spermatogonial chromosomes side by side in synapsis, that division must be conceived as a true reduction. A number of observations supporting this view will be brought forward in my forthcoming work on *Brachystola*.

Again, if the normal course of inheritance depends upon the accurate chromatin division accomplished by mitosis, it would appear that the interjection, into any part of the germ cycle, of the gross processes of amitosis could result only in a radical deviation from that normal course. Such an occurrence has actually been described by Meves, McGregor and others in the primary spermatogonia of amphibians. In these cases, however, it appears that fission of the cell body does not necessarily follow amitotic division of the nucleus. I would suggest, therefore, the possibility that the process may be of no significance in inheritance, since by the disappearance of the nuclear membranes in preparation for the first mitotic division, the original condition is restored, and the chromosomes may enter the equatorial plate as if no amitotic process had intervened.[20]

There is one observation in connection with the accessory chromosome which deserves mention in any treatment of the chromosomes as agents in heredity. This element always divides longitudinally and hence probably equationally. It fails to divide in the first maturation mitosis, in which the ordinary chromosomes are divided equationally, but passes entire to one of the resulting cells. In the second maturation division, by which the reduction of the ordinary chromosomes is effected, the accessory divides longitudinally.*

My observations in regard to the accessory chromosome lend support to the hypothesis of McClung[21] that of the four spermatozoa arising from a single primary spermatocyte, those two which contain this element enter into the formation of male offspring, while the other two, which receive only ordinary chromosomes take part in the production of females. If this hypothesis be true, then it is plain that in the character of sex the reduction occurs in the first maturation mitosis, since it is this division which separates cells capable of producing only males from those capable of producing only females. Thus we are confronted with the probability that reduction in the field of one character occurs in one of the maturation divisions and that of all the remaining characters in the other division. The significance of such an arrangement, though not easy of perception, is nevertheless great. As regards their chromosome groups, the two cells resulting from each reduction mitosis are conjugates and, therefore, opposites from the standpoint of any individual character. Thus if we consider a hypothetical form having eight chromosomes comprising the paternal series A, B, C, D and the maternal series a, b, c, d, one of

the cells resulting from the reduction division might contain the series A, b, c, D, in which case its sister cell would receive the conjugate series a, B, C, d. It is plain that these conjugates, differing from each other in every possible character, represent the most widely different sperms the organism can produce. Now if reduction in the sex-determining chromatin also took place in this division it is apparent that these two diametrically opposite series would enter into individuals of different sexes; but if the sex reduction is previously accomplished by the asymmetrical distribution of the accessory in the first division, then both the members of each conjugate pair must take part in the production either of males or of females and thus all extremes of chromosome combination are provided for within the limits of each sex.

POSTSCRIPT

The interesting and important communication of Guyer[22] on "Hybridism and the Germ Cell" is received too late for consideration in the body of this paper. This investigator also has applied conclusions from cytological data to the explanation of certain phenomena of heredity, and his comparative observations on the spermatogenesis of fertile and infertile hybrids are an important contribution to the cytological study of the subject. The conclusions drawn are of great interest but, I think, in some cases, open to criticism. In assuming that there is a "segregation of maternal and paternal chromosomes into separate cells, which may be considered 'pure' germ cells containing qualities of only one species" (p. 19), he repeats the error of Cannon which has already been dealt with in the early part of this paper. No mention is made in the paper of Mendel's law but in considering the inbred pigeon hybrids from which his material was obtained, the author expresses his familiarity with manifestations of the Mendelian principle by the statement that "in the third generation there is generally a return to the original colors of the grandparents." In cases which seem to resemble one grandparent in all particulars it is clear that the conception of pure germ cells may be strictly applied, but the author was familiar with cases of inbred hybrids which plainly show mixtures. These he is inclined to explain in two ways as follows: (1) "Union of two cells representing each

---

*The chromosome x of *Protenor*, which of all chromosomes in nonorthopteran forms most closely resembles the accessory, is also described by Montgomery (1901) as dividing in the reducing division, and failing to divide in the equation division — a fact which is the more remarkable because in *Protenor*, as in all Hemiptera-Heteroptera thus far described, reduction is accomplished in the *first* maturation division.

of the two original species would yield an offspring of the mixed type." (2) "Besides through the mixing just indicated, variability may be due also in some cases to the not infrequent inequalities in the division of individual chromosomes, through which varying proportions of the chromatin of each species may appear in certain of the mature germ cells" (p. 20).

The first of these explanations would accord with the result of Mendelian experiment but for the fact that it is erroneously applied (and without cytological grounds) to *all* the characters or chromosomes instead of to individuals. As for the second passage quoted, there can be little doubt that irregular division of chromosomes would be likely to produce marked variation, but as Guyer himself observes, *these irregularities increase with the degree of infertility*. It seems natural to conclude, therefore, that they are not only pathological but perhaps in part the cause of the infertile condition. Furthermore, on the hypothesis of individuality of chromosomes, which Guyer accepts, the loss of a portion of a chromosome by irregular division would be permanent and the effect of repetitions of the operation upon the descendants of a single chromosome group (which he regards as transmitted as a whole) would be so marked a depletion of chromatic substance as must lead soon to malfunction and ultimately to sterility.

As already noted the first of these two explanations of the causes of variation would allow only four possible combinations of chromosomes in the offspring of a single pair. But we know that except in the case of identical twins, duplicates practically never appear in the offspring of a pair however numerous the progeny. Therefore, whatever the number of the offspring, the variations of all except the few provided for by the four normal chromosome combinations must be accounted for by obviously pathological division processes, which tend strongly in the direction of sterility. But in the report of Bateson and Saunders to the Evolution Committee we find the statement: "We know no Mendelian case in which fertility is impaired" (p. 148). When we reflect that the vast majority of cases studied by these observers were Mendelian and connect this piece of evidence with the testimony of Cannon[23] that the maturation processes of variable cotton hybrids are either normal or so distinctly abnormal as to entail sterility and with Guyer's own

admission that the abnormalities in mitosis increase with the degree of sterility, the balance is strongly against the efficacy of pathological mitoses as factors in normal hybrid variation.

I take pleasure in acknowledging my indebtedness to Professor E. B. Wilson for invaluable counsel in the presentation of a subject offering many difficulties.

REFERENCES

1. Sutton, Walter S., "On the Morphology of the Chromosome Group in Brachystola magna," *Biol. Bull.*, IV., 1, 1902.
2. Mendel, Gregor Johann, "Vesuche über Pflanzen-Hybriden," *Verh. naturf. Vers. in Brünn* IV., and in Osterwald's *Klassiker der exakten Wissenschaft*. English translation in *Journ. Roy. Hort. Soc.*, XXVI., 1901. Later reprinted with modifications and corrections in Bateson's "Mendel's Principles of Heredity," Cambridge, 1902, p. 40.
3. Wilson, E. B., "Mendel's Principles of Heredity and the Maturation of the Germ-Cells," *Science*, XVI., 416.
4. Bateson, W., "Mendel's Principles of Heredity," Cambridge, 1902, p. 30.
5. Sutton, W. S., *loc. cit.*
6. The conclusion that synapsis involves a union of paternal and maternal chromosomes in pairs was first reached by Montgomery in 1901. Montgomery, T. H., Jr., "A Study of the Chromosomes of the Germ-Cells of Metazoa," *Trans. Amer. Phil. Soc.*, XX.
7. Cannon, W. A., "A Cytological Basis for the Mendelian Laws," *Bull. Torrey Botanical Club*, 29, 1902.
8. Absolute proof is impossible in a pure-bred form on account of the impossibility of distinguishing between maternal and paternal members of any synaptic pair. If, however, such hybrids as those obtained by Moenkhaus (Moenkhaus, W. J., "Early Development in Certain Hybrid Species," Report of Second Meeting of Naturalists at Chicago, *Science*, XIII., 323), with fishes can be reared to sexual maturity absolute proof of this point may be expected. This observer was able in the early cells of certain fish hybrids to distinguish the maternal from the paternal chromosomes by differences

in form, and if the same can be done in the maturation divisions the question of the distribution of chromosomes in reduction becomes a very simple matter of observation.

9. Boveri, Th., "Ueber Mehrpolige Mitosen als Mittel zur Analyse des Zellkerns," *Verb. d. Phys.-Med. Ges. zu Würzburg*, N. F., Bd. XXXV., 1902. It appears from a personal letter that Boveri had noted the correspondence between chromosomic behavior as deducible from his experiments and the results on plant hybrids — as indicated also in reference 1, *l. c., p.* 81.

10. Cannon, W. A., *loc. cit.*

11. Bateson and Saunders, Experimental Studies in the Physiology of Heredity (Reports to the Evolution Committee, I., London, 1902) p. 81, paragraphs 11 and 12.

12. *Cf.* Bateson and Saunders, *loc. cit.*

13. *Ibid.*

14. *Cf.* Mendel's experiments on *Hieracium.*

15. Bateson and Saunders, *loc. cit.*, p. 154.

16. *Cf.* Bateson and Saunders, pp. 135, 136.

17. Bateson and Saunders, p. 156.

18. Fick, R., "Mittheilung ueber Eireifung bei Amphibien," *Supp. Anat. Anz.*, XVI.

19. Montgomery, T. H., Jr., *loc. cit.*

20. It is of interest in connection with this question that there occurs regularly in each of the spermatogonial generations in *Brachystola* a condition of the nucleus which suggests amitosis but which in reality is nothing more than the enclosure of the different chromosomes in partially separated vesicles. *Cf.* Sutton, W. S., "The Spermatogonial Divisions in Brachystola Magna," *Kans. Univ. Quart.*, IX., 2.

21. McClung, C. E., "The Accessory Chromosome — Sex Determinant?" *Biol. Bull.*, III., 1 and 2, 1902. "Notes on the Accessory Chromosome," *Anat. Anz.*, XX., pp. 220–226.

22. Guyer, M. F., "Hybridism and the Germ Cell," *Bulletin of the University of Cincinnati*, No. 21, 1902.

23. Cannon, W. A., *loc. cit.*

Sex-limited inheritance in *Drosophila*

THOMAS H. MORGAN

Reprinted with permission from Science, *vol. 32, 1910, pp. 120–122.*

Much of the experimental support for the early conceptual foundations of genetics came out of a small laboratory at Columbia University, under the direction of Thomas Hunt Morgan. Because of the experimental organism employed in the laboratory, the room became known as the "fly room." Morgan was an embryologist-turned-geneticist, viewing the conceptual evolution in genetics with a developmentalist's eye. He initially rejected the idea of the discrete nature of the units of heredity. He felt that theory too closely approached the ideas of the early preformationists. He viewed the organism's phenotype as the result of the action of the total genetic content working in concert; and, he felt that the appearance of an organism could not be subdivided into discrete traits. However, when the increase in data supporting the discrete nature of the units of heredity gradually forced his allegiance to the Mendelian camp, he became one of its strongest supporters. Due to this transition, Morgan found the suggestion of a physical tie between gene and chromosome unconvincing. He felt there was no evidence to support the prediction that certain units should assort together if they were part of chromosomes. It seems paradoxical, then, that the first experimental evidence to support the chromosome theory should have been provided by Morgan.

The following paper presents his observations of the pattern of transmission of eye color in respect to the sex of the fly. As he prepared this paper, Morgan remained biased against the chromosome theory of inheritance. Initially he attempted to explain his results while maintaining his belief in the separate existence of the hereditary units controlling eye color and the chromosomes controlling sex determination (see Wilson's paper, page 191). The struggle between the experimental argument and Morgan's personal viewpoint is evident in the paper. Ultimately, he was driven to conclude that the unit for eye color was coupled to the sex-determining factor (*chromosome*). Although he attempted to differentiate between the unit of inheritance and the X chromosome, his subsequent finding of additional mutant phenotypes which displayed sex-linked patterns of inheritance led Morgan to accept the hypothesis that the units of inheritance were portions of the chromosome.

In a pedigree culture of *Drosophila* which had been running for nearly a year through a considerable number of generations, a male appeared with white eyes. The normal flies have brilliant red eyes.

The white-eyed male, bred to his red-eyed sisters, produced 1,237 red-eyed offspring, (F$_1$), and 3 white-eyed males. The occurrence of these three white-eyed males (F$_1$) (due evidently to further sporting) will, in the present communication, be ignored.

The F$_1$ hybrids, inbred, produced:

2,459 red-eyed females
1,011 red-eyed males
782 white-eyed males

No white-eyed females appeared. The new character showed itself therefore to be sex-limited in the sense that it was transmitted only to the grandsons. But that the character is not incompatible with femaleness is shown by the following experiment. The white-eyed male (mutant) was later crossed with some of his daughters (F$_1$), and produced:

187

129 red-eyed females
132 red-eyed males
88 white-eyed females
86 white-eyed males

The results show that the new character, white eyes, can be carried over to the females by a suitable cross, and is in consequence in this sense not limited to one sex. It will be noted that the four classes of individuals occur in approximately equal numbers (25 percent).

An Hypothesis to Account for the Results The results just described can be accounted for by the following hypothesis. Assume that all of the spermatozoa of the white-eyed male carry the "factor" for white eyes "W"; that half of the spermatozoa carry a sex factor "X," the other half lack it, i.e., the male is heterozygous for sex. Thus the symbol for the male is "WWX," and for his two kinds of spermatozoa WX–W.

Assume that all of the eggs of the red-eyed female carry the red-eyed "factor" R; and that all of the eggs (after reduction) carry one X each, the symbol for the red-eyed female will be therefore RRXX and that for her eggs will be RX–RX.

When the white-eyed male (sport) is crossed with his red-eyed sisters, the following combinations result:

WX–W (male)
RX–RX (female)
———————————
RWXX (50%)–RWX (50%)
Red female   Red male

When these F_1 individuals are mated, the following table shows the expected combinations that result:

RX–WX (F_1 female)
RX–W (F_1 male)
—————————————————————
RRXX–RWXX–RWX–WWX
(25%)   (25%)   (25%)   (25%)
Red   Red   Red   White
female female male   male

It will be seen from the last formulæ that the outcome is Mendelian in the sense that there are three reds to one white. But it is also apparent that all of the whites are confined to the male sex.

It will also be noted that there are two classes of red females — one pure RRXX and one hybrid RWXX — but only one class of red males (RWX). This point will be taken up later. In order to obtain these results it is necessary to assume, as in the last scheme, that, when the two classes of the spermatozoa are formed in the F_1 red male (RWX), R and X go together — otherwise the results will not follow (with the symbolism here used). This all important point can not be fully discussed in this communication.

The hypothesis just utilized to explain these results first obtained can be tested in several ways.

VERIFICATION OF HYPOTHESIS

First Verification If the symbol for the white male is WWX, and for the white female WWXX, the germ cells will be WX–W (male) and WX–WX (female), respectively. Mated, these individuals should give

WX–W (male)
WX–WX (female)
——————————————————————
WWXX (50%)–WWX (50%)
White female   White male

All of the offspring should be white, and male and female in equal numbers; this in fact is the case.

Second Verification As stated, there should be two classes of females in the F_2 generation, namely, RRXX and RWXX. This can be tested by pairing individual females with white males. In the one instance (RRXX) all the offspring should be red —

RX–RX (female)
WX–W (male)
——————————————
RWXX–RWX

and in the other instance (RWXX) there should be four classes of individuals in equal numbers, thus:

RX–WX (female)
WX–W (male)
———————————————————————
RWXX–WWXX–RWX–WWX

Tests of the F_2 red females show in fact that these two classes exist.

Third Verification The red F_1 females should all be RWXX, and should give with any white male the four combinations last described. Such in fact is found to be the case.

Fourth Verification The red F_1 males (RWX) should also be heterozygous. Crossed with white females (WWXX) all the female offspring should be red-eyed, and all the male offspring white-eyed, thus:

$$RX-W \text{ (red male)}$$
$$WX-WX \text{ (white female)}$$
$$\overline{RWXX-WWX}$$

Here again the anticipation was verified, for all of the females were red-eyed and all of the males were white-eyed.

CROSSING THE NEW TYPE WITH WILD MALES AND FEMALES

A most surprising fact appeared when a white-eyed female was paired to a wild, red-eyed male, i.e., to an individual of an unrelated stock. The anticipation was that wild males and females alike carry the factor for red eyes, but the experiments showed that all wild males are heterozygous for red eyes, and that all the wild females are homozygous. Thus when the white-eyed female is crossed with a wild red-eyed male, all of the female offspring are red-eyed, and all the male offspring white-eyed. The results can be accounted for on the assumption that the wild male is RWX. Thus:

$$RX-W \text{ (red male)}$$
$$WX-WX \text{ (white female)}$$
$$\overline{RWXX (50\%)-WWX (50\%)}$$

The converse cross between a white-eyed male RWX and a wild, red-eyed female shows that the wild female is homozygous both for X and for red eyes. Thus:

$$WX-W \text{ (white male)}$$
$$RX-RX \text{ (red female)}$$
$$\overline{RWXX (50\%)-RWX (50\%)}$$

The results give, in fact, only red males and females in equal numbers.

GENERAL CONCLUSIONS

The most important consideration from these results is that in every point they furnish the converse evidence from that given by Abraxas as worked out by Punnett and Raynor. The two cases supplement each other in every way, and it is significant to note in this connection that in nature only females of the sport *Abraxas lacticolor* occur, while in *Drosophila* I have obtained only the male sport. Significant, too, is the fact that analysis of the result shows that the wild female *Abraxas grossulariata* is heterozygous for color and sex, while in *Drosophila* it is the male that is heterozygous for these two characters.

Since the wild males (RWX) are heterozygous for red eyes, and the females (RXRX) homozygous, it seems probable that the sport arose from a change in a single egg of such a sort that instead of being RX (after reduction) the red factor dropped out, so that RX became WX or simply OX. If this view is correct it follows that the mutation took place in the egg of a female from which a male was produced by combination with the sperm carrying no X, no R (or W in our formulæ). In other words, if the formula for the eggs of the normal female is RX-RX, then the formula for the particular egg that sported will be WX; i.e., one R dropped out of the egg leaving it WX (or no R and one X), which may be written OX. This egg we assume was fertilized by a male-producing sperm. The formula for the two classes of spermatozoa is RX-O. The latter, O, is the male-producing sperm, which combining with the egg OX (see above) gives OOX (or WWX), which is the formula for the white-eyed male mutant.

The transfer of the new character (white eyes) to the female (by crossing a white-eyed male, OOX to a heterozygous female (F_1)) can therefore be expressed as follows:

$$OX-O \text{ (white male)}$$
$$RX-OX \text{ (F_1 female)}$$

RXOX	RXO	OOXX	OOX
Red	Red	White	White
female	male	female	male

It now becomes evident why we found it necessary to assume a coupling of R and X in one of the spermatozoa of the red-eyed F_1 hybrid (RXO). The fact is that this R and X are combined, and have never existed apart.

It has been assumed that the white-eyed mutant arose by a male-producing sperm (O) fertilizing an egg (OX) that had mutated. It may be asked what would have been the result if a female-producing sperm (RX) had fertilized this egg (OX)? Evidently a heterozygous female RXOX would arise, which, fertilized later by any normal male (RX–O) would produce in the next generation pure red females RRXX, red heterozygous females RXOX, red males RXO, and white males OOX (25 percent). As yet I have found no evidence that white-eyed sports occur in such numbers. Selective fertilization may be involved in the answer to this question.

The chromosomes in relation to the determination of sex in insects

E. B. WILSON

Reprinted with permission from Science, *vol. 22, 1905, pp. 500–502.*

In the last paper, Morgan incorporated a major premise: the sex of an individual was controlled by its chromosomes. Cytological evidence for this belief had accumulated since the latter part of the nineteenth century. Edmund B. Wilson's work was very important to both the empirical investigations and the theoretical considerations concerning the determination of sex. In 1905, in the short paper which follows, he clearly stated the level of understanding cytologists had attained of the chromosomal contribution to the determination of sex in insects. He qualified his review, by prophetically suggesting, at the end, that it probably was not only the qualitative distribution of chromosomes that determined sex; but, it was also the quantitative composition of the entire chromosomal complement. Experimental support for this statement would not be provided for another 30 years.

Material procured during the past summer demonstrates with great clearness that the sexes of Hemiptera show constant and characteristic differences in the chromosome groups, which are of such a nature as to leave no doubt that a definite connection of some kind between the chromosomes and the determination of sex exists in these animals. These differences are of two types. In one of these, the cells of the female possess one more chromosome than those of the male; in the other, both sexes possess the same number of chromosomes, but one of the chromosomes in the male is much smaller than the corresponding one in the female (which is in agreement with the observations of Stevens on the beetle *Tenebrio*). These types may conveniently be designated as *A* and *B*, respectively. The essential facts have been determined in three genera of each type, namely, (type *A*) *Protenor belfragei, Anasa tristis* and *Alydus pilosulus,* and (type *B*) *Lygaeus turcicus, Euschistus fissilis* and *Cœnus delius.* The chromosome groups have been examined in the dividing oogonia and ovarian follicle cells of the female and in the dividing spermatogonia and investing cells of the testis in case of the male.

Type *A* includes those forms in which (as has been known since Henking's paper of 1890 on *Pyrrochoris*) the spermatozoa are of two classes, one of which contains one more chromosome (the so-called accessory or heterotropic chromosome) than the other. In this type the somatic number of chromosomes in the female is an even one, while the somatic number in the male is one less (hence an odd number) the actual numbers being in *Protenor* and *Alydus* ♀ 14, ♂ 13, and in *Anasa* ♀ 22, ♂ 21. A study of the chromosome groups in the two sexes brings out the following additional facts. In the cells of the female all the chromosomes may be arranged two by two to form pairs, each consisting of two chromosomes of equal size, as is most obvious in the beautiful chromosome groups of *Protenor*, where the size differences of the chromosomes are very marked. In the male all the chromosomes may be thus symmetrically paired with the exception of one which is without a mate. This chromosome is the "accessory" or heterotropic one; and it is a consequence of its unpaired character that it passes into only one half of the spermatozoa.

In type *B* all of the spermatozoa contain the same number of chromosomes (half the somatic number in both sexes), but they are, nevertheless, of two classes, one of which contains a large and one a small "idiochromosome." Both sexes have the same somatic number of chromosomes (fourteen in the three examples mentioned above), but differ as follows: In the cells of the female (oogonia and follicle-cells) all of the chromosomes may, as in

191

type A, be arranged two by two in equal pairs, and a small idiochromosome is not present. In the cells of the male all but two may be thus equally paired. These two are the unequal idiochromosomes, and during the maturation process they are so distributed that the small one passes into one half of the spermatozoa, the large one into the other half.

These facts admit, I believe, of but one interpretation. Since all of the chromosomes in the female (oogonia) may be symmetrically paired, there can be no doubt that synapsis in this sex gives rise to the reduced number of symmetrical bivalents, and that consequently all of the eggs receive the same number of chromosomes. This number (eleven in *Anasa*, seven in *Protenor* or *Alydus*) is the same as that present in those spermatozoa that contain the "accessory" chromosome. It is evident that both forms of spermatozoa are functional, and that in type A females are produced from eggs fertilized by spermatozoa that contain the "accessory" chromosome, while males are produced from eggs fertilized by spermatozoa that lack this chromosome (the reverse of the conjecture made by McClung). Thus if n be the somatic number in the female $n/2$ is the number in all of the matured eggs, $n/2$ the number in one half of the spermatozoa (namely, those that contain the "accessory"), and $n/2 - 1$ the number in the other half. Accordingly:

In fertilization

$$\text{Egg } \frac{n}{2} + \text{spermatozoon } \frac{n}{2} = n(\text{female}).$$

$$\text{Egg } \frac{n}{2} + \text{spermatozoon } \frac{n}{2} - 1 = n - 1(\text{male}).$$

The validity of this interpretation is completely established by the case of *Protenor*, where, as was first shown by Montgomery, the "accessory" is at every period unmistakably recognizable by its great size. The spermatogonial divisions invariably show but one such large chromosome, while an equal pair of exactly similar chromosomes appear in the oogonial divisions. One of these in the female must have been derived in fertilization from the egg nucleus, the other (obviously the "accessory") from the sperm nucleus. It is evident, therefore, that all of the matured eggs must before fertilization contain a chromosome that is the maternal mate of the

"accessory" of the male, and that females are produced from eggs fertilized by spermatozoa that contain a similar group i.e., those containing the "accessory"). The presence of but one large chromosome (the "accessory") in the somatic nuclei of the male can only mean that males arise from eggs fertilized by spermatozoa that lack such a chromosome, and that the single "accessory" of the male is derived in fertilization from the egg nucleus.

In type B all of the eggs must contain a chromosome corresponding to the large idiochromosome of the male. Upon fertilization by a spermatozoon containing the large idiochromosome a female is produced, while fertilization by a spermatozoon containing the small one produces a male.

The two types distinguished above may readily be reduced to one; for if the small idiochromosome of type B be supposed to disappear, the phenomena become identical with those in type A. There can be little doubt that such has been the actual origin of the latter type, and that the "accessory" chromosome was originally a large idiochromosome, its smaller mate having vanished. The unpaired character of the "accessory" chromosome thus finds a complete explanation, and its behavior loses its apparently anomalous character.

The foregoing facts irresistibly lead to the conclusion that a causal connection of some kind exists between the chromosomes and the determination of sex; and at first thought they naturally suggest the conclusion that the idiochromosomes and heterotropic chromosomes are actually sex determinants, as was conjectured by McClung in case of the "accessory" chromosome. Analysis will show, however, that great, if not insuperable, difficulties are encountered by any form of the assumption that these chromosomes are specifically male or female sex determinants. It is more probable, for reasons that will be set forth hereafter, that the difference between eggs and spermatozoa is primarily due to differences of degree or intensity, rather than of kind, in the activity of the chromosome groups in the two sexes; and we may here find a clue to a general theory of sex determination that will accord with the facts observed in hemiptera. A significant fact that bears on this question is that in both types the two sexes differ in respect to the behavior of the idiochromosomes or "accessory" chromosomes during the synaptic and growth periods, these chromosomes assuming in the male the form of condensed chromosome nucleoli, while in the female they remain, like the other chromosomes, in a diffused condition. This indicates that during these periods these chromosomes play a more active part in the metabolism of the cell in the female than in the male. The primary factor in the differentiation of the germ cells may, therefore, be a matter of metabolism, perhaps one of growth.

In the seven years following Morgan's discovery of the initial white-eyed mutant, a vast quantity of experimental data in support of the chromosome theory of inheritance emerged from the Columbia University fly room. Much of this data was the work of an exceptional group of graduate students. One, A. H. Sturtevant, was still an undergraduate when, in a discussion with Morgan, he was struck with the possibility of mapping the positions of linked genes by examining the amount of recombination taking place between them. Six sex-linked characters had been detected. Using these characters and assuming they were controlled by regions of the X chromosome, he carried out the recombination studies detailed in the next paper. His results — with the exception of the recombination values used for establishing additivity between three relatively distant genes — supported a linear ordering of genes along the chromosome. H. J. Muller, another graduate student, successfully explained the deviations from additivity observed by Sturtevant. Muller proposed that the occurrence of double crossovers would reduce the apparent number of detectable recombinants between linked genes. Since double crossovers would be more likely to occur as the distance between linked genes increased, the reduction in recombinant values would occur almost exclusively in studies dealing with widely separated genes. He derived a formula which would compensate for the presence of double crossovers and when it was applied to Sturtevant's data, additivity was obtained — even for three relatively distant genes. Since the phenomenon of double crossovers assumes that genes are located on the chromosome in a linear sequence, Sturtevant's and Muller's investigations, together, provided experimental support for the chromosome theory of inheritance.

The second paper here is also the product of one of Morgan's graduate students, Calvin Bridges. Bridges was interested in the flies that exhibited exceptional phenotypes as a result of Morgan's crosses with sex-linked characters. Morgan was satisfied to explain these rare flies as the result of spontaneous changes which he called *sports.* Normally, if one crosses a female fly, homozygous for the recessive expression of a sex-linked character, with a male, hemizygous for the dominant expression for the same character, the next generation displays progeny in which the females now possess the parental male's dominant expression and the males possess the parental female's recessive expression. This has been named *crisscross inheritance.* Assuming the genes are portions of chromosomes, the chromosomal basis for this pattern of inheritance is due to the normal disjunction of homologous chromosomes during meiosis. Bridges observed that Morgan's rare flies exhibited phenotypes in which the male progeny resembled their male parents and the female progeny resembled their female parents. He suggested that a possible explanation for this occurrence was the breakdown of normal disjunction in the female parent. If this breakdown occurred, the rare eggs would either contain two X chromosomes or no sex chromosomes at all. Following fertilization, the gamete with the two X chromosomes would unite with a Y-chromosome-bearing gamete from the male — producing the rare, white-eyed female progeny; the nullo-X eggs would fuse with the X-chromosome-bearing sperm yielding a rare, red-eyed male. Had Bridges stopped at this point, his would have been an interesting hypothesis only. However, he formulated testable predictions as to what the chromosomal constitution of these rare flies should be if his explanation were valid. In addition, he predicted that the rare females, when analyzed genetically, should produce nondisjunctional events at an elevated frequency of occurrence, due to the presence of the extra Y chromosome. The results of the cytological examination of the flies and the genetic studies of the rare females correlated with his predictions — providing an elegant proof for the validity of the chromosome theory of inheritance.

The linear arrangement of six sex-linked factors in *Drosophila,* as shown by their mode of association

A. H. STURTEVANT

Reprinted with permission from Journal of Experimental Zoology, *vol. 14, 1913, pp. 43–59.*

HISTORICAL

The parallel between the behavior of the chromosomes in reduction and that of Mendelian factors in segregation was first pointed out by Sutton ('02) though earlier in the same year Boveri ('02) had referred to a possible connection. In this paper and others Boveri brought forward considerable evidence from the field of experimental embryology indicating that the chromosomes play an important role in development and inheritance. The first attempt at connecting any given somatic character with a definite chromosome came with McClung's ('02) suggestion that the accessory chromosome is a sex-determiner. Stevens ('05) and Wilson ('05) verified this by showing that in numerous forms there is a sex chromosome, present in all the eggs and in the female-producing sperm, but absent, or represented by a smaller homologue, in the male-producing sperm. A further step was made when Morgan ('10) showed that the factor for color in the eyes of the fly *Drosophila ampelophila* follows the distribution of the sex chromosome already found in the same species by Stevens ('08). Later, on the appearance of a sex-linked wing mutation in *Drosophila,* Morgan ('10a, '11) was able to make clear a new point. By crossing white-eyed, long-winged flies to those with red eyes and rudimentary wings (the new sex-linked character) he obtained, in F_2, white-eyed, rudimentary-winged flies. This could happen only if "crossing over" is possible; which means, on the assumption that both of these factors are in the sex chromosomes, that an interchange of materials between homologous chromosomes occurs (in the female only, since the male has only one sex chromosome). A point not noticed at this time came out later in connection with other sex-linked factors in *Drosophila* (Moran '11d). It became evident that some of the sex-linked factors are associated, i.e., that crossing over does not occur freely between some factors, as shown by the fact that the combinations present in the F_1 flies are much more frequent in F_2 than are new combinations of the same characters. This means, on the chromosome view, that the chromosomes, or at least certain segments of them, are more likely to remain intact during reduction than they are to interchange materials.[*] On the basis of these facts Morgan ('11 c, '11 d) has made a suggestion as to the physical basis of coupling. He uses Janssens' ('09) chiasmatype hypothesis as a mechanism. As he expresses it (Morgan '11 c):

If the materials that represent these factors are contained in the chromosomes, and if those that "couple" be near together in a linear series, then when the parental pairs (in the heterozygote) conjugate like regions will stand opposed. There is good evidence to support the view that during the strepsinema stage homologous chromosomes twist around each other, but when the chromosomes separate (split) the split is in a single plane, as maintained by Janssens. In consequence, the original materials will, for short distances, be more likely to fall on the same side of the split, while remoter regions will be as likely to fall on the same side as the last, as on the opposite side. In consequence, we find coupling in certain characters, and little or no evidence at all of coupling in other characters, the difference depending on the linear distance apart of the chromosomal materials that represent the factors. Such an explanation will account for all the many phenomena that I have observed and will explain equally, I think, the other cases so far described. The results are a simple mechanical result of the location of the materials in the chromosomes, and of the method of union of homologous chromosomes, and the proportions that result are not so much the expression of a numerical system as of the relative location of the factors in the chromosomes.

---

[*]It is interesting to read, in this connection, Lock's ('06, p. 248–253) discussion of the matter.

SCOPE OF THIS INVESTIGATION

It would seem, if this hypothesis be correct, that the proportion of "crossovers" could be used as an index of the distance between any two factors. Then by determining the distances (in the above sense) between A and B and between B and C, one should be able to predict AC. For, if proportion of crossovers really represents distance, AC must be approximately, either AB plus BC, or AB minus BC, and not any intermediate value. From purely mathematical considerations, however, the sum and the difference of the proportion of crossovers between A and B and those between B and C are only *limiting* values for the proportion of crossovers between A and C. By using several pairs of factors one should be able to apply this test in several cases. Furthermore, experiments involving three or more sex-linked allelomorphic pairs together should furnish another and perhaps more crucial test of the view. The present paper is a preliminary report of the investigation of these matters.

   I wish to thank Dr. Morgan for his kindness in furnishing me with material for this investigation, and for his encouragement and the suggestions he has offered during the progress of the work. I have also been greatly helped by numerous discussions of the theoretical side of the matter with Messrs. H. J. Muller, E. Altenburg, C. B. Bridges, and others. Mr. Muller's suggestions have been especially helpful during the actual preparation of the paper.

THE SIX FACTORS CONCERNED

In this paper I shall treat of six sex-linked factors and their interrelationships. These factors I shall discuss in the order in which they seem to be arranged.

   B stands for the black factor. Flies recessive with respect to it (b) have yellow body color. The factor was first described and its inheritance given by Morgan ('11a).

   C is a factor which allows color to appear in the eyes. The white-eyed fly (first described by Morgan '10) is now known to be always recessive with respect both to C and to the next factor.

   O. Flies recessive with respect to O (o) have eosin eyes. The relation between C and O has been ex-

plained by Morgan in a paper now in print and about to appear in the Proceedings of the Academy of Natural Sciences in Philadelphia.

   P. Flies with p have vermilion eyes instead of the ordinary red (Morgan '11 d).

   R. This and the next factor both affect the wings. The normal wing is RM. The rM wing is known as miniature, the Rm as rudimentary, and the rm as rudimentary-miniature. This factor R is the one designated L by Morgan ('11 d) and Morgan and Cattell ('12). The L of Morgan's earlier paper ('11) was the next factor.

   M. This has been discussed above, under R. The miniature and rudimentary wings are described by Morgan ('11 a).

   The relative position of these factors is

$$B, \frac{C}{O}, P, R, M$$

C and O are placed at the same point because they are completely linked. Thousands of flies had been raised from the cross CO (red) by co (white) before it was known that there were two factors concerned. The discovery was finally made because of a mutation and not through any crossing over. It is obvious, then, that unless coupling strength be variable, the same gametic ratio must be obtained whether, in connection with other allelomorphic pairs, one uses CO (red) as against co (white), Co (eosin) against co (white), or CO (red) against Co (eosin) (the cO combination is not known).

METHOD OF CALCULATING STRENGTH OF ASSOCIATION

In order to illustrate the method used for calculating the gametic ratio I shall use the factors P and M. The cross used in this case was, long-winged, vermilion-eyed female by rudimentary-winged, red-eyed male. The analysis and results are seen in Table 1.

   It is of course obvious from the figures that there is something peculiar about the rudimentary-winged flies, since they appear in far too small numbers. This point need not detain us here, as it always comes up in connection with rudimentary crosses, and is being investigated by Morgan. The point of

Table 1

| Long-vermilion | ♀—MpX MpX |
| Rudimentary-red | ♂—mPX |

F_1

| MpX mPX—long-red | ♀ |
| MpX   —long-vermilion | ♂ |

Gametes F_1

| Eggs —MPX mPX MpX mpX |
| Sperm—MpX |

F_2

MPX MpX ⎫	—long-red ♀—451
mPX MpX ⎭	
MpX MpX ⎫	—long-vermilion ♀—417
mpX MpX ⎭	
MPX	—long-red ♂—105
mPX	—rudimentary-red ♂—33
MpX	—long-vermilion ♂—316
mpX	—rudimentary-vermilion ♂—4

interest at present is the linkage. In the F_2 generation the original combinations, red–rudimentary and vermilion–long, are much more frequent in the males (allowing for the low viability of rudimentary) than are the two new or crossover combinations, red–long and vermilion–rudimentary. It is obvious from the analysis that no evidence of association can be found in the females, since the M present in all female-producing sperm masks m when it occurs. But the ratio of crossovers in the gametes is given without complication by the F_2 males, since the male-producing sperm of the F_1 male bore no sex-linked genes. There are in this case 349 males in the noncrossover classes and 109 in the crossovers. The method which has seemed most satisfactory for expressing the relative position of factors, on the theory proposed in the beginning of this paper, is as follows. The unit of "distance" is taken as a portion of the chromosome of such length that, on the aver-

age, one crossover will occur in it out of every 100 gametes formed. That is, percent of crossovers is used as an index of distance. In the case of P and M there occurred 109 crossovers in 405 gametes, a ratio of 26.9 in 100; 26.9, the percent of crossovers, is considered as the "distance" between P and M.

THE LINEAR ARRANGEMENT OF THE FACTORS

Table 2 shows the proportion of crossovers in those cases which have been worked out. The detailed results of the crosses involved are given at the end of this paper. The 16287 cases of B and CO are from Dexter ('12). Inasmuch as C and O are completely linked I have added the numbers for C, for O, and for C and O taken together, giving the total results in the lines beginning (C, O) P, B (C, O), etc., and have used these figures, instead of the individual C, O, or CO results, in my calculations. The fractions in the column marked "proportion of crossovers" represent the number of crossovers (numerator) to total available gametes (denominator).

As will be explained later, one is more likely to obtain accurate figures for distances if those distances are short, i.e., if the association is strong. For this reason I shall, in so far as possible, use the percent of crossovers between adjacent points in mapping out the distances between the various factors. Thus, B (C, O), (C, O) P, PR, and PM form the basis of the diagram. The figures on the diagram at the bottom of the page represent calculated distances from B.

Of course there is no knowing whether or not these distances as drawn represent the actual relative spatial distances apart of the factors. Thus the distance CP may in reality be shorter than the distance BC, but what we do know is that a break is far more likely to come between C and P than between B and C. Hence, either CP is a long space, or else it is for some reason a weak one. The point I wish to make here is that we have no means of knowing that the

O		P R		M
B C				
├──┼──	────────────	├──┼──	────────	──┤
0.0 1.0		30.7 33.7		57.6

Table 2

Factors concerned	Proportion of crossovers	Percent of crossovers
BCO	$\dfrac{193}{16287}$	1.2
BO	$\dfrac{2}{373}$	0.5
BP	$\dfrac{1464}{4551}$	32.2
BR	$\dfrac{115}{324}$	35.5
BM	$\dfrac{260}{693}$	37.6
COP	$\dfrac{224}{748}$	30.0
COR	$\dfrac{1643}{4749}$	34.6
COM	$\dfrac{76}{161}$	47.2
OP	$\dfrac{247}{836}$	29.4
OR	$\dfrac{183}{538}$	34.0
OM	$\dfrac{218}{404}$	54.0
CR	$\dfrac{236}{829}$	28.5
CM	$\dfrac{112}{333}$	33.6
B(C, O)	$\dfrac{214}{21736}$	1.0
(C, O)P	$\dfrac{471}{1584}$	29.7
(C, O)R	$\dfrac{2062}{6116}$	33.7
(C, O)M	$\dfrac{406}{898}$	45.2
PR	$\dfrac{17}{573}$	3.0
PM	$\dfrac{109}{405}$	26.9

chromosomes are of uniform strength, and if there are strong or weak places, then that will prevent our diagram from representing actual relative distances — but, I think, will not detract from its value as a diagram.

Just how far our theory stands the test is shown by Table 3, giving observed percent of crossovers, and distances as calculated from the figures given in the diagram of the chromosome. Table 3 includes all pairs of factors given in Table 2 but not used in the preparation of the diagram.

It will be noticed at once that the long distances, BM, and (C, O) M, give smaller percent of crossovers, than the calculation calls for. This is a point which was to be expected, and will be discussed later. For the present we may dismiss it with the statement that it is probably due to the occurrence of two breaks in the same chromosome, or "double crossing over." But in the case of the shorter distances the correspondence with expectation is perhaps as close as was to be expected with the small numbers that are available. Thus, BP is 3.2 less than BR, the difference expected being 3.0. (C, O) R is less than BR by 1.8 instead of by 1.0. It has actually been found possible to predict the strength of association between two factors by this method, fair approximations having been given for BR and for certain combinations involving factors not treated in this paper, before the crosses were made.

DOUBLE CROSSING OVER

On the chiasmatype hypothesis it will sometimes happen, as shown by Dexter ('12) and intimated by Morgan ('11d) that a section of, say, maternal chromosome will come to have paternal elements at both ends, and perhaps more maternal segments

Table 3

Factors	Calculated distance	Observed percent of crossovers
BP	30.7	32.2
BR	33.7	35.5
BM	57.6	37.6
(C, O)R	32.7	33.7
(C, O)M	56.6	45.2

beyond these. Now if this can happen it introduces a complication into the results. Thus, if a break occurs between B and P, and another between P and M, then, unless we can follow P also, there will be no evidence of crossing over between B and M, and the fly hatched from the resulting gamete will be placed in the noncrossover class, though in reality he represents two crossovers. In order to see if double crossing over really does occur it is necessary to use three or more sex-linked allelomorphic pairs in the same experiment. Such cases have been reported by Morgan ('11d) and Morgan and Cattell ('12) for the factors B, CO, and R. They made such crosses as long–gray–red by miniature–yellow–white, and long–yellow–red by miniature–gray–white, etc. The details and analyses are given in the original papers, and for our present purpose it is only the flies that are available for observations on double crossing over that are of interest. Table 4 gives a graphical representation of what happened in the 10495 cases.

Double crossing over does then occur, but it is to be noted that the occurrence of the break between B and CO tends to prevent that between CO and R (or vice versa). Thus where B and CO did not separate, the gametic ratio for CO and R was about 1 to 2, but in the cases where B and CO did separate it was about 1 to 6.5. Three similar cases from my own results, though done on a smaller scale, are given in the table at the end of this paper. The results are represented in Tables 5, 6 and 7.

It will be noted that here also the evidence, so far as it goes, indicated that the occurrence of one cross-over makes another one less likely to occur in the same gamete. In the case of BOPR there was an opportunity for triple crossing over, but it did not occur. Of course, on the view here presented there is no reason why it should not occur, if enough flies were raised. An examination of the figures will

show that it was not to be expected in such small numbers as are here given. So far as I know there is, at present, no evidence that triple crossing over takes place, but it seems highly probable that it will be shown to occur.[*]

Unfortunately, in none of the four cases given above are two comparatively long distances involved, and in only one are there enough figures to form a fair basis for calculation, so that it seems as yet hardly possible to determine how much effect double crossing over has in pulling down the observed percent of crossovers in the case of BM and (C, O) M. Whether or not this effect is partly counter-balanced by triple crossing over must also remain unsettled as yet. Work now under way should furnish answers to both these questions.

---

[*] A case of triple crossing over within the distance CR was observed after this paper went to press.

Table 4

No crossing over	Single crossing over		Double crossing over
B CO R	B CO R	B CO R	B CO R
6972	3454	60	9

Table 5

No crossing	Single crossing over		Double crossing over
O P R	O P R	O P R	O P R
194	102	11	1

Table 6

No crossing	Single crossing over		Double crossing over
B O M	B O M	B O M	B O M
278	160	1	0

Table 7

B O P R	B O P R	B O P R	B O P R	B O P R	B O P R	B O P R	B O P R
393	203	19	6	2	1	1	0

Table 8

(The meaning of the phrase 'proportion of crossovers' is given on p. 196)

BO. P_1: gray-eosin ♀ X yellow-red ♂
      F_1: gray-red ♀ X gray-eosin ♂
F_2: ♀♀, g.r. 241, g.e. 196
      ♂♂, g.r. 0, g.e. 176, y.r. 195, y.e. 2

Proportion of crossovers, $\dfrac{2}{373}$

BP. P_1: gray-red ♀ X yellow-vermilion ♂
      F_1: gray-red ♀ X gray-red ♂
F_2: ♀♀, g.r. 98;
      ♂♂, g.r. 59, g.v. 16, y.r. 24, y.v. 33
      Black cross, F_1 gray-red ♀♀ from above X yellow-vermilion ♂♂
F_2: ♀♀, g.r. 31, g.v. 11, y.r. 12, y.v. 41
      ♂♂, g.r. 23, g.v. 13, y.r. 8 y.v. 21
      P_1: gray-vermilion ♀ X yellow-red ♂
      F_1: gray-red ♀ X gray-vermilion ♂
F_2: ♀♀, g.r. 199, g.v. 182
      ♂♂, g.r. 54, g.v. 149, y.r. 119, y.v. 41
      P_1: yellow-vermilion ♀ X gray-red ♂
      F_1: gray-red ♀ X yellow-vermilion ♂
F_2: ♀♀, g.r. 472, g.v. 240, y.r. 213, y.v. 414
      ♂♂, g.r. 385, g.v. 186, y.r. 189, y.v. 324
      F_1: gray-vermilion X yellow-red (sexes not recorded)
      F_1: gray-red ♀♀. These were mated to yellow-vermilion ♂♂ of other stock
F_2: ♀♀, g.r. 50, g.v. 96, y.r. 68, y.v. 41
      ♂♂, g.r. 44, g.v. 105, y.r. 86, y.v. 47

Proportion of crossovers, adding ♀♀ from BOPR (below), $\dfrac{1464}{4551}$

BR. P_1: miniature-yellow ♀ X long-gray ♂
      F_1: long-gray ♀ X miniature-yellow ♂
F_2: ♀♀, l.g. 14, l.y. 2, m.g. 7, m.y. 6
      ♂♂, l.g. 10, l.y. 1, m.g. 6, m.y. 8
      P_1: long-yellow ♀ X miniature-gray ♂
      F_1: long-gray ♀ X long-yellow ♂
F_2: ♀♀, l.g. 148, l.y. 130
      ♂♂, l.g. 51, l.y. 82, m.g. 89, m.y. 48

Proportion of crossovers, $\dfrac{115}{324}$

BM. P_1: long-yellow ♀ X rudimentary-gray ♂
      F_1: long-gray ♀ X long-yellow ♂
F_2: ♀♀, l.g. 591, l.y. 549
      ♂♂, l.g. 228, l.y. 371, r.g. 20, r.y. 3
      P_1: long-gray ♀ X rudimentary-yellow ♂

      F_1: long-gray ♀ X long-gray ♂
F_2: ♀♀, l.g. 152
      ♂♂, l.g. 42, l.y. 29, r.g. 0, r.y. 0

Proportion of crossovers, $\dfrac{260}{693}$

COP. P_1: vermilion ♀ X white ♂
       F_1: red ♀ X vermilion ♂
F_2: ♀♀, r. 320, v. 294
      ♂♂, r. 86, v. 206, w. 211
(7 of the vermilion ♀♀ known from tests to be CC, 2 known to be Cc. 7 white ♂♂ Pp, 2 pp.)
      Back cross. F_1 red ♀♀ from above X white ♂♂, gave
F_2: ♀♀, r. 195, w. 227,
      ♂♂, r. 66, v. 164, w. 184
      Our cross, F_1 ♀♀ as above X white ♂♂ recessive in P, gave
F_2: ♀♀, r. 35, v. 65, w. 98
      ♂♂, r. 33, v. 75, w. 95

Proportion of crossovers, $\dfrac{224}{748}$

COR. P_1: miniature-white ♀ X long-red ♂
       F_1: long-red ♀ X miniature-white ♂
F_2: ♀♀, l.r. 193, l.w. 109, m.r. 124, m.w. 208
      ♂♂, l.r. 202, l.w. 114, m.r. 123, m.w. 174
      P_1: long-white ♀ X miniature-red ♂
      F_1: long-red ♀ X long-white ♂
F_2: ♀♀, l.r. 194, l.w. 160
      ♂♂, l.r. 52, l.w. 124, m.r. 97, m.w. 41

Proportion of crossovers, $\dfrac{563}{1561}$; or, adding such available figures from

Morgan ('11 d) and Morgan and Cattell ('12) as are not complicated by the presence of yellow or brown flies, $\dfrac{1643}{4749}$

COM. P_1: long-white ♀ X rudimentary-red ♂
       F_1: long-red ♀ X long-white ♂
F_2: ♀♀, l.r. 157, l.w. 127
      ♂♂, l.r. 74, l.w. 82, ru.r. 3, ru.w. 2

Proportion of crossovers, $\dfrac{76}{161}$

OP. P_1: black-red ♀ X black eosin-vermilion ♂
      F_1: black-red ♀ X black-red ♂
F_2: (all black), ♀♀, r. 885
             ♂♂, r. 321, v. 125, e. 122, e.-v. 268

Proportion of crossovers, $\dfrac{247}{836}$

Table 8 continued

OR. P_1: long–red ♀ X miniature–eosin ♂
       F_1: long–red ♀ X long–red ♂
F_2:   ♀♀, l.r. 408
       ♂♂, l.r. 145, l.e. 67, m.r. 70, m.e. 100
       P_1: long–eosin ♀ X miniature–red ♂
       F_1: long–red ♀ X long–eosin ♂
F_2:   ♀♀, l.r. 100, l.e. 95
       ♂♂, l.r. 27, l.e. 54, m.r. 56, m.e. 19

       Proportion of crossovers, $\dfrac{183}{538}$

OM. P_1: long–eosin ♀ X rudimentary–red ♂
       F_1: long–red ♀ X long–eosin ♂
F_2:   ♀♀, l.r. 368, l.e. 266
       ♂♂, l.r. 194, l.e. 146, ru.r. 40, ru.e. 24

       Proportion of crossovers, $\dfrac{218}{404}$

CR. P_1: long–white ♀ X miniature–eosin ♂
       F_1: long–eosin ♀ X long–white ♂
F_2:   ♀♀, l.e. 185, l.w. 205
       ♂♂, l.e. 54, l.w. 147, m.e. 149, m.w. 42
       P_1: long–eosin ♀ X miniature–white ♂
       F_1: long–eosin ♀ X long–eosin ♂
F_2:   ♀♀, l.e. 527
       ♂♂, l.e. 169, l.w. 85, m.e. 55, m.w. 128

       Proportion of crossovers, $\dfrac{236}{829}$

CM. P_1: long–white ♀ X rudimentary–eosin ♂
       F_1: long–eosin ♀ X long–white ♂
F_2:   ♀♀, l.e. 328, l.w. 371
       ♂♂, l.e. 112, l.w. 217, ru.e. 4, ru.w. 0

       Proportion of crossovers, $\dfrac{112}{333}$

PR. P_1: long–vermilion (yellow) ♀ X miniature–red
       (yellow) ♂
       F_1: long–red–yellow ♀ X long–vermilion–yellow ♂
F_2:   (all y.) ♀♀, l.r. 138, l.v. 110
             ♂♂, l.r. 8, l.v. 117, m.r. 97, m.v. 1

P_1: long–vermilion (gray) ♀ X miniature–red ♂
   F_1: long–red ♀ X long–vermilion ♂
F_2:   ♀♀, l.r. 116, l.v. 110
       ♂♂, l.r. 2, l.v. 81, m.r. 96, m.v. 1
       P_1: miniature–red ♀ X long–vermilion ♂
       F_1: long–red ♀ X miniature–red ♂
F_1:   ♀♀, l.r. 45, m.r. 49
       ♂♂, l.r. 1, l.v. 27, m.r. 26, m.v. 0
       F_1: long–red ♀♀ from above X miniature–red ♂♂
       of other stock, gave
F_2:   ♀♀, l.r. 74, m.r. 52
       ♂♂, l.r. 8, l.v. 66, m.r. 46, m.v. 1

       Proportion of crossovers, $\dfrac{17}{573}$

PM. P_1: long–vermilion ♀ X rudimentary–red ♂
       F_1: long–red ♀ X long–vermilion ♂
F_2:   ♀♀, l.r. 451, l.v. 417
       ♂♂, l.r. 105, l.v. 316, ru.r. 33, ru.v. 4

       Proportion of crossovers, $\dfrac{109}{405}$

OPR. P_1: long–vermilion ♀ X miniature–eosin ♂
       F_1: long–red ♀ X long–vermilion ♂
F_2:   ♀♀, l.r. 205, l.v. 182
       ♂♂, l.r. 1, l.v. 109, l.e. 8, l.e.–v. 53, m.r. 49, m.v. 3,
       m.e. 85, m.e.–v. 0

BOM. P_1: long–red–yellow ♀ X rudimentary–eosin–gray ♂
       F_1: long–red–gray ♀ X long–red–yellow ♂
F_2:   ♀♀, l.r.g. 530, l.r.y. 453
       ♂♂, l.r.g. 1, l.r.y. 274, l.e.g. 156, l.e.y. 0, ru.r.g. 0,
       ru.r.y. 4, ru.e.g. 4, ru.e.y. 0

BOPR. P_1: long–vermilion–brown ♀ X miniature–eosin–
       black ♂
       F_1: long–red–black ♀ X long–vermilion–brown ♂
F_2:   ♀♀, l.r.bl. 305, l.r.br. 113, l.v.bl. 162, l.v.br. 256
       ♂♂, l.r.bl. 0, l.r.br. 2, l.v.bl. 3, l.v.br. 185, l.e.bl. 9,
       l.e.br. 0, l.e.–v.bl. 127, l.e.–v.br. 0, m.r.bl. 1, m.r.br.
       76, m.v.bl. 1, m.v.br. 10, m.e.bl. 208, m.e.br. 3,
       m.e.–v.bl. 0, m.e.–v.br. 0

POSSIBLE OBJECTIONS TO
THESE RESULTS

It will be noted that there appears to be some variation in coupling strength. Thus, I found (CO) R to
be 36.7; Morgan and Cattell obtained the result
33.9; for OR I got 34.0, and for CR, 28.5. The
standard error for the difference between (CO) R

(all figures) and CR is 1.84 percent, which means
that a difference of 5.5 percent is probably significant (Yule '11, p. 264). The observed difference is
6.1 percent, showing that there is some complication
present. Similarly, BM gave 37.6, while OM gave
54.0 — and BOM gave 36.7 for BM, and 36.5 for
OM. There is obviously some complication in these

cases, but I am inclined to think that the disturbing factor discussed below (viability) will explain this. However, experiments are now under way to test the effect of certain external conditions on coupling strength. It will be seen that on the whole when large numbers are obtained in different experiments and are averaged, a fairly consistent scheme results. Final judgment on this matter must, however, be withheld until the subject can be followed up by further experiments.

Another point which should be considered in this connection is the effect of differences in viability. In the case of P and M, used above as an illustration, the rudimentary-winged flies are much less likely to develop than are the longs. Now if the viability of red and vermilion is different, then the longs do not give a fair measure of the linkage, and the rudimentaries, being present in such small numbers, do not even up the matter. It is probable that there is no serious error due to this cause except in the case of rudimentary crosses, since the two sides will tend to even up, unless one is very much less viable than the other, and this is true only in the case of rudimentary. It is worth noting that the only serious disagreements between observation and calculation occur in the case of rudimentary crosses (BM, and (CO) M). Certain data of Morgan's now in print, and further work already planned, will probably throw considerable light on the question of the position and behavior of this factor M.

SUMMARY

It has been found possible to arrange six sex-linked factors in *Drosophila* in a linear series, using the number of crossovers per 100 cases as an index of the distance between any two factors. This scheme gives consistent results, in the main.

A source of error in predicting the strength of association between untried factors is found in double crossing over. The occurrence of this phenomenon is demonstrated, and it is shown not to occur as often as would be expected from a purely mathematical point of view, but the conditions governing its frequency are as yet not worked out.

These results are explained on the basis of Morgan's application of Janssens' chiasmatype hypothesis to associative inheritance. They form a new

argument in favor of the chromosome view of inheritance, since they strongly indicate that the factors investigated are arranged in a linear series, at least mathematically.

REFERENCES

Boveri, T., 1902. "Ueber mehrpolige Mitosen als Mittel zur Analyse des Zellkerns." *Verh, Phys.-Med. Ges Würzburg.*, N.F., Bd. 35, p. 67.

Dexter, J. S., 1912. On coupling of certain sex-linked characters in Drosophila. *Biol. Bull.*, vol. 23, p. 183.

Janssens, F. A., 1909. La théorie de la chiasmatypie. *La Cellule,* tom. 25, p. 389.

Lock, R. H., 1906. Recent progress in the study of variation, heredity, and evolution. London and New York.

McClung, C. E., 1902. The accessory chromosome — sex determinant? *Biol. Bull.*, vol. 3, p. 43.

Morgan, T. H., 1910. Sex-limited inheritance in Drosophila. *Science,* n.s., vol. 32, p. 1.
1910a. The method of inheritance of two sex-limited characters in the same animal. *Proc. Soc. Exp. Biol. Med.,* vol. 8, p. 17.
1911. The application of the conception of pure lines to sex-limited inheritance and to sexual dimorphism. *Amer. Nat.,* vol. 45, p. 65.
1911a. The origin of nine wing mutations in Drosophila. *Science,* n.s., vol. 33, p. 496.
1911b. The origin of five mutations in eye color in Drosophila and their modes of inheritance. *Science,* n.s., vol. 33, p. 534.
1911c. Random segregation versus coupling in Mendelian inheritance. *Science,* n.s., vol. 34, p. 384.
1911d. An attempt to analyze the constitution of the chromosomes on the basis of sex-limited inheritance in Drosophila. *Jour. Exp. Zoöl.,* vol. 11, p. 365.

Morgan, T. H. and Cattell, E., 1912. Data for the study of sex-limited inheritance in Drosophila. *Jour. Exp. Zoöl.,* vol. 13, p. 79.

Stevens, N. M., 1905. Studies in spermatogenesis with special reference to the "accessory chromosome." *Carnegie Inst. Washington Publ.,* 36.
1908. A study of the germ-cells of certain Diptera. *Jour. Exp. Zoöl.,* vol. 5, p. 359.

Sutton, W. S., 1902. On the morphology of the chromosome group in Brachystola magna. *Biol. Bull.*, vol. 4, p. 39.

Wilson, E. B., 1905. The behavior of the idiochromosomes in Hemiptera. *Jour. Exp. Zoöl.*, vol. 2, p. 371.

1906. The sexual differences of the chromosome-groups in Hemiptera, with some considerations on the determination and inheritance of sex. *Jour. Exp. Zoöl.*, vol. 3, p. 1.

Yule, G. U., 1911. *An introduction to the theory of statistics.* London.

Excerpt from: Nondisjunction as proof of the chromosome theory of heredity

CALVIN B. BRIDGES

Reprinted with permission from Genetics, vol. 1:1-52, 1916, pp. 107–163.

PRIMARY NONDISJUNCTION IN THE FEMALE

Ordinarily in a cross to a male with the dominant character all the sons and none of the daughters show the recessive sex-linked characters of the mother. Similarly, all the daughters and none of the sons show the dominant sex-linked characters of the father. The peculiarity of nondisjunction is that sometimes a female transcends these rules and produces a daughter like herself (a matroclinous daughter) or a son like the father (a patroclinous son), while the rest of the offspring are perfectly regular, showing the expected crisscross inheritance. Such exceptions, produced by a normal XX female, may be called primary.

The production of primary exceptions by a normal XX female may be supposed to result from an aberrant reduction division at which *the two X chromosomes fail to disjoin from each other. In consequence both remain in the egg or both pass out into the polar body.* In the former case the egg will be left with two X chromosomes and in the latter case with no X.

If the genes for sex-linked characters are carried by the X chromosomes, then each of the X chromosomes of the XX egg of a vermilion female will carry the gene for vermilion. The fertilization of such XX and zero eggs by the X and by the Y spermatozoa of a wild male will result in four new types of zygotes.

1. The XX egg fertilized by the X sperm gives an XXX zygote which might be expected to develop into a female. No females of this class have been found, and it is certain that they die.

2. The fertilization of the XX egg by the Y sperm gives rise to a female having an extra Y chromosome (XXY). Since both of the X chromosomes came from the vermilion-eyed mother, this daughter must be a vermilion *matroclinous exception.*

3. The fertilization of the zero egg by the X sperm gives rise to a male which has no Y chromosome (XO), and whose X coming from his red-eyed father brings in the red gene which makes the son a *patroclinous exception.* These XO males are viable but are completely sterile.

4. The zero egg by the Y sperm gives a zygote (OY) which is not viable.

Perhaps the cause of the initial aberrant reduction which constitutes primary nondisjunction is a mechanical entanglement (an incomplete untwisting from a strepsinema stage) of the two X chromosomes, resulting in a delayed reduction. In such cases the formation of the cell boundaries would catch the lagging X's and include them in one or the other cell, and perhaps very often (as in certain nematodes) would prevent their leaving the middle of the spindle to join either daughter nucleus. If such an occurrence were common there should be more zero than XX eggs and consequently more primary exceptions should be males than females.

In studying primary nondisjunction we are dependent on what material chance offers, since we know of no means of controlling the process. It is equally as likely that an egg produced by primary nondisjunction will become a nonviable zygote (XXX and OY) as that it shall be viable (XXY and XO). For this reason it is impossible to detect half of those cases of primary nondisjunction which really occur.

The XO male is viable and should offer an interesting field for further work, but — he is sterile. The direct opening offered for further work is through the matroclinous XXY daughter, which is perfectly fertile and which produces further exceptions which we may call secondary.

PRIMARY NONDISJUNCTION IN THE MALE

If primary nondisjunction occurred in the male, XY and zero sperm would be formed, but the zygotes from them would not differ in their sex-linked characters from regular offspring, so that such an occurrence could not be detected immediately. However, the XY sperm would give rise to XXY daughters, and these in turn would produce secondary exceptions which could be observed.

Primary nondisjunction has been actually seen to occur in the male of *Metapodius*. Wilson found three spermatocytes in which X followed Y to one pole at the reduction division (Wilson 1909).

SECONDARY NONDISJUNCTION IN THE FEMALE

It has been shown that matroclinous daughters of the constitution XXY may arise as the result of primary nondisjunction. The results from the outcrossing of several matroclinous daughters to males having other sex-linked characters were given in 1913. Of unusual interest was the appearance in F_1 of about four percent of further exceptions (secondary exceptions). That is, about four percent of the daughters were like the mother and four percent of the sons were like the father. The remaining sons and daughters were of the kinds expected.

The explanation given at first for the fact that exceptional daughters inherit from their mother the power of producing exceptions, was that each X of the exceptional female carried a gene which caused these chromosomes to undergo reduction abnormally in a small percentage of cases. Since these chromosomes descend directly to their exceptional daughters, they would transmit to those daughters the same gene and consequently the same power of producing exceptions.

Later work has provided data which can not be explained by appealing to the action of a gene in the X chromosome, and which prove that these secondary exceptions are due to the presence of the extra Y. In an XXY female there are three homologous sex chromosomes, between any two of which synapsis may occur, that is, synapsis may be of the XX or the XY type (homo- and heterosynapsis). In only about sixteen percent of cases does heterosynapsis occur, while about 84 percent of cases are homosynaptic and the Y remains unsynapsed. At the reduction division the two chromosomes that have synapsed, disjoin, one going to each pole, and the free chromosome goes to one pole as often with the one as with the other of the disjoined chromosomes. Thus, after heterosynapsis the reduction divisions are of two kinds, the XX-Y and the X-XY types. Half the eggs that come from the XX-Y type of reduction are XX and the other half are Y. For the X-XY type the eggs are X and XY, as many of one kind as of the other. After homosynapsis all the reductions are X-XY. As a result of reduction of these two types there are four classes of eggs — two of which, X and XY, are composite and large (46 percent), and two of which, XX and Y are of single origin and small (4 percent). If these eggs are produced by a vermilion-eyed female, both of whose X chromosomes carry the recessive gene for vermilion, then the eight classes of zygotes will result upon fertilization by a wild male, which produces X and Y sperm.

The XX eggs fertilized by X sperm give XXX individuals which are unable to live.

The XX eggs fertilized by the Y sperm give individuals which are exact duplicates of their mother in their sex chromosomes, and like her are females each containing an extra Y chromosome. Since the gene for vermilion is carried by the X chromosome, these females have vermilion eyes and hence are matroclinous exceptions. Since they have not received an X from their father, they can neither show nor transmit his sex-linked characters. If in the mother the presence of the extra Y led to the production of secondary exceptions, then these XXY daughters should also give exceptions, and this is in fact the case.

The Y eggs fertilized by the X sperm give males. These males have received their X from their father and they show his sex-linked characters, that is, they are patroclinous exceptions. Since in chromosome constitution (XY) these males are not different from ordinary males, they should have no power of producing exceptions. This has been shown to be the case.

The Y eggs fertilized by Y sperm give YY individuals which are unable to live.

The X eggs by X sperm give regular XX females, and by Y sperm give regular XY males. Neither of these two classes is able to produce secondary exceptions or to transmit nondisjunction.

The XY eggs by X sperm give XXY females *which, because of the extra Y, possess the power of producing secondary exceptions, though they themselves are not exceptions.*

The XY eggs by Y sperm give XYY males. These males do not give rise to genetic exceptions in F_1, but they endow some of their daughters with an extra Y (XY sperm and X egg) which enables these daughters to produce secondary exceptions.

By breeding in each generation the exceptional daughters with the exceptional sons a line was maintained in which the entire set of sex-linked genes of the mother was handed down to the exceptional daughters and the entire set of the father to the exceptional sons.

That the XX eggs which developed into matroclinous exceptions had really been fertilized by normal sperm of the father was proved by the introduction into such daughters of autosomal genes from the father. The inheritance was uniparental with respect to the sex-linked genes, but biparental and wholly normal with respect to the autosomal genes. The fact that exceptional offspring inherit sex-linked characters from only one parent, but at the same time inherit the autosomal characters from both parents is explained if the sex chromosomes are the only chromosomes which have undergone nondisjunction, the ordinary chromosomes disjoining normally. . . .

CYTOLOGICAL PROOF OF THE OCCURRENCE OF XXY FEMALES

The breeding work presented in the last two [not included here] sections has furnished data which show that the cause of secondary nondisjunction cannot be a gene carried by the X chromosome, while the same data are consistent with the assumption that a female which produces secondary exceptions does so because of the presence of an extra Y chromosome. Likewise the data from the tests of the constitution of the regular daughters,

the regular sons, the exceptional daughters, and the exceptional sons, all lead to this same conclusion.

Accordingly, the prediction was made that cytological examination of the daughters of an exceptional female would demonstrate the presence of an extra chromosome in half of the daughters while the other half would show only normal figures. This prediction has been completely verified.

The ovaries at the mid-pupal stage of development offer the best material for examination. Nearly a hundred pairs of ovaries were dissected from pupae of cultures the mother of which was an exceptional female. Eighteen of these gave oogonial metaphases which were clear enough to give a reliable count of the chromosomes.

Nine of the females showed figures like those published by Miss Stevens (1908) in which the X chromosomes are a pair of straight rods. (See Fig. 3, No. 1–3).

The other nine females showed these two X chromosomes and in addition a chromosome which differed from both the X's in that it had the shape of a V with one arm shorter than the other. This chromosome was identified as the Y from the following considerations.

No figures which showed this extra chromosome were found by Metz (1914) when he examined the chromosomes of several wild stocks of *Drosophila ampelophila [D. melanogaster]*. From the work of Stevens (1908), Metz (1914) and myself (Fig. 3, No. 1–3) there can be no doubt that the normal condition of the female is that shown in the figure. The new type of figure which I found differs from the normal only by the addition of this chromosome. Fortunately, there are several good figures in each of four or five of the XXY females, and all the figures in any one female show the same condition.

The figures given by Stevens, Metz, and myself, show that homologous chromosomes usually lie together as actual pairs. In the figures showing the extra chromosome, this chromosome is usually found in company with the two straight chromosomes so that it behaves as a homologue to them.

Very recently I have found several excellent spermatogonial figures in the testes of larval males; these show beyond question that the identification of the Y has been correct, for the Y has in these males the same characteristics as the supernumerary chromosome of the XXY females.

Breeding tests with sex-linked characters have shown that half the regular daughters produce exceptions to the inheritance of sex and sex-linked characters; parallel with this is the fact that half of the regular daughters possess this extra Y chromosome. Normal females do not possess this chromosome and do not produce exceptions, so that the exceptions must be produced by the daughters with the extra chromosome.

Recently over forty freshly hatched females which were first classified as exceptions were dissected and a cytological examination made of their chromosomes. In over a dozen of these individuals sufficiently clear figures were found to be sure of the number and character of the chromosomes. *In every case the exceptional female was found to be XXY.* This direct examination of the exceptions gives entirely conclusive proof that the cause of the production of secondary exceptions is the presence of the Y.

Miss Stevens's (1908) work upon the male showed a pair of unequal chromosomes in place of the pair of equal straight rods of the female. The longer of these two chromosomes seems to have the shape of a J in some of Stevens's figures. The perpetuation of this longer chromosome in the male line can only be explained if these chromosomes have a causal connection with the differentiation of sex.

Of special interest is the condition shown in figures 20 to 24 (Fig. 3) which are from a single female with two X and two Y chromosomes, XXYY. This female was from a stock culture in which about half the females were expected to be XXY and half the males XYY. As we have seen, nearly half the eggs of an XXY female are XY, and thirty-three percent of the spermatozoa of an XYY male are XY. In the next generation therefore XXYY females should not be at all rare in such a stock. This female gave an unusual number of good figures, there being ten figures in which the identification of every chromosome is fairly certain.

This female has an additional value as evidence since the increase in the number of chromosomes is more striking, and since the occurrence of such a female gives indirect cytological proof of the occurrence of XYY males.

The ratio between XXY and XX daughters was much more easily determined by breeding tests than by cytological examination; accordingly only enough females were examined to prove that XXY females do actually exist. The proof of this point is beyond question. Also it chanced that the ratio of nine XXY to nine XX females was the equality expected from the breeding tests.

It will be noticed that often the figures show chromosomes split in preparation for the coming division. The difficulty in understanding the figures published by Stevens disappears if it is assumed that in the male such a split in the long arm of the Y chromosomes appears relatively early while the short arm splits later. With this interpretation, practically every figure given by Stevens falls into line with the evidence which the XXY females furnish, namely, that Y is the longer member of the XY pair, that it is attached subterminally, and has a J shape. . . .

SUMMARY AND CONCLUSIONS

Evidence has been presented which proves that the occasional (1 in 1700) matroclinous daughter or patroclinous son produced by females known to be XX in composition is due to primary nondisjunction, that is, the X's fail to disjoin and are both included in the egg or both extruded to the polar cell.

The fertilization of the zero egg by an X sperm of a normal male results in a patroclinous XO son. He is entirely unaltered in somatic appearance, both as to sex-linked characters and as to sexual characters, but he is absolutely sterile. This difference between XO and XY males proves that the Y has some normal function in *Drosophila*.

The fertilization of an XX egg by a Y sperm of a normal male gives rise to a matroclinous daughter of the constitution XXY. The constitution of a matroclinous female as XXY has been proved by direct cytological examination and by conclusive genetic tests.

Matroclinous females always produce further exceptions which we may call secondary, to the extent of 4.3 percent. The cause of this production is the fact that the presence of the extra Y forces both X's to enter the same cell in a certain percent of reductions.

In an XXY female the sex chromosomes do not synapse as a triad, but two synapse, leaving the

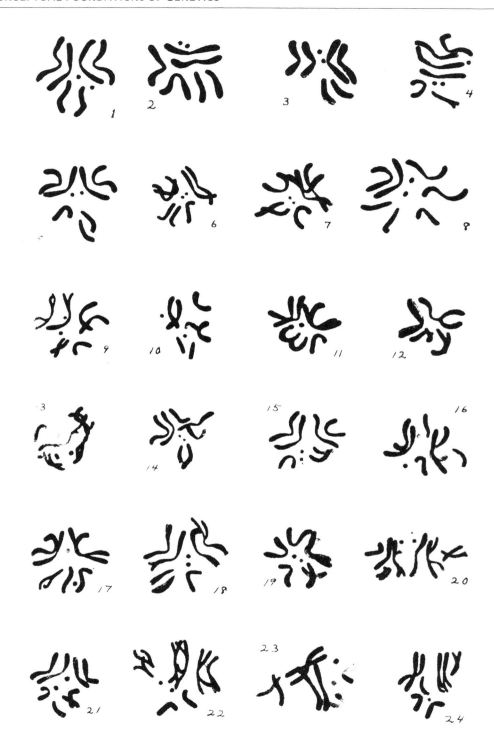

third unsynapsed. Synapses are not at random, but are highly preferential; in 16.5 percent of cases Y synapses with one or the other X (heterosynapsis) and the remaining X is unsynapsed; in 83.5 percent of cases synapsis is between X and X (homo-synapsis) and the Y is unsynapsed.

At the reduction division the two synapsed chromosomes disjoin and pass to opposite poles. The free chromosome goes with one or the other at random.

Reductions are not preferential; the polar spindle delivers two chromosomes to the polar cell as often as to the egg.

After XX synapsis the amount of crossing over is slightly increased (13.5 percent) in some manner by the presence of the extra Y.

After XY synapsis there is no crossing over — either between the X and Y or between the synapsed X and the free X.

After XY synapses the eggs are XX and Y, and X and XY. These four classes of eggs are in equal numbers and are noncrossovers.

The XX eggs by Y sperm give matroclinous daughters which are exact reproductions of the mother in all respects.

The Y eggs by X sperm give patroclinous sons which can give nondisjunctional effects neither in F_1 nor in F_2.

The X and XY eggs from XY synapses are indistinguishable from the noncrossover classes of the X and XY eggs which are from XX synapses. As a result the linkage values must be corrected.

The XY egg by X sperm gives an XXY regular daughter which nevertheless gives 4.3 percent of secondary exceptions by virtue of the extra Y.

YO, YY, and XXX zygotes are unable to live.

The XY egg by Y sperm gives XYY males. These males produce no exceptions in F_1 but produce XY sperm which, fertilizing X eggs, give rise to XXY daughters, and these produce secondary exceptions.

Synapses in an XYY male are probably at random.

The source of chromosomes, whether maternal or paternal, is without effect upon their subsequent behavior at synapsis and reduction.

The predominant type of nondisjunction has been shown to be preceded by XY synapsis and to take place at the reduction division. A rare type of nondisjunction takes place at an equational division. Equational nondisjunction is apparently always

---

◀ **Fig. 3.** The figures in this plate were drawn at table level; tube length 160 mm; Zeiss compensating ocular 12X; and Zeiss apochromatic 1.5 mm oil immersion objective, N.A. 1.30. The figures were then enlarged 2¼ diameters, and in reproduction were reduced in the ratio 3:2. The resulting magnification is 5:115 diameters.

Figures 1–3 are oogonial plates for wild females. Figures 2 and 3 are from the same cyst; Figure 1 is from another individual. These figures are from freshly hatched mature flies; the rest of the figures are from pupae.

Figure 4 is a spermatogonial plate of a wild male (the extra granule is probably of no significance). I now have several good figures from the testes of larvae; four of these are diagrammatic in clearness, and show that the Y has the same character in the male as it has when transferred to the female.

Figures 5–19 are from XXY daughters of an exceptional mother. The plates are oogonial with the exception of 6 and 10 which are from other ovarian cells. Figures 5–10 are from one individual, as are 11–13, 14–16, 17, and 18–19. Figure 5 is of unusual clearness.

Figure 13 shows a typical late prophase in which the greatly elongated chromosomes are arranged about the periphery of the nucleus.

Figure 6 shows the X chromosomes just drawing into the equatorial plate; the outer ends are still curved in contact with the nuclear wall; the same is true of figure 8.

Figures 5, 7, 10–12, 13–16 and 19 are full metaphase groups.

Figure 9 shows a late metaphase group in which the chromosomes are already split; the same condition is seen in 4, 17, 18, and especially well in 20, 22, and 24. It is characteristic of this split that it begins at the free ends of the chromosomes and proceeds toward the spindle attachment; the separation at the point of attachment does not take place until the elongation of the cell.

Figures 20–24 are from an XXYY individual which was found in a stock mass culture in which half the parental flies were XXY females and half were XYY males.

In general the Y is the most sharply defined of all the chromosomes; this is seen especially well in early metaphases such as Figure 6, or late metaphases such as Figure 22.

preceded by XX synapsis and crossing over. Equational nondisjunction offers the possibility of determining whether crossing over in *Drosophila* takes place by the chiasmatype (four-strand stage) method or at a two-strand stage.

Somatic nondisjunction explains the occurrence of gynandromorphs and mosaics in *Drosophila*.

Unusually high percentages of exceptions occur and are irregularly inherited. It is suggested that the cause of the high nondisjunction is a mutation in the Y chromosome.

The occurrence of nondisjunction in Abraxas is shown both by cytological and by breeding tests. Various other forms, namely, canaries, fowls, pigeons, and the moth, *Aglia tau*, show exceptions to sex-linkage which may be explained as due to nondisjunction.

In *Oenothera lata*, autosomal nondisjunction has been studied. A single triallelomorphic individual in Paratettix may be due to autosomal nondisjunction.

The occurrence of supernumerary Y chromosomes and of triploid or multiple chromosomes in various forms can likewise be explained as the result of nondisjunction.

The genetic and cytological evidence in the case of nondisjunction leaves no escape from the conclusion that the X chromosomes are the carriers of the genes for the sex-linked characters. The distribution of sex-linked genes (as tested by experimental breeding methods) has been demonstrated to be identical, through all the details of a unique process, with the distribution of the X chromosomes (as tested by direct cytological examination). The argument that the cell as a whole possesses the tendency to develop certain characters is completely nullified by the fact that in these cases the cells that produce exceptions are of exactly the same parentage as those which do not produce exceptions, the only difference being the parentage of a particular chromosome, the X. Those eggs which have lost nothing but the X chromosome have completely lost therewith the ability to produce any of the maternal sex-linked characters, and with the introduction

of an X from the father these eggs have developed all of the sex-linked characters of the father. Conversely, those eggs which have retained both X's of the mother and have received no X from the father show all of the sex-linked characters of the mother and none from the father. The breach which Gregory found in his case of the tetraploid Primula, namely, that the doubling of the genes is only an expression of the doubleness of the *cell-as-a-whole*, becomes in this case the strongest bulwark; for here the cell as a whole remains constant and the issue is restricted to *particular* chromosomes and a *particular* class of genes.

Experimental proof is given that particular chromosomes, the X chromosomes, are the differentiators of sex; the X chromosome constitution of an individual is the cause of the development by that individual of a particular sex, and is not the result of sex already determined by some other agent. The sex is not determined in the egg or the sperm as such, but is determined at the moment of fertilization; for the X sperm of a male gives rise to a female when it fertilizes an egg containing an X, but to a male if it fertilizes an egg containing a Y or no sex chromosome at all. Likewise the Y sperm of a male gives rise to a female when fertilizing an XX egg and to a male when fertilizing an X egg. These facts in connection with the fact that an X egg of a female produces a male if fertilized by an X sperm prove that the segregation of the X chromosomes is the segregation of the sex differentiators. The presence of two X chromosomes determines that an individual shall be a female, the presence of one X that the individual shall be a male. The origin of these chromosomes whether maternal or paternal is without significance in the production of sex.

The Y chromosome is without effect upon the sex or the characters of the individual, for males may have one Y, two Y's, or may lack Y entirely (males lacking Y are sterile); and females may have one or two supernumerary Y's with no change in appearance in any case.

A correlation of cytological and genetical crossing over in *Zea mays*

HARRIET B. CREIGHTON and BARBARA MCCLINTOCK

Reprinted by authors' and publisher's permission from Proceedings of the
National Academy of Sciences, U.S., *vol. 17, 1931, pp. 492–497.*

Although, by 1920, the chromosome theory of inheritance was firmly entrenched in genetic thought, there was still no direct proof of its validity. All of the work emerging from the Columbia University fly room experimentally supported predictions based upon the hypothesis that genes resided on the chromosomes. However, none of this work demonstrated a direct physical correlation between a certain portion of a chromosome and an established genetic trait. To do this, regions of chromosomes had to be morphologically detected and shown to be affiliated with the control of a transmissible character. In 1931, Harriet Creighton and Barbara McClintock presented the following paper. Again, it called for the combining of cytological and genetic techniques to provide evidence which demonstrated that a certain region of one of the *Zea mays* chromosomes was responsible for the control of a pair of Mendelizing traits: the color of the aleurone and the substantive nature (starchy versus waxy) of the endosperm.

A requirement for the genetical study of crossing over is the heterozygous condition of two allelomorphic factors in the same linkage group. The analysis of the behavior of homologous or partially homologous chromosomes, which are morphologically distinguishable at two points, should show evidence of cytological crossing over. It is the aim of the present paper to show that cytological crossing over occurs and that it is accompanied by genetical crossing over.

In a certain strain of maize the second smallest chromosome (chromosome 9) possesses a conspicuous knob at the end of the short arm. Its distribution through successive generations is similar to that of a gene. If a plant possessing knobs at the ends of both of its second smallest chromosomes is crossed to a plant with no knobs, cytological observations show that in the resulting F_1 individuals only one member of the homologous pair possesses a knob. When such an individual is backcrossed to one having no knob on either chromosome, half of the offspring are heterozygous for the knob and half possess no knob at all. The knob, therefore, is a constant feature of the chromosome possessing it. When present on one chromosome and not on its homologue, the knob renders the chromosome pair visibly heteromorphic.

In a previous report[*] it was shown that in a certain strain of maize an interchange had taken place between chromosomes 8 and 9. The interchanged pieces were unequal in size; the long arm of chromosome 9 was increased in relative length, whereas the long arm of chromosome 8 was correspondingly shortened. When a gamete possessing these two interchanged chromosomes meets a gamete containing a normal chromosome set, meiosis in the resulting individual is characterized by a side-by-side synapsis of homologous parts. Therefore, it should be possible to have crossing over between the knob and the interchange point.

In the previous report it was also shown that in such an individual the only functioning gametes are those which possess either the two normal chromosomes (N, n) or the two interchanged chromosomes (I, i), i.e., the full genome in one or the other arrangement. The functional gametes therefore possess either the shorter, normal, knobbed chromosome (n) or the longer, interchanged, knobbed chromosome

[*]McClintock, B., *Proc. Nat. Acad. Sci.,* 16:791–796, 1930.

(*I*). Hence, when such a plant is crossed to a plant possessing the normal chromosome complement, the presence of the normal chromosome in functioning gametes of the former will be indicated by the appearance of ten bivalents in the prophase of meiosis of the resulting individuals. The presence of the interchanged chromosome in other gametes will be indicated in other F_1 individuals by the appearance of eight bivalents plus a ring of four chromosomes in the late prophase of meiosis.

If a gamete possessing a normal chromosome number 9 with no knob, meets a gamete possessing an interchanged chromosome with a knob, it is clear that these two chromosomes which synapse along their homologous parts during prophase of meiosis in the resulting individual are visibly different at each of their two ends. If no crossing over occurs, the gametes formed by such an individual will contain either the knobbed, interchanged chromosome (*a,* Fig. 1) or the normal chromosome without a knob (*d,* Fig. 1). Gametes containing either a knobbed, normal chromosome (*c,* Fig. 1) or a knobless, interchanged chromosome (*b,* Fig. 1) will be formed as a result of crossing over. If such an individual is crossed to a plant possessing two normal knobless chromosomes, the resulting individuals will be of four kinds. The noncrossover gametes would give rise to individuals which show either (1) ten bivalents at prophase of meiosis and no knob on chromosome 9, indicating that a gamete with a chromosome of type *d* has functioned or (2) a ring of four chromosomes with a single conspicuous knob, indicating that a gamete of type *a* has functioned. The crossover types will be recognizable as individuals which possess either (1) ten bivalents and a single knob associated with bivalent chromosome 9 or (2) a ring of four chromosomes with no knob, indicating that crossover gametes of types *c* and *b*, respectively, have functioned. The results of such a cross are given in culture 337, Table 1. Similarly, if such a plant is crossed to a normal plant possessing knobs at the ends of both number 9 chromosomes and if crossing over occurs, the resulting individuals should be of four kinds. The noncrossover types would be represented by (1) plants homozygous for the knob and possessing the interchanged chromosome and (2) plants heterozygous for the knob and possessing two normal chromosomes. The functioning of gametes which had been produced as the result of crossing over

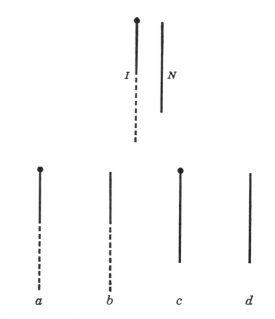

Fig. 1. (*above*) Diagram of the chromosomes in which crossing over was studied.
(*below*) Diagram of chromosome types found in gametes of a plant with the constitution shown above.

a—Knobbed, interchanged chromosome
b—Knobless, interchanged chromosome
c—Knobbed, normal chromosome
d—Knobless, normal chromosome

a and d are noncrossover types.
b and c are crossover types.

between the knob and the interchange would give rise to (1) individuals heterozygous for the knob and possessing the interchanged chromosome and (2) those homozygous for the knob and possessing two normal chromosomes. The results of such crosses are given in culture A125 and 340, Table 1. Although the data are few, they are consistent. The amount of crossing over between the knob and the interchange, as measured from these data, is approximately 39%.

In the preceding paper it was shown that the knobbed chromosome carries the genes for colored aleurone (*C*), shrunken endosperm (*sh*) and waxy endosperm (*wx*). Furthermore, it was shown that the order of these genes, beginning at the interchange point is *wx-sh-C*. It is possible, also, that

Table 1

	Knob-interchanged / Knobless-normal X Knobless-normal, culture 337 and Knobbed-normal cultures A125 and 340			
Culture	**Plant possessing 2 normal chromosomes**		**Plants possessing an interchanged chromosome**	
	Noncrossovers	*Crossovers*	*Noncrossovers*	*Crossovers*
337	8	3	6	2
A125	39	31	36	23
340	5	3	5	3
Totals	52	37	47	28

Table 2

	Knob-C-wx / Knobless-c-Wx X Knobless-c-wx							
C-wx		**c-Wx**		**C-Wx**		**c-wx**	
Knob	*Knobless*	*Knob*	*Knobless*	*Knob*	*Knobless*	*Knob*	*Knobless*
12	5	5	34	4	0	0	3

these genes all lie in the short arm of the knobbed chromosome. Therefore, a linkage between the knob and these genes is to be expected.

One chromosome number 9 in a plant possessing the normal complement had a knob and carried the genes C and wx. Its homologue was knobless and carried the genes c and Wx. The noncrossover gametes should contain a knobbed-C-wx or a knob-less-c-Wx chromosome. Crossing over in region 1 (between the knob and C) would give rise to knob-less-C-wx and knobbed-c-Wx chromosomes. Crossing over in region 2 (between C and wx) would give rise to knobbed-C-Wx and knobless-c-wx chromosomes. The results of crossing such a plant to a knobless-c-wx type are given in Table 2. It should be expected on the basis of interference that the knob and C would remain together when a crossover occurred between C and wx; hence, the individuals arising from colored starchy (C-Wx) kernels should possess a knob, whereas those coming from colorless, waxy (c-wx) kernels should be knobless. Although the data are few they are convincing. It is obvious that

there is a fairly close association between the knob and C.

To obtain a correlation between cytological and genetic crossing over it is necessary to have a plant heteromorphic for the knob, the genes c and wx and the interchange. Plant 338 (17) possessed in one chromosome the knob, the genes C and wx and the interchanged piece of chromosome 8. The other chromosome was normal, knobless and contained the genes c and Wx. This plant was crossed to an individual possessing two normal, knobless chromosomes with the genes c-Wx and c-wx, respectively. This cross is diagrammed as follows:

The results of the cross are given in Table 3. In this case all the colored kernels gave rise to individuals possessing a knob, whereas all the colorless kernels gave rise to individuals showing no knob.

Table 3

	Knob-*C*-*wx*-interchanged / Knobless-*c*-*Wx*-normal	× Knobless-*c*-*Wx*-normal / Knobless-*c*-*wx*-normal	

Plant number	Knobbed or Knobless	Interchanged or normal	
Class I, *C*-*wx* kernels			
1	Knob	Interchanged	
2	Knob	Interchanged	
3	Knob	Interchanged	
Class II, *c*-*wx* kernels			
1	Knobless	Interchanged	
2	Knobless	Interchanged	
Class III, *C*-*Wx* kernels			*Pollen*
1	Knob	Normal	*WxWx*
2	Knob	Normal	. . . . .
3	. . . . .	Normal	*WxWx*
5	Knob	Normal	. . . . .
6	Knob	. . . . . .	. . . . .
7	Knob	Normal	. . . . .
8	Knob	Normal	. . . . .
Class IV, *c*-*Wx* kernels			
1	Knobless	Normal	*Wxwx*
2	Knobless	Normal	*Wxwx*
3	Knobless	Interchanged	*Wxwx*
4	Knobless	Normal	*Wxwx*
5	Knobless	Interchanged	*WxWx*
6	Knobless	Normal	*WxWx*
7	Knobless	Interchanged	*Wxwx*
8	Knobless	Interchanged	*WxWx*
9	Knobless	Normal	*WxWx*
10	Knobless	Normal	*WxWx*
11	Knobless	Normal	*Wxwx*
12	Knobless	Normal	*Wxwx*
13	Knobless	Normal	*WxWx*
14	Knobless	Normal	*WxWx*
15	Knobless	Normal	*Wx. . . .*

The amount of crossing over between the knob and the interchange point is approximately 39% (Table 1), between *c* and the interchange approximately 33%, between *wx* and the interchange, 13% (preceding paper). With this information in mind it is possible to analyze the data given in Table 3. The data are necessarily few since the ear contained but few kernels. The three individuals in class I are clearly noncrossover types. In class II the individuals have resulted from a crossover in region 2, i.e., between *c* and *wx*. In this case a crossover in region 2 has not been accompanied by a crossover in region 1 (between the knob and *C*) or region 3 (between *wx* and the interchange). All the individuals in class III had normal chromosomes. Unfortunately, pollen was obtained from only 1 of the 6 individuals examined for the presence of the knob. This one individual was clearly of the type expected to come from a gamete produced through crossing over in region 2. Class IV is more difficult to analyze. Plants 6, 9, 10, 13, and 14 are normal and *WxWx*; they therefore represent noncrossover types. An equal number of noncrossover types are expected among the normal *Wxwx* class. Plants 1, 2, 4, 11 and 12 may be of this type. It is possible but improbable that they have arisen through the union of a *c*-*Wx* gamete with a gamete resulting from a double crossover in region 2 and 3. Plants 5 and 8 are single crossovers in region 3, whereas plants 3 and 7 probably represent single crossovers in region 2 or 3.

The foregoing evidence points to the fact that cytological crossing over occurs and is accompanied by the expected types of genetic crossing over.

Conclusions Pairing chromosomes, heteromorphic in two regions, have been shown to exchange parts at the same time they exchange genes assigned to these regions.

The authors wish to express appreciation to Dr. L. W. Sharp for aid in the revision of the manuscripts of this and the preceding paper. They are indebted to Dr. C. R. Burnham for furnishing unpublished data and for some of the material studied.

Part 3. Transmission and recombination of the genetic material in prokaryotes

RECOMBINATION IN VIRUSES

Studies on the transmission and recombination of the genetic material in viruses began long after similar studies in eukaryotic organisms. In the late 1940s, Bailey, Delbrück, Hershey, and Rotman published thorough studies[2,6,7] of recombination in the T2 bacteriophage and established, in skeletal form, the basic features of recombination in this organism. They showed that the study of phage genetics conformed in many respects to the principles established in genetic studies of higher organisms. Thus, linkage maps could be constructed on the basis of recombination frequencies.

But, some features of phage recombination did differ from those of recombination in higher organisms. For example, Hershey and Rotman[7] demonstrated that genetic exchange was not necessarily a reciprocal event. In 1953, Visconti and Delbrück[15] pointed out that phage recombination involved repeated interactions and exchanges between phage chromosomes; and, that any one phage chromosome could exchange genetic information with several other chromosomes. All this occurred during the course of a single life cycle — in contrast to genetic exchange during meiosis in higher organisms, where the phenomenon is restricted in both time and space.

Crucial to studies of genetic recombination in viruses was the work on the bacteriophage life cycle begun in earnest by Ellis and Delbrück in 1938.[3] Their studies showed that a bacterium infected with a single phage released approximately 100 progeny phage about 30 minutes after the initial infection. The liberation of the progeny phage occurred through lysis of the infected bacterium. Their experiment, called *the one-step growth experiment,* defined (1) a latent period in which the virus particles reproduce inside the host cell; (2) a rise period in which cells begin to lyse and release progeny phage; and (3) a plateau period in which lysis is completed and the number of free progeny phage particles is stabilized. This study led to Hershey and Chase's[5] work in 1952 which revealed that the phage DNA entered the host cell and directed the synthesis of progeny phage — confirming the notion that DNA was the genetic material. In 1953, Watson and Crick presented their model of DNA's structure. The problem of genetic exchange in bacteriophage now became one of describing the interaction between DNA molecules inside of a host cell.

There were two general ways of viewing recombination in bacteriophage. The copy-choice view suggested that as DNA molecules were replicated, the newly synthesized strand would switch templates. This resulted in a new DNA molecule (*phage chromosome*) with genes from two templates:

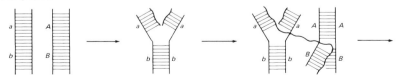

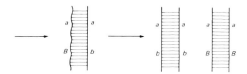

The breakage-reunion view suggested a mechanism similar to the one established in higher organisms. That is, DNA molecules would break and rejoin, giving recombinant molecules:

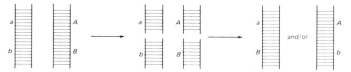

Both of the genetic exchange hypotheses lent themselves to experimental testing. The copy-choice idea predicted that recombinant molecules could only be formed by DNA synthesis; the breakage-reunion viewpoint argued that DNA synthesis was not necessary to the formation of recombinants. In 1961, Meselson and Weigle presented data in support of the breakage-reunion viewpoint[13] and, in 1964, Meselson confirmed it.[12]

Today our concept of recombination in phage is somewhat different. Breakage and reunion is undoubtedly the general mechanism by which phage DNA molecules recombine — as it probably is in higher organisms — but DNA synthesis is probably an integral part of the process, as can be seen in the following model:

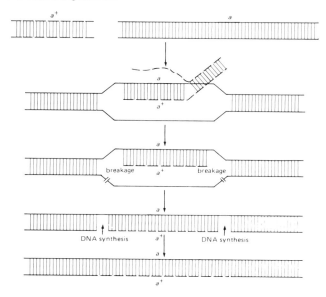

A model describing the formation of recombinant DNA molecules. Solid lines represent strands of the host DNA; dashed strands represent strands of the donor DNA. (*Modification of a model presented by Fox, 1966.* J. Gen. Physiol. *49:183-96.)*

In summary, we can point to many similarities in the exchange of genetic information in viruses and higher organisms. We must be careful, however, to emphasize the superficial nature of these similarities and to point out the many vast differences which exist.

RECOMBINATION IN BACTERIA

At about the same time that genetic recombination was discovered in viruses, processes of genetic recombination were discovered in bacteria. In 1946, Lederberg and Tatum found a mechanism of genetic exchange in the bacterium *Escherichia coli.*[10,11] The mechanism was called *conjugation* and involved a joining of cells of different mating types and the passage of genetic information from one cell to the other. In 1952, Zinder and Lederberg[16] showed that in *Salmonella,* viruses could (1) infect the cells; (2) pick up *Salmonella* genes; and (3) introduce those genes into other *Salmonella* cells. This process is called *transduction* and is unusual in that it employs a viral intermediary to carry bacterial genes from one cell to the next. The first mechanism found by which genes from one bacterium are transferred to another was *transformation.* First discovered by Griffith[4] in 1928, transformation is discussed in connection with the identification of the genetic material in Part 1. In the remainder of this introduction, we would like to discuss these three bacterial recombination processes.

TRANSFORMATION   There is no need to repeat the history of the transformation phenomenon; it was discussed at length in Part 1. We might, however, profit from some insight into its current level of understanding.
   The early interpretation of transformation was that bacterial DNA, free in the medium, was picked up by bacterial cells growing in the same liquid medium, transported to the cell's interior, and incorporated into that recipient cell's own genome. Today the details of this process are better known. To begin with, the recipient cell must be physiologically competent to pick up the free DNA. If it is, the DNA attaches to specific sites on the bacterial cell wall, is actively transported to the cell's interior, is converted to single-stranded DNA, and then this single strand is incorporated into the host cell's genome. The single-stranded incorporation is a replacement phenomenon rather than an addition phenomenon. That is to say, the single-stranded fragment of donor DNA replaces a homologous segment of recipient DNA; it does not add to the recipient DNA already present.[8] Transformation has been an important tool for use in genetic recombination studies in organisms that have demonstrated no other means of generating recombinants. Using the process, we have been able to construct genetic maps of bacterial chromosomes based on frequencies of genetic exchange.

CONJUGATION   In 1946, Lederberg and Tatum published a report[11] that began: "Hershey has reported the occurrence of novel combinations of inherited characters in a bacterial virus. It may not be amiss to describe briefly some experimental fragments, relating to a situation in the bacterium *Escherichia coli,* which may be similar in some respects." Bacterial conjugation thus was announced. Of course, the only similarities in the two phenomena (conjugation and genetic recombination in viruses) are found in the ends and not the means — both can generate new combinations of genes. In his paper of 1947, included in this part, Lederberg showed how conjugation could be employed to map genes.

Conjugation operates with three major bacterial cell types. Two of these types can donate genetic information; the third type can serve as a recipient of that genetic information. The donor strains have an additional segment of DNA called a fertility factor (*F factor*), and the two differ from each other in the physical state of that F factor; in F$^+$ cells, the F factor is an autonomous circular piece of DNA that is not integrated into the bacterial chromosome; in Hfr cells, the F factor is integrated into the bacterial chromosome. The recipient strain of cells lacks the F factor altogether, and is referred to as F$^-$. Conjugation occurs when an F$^-$ cell meets either an F$^+$ cell or an Hfr cell. In an F$^-$ × F$^+$ mating, the cells join and a tube connects them. Then the F factor, alone, passes from the F$^+$ to the F$^-$ cell. No other F$^+$ bacterial genes are involved in the transfer. In an Hfr × F$^-$ mating, bacterial genes are transferred. The circular *E. coli* chromosome breaks at the point of F-factor integration, and a single strand of the bacterial chromosome passes into the F$^-$ cell behind a small segment of the F factor. The remaining F-factor DNA is the last to enter the F$^-$ cell and rarely does so because the two conjugating cells usually separate before conjugation is complete. The incorporation of single-stranded DNA in the F$^-$ chromosome is reminiscent of the mechanism of transformation. A diagrammatic summary of conjugation is presented below:

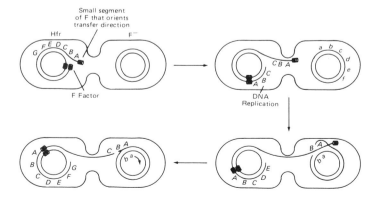

An Hfr cell can revert to an F$^+$ cell when the F factor leaves its integrated state to become free and autonomous. Occasionally, as this happens, the F factor incorporates some chromosomal genes into its structure. Such a cell is called F′ and mates like F⁺ cells except that it transfers those few chromosomal genes along with the F-factor DNA.

Conjugation, like transformation, has enabled us to study the linear relationship of genes on the *E. coli* chromosome in detail. Map distances are not expressed in terms of genetic exchange frequencies but in terms of time. Under optimal conditions, the entire Hfr chromosome can be transferred to an F⁻ cell in 90 minutes; and, each gene's position is a function of when it enters the recipient cell. For example, the first gene to enter the F⁻ cell maps at point 0; a gene entering 10 minutes later at point 10, and so on up to point 90. A general map of the *E. coli* chromosome is seen below:

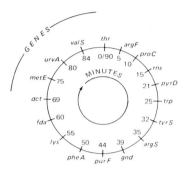

TRANSDUCTION In 1952, Zinder and Lederberg discovered that a bacteriophage which infects *Salmonella* can transfer bacterial genes from one bacterium to another. The process of phage-mediated bacterial gene transfer is called *transduction.* In this particular system, the phage could transduce any of the *Salmonella* genes, therefore the process was called *generalized transduction.* Two years later (1954), a similar process was discovered in *E. coli,* by Morse. His studies were expanded by himself and the Lederbergs[14] in 1956. At that time they established that the bacteriophage lambda (λ) transduced only *E. coli* genes in the *Gal* and *Bio* regions — therefore the process was termed *specialized transduction.*

Generalized transduction is less complex than specialized transduction. In generalized transduction, the bacterial cell chromosome breaks up and, as mature phage particles form, some phage protein condenses around fragmented bacterial DNA to produce a transducing bacteriophage. This transducing phage, carrying bacterial DNA, infects another bacterium and injects the bacterial genes into the cell — setting up a recombination between the host cell's genes and the genes on the transduced fragment.[9]

Specialized λ transduction requires the integration of the λ DNA into the host cell's chromosome. This integration occurs at a specific site between the *Gal* and the *Bio* regions of the *E. coli* chromosome and is remarkably similar to the formation of Hfr from F⁺. A circular λ chromosome (*prophage*) integrates into the circular *E. coli* chromosome, as seen on page 224.

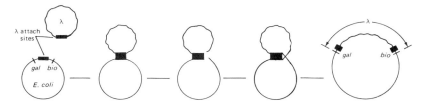

A transducing λ occurs when the prophage is released from its integrated state and, in the process of being released, picks up the *Gal* or *Bio* genetic region on the bacterial chromosome. Often when the bacterial *Gal* genes are picked up, an equivalent segment of λ DNA is left behind in the *E. coli* chromosome. The λ thus formed is defective and is called λ *dg* (*d* = defective, *g* = *Gal*). It is unable to complete a life cycle because it has lost some of its genes. In these instances a normal, nontransducing λ phage (*helper phage*) is required if λ *dg* is to produce progeny λ *dg* particles. The helper phage supplies the gene products that are required for the synthesis of mature phage particles, which λ *dg* cannot synthesize because of its missing genes.

When λ *dg* (carrying *Gal*⁺) infects a *Gal*⁻ cell, a number of things can happen. (1) The λ *dg* DNA can attach to or integrate with the host chromosome, producing a *Gal*⁺/*Gal*⁻ heterogenote; (2) the *Gal*⁺ gene can replace the *Gal*⁻ gene; or (3) the λ *dg* DNA could be degraded.[1] Each of these possibilities has specific consequences, some of which are discussed in the papers included in this part.

Transduction, like transformation and conjugation, provides a means of generating new combinations of genes in bacteria. Generating new genetic combinations is a major force in evolution. Also, all three of these processes have provided extremely valuable information on the structure and function of genes in prokaryotic systems.

REFERENCES

1. Borek, E., and A. Ryan 1973. Lysogenic induction. *Prog. in Nuc. Acid Res. and Mol. Biol.* 13:249–300.

2. Delbrück, M., and W. T. Bailey 1946. Induced mutations in bacterial viruses. *Cold Spring Harbor Symp. Quant. Biol.* 11:33.

3. Ellis, E. L., and M. Delbrück 1939. The growth of bacteriophage. *J. Gen. Physiol.* 22:365–84.

4. Griffith, F. 1928. The significance of pneumococcal types. *J. Hyg.* 27:113–59.

5. Hershey, A. D., and M. Chase 1952. Independent functions of viral protein and nucleic acid in growth of bacteriophage. *J. Gen. Physiol.* 36:39–56.

6. Hershey, A. D., and R. Rotman 1948. Linkage among genes controlling inhibition of lyses in bacterial viruses. *Proc. Nat. Acad. Sci. U.S.* 34:89.

7. Hershey, A. D., and R. Rotman 1949. Genetic recombination between host-

range and plaque-type mutants of bacteriophage in single bacterial cells. *Genetics* 34:44.

8. Hotchkiss, R. D., and M. Gabor 1970. Bacterial transformation with special reference to recombination process. *Ann. Rev. Genetics* 4:193–224.

9. Ikeda, H., and J. Tomizawa 1965. Transducing fragments in generalized transduction by phage PI. III studies with small phage particles. *J. Mol. Biol.* 14: 120–29.

10. Lederberg, J., and E. L. Tatum 1946. Gene recombination in *E. Coli. Nature* 158:558.

11. Lederberg, J., and E. L. Tatum 1946. Novel genotypes in mixed cultures of biochemical mutants of bacteria. *Cold Spring Harbor Symp. Quant. Biol.* 11:113–14.

12. Meselson, M. 1964. On the mechanism of genetic recombination between DNA molecules. *J. Mol. Biol.* 9:734–45.

13. Meselson, M., and J. J. Weigle 1961. Chromosome breakage accompanying genetic recombination in bacteriophage. *Proc. Nat. Acad. Sci. U.S.* 47:857.

14. Morse, M. L., E. M. Lederberg, and J. Lederberg 1956. Transduction in *Escherichia coli* K-12. *Genetics* 41:142–56.

15. Visconti, N., and M. Delbrück 1953. The mechanism of recombination in phage. *Genetics* 38:5.

16. Zinder, N., and J. Lederberg 1952. Genetic exchange in *Salmonella. J. Bacteriol.* 64:679–99.

The growth of bacteriophage*

EMORY L. ELLIS and MAX DELBRÜCK[†]

Reprinted by permission of the authors and Rockefeller University Press from
Journal of General Physiology, vol. 22(3), 1939, pp. 365-384. (Accepted for
publication, September 7, 1938)

This paper was a landmark in the field of bacterial virus genetics; in fact, it marked the beginning of modern phage research. The paper describes the "one-step growth experiment," an experiment that demonstrated that when a parent virus infects a cell, there is a latent period of about 20 minutes, followed by the bursting open of the cell and the release of progeny viruses. The paper also describes the "single-burst experiment," wherein individually infected bacteria rather than mass cultures of infected cells are studied. The major questions raised by this work centered on the intracellular phases of the viral development. The virus attaches to a susceptible *E. coli* cell and about 30 minutes later the infected cells break open, releasing hundreds of progeny phage. What are the mechanisms for this remarkable process?

This question became a central issue to the members of the "phage group": Max Delbrück, Alfred D. Hershey, and Salvador E. Luria. During the 1940s, this group, later joined by the pioneer electron microscopist, T. F. Anderson, attacked the problem of phage replication and paved the way to the pivotal years of 1952 and 1953. In 1952, Hershey and Chase showed that it was DNA that entered the cell and directed the intracellular process of phage replication; and, in 1953, Watson and Crick presented their model of DNA's structure.

INTRODUCTION

Certain large protein molecules (viruses) possess the property of multiplying within living organisms. This process, which is at once so foreign to chemistry and so fundamental to biology, is exemplified in the multiplication of bacteriophage in the presence of susceptible bacteria.

Bacteriophage offers a number of advantages for the study of the multiplication process not available with viruses which multiply at the expense of more complex hosts. It can be stored indefinitely in the absence of a host without deterioration. Its concentration can be determined with fair accuracy by several methods, and even the individual particles can be counted by d'Herelle's method. It can be concentrated, purified, and generally handled like nucleoprotein, to which class of substances it apparently belongs (Schlesinger[1] and Northrop).[2] The host organism is easy to culture and in some cases can be grown in purely synthetic media, thus the conditions of growth of the host and of the phage can be controlled and varied in a quantitative and chemically well defined way.

Before the main problem, which is elucidation of the multiplication process, can be studied, certain information regarding the behavior of phage is needed. Above all, the "natural history" of bacteriophage, *i.e.*, its growth under a well-defined set of cultural conditions, is as yet insufficiently known, the only extensive quantitative work being that of Krueger and Northrop[3] on an anti-*staphylococcus* phage. The present work is a study of this problem, the growth of another phage (anti-*Escherichia coli* phage) under a standardized set of culture conditions.

*(From the William G. Kerckhoff Laboratories of the Biological Sciences, California Institute of Technology, Pasadena)
[†]Fellow of The Rockefeller Foundation

EXPERIMENTAL

Bacteria Culture Out host organism was a strain of *Escherihia coli*, which was kindly provided by Dr. C. C. Lindegren. Difco nutrient broth (pH 6.6–6.8) and nutrient agar were selected as culture media. These media were selected for the present work because of the complications which arise when synthetic media are used. We thus avoided the difficulties arising from the need for accessory growth factors.

Isolation, Culture, and Storage of Phage A bacteriophage active against this strain of *coli* was isolated in the usual way from fresh sewage filtrates. Its homogeneity was assured by five successive single plaque isolations. The properties of this phage remained constant throughout the work. The average plaque size on 1.5 percent agar medium was 0.5 to 1.0 mm.

Phage was prepared by adding to 25 cc of broth, 0.1 cc of a 20 hour culture of bacteria, and 0.1 cc of a previous phage preparation. After 3½ hours at 37° the culture had become clear, and contained about 10^9 phage particles.

Such lysates even though stored in the ice box, decreased in phage concentration to about 20 percent of their initial value in 1 day, and to about 2 percent in a week, after which they remained constant. Part of this lost phage activity was found to be present in a small quantity of a precipitate which had sedimented during this storage period.

Therefore, lysates were always filtered through Jena sintered glass filters (5 on 3 grade) immediately after preparation. The phage concentration of these filtrates also decreased on storage, though more slowly, falling to 20 percent in a week. However, 1:100 dilutions in distilled water of the fresh filtered lysates retained a constant assay value for several months, and these diluted preparations were used in the work reported here, except where otherwise specified.

This inactivation of our undiluted filtered phage suspensions on standing is probably a result of a combination of phage and specific phage inhibiting substances from the bacteria, as suggested by Burnett.[4,5] To test this hypothesis we prepared a polysaccharide fraction from agar cultures of these bacteria, according to a method reported by Heidelberger *et al.*[6] Aqueous solutions of this material, when mixed with phage suspensions, rapidly inactivated the phage.

Method of Assay We have used a modification of the plaque counting method of d'Herelle[7] throughout this work for the determination of phage concentrations. Although the plaque counting method has been reported unsatisfactory by various investigators, under our conditions it has proven to be entirely satisfactory.

Phage preparations suitably diluted in 18-hour broth cultures of bacteria to give a readily countable number of plaques (100 to 1000) were spread with a bent glass capillary over the surface of nutrient agar plates which had been dried by inverting on sterile filter paper overnight. The plates were then incubated 6 to 24 hours at 37°C at which time the plaques were readily distinguishable. The 0.1 cc used for spreading was completely soaked into the agar thus prepared in 2 to 3 minutes, thus giving no opportunity for the multiplication of phage in the liquid phase. Each step of each dilution was done with fresh sterile glassware. Tests of the amount of phage adhering to the glass spreaders showed that this quantity is negligible.

The time of contact between phage and bacteria in the final dilution before plating has no measurable influence on the plaque count, up to 5 minutes at 25°C. Even if phage alone is spread on the plate and allowed to soak in for 10 minutes, before seeding the plate with bacteria, only a small decrease in plaque count is apparent (about 20 percent). This decrease we attribute to failure of some phage particles to come into contact with bacteria.

Under parallel conditions, the reproducibility of an assay is limited by the sampling error, which in this case is equal to the square root of the number of plaques (10 percent for counts of 100; 3.2 percent for counts of 1000). To test the effect of phage concentration on the number of plaques obtained, successive dilutions of a phage preparation were all plated, and the number of plaques enumerated. Over a 100-fold range of dilution, the plaque count was in linear proportion to the phage concentration. (See Fig. 1.)

Fig. 1. Proportionality of the phage concentration to the plaque count. Successive twofold dilutions of a phage preparation were plated in duplicate on nutrient agar; 0.1 cc on each plate. The plaque counts from two such series of dilutions are plotted against the relative phage concentration, both on a logarithmic scale.

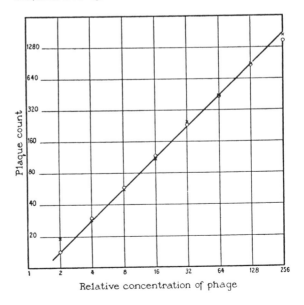

Dreyer and Campbell-Renton[8] using a different anti-*coli* phage and an anti-*staphylococcus* phage, and a different technique found a complicated dependence of plaque count on dilution. Such a finding is incompatible with the concept that phage particles behave as single particles, *i.e.*, without interaction, with respect to plaque formation. Our experiments showed no evidence of such a complicated behavior, and we ascribe it therefore to some secondary cause inherent in their procedure.

Bronfenbrenner and Korb[9] using a phage active against *B. dysenteriae* Shiga, and a different plating technique found

that when the agar concentration was changed from 1 percent to 2.5 percent, the number of plaques was reduced to 1 percent of its former value. They ascribed this to a change in the water supplied to the bacteria. With the technique which we have employed, variation of the agar concentration from 0.75 percent to 3.0 percent, had little influence on the number of plaques produced, though the size decreased noticeably with increasing agar concentration. (See Table 1.)

Changes in the concentration of bacteria spread with the phage on the agar plates had no important influence on the

Table 1 Independence of plaque count on plating method

Agar concentration

Plates were prepared in which the agar strength varied, and all spread with 0.1 cc of the same dilution of a phage preparation. There is no significant difference in the numbers of plaques obtained.

Agar concentration, *percent*	0.75	1.5	3.0
Plaque counts	394	373	424
	408	430	427
	376	443	455
	411	465	416
	373	404	469
Average	392	423	438
Plaque size, *mm*	2	0.5	0.2

Concentration of plating *coli*

A broth suspension of bacteria (10^9 bacteria/cc) was prepared from a 24 hour agar slant and used at various dilutions, as the plating suspension for a single phage dilution. There are no significant differences in the plaque counts except at the highest dilution of the bacterial suspension, where the count is about 15 percent lower.

Concentration	Plaque count
1	920
1/5	961
1/25	854
1/125	773

Temperature of plate incubation

Twelve plates were spread with 0.1 cc of the same suspension of phage and bacteria, divided into three groups, and incubated at different temperatures. There were no significant differences in the plaque counts obtained.

Temperature, °C	37	24	10
Plaque Count	352	384	405
	343	405	377
	386	403	400
	422	479	406
Average	376	418	397

number of plaques obtained. (See Table 1.) The temperature at which plates were incubated had no significant effect on the number of plaques produced. (See Table 1.)

In appraising the accuracy of this method, several points must be borne in mind. With our phage, our experiments confirm in the main the picture proposed by d'Herelle, according to which a phage particle grows in the following way: it becomes attached to a susceptible bacterium, multiplies upon or within it up to a critical time, when the newly formed phage particles are dispersed into the solution.

In the plaque counting method a single phage particle and an infected bacterium containing any number of phage particles will each give only one plaque. This method therefore, does not give the number of phage particles but the number of loci within the solution at which one or more phage particles exist. These loci will hereafter be called "infective centers." The linear relationship between phage concentration and plaque count (Fig. 1) does not prove that the number of plaques is equal to the number of infective centers, but only that it is proportional to this number. We shall call the fraction of infective centers which produces plaques the "efficiency of plating." With the concentrations of phage and bacteria which we have used this coefficient is essentially the fraction of infected bacteria in the suspension spread on the plate, which goes through to lysis under our cultural conditions on the agar medium. After plating, the phage particles released by this lysis infect the surrounding bacteria, increasing only the size, and not the number of plaques.

The experimental determination of the efficiency of plating is described in a later section (see p. 234). The coefficient varies from 0.3 to 0.5. This means that three to five out of every ten infected bacteria produce plaques. The fact that the efficiency of plating is relatively insensitive to variations in the temperature of plate incubation, density of plating *coli*, concentration of agar, etc. indicates that a definite fraction of the infected bacteria in the broth cultures do not readily go through to lysis when transferred to agar plates. For most experiments only the relative assay is significant; we have therefore, given the values derived directly from the plaque counts without taking into account the efficiency of plating, unless the contrary is stated.

Growth Measurements

The main features of the growth of this phage in broth cultures of the host are shown in Fig. 2. After a small initial increase (discussed below) the number of infective centers (individual phage particles, plus infected bacteria) in the suspension remains constant for a time, then rises sharply to a new value, after which it again remains constant. Later, a second sharp rise, not as clear-cut as the

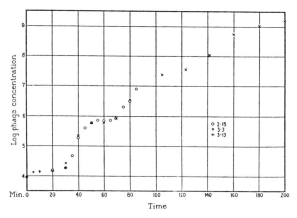

Fig. 2. Growth of phage in the presence of growing bacteria at 37°C. A diluted phage preparation was mixed with a suspension of bacteria containing 2 X 10⁸ organisms per cc, and diluted after 3 minutes 1 to 50 in broth. At this time about 70 percent of the phage had become attached to bacteria. The total number of infective centers was determined at intervals on samples of this growth mixture. Three such experiments, done on different days, are plotted in this figure. The same curve was easily reproducible with all phage preparations stored under proper conditions.

first, and finally a third rise occur. At this time visible lysis of the bacterial suspension takes place. A number of features of the growth process may be deduced from this and similar experiments, and this is the main concern of the present paper.

The Initial Rise

When a concentration of phage suitable for plating was added to a suspension of bacteria, and plated at once, a reproducible plaque count was obtained. If the suspension with added phage was allowed to stand 5 minutes at 37°C (or 20 minutes at 25°C) the number of plaques obtained on plating the suspension was found to be 1.6 times higher. This initial rise is not to be confused with the first "burst" which occurs later and increases the plaque count 70-fold. After the initial rise, the new value is readily duplicated and remains constant until the start of the first burst in the growth curve (30 minutes at 37° and 60 minutes at 25°).

This initial rise we attribute not to an increase in the number of infective centers, but to an increase in the probability of plaque formation (*i.e.*, an increase in the efficiency of plating) by infected bacteria in a progressed state; that is, bacteria in which the phage particle has commenced to multiply. That this rise results from a change in the efficiency of plating and not from a quick increase in the number of infective centers is evident from the following experiment. Bacteria were grown for 24 hours at 25°C on agar slants, then suspended in broth. Phage was added to this suspension and to a suspension of bacteria grown in the usual way, and the concentration of infective centers was determined on both. The initial value was 1.6 times higher in the agar grown bacteria than in the control experiment, and remained constant until actual growth occurred. The initial rise was therefore absent in this case, clearly a result of an increase in the efficiency of plating. A sufficient number of experiments were performed with bacteria grown on agar to indicate that in other respects their behavior is similar to that of the bacteria grown in broth. The bacteria grown in this way on agar slants are in some way more susceptible to lysis than the broth cultured bacteria.

Adsorption

The first step in the growth of bacteriophage is its attachment to susceptible bacteria. The rate of this attachment can be readily measured by centrifuging the bacteria out of a suspension containing phage, at various times, and determining the amount of phage which remains unattached in the supernatant (cf. Krueger[10]).*

According to the picture of phage growth outlined above, phage cannot multiply except when attached to bacteria; therefore, the rate of attachment may, under certain conditions, limit the rate of growth. We wished to determine the rate of this adsorption so that it could be taken into account in the interpretation of growth experiments, or eliminated if possible, as a factor influencing the growth rate.

Our growth curves show that there is no increase in the number of infective centers up to a critical time; we could therefore, make measurements of the adsorption on living bacteria suspended in broth, so long as the time allowed for attachment was less than the time to the start of the first burst in the growth curve. The adsorption proved to be so rapid that this time interval was ample to obtain adsorption of all but a few percent of the free phage if the bacteria concentration was above 3×10^7. The number of bacteria remained constant; the lag phase in their growth was longer than the experimental period.

The rate of attachment was found to be first order with respect to the concentration of free phage (P_f) and first order with respect to the concentration of bacteria (B) over a wide range of concentrations, in agreement with the results reported by Krueger.[10] That is, the concentration of free phage followed the equation

$$-\frac{d(P_f)}{dt} = k_a(P_f)(B)$$

in which k_a was found to be 1.2×10^{-9} cm^3/min at 15° and 1.9×10^{-9} cm^3/min at 25°C. These rate constants are about five times greater than those reported by Krueger.[10] With our ordinary 18-hour bacteria cultures (containing 2×10^8 B. coli/cc) we thus obtain 70 percent attachment of phage in 3 minutes and 98 percent in 10 minutes. The adsorption follows the equation accurately until more than 90 percent attachment has been accomplished, and then slows down somewhat, indicating either that not all the phage particles have the same affinity for the bacteria, or that equilibrium is being approached. Other experiments not recorded here suggest that, if an equilibrium exists, it lies too far in favor of adsorption to be readily detected. This equation expresses the rate of adsorption even when a tenfold excess of phage over bacteria is present, indicating that a single bacterium can accommodate a large number of phage particles on its surface, as found by several previous workers.[5, 10]

Krueger[10] found a tree equilibrium between free and adsorbed phage. The absence of a detectable desorption in our case may result from the fixation of adsorbed phage by growth processes, since our conditions permitted growth, whereas Krueger's experiments were conducted at a temperature at which the phage could not grow.

---

*A very careful study of the adsorption of a coli-phage has also been made by Schlesinger (Schlesinger, M., Z. Hyg. u. Infektionskrankh., 1932, 114:136, 149). Our results, which are less accurate and complete, agree qualitatively and quantitatively with the results of his detailed studies.

Growth of Phage

Following adsorption of the phage particle on a susceptible bacterium, multiplication occurs, though this is not apparent as an increase in the number of plaques until the bacterium releases the resulting colony of phage particles into the solution. Because the adsorption under proper conditions is so rapid and complete (as shown above) experiments could be devised in which only the influence of the processes following adsorption could be observed.

The details of these experiments were as follows: 0.1 cc of a phage suspension of appropriate concentration was added to 0.9 cc of an 18 hour broth bacterial culture, containing about 2×10^8 *B. coli* / cc. After standing for a few minutes, 70 to 90 percent of the phage was attached to the bacteria. At this time, the mixture was diluted 50-fold in broth (previously adjusted to the required temperature) and incubated. Samples were removed at regular intervals, and the concentration of infective centers determined.

The results of three experiments at 37°C are plotted in Fig. 2, and confirm the suggestion of d'Herelle that phage multiplies under a spatial constraint, *i.e.*, within or upon the bacterium, and is suddenly liberated in a burst. It is seen that after the initial rise (discussed above) the count of infective centers remains constant up to 30 minutes, and then rises about 70-fold above the initial value. The rise corresponds to the liberation of the phage particles which have multiplied in the initial constant period. This interpretation was verified by measurements of the free phage by centrifuging out the infected bacteria, and determining the number of phage particles in the supernatant liquid. The free phage concentration after adsorption was, of course, small compared to the total and remained constant up to the time of the first rise. It then rose steeply and became substantially equal to the total phage.

The number of bacteria lysed in this first burst is too small a fraction of the total bacteria used in these experiments to be measured as a change in turbidity; the ratio of uninfected bacteria to the total possible number of infected bacteria before the first burst is 400 to 1, the largest number of bacteria which can disappear in the first burst is therefore only 0.25 percent of the total.

The phage particles liberated in the first burst are free to infect more bacteria. These phage particles

then multiply within or on the newly infected bacteria; nevertheless, as before, the concentration of infective centers remains constant until these bacteria are lysed and release the phage which they contain into the medium. This gives the second burst which begins at about 70 minutes from the start of the experiment. Since the uninfected bacteria have been growing during this time, the bacteria lysed in the second burst amount to less than 5 percent of the total bacteria present at this time. There is again therefore, no visible lysis.

This process is repeated, leading to a third rise of smaller magnitude starting at 120 minutes. At this time, inspection of the culture, which has until now been growing more turbid with the growth of the uninfected bacteria, shows a rapid lysis. The number of phage particles available at the end of the second rise was sufficient to infect the remainder of the bacteria.

These results are typical of a large number of such experiments, at 37°, all of which gave the 70-fold burst size, *i.e.*, an average of 70 phage particles per infected bacterium, occurring quite accurately at the time shown, 30 minutes. Indeed, one of the most striking features of these experiments was the constancy of the time interval from adsorption to the start of the first burst. The magnitude of the rise (70-fold) was likewise readily reproducible by all phage preparations which had been stored under proper conditions to prevent deterioration (see above).

Multiple Infection

The adsorption measurements showed that a single bacterium can adsorb many phage particles. The subsequent growth of phage in these "multiple infected" bacteria might conceivably lead to (*a*) an increase in burst size; (*b*) a burst at an earlier time, or (*c*) the same burst size at the same time, as if only one of the adsorbed particles had been effective, and the others inactivated. In the presence of very great excesses of phage, Krueger and Northrop[3] and Northrop[2] report that visible lysis of the bacteria occurs in a very short time. It was possible therefore, that in our case, the latent period could be shortened by multiple infection. To determine this point, we have made several experiments of which the following is an example. 0.8 cc of a freshly prepared phage suspension containing 4×10^9 particles per cc (assay corrected for efficiency of plating) was added to 0.2 cc of bacterial suspension containing 4×10^9 bacteria per cc. The ratio of phage to bacteria in this mixture was 4 to 1. 5 minutes were allowed for adsorption, and then the mixture was diluted 1 to 12,500 in broth, incubated at 25°, and the growth of the phage followed by plating at 20 minute intervals, with a control growth curve in which the phage to bacteria ratio was 1 to 10. No significant difference was found either in the latent period or in the size of the burst. The bacteria which had adsorbed several phage particles behaved as if only one of these particles was effective.

Effect of Temperature on Latent Period and Burst Size

A change in temperature might change either the latent period, *i.e.*, the time of the burst, or change the size of the burst, or both. In order to obtain more accurate estimates of the burst size it is desirable to minimize reinfection during the period of observation. This is obtained by diluting the phage-bacteria mixture (after initial contact to secure adsorption) to such an extent that the rate of adsorption then becomes extremely small. In this way, a single "cycle" of growth, (infection, growth, burst) was obtained as the following example shows. 0.1 cc of phage of appropriate and known concentration was added to 0.9 cc of an 18-hour culture and allowed to stand in this concentrated bacterial suspension for 10 minutes at the temperature of the experiment. This mixture was then diluted $1:10^4$ in broth and incubated at the temperature chosen. Samples of this diluted mixture were withdrawn at regular intervals and assayed. The results of three such experiments are plotted in Fig. 3. The rise corresponds to the average number of phage produced per burst, and its value can be appraised better in these experiments than in the complete growth curve previously given (Fig. 2) where there is probably some overlapping of the steps. In these experiments the rise is seen to be practically identical at the three temperatures, and equals about sixty particles per infected bacterium, but the time at which the rise occurred was 30 minutes at 37°, 60 minutes at 25°, and 180 minutes at 16.6°. This shows that the effect of temperature is solely on the latent period.

We have also made separate measurements of the rate of bacterial growth under the conditions of these experiments. They show that the average division period of the bacteria in their logarithmic growth phase varies in the same way with temperature, as the length of the latent period of phage growth. The figures are:

Temperature	Division period of *B*	Latent period of *P* growth
°C	*min*	*min*
16.6	About 120	180
25	42	60
37	21	30

There is a constant ratio (3/2) between the latent period of phage growth and the division period of the bacteria. This coincidence suggests a connection between the time required for division of a bacterium under optimum growth conditions, and the time from its infection by phage to its lysis.

Individual Phage Particle

The growth curves described above give averages only of large numbers of bursts. They can, however, also be studied individually, as was first done by Burnett.[11]

If from a mixture containing many particles very small samples are withdrawn, containing each on the average only about one or less particles, then the fraction p_r of samples containing r particles is given by Poissons'[12] formula,

$$p_r = \frac{n^r e^{-n}}{r!} \tag{1}$$

where n is the average number of particles in a sample and e is the Napierian logarithm base. If the average number n is unknown, it can be evaluated from an experimental determination of any single one of the p_r, for instance from a determination of p_0, the fraction of samples containing no particles:

$$n = -\ln p_0 \tag{2}$$

Let us now consider the following experiment. A small number of phage particles is added to a sus-

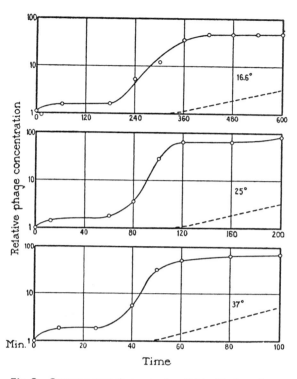

Fig. 3. One-step growth curves. A suitable dilution of phage was mixed with a suspension of bacteria containing 2×10^8 organisms per cc and allowed to stand at the indicated temperature for 10 minutes to obtain more than 90 percent adsorption of the phage. This mixture was then diluted $1:10^4$ in broth, and incubated. It was again diluted 1:10 at the start of the first rise to further decrease the rate of adsorption of the phage set free in the first step. The time scales are in the ratio 1:2:6 for the temperatures 37, 25, and 16.6°C. Log P/P_0 is plotted, P_0 being the initial concentration of infective centers and P the concentration at the time t. The broken line indicates the growth curve of the bacteria under the corresponding conditions.

pension containing bacteria in high concentration. Within a few minutes each phage particle has attached itself to a bacterium. The mixture is then diluted with a large volume of broth, in order to have the bacteria in low concentration so that after the first burst a long time elapses before reinfection, as in the one step growth curves. Samples (0.05 cc) are removed from this mixture to separate small vials and incubated at the desired temperature. If these samples are plated separately (after adding a drop of bacterial suspension to each vial) before the

occurrence of bursts, the fraction of the plates containing 0, 1, 2, *etc.*, plaques is found to conform to formula (1) (see Table 2). In this experiment we could also have inferred the average number of particles per sample, using formula (2), from the fraction of the plates showing no plaques (giving 0.93 per sample) instead of from the total number of plaques (27/33 = 0.82 per sample).

Experimental Measure of Efficiency of Plating

If the samples are incubated until the bursts have occurred, and then plated, the samples which had no particles will still show no plaques, those with one or more particles will show a large number, depending on the size of the burst, and on the efficiency of plating. In any case, if we wait until all bursts have occurred, only those samples which really contained no particle will show no plaques, quite independent of any inefficiency of plating. From this fraction of plates showing no plaques we can therefore evaluate the true number of particles originally present in the solution, and by comparison with the regular assay evaluate the efficiency of plating. In this way we have determined our efficiency of plating to be about 0.4. For instance, one such experiment gave no plaques on 23 out of 40 plates, and many plaques on each of the remaining plates. This gives

Table 2 Distribution of individual particles among small samples

	p_r (experimental)	p_r (calculated)
0 plaques on 13 plates	0.394	0.441
1 plaque on 14 plates	0.424	0.363
2 plaques on 5 plates	0.151	0.148
3 plaques on 1 plate	0.033	0.040
4 plaques on 0 plates	0.000	0.008
27 plaques on 33 plates	1.002	1.000

A suitably diluted phage preparation was added to 5 cc of 18-hour bacteria culture and 0.1 cc samples of this mixture were plated. The distribution of particles among the samples is that predicted by formula (1).

$$p_0 = \frac{23}{40} \text{ or } 0.57$$

from which $n = 0.56$ particles per sample. A parallel assay of the stock phage used indicated 0.22 particles per sample; the plating efficiency was therefore

$$\frac{0.22}{0.56} = 0.39$$

This plating efficiency remains fairly constant under our standard conditions for assay. The increase in probability of plaque formation which we suppose to take place following the infection of a bacterium by the phage particle, *i.e.*, the initial rise, brings the plating efficiency up to 0.65.

The Burst Size

Single particle experiments such as that described above, revealed a great fluctuation in the magnitude of individual bursts, far larger than one would expect from the differences in size of the individual bacteria in a culture; indeed, they vary from a few particles to two hundred or more. Data from one such experiment are given in Table 3.

We at first suspected that the fluctuation in burst size was connected with the time of the burst, in that early bursts were small and late bursts big, and the fluctuation was due to the experimental superposition of these. However, measurement of a large number of bursts, plated at a time when only a small fraction of the bursts had occurred, showed the same large fluctuation. We then suspected that the particles of a burst were not liberated simultaneously, but over an interval of time. In this case one might expect a greater homogeneity in burst size, if measurements were made at a late time when they are at their maximum value. This view also was found by experiment to be false.

The cause of the great fluctuation in burst size is therefore still obscure.

DISCUSSION

The results presented above show that the growth of this strain of phage is not uniform, but in bursts. These bursts though of constant average size, under

Table 3 Fluctuation in individual burst size

	Bursts
25 plates show 0 plaques	
1 plate shows 1 plaque	
14 plates show bursts	130
Average burst size, taking account of	58
probable doubles = 48	26
	123
	83
	9
	31
	5
	53
	48
	72
	45
	190
	9
Total	*882 plaques*

97.9 percent of phage attached to bacteria in presence of excess bacteria (10 minutes), this mixture diluted, and samples incubated 200 minutes, then entire sample plated with added bacteria.

our conditions, vary widely in individual size. A burst occurs after a definite latent period following the adsorption of the phage on susceptible bacteria, and visible lysis coincides only with the last stepwise rise in the growth curve when the phage particles outnumber the bacteria present. It seemed reasonable to us to assume that the burst is identical with the lysis of the individual bacterium.

Krueger and Northrop,[3] in their careful quantitative studies of an anti-*staphylococcus* phage came to an interpretation of their results which differs in some important respects from the above:

1. Their growth curves were smooth and gave no indication of steps; they concluded therefore that the production of phage is a continuous process.

2. In their case, the free phage during the logarithmic phase of a growth curve was an almost constant small fraction of the total phage. This led them to the view that there is an equilibrium between intracellular and extracellular phage. With an improved technique, Krueger[10] found that the fraction of free

phage decreased in proportion to the growth of the bacteria, in conformity with the assumption of an equilibrium between two phases.

3. Krueger and Northrop[3] found that visible lysis occurred when a critical ratio of total phage to bacteria had been attained, and they assumed that there was no lysis in the earlier period of phage growth.

To appreciate the nature of these differences it must be born in mind that their method of assay was essentially different from ours. They used, as a measure of the "activity" of the sample of phage assayed, the time required for it to lyse a test suspension of bacteria under standard conditions. This time interval, according to the picture of the growth process given here, is the composite effect of a number of factors: the average time required for adsorption of free phage, its rate of growth in the infected bacteria, the time and size of burst, and the average time required for repetition of this process until the number of phage particles exceeds the number of bacteria and infects substantially all of them. Then, after a time interval equal to the latent period, lysis occurs.

This lysis assay method tends to measure the total number of phage particles rather than the number of infective centers as the following considerations show. Let us take a sample of a growth mixture in which is suspended one infected bacterium containing fifty phage particles. If this sample is plated, it can show but a single plaque. However, if the sample is assayed by the lysis method, this single infective center soon sets free its fifty particles (or more, if multiplication is still proceeding) and the time required to attain lysis will approximate that for fifty free particles rather than that for a single particle.

Since the burst does not lead to an increase in the number of phage particles, but only to their dispersion into the solution, the lysis method cannot give any steps in the concentration of the *total* phage in a growth curve. On the other hand one might have expected a step-wise increase in the concentration of *free* phage. However, the adsorption rate of the phage used by Krueger[10] is so slow that the infection of the bacteria is spread over a time longer than the presumed latent period, and therefore the bursts would be similarly spread in time, smoothing out any steps which might other-

wise appear. Moreover, their measurements were made at 30 minute intervals, which even in our case would have been insufficient to reveal the steps.

The ratio between intracellular and extracellular phage would be determined, according to this picture of phage growth, by the ratio of the average time of adsorption to the average latent period. The average time of adsorption would decrease as the bacteria increased, shifting the ratio of intracellular to extracellular phage in precisely the manner described by Krueger.[10]

As we have indicated in the description of our growth curves, lysis of bacteria should become visible only at a late time. Infection of a large fraction of the bacteria is possible only after the free phage has attained a value comparable to the number of bacteria, and visible lysis should then set in after the lapse of a latent period. At this time the total phage (by activity assay) will be already large compared to the number of bacteria, in agreement with Krueger and Northrop's findings.

It appears therefore that while Krueger and Northrop's picture does not apply to our phage and bacteria, their results do not exclude for their phage the picture which we have adopted. It would be of fundamental importance if two phages behave in such a markedly different way.

SUMMARY

1. An anti-*Escherichia coli* phage has been isolated and its behavior studied.

2. A plaque counting method for this phage is described, and shown to give a number of plaques which is proportional to the phage concentration. The number of plaques is shown to be independent of agar concentration, temperature of plate incubation, and concentration of the suspension of plating bacteria.

3. The efficiency of plating, *i.e.*, the probability of plaque formation by a phage particle, depends somewhat on the culture of bacteria used for plating, and averages around 0.4.

4. Methods are described to avoid the inactivation of phage by substances in the fresh lysates.

5. The growth of phage can be divided into three

periods: adsorption of the phage on the bacterium, growth upon or within the bacterium (latent period), and the release of the phage (burst).

6. The rate of adsorption of phage was found to be proportional to the concentration of phage and to the concentration of bacteria. The rate constant k_a is 1.2×10^{-9} cm^3/min at 15°C and 1.9×10^{-9} cm^3/min at 25°.

7. The average latent period varies with the temperature in the same way as the division period of the bacteria.

8. The latent period before a burst of individual infected bacteria varies under constant conditions between a minimal value and about twice this value.

9. The average latent period and the average burst size are neither increased nor decreased by a four-fold infection of the bacteria with phage.

10. The average burst size is independent of the temperature, and is about 60 phage particles per bacterium.

11. The individual bursts vary in size from a few particles to about 200. The same variability is found when the early bursts are measured separately, and when all the bursts are measured at a late time.

One of us (E. L. E.) wishes to acknowledge a grant in aid from Mrs. Seeley W. Mudd. Acknowledgment is also made of the assistance of Mr. Dean Nichols during the preliminary phases of the work.

REFERENCES

1. Schlesinger, M., *Biochem. Z.,* Berlin, 1934, 273:306.
2. Northrop, J. H., *J. Gen. Physiol.*, 1938, 21:335.
3. Krueger, A. P., and Northrop, J. H., *J. Gen. Physiol.*, 1930, 14:223.
4. Burnet, F. M., *Brit. J. Exp. Path.*, 1927, 8:121.
5. Burnet, F. M., Keogh, E. V., and Lush, D., *Australian J. Exp. Biol. and Med. Sc.,* 1937, 15:suppl. to part 3, p. 227.
6. Heidelberger, M., Kendall, F. E., and Scherp, H. W., *J. Exp. Med.*, 1936, 64:559.
7. d'Hérelle, F., The bacteriophage and its behavior, Baltimore, The Williams & Wilkins Co., 1926.
8. Dreyer, C., and Campbell-Renton, M. L., *J. Path. and Bact.*, 1933, 36:399.
9. Bronfenbrenner, J. J., and Korb, C., *Proc. Soc. Exp. Biol. and Med.*, 1923, 21:315.
10. Krueger, A. P., *J. Gen. Physiol.*, 1931, 14:493.
11. Burnet, F. M., *Brit. J. Exp. Path.*, 1929, 10:109.
12. Poisson, S. D., Recherches sur la probabilité des jugements en matière criminelle et en matière civile, précédées des règles générales du calcul des probabilités, Paris, 1837.

On the mechanism of genetic recombination between DNA molecules*

MATTHEW MESELSON†

Reprinted by author's and publisher's permission from Journal of Molecular
Biology, *vol. 9, 1964, pp. 734–745.*

In the late 1940s, Delbrück and Bailey, with Hershey and Rotman, demonstrated that virus chromosomes could undergo genetic exchange during intracellular replication. Their demonstration raised the important question: how does the exchange occur? Two important ideas had to be tested: First, recombination by copy-choice, and second, the recombination by breakage-fusion (see the introduction to this part for a discussion of these terms). When parental DNA could be distinguished from newly synthesized DNA (see the paper by Meselson and Stahl in Part 1), a basis was established for distinguishing between the two hypotheses. In 1961, Meselson and Weigle (*Proc. Nat. Acad. Sci. U.S.* 47:857:868) published a report that suggested that chromosomes recombined by a process of breakage and fusion, not copy-choice. However, their paper could not eliminate the possibility of a limited amount of synthesis accompanying the recombinational process.

Meselson followed up on his work in 1961 with this paper. The paper clearly demonstrated that recombination occurs in the absence of significant DNA synthesis. Meselson accomplished this by using DNA molecules that were *both* density-labeled. He found recombinant chromosomes that contained *all* parental DNA — no synthesis had occurred at all — thus establishing breakage-fusion as the most reasonable method of phage chromosome recombination. Although some DNA synthesis was probably required, it was involved solely with the repair of minor gaps in the molecule.

A two-factor cross was performed between bacteriophages labeled with heavy isotopes. Recombinants were found with chromosomes formed entirely or almost entirely of parental DNA. This and other features of the distribution of parental DNA among recombinant phages and among their descendants show that genetic recombination occurs by breakage and joining of double-stranded DNA molecules. Also, there is some indication that a small amount of DNA is removed and resynthesized in the formation of recombinant molecules.

1. INTRODUCTION

Genetic recombination between bacteriophages was discovered by Delbrück & Bailey (1946) in the course of experiments designed to determine whether more than one infecting phage can multiply in a single bacterium. They examined phages produced by cells jointly infected with two types of phage differing from one another in two genetic characters and found not only the two parental phage types but also two types with combinations of the parental characters. For a time, it was suspected that the new types resulted from a novel process more akin to mutation than to recombination. However, Hershey & Rotman (1948) showed that the phenomenon was indeed genetic recombination, for its operation closely followed rules already worked out for recombination in higher organisms.

Although the details of the mechanism are not known, there is strong evidence in the case of bacteriophage λ that recombination results from the breakage and joining of DNA molecules. The initial evidence for breakage and joining came from measurements of the amount of labeled DNA in selected recombinants yielded by cells infected jointly with isotopically labeled and unlabeled phages (Meselson & Weigle 1961; Kellenberger, Zichichi & Weigle

*Received 24 April 1964.
†Biological Laboratories, Harvard University, Cambridge, Massachusetts, U.S.A.

1961). Recombinant phages were found to contain labeled DNA in discrete proportions, corresponding to the proportion of the λ genetic map lying beyond the selected site of recombination in the direction of the allele contributed by the labeled parent phage. Since the chromosome of λ is a single DNA molecule along which hereditary determinants are spaced at least approximately as they are along the λ genetic map (Kaiser 1962 and personal communication), this simple result may be explained as the result of breakage and joining of chromosomes at the selected site of recombination. Alternatively, it could reflect the operation of a rather less plausible mechanism whereby a fragment of a parental chromosome is rebuilt by synthesis of its missing length on a template provided by the homologous portion of a chromosome of different parentage. The latter mechanism has been called breakage and copying.

Breakage and joining may be distinguished from breakage and copying by an examination of the amount of parental DNA in recombinant phages from a cross between parents *both* of which contain labeled DNA. Only the former of the two mechanisms can produce a recombinant containing label from both parents. Using host-induced modification as a label, Ihler & Meselson (1963) obtained evidence that both parents do contribute lengths of DNA which become joined together to form recombinant chromosomes.

The mechanism of recombination is specified more conclusively and in greater detail by the present results, a preliminary report of which has appeared elsewhere (Meselson 1962). The occurrence of recombination by breakage and joining of λ chromosomes is confirmed, and the process is shown to be independent of chromosome replication in the sense that recombinants may be formed entirely, or almost entirely, of unreplicated chromosome fragments. However, there is some indication that a small amount of DNA is removed and resynthesized in the formation of recombinant chromosomes.

2. MATERIALS AND METHODS

(a) Bacteriophages

Phage λ, the wild type of Kaiser (1957) and λhc were used as parents for all crosses. Phage λhc is a recombinant isolated from a cross of the λh of Kaiser (1962) by λc$_{26}$, a nitrous acid-induced c$_1$ mutant of λ.

All of these phages have buoyant density 1.508 g cm^{-3}

in pure CsCl colution of 20°C. To insure homology of the two parental phages, the h allele was passed through five consecutive crosses alternately to λc$_{26}$ and λ, selecting in the first case for λhc and in the second for λh. Although many thousands of plaques were examined, there was no indication of segregation of characters other than h and c.

Stocks of λ were prepared by the induction of lysogenic bacteria with ultraviolet light. Stocks of λhc were produced by lytic multiplication. Phages uniformly labeled with ^{13}C and ^{15}N were prepared and purified according to Meselson & Weigle (1961). The uniformity of labeling of each stock was verified by density-gradient centrifugation. The frequency of λh was found to be less than 0.01% in each stock.

(b) Bacteria

Escherichia coli K12 strain 3110 (E. M. Lederberg) was used for preparation of all parental phage stocks. Strain C600.5, a mutant of *E. coli* K12 C600 (Appleyard 1954), was used as host for all crosses. The host modification and restriction properties of this mutant are identical to those of *E. coli* strain C (Bertani & Weigle 1953). Strain C600.5 was used in preference to C, because λ is rapidly inactivated in lysates of our strain C. Strains C600, C600.5, and *E. coli* K12/λ (Kaiser 1962) were used as indicators.

(c) Methods

Crosses and low-multiplicity growth cycles were performed on strain C600.5 following Ihler & Meselson (1963). In crosses, the multiplicity of infection was approximately three of each parental type. Recombinants λh were scored as turbid plaques on a mixed indicator containing equal concentrations of C600.5 and ultraviolet-irradiated K/λ. On this indicator λhc produces clear plaques whereas λ and λc give rise to barely discernable ghost plaques. Irradiation removes the barrier which otherwise would restrict phages produced on strain C600.5 from infecting K/λ. For the assay of unrestricted phages, a mixture of C600 and nonirradiated K/λ was used and is referred to as restricting indicator. The validity of both scoring procedures was established by appropriate reconstruction experiments. Only one parental type and one recombinant type, λhc and λh were scored. Density-gradient centrifugation was performed according to Meselson & Weigle (1961).

3. OUTLINE OF THE EXPERIMENTS

When several λ phages simultaneously infect a cell, some of the chromosomes replicate semiconservatively while others remain unreplicated (Meselson & Weigle 1961). Accordingly, a cell infected with several isotopically labeled phages will contain chromosomes with both polynucleotide strands labeled, others with only one labeled strand, and still others with no label. If genetic recombination occurs by simple breakage and joining, then recombination at a given site followed by phage maturation could yield altogether nine discrete phage types with respect to the proportion of labeled DNA they contain. If the site of recombination is chosen at the center of the chromosome, the number of possible classes is reduced to five: fully labeled, three-quarter, one-half, one-quarter and unlabeled. The fully-labeled and three-quarter-labeled classes deserve special attention, because they are expected to result from breakage and joining but not from breakage and copying. An important result of the experiments described below is the discovery of these two classes among the recombinant phages from a suitably designed cross. However, this result does not prove that segments from different chromosomes have become permanently joined to one another. Instead, they could be only temporarily associated. That this is not the case is shown by the finding of half-labeled recombinant phages among the progeny yielded by cells singly infected with fully-labeled, three-quarter-labeled or half-labeled recombinant phages. This result shows that recombination results in an association between polynucleotide chains from different parent chromosomes which is not disrupted by the events of chromosome injection, replication or maturation.

The genetic markers h (host range) and c (clear plaque) were chosen for these experiments, because of their symmetrical location about the center of the linkage map of λ. That they also straddle the center of the λ chromosome is shown both by the present results and by the behavior of infectious molecules of λ DNA when they are broken in half by hydrodynamic shear (Kaiser 1962). The frequency of recombination between h and c is approximately 7%, and the hc interval comprises about one third of the λ genetic map.

In order to improve the resolution of certain classes of labeled phages in the presence of a great excess of unlabeled phages, several gradients were assayed on restricting indicator as described under Materials and Methods. On this indicator, phages from strain C600.5 plate with very low efficiency unless they contain at least one strand composed of parental DNA (i.e., DNA synthesized on strain 3110). This is an example of the phenomena of host-induced modification and restriction elucidated by Arber & Dussoix (1962). It should be emphasized that we have relied on host-induced modification only as a means of suppressing the appearance of unlabeled phages, and not as a label for parental DNA.

4. RESULTS

(a) A Cross with Isotopically Unlabeled Phages

Figure 1 shows the density distribution of phages from a cross of unlabeled λ with unlabeled λhc. It is seen that both λhc and λh form a single sharp band in the gradient. Bands of exactly the same shape and location were found from assays performed on restricting indicator bacteria. The complete absence of phages with atypically high density provides assurance that variations in the amount of isotopic label are the only cause of the density variations observed in the remainder of these experiments. The light tail seen on each band results from mixing caused by the method of collecting fractions.

(b) Crosses with Isotopically Labeled Phages

The density distributions of λhc and λh from a cross of $^{13}C^{15}N$-labeled λhc by $^{13}C^{15}N$-labeled λ are shown in Fig. 2. The parental type λhc is seen to occur primarily in three discrete modes, corresponding to phages with conserved, semiconserved and completely new chromosomes (Meselson & Weigle 1961). The distribution of the recombinant type λh shows inflections at the positions expected for phages with fully-labeled, three-quarter-labeled, and one-half-labeled chromosomes; but lack of resolution prevents the positive identification of these categories.

An attempt was made to obtain increased resolution by assaying the distribution on restricting indicator bacteria, so as to suppress the appearance of phages lacking a completely labeled polynucleotide chain. Figure 3 shows that this procedure serves to reveal a discrete mode of half-labeled recombinants but that fully-labeled and three-quarter-labeled λh remain unresolved. The slight peaks at fraction 30

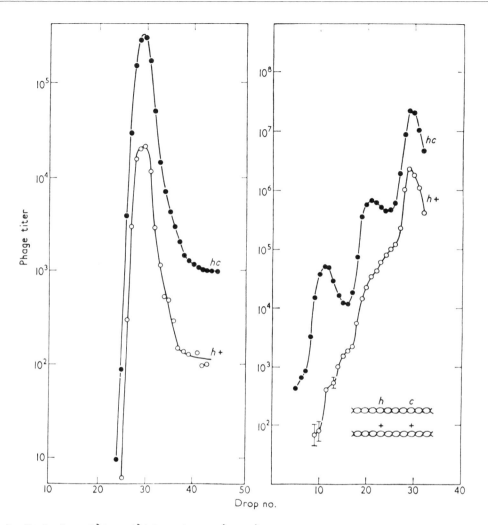

Fig. 1. Density distributions of λhc and λh from the cross $\lambda hc \times \lambda$.

Fig. 2. Density distributions of λhc and λh from the cross $[^{13}C^{15}N]\lambda hc \times [^{13}C^{15}N]\lambda$. Fractions 12 and 13 were pooled for the assay of λh. Error tags indicate 90% confidence limits. The three modes in the distribution of λhc are formed by phages with conserved, semiconserved, and newly synthesized chromosomes, in order of decreasing density. The location of mutations h and c on the chromosome is indicated schematically at the lower right of the figure.

are formed by unlabeled phages plating at the low efficiency with which λ from strain C600.5 infects most K12 strains.

The difficulty caused by insufficient resolution was overcome by subjecting fractions from each region of interest in the original gradient to further density-gradient analysis. The distributions of phages

in pooled fractions taken from the fully-, three-quarter-, and one-half-labeled regions of the original gradient are shown in Figs. 4, 5 and 6 respectively. Discrete bands of recombinant phages are found in all three cases. The distributions of Fig. 5 are of special interest. The contrast between the distinct unimodal band of three-quarter-labeled recombinants

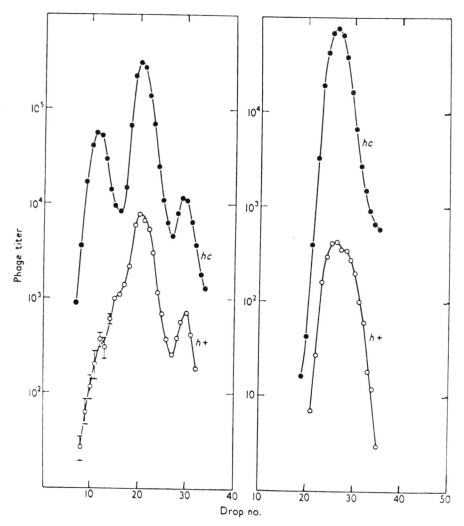

Fig. 3. Density distributions of λ*hc* and λ*h* in the gradient depicted in Fig. 2 found by assays on restricting indicator bacteria. Error tags indicate 90% confidence limits.

Fig. 4. Density distributions of λ*hc* and λ*h* from the fully-labeled region (fractions 10 to 13) of the gradient shown in Fig. 2.

and the broad bimodal distribution of the accompanying nonrecombinant phages shows most strikingly that the labeled recombinants are indeed a discrete density species formed by the breakage and joining of double-stranded chromosomes. (It may be noted that the band of three-quarter-labeled recombinants appears broadened by about two drops, probably as a result of the considerable length of the *hc* interval,

anywhere within which recombination presumably may occur. If this explanation is correct, then the sharpness of the band of half-labeled recombinants would suggest that most of them arose when the composition of the vegetative phage pool favored recombination between two conserved or two hybrid chromosomes more than recombination between a conserved and a completely new chromosome.

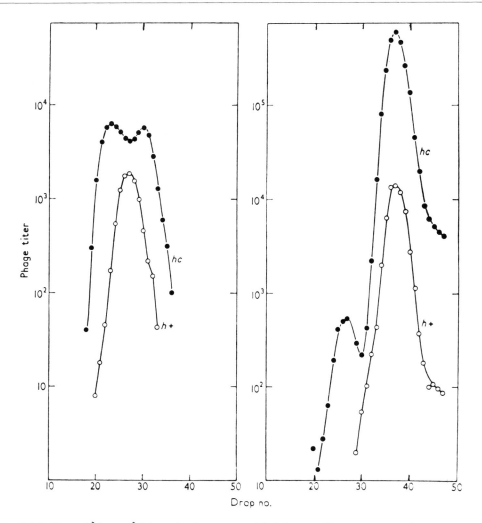

Fig. 5. Density distributions of λ*hc* and λ*h* from the three-quarter-labeled region (fractions 14 to 17) of the gradient shown in Fig. 2. Conserved and semiconserved λ*hc* are seen to be barely resolved.

Fig. 6. Density distributions of λ*hc* and λ*h* from the one-half-labeled region (fractions 19 to 21) of the gradient shown in Fig. 2.

In this regard, the relative amounts of the various labeled classes of parental and recombinant phages found in these experiments are compatible with the generally accepted view that recombination occurs throughout the latent period.)

Certain details of the distributions of Figs. 4 to 6 should be noted. A light shoulder appears on the band of fully-labeled recombinants and possibly also on the band of three-quarter-labeled recombinants, whereas the half-labeled recombinants form a band as sharp as do nonrecombinant phages. Upon repe-

tition of the entire experiment, these features of the various distributions were found to recur.

(c) One-step Growth of the Cross Progeny

Phages in reserve portions of the fully-, three-quarter-, and one-half-labeled fractions described above were allowed to multiply (one cycle) on strain C600.5. The total multiplicity of infection was 0.01 phage per bacterium. Each of the three lysates was analysed

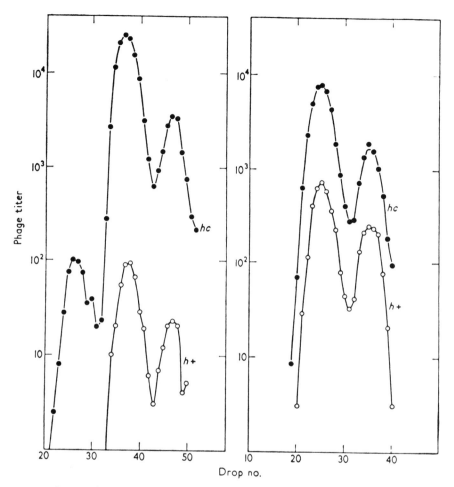

Fig. 7. Distributions of λ*hc* and λ*h* yielded by cells singly infected with fully-labeled phages from the gradient of Fig. 2. Assays were performed on restricting indicator bacteria. The most dense band of λ*hc* contains unadsorbed phages with fully-labeled chromosomes. The bands at fractions 37 and 47 contain phages with one-half-labeled and unlabeled chromosomes respectively.

Fig. 8. Distributions of λ*hc* and λ*h* yielded by cells singly infected with three-quarter-labeled phages from the gradient of Fig. 2. Assays were performed on restricting indicator bacteria. The bands at fractions 25 and 35 are formed by phages with half-labeled and unlabeled chromosomes respectively. One-quarter-labeled recombinant phages are not expected to plate on the restricting indicator and, accordingly, no indication of their presence is seen.

by density-gradient centrifugation with results shown in Figs. 7, 8 and 9. Assays were performed on restricting indicator bacteria in order to suppress the appearance of unlabeled phages, which otherwise would obscure the discrete bands of half-labeled λ*hc* and λ*h* seen in each gradient. Comparison of Figs. 7 to 9 with Figs. 4 to 6 shows that, in each case, the recombinant phages give rise to half-labeled progeny with approximately the same efficiency as do the parent type phages.

(d) Crosses of Labeled and Unlabeled Phages

Two crosses were performed in which only one of the parental types was labeled with heavy isotopes.

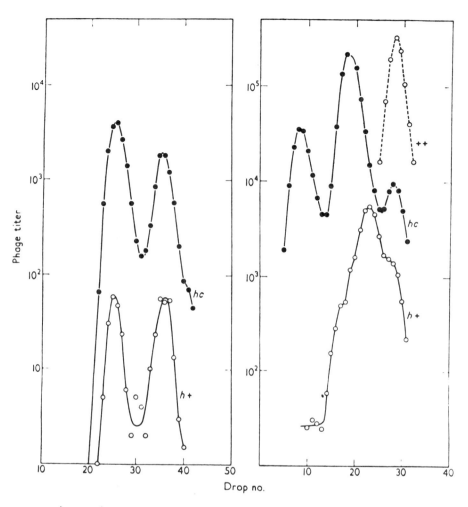

Fig. 9. Distributions of λ*hc* and λ*h* yielded by cells singly infected with one-half-labeled phages from the gradient of Fig. 2. Assays were performed on restricting indicator bacteria. The bands at fractions 26 and 36 are formed by phages with half-labeled and unlabeled chromosomes respectively.

Fig. 10. Distributions of λ, λ*hc*, and λ*h* from the cross [^{13}C^{15}N] λ*hc* X λ. Assays were performed on restricting indicator bacteria.

The results provide additional assurance that the fully-, three-quarter- and one-half-labeled recombinants examined above do in fact derive their label jointly from chromosomes of both parental types, as expected for recombination by simple breaking and joining. Figures 10 and 11 show the density distributions of progeny from the crosses [^{13}C^{15}N] λ*hc* X λ and λ*hc* X [^{13}C^{15}N]λ respectively. Assays were per-

formed on restricting indicator bacteria in order to suppress the appearance of unlabeled phages. The prominent band of unrestricted λ*h* found in both gradients occurs near the position expected for phages with one-quarter-labeled chromosomes. The exact location of the band corresponds to slightly less than one-quarter label when λ is the labeled parent, and slightly more when λ*hc* is labeled. If the

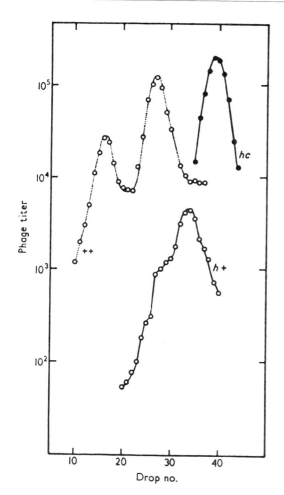

Fig. 11. Distributions of λ, λ*hc*, and λ*h* from the cross λ*hc* × [^{13}C^{15}N] λ. Assays were performed on restricting indicator bacteria. The shoulder of λ*h* at fractions 27 to 30 probably results from multiple recombination between hybrid and light chromosomes.

probability of recombination is relatively uniform in the region between *h* and *c*, then we may conclude that these markers are equidistant from a point about 55% of the way in from the "left" end of the chromosome.

Apart from the prominent band of one-quarter-labeled λ*h*, there is an inflection in the distributions of Figs. 10 and 11 suggesting the presence of half-labeled λ*h*. No three-quarter- or fully-labeled recom-

binants are expected, nor is there any indication of their presence.

5. DISCUSSION

Parental DNA is found in recombinant phages in discrete amounts which cannot be accounted for by breakage and copying, copy-choice, or any process which would alter the genotype of a chromosome without substantial replacement of chromosomic material. Instead, the observed distribution of parental DNA among the progeny of crosses and among their descendants shows that genetic recombination in λ occurs by the breakage and joining of double-stranded phage chromosomes. However, a reservation must be placed on the interpretation of the evidence for the joining of chromosomes. The discovery of phages with half-labeled chromosomes among the descendants of labeled recombinants shows only that the two parental contributions to at least one strand of the recombinant chromosome are able to remain associated throughout the processes of injection, replication and phage maturation. Whether both strands retain their integrity in this fashion and whether the association results from the formation of a covalent bond can not be decided.

It may be inquired whether breakage and joining are confined to a site or sites between *h* and *c*, while other mechanisms of recombination operate elsewhere on the λ chromosome. This possibility is rendered very unlikely by a number of observations, including the finding that chromosome breakage is associated with recombination in regions not overlapping the *hc* interval (Meselson & Weigle 1961; Jordan & Meselson unpublished experiments).

Considerable evidence for recombination by breakage and joining of DNA molecules has been obtained from experiments with T-even bacteriophages and with several bacteria. In the former case, it is well established that fragments of parental chromosomes appear in progeny phages (see Delbrück & Stent 1957; Hershey & Burgi 1956; Levinthal & Thomas 1957; Roller 1961; Kosinski 1961; Kahn 1964). Although other explanations can be devised, the implication is that recombination by breakage and joining is occurring in these cases. This interpretation is strengthened by the recent findings of Tomizawa & Anraku (1964), who have extracted from T4-infected cells structures containing DNA fragments from different parental chromosomes. In recombinants produced by bacterial conjugation, Siddiqi (1963) found an association between a paternal genetic character and radioisotope which had been used to label the female cells before mating. In bacterial transformation, structures which are either true recombinants or zygotes may be recovered very soon after the uptake of transforming DNA and under conditions which considerably inhibit DNA synthesis (Fox 1960; Voll & Goodgal 1961). It appears that these structures are inactivated at the same initial rate as in donor DNA if the latter is extensively labeled with ^{32}P (Fox 1962). These various observations have been interpreted as evidence that double-stranded donor DNA is integrated directly into the recipient chromosome. This would be in accord with the finding of recombination by breakage and joining between unreplicated double-stranded λ DNA molecules. More recently, however, Lacks (1962), Bodmer & Ganesan (personal communication) and Fox (personal communication) have been led by studies of the fate of donor DNA after uptake by recipient cells to favor the possibility that only one strand of donor DNA is integrated during transformation. These apparently conflicting views might be reconciled with each other, and with our knowledge of the mechanism of recombination in λ, if joining were to take place through the agency of relatively extensive hybrid regions as described in Figure 12.

Fig. 12. Possible structure of an unreplicated recombinant DNA molecule. At least in mature phages, the possibility that the region of joining contains more than two DNA strands is made unlikely by the absence of heavy shoulders on all bands of recombinant phages and by the observation that λ heterozygotes are not more dense than other λ particles (Kellenberger, Zichichi & Epstein 1962; Meselson, unpublished experiments).
Light solid and dashed lines represent polynucleotide strands contributed by two different parent molecules. Heavy lines represent either regions from which DNA has been removed and resynthesized along the opposing strand, or else regions in which missing DNA is not yet replaced. In the latter case, replacement is thought of as occurring during replication. The removal and repair of DNA may take place on only one strand rather than on both, as depicted in the figure.

There is some suggestion in the present experiments that joining is accompanied by a small amount of DNA synthesis. A light shoulder appears with the band of fully-labeled recombinants; the band of three-quarter-labeled recombinants exhibits a slight and possibly significant light shoulder; whereas the one-half-labeled and unlabeled recombinants form bands as sharp in appearance as those of nonrecombinant phages. These various features may be explained by assuming that up to five or ten percent of the DNA of the λ chromosome is removed and resynthesized in the course of genetic recombination. The replacement of labeled DNA with unlabeled DNA would explain why a light shoulder appears with bands of fully-labeled recombinants, but not with unlabeled recombinants. The bands of three-quarter- and one-half-labeled recombinants would possess light shoulders with intermediate displacements, which in the case of the latter may be so slight as to have escaped detection. An alternative explanation of the light shoulders is that they contain phages from which up to one or two percent of the DNA has been removed without replacement. If most of the one-half-labeled and unlabeled recombinants replicate at least once before maturation, the missing DNA might usually be restored, or less likely, the deficient strands might be lost, explaining the absence of substantial light shoulders in these cases. Although neither explanation of the light shoulders can be substantiated without additional information, both entail the removal and probably the resynthesis of a small amount of DNA in recombinant chromosomes.

If we consider (1) the present demonstration that recombination occurs by breakage and joining of double-stranded DNA molecules, (2) the indication that some DNA is removed and resynthesized in the course of recombination, and (3) the extensive evidence that bacteriophage recombinants arise through the formation of partially heterozygous structures (see Luria 1962), we are led to imagine that the unreplicated recombinant chromosome resembles the structure depicted in Fig. 12. The possibility that such structures are responsible for recombination in higher organisms as well as in bacteria and viruses has been discussed by Whitehouse (1963) and by the author (Meselson 1963).

I am grateful to Mrs. Miriam Wright for superb technical assistance in performing these experiments.

This work was supported by grants from the U.S. National Science Foundation.

REFERENCES

Appleyard, R. K. (1954). *Genetics,* 39:440.

Arbor, W. & Dussoix, D (1962). *J. Mol. Biol.* 5:18.

Bertani, G. & Weigle, J. (1953). *J. Bact.* 65:113.

Delbrück, M. & Bailey, W. T. (1946). *Cold Spr. Harb. Symp. Quant. Biol.* 11:33.

Delbrück, M. & Stent, G. S. (1957). In *The Chemical Basis of Heredity,* ed. by W. D. McElroy & B. Glass, p. 699. Baltimore: Johns Hopkins Press.

Fox, M. S. (1960). *Nature,* 187:1004.

Fox, M. S. (1962). *Proc. Nat. Acad. Sci., Wash.* 48: 1043.

Hershey, A. D. & Burgi, E. (1956). *Cold Spr. Harb. Symp. Quant. Biol.* 21:91.

Hershey, A. D. & Rotman, R. (1948). *Proc. Nat. Acad. Sci., Wash.* 34:89.

Ihler, G. & Meselson, M. (1963). *Virology,* 21:7.

Kahn, P. L. (1964). *J. Mol. Biol.* 8:392.

Kaiser, A. D. (1957). *Virology,* 3:42.

Kaiser, A. D. (1962). *J. Mol. Biol.* 4:275.

Kellenberger, G., Zichichi, M. L. & Epstein, H. T. (1962). *Virology,* 17:44.

Kellenberger, G., Zichichi, M. L. & Weigle, J. J. (1961). *Proc. Nat. Acad. Sci., Wash.* 47:869.

Kosinski, A. W. (1961). *Virology,* 13:124.

Lacks, S. (1962). *J. Mol. Biol.* 5:119.

Leventhal, C. & Thomas, C. A. Jr. (1957). *Biochem. biophys. Acta,* 23:453.

Luria, S. E. (1962). *Ann. Rev. Microbiol.* 16:205.

Meselson, M. (1962). *Pontificiae Acad. Sci. Scripta Varia,* 22:173.

Meselson, M. (1963). *Symposium of 16th International Congress of Zoology.* New York: Doubleday & Co.

Meselson, M. & Weigle, J. J. (1961). *Proc. Nat. Acad. Sci., Wash.* 47:857.

Roller, A. (1961). Ph.D. Thesis, California Institute of Technology.

Siddiqi, O. H. (1963). *Proc. Nat. Acad. Sci., Wash.* 49:589.

Tomizawa, J. & Anraku, N. (1964). *J. Mol. Biol.* 8:516.

Voll, M. J. & Goodgal, S. H. (1961). *Proc. Nat. Acad. Sci., Wash.* 47:505.

Whitehouse, H. L. K. (1963). *Nature,* 199:1034.

Conjugation and genetic recombination in *Escherichia coli* K-12

E. L. WOLLMAN, F. JACOB and W. HAYES*

Reprinted by authors' and publisher's permission from Cold Spring Harbor Symposia on Quantitative Biology, *vol. 21, 1956, pp. 141–162.*

In 1946, Lederberg and Tatum demonstrated genetic recombination by conjugation in *E. coli*. Unfortunately the mechanism was not well understood; so, in 1947, Lederberg did further analysis. The theory of one-way transfer of genetic information from donor to recipient, with the enhancement of donor fertility by ultraviolet light under conditions similar to prophage induction, led to the conclusion that the donor state might be a function of a viruslike segment of DNA in donor cells. Recipient cells would lack this extra DNA.

In 1953, Hayes hypothesized that the donor state was caused by an extrachromosomal factor called F (for "fertility"): cells having F were called F$^+$ and those lacking it were F$^-$. Hayes, Cavalli, and the Lederbergs, after extensive analysis of F, suggested that the F factor was freely transmissible to F$^-$ cells during the process of conjugation.

A debate ensued between Cavalli and the Lederbergs on the one hand, and Hayes on the other. Cavalli and the Lederbergs felt that F determined mating compatability with the F factor somehow directing chromosome transfer. Hayes believed that F was directly responsible for mediating chromosome transfer from donor to recipient by effectively associating with the chromosome. During the debate, an Hfr strain was discovered and problems began to clear up: unidirectional transfer, partial transfer, F$^+$, and Hfr began to fall into a pattern. All of the work during the early 1950s was a necessary prelude to the innovative and vital research carried out by Wollman, Jacob, and Hayes in 1956, wherein conjugation was investigated definitively for the first time. This paper describes in detail the process by which genetic material is passed from one cell to a recipient cell via conjugation.

"... it is also a good rule not to put overmuch confidence in the observational results that are put forward *until they have been confirmed by theory*." — Sir Arthur Eddington, "New Pathways in Science."

It is just ten years since the first report of genetic recombination in *Escherichia coli* K-12 was made to this Symposium by Lederberg and Tatum. Since then a great deal of work has been devoted to study of the genetics of recombination but, in general, such studies have tended to stress the complexity of the genetic process by bringing to light the anomalies inherent in it rather than to elucidate its mechanism. We do not intend to summarise all this work here since it has already been covered by recent reviews (Lederberg, Lederberg, Zinder and Lively 1951; Hayes 1953b; Lederberg 1955; Cavalli and Jinks 1956) but only to recall such facts as are strictly relevant to our present purpose. The object of this paper is to define the different steps involved in the processes of conjugation and recombination in *E. coli* K-12, to offer a systematic analysis of what we know about these steps, based primarily on work done in the authors' laboratories in Paris and London, and from this analysis to construct a simple model of the mating process. While we are very conscious that many gaps remain in our knowledge

---

*Institut Pasteur, Paris, France and The Postgraduate Medical School of London, London, England.

of this subject and that alternative hypotheses may be advanced to explain some of the experimental results, we propose to adopt a rather didactic approach in order to clarify the presentation of this work.

I. THE NATURE OF THE SYSTEM

Since Lederberg and Tatum's discovery, most of the work on genetic recombination in *E. coli* has involved crosses in which the observed frequency of appearance of recombinants was only 10^{-5} to 10^{-6} that of the parental population. This necessitated the use of the original experimental design of Lederberg and Tatum in which a mixture of two doubly auxotrophic mutants of the K-12 strain of *E. coli* was plated on a minimal medium on which only prototrophs, that is, bacteria in which the nutritional deficiencies of both parents had been eliminated by recombination, could grow. Under such conditions it is impossible directly to investigate events occurring at the cellular level, while interpretation of genetic results is difficult since selection permits the recovery of only one class of the progeny of numerous mating events.

When the auxotrophic parental strains were further marked by a series of differential characters which were not selected by the minimal medium used for prototroph detection, these unselected markers were found to have undergone reassortment among the prototrophic progeny of the cross. Analysis of such reassortments revealed linkage of many of the unselected markers to one another as well as to the nutritional selective markers, and four groups of clearly linked genes were revealed. However, although all the characters studied appeared to show some degree of interdependence in their inheritance, they could not be mapped on a single chromosome. Difficulties were even encountered in the analysis of recombination data between clearly linked genes, while the results of backcrosses between recombinants and parental strains frequently diverged grossly from those anticipated by genetical theory. It seemed likely that pure genetic analysis would lead only to increasing complication, rather than to clarification, of the picture and that further progress awaited the discovery of some fundamental aspect of the mating process which had not hitherto been taken into account.

A. One-way Genetic Transfer and F Polarity

It was first demonstrated that the two parental strains of a fertile cross played different roles in mating. Although both strains were equally sensitive to streptomycin, treatment of one of them with this drug invariably resulted in sterility of the cross, whereas treatment of the other did not have this effect although the fertility of the cross was often greatly reduced. It therefore appeared that while the role of one parent was transient and independent of the capacity for further multiplication, the continued viability of the other parent was vital. The function of the first parent or gene donor was simply to fertilise the other parent or gene recipient, which thus became the zygote cell (Hayes 1952a). In the light of subsequent work it is now believed that this differential effect was revealed because the bactericidal action of streptomycin was sufficiently slow to allow the donor population, or a moiety of it, to initiate and complete its fertilising function although further cell division was suppressed.

Another difference between the two strains is that irradiation of donor strains with small doses of ultraviolet light markedly enhances the fertility of crosses while similar irradiation of recipient strains does not (Hayes 1952b).

This functional distinction between the two strains was independently confirmed, at the same time, by the observation that crosses between recipient strains were always sterile. Donor strains were called F^+ (for fertility) and recipient strains F^- (Lederberg, Cavalli and Lederberg 1952). It was further shown that F^- cells could be converted to F^+ by contact with F^+ cells, with an efficiency about 10^5 times higher than that of recombination. The F^+ character acquired in this way by an F^- cell is inheritable and very stable in cultures, but a chromosomal locus for it could not be determined (Lederberg *et al.* 1952; Cavalli, Lederberg and Lederberg 1953; Hayes 1953a, b).

B. Partial Transfer of Genetic Material

Comparison of the results of crosses of reciprocal F polarity between the same parental strains showed that the majority of the unselected markers present among recombinants were those of the F^- (recipient) bacteria, irrespective of the cross or of the selection employed. The hypothesis was therefore made

(Hayes 1953a) that only a part of the donor chromosome was transferred to the recipient cell to form an incomplete zygote. An alternative hypothesis, based on results obtained with diploid heterozygotes (Lederberg 1949), was that a complete zygote was formed but that part of the genetic material, usually contributed by the donor cell, was subsequently eliminated after crossing over (see Lederberg 1955). Evidence strongly favouring the hypothesis of partial transfer will be presented later.

C. Hfr Systems

The usual frequency of recombinants arising from $F^+ \times F^-$ crosses is about 10^{-5} to 10^{-6} of the parental population. Two mutant strains of the same F^+ strain were independently isolated which showed a very much higher frequency of recombination in crosses with F^- strains. Such strains were called Hfr (high frequency of recombination). One of these strains (HfrC) was isolated after treatment of the parent F^+ culture with nitrogen mustard (Cavalli 1950); the other (HfrH) arose spontaneously (Hayes 1953b). It is this latter strain, HfrH, of Hfr derivatives from it, which has predominantly been employed in the work to be described.

Hfr strains behave as typical donor strains but differ strikingly from F^+ strains in several important respects:

1. When crossed with F^- strains under the conditions generally used throughout this work, the frequency of recombinants is about 2×10^4 times higher than in the equivalent $F^+ \times F^-$ cross (Hayes 1957).

2. Hfr behaviour is only manifest when selection is made for inheritance of an Hfr marker or markers (T^+L^+ or Lac^+) situated on a particular linkage group. Selection for markers on other linkage groups (M, B_1 or S, Mal) yields only a low frequency of recombinants (Lfr) comparable to that found in $F^+ \times F^-$ crosses (Hayes 1953b). Since only a single group of linked characters is therefore transmitted at high frequency to the recombinants, this system offers more dramatic evidence for the hypothesis, suggested by the results of $F^+ \times F^-$ crosses, that only this group of characters is usually transferred

from the Hfr to the F^- cell to form a partial zygote. The Hfr chromosome can thus provisionally be visualised (Fig. 1) as having a preferential region of rupture, such that only characters located on region A will be transferred at high frequency. It is evident that the terms "high" or "low" frequency of recombination have meaning only when the selective conditions of the cross are defined. As will be seen later, there are other conditions which can profoundly influence the productivity of a cross.

3. Hfr strains, unlike F^+, do not convert F^- strains to either F^+ or Hfr at high frequency, nor is the F^+ character inherited by recombinants from Hfr $\times$ F^- crosses. Recombinants from such crosses are F^- when markers on region A (Fig. 1) are selected, but when markers on region B are selected they are either F^- or Hfr depending on the particular selection employed. Nevertheless Hfr strains can revert at a low rate to an F^+ state indistinguishable from that of the parent F^+ strain.

4. Irradiation of Hfr strains, unlike that of F^+ strains, does not increase the frequency of recombinants when markers on region A are selected, but markedly enhances Hfr fertility *at low frequency* when selection is made for markers on region B.

The use of Hfr strains has provided proof of one-way transfer in this system. For example when exconjugants from visually observed pairs of Hfr and F^- cells are isolated with a micromanipulator and cultured, only the F^- partners yield recombinants (Lederberg 1955). Furthermore, the discovery of zygotic induction, now to be recounted, offers not only independent proof of one-way transfer but also strong evidence in support of partial transfer as well.

Fig. 1. The *E. coli* K–12 linkage group (after Cavalli and Jinks 1956). Symbols refer to threonine (*T*), leucine (*L*), methionine (*M*) and thiamine (*B₁*) synthesis; resistance to sodium azide (*Az*), bacteriophage T1, bacteriophage T6, streptomycin (*S*); fermentation of lactose (*Lac*), galactose (*Gal*), maltose (*Mal*), xylose (*Xyl*) and mannitol (*Mtl*).

D. Zygotic Induction

The wild-type strain of *E. coli* K-12 and most of its derivatives are lysogenic for the temperate phage λ, but nonlysogenic strains have been isolated from it (E. Lederberg 1951) and can be obtained at will. In crosses between lysogenic and nonlysogenic strains, lysogeny segregates and is linked to certain galactose markers (Lederberg and Lederberg 1953; Wollman 1953). Reciprocal crosses between lysogenic (ly^+) and nonlysogenic (ly^-) bacteria do not give symmetrical results, however, so far as inheritance of lysogeny is concerned (Wollman 1953; Appleyard 1954). In a cross between nonlysogenic HfrH and a lysogenic F⁻ strain, lysogeny segregates among the recombinants and can thus be shown to occupy a genetic locus on the chromosome near to the *Gal* locus. Similarly when HfrH and F⁻ parental strains, each of which is lysogenic for a different mutant of λ phage, are crossed, the two prophages segregate among recombinants in precisely the same way as did lysogeny and nonlysogeny in the previous cross, thus demonstrating that it is the prophage which is the genetic determinant of lysogeny (Wollman and Jacob 1954, 1957).

On the other hand when the Hfr parent is characterised by lysogeny and the F⁻ parent by nonlysogeny, the outcome of the cross is quite different, for lysogeny is no longer found to be inherited by any recombinant. Instead, the transfer of λ prophage is invariably followed by the *immediate* induction of its development within the fertilised F⁻ cells with consequent lysis of the zygotes and liberation of λ phage. When the F⁻ strain used is one which imposes a phenotypic modification on λ phage, this modification is found to characterise the liberated phage (Jacob and Wollman 1954, 1956b). This is a clear example of the one-way transfer of genetic characters in Hfr X F⁻ crosses.

Lambda prophage is therefore a genetic character whose *transfer to the F⁻ cell is immediately expressed* by zygotic induction, before genetic recombination has taken place. The fact that some recombinants are formed in HfrH(ly)⁺ X F⁻(ly)⁻ crosses but that these recombinants do not contain λ prophage, indicates that in these cases λ prophage has not been transferred to the F⁻ recipient cells. This demonstrates that the genetic locus, λ prophage, at least, is not transferred to many zygotes which yield recombinants, that is that these zygotes are not complete. This method of approach to the problem of whether the partial inheritance of Hfr characters among recombinants is entirely due to partial transfer from donor to recipient cells will be amplified later in this paper.

E. The Presumptive States of Mating

Despite the restriction that only a limited group of Hfr genetic characters are inherited, the proportion of recombinants issuing from Hfr X F⁻ crosses is sufficiently high to allow a quantitative study of the mating process. When Hfr and F⁻ cells are mixed one can, *a priori*, distinguish the following presumptive stages of mating:

1. Conjugation   This stage is synonymous with that of zygote formation and includes the entire sequence of events from the first encounter between donor and recipient cells to the completion of genetic transfer. It can be divided into the following substages:

  (a) *Collision*. This is obviously the essential first step and is a process determined by chance.

  (b) *Effective contact*. This step follows collision and involves a specific attachment between a donor and a recipient cell which probably depends on the surface properties of the two cells.

  (c) *Genetic transfer*. The genome, or part of the genome, of the donor cell is transferred to the recipient cell which thus becomes the zygote. When transfer is accomplished the donor cell has discharged its function and is no longer necessary.

2. Formation of Recombinants   This process comprises two important steps:

  (a) *Integration*. The process at the genetic level whereby the chromosome of the recombinant cell is evolved in the zygote from the chromosomal contributions of the two parents.

  (b) *Expression of the recombinants*. This stage comprises *segregation* of the haploid recombinant cell from the zygote cell and the *phenotypic expression* of the new recombinant chromosome in the cell which inherits it.

Before presenting an experimental analysis of these stages by means of Hfr X F⁻ crosses, we must first examine the methodology of such crosses and the kind of information which can be derived from them.

F. Methods of Approach to Hfr X F⁻ Crosses

Until recently the occurrence of mating could be measured only by the formation of recombinants. Since recombinants are the end products of the reaction between mating cells, so that their number and constitution can potentially be modified at any stage of the process, the analytical information which they can yield is limited. What is needed in analysis is to measure the different available expressions of recombination so that information concerning different stages of the process is obtained which can then be evaluated and equated. The phenomenon of zygotic induction answers this need by giving a direct indication of the extent of genetic transfer from the Hfr cells to the zygotes, since ultraviolet-inducible prophages, located on the chromosome segment transferred at high frequency by HfrH, are expressed immediately upon transfer to the recipient cell, *before* the formation of recombinants, by the liberation of phage. Thus comparison of the results of zygotic induction and of recombination from the same cross has yielded valuable information about the nature of both transfer and integration (Jacob and Wollman 1955a; Wollman and Jacob 1957).

Conditions of mating: For comparative quantitative studies, crosses are made under the following basic conditions. Exponentially growing cultures of the Hfr and F⁻ strains are mixed in liquid medium and aerated, the population density of one of the strains, usually the F⁻ recipient, generally being in 20-fold excess in order to increase the efficiency of specific collisions. The frequency of the effect to be measured is then expressed as a percentage of the number of minority parental cells initially present. Samples of the mixture are removed at suitable times, diluted to prevent further contacts, and plated either on a selective minimal medium if recombination is being studied or, in the case of zygotic induction, with a phage-sensitive indicator strain on nutrient agar. It is clear that lysogenic Hfr parental cells, if present, will interfere with results by themselves producing infectious centres if they are permitted to grow on the latter medium. Moreover, in many recombination experiments a prototrophic Hfr strain, capable of growth on mini-

mal agar, was used. It is therefore necessary either to prevent further growth of the Hfr parent after plating, or to eliminate it from the mixture, once genetic transfer has been effected and it is no longer required. When the Hfr parent is streptomycin-sensitive and the F⁻ parent resistant, the simplest and most efficient way of inhibiting Hfr growth is to incorporate streptomycin in the agar on which the mixture is plated. As one might expect from the fact that streptomycin sensitivity from strain HfrH is never inherited among recombinants formed at high frequency, the use of streptomycin in this way has no effect on the outcome of the cross (Hayes 1953b).

Elimination of Hfr cells is achieved by treating the undiluted mixture with a high titre preparation of virulent phage T6 to which only the Hfr parent is susceptible. After subsequent dilution, the concentration of phage on the plate is not such as to affect the frequency of appearance of recombinants inheriting sensitivity to the phage from the Hfr parent: the use of antiphage serum is, therefore, unnecessary. In most of the recombination experiments in which the phage technique was used, the Hfr strain was auxotrophic and streptomycin-resistant so that streptomycin was not added to the minimal agar (Hayes 1957).

Expressions of Mating 1. Formation of Recombinants Selection of recombinants from an Hfr X F⁻ cross means that selection is made for inheritance by the F⁻ cells of characters derived from the Hfr donor cells (Hayes 1953b). Accordingly, Hfr and F⁻ strains are chosen which differ in as many as possible of those markers which are inherited at high frequency by the F⁻ cells (Fig. 1). The F⁻ recipient strain is marked by the minus alleles of those characters which can be used for selection, for example threonine⁻, leucine⁻, lactose⁻, galactose⁻ ($T^-L^-Lac^-Gal^-$), and is resistant to streptomycin (S^r). The Hfr strain has the wild alleles of these characters and is sensitive to streptomycin ($T^+L^+Lac^+Gal^+S^s$). In addition, the two strains can be made to differ in markers such as resistance or sensitivity to sodium azide (Az^r, Az^s) or to virulent phages ($T1^r$, $T1^s$) which are not usually used for selection. Thus by the use of suitable selective media one can select for different recombinant classes issuing from the same cross. For instance in a cross between Hfr·$T^+L^+Az^sT1^sLac_1{}^+Gal_b{}^+S^s$ and F⁻·$T^-L^-Az^rT1^rLac_1{}^-Gal_b{}^-S^r$, $T^+L^+S^r$, $T^+L^+S^r$

or Gal^+S^r recombinants can be specifically selected and the inheritance of the other, unselected, markers among them studied. Two kinds of information can be obtained from such crosses:

(a) *The frequency of recombination:* the number of recombinants of any type (e.g., $T^+L^+S^r$ or Gal^+S^r) per 100 initial Hfr Cells.

(b) *The genetic constitution* of these recombinants, that is, the distribution among them of unselected Hfr markers.

2. Formation of Infectious Centres   What is looked for here is the proportion of lysogenic Hfr cells which, on mating with nonlysogenic F⁻ cells, transfer their prophage to the recipient F⁻ cells which thereby liberate phage as a result of zygotic induction. The cross is therefore made between $Hfr(ly)^+$ S^s and $F^-(ly)^-S^r$ parental strains and the mixture plated for infectious centres with a streptomycin-resistant, phage-sensitive indicator strain, on streptomycin-agar. Results are expressed as the *frequency of induction* which is the number of infectious centres (plaques) per 100 initial Hfr cells.

In the following two sections the various stages of mating in Hfr × F⁻ crosses will be discussed in detail against the background of our present knowledge. Our immediate aim is not to offer a definitive answer to the many questions which can be posed about the problem but rather to report the facts as we know them in the light of our own experience. These facts relate mainly to the experimental definition of each stage and substage, and to the analysis of some of them in kinetic terms by means of the methods we have already outlined.

II. CONJUGATION IN Hfr × F⁻ CROSSES

Conjugation connotes the sequence of events from the initial collision to the completion of fertilisation of the F⁻ cell, that is up to the time at which further participation of the Hfr donor cell is no longer necessary.

A. The Kinetics of Conjugation

1. Effective Contacts   A standard cross is made between an Hfr·S^s and an F⁻·S^r strain, both strains being either lysogenic or nonlysogenic. At intervals

after mixing in broth, samples are removed, diluted and plated for recombinants. Selection is made for inheritance of either T^+L^+ or of Gal^+ from the Hfr parent, these being the most widely separated of the known selective markers on that part of the Hfr chromosome which is inherited at high frequency (Wollman and Jacob 1955, 1957; Jacob and Wollman 1954, 1956b). The results of such an experiment are shown in Figure 2A.

Figure 2B shows the results obtained under the same conditions when the Hfr strain is lysogenic for λ prophage and the F⁻ strain is nonlysogenic, so that zygotic induction occurs, the samples being plated for infectious centres as well as for $T^+L^+S^r$ and Gal^+S^r recombinants.

Inspection of Figures 2A and 2B shows: 1. In all cases the curves start from the origin (i.e., from the

Fig. 2. The kinetics of conjugation. Exponential broth cultures of HfrH S^s and P678 F⁻S^r are mixed at time 0 (1 to 2.10^7 Hfr/ml and 2 to 5.10^8 F⁻/ml) and aerated in broth. The differential characters of the two strains are: HfrH $S^s T^+ L^+$ $Az^s T1^s$ $Lac_1^+ Gal_b^+$ (for symbols see Fig. 1) and: P678F⁻$S^r T^- L^- Az^r T1^r Lac_1^- Gal_b^-$. At different intervals of time samples are immediately plated, either for recombinants on glucose minimal medium + streptomycin ($T^+L^+S^r$ recombinants, curves 1) and on galactose minimal medium + threonine (*T*), leucine (*L*), and streptomycin (Gal^+S^r recombinants, curves 2), or for infectious centers on nutrient agar + streptomycin resistant indicator bacteria for phage λ (curve 3). The numbers of recombinants or of infectious centers obtained are expressed as a percentage of the number of initial Hfr cells. 2A, cross between HfrH S^s (λ)⁺ and P678 F⁻S^r(λ)⁺, both lysogenic (λ)⁺ for phage λ. 2B, cross between HfrH S^s (λ)⁺ and P678 F⁻S^r(λ)⁻. *(From Wollman and Jacob 1957.)*

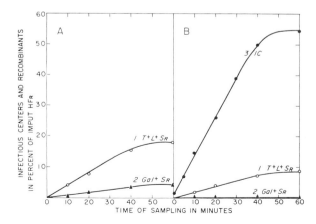

time of mixing) and then continue to rise until a plateau is reached, irrespective of whether the curves express the formation of recombinants or of infectious centres. 2. All the curves reach their plateaux at about the same time although the maximum levels represented by the plateaux vary from 58 percent for frequency of induction (Fig. 2B) to 5 percent for frequency of Gal^+S^r recombination (Fig. 2A). Since the slope of each curve, from its inception to its maximum, represents the same rate of increase (about 2.5% per minute), all must be an expression of the same phenomenon, the conjugation of Hfr and F⁻ cells. The fact that all the plateaux are attained at about the same time indicates that at this time the maximum frequency of conjugation has been achieved.

2. Zygote Formation Crosses are made in the same manner as before but the samples, instead of being plated directly, are treated with a high titre preparation of phage T6, to which only the Hfr cells are susceptible, before dilution and plating (Hayes 1957). The results of such an experiment are shown in Figure 6A (first curve).

The striking difference between curve 1 of Figure 2A and the first curve of Figure 6A is that whereas, in the first case, the number of T^+L^+ recombinants begins to rise from the time of mixing, when the samples are treated with phage there is a lag of eight minutes before recombinants begin to appear. This difference is explained in the following way. The curves constructed from untreated samples represent the kinetics of those contacts between Hfr and F⁻ cells which survive dilution and plating and can therefore continue through the succeeding steps leading to formation of recombinants or infectious centres on the plate; they thus express the *kinetics of the formation of effective contacts*. On the other hand, the curves constructed from samples treated with phage indicate the rate of development of mating pairs in which the markers T^+L^+ have already been transferred from donor to recipient cell at the time of treatment, so that the Hfr cell is no longer needed to form T^+L^+ recombinants: that is, the *kinetics of zygote formation*. From this two inferences can be drawn:

1. Effective contacts are formed very quickly, since the kinetic curve arises from the zero point of the time axis.

2. Transfer of the markers T^+L^+ from the Hfr to the F⁻ parent occupies about eight minutes from the time of contact.

B. Genetic Transfer

1. The Kinetics of Transfer In an attempt to distinguish actual genetic transfer from conjugation as a whole, samples removed at intervals after mixing the Hfr and F⁻ cells were either plated directly as before, or subjected to violent agitation in a Waring blender before dilution and plating in order to separate bacteria in the act of conjugation (Wollman and Jacob 1955, 1957). Treatment in a blender affects neither the viability of bacterial cells (Anderson 1949) or recombinants, nor the ability of phage-infected cells to form plaques (Hershey and Chase 1952). The results of two experiments of this kind are given in Figure 3. Curves 1a and 2a represent the kinetics of the formation of T^+L^+ and Gal^+ recombinants respectively in an Hfr × F⁻ cross in which both parents are lysogenic for λ phage, samples being diluted and plated without treatment. Curves 1b and 2b are derived from equivalent samples after treatment in the blender. Similarly, curves 3a and 3b express the formation of infectious centres, without treatment and after treatment respectively, in an Hfr × F⁻ cross in which the Hfr parent is lysogenic for λ phage and the F⁻ parent nonlysogenic.

It will be seen that while the curves obtained from the untreated samples all arise from the origin and reach a plateau at about 50 minutes, as in Figure 2A and 2B, the corresponding curves obtained after treatment do *not* pass through the origin but nevertheless reach the same plateaux at about the same time. Moreover the times at which the various genetic markers appear to enter the zygote vary widely. For instance, T^+L^+ recombinants begin to appear at about nine minutes, as after phage treatment, while the appearance of Gal^+ recombinants and λ infectious centres is delayed until 25 minutes. Since the time of appearance of markers widely separated on the chromosome, such as T^+L^+ on the one hand and Gal_b^+ on the other, is very different whereas closely linked markers like Gal_b and λ prophage appear at the same time despite the great disparity in their mode of expression, the conclusion is inescapable that *there exists a definite relationship between the*

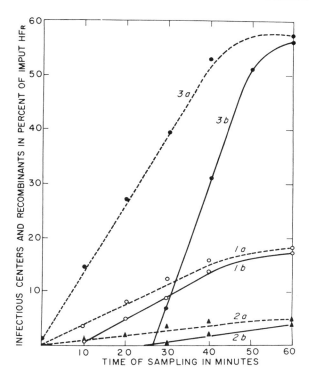

Fig. 3. The kinetics of transfer. The experiments are the same as illustrated on Figure 2. At different time intervals samples of the mating mixtures are either plated directly on indicator media (dotted lines) or treated for 2 minutes in a Waring blender and then immediately plated (solid lines). Curves 1 and 2 refer to the kinetics of formation of $T^+L^+S^r$ recombinants (curves 1a and 1b) and of Gal^+S^r recombinants (curves 2a and 2b) in a cross between Hfr$(\lambda)^+$ X F$^-(\lambda)^+$. Curves 3a and 3b represent the kinetics of formation of infectious centers in a cross between Hfr$(\lambda)^+$ X F$^-(\lambda)^-$. *(From Wollman and Jacob 1957.)*

time at which a given marker is transferred from the Hfr to the F$^-$ cell and the location of that marker on the Hfr chromosome.

2. Transfer as an Oriented Process If this point of view is correct, then the frequency distribution among T^+L^+ recombinants of markers situated between TL and Gal_b on the Hfr chromosome should vary widely according to the time after mixing at which samples are treated in the blender. T^+L^+ recombinants from untreated samples (Figure 3, curve 1a) and from treated samples (Figure 3, curve

1b) were therefore analysed for inheritance of unselected markers as a function of the time of sampling. The genetic constitution of T^+L^+ recombinants derived from untreated samples is constant irrespective of the time of sampling and, therefore, of the proportion of the population that have formed effective contacts. As can be seen from the table and Figure 4, the percentage of various Hfr markers found among $T^+L^+S^r$ recombinants is approximately: Az^s 90 percent, $T1^s$ 75 percent, Lac^+ 45 percent and Gal_b^+ 28 percent. On the contrary, the frequency of inheritance of these same markers among $T^+L^+S^r$ recombinants stemming from the *treated* samples varies in a striking way according to the time of sampling, as the curves in Figure 4 show. The time at which any given character begins to appear among T^+L^+ recombinants is directly related to its distance from TL on the chromosome map (Fig. 5), the further the distance the greater being the delay in its appearance. Once a character begins to be found among T^+L^+ recombinants, the frequency of its inheritance increases rapidly as a function of time until its

Fig. 4. Time sequence of transfer of different characters. $T^+L^+S^r$ recombinants formed at different times on curves 1a and 1b, Figure 3, have been analysed as to their genetic constitution as a function of time. The distribution among $T^+L^+S^r$ recombinants of characters originating from Hfr donor is represented as a percentage of the $T^+L^+S^r$ recombinants arising from the same sample. In dotted lines: from untreated samples. In solid lines: from treated samples. *(From Wollman and Jacob, 1954 and 1957.)*

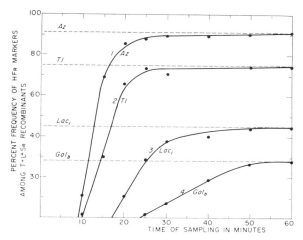

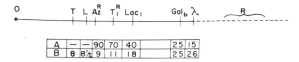

Fig. 5. Genetic map of O–R the chromosomal segment of Hfr H. The location of the different characters as measured in HfrH X P678 F⁻ crosses by A: the percentage of $T^+L^+S^r$ recombinants which have inherited the different Hfr alleles; B: the time at which individual Hfr characters start penetrating into the F⁻ recipient cells in a Waring blender experiment. *(From Wollman and Jacob 1957.)*

normal level is attained. Thus *Az* first appears at about nine minutes, *T1* at ten minutes, *Lac* at 18 minutes and *Gal* at 25 minutes. At 50 minutes after mixing, all the markers transferred from the Hfr to the F⁻ cell have reached their normal frequency of inheritance. It is therefore clear that the Hfr chromosome penetrates the F⁻ cell in a specifically orientated way. The extremity, O (for "origin") enters first, to be followed by T^+L^+ eight to nine minutes later, and then by the other markers in the order of their arrangement on the chromosome and at intervals of time proportional to the distance between them until, at about 35 minutes after contact, the whole segment $O - R$ has been transferred (Fig. 5). The effect of treatment in the blender thus seems to cut this chromosome segment during transfer and thus to determine the length of segment which enters the zygote and the markers which will be available for subsequent incorporation into recombinants (Wollman and Jacob 1955, 1957).

Two kinds of evidence indicate that the agitation intervenes in *transfer* rather than in subsequent stages:

1. The very closely linked markers Gal_b and λ prophage appear at about the same time (25 minutes) after mixing despite the fact that transfer of prophage to the zygote is expressed immediately after transfer while the Gal_b marker is involved in the process of recombination.

2. Results similar to those obtained with the blender are given by treatment of samples from a mating mixture with virulent phage to which only the Hfr cells are susceptible (see Fig. 6). Destruction of the

Hfr parent in this way allows an analysis of events occurring solely within the F⁻ cells which are unaffected by the phage (Hayes 1957).

C. Physiology of Conjugation

1. Nutritional Requirements Zygote formation does not occur at a significant rate if parental cultures grown aerobically in broth are washed and mixed in buffer. This is especially the case if the parental suspensions are starved by aeration in buffer before mixing. Addition to the buffer of both glucose and sodium aspartate, but neither alone, allows zygote formation to proceed at a rate comparable to that found in broth. Krebs cycle dicarboxylic acids and glutamic acid, but no other dicarboxylic or amino acids, can replace aspartic acid in promoting the formation of zygotes while such metabolic inhibitors as fluoracetic acid, malonic acid, and cyanide, as well as 2:4 dinitrophenol

Fig. 6. The kinetics of zygote formation, using phage T6 to eliminate the Hfr parental cells. Cultures of HfrHM⁻T6ˢ and F⁻ TLB₁⁻T6ʳ were washed, resuspended in buffer + glucose + sodium aspartate at the appropriate temperature, mixed and aerated at the same temperature. At intervals samples were treated with phage T6, diluted and planted on: a. minimal agar + B₁ + glucose (selection for inheritance of T^+L^+ only from the Hfr parent); b. minimal agar + B₁ + lactose (selection for inheritance of T^+L^+ and Lac^+). A shows the effect of varying the temperature at which zygote formation occurs. B shows the effect of varying the concentration of the parental cells or the pH of the medium. *(From Hayes 1957, J. Gen. Microbiol., in press.)*

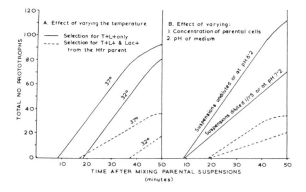

(DNP), inhibit the process. Zygote formation does not occur in buffer + glucose + sodium aspartate under anaerobic conditions. It thus appears that energy liberated by the Krebs cycle is required (Fisher 1957a). Moreover, differential starvation experiments show that only the Hfr parent requires energy, the F⁻ cells playing an entirely passive part until the completion of genetic transfer.

For what purposes do the Hfr cells require their energy? If a series of identical mixtures of Hfr and F⁻ cells, previously grown aerobically in broth, are made under anaerobic conditions and, at different intervals after mixing, oxygen is then admitted to each mixture in turn, it is found that, in each case, zygote formation (*i.e.*, transfer of T^+L^+ from the Hfr to the F⁻ cells) begins at eight minutes after the commencement of oxygenation and then progresses at precisely the same rate *irrespective of the duration of the preceding anaerobiosis*. This shows that, under anaerobic conditions, nothing happens which affects the efficiency of subsequent conjugation although the probability of chance contacts is the same as in the presence of oxygen. Thus the establishment of effective contacts between parental cells is an energy-requiring act performed by the donor cells (Fisher 1957b).

The kinetics of genetic transfer, as demonstrated by the use of the Waring blender, can be reproduced with great precision by arresting the transfer of the chromosome with DNP and then killing the Hfr cells with virulent phage. The use of DNP alone does not kill the parental cells so that transfer is resumed when the effective concentration of the drug is reduced by dilution. Similarly, if an Hfr X F⁻ mixture at 37°, in which chromosome transfer is proceeding, is diluted to prevent further contacts at about 20 minutes after mixing and then rapidly cooled to 4°, analysis of subsequent samples reveals no change in the inheritance of the Hfr marker Lac^+ among T^+L^+ recombinants, showing that chromosome transfer has been arrested. When, however, the diluted mixture is warmed again to 37°, the inheritance of Lac^+, which had only begun to enter at the time of cooling, immediately commences to rise again at its initial rate until its normal frequency is attained. Since the formation of new contacts was prevented by dilution, the secondary increase in Lac^+ inheritance among T^+L^+ recombinants must have been due to subsequent entry of

this locus into cells which had previously acquired T^+L^+. Such experiments not only confirm the continuity of chromosomal movement into the F⁻ cell but show that energy is required throughout the period of chromosome transfer. However, energy is not needed to hold mating cells together once effective contacts have been established (Fisher 1957b).

2. The Effect of Temperature As might be expected from the fact that the Hfr cells require energy for chromosomal transfer, the efficiency of zygote formation falls rapidly as the temperature diverges from the optimal at 37°. Thus at 44° the number of F⁻ cells which have received the T^+L^+ markers from the Hfr parent in a given time is only 25 percent, and at 25° only 6 percent that at 37° (Hayes 1956). Figure 6A shows what happens when the kinetics of chromosome transfer are studied in identical Hfr X F⁻ crosses in buffer + glucose + sodium aspartate, at 37° and at 32°. The reduction of temperature will lower the overall capacity of the Hfr cells to produce energy by carbohydrate oxidation. At 37° the T^+L^+ markers (continuous line) begin to enter the F⁻ cells at eight minutes and the marker Lac^+ (interrupted line) at 17 minutes after mixing, so that transfer from *TL* to *Lac* occupies nine minutes at this temperature. At 32°, on the other hand, T^+L^+ enters at 18 minutes and Lac^+ at 38 minutes so that transfer of the same piece of chromosome takes 20 minutes, that is about twice as long, at the lower temperature. It has been mentioned that since the curve expressing the kinetics of formation of effective contacts commences at zero time, effective contacts must be made very quickly so that it may be assumed that transfer of the O — *TL* segment occupies virtually the whole of the lag period preceding entry of T^+L^+ into the F⁻ cell. Thus at 37°, transfer of O — *TL* takes eight minutes and of O — *TL* — *Lac* 17 minutes, while at 32° transfer of these same segments takes 18 and 38 minutes respectively. The proportionality between these times strongly suggests that that part of the chromosome from O to *Lac*, at least, enters the F⁻ cell at a uniform rate which depends upon the available energy, so that measurement of the relative distances between chromosomal markers in terms of time is a valid procedure (Hayes 1957).

The fact that the curves shown in Figure 6A are

parallel, although displaced on the time axis, indicates that the only effect of this degree of limitation of energy, in so far as conjugation is concerned, is to slow the rate of chromosome transfer.

3. The Effect of Population Density   When the frequency with which chance contacts occur in Hfr × F⁻ mixtures is altered by mixing the same number of cells of each parent in a different volume of fluid, the relationship of the resulting curves (shown in Fig. 6B) is quite different from that characterising alterations in temperature. It will be seen that the time of entry of various markers into the F⁻ cells remains unchanged when the population density of the parental cells is reduced; the curves arise from the same point on the time axis but their slopes differ (Hayes 1957). This same effect of population density is observed whether the manifestation studied is formation of recombinants or infectious centres. It is thus evident that the *slope* of the curves expressing the rate of zygote formation is a simple function of the frequency of chance contacts between Hfr and F⁺ cells, within a certain population range.

4. The Effect of pH   The rate of zygote formation in buffer + glucose + sodium aspartate is doubled when the pH is reduced from 7.2 to 6.2. Since this effect is also apparent when the cross is made in unsupplemented buffer it must be unrelated to energy requirements, because, under these conditions, the Hfr cells are limited to their endogenous resources (Fisher 1957b). The effect is not found, or is not marked, when crosses are made in broth. When the kinetics of chromosomal transfer are compared for identical Hfr × F⁻ crosses in buffer + glucose + sodium aspartate at pH 7.2 and at pH 6.2, the same result is obtained as when the population density is varied, as Figure 6B demonstrates (Hayes 1957). The effect of lowering the pH must therefore be ascribed to physical action on the cell surfaces which facilitates the intimacy of chance contacts so that a higher proportion of such contacts can become effective in a given time. This is in keeping with the surface differences, probably attributable to differences of charge, between F⁺ and F⁻ cells which have previously been described (Maccacaro 1955).

A difference between the surface antigenic properties of donor and recipient strains is suggested by the results of preliminary experiments by L. and S. Le Minor at the Pasteur Institute.

D. The Frequency of Conjugation

When Hfr cells are mixed in broth at 37° with an excess of F⁻ cells, what proportion of the Hfr cells have conjugated at about 50 minutes after mixing, when the curves defining the kinetics of effective contact formation have reached their plateaux (Fig. 2)? If one were to examine only one expression of conjugation, such as the frequency of T^+L^+ recombinants, no absolute answer could be given to this question since the proportion of zygotes which do not yield recombinants of this class is not known. Since the frequency of T^+L^+ recombinants in Hfr $(\lambda)^+$ × F⁻$(\lambda)^+$ crosses is about 20 percent of the Hfr cells initially present, it could only be asserted that not less than 20 percent of the Hfr cells must have conjugated. By correlating the frequency of recombination with that of zygotic induction, however, it can be estimated that approximately 100 percent of the Hfr population conjugate (Jacob and Wollman 1955a; Wollman and Jacob 1957).

As has been said, in the cross Hfr $(\lambda)^+$ × F⁻$(\lambda)^+$, 20 percent of the Hfr cells form zygotes which yield T^+L^+ recombinants (Fig. 2A). When the same Hfr $(\lambda)^+$ strain is crossed under the same conditions with F⁻$(\lambda)^-$, about 50 percent of the Hfr cells transmit λ prophage to the F⁻ cells with the result that about 50 percent of the maximum number of possible zygotes is destroyed by zygotic induction. But, as reference to Figure 2B will show, this destruction of zygotes only reduces the frequency of T^+L^+ recombinants by a factor of 2, from 20 percent to 10 percent. If it is assumed that those zygotes which yield T^+L^+ recombinants in Hfr $(\lambda)^+$ × F⁻$(\lambda)^-$ crosses did not inherit λ prophage, then since only half of the potential T^+L^+ recombinants are removed when 50 percent of the total possible zygotes are destroyed, the total number of zygotes formed must account for 100 percent of the Hfr cells present.

When Hfr × F⁻ cells are mixed together and the mixture examined at intervals under a light microscope, an increasing number of pairs is observed which remain in contact for long periods. Such pairs are particularly noticeable when the two parental strains differ morphologically as in the case of HfrH and F⁻ *E. coli* strain C (Lieb, Weigle and Kellenberger 1955). Electron microphotographs obtained by Dr. T. F. Anderson (Anderson, Wollman and Jacob 1957) are presented in Figure 7. The top picture shows conjugation between HfrH and the F⁻ strain P678 which has been used in most of our studies. HfrH has long flagella whereas P678 is covered with bristles. As an additional distinguishing feature, HfrH has had ultraviolet-killed λc phage adsorbed to its surface: P678 is resistant to this phage. The middle picture demonstrates conjugation between HfrH and *E. coli* C. In this case the two strains are so different in their morphology that additional labeling is not required. In both crosses the conjugating bacteria appear to be united by a cellular bridge.

E. The Evidence for Partial Genetic Transfer

The experiments reported above demonstrate that, by use of the blender or the phage technique, it is possible to regulate at will the length of chromosome segment transferred from the Hfr to the F⁻ cell. The partial zygote thus formed is perfectly operative and recombinants will be formed in exactly the same fashion as when transfer is allowed to proceed normally. Interruption of the mating process at about 50 minutes after mixing when Gal_b, the last known biochemical marker on the $O — R$ segment, has been transferred (Fig. 5), has no effect either on the number or on the genetic constitution of the recombinants formed.

It can be shown that in a normal cross, where only those characters situated on the $O — R$ segment *appear* to be transferred from donor to recipient cells, the whole of this segment is not, in fact, transferred to every zygote (Jacob and Wollman 1955a; Wollman and Jacob 1957). As the leading locus, O, followed in order by the other loci, penetrates the F⁻ cell, spontaneous breakages may occur in certain cases. The occurrence of zygotic induction in certain crosses makes it possible to estimate some of these breakages. Penetration of λ prophage into a nonlysogenic recipient cell leads to destruction of the zygote. If λ prophage entered every zygote no recombinants would ever be formed: an Hfr $(λ)^+$ × F⁻ $(λ)^-$ cross would be sterile. This, however, is not the case since half the usual number of T^+L^+ recombinants (10% instead of 20%) is re-

covered while only about one-half the number of possible zygotes are scored as infectious centres (see *The Frequency of Conjugation*, above). These results are readily interpretable in the light of the mechanism of transfer which has been described. About half the zygotes inherit the Hfr segment $O — Gal — λ$ and are thus destroyed by zygotic induction. The other half receives a segment *shorter* than $O — Gal$ and will not be destroyed, so that about 20 percent of them will yield T^+L^+ recombinants. These T^+L^+ recombinants constitute a special group in so far as they are derived from zygotes which have been selected by acquiring a shorter normal piece of Hfr chromosome. Thus, if the theory is right, they should show a lower frequency of inheritance of those unselected markers least linked to TL than does the *whole* group of T^+L^+ recombinants from a cross in which zygotic induction does not occur (Wollman and Jacob 1954, 1957). That this is indeed so can be seen by comparing the results of crosses 3 and 4 in Table 1.

In the case of zygotic induction (cross 4) the frequency among T^+L^+ recombinants of those markers located close to λ prophage is greatly reduced, the closer the linkage to λ the greater being the reduction. In particular, the marker Gal_b, which is most closely linked to λ, is almost absent as can also be inferred from the $T^+L^+:Gal^+$ ratios in the table. On the other hand there is no significant difference in the inheritance of unselected markers between crosses 3 and 4 (table) when Gal^+ is selected instead of T^+L^+, showing that if Gal^+ enters the zygote and the zygote survives, the outcome of the cross is the same irrespective of the lysogenic relationships. As a matter of fact the orientated nature of chromosomal transfer from the Hfr to the F⁻ cell can be deduced directly from comparison of the data of Figures 2A and 2B and of the table. The blender experiment is only a direct demonstration of its reality.

It can also be shown by the study of a series of inducible prophages (Jacob and Wollman 1955a, 1956b) that the probability of chromosomal rupture increases rapidly as R is approached. All these prophages are located on the $O — R$ segment between Gal_b and R. The frequency of zygotic induction which they evoke is a measure of the frequency of their transfer from Hfr$(ly)^+$ to F⁻$(ly)^-$. For the seven different prophages studied, this ranges from 65 percent to 5 percent. When this information is

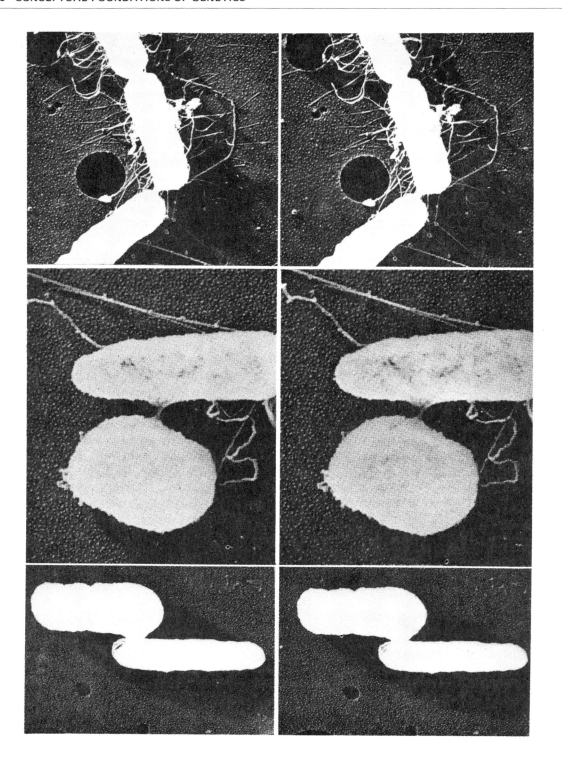

Table 1 Genetic analysis of $T^+L^+S^r$ and Gal^+S^r recombinants obtained in various crosses between HfrH–S^s and P678 F⁻–S^r

No.	Crosses	Ratio $T^+L^+S^r/Gal^+S^r$	Genetic constitution of recombinants $T^+L^+S^r$					Gal^+S^r				
			Az	T1	Lac	Gal_b	(λ)	TL	Az	T1	Lac	(λ)
1	Hfr(λ)⁻ X F⁻(λ)⁻	4.2	91	72	48	27	—	83	78	79	81	—
2	Hfr(λ)⁻ X F⁻(λ)⁺	3.7	92	73	49	31	15	75	75	74	74	84
3	Hfr(λ)⁺ X F⁻(λ)⁺	4.2	90	70	47	29	14	80	78	78	82	82
4	Hfr(λ)⁺ X F⁻(λ)⁻	54	86	60	21	2.5	<0.1	82	79	78	74	1
5	Hfr(λ)⁻ U.V. X F⁻(λ)⁻	4.5	50	32	9	1	—	17	24	29	41	—

The strains used are, as in Figure 2, HfrH · $S^sT^+L^+Az^sT_1^sLac_1^+Gal_b^+$ and P678 F⁻$S^rT^-L^-Az^rT1^rLac^-Gal_b^-$ (for symbols see Figure 1).

In column 1 are the different crosses reported (for details see text).

In column 2 the ratio of the numbers of $T^+L^+S^r$ to Gal^+S^r recombinants obtained.

In column 3 the distribution of the different alleles contributed by the Hfr donor to $T^+L^+S^r$ recombinants on one side, to Gal^+S^r recombinants on the other side, as a percentage of recombinants having inherited those characters. *(From Wollman and Jacob, 1957.)*

correlated with the order of arrangement of these prophages on the chromosome, established by genetic analysis of Hfr$(ly)^-$ X F⁻$(ly)^+$ crosses in which there is no zygotic induction, it is found that the further a prophage is located from Gal_b the lower is the frequency of prophage transfer. The relative positions of these prophages can also be assessed by the blender technique, their times of entry ranging from 25 to 33 minutes.

The point of rupture, R, beyond which no markers enter the F⁻ cell can formally be considered as genetically equivalent, although conceptually opposed, to the elimination locus, E, which Cavalli and Jinks (1956), working with F⁺ X F⁻ systems, have located at, or very close to, the λ locus.

III. THE FORMATION OF RECOMBINANTS

As a result of conjugation a zygote is formed. The experiments recounted in the previous sections indicate unambiguously that the mechanisms involved in bacterial conjugation are quite different from what is found in other organisms. The two main differences are the following:

1. Bacterial conjugation involves a one-way transfer of genetic material from a donor to a recipient strain.

2. This transfer is partial since only a piece of the Hfr chromosome of variable size is injected into the

Fig. 7. Stereoscopic electron micrographs showing connections between cells of opposite mating type. Top. Mixture of *E. coli* K-12 Hfr and K-12 F⁻/λ. Bacteriophage λc (which had previously been killed with ultraviolet light) was adsorbed on the Hfr cells to identify them. The F⁻/λ which failed to adsorb the phage could also be recognized by their numerous "bristles." In this view, a bridge 1500 Å wide can be seen connecting the end of a dividing F⁻ bacterium with the end of a Hfr bacterium to which bacteriophage particles are attached. EMG 5.IV.56 BI, 2. X 21,000. Middle. Mixture of *E. coli* K-12 Hfr and *E. coli* CF⁻. At the top is a typically long and narrow K–12 Hfr cell connected by a 500 Å bridge to a typically short and plump CF⁻ cell. EMG 20.IV.56 C 3, 4. X 40,000. Bottom. Mixture of *E. coli* K-12 F⁺ and *E. coli* CF⁻. Here the plump CF⁻ cell at the top is in intimate contact with the narrow K–12 F⁺ cell in the lower part of the picture. EMG 3.V.56 D, 4, 5. X 20,000. It seems probable that during conjugation the genetic material from Hfr or F⁺ donor cells passes to F⁻ recipient cells through connections like the ones shown here.

Twenty minutes after mixing, specimens were fixed in the vapor over 2% OsO₄ and then dried by the critical point method to conserve the three dimensional structure of the specimen. Specimens were lightly shadowed with a An-Pd mixture and coated with 10 Å of carbon to reduce charging effects in the electron microscope. *(From Anderson, Wollman and Jacob 1957).*

recipient cell. The resulting zygote is therefore a rather peculiar one for which the term *merozygote* has been proposed (Wollman and Jacob, 1957).

Once the merozygote is formed, a series of processes is initiated which ultimately result in the appearance of a recombinant bacterium. It is unlikely that these processes differ fundamentally from those which mediate the expression of other kinds of partial genetic transfer among bacteria, such as transformation and transduction. In all three cases a recombinant chromosome is evolved which is essentially that of the recipient parent save that a segment or segments of variable size have been replaced by homologous fragments of the donor genome. Once this integration has been achieved, the recombinant cell will emerge through segregation and phenotypic expression of the character or characters contributed by the donor parent.

We will now discuss in turn the available information relating to each of these steps.

A. Chromosomal Integration

There is at present little understanding the processes which lead to the formation of a recombinant chromosome from the two parental constituents of a merozygote. Since there is insufficient evidence on which to base a definite hypothesis we will simply examine briefly the facts that are known to us.

1. Genetic Patterns of Integration The segment of Hfr chromosome which enters the zygote may or may not be incorporated into a new recombinant chromosome. When, however, integration does occur, the asymmetry of the chromosomal contributions from the two parents may be expected to impose certain patterns of recombination.

(a) Efficiency of integration. It is possible to calculate the probability that a character derived from the donor parent will appear in recombinants, and this probability is found to be very constant for any given cross. It has already been shown, by comparison of the frequency of recombination with that of zygotic induction, that whereas the Hfr markers T^+L^+ are transferred to about 100 percent of zygotes, only some 20 percent of these zygotes yield T^+L^+ recombinants. Thus the efficiency of integration of these markers together is about 1/5.

Similarly we know that the Gal_b marker is transferred to about 60 percent of zygotes but is only found among recombinants issuing from five percent of the total zygotes, so that it is integrated with an efficiency of about 1/12. There is thus rather a small probability of integration for an Hfr gene which has entered the zygote, and this probability varies with the location of the gene on the chromosome (Jacob and Wollman 1955a; Wollman and Jacob 1957).

(b) Polarity of integration. From the example just given it would appear that the closer a gene is situated to O the greater is its chance of integration. The O — R segment therefore seems to be characterized by an integration gradient as well as by a gradient of transfer. These two gradients are exemplified by comparison of the genetic constitution of $T^+L^+S^r$ and Gal^+S^r recombinants from crosses 1, 2 and 3 in the table. While the frequency of inheritance of markers among T^+L^+ recombinants is strongly dependent on their distances from O, in the Gal^+ recombinants all the unselected markers on the O — R region appear with the same high frequency (Jacob and Wollman 1955a; Wollman and Jacob 1957). Figure 8 illustrates the types of recombinants observed in the Gal^+S^r class, which have issued from zygotes which certainly received the whole of the O — Gal_b segment from the Hfr parent. It will be seen that almost 70 percent have inherited the TL — Gal_b segment as a whole. If O

Fig. 8. Patterns of integration among Gal^+S^r recombinants in a cross between HfrH S^s and P678 F⁻S^r. Different classes observed, expressed as a percentage of total Gal^+S^r recombinants

Percent	Alleles from Hfr donor				Crossing over in region
68	TL	T1	Lac	Gal	1
5		T1	Lac	Gal	1 and 2
1.7			Lac	Gal	1 and 3
11.6				Gal	1 and 4
7.5	TL		Lac		1, 2 and 3
2.5	TL	T1		Gal	1, 3 and 4
1.7	TL			Gal	1, 2 and 4
2		T1		Gal	1, 2, 3 and 4

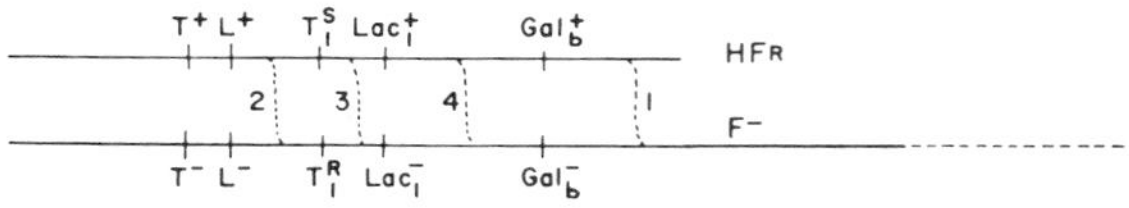

represents the extremity of the chromosome, as seems reasonable, these recombinants would result from a single crossover in position 1. The other recombinants would be the result of two or more crossings over. A certain amount of negative interference is therefore apparent.

The table clearly shows that the combined effect of a gradient of integration as well as of transfer is to facilitate mapping of the *order* of unselected markers when characters close to O (such as T^+L^+) are selected but to render this mapping almost impossible when selection is made for characters (such as Gal_b) situated close to R since, in this latter case, all the unselected markers appear with the same high frequency among the recombinants. The possibility of simple genetic mapping is precluded even following T^+L^+ selection, since the genetic outcome also depends in part on variations in transfer. Thus the only reliable way of mapping the *distances* between genes as well as their order is to measure their relative times of appearance in the F$^-$ cell by means of the blender or some equivalent technique. The only assumption which must be made to establish the validity of this mode of measurement is that injection of the Hfr chromosome into the F$^-$ cell proceeds at a constant speed, and evidence has been provided that this is so as long as the available energy is kept constant. The value of the blender technique is well shown by attempts to map some of the inducible prophages located on the *Gal — R* segment. Whereas some of the phages situated farthest from *Gal* can scarcely be located by analysis of a cross between lysogenic and nonlysogenic cells, the blender technique allows an accurate determination of the time at which the prophage enters the F$^-$ cell (Wollman and Jacob 1957).

2. The Process of Integration   The following experiments may throw some light on how integration is accomplished.

(a) The effect of ultraviolet light on integration. If, in a nonlysogenic system, the Hfr parent is exposed to small doses of ultraviolet light prior to mating, the genetic constitution of recombinants is drastically modified although their total number is only slightly decreased, irrespective of whether T^+L^+ or Gal^+ is selected (Jacob and Wollman 1955a and unpub.). The table (cross 5) reveals that the frequency with which unselected markers are

found among both $T^+L^+S^r$ and Gal^+S^r recombinants is strikingly reduced by ultraviolet treatment. This means that radiation results in a loosening of the linkage observed among the markers lying between *TL* and *Gal* or, to put it another way, that the probability of a crossover occurring between two markers is increased. Since the frequency of recombination is decreased proportionately for both classes, it seems likely that radiation affects the processes of integration within the merozygote rather than genetic transfer from Hfr to F$^-$ cell.

If a mating mixture is diluted in nutrient broth at 37° 15 minutes after mixing, in order to prevent further collisions, and samples are thereafter irradiated with ultraviolet at intervals, it is observed that irradiation affects the genetic constitution of recombinants up to 70 minutes after the time of mixing. Thus ultraviolet irradiation evokes the same genetic outcome whether the Hfr cells alone or the formed zygotes are treated. After 70 minutes the effect diminishes until, at 120 minutes, it is no longer found. This suggests that the processes involved in the formation of the recombinant chromosome begin to be completed at about 70 minutes and are complete by 120 minutes.

This action of ultraviolet light on bacterial recombination recalls the similar effect which the same ultraviolet doses exert on recombination in λ phage (Jacob and Wollman 1955b) which was thought to be best interpreted according to Levinthal's hypothesis of replication and recombination (Levinthal 1954). According to this view recombination does not take place by breakage and reunion of already formed vegetative particles, but through partial replicas commenced on one phage type and finished on another. In an elementary act of recombination, only one recombinant type would therefore be formed. Small doses of ultraviolet light would interfere with the process of replication so that the replicas formed would be smaller. The same interpretation could account for the results found with bacteria, recombination in them, too, resulting from the integration of replicas formed along the transferred chromosome segment during replication of the recipient's chromosome.

(b) Labeling of the Hfr parent with P^{32}. Another way of studying both the kinetics and the mechanism of integration is to label the DNA of the Hfr parental cells with P^{32}. After conjugation with a nonradioactive F$^-$ strain, the behaviour of the radio-

active material transferred to the zygote can then be followed in a relatively simple manner, since it is known that bacteria containing a high specific radioactivity lose their ability to multiply as a function of P^{32} decay (Fuerst and Stent 1956). Once the Hfr cells have injected their radioactive genetic material into the nonradioactive F^- cells, it is possible to arrest the functions of the zygotes by freezing them at various intervals thereafter, and then subsequently to measure the survival of various classes of recombinants as a function of time (Fuerst, Jacob and Wollman, in preparation). Results so far obtained have provided two kinds of information. First of all, when the mating process is stopped early (40 minutes after mixing) the final rate of inactivation observed among the recombinants varies according to the selection. $T^+L^+Gal^+$ recombinants, on which inheritance of both T^+L^+ and Gal^+ from the Hfr parent has been imposed, are inactivated at a faster rate than are T^+L^+ recombinants. At first approximation the rate of inactivation appears to be correlated with the length of the segment selected for ($O — TL$ or $O — TL — Gal$), as estimated by the blender technique and recombination data. The second point is that, for each class of recombinants, the rate of inactivation decreases between 40 and 80 minutes after mixing and is no longer measurable after 100 minutes. Whatever the mechanism of this stabilisation may be, it implies that after 100 minutes the genetic information of the Hfr fragment has been transmitted to material which is not susceptible to P^{32} decay.

B. Expression of Recombinants

This is the final stage in the sequence of events which follows fertilisation of the F^- cell. The recombinant chromosome has been formed within the zygote. The process whereby it attains an autonomous position within a haploid cell is known as *segregation*. Each bacterial cell normally possesses 2 to 4 chromatinic bodies which are justifiably regarded as nuclear analogues. The fact that recombinants are usually clonal indicates that only one of the nuclei of either donor or recipient cell normally participates in zygote formation. Thus after conjugation the recipient cell (zygote) will contain at least one haploid F^- nucleus and one heterozygous or hemizygous diploid nucleus. It therefore resembles

a heterokaryon rather than a simple zygote so that segregation will yield at least one F^- cell in addition to the products of the heterozygote itself. On the other hand, partial transfer would exclude the appearance of the Hfr parental type among the progeny of the zygote cell. This has been shown to be the case in the HfrH × F^- cross (Hayes 1957).

For one reason or another all the characters inherited from the donor parent may not be able to operate functionally at the time of segregation, that is, their *phenotypic expression* may be delayed. The time required for expression will depend not only on whether a gene is a dominant or recessive allele but also on whether the particular character a gene controls manifests itself directly, for example through synthesis of an enzyme, or requires reorganization of some complex structure such as the cell wall. In this section some facts concerning the kinetics of segregation and phenotypic expression will be discussed.

1. The Kinetics of Segregation The time at which the recombinant segregant cells undergo their first division, either on minimal agar or in broth, is easily assessed. The first step is to allow zygotes to form for about 30 minutes in an Hfr × F^- cross in broth and then to eliminate the Hfr parent by treatment with phage so that only zygotes and F^- cells remain. To assess the time of segregation on minimal agar, the zygotes are spread on a series of plates selective for T^+L^+ recombinants and, at intervals thereafter, plates are removed in turn and their surfaces vigorously rubbed with distilled water by means of a glass spreader. This separates the progeny of any T^+L^+ recombinants that have already divided so that the subsequent colony count is doubled for each generation (Hayes 1957). An example of the results obtained by this kind of experiment is illustrated by the curve (continuous line) in Figure 9A. The T^+L^+ recombinants begin to divide about 120 minutes after plating and thereafter multiply with a generation time of about 60 minutes.

To follow segregation in broth, the zygotes are diluted into fresh broth. Samples are then removed as a function of time, appropriately diluted and plated on minimal agar for T^+L^+ recombinants (Hayes 1957). A representative result is given by the curve (continuous line) in Figure 9B. The recombinants begin to divide at about 100 minutes after dilution (i.e., about 140 minutes after mixing

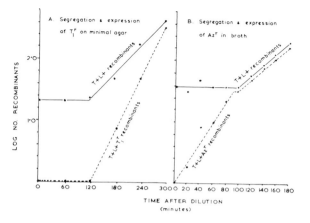

Fig. 9. The kinetics of segregation and phenotypic expression. A. On minimal agar (+B₁). Young broth cultures of HfrH $M^-Az^rT1^rT6^s$ and F^- $TLB_1^-Az^sT1^sT6^r$ were mixed and aerated at 37° for 30 minutes. The mixture was then treated with phage T6 for 10 minutes to eliminate the Hfr cells, diluted and plated on each of two series of warm plates of minimal agar + B₁, and incubated. At intervals thereafter the surfaces of the plates of one series were rubbed in turn with distilled water to separate the progeny of T^+L^+ recombinants; plates of the other series were similarly rubbed with a washed, high titre T1 suspension so as to select only T^+L^+ recombinants in which $T1^r$ had become phenotypically expressed. B. *In nutrient broth.* After treatment with phage T6 as above, the mixture was diluted into fresh broth at 37° and aerated at 37°. At intervals thereafter samples were further diluted and plated on: 1. minimal agar + B₁, for T^+L^+ recombinants; 2. minimal agar + B₁ + sodium azide (M/1500) for T^+L^+ recombinants in which Az^r had become phenotypically expressed. *(From Hayes 1957, J. Gen. Microbiol. in press.)*

the parental cultures) with a generation time of 20 minutes.

Assuming that timing starts when the parental cultures are mixed, it may be profitable to discuss a very tentative correlation of our findings if only to indicate the potentialities of the kinetic approach in analysis.

There are three sets of data. The first in order of timing is the stabilisation of the inactivation of zygotes due to decay of P³² introduced in the donor chromosome fragment: this becomes complete about 100 minutes after mixing. It suggests that, at this time, a P³²-free replica or template of the donor genetic contribution has been formed in all the zygotes. Secondly, ultraviolet irradiation ceases to have any effect on the genetic constitution of recombinants at about 120 minutes implying that the process of integration has by then been completed in all the zygotes. Twenty minutes later (i.e., the generation time of both parental and recombinant cells), at 140 minutes after mixing, the first division of the recombinant segregants is initiated.

2. The Kinetics of Phenotypic Expression The Hfr genes determining resistance or sensitivity to sodium azide and to phage T1 are linked to the selective markers T^+L^+ and are inherited respectively by about 90 percent and 75 percent of T^+L^+ recombinants (Fig. 5). For this reason, and because they manifest their functions in quite different ways, these two markers are admirably suited to the study of phenotypic expression. Only the expression of *resistance* to sodium azide (Az^r) and phage T1($T1^r$), inherited from the Hfr parent, has so far been examined (Hayes 1957).

The experiments are conducted in parallel with those for determining the kinetics of segregation either on minimal agar or in nutrient broth. In the cross Hfr·Az^rT1^r × F⁻·Az^sT1^s selection is made for $T^+L^+Az^r$ or $T^+L^+T1^r$ recombinants, as a function of time, as well as for the whole T^+L^+ recombinant class. The kinetics of expression, in terms of generation time, are about the same whether assessed on minimal agar, nutrient agar or in broth. Figure 9A correlates the kinetics of segregation (continuous line) and expression of the character T1^r (interrupted line) on minimal agar. In Figure 9B, segregation in broth (continuous line) is similarly related to phenotypic expression of the character Az^r (interrupted line). The patterns of expression of the two markers inherited from the Hfr parent are quite different. Expression of the character Az^r commences at the time of dilution (or of plating) of the zygotes and then rises exponentially to become complete just before the segregants which inherit it start to divide (Figure 9B). In contrast, the character T1^r does not begin to be expressed until *after* segregation while full expression (i.e., in 75% of all T^+L^+ recombinants) is delayed until the fourth generation (Figure 9A).

The gene $T1^r$ has been reported, from work on *E. coli* diploids, to be recessive to $T1^s$ (Lederberg 1949) so that its expression is not to be expected until after segregation; the further extension of its phenotypic lag is a natural consequence of the dependence of the development of phage resistance on structural alteration of the cell wall. Since expression of the character Az^r rises rapidly during a period when both alleles must be present together in the diploid merozygote, it follows that Az^r is dominant to Az^s; this is an exception to the rule that wild-type alleles are dominant to their mutant alleles in *E. coli*.

Knowledge of the kinetics of phenotypic expression following recombination in bacteria has a twofold importance. Firstly it offers a normal control against which the results of mutation kinetics can be studied. Secondly, it is necessary for the rational design of recombination experiments, whether in *E. coli* or in other bacterial species. For example, the choice of selective markers which are recessive, or whose expression is normally delayed, is unlikely to yield recombinants from an otherwise fertile cross unless application of the selective agent is withheld until expression has occurred. This is well exemplified by the fact that virtually no Az^r recombinants appear if recently formed zygotes which have inherited the gene Az^r from the donor parent are plated directly on medium containing sodium azide, despite the fact that Az^r appears to be dominant and is rapidly expressed.

IV. THE RELATIONSHIP BETWEEN F⁺ AND Hfr SYSTEMS

The discussion so far has dealt exclusively with the features of crosses involving an Hfr donor (HfrH) and F⁻ recipient strains. In this section we propose to examine the properties of the equivalent F⁺ × F⁻ cross and hope to show that a comparative analysis of the two systems may lead to a unified interpretation of the mating process in *E. coli*. The principal points of difference between F⁺ and Hfr strains have already been summarized and will not be reiterated here. The main aim of this analysis is to attempt to equate the two systems. It will be found that some of the paradoxes which seem to exist are resolved as the analysis proceeds: others remain and will be discussed in their proper context.

In the light of the results of analysis of the Hfr system, it may be asked whether the low frequency of recombination found in F⁺ × F⁻ crosses is due to low frequency of conjugation, of transfer, or of

integration. The evidence relevant to this question will now be presented.

A. Analysis of F⁺ × F⁻ Crosses

1. Conjugation There is one character of F⁺ cells which is transferred to F⁻ cells at high frequency. This is the F⁺ character itself, and its transfer requires cellular contact (Cavalli *et al.* 1953; Hayes 1953a). The kinetics of F⁺ transfer can be studied by the same methods as those used for the analysis of Hfr systems. A mixture of F⁺ and F⁻ cells is made in broth, the F⁺ cells being in excess. Samples are removed at intervals and plated, both directly and after treatment in a Waring Blender, on a medium on which only the F⁻ recipient cells can grow. The resulting F⁻ clones are then scored for inheritance of F⁺ as a function of time after mixing. In the case of the untreated samples it is found that the curve expressing the proportion of F⁻ cells which have inherited the character F⁺ arises from the origin and increases linearly until a plateau is reached about 50 minutes after mixing, when some 75 percent of the F⁻ cells have been converted. In the case of the treated samples, the F⁺ character begins to appear at about five minutes after mixing (Jacob and Wollman 1955a). It thus turns out that transfer of F⁺ from F⁺ donor to F⁻ recipient cells is very similar to the transfer of such characters as T^+L^+ from Hfr to F⁻ cells (Fig. 3) except that F⁺ transfer occurs several minutes earlier than any of the hitherto known markers on the $O — R$ segment of HfrH. It therefore seems probable that F⁺ is located on the F⁺ chromosome within a region $O' — R'$ which is transferred at high frequency to the F⁻ cell. Another kind of character which appears to be transferred at high frequency from F⁺ to F⁻ cells is the ability to synthesise certain colicines such as colicine E, originally a character of *E. coli* strain K-30 (Frédéricq and Betz-Bareau 1953). Kinetic analysis of the transfer of this ability shows that production of colicine E is transferred from a K-12 F⁺ strain to an F⁻ recipient strain at about two and a half minutes after mixing. The ability to produce colicine E would therefore seem to be controlled by a locus situated between $0'$ and F⁺ on the F⁺ chromosome (Jacob and Wollman, unpub.).

One can thus conclude that conjugation between F⁺ and F⁻ bacteria is no different from Hfr × F⁻ conjugation and occurs with a frequency of the order of 100 percent. In substantiation of this, conjoint pairs of cells, and cellular bridges uniting them, similar to those observed in Hfr X F⁻ crosses, can also be seen in F⁺ X F⁻ crosses by means of phase contrast and electron microscopy (Fig. 7, bottom picture) (Anderson, Wollman and Jacob 1957).

2. Transfer and Integration Since the frequency of conjugation, and of transfer of certain markers, is found to be high in F⁺ X F⁻ crosses, the question arises whether those F⁺ markers which are not inherited at high frequency are absent because they are not transferred, or because they are transferred but not integrated. The most significant evidence is derived from zygotic induction. Whereas this phenomenon can be directly demonstrated in the cross HfrH (λ)⁺ X F⁻ (λ)⁻, it is hardly detectable in the equivalent F⁺ (λ)⁺ X F⁻ (λ)⁻ cross (Jacob and Wollman 1956b). Since the transfer of λ prophage is immediately expressed it must be concluded that prophage is not *transferred* at high frequency from an F⁺ lysogenic donor to an F⁻ nonlysogenic recipient cell. It must therefore be assumed that the failure of at least the *Gal* − λ region of the F⁺ chromosome to be inherited at high frequency is due to lack of transfer. The hypothesis can therefore be made that the two systems are similar in every respect save one; that the location of the region of preferential rupture is different. Accordingly, the chromosome of F⁺ bacteria also comprises two segments separated by a new region of rupture, R', such that the segment $O' — R'$, carrying the markers colicine E and F⁺, is transferred at high frequency. The other known genetic markers are all located on the other segment which is not transferred at high frequency.

B. The Mechanism of Low Frequency of Recombination

HfrH cells can only transfer, at high frequency, markers located on the $O — R$ segment of the chromosome. In the case of the equivalent F⁺ cells, only markers on the chromosome segment $O' — R'$ can similarly be transferred to F⁻ cells. Let us consider how, in either case, a marker situated on neither of these segments may be transferred at low frequency. There are two possible hypotheses;

either each Hfr or F⁺ cell has a small but equal probability of transferring a chromosome segment on which the marker is located, or else there exists in each Hfr or F⁺ population a small fraction of mutant cells which can transfer the marker with high efficiency. In this latter event the number of recombinants formed by independent cultures of the same F⁺ strain should be extremely variable as compared with those formed by samples of the same culture. The results of such fluctuation tests (Luria and Delbrück 1943) in which the F⁺ population was standardised so as to contain only a few presumptive mutant clones in each independent culture, are in agreement with this prediction (Jacob and Wollman 1956a). Genetic analysis of recombinants isolated from fluctuation tests of this kind affords striking confirmation of the mutation hypothesis. When the genetic patterns of the progeny of different independent cultures are compared they are found to vary widely between cultures, both in the characters of the F⁺ parent which are transferred and the proportion of these characters found among recombinants. On the other hand, the genetic constitution of recombinants arising from different samples of the same culture is very homogeneous.

It is therefore probable that most, if not all, of the recombinants formed in an F⁺ X F⁻ cross must be due to preexisting Hfr mutants in the F⁺ population. By means of the method of indirect selection known as "replica plating" (Lederberg and Lederberg 1952), it is in fact possible to isolate different types of Hfr strains from F⁺ cultures. When the means of selection are adequate, the efficiency of isolation of Hfr mutants responsible for particular recombinant colonies is close to 100 percent. By selecting for recombinants which have inherited different markers from the F⁺ parent it is possible, by the replica plating method, to isolate a variety of different Hfr strains capable of transmitting the selected marker to recombinants at high frequency. These Hfr strains differ from one another not only in the characters they can transfer to the zygote but also in the *sequence* in which these characters are transferred (Jacob and Wollman 1956a). Each of the known markers of *E. coli* K-12 can be transferred at high frequency by one or another Hfr strain but no single Hfr strain has so far been isolated which alone can transfer all these markers. That low frequency of appearance of a given character among recombinants is indeed due to the fact that the locus determining it is not situated on the chromosome segment transferred at high frequency, is demonstrated by the fact that zygotic induction is not observed in crosses between strains of Hfr $(\lambda)^+$ which do not transmit the Gal_b character, and nonlysogenic F$^-$ strains. This situation is found in crosses involving HfrC. This strain has been otherwise shown by Skaar and Garen (personal communication) to transfer the TL — Lac segment in a sequence which is the reverse of that exhibited by HfrH.

The hypothesis therefore appears well substantiated that characters transmitted to recombinants at low frequency, either in F$^+$ × F$^-$ or in Hfr × F$^-$ crosses, are transferred by spontaneous Hfr mutants. Among the Hfr strains isolated there seem to exist certain preferential patterns of transfer but the possibility has not yet been excluded that this is due to a bias introduced by the use of streptomycin as a contraselective agent. The properties of these different Hfr strains is still under investigation.

Despite the low frequency with which recombinants are found, F$^+$ × F$^-$ mixtures in broth can be analysed kinetically by the use of virulent phage to eliminate the F$^+$ parent, in the same way as the Waring blender has been used for analysis of Hfr × F$^-$ crosses (Hayes 1957). In comparative experiments of this kind the cross HfrH × F$^-$ is found to be about 2×10^4 times more productive of T^+L^+ recombinants than the equivalent F$^+$ × F$^-$ cross. When selection is made for the T^+L^+, $T^+L^+Lac^+$ or $T^+L^+B_1^+$ markers of the F$^+$ parent (see Fig. 1), it is found that the transfer of T^+L^+ starts at about eight minutes and of Lac^+ at 18 minutes after mixing, exactly as in the equivalent Hfr × F$^-$ cross. Unlike the Hfr × F$^-$ cross, however, the F$^+$ marker B_1^+ is transferred at about 45 minutes after mixing. No recombinants inheriting the xylose, maltose or streptomycin markers of the F$^+$ parent are found up to 90 minutes after mixing when selection is made for them. This implies that among those Hfr mutants which can transfer T^+L^+ at high frequency, the most common types are broadly similar to HfrH. The late transfer of the F$^+$ marker B_1^+, and its appearance among T^+L^+ recombinants with a frequency of five percent, indicates, however, that at least five percent of these Hfr mutants can transfer at high frequency a different chromosome segment.

V. DISCUSSION

A. Sexual Differentiation

Sexual conjugation in *E. coli* takes place between cells of opposite mating type which can, for convenience, be referred to as *donor* and *recipient* since, after cellular contact, genetic material is transferred from the former to the latter. At the present time these two mating types can only be defined in terms of the properties expressed by the donor. These properties comprise a surface configuration which enables effective contact with the recipient cell's surface to be established, the ability to inject a chromosomal segment, O — R, which differs from donor to donor and finally, the capacity to mutate from one donor type to another. Upon high frequency transfer of a small chromosome segment from an F$^+$ donor to an F$^-$ recipient, the recipient is thereby converted into an F$^+$ donor in which these three properties are simultaneously expressed. It is not known, however, whether these properties are all expressions of the same genetic factor or whether they may exist independently.

The distinction between F$^+$ and Hfr donor strains is more artificial than real since the frequency of recombination observed in practice will depend only upon whether the selective markers used happen to lie on the O — R segment or beyond it. Since the low frequency of recombination found in the latter case appears to be mediated by different Hfr mutants preexisting in the donor culture, use of the terms "Hfr" and "F$^+$" to denote a fundamental distinction in the capacity of the two strains to yield recombinants loses its validity. The term "F$^+$ strain" has also been used to imply the ability of the strain to transmit the donor state to F$^+$ cells at high frequency. In contrast, neither of the Hfr strains hitherto isolated could transmit either the F$^+$ or Hfr donor state at high frequency. However, when selection is made for certain markers of either HfrH or HfrC, the Hfr donor state is inherited by a high proportion of recombinants and, in the case of HfrC at least, appears to be controlled by a specific locus closely linked to one of the Gal loci (Hayes 1953b; Cavalli and Jinks 1956).

In its transmissibility and inheritance, F$^+$ displays the properties of a gene. In strain HfrH, F$^+$ is no longer found as such but appears to be replaced by a new locus determining the donor state, situated distal to a new region of rupture. Since this strain can revert to the original F$^+$ donor state, the most likely hypothesis is that the *position* of F$^+$ determines the region of preferential rupture and that mutation to Hfr is a function of the shift of F$^+$ to a new position on the chromosome. If, in a Hfr strain, F$^+$ happened to be located on the $O - R$ segment proximal to the region of rupture it determined, then the donor state would be transferred at high frequency to F$^-$ cells as in the case of F$^+$ strains. The range of Hfr strains which have now been isolated has not yet been investigated from this point of view. It should be pointed out, however, that in F$^+$ × F$^-$ crosses all, or nearly all, the recombinants may be F$^+$ even when the cross is made in broth and the F$^+$ parent is eliminated by phage to prevent secondary transfer of the F$^+$ character to the recombinant cells after plating. If, as is supposed, these recombinants result from Hfr mutants one would expect them to inherit either the Hfr or the F$^-$ character, not the F$^+$ (Hayes 1957).

Different donor strains differ in the chromosomal location of the region of rupture, R, which determines the length of chromosome they are able to transfer. Any donor type apparently possesses the potentiality to mutate to other donor types, with a consequent shift in the position of

$$R \ (F^+ \rightleftharpoons Hfr_1 \rightleftharpoons Hfr_2 \ etc.)$$

as well as to the recipient type (Lederberg *et al.* 1952; Hayes 1953a). When a recipient F$^-$ cell inherits that region of a donor chromosome controlling the donor properties of the cell (F$^+$ or Hfr) it also inherits the capacity to mutate to other donor types. Thus it has been possible to obtain F$^+$ and various Hfr derivatives of *E. coli* strain C which was originally F$^-$ (Jacob and Wollman, unpub.). The occurrence of mutation from the recipient to the donor state has not yet been reported.

A population of donor cells is therefore heterogeneous and contains, in addition to a majority of the original type which, on conjugation, can transfer a particular chromosomal segment, a variety of other donor types which are responsible for the recombinants formed at low frequency. Recombinants obtained by crossing an F$^+$ population of this kind with an F$^-$ population would be formed by different Hfr mutants transferring a different range of characters. Their number and genetic make-up

would depend on the relative frequencies of the different Hfr types present in the donor population. Moreover it may be expected that the results of crosses involving the same F$^-$ strain but F$^+$ strains of different origin will not be the same but will differ according to the "Hfr mutability pattern" of the F$^+$ strain employed. The difficulties encountered in the genetic analysis of F$^+$ × F$^-$ crosses are readily understandable in such a model (Jacob and Wollman 1956a).

The mechanism of the mutation from one donor type to another is still a matter for speculation. It may be an example of chromosomal mutation since it seems to involve breakage and rearrangement within the chromosome which can only be expressed when the donor character is present. Such mutation appears to be associated with a change in the location of the donor character although it is not known whether the alterations in position are the cause or the effect of the rearrangements. This behaviour is reminiscent of the phenomena found in maize by McClintock (1956). The "donor character" could be conceived as being a "controlling element" whose location commands, in an unknown fashion, the region of chromosome breakage, R, as well as the pattern of chromosome rearrangements. Whether or not the "donor character" will be transferred on conjugation will depend on whether it is located between O and R or distal to R.

B. Meromixis

Once the genetic contribution of the donor cell has been transferred to the recipient cell a partial zygote or *merozygote* is formed which comprises the entire genome of the recipient cell together with a chromosomal segment of variable size from the donor cell. Thus a comparable situation is attained in transformation, transduction and conjugation, the three varieties of genetic transfer found among bacteria, for in each only a fragment of the genome of a donor cell is transferred to a recipient cell. The difference between them lies in the method whereby genetic transfer is accomplished, transformation being effected by the DNA extracts of donor cells, transduction by a phage vector and conjugation by direct injection of a chromosomal segment into the recipient cell by the donor cell. In some strains like

E. coli K-12 transfer of genetic material can be performed by either conjugation or transduction (Lennox 1955; Jacob 1955). We therefore propose to unite, under the name of *meromixis*, these three known processes which are characterised by partial genetic transfer.

As to the mechanisms whereby characters transferred from the donor cell are integrated into a recombinant chromosome, the same problems are clearly raised by the three types of meromixis, and have already been considered in the case of transformation (Ephrussi-Taylor 1955; Hotchkiss 1955) and transduction (Demerec and Demerec 1955; Lederberg 1955). In conjugation it is evident that a segment of donor chromosome, although present in a merozygote, does not always participate in the formation of a recombinant chromosome and, when it does, that it is not necessarily incorporated as a whole. A mechanism analogous to crossing over must therefore be assumed although its physical basis is likely to differ from one of simple breakage and reunion in view of the asymmetry of the parental components, one of which may be very small. The question therefore arises whether the transferred fragment itself enters into the constitution of the recombinant chromosome, or whether some mechanism in which replication and recombination are combined, such as that proposed by Levinthal for recombination in phage, is not more likely. The experiments briefly reported here on the effects of ultraviolet light on recombination, as well as those involving the transfer of a radioactive chromosome segment to the merozygote, favour such an hypothesis although their interpretation cannot yet be regarded as unequivocal.

Recombination of genetic characters in bacteria is thus accomplished by means of mechanisms very different from those known in other organisms, since chromosomal fragments of variable size are transferred from one bacterium to another. Of the three known processes by which this transfer is accomplished, that of conjugation in *E. coli* is the most highly developed and most closely resembles a sexual mechanism with sex differentiation, the donor's chromosome segment playing the role of an incomplete male gamete while the recipient cell is analogous to a female gamete which contributes both its intact genome and its cytoplasm to the zygote.

SUMMARY

1. When Hfr and F⁻ bacteria are mixed, effective pairing between donor and recipient cells follows chance collisions. The efficiency of this pairing is modifiable by environmental influences such as pH, and is an energy requiring process. An actual bridge of cellular material can be seen to join the paired cells in electron microphotographs. Under optimal conditions, 100 percent of Hfr cells conjugate.

2. Effective contact is immediately followed by the orientated transfer of a segment of Hfr chromosome, $O — R$, to the F⁻ cell which thus becomes a partial zygote. Loci situated on this segment always penetrate the F⁻ cell in the same order as their arrangement on the chromosome. Transfer of the whole $O — R$ segment takes about 35 minutes in broth at 37°. Markers located beyond R do not enter the F⁻ cell at all and are therefore not inherited among recombinants.

3. During chromosome transfer spontaneous breaks occur and the probability of breaking increases towards its distal end. This distal end, R, is therefore thought to be a region of increasing fragility rather than a well defined locus.

4. Chromosome transfer is an energy-requiring process carried out solely by the Hfr cell. The speed at which the chromosome proceeds appears to be constant at any given temperature but can be modified by changes in the energy supply.

5. The $O — R$ segment of chromosome can be ruptured at will during transfer, either by mechanical separation of the mating bacteria or by selectively killing the donor parent with phage. Only those Hfr markers which have already entered the F⁻ cell at the time of treatment will participate in recombination.

6. Hfr markers which have entered the zygote have a probability of being integrated into the recombinant chromosome which becomes less the further the marker is situated from O on the $O — R$ segment, as might be deduced from the partial nature of the zygote if it is assumed that O is an extremity of the donor chromosome.

7. The time at which a marker inherited from the donor cell is phenotypically expressed in the recombinant segregant cell or its progeny depends on whether it is dominant or recessive, as well as on its manner of expression. Expression of the Hfr gene determining resistance to sodium azide commences in the zygote and is complete before the segregants begin to divide: resistance to phage T1 is not expressed until after segregation and requires four generations for completion.

8. $F^+ \times F^-$ crosses show a high frequency of conjugation and of F^+ transfer. The kinetics of F^+ transfer are compatible with the view that F^+ donor cells transfer a segment of chromosome $O' - R'$, on which F^+ is located, to F^- cells. F^+ markers situated distal to R' are not transferred.

9. Transfer of any particular marker situated distal to R on the Hfr chromosome, or to R' on the F^+ chromosome, is mediated by Hfr mutants in the donor population which can transfer the marker to F^- cells at high frequency. These Hfr mutants are very heterogeneous both in the range of markers they can transfer and in the order of transference.

10. The mechanisms by which donor strains may mutate from one type to another are discussed.

REFERENCES

Anderson, T. F., 1949, The reactions of bacterial viruses with their host cells. Bot. Rev. 15:477.

Anderson, T. F., Wollman, E. L., and Jacob, F., 1957, Sur les processus de conjugaison et de recombinaison génétique chez *E. coli*. III. Aspects morphologiques en microscopie électronique. Ann. Inst. Pasteur (in press).

Appleyard, R. K., 1954, Segregation of λ lysogenicity during bacterial recombination in *Escherichia coli* K-12. Genetics 39:429–439.

Cavalli, L. L., 1950, La sessualita nei batteri. Boll. Ist. sierotera. Milano 29:1–9.

Cavalli-Sforza, L. L., and Jinks, J. L., 1956, Studies on the genetic system of *Escherichia coli* K-12. J. Genet. 54:87–112.

Cavalli, L. L., Lederberg, J., and Lederberg, E. M., 1953, An infective factor controlling sex compatibility in *Bacterium coli*. J. Gen. Microbiol. 8:89–103.

Demerec, M., and Demerec, Z. E., 1955, Analysis of linkage relationships in *Salmonella* by transduction techniques. Brookhaven Symp. Biol. 8:75–87.

Eddington, A., 1935, New Pathways in Science. Cambridge, England, Cambridge Univ. Press.

Ephrussi-Taylor, H., 1955, Current status of bacterial transformations. Adv. Virus Res. 275–307.

Fisher, K., 1957a, The role of the Krebs cycle in conjugation in *Escherichia coli* K-12. Gen. Microbiol. 16:No. 1 (in press).

1957b, The nature of the endergonic processes in conjugation in *Escherichia coli* K-12 J. Gen. Microbiol. 16:No. 1 (in press).

Fredericq, P., and Betz-Bareau, M., 1953, Transfert génétique de la propriété colicinogène en rapport avec la polarité F des parents. C. R. Soc. Biol. 147:2043–2045.

Fuerst, C. R., and Stent, G. S., 1956, Inactivation of bacteria by decay of incorporated radioactive phosphorus. J. Gen. Physiol. 40:73–90.

Fuerst, C. R., Jacob, F., and Wollman, E. L., 1957, Sur les processus de conjugaison et de recombinaison génétique chez *E. coli*. Etude de la recombinaison à l'aide du phosphore radioactif (in preparation).

Hayes, W., 1952a, Recombination in *Bact. coli* K-12: unidirectional transfer of genetic material. Nature, Lond. 169:118–119.

1952b, Genetic recombination in *Bact. coli* K-12: analysis of the stimulating effect of ultraviolet light. Nature, Lond. 169:1017–1018.

1953a, Observations on a transmissible agent determining sexual differentiation in *Bact. coli*. J. Gen. Microbiol. 8:72–88.

1953b, The mechanism of genetic recombination in *Escherichia coli*. Cold Spring Harb. Symp. Quant. Biol. 18:75–93.

1957, The kinetics of the mating process in *Escherichia coli*. J. Gen. Microb. 16: No. 1 (in press).

Hershey, A. D., and Chase, M., 1952, Independent functions of viral protein and nucleic acid in growth of bacteriophage. J. Gen. Physiol. 36:39–56.

Hotchkiss, R. D., 1955, Bacterial transformation. J. Cell. Comp. Physiol. 45:suppl. 2:1–22.

Jacob, F., 1955, Transduction of lysogeny in *Escherichia coli.* Virology 1:207–220.

Jacob, F., and Wollman, E. L., 1954, Induction spontanée du développement du bactériophage λ au cours de la recombinaison génétique chez *E. coli* K-12. C. R. Acad. Sci. 239:317–319.

1955a, Etapes de la recombinaison génétique chez *E. coli* K-12. C. R. Acad. Sci. 240:2566–2568.

1955b, Etude génétique d'un bactériophage tempéré d'*E. coli.* III Effet du rayonnement ultraviolet sur la recombinaison génétique. Ann. Inst. Pasteur 88:724–749.

1956a, Recombinaison génétique et mutants de fertilité. C. R. Acad. Sci., 242:303–306.

1956b, Sur les processus de conjugaison et de recombinaison génétique chez *E. coli.* I. L'induction par conjugaison ou induction zygotique. Ann. Inst. Pasteur 91:486–510.

Lederberg, E. M., 1951, Lysogenicity in *E. coli* K-12. Genetics 36:560.

Lederberg, E. M., and Lederberg, J., 1953, Genetic studies of lysogenicity in *E. coli.* Genetics 38:51–64.

Lederberg, J., 1947, Gene recombination and linked segregations in *E. coli.* Genetics 32:505–525.

1949, Aberrant heterozygotes in *Escherichia coli.* Proc. Nat. Acad. Sci. Wash. 35:178–184.

1955, Recombination mechanisms in bacteria. J. Cell. Comp. Physiol. 45, suppl. 2:75–107.

Lederberg, J., and Lederberg, E. M., 1952, Replica plating and indirect selection of bacterial mutants. J. Bact. 63:399–406.

Lederberg, J., Lederberg, E. M., Zinder, N. D., and Lively, E. R., 1951, Recombination analysis of bacterial heredity. Cold Spring Harb. Symp. Quant. Biol. 16:413–441.

Lederberg, J., and Tatum, E. L., 1946, Novel genotypes in mixed cultures of biochemical mutants of bacteria. Cold Spring Harb. Symp. Quant. Biol. 11:113–114.

Lennox, E. S., 1955, Transduction of linked genetic characters of the host by bacteriophage P_1. Virology 1:190–206.

Levinthal, C., 1954, Recombination in phage T2: its relationship to heterozygosis and growth. Genetics 39:169–184.

Lieb, M., Weigle, J. J. and Kellenberger, E., 1955, A study of hybrids between two strains of *E. coli.* J. Bact. 69:468–471.

Maccacaro, G. A., 1955, Cell surface and fertility in *E. coli.* Nature, Lond. 176:125–126.

McClintock, B., 1956, Controlling elements and the gene. Cold Spring Harb. Symp. Quant. Biol. 21:197–216.

Newcombe, M. B., and Nyholm, M. H., 1950, Anomalous segregation in crosses of *E. coli.* Amer. Nat. 84:457–465.

Rothfels, K. H., 1952, Gene linearity and negative interference in crosses of *E. coli.* Genetics 37:297–311.

Wollman, E. L., 1953, Sur le determinisme génétique de la lysogénie. Ann. Inst. Pasteur 84:281–293.

Wollman, E. L., and Jacob, F., 1954, Lysogénie et recombinaison génétique chez *E. coli* K-12. C. R. Acad. Sci. 239:455–456.

1955, Sur le mécanisme du transfert de matériel génétique au cours de la recombinaison chez *E. coli* K-12. C. R. Acad. Sci. 240:2449–2451.

1957, Sur les processus de conjugaison et de recombinaison génétique chez *E. coli.* II. Polarité du transfert et de la recombinaison génétique. Ann. Inst. Pasteur (in press).

Transduction in *Escherichia coli* K-12*

M. L. MORSE,[†] ESTHER M. LEDERBERG, and JOSHUA LEDERBERG[‡]

Reprinted by authors' and publisher's permission from Genetics, *vol. 41, 1956, pp. 142–156.*

Morse, like Zinder, was a student of the Lederbergs when he discovered specialized transduction in *E. coli*. In 1954, he found that λ-transduced *Gal* genes only. In this paper, written in 1956, Morse and the Lederbergs followed through on the issue of specialized transduction and described other remarkable properties of this phenomenon as well. They showed, for example, that λ-transduced bacteria carry both the *Gal*⁺ gene, brought in by λ, and the mutant (*Gal*⁻) gene. Furthermore, the *Gal*⁺/*Gal*⁻ heterogenotes were lysogenic and when induced by ultraviolet irradiation produced both transducing λ and normal, nontransducing λ in roughly equal proportions.

A system of genetic transduction has been discovered in the sexually fertile K-12 strain of *Escherichia coli*. This transduction is mediated by lambda, a temperate phage for which K-12 is normally lysogenic.

The distinctive features of the lambda-K-12 system include the following: (1) The transductions are limited to a cluster of genes for galactose fermentation. The *Gal* loci are closely linked to each other and to *Lp*, the locus for lambda-maintenance. (2) The transducing competence of lambda depends on how it is prepared. Competent lambda is produced by induction of lysogenic bacteria; lambda harvested from infected, sensitive hosts is incompetent. (3) The transduction clones are often heterogenotic, that is, heterozygous for the *Gal* genes which they continue to segregate. Technical advantages of the

lambda system include recombinational analysis by the sexual cycle and the availability of lysates in which nearly every lambda particle is competent.

MATERIALS AND METHODS

Cultures

The origin and history of the *Escherichia coli* K-12 cultures studied have already been described (E. Lederberg 1950, 1952; Lederberg and Lederberg 1953). The emphasis will be placed here on the *Gal* loci (+ = fermenting galactose; – = nonfermenting) and on the locus which controls the maintenance of lambda (Lp_1).

The phenotypes of cultures with different alleles of Lp_1 are as follows:

	Lysed by lambda	Lyses Lp_1[8] culture
$Lp_1{}^s$ culture (sensitive)	yes	no
$Lp_1{}^+$ culture (lysogenic)	no	yes
$Lp_1{}^r$ culture (immune)	no	no

Regardless of their Lp_1 genotype, cultures have been found to adsorb lambda. Thus $Lp_1{}^+$ and $Lp_1{}^r$ are resistant to lysis by lambda in spite of their ability to adsorb the phage. In contrast with this,

*Paper No. 589 of the Department of Genetics. This work has been supported at various times by the Atomic Energy Commission, Contract AT(11-1)-64, Proj. 10; a research grant (C2157) from the National Cancer Institute, Public Health Service and grants from the Research Committee, University of Wisconsin, with funds allotted by the Wisconsin Alumni Research Foundation. Received August 26, 1955.
[†]Predoctoral research fellow of the National Science Foundation, 1953–54.
[‡]Department of Genetics, University of Wisconsin, Madison, Wisconsin.

mutants resistant to lambda-2, a virulent mutant of lambda, are resistant because they do not adsorb either lambda or lambda-2 under the experimental conditions used here.

Media

The media used include: broth, Difco penassay; agar for phage assay, Difco nutrient agar with 0.5 percent NaCl; indicator medium, EMB agar plus one percent sugar; minimal agar, D(O); and minimal indicator agar, EMS (J. Lederberg 1950). Special supplements were added where indicated. All dilutions of phage lysates were made in either penassay or nutrient saline broth, and cell suspensions were diluted in either 0.5 percent saline or penassay broth.

General Methods

Plates and tubes were incubated at 37°C. When high cell densities were desired, broth cultures were aerated by bubbling filtered air through them. Propylene glycol monolaurate (Glyco Products Co., Inc.) at a final concentration of 0.01 percent was added to bubbled cultures to lessen foaming. Phage assays were made either in agar layer or by spreading a portion of dilute lysate with Gal^- cells on EMB galactose agar.

Lysates containing lambda in high titer were prepared by two methods: (1) "Induced lambda" was liberated from lysogenic bacteria after treatment with ultraviolet (UV) (Weigle and Delbrück 1951); (2) "Lytic lambda" was harvested from sensitive bacteria infected with free lambda. The induced lambda was prepared as follows: aerated, penassay grown cultures of an Lp^+ strain (ca. 10^9 cells per ml) were sedimented in the centrifuge, the broth discarded, and the cells resuspended in 0.5 percent saline. The cell suspensions (10 ml) were exposed to the radiation from a GE Sterilamp (45 seconds at 50 cm) in open petri dishes on a platform shaker. After irradiation the suspensions were diluted with an equivalent volume of double strength penassay broth and aerated at 37° C until maximal clearing was obtained. This usually required from 2 to 3 hours. To produce lytic lambda, an inoculum of induced lambda was adsorbed on to penassay grown

sensitive cells. After the adsorption period the cells were sedimented to separate them from the penassay broth and resuspended in nutrient saline broth. The suspension was then aerated until maximal clearing was obtained (4–5 hours). Induced lysates have phage plaque titers of about 3×10^{10} particles per ml, while lytic lysates have about 10^{10}.

Induced lambda was used in all experiments unless otherwise stated.

Methods for Testing for Transduction

In order to detect infrequent genetic changes, selective agar media were used: EMB agar for fermentation markers; EMB agar plus 100 micrograms per ml streptomycin for streptomycin resistance; minimal agar for nutritional markers. About 10^8 mutant cells in 0.1 ml broth or saline, and 0.1–0.2 ml of lysate were added to the surface of each agar plate and then spread with a bent glass rod. The plates were incubated 2–3 days before being scored.

Transduction clones selected by these methods develop in a heavy background of unchanged cells. On EMB medium, negative cells grow at the expense of the peptone; by using sugar as well, positive clones form papillate outgrowths from the negative background. EMB agar serves as an indicator as well as a selective medium; isolated positive colonies are deeply colored, whilte negative colonies remain translucent (illustrated in Fig. 3).

The transduction clones were purified by the following procedure. Papillae were picked with a needle and suspended in 1 ml of sterile water. A loopful (ca. 0.001 ml) of this suspension was then streaked upon a portion of another plate of the EMB agar. These primary dispersals of the transduction clones were nearly always mixed. Direct picking and streaking, or spotting without any purification cannot be trusted. From the primary streaks a single colony that looked pure was picked to water and streaked as before. This operation was repeated once again, and a single colony from thè last streaking was taken to represent the transduction clone. In addition to freeing the transduction clone from unchanged background cells, this method of purification may also act selectively within an unstable clone. Picking apparently pure colonies leads to an overestimate of the fraction of nonsegregating clones.

RESULTS

The Transductions

Although a number of different loci affecting diverse portions of the genotype were tested, only genes of a cluster of loci for galactose fermentation were transduced by lambda lysates (Morse 1954). The *Gal* loci, of which about seven have been investigated thus far, are closely linked to one another (less than one percent recombination) and to Lp_1, the locus for lambda maintenance (Lederberg and Lederberg 1953, and unpublished).

The transformation of Gal^- cells to Gal^+ by induced lambda is illustrated in Figure 1. Each papilla is a clone of galactose fermenting cells; on the area of the plate to which lysate was added, most of the Gal^+ papillae are transduction clones. The quantitative relationships are illustrated in Figure 2. The data can be summarized: (1) Regardless of the Lp_1 genotype of the recipient, transductions were obtained; (2) with each genotype the number of transductions was proportional to the amount of lysate plated; (3) Lp^s recipient cultures gave 5- to 10-fold more papillae per unit of lysate than either Lp_1^+ or Lp_1^r. Fur-

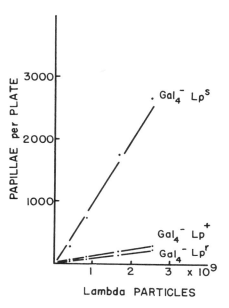

Fig. 2. Proportionality between amount of lambda lysate (LFT) plated and number of papillae formed from Lp^s, Lp^+ and Lp^r Gal^- cultures. The ratio of papillae to lambda particles is 10^{-6} for an Lp^s culture, 10^{-7} for Lp^+ and Lp^r cultures.

Fig. 1. The production of Gal^+ papillae from a Gal^- background of cells by a lambda lysate. Left, the control, no lysate added. Right, 0.1 ml of lysate from a Gal^+ culture. Some of the papillae have been picked with a needle.

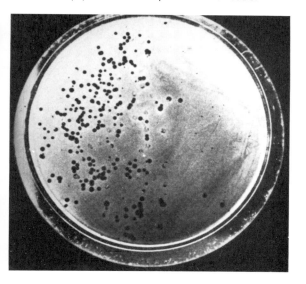

ther, the transducing activity of lysates (which contain 10^{10} lambda per ml) varies according to the number of cells plated: (1) with Lp_1^+ Gal^- cultures there is a twofold increase between 10^6–10^9 cells per plate, with a plateau of maximal yield around 10^8 cells per plate; (2) Lp^s Gal^- recipient cultures show about a sixfold increase over a similar range of cell platings, with the highest yield at the highest cell density.

The transducing activity of lysates is specific; that is, a lysate of a Gal_x^- culture will not transform Gal_x^- cultures (Table 1) but Gal^+ papillae were found with a Gal_y^- culture. The specificity is extended further in that some galactose positive phenotypic reversals of Gal^- culture give lysates with transducing activity on the original Gal_x^- indicator (Table 2). The different types of phenotypic reversals may be understood under the following hypothesis: (1) reverse mutations (Gal_x^- to Gal_x^+) yield cultures that give active lysates, and (2) suppressor mutations (Gal_x^- Suppressor$^-$ to Gal_x^- Suppressor$^+$)

Table 1 Transformation of Gal^- cultures by lysates of Gal^+

Recipient culture	Lysate of:	Lambda titer in 10^{10} per ml	Number of $Gal+$ papillae		$Gal+$ papillae per 10^7 lambda
			Control (no lysate)*	0.1 ml lysate	
$Gal_1^-Lp^+$	Gal^+	1.4	2	405	2.9
	Gal_1^-	2.4	2	2	—
$Gal_2^-Lp^+$	Gal^+	1.4	20	356	2.4
	Gal_2^-	4.9	20	10	—
$Gal_4^-Lp^+$	Gal^+	1.4	47	394	2.5
	Gal_4^-	1.7	47	50	—
$Gal_4^-Lp^s$	Gal^+	1.4	4	2112	15.1
	Gal_4^-	1.7	163	86	—
$Gal_4^-Lp^s$†	Gal^+	2.3	10	3020	13.1
	Gal_4^-	1.7	10	18	—
$Gal_4^-Lp^s$†	Gal^+	2.3	5	1296	5.6
$Gal_4^-Lp^r$	Gal^+	2.3	40	161	0.5
$Gal_4^-Lp^r$‡	Gal^+	1.4	29	129	0.7
$Gal_4^-Lp^r$‡	Gal^+	1.6	28	92	0.4

*The Gal^+ papillae on the control are spontaneous reversals of phenotype.
†Different stocks.
‡Different experiments.

will yield incompetent cultures when the suppressor lies outside the region transduced.

From the data in table 1 and figure 2 the ratio of the transducing particles to the lambda particles in a lysate may be obtained. Lp^s recipient cultures give about one transduction per 10^6 lambda; Lp^+ recipients, one per 10^7. One per 10^6–10^7 lambda will be referred to as LFT (low frequency of transduction).

Examination of the Gal⁺ Clones Formed by Transduction

After purification the transduction clones were examined for changes at loci other than the Gal

series. A number of markers were examined, including fermentative, nutritional, and phage and drug resistance mutations. The only changes at other loci were Lp^s to Lp^+ in lambda sensitive recipients, and occasionally Lp^r to Lp^+ in lambda immune cultures. Transduction clones from Lp^+ recipients were invariably Lp^+.

To determine whether lysogeny was causally related to transduction, a reconstruction experiment was done. To a mixture of lysate and $Gal^- Lp^s$ cells, $Gal^+ Lp^s$ cells labeled with a mutant character were added to estimate the frequency of chance lysogenization in the untransformed cells in a transduction mixture. After papillae had formed, they were picked, purified, and on the basis of the differential

Table 2 The action of lysates of reverse mutants

Lp^+ recipient culture	Lysate of reversion	Numbers of Gal^+ papillae observed	
		Control (no lysate)	0.1 ml lysate
Gal_1^-	Gal_1^+ #1	0	648
Gal_2^-	Gal_2^+ #1	10	96
	Gal_2^+ #2	6	552
Gal_4^-	Gal_4^+ #5	39	204
	Gal_4^+ #8	25	291

label, divided into: (1) the inserted Gal^+, and (2) the transductions. The frequency of lysogeny was determined in the two classes, and in the Gal^- background. Whereas unchanged Gal^- cells and the inserted Gal^+ were each only 70 percent lysogenized, the transduction clones were 100 percent lysogenized (Table 3).

When Lp^r cultures were used as recipients, 14/112 (12 percent) of transduction clones formed were Lp^+. Although the fraction is small, all previous attempts to lysogenize these cultures have been unsuccessful. The isolation of transduction clones evidently selects for these cells that have been infected with lambda particles from the input lysate.

The original Gal^+ strain and spontaneous reversions of the Gal^- mutants have all been stable in ordinary culture. However, the Gal^+ clones formed by transduction are unstable for galactose fermentation as shown by the recurrence of negative and mosaic colonies (Fig. 3). Despite many serial single colony isolations the galactose-positives continue to segregate galactose negative progeny. They behave as if heterozygous for a single gene (or short chromosome segment) and may be designated as heterogenotes. Instability among the transduction clones is quite frequent; 484 of 609 clones (70 percent) were found unstable (representative data are given in Table 4). This estimate is probably low because the purification procedure acts selectively against unstable clones.

The frequency of segregation has been estimated from the incidence of Gal^- in small clones of heterogenotes. The probability of segregation per bacterial division is about 2×10^{-3} (Table 5). By repeated reisolation, however, heterogenotic lines can be maintained indefinitely.

The segregants from the heterogenotic clones were examined with regard to their Gal and Lp character.

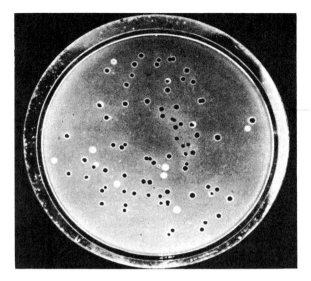

Fig. 3. EMB galactose agar plate spread with cells form a culture of a heterogenote, showing Gal^+, Gal^- and sectoring colonies.

Table 4 Frequency of instability for galactose fermentation among the transduction clones

Recipient cells	Unstable clones/ total examined	Percent unstable
$Gal_1^- Lp^s$	9/22	41
$Gal_1^- Lp^+$	40/48	83
$Gal_2^- Lp^+$	22/24	92
$Gal_4^- Lp^s$	13/24	54
Lp^+	20/24	83
Lp^r	29/48	60
$Gal_6^- Lp^s$	6/8	75
$Gal_8^- Lp^s$	28/48	58
Lp^+	16/24	67

Table 3 Comparison of the lysogenization of transformed and nontransformed sensitive strains: reconstruction experiment

Types recovered from mixture* of Lp^s bacteria and LFT lysate	Number of clones examined	Percent of clones lysogenized
Inserted $Gal^+Lac^+S^s$	46	68.5
Recipient $Gal^-Lac^-S^r$ (nontransformed)	40	72.5
Transduction $Gal^+Lac^-S^r$	103	100.

*10^8 $Gal^-Lac^-S^r$, 100 $Gal^+Lac^+S^s$, 10^9 lambda particles.

Table 5 Frequency of segregation from the heterogenotes

Heterogenote	Test clones*			Probability of segregation per 10^3 bacterial divisions†
	Number of cells in inoculum	Number of Gal⁻ cells	Total cells	
Gal_1^- _/|Gal^+	2.1‡	6	1169	1
		3	595	1
		4	251	4
		23	1252	3.6
		9	1113	2
		19	897	4.3
		103	2750	6.6
		319	1622	36.8
		22	1966	2.0
		0	237	0
Gal_2^- _/|Gal^+	1.5§	11	323	8.1
		2	176	3
		8	1669	0.9
		3	317	2
		52	1236	8.2
		0	10	0
		36	1055	6.7
		3	299	2
		6	386	4
		55	1965	5.1

*A fully grown culture in penassay broth was diluted to give about 10 cells per ml. Twenty samples of 0.1 ml were taken up in 0.2 ml serological pipettes which were supported in a horizontal position on a tray. The pipettes were incubated at 37°C for 4.5 hours. Each pipette was then blown out on to an EMB galactose agar plate, and the inside of the pipette washed with 0.1 ml of broth. The washing was added to the plate, and the inoculum spread with a glass rod. After 18 hours incubation at 37°C the number Gal^+ and Gal^- colonies on the plates was determined.

†Using the equation $a = 0.602r/N \log N$, (modified for the indicated units from Luria and Delbrück 1943) where r = the number of Gal^- segregants and N = the clone size. The probability of segregation is also estimated by the fraction of cultures containing no segregants.

$$a = \frac{2.3}{N} \log \frac{1}{P_0} \quad (P_0 = \text{fraction of cultures with no segregants.})$$

In the first experiment, using $N = 2^{10}$

$$a = \frac{2.3}{1024} \log 1/1/19 = 2.8 \times 10^{-3}$$

In the second experiment, using $N = 2^{10}$

$$a = \frac{2.3}{1024} \log 1/1/11 = 2.6 \times 10^{-3}$$

‡The assay plates showed this culture to have $Gal^+:Gal^-$ in the ratio 106:4. Of the twenty samples in this experiment, one contained only Gal^+, one contained only Gal^-, and 18, both Gal^+ and Gal^-. Only the plates that were counted are given. Nine plates were too crowded to be counted accurately.

§The ratio of $Gal^+:Gal^-$ in the parent culture was 128:19. The twenty cultures were distributed as follows: failed to grow, 9; contained only Gal^+, 1; contained both Gal^+ and Gal^-, 10. One plate had approximately equal numbers of Gal^+ and Gal^- and was assumed to have come from a mixed inoculum.

Lysates of the segregants have no transducing activity on the Gal^- culture that was used as the recipient in forming the transduction clone and are therefore allelic to it. The same lysates continue to give one transduction per 10^6–10^7 lambda (LFT ratio) on nonallelic Gal^- cultures. With different recipient cultures the Lp alleles of the segregants were (1) Lp^+ recipient, all segregants Lp^+; (2) Lp^s recipients, all segregants Lp^+; (3) Lp^r cells, the segregants were usually Lp^r. In one instance, a heterogenote segregated both Lp^+/Lp^r and Gal^+/Gal^-.

Lysates prepared from the heterogenotes have two outstanding features: (1) instead of containing 10^{10} lambda particles per ml, they seldom have titers higher than 5×10^8, particularly if they originate from cultures containing few Gal^- segregants; (2) the number of transducing particles in these lysates is often nearly equal to the number of lambda particles in the lysate (Table 6). These lysates will be referred to as HFT (giving a high frequency of transduction).

Transductions with Lysates of Heterogenotes

Platings of highly diluted HFT lysate with Lp^s and Lp^+ bacteria give a number of papillae. The number of papillae obtained with Lp^s cells is, however, less than that obtained with Lp^+. The lower yield with Lp^s recipients may result at least in part from the loss of potential transductions through lysis of the recipient cell or of some of its early progeny.

With HFT lysates it is possible to transform a large fraction of a cell population, and to observe transduction without strong selection. By adsorbing HFT lambda onto cells, diluting and plating on EMB galactose to obtain well-isolated colonies it is possible to study individual transduction clones derived from single particle infections of isolated bacteria. At the optimal ratio of about 10 lambda particles per cell, the fraction of cells transformed ranged from 5 to 15 percent.

Evidence that Lambda is the Vector of Transduction

That lambda is the vector of Gal transduction is suggested by previous experiments: (1) the 100 percent lysogenization of Lp^s recipients by LFT lysate transductions; (2) the incidence of lysogenicity in transduction to Lp^r recipients. In addition, lambda and the transducing agent are adsorbed to about the same degree by Lp^s cells, and both are inactivated by crude anti-lambda serum. More definite evidence was the failure of lambda-2 resistant cells to adsorb either lambda or transducing activity or to be transformed even by HFT lysates (Table 7).

Conclusive evidence that lambda is the vector of transduction is found from the behavior of single transduction clones: (1) Heterogenotes formed from HFT lysate and Lp^s cells at low lambda multiplicity are always either overtly lysogenic (Lp^+) or carry a defective prophage (Lp^r) (Table 8). (2) Pro-

Table 6 The high frequency of transduction (HFT) given by lysates of heterogenotes

Heterogenote	Lysate of the heterogenote		
	Lambda particles per ml	Transductions per ml	Transductions per lambda particle
Gal_1^- $//Gal^+$	1.2×10^8	2.1×10^7	1/5.7
Gal_1^- $//Gal^+$	5.8×10^8	1.8×10^7	1/32*
Gal_2^- $//Gal^+$	5.4×10^7	3.6×10^7	1/1.5
Gal_2^- $//Gal^+$	7.6×10^7	4.2×10^7	1/1.8
Gal_4^- $//Gal^+$	1.5×10^8	7.4×10^7	1/2.0
Gal_4^- $//Gal^+$	7.3×10^8	2.5×10^7	1/29*

*With the exception of these cases, the cultures used for making the lysates were started from a single apparently pure Gal^+ colony on EMB galactose. The lower ratio in the exceptional cases, and the higher lambda titer is probably the result of the presence in the source cultures of a larger number of Gal^- segregants. Assay of the transductions was made with Lp^+ cells.

portionality between number of transductions and amount of lysate at high dilution (Fig. 4). For a two-factor system to be invoked at these dilutions, the accessory factor would have to exceed the lambda by at least 10^{10}, which would imply a concentration of this fancied element in undiluted HFT lysate of 10^{18} per ml, which should be compared with Avogadro's number.

Early Segregation of Lp and Gal in Transduction Clones

HFT Gal^+ lambda was mixed with a culture of Gal_4^- bacteria to give 2.6×10^7 lambda and 7×10^8 cells per ml, a multiplicity ratio of 0.04. The suspension was then diluted and plated on EMB Gal to give about 100 cells per plate. After 24 hours incu-

Table 7   Failure of transduction to lambda-2 resistant mutants

Recipient cells (Lp^+)	Lambda-2 reaction*	Number of fermenting clones	
		No lysate added	0.1 ml of Gal^+-lysate
Gal_1^-	sensitive	1	426 (LFT)
	resistant	1	2
Gal_2^-	sensitive	20	356
	resistant	14	14
Gal_4^-	sensitive	89	296
	resistant	50	57
Gal_1^-	sensitive	2	10^7 (HFT)
	resistant	3	4

*Lambda-2 resistant mutants do not adsorb lambda or lambda-2.

Table 8   Incidence of lysogenicity in isolated heterogenotes

1. The transductions

Gal^- cells exposed to:	Number of colonies observed		
	Unaltered Gal^-	With Gal^+	Gal^- partially lysed
Broth	3280	0	0
HFT lysate*	2801	31	54

2. Examination of the colonies after exposure to HFT lysate

Colony type	Number of colonies examined	Number of colonies		
		Lp^s	Lp^+	Lp^r
Unaltered Gal^-	31	31	0	0
With Gal^+	26	0	23	3

*One ml of cell suspension (4.1×10^9 cells) was added to one ml of HFT lysate (1.2×10^9 plaques per ml, 3.0×10^8 transducing particles per ml) and the mixture incubated at 37°C for 10 minutes. The cells were then centrifuged down, the supernatant discarded, and the cells resuspended in one ml of broth. The suspension was then diluted and plated on EMB galactose agar. The tube contained 3.5×10^9 cells after HFT lysate exposure. 1.1 percent of exposed cells were transformed, and 1.8×10^7 transductions per ml were accomplished.

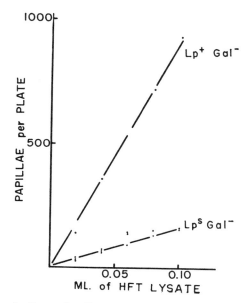

Fig. 4. Proportionality between amount of HFT lysate and number of papillae formed from Lp^s and $Lp^+ Gal^-$ cultures. For the assays, a lysate containing 1.6×10^8 phage/ml was diluted a thousandfold.

bation, on 7 plates, a total of 8 colonies with Gal^+ sectors was noted. Each of these colonies was sectored, with a large Gal^- component. Each colony was restreaked, and 20 to 30 Gal^- reisolated from each line. Of the 8 lines, the Gal^- from 3 gave only Lp^+, from 5 gave mainly Lp^s with a few Lp^+. Ten Gal^+ (heterogenote) colonies were also picked from each line. All of them were Lp^+ and of a total of 297 Gal^- segregants subsequently reisolated from these 60 heterogenotic colonies, all were Lp^+ also. The frequent segregation of Lp^+/Lp^s subclones from lambda-infected Lp^s cells has been noted previously (Lederberg and Lederberg 1953; Lieb 1953). The correlation of Lp^+ and Gal^+ evidently extended, in the 5/8 clones that segregated both markers, to the early intraclonal progeny. Since the heterogenotes do not continue to segregate Lp^s, these results are economically interpreted on the basis of the multinucleate character of the bacterial cells. The early segregation would represent the separation of unaltered $Lp^s Gal^-$ nuclei from the nucleus with which the prophage-Gal^+ complex has associated. The segregation of Gal and stability of Lp in the

heterogenotic subclones will be taken in later communications.

The Failure to Observe Transduction with Lytic Lambda

The experiments described above employed UV induced lysates. That lytic lambda, prepared by the growth of lambda on sensitive cells is incompetent in transduction is evident from the following: (1) lytic lambda failed to augment the number of papillae when added to Gal^- cells on EMB galactose agar; (2) the occasional Gal^+ clones that were found on plates to which lytic lambda was added were all stable and were presumably spontaneous reversions. The lysates used in these experiments were made by growing induced lambda from a Gal_4^- culture on a Gal^+ culture, and the initial tests of competence of the lysates were made on Gal_4^- cultures. In this way, confusion by "carry over" of the inoculum phage was avoided. The experiments were executed on a scale that should have detected as little as 3% of the activity per phage of LFT induced lysates.

Failure to Observe Transduction at Loci Other than Gal

Attempts with LFT, HFT, or lytic lysates to transduce genes at other loci were unsuccessful.

The unsuccessful tests for transductions of prototrophy to auxotrophic cultures involved: histidine; leucine (two loci); methionine; proline; glycine or serine; tryptophane.

The fermentation markers that were not transduced included: lactose (Lac_1); maltose (two loci); arabinose (two loci); xylose; glucose.

The attempt to transduce streptomycin resistance to sensitive cells was unsuccessful.

In the *E. coli* compatibility system, failure to transduce the following was noted: (1) by lysates of Hfr cultures, F^+ and F^- recipients to Hfr; and F^- recipients to F^+; (2) by lysates of F^+, F^- recipients to F^+.

The most extensive tests were made on genes at loci known to be linked to the Gal series (Hfr, histidine; Cavalli-Sforza, personal communication, and proline), or mutations other than *Gal* (W435, *Lac₃⁻*, Lederberg and Lederberg 1953 and some auxotrophs) which had occurred coincidentally with changes of Lp^+ cultures to Lp^s.

In considering the transduction of specific loci, interactive effects should be kept in mind. For example, papillae were observed on EMB lactose, arabinose, and xylose, respectively, in tests with multiple marker stocks. When purified, however, these papillae were negative for the indicated sugar, but gave galactose-positive colonies. Historically, transduction papillae were first observed in platings of a treated $Gal^- Lac^-$ culture on EMB lactose. The papillae proved to be $Gal^+ Lac^-$ rather than $Gal^- Lac^+$. Evidently, all these sugars have slight selective potentials for Gal^+ clones.

Other Observations

Most lambda lysates are viscous when first obtained. The viscosity is destroyed: (1) by DNAase, an indication that DNA is the cause of viscosity; (2) spontaneously at a slow rate. Exposure of lambda lysates to DNAase has not affected either transduction or plaque titers.

Transduction of the *Gal* gene is not restricted when either the donor or the recipient culture is (1) a prototroph or any of a variety of auxotrophs; (2) Hfr, F^+ or F^-, in any combination. Transduction is controlled (1) by the method of lysate production, and (2) the ability of the recipient cells to adsorb lambda. The only genes transduced are the *Gal* loci.

Gal^- mutants in *E. coli* strains other than K-12 that adsorb lambda can be transformed. As in strain K-12 the transformation does not require that the recipient be sensitive; among the susceptible strains are lambda sensitives, lambda immunes, and host modifiers of K-12 lambda (E. Lederberg 1954). However, lambda was incompetent when tested on galactose negative mutants of Salmonella, and transducing Salmonella phage (Zinder and Lederberg 1952) failed to transform *E. coli*.

DISCUSSION

Galactose-negative cultures of *E. coli* are transformed to galactose-positive by certain lysates containing the phage lambda. That this process is genetic transduction by lambda particles is established by the following: (1) Gal_y^- cells are transformed to Gal^+ by lysates of Gal_y^+ cultures but not by Gal_y^-. (2) However, Gal_y^+ obtained by reversion regains its ability to transform the Gal_y^-, which emphasizes the role of the donor genotype in effective transformation. (3) The transformed positives are unstable, and segregate Gal_y^- and not other galactose types. The various "Gal_y" used for these experiments include Gal_1, Gal_2, Gal_3, Gal_4, Gal_6, Gal_7 and Gal_8. (4) All transduction clones obtained from Lp^s recipients become lysogenic for lambda (either Lp^+ or Lp^r). (5) Transduction is not obtained with cells unable to adsorb lambda.

The contrasting features of the *E. coli*-lambda and the Salmonella systems of transduction are summarized below.

The two systems are alike in the following respects: (1) Genetic factors are carried by phage particles; (2) The specificity of the transducing particles is determined by the genetic content of the donor bacteria, in contrast to lysogenic conversions (Uetake, et al. 1955); (3) The genetic material is inaccessible to DNAase and other enzymes; (4) Trans-

Range of genes transduced	*E. coli* K-12 phage lambda only *Gal*	Salmonella phage PLT 22 any selectable marker
Localization of prophage	*Lp* locus linked to *Gal*	Unknown
Competence of lytic phage	No	yes
Transduction clones	unstable heterogenotes	stable
Efficiency of transduction, per phage	LFT 10^{-6} HFT 10^{-1}	10^{-5}–10^{-6}
Sexual fertility of the host	Fertile, subject to F compatibility system	Unknown

duction occurs without regard (except for quantitative changes in yield) to the lysogenic or sensitive status of the recipient cells. In both systems UV induced phage is competent, but lytic phage is competent only in Salmonella.

However, the two systems evidently do not cross-react; lambda does not transform Salmonella and conversely, probably because of the specificity of phage adsorption.

Several of these features may be related in origin. For example, the limitation both on the mode of inclusion in the phage (i.e., only after induction of a lysogenic bacterium), and on the genetic material that can be transduced suggest that the physical proximity of the *Gal* loci to the prophage site determines transduction competence of lambda. This is supported by the linkage observed in crosses of *Lp* to *Gal*. Presumably the linked *Gal* genes may sometimes accompany the prophage into the maturing lambda particle when lysogenic bacteria are irradiated. The failure to obtain lambda particles with transducing activity when the phage is grown lytically on sensitive cells would be explained on this hypothesis, since the lambda may have no specific association with the *Lp-Gal* chromosomal segment during lytic growth.

The heterogenotic clones which result from transduction are isolated through the effectiveness of the *Gal* genes that accompany the prophage. In LFT transductions, this is a rare event; the HFT quality of lysates from heterogenotes may result in part from the prior selection of an effective fragment and its reproduction as such in the growth of the clone.

The persistence of the fragment in transduction clones requires an *ad hoc* explanation, possibly related to the presence of an *Lp* region in the fragment. For example, *Lp* might be closely linked to a centromere; it may function as a centromere itself; it may be adapted to synapse with the homologous site of an intact chromosome.

At any rate, the *Lp* region is singular in at least two respects: it is close to a regular point of breakage in crosses determined by *F* polarity (Lederberg and Lederberg 1953; Nelson and Lederberg 1954; Cavalli-Sforza and Jinks 1956) and the *Lp* segment (considered as prophage) is capable of independent replication as a phage. If comparable singular regions exist in Salmonella, they have not yet been revealed in the occurrence of heterogenotes.

The occurrence of sexual recombination and trans-

duction in the same organism raises the technical question of their experimental confusion. Since sexual recombination requires intact cells, and transduction is accomplished with a cell-free lysate, sexual recombination can have no direct bearing on transduction experiments. Furthermore, although crossing is completely blocked between *F⁻* cultures, the compatibility status has no effect on transduction. On the other hand, the rarity of LFT transduction makes it *a priori* unlikely that transduction will significantly interfere with segregation ratios in crosses.

Crosses of the various combinations of cultures carrying different *Lp* alleles will be presented in detail in further reports. However, they have indicated that combinations involving *Lp⁺* (where transduction could occur) do not give appreciably different frequencies of *Gal⁺* than *Lpˢ* × *Lpˢ* crosses (where lambda transduction is not possible). In addition, the *Gal⁺* prototrophic recombinants obtained are stable for galactose fermentation. Even crosses of known heterogenotes (capable of HFT lambda) have not given increased frequencies of *Gal⁺*. These observations suggest that lambda transduction has not significantly affected results obtained by crossing.

The mosaic colonies of heterogenotic cultures (Fig. 3) are reminiscent of those formed by segregating diploids of E. *coli* K-12. The latter, of course, are segregating blocks of many linked markers, not merely the *Gal* genes. Diploids are, however, more difficult to maintain without the benefit of balanced selective markers. They segregate twenty times as frequently as heterogenotes, as can be judged from the appearance of the colonies and from rates calculated from cell pedigrees (Zelle and Lederberg 1952 and unpublished).

Further studies involving the use of two or more *Gal* markers, and relating transduction to sexual recombination analysis will be presented shortly, together with further consideration of the genetics of the prophage.

SUMMARY

Transduction of several *Gal⁺* genes from galactose positive (*Gal⁺*) to galactose negative cells (*Gal⁻*) by the bacteriophage lambda has been demonstrated. The resultant galactose positive clones have been found to be heterozygous for the *Gal* region and have been designated as heterogenotes ($Gal^- \_/\!\!/ Gal^+$). Segregation and the reappearance of Gal^- from the heterogenotes occurs about once per 10^3 bacterial divisions. The low frequency of lambda particles with *Gal* genes ($1/10^6$) from haploid cultures resembles other transduction systems. However, heterogenotic cultures produce lysates in which nearly every lambda particle carries *Gal* genes. No other markers have been transduced by lambda, and the competence of lambda in transduction depends upon its production from lysogenic cells, rather than by lytic growth on sensitive bacteria.

LITERATURE CITED

Cavalli-Sforza, L. L., and J. L. Jinks, 1956. Studies on the genetic system of *E. coli* K-12. J. Genet. In press.

Lederberg, E., 1950. Genetic control of mutability in the bacterium *Escherichia coli*. Doctoral Dissertation, University of Wisconsin.

———— 1952. Allelic relationships and reverse mutation in *Escherichia coli*. Genetics 37:469–483.

———— 1954. The inheritance of lysogenicity in interstrain crosses of *Escherichia coli*. Genetics 39:978.

Lederberg, E., and J. Lederberg, 1953. Genetic studies of lysogenicity in *Escherichia coli*. Genetics 38:51–64.

Lederberg, J., 1950. Isolation and characterization of biochemical mutants of bacteria. Methods in Medical Research 3:5–22.

Lieb, M., 1953. The establishment of lysogenicity in *Escherichia coli*. J. Bacteriol. 65:642–651.

Luria, S. E., and M. Delbrück, 1943. Mutations of bacteria from virus sensitivity to virus resistance. Genetics 28:491–511.

Morse, M. L., 1954. Transduction of certain loci in *Escherichia coli* K-12. Genetics 39:984.

Nelson, T. C., and J. Lederberg, 1954. Postzygotic elimination of genetic factors in *Escherichia coli*. Proc. Nat. Acad. Sci. U.S. 40:415–419.

Uetake, H., T. Nakagawa, and T. Akiba, 1955. The relationship of bacteriophage to antigenic changes in salmonellas of Group E. J. Bacteriol. 69:571–579.

Weigle, J. J., and M. Delbrück, 1951. Mutual exclusion between an infecting phage and a carried phage. J. Bacteriol. 62:301–318.

Zelle, M. R., and J. Lederberg, 1951. Single-cell isolations of diploid heterozygous *Escherichia coli*. J. Bacteriol. 61:351–355.

Zinder, N., and J. Lederberg, 1952. Genetic exchange in Salmonella. J. Bacteriol. 64:679–699.

Genetic recombination between λ prophage and irradiated λ *dg* phage *

ALLAN CAMPBELL†

Reprinted by permission of the author and Academic Press Inc. from Virology, *vol. 23, 1964, pp. 234–251.*

Campbell with Arber, Kellenberger, Weigle, and others, was a pioneer in λ transduction studies. By the time this paper appeared, the structure of λ *dg* was fairly well understood. At least it was known that λ genes had been replaced by *E. coli* genes. The issue now became one of describing how λ *dg* formed and how it integrated with the chromosome of the host cell.

Our understanding of λ *dg* formation and its association with the *E. coli* chromosome comes in large part from Campbell's work. In this paper, Campbell explores an earlier finding that the gene sequence in a λ prophage differs from the gene sequence of a vegetative λ chromosome by a circular permutation. This led to the idea that λ has a linear chromosome that circularizes upon entering an *E. coli* cell. Campbell notes how the interaction of two circular chromosomes (*E. coli* and λ) lead to lysogeny and λ *dg* formation.

Galactose-negative cells of *Escherichia coli* made lysogenic for multiple mutants of phage λ have been exposed to λ *dg* irradiated at various doses up to about 8 phage lethal hits. The pattern of rescue of phage markers among the heterogenotic transductants indicates that recombination occurs according to the genetic map of the prophage rather than that of the vegetative phage. The proportion of transductants that are heterogenotic rather than stable depends on the extent of the *dg* region, especially at high doses of UV. Heterogenote formation is lower and more highly UV sensitive with λ *dg* mutants that have few or no known genes at the left end of the lineom. From this we conclude that the free λ *dg* particle probably has physical ends corresponding to the ends of the vegetative map, and that the ends become joined prior to lysogenization and marker rescue.

INTRODUCTION

Results obtained in several laboratories during the last few years (Calef and Licciardello 1960; Campbell 1963a; Rothman, J., personal communication) indicate that the genetic order of λ prophage differs from that of vegetative λ phage by a circular permutation. In order to get one map from the other, one must join the two ends and then cut open the resulting circle at another place. It is simplest to imagine that this formal genetic result reflects an actual joining and cutting of a physical structure. We know nothing about the mechanism of either the joining or the cutting process.

The present paper will describe some observations that may be relevant. The experiments are an extension of the findings of Arber (1958) on the effect of ultraviolet light (UV) on transduction of the galactose genes by phage λ.

A high frequency transducing (HFT) lysate contains two types of particles, active (plaque-forming) λ and transducing λ (λ *dg*). On infecting a sensitive recipient, λ *dg* can produce two kinds of transductants: (1) stable types, in which the galactose genes of the λ *dg* have replaced their homologues in the

*Supported by grant E-2862 of the United States Public Health Service, Division of Allergy and Infectious Diseases. Accepted February 14, 1964.
†Research Career Awardee of the U. S. Public Health Service. Department of Biology, University of Rochester, Rochester, New York.

lineom of the bacterial recipient; and (2) hetero-genotes, in which the λ dg phage has lysogenized the recipient and constitutes an addition to the genome rather than a replacement. Operationally, the two types are distinguished by the presence of a few Gal^- segregants in any culture of a heterogenote.

When λ dg is irradiated, it rapidly loses its ability to form heterogenotes. The data are compatible with the notion that a single UV hit which would inactivate the plaque-forming ability of an active λ particle will render λ dg incapable of lysogenizing. The formation of stable transductants is much less UV sensitive. When an HFT lysate is irradiated for various doses, the number of transductants first increases by a large factor at low doses and then decreases with a slope about one-third that for active λ phage (see Fig. 2). The reason for the increase is not clearly understood. Under conditions of single infection, the efficiency of transduction by λ dg is very low, and UV presumably enhances this efficiency by reducing the number of alternative fates (e.g., lysis) of the infected cell. In any case, essentially all the transductants found after the increase are stable rather than heterogenotic. We can reasonably imagine that once a λ dg particle has been hit it can neither lyse the cell nor lysogenize it, and therefore forms a stable transductant by default.

This idea is supported by Arber's results when he used as recipient a culture already lysogenic for λ. The curve for total transductants versus dose was almost the same as for a nonlysogenic recipient. However, the large majority of transductants, even at fairly high doses, were then heterogenotic rather than stable. Somehow, the presence of the intact phage genome of the prophage helps to rescue the transducing phage as a complete structure, which can not only transduce, but also lysogenize. This is apparently the result of a recombination that rescues the Gal part of the λ dg away from the radiation damages.

The present investigation has extended Arber's experiment by using different λ dg stocks of known constitution and employing a recipient whose prophage carries several genetic markers, thus showing which parts of the λ genome are most frequently rescued in conjunction with the Gal region. It will be seen that the genetic map constructed from such marker rescue data is that of the prophage rather than that of the vegetative phage.

MATERIALS

The biological materials are diagrammed in Fig. 1. They consist of a collection of donor and recipient stocks.

Donor Stocks   These are stocks of the transducing phage λ dg. It is generally preserved as a prophage, HFT lysates being made by induction and super-infection. λ dg is a derivative of λ that has picked up the galactose genes of the host and has simultaneously lost a connected block of phage genes. The genes in this block comprise the dg region. λ dg stocks differ one from another in the extent of the dg region. The position of the right end has not been observed to vary, but the left end may be in any one of at least 30 different places. As some of the properties of λ dg will be seen to depend on the position of this left end, we have in this paper designated each λ dg by a number corresponding to the region in which its end point falls (Campbell 1963b). λ dg_1 is missing all known genes on the left end of the vegetative map of λ (Campbell 1964). As all the λ dg mutants were made from wild-type λ, they contain the wild-type alleles of all mutants outside the dg region. As the genes of the dg region are missing, it is operationally equivalent to a deletion, and it will sometimes be convenient to refer to it as one.

Recipient Stocks   These are derivatives of the $Gal_1 Gal_2$ strain W3350 that are lysogenic for multiple mutants of λ. The multiple mutants were generally synthesized as prophages by transduction and segregation (Campbell 1963a). We have used throughout suppressor-sensitive mutants, recognizable by their ability to form plaques on strain C600 but not on strain W3350. The mutants score as recessive to their wild-type alleles in a double lysogen, so that a transductant can be analyzed for all those phage genes it has acquired from the donor. The composition of the prophages and the position of the mutants on the genetic map of vegetative λ are shown in Fig. 1.

The first two recipients shown differ in the one marker sus_7. The quadruple mutant was synthesized first and used in the earlier experiments (Tables 2, 4, 6, and 10). The quintuple mutant was used later with λ dg_1, λ dg_2, λ dg_3, and λ dg_4 as donors (Ta-

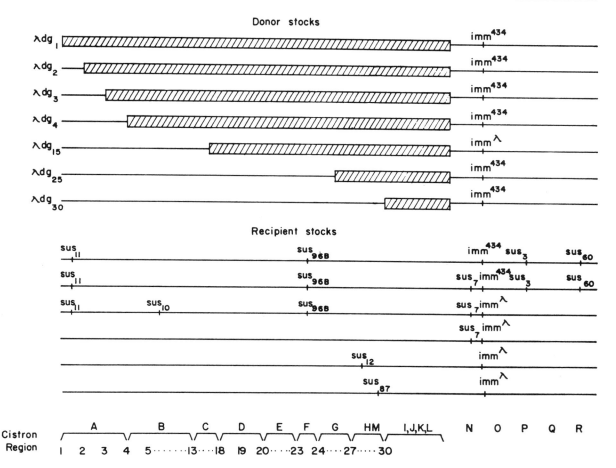

Fig. 1. Diagrammatic representation of donor and recipient stocks. Each donor carries a different type of λ *dg* prophage, all originally derived from wild-type λ or λ *imm*^{434}. The extent of the *dg* region is shown by crosshatching. The recipients carry multiple mutant prophages of the indicated constitution. Cistrons and regions are defined as previously (Campbell 1961, 1963b). The map is drawn with the vegetative order. Only order is indicated; distances are not drawn to scale.

bles 7, 8, 9, and 11). With these donors, sus_{96B} is unusable because it falls within the *dg* region. The third recipient (sus_{11} sus_{10} sus_{96B} sus_7 *imm*λ), kindly provided to us by June Rothman, was used with λ dg_{15} as donor (Table 5). The single marker stocks sus_7, sus_{12}, and sus_{87} were employed in one experiment with λ dg_{30} (Table 3). Except for this one experiment, the immunity specificity was the same in donor and recipient. This avoids the many complications of heteroimmune transductions.

METHODS AND CONTROLS

For general procedures, composition of media, scoring of mutant genotypes, etc., the reader is referred to earlier papers (Campbell 1959, 1961). Here we shall confine ourselves to certain technical points about the present experiment.

Double Lysogeny of Heterogenotic Transductants
Arber (1958) showed that transduction of a lyso-

genic recipient by a UV-irradiated λ *dg* generally produced lysogenic heterogenotes. As they are heterogenotes, they have become lysogenized by λ *dg*; as they are overtly lysogenic, they still carry the original prophage. They are thus double lysogens. If the λ *dg* were to replace the prophage rather than to add to the recipient genome, the resulting heterogenote would be a defective lysogen.

In our system, the recipients are lysogenic, but the prophage carries mutations that render it incapable of growth on the strain by which it is carried. Except where the wild alleles of all these markers are brought in by the donor, the transductants are thus defective lysogens in any event. To test whether the original prophage has been lost, we look in this case for the wild alleles of genes from the *dg* region. These cannot derive from the donor, because they are not represented in λ *dg* at all. We have therefore tested all transductants and segregants for the marker $sus_6{}^+$, which is in the *dg* region of all known λ *dg* mutants. In all the homoimmune transductions performed, 5/3065 heterogenotic transductants and 5/1776 stable transductants had lost the $sus_6{}^+$ gene. This compares with 2/601 and 0/131 in the unirradiated controls. The nature of these exceptional transductants was verified by further testing of them and their segregants for immunity and for other genes in the *dg* region. Thus, substitutions of λ prophage by λ *dg* can occur, but they are rare and are not significantly increased by UV.

Such substitutions are more common in heteroimmune transductions where the immunity specificity of λ *dg* differs from that of the prophage (Campbell and Balbinder 1959). In the present series, some of the transductions shown in Table 3 were heteroimmune. In the non-UV'd controls, we found that 3/3 stable and 18/57 heterogenotic transductants had lost the λ prophage; in the UV'd samples (all exposed to 8 minutes of UV), the figures were 1/112 stables and 0/188 heterogenotes. So, even in heteroimmune transductions, substitution scarcely occurs with UV'd λ *dg*.

Thus, although we shall show that the λ *dg* appearing in the transductants from a UV'd lysate is usually a recombinant, the parent prophage which contributed to its formation is not consumed in the process but remains.

Transductions   For all transductions, an HFT lysate

was diluted 1:10 in 0.01 *M* MgSO₄, and irradiated the desired length of time. Then 0.1–0.4 ml of an appropriate dilution was added to 0.9–0.6 ml of recipient cells which had been grown with aeration to near saturation and resuspended at a 1:3 dilution in 0.01 *M* MgSO₄. After 20 minutes' adsorption, 0.1–0.25 ml was plated together with 0.1 ml anti-λ serum ($K = 30$–200 min⁻¹ on eosin methylene blue agar. Transductants were counted after 2–4 days and purified by two restreakings. From the second restreaking, the transductant could be classified as stable or heterogenotic. From each heterogenotic transductant, a *Gal⁻* segregant, as well as a *Gal⁺* colony, was picked from the second plate. (See Results: Marker Rescue Experiments.)

The multiplicity of active phage (assayed at zero

Fig. 2. Survival of transducing and of plaque-forming ability as a function of UV dose. A GE germicidal lamp was used at a distance of 50 cm, giving a dose rate of about 966 erg min⁻¹ mm⁻². After irradiation, one aliquot was adsorbed 15 minutes to sensitive bacteria and plated for plaque count; a second aliquot was adsorbed 20 minutes to *Gal⁻* bacteria and spread on EMB galactose agar with anti-λ serum. Transduction plates were counted in 2–4 days.

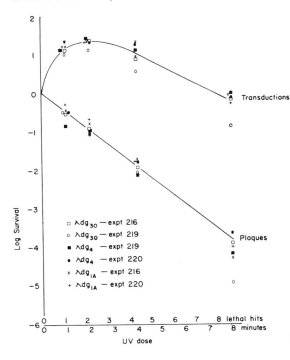

dose) added to the recipients was in no case greater than 10^{-2}. With a homoimmune recipient, there is no helping effect, with or without UV, and we have no reason to expect multiplicity reactivation here. However, if there were some multiplicity reactivation, or if the serum were not sufficiently strong or well mixed to completely prevent phage growth on the assay plates, there is the possibility that the active phage in the donor lysate might be contributing genetic markers to a small fraction of the transductants, and this might color our results considerably in certain cases.

The best control on this point comes from experiments in which the recipient carries one or more mutations within the dg region of the donor; e.g., $\lambda\ dg_4$ on a recipient whose prophage is sus_{96B} or $\lambda\ dg_1$ on one that is sus_{11}. If active phage particles from the donor lysate are contributing phage genes to the transductants, they should also contribute genes from the dg region, which $\lambda\ dg$ cannot. Of 200 transductants produced at a high UV dose (8 minutes), none had picked up any donor alleles from the dg region. If a transductant is unlikely to have been mixedly infected by λ and $\lambda\ dg$, it is likewise unlikely to have been multiply infected with two $\lambda\ dg$ particles, and we conclude that most (probably all) of the transductants we observe are each derived from a single infection by one irradiated $\lambda\ dg$ particle.

RESULTS

UV Sensitivity of Transduction

The effect of UV dose on the transducing ability of some HFT lysates is shown in Fig. 2. The general shape of these curves is similar to that found by Arber (1958). We have included three $\lambda\ dg$ mutants, each of which was measured in two separate experiments. These include the extreme types $\lambda\ dg_1$ and $\lambda\ dg_{30}$. There is some variability between experiments, but no consistent difference between one $\lambda\ dg$ and another. If one examines such curves for other $\lambda\ dg$ mutants, one sees no trend to indicate that the length of the dg region influences the shape of this curve.

In the bottom part of Fig. 2 is shown the survival of free phage in each lysate. These curves should be identical and provide an internal standard on the

constancy of the dose rate. As precise dosimetry is not critical to our argument, the agreement observed is quite adequate. From the slope of this curve, the UV dose can be converted from minutes to phage-lethal hits, as shown on the abscissa below.

Marker Rescue Experiments

The first experiments were done with the shortest deletion available ($\lambda\ dg_{30}$), the recipient strain W3350 ($\lambda\ sus_{11}\ sus_{96B}\ imm^{434} - sus_3\ sus_{60}$) and a high dose (8 minutes) of UV. The results are shown in the first column of Table 1. Only 2% of the heterogenotic transductants have a complete donor-type phage genome, and 63% have picked up from the donor none of the four genes tested.

Our aim at the outset was to construct a linkage map from such marker rescue data. If the rescue is recombinational, it will be a region which is rescued rather than a point. As we are selecting for rescue of Gal, the probability that a phage marker will be in the rescued region will depend on its proximity to Gal. Furthermore, the joint rescue of a phage marker and Gal should entail the simultaneous rescue of all other phage markers between it and Gal on the linkage structure.

The results of Table 1 were discouraging in that they suggested only a feeble linkage between Gal and one phage marker (sus_{96B}). Also, single marker rescues are much more common than multiple marker rescues, regardless of the single marker concerned. This suggests that a large fraction of the observed marker rescues were not true joint rescues with Gal, in the sense of a single event that rescues both Gal and the marker in question together with all intervening material, but rather resulted from a fairly high probability of independent cross reactivation. To verify this interpretation, we picked from each transductant one Gal^- segregant to see whether the donor markers were lost along with the Gal^+ gene. If they have been acquired as a true joint rescue, they should comprise part of a $\lambda\ dg$ prophage, which is usually lost as a unit in segregation (Campbell 1963a). If they have come from an independent cross reactivation, they may have become part of the nontransducing phage (or of a second active phage produced by the cross reactivation), in which case they should seldom be lost at segregation. The results (Table 1, column 2) show

Table 1   Transduction of W3350 ($\lambda sus_{11}sus_{96B}imm^{434}sus_3sus_{60}$) by λdg_{30} irradiated 8 minutes

Gal+ transductants		Gal− segregants		Corrected* No.
Type	No.	Type	No.	
$+ + + +$	8	$+ + + +$	5	
		$+ sus_{96B}sus_3 +$	1	
		$+ sus_{96B}sus_3sus_{60}$	1	
		$sus_{11}sus_{96B}sus_3sus_{60}$	1	1.5
$+ + + sus_{60}$	2	$+ + + sus_{60}$	1	
		$+ sus_{96B} + sus_{60}$	1	
$+ + sus_3 +$	2	$+ + sus_3sus_{60}$	1	
		$sus_{11} + sus_3sus_{60}$	1	
$+ sus_{96B} + +$	1	$+ sus_{96B} + +$	1	
$+ + sus_3sus_{60}$	12	$+ + sus_3sus_{60}$	7	
		$+ sus_{96B}sus_3sus_{60}$	3	
		$sus_{11} + sus_3sus_{60}$	1	
		Not scored	1	
$+ sus_{96B} + sus_{60}$	5	$+ sus_{96B} + sus_{60}$	4	
		$+ sus_{96B}sus_3sus_{60}$	1	
$sus_{11} + + sus_{60}$	10	$sus_{11} + + sus_{60}$	5	
		$sus_{11} + sus_3sus_{60}$	1	
		$sus_{11}sus_{96B} + sus_{60}$	2	
		$sus_{11}sus_{96B}sus_3sus_{60}$	1	1.4
		Not scored	1	
$sus_{11} + sus_3 +$	2	$sus_{11}sus_{96B}sus_3 +$	1	
		$sus_{11}sus_{96B}sus_3sus_{60}$	1	1.3
$sus_{11}sus_{96B} + +$	1	$sus_{11}sus_{96B}sus_3 +$	1	
$+ sus_{96B}sus_3sus_{60}$	12	$+ sus_{96B}sus_3sus_{60}$	7	
		$sus_{11}sus_{96B}sus_3sus_{60}$	5	6.0

*Corrected for bias introduced by selecting only those individuals which produced a recipient type Gal− segregant.

clearly that much independent cross reactivation occurs. For example, of 41 transductants that have become $sus_{11}{}^+$, only 9 have lost this allele on becoming Gal−. From previous experience (Campbell 1963a), we know that from an individual carrying the two prophages $\lambda dg sus_{11}{}^+$ and λsus_{11} (the expected product of a true joint rescue) 83% of the segregants should be sus_{11}.

To detect genetic linkages from marker rescue data, it is desirable to reduce the noise by tabulating only those individuals where a true joint rescue has occurred. It is not feasible, however, to analyze each transductant completely, i.e., to look at many segregants from it, and a single segregant may have failed by chance to segregate a particular marker.

The most workable procedure we have found is the following:

For every transduction performed, we proceed as in Table 1 to study a number of transductants and to pick one Gal− segregant from each heterogenotic transductant. We then tabulate only those transductants which have segregated on the first try all donor alleles and again carry a pure recipient-type prophage. This eliminates much of the noise due to independent cross reactivation. It also introduces some bias into the totals, because we have thrown away some individuals which did arise by true joint rescue but which happened not to segregate one or more of the donor alleles acquired. The probability that this will occur depends on the gene or genes

Table 1 continued

Gal$^+$ transductants		Gal$^-$ segregants		Corrected* No.
Type	No.	Type	No.	
sus_{11} + sus_3sus_{60}	39	sus_{11} + sus_3sus_{60}	11	
		$sus_{11}sus_{96B}sus_3sus_{60}$	25	28.5
		Not scored	3	
$sus_{11}sus_{96B}$ + sus_{60}	19	$sus_{11}sus_{96B}$ + sus_{60}	14	
		$sus_{11}sus_{96B}sus_3sus_{60}$	4	5.6
		Not scored	1	
$sus_{11}sus_{96B}sus_3$ +	8	$sus_{11}sus_{96B}sus_3$ +	6	
		$sus_{11}sus_{96B}sus_3sus_{60}$	1	1.3
		Not scored	1	
$sus_{11}sus_{96B}sus_3sus_{60}$	212	$sus_{11}sus_{96B}sus_3sus_{60}$	209	212
		Not scored	3	
Not scored	1	$sus_{11}sus_{96B}sus_3sus_{60}$	1	
$sus_{11}sus_{96B}sus_3sus_{60}$ (defective)	1	Sensitive	1	
+ + + +	1			
+ + + sus_{60}	1			
+ sus_{96B} + +	1			
+ + sus_3sus_{60}	1			
sus_{11} + sus_3 +	1			
$sus_{11}sus_{96B}$ + +	2			
+ $sus_{96B}sus_3sus_{60}$	6			
sus_{11} + sus_3sus_{60}	11			
$sus_{11}sus_{96B}$ + sus_{60}	8			
$sus_{11}sus_{96B}sus_3$ +	1			
$sus_{11}sus_{96B}sus_3sus_{60}$	127			
Sensitive	2			
Total	497			

(Gal$^+$ transductant rows from "+ + + +" through "Sensitive 2" bracketed as **stable**)

concerned and can be calculated from segregation data (Campbell 1963a). The corrected numbers are shown in column 3 of Table 1. The correction is not very accurate. Some of the segregation data come from different λ dg mutants, and we will have occasion below to use some markers that were not employed in the segregation experiments, and for which we have estimated the correction factor from the behavior of other linked markers. It will turn out that the conclusions do not depend very much on the precise values of these correction factors.

From column 3 of Table 1, we see that the rescue of sus_{96B} is more common than that of any other marker. The linkage is weak, and no linkage among other markers is ascertainable. More positive information on linkages could in principle be obtained by (a) lowering the UV dose, (b) employing a recipient with markers closer to the left end of the dg region, or (c) using a different λ dg whose dg region terminates closer to sus_{96B}.

All three approaches have been employed; the data, sorted and corrected as indicated above, are tabulated in Tables 2–11. The data extracted from Table 1 are repeated in Table 2, column 5. Table 2 shows the effect of different UV doses with the same donor and recipient used in Table 1. The second row of Table 2 gives the total number of individuals counted. Of these, only that percentage

Table 2 Transduction of W3350 ($\lambda\, sus_{11}sus_{96B}imm^{434}\, sus_3sus_{60}$) by $\lambda\, dg_{30}$

UV dose (minutes)	0	1	2	4	8	Total	Corrected† total	X-over‡
Number	249	250	250	250	500	1499		
% Hets	92%	96%	89%	79%	68%			
% Hets comp. seg.*	36%	42%	42%	55%	76%			
$+ + + +$	71	40	28	3	1	143	215	Parental
$+ + + sus_{60}$	0	0	1	0	0	1	1.5	CD
$+ + sus_3 +$	1	13	3	4	0	21	30.3	D
$+ sus_{96B} + +$	0	4	0	1	0	5	7.4	AB
$sus_{11} + + +$	0	0	1	0	0	1	1.5	BC
$+ + sus_3 sus_{60}$	2	2	5	1	0	10	13.7	C
$+ sus_{96B} + sus_{60}$	0	0	0	1	0	1	1.4	ABCD
$sus_{11} + + sus_{60}$	1	0	4	3	1	9	12.8	BD
$sus_{11} + sus_3 +$	0	0	0	0	1	1	1.3	BCD
$sus_{11}\, sus_{96B} + +$	1	1	0	0	0	2	2.8	AC
$+ sus_{96B} sus_3 sus_{60}$	0	1	5	4	5	15	18.0	ABC
$sus_{11} + sus_3 sus_{60}$	3	9	10	17	25	64	72.9	B
$sus_{11} sus_{96B} + sus_{60}$	0	4	4	5	4	17	24.8	AD
$sus_{11} sus_{96B} sus_3 +$	0	0	0	0	1	1	1.3	ACD
$sus_{11} sus_{96B} sus_3 sus_{60}$	1	26	29	67	212	335	335	A
	80	100	90	106	250	626	740	

*Percentage, among the heterogenotes, of individuals from which the Gal^- segregant examined was pure recipient type.

†Corrected for bias introduced by selecting only those individuals which produced a recipient type Gal^- segregant.

‡Crossovers on basis of map with prophage order:

$$Gal^A sus_{96B}{}^B sus_{11}{}^C sus_{60}{}^D sus_3$$

shown in the third row were heterogenotes. The properties of the stable transductants are not included in the table. Among the heterogenotes, only the fraction stated in the fourth row segregated out all the donor markers on the first try. In column 6, the data from all doses have been totaled, and the corrected totals are given in column 7.

Table 2 shows clearly the effect of UV dose on marker rescue. At zero dose, almost all transductants have acquired the whole $\lambda\, dg$ genome. At the highest dose used, most of the transductants have acquired only the Gal region from the donor phage. The experiment can be considered as a cross between the $\lambda\, dg$ and the λ prophage, where the effect of increasing UV dose is to augment the recombination frequency. In other words, we have a cross in which the recombination frequency can be varied at will. It is clear that for ordering a given group of genes a certain level of recombination will be optimal, i.e., will give the most information for a given number of individuals scored. At other levels the same gene

order will be obtained, but with more work. It is therefore permissible to pool the data from different doses, as we have done in column 6 of Table 2, a strategy which may even be superior in some cases to working entirely at a single optimal dose. We shall work with pooled data wherever we are concerned only with the determination of gene order. It must be emphasized that one cannot calculate from such pooled data values for genetic distance that have any meaning in the comparison of one $\lambda\, dg$ with another.

The data of Table 2 indicate the genetic map

$$Gal\text{-}sus_{96B}\text{-}sus_{11}\text{-}sus_{60}\text{-}sus_3$$

the genes appearing in the *prophage* rather than in the *vegetative* order. In the last column, the transductants are classified into single or multiple crossover types according to this map. It is seen that the single crossover types are consistently more numerous than the multiple crossover types.

Table 3 shows the results obtained with the same

Table 3 Survival of single markers from $\lambda \, dg_{30}$ irradiated 8 minutes

Marker	Frequency*		Corrected frequencies‡		Survival§
	0 UV	8 min. UV	0 UV	8 min. UV	
sus_{87}	−†	0.63 (35/51)	—	0.66	0.66
sus_{12}	0.73 (19/26)	0.44 (25/57)	0.76	0.47	0.62
sus_{96B}	0.98 (78/80)	0.14 (36/266)	0.98	0.16	0.16
sus_7	0.45 (9/20)	0 (0/70)	0.48	0	0

*Frequency of acquisition of wild-type allele of a marker, among those individuals from which the Gal^- segregant examined was mutant for that marker.
†Not measured.
‡Defined as $fc/(1 - f + fc)$ where f is the measured frequency and c, the correction factor, is the reciprocal of the segregation frequency for the marker in question.
§(Corrected value at 8 minutes):(corrected value at 0 minutes).

Table 4 Transduction of W3350 ($\lambda \, sus_{11}sus_{96B}imm^{434}sus_3sus_{60}$) by $\lambda \, dg_{25}$

UV dose (minutes)	0	1	2	4	8	Total	Corrected† total	Cross-over‡
Number	100	93	100	100	160	553		
% hets	93%	95%	85%	81%	71%			
% hets comp. seg.*	56%	65%	54%	42%	73%			
$+\ +\ +\ +$	49	39	20	2	2	112	168	Parental
$+\ +\ +\ sus_{60}$	0	0	0	1	1	2	3.0	CD
$+\ +\ sus_3\ +$	2	5	2	2	1	12	16.8	D
$+\ sus_{96B}\ +\ +$	0	1	0	1	0	2	2.9	AB
$sus_{11}\ +\ +\ +$	0	1	2	0	0	3	4.4	BC
$+\ +\ sus_3sus_{60}$	0	0	2	0	1	3	4.1	C
$sus_{11}\ +\ +\ sus_{60}$	0	1	2	2	1	6	8.5	BD
$+\ sus_{96B}sus_3sus_{60}$	0	0	1	2	1	4	4.8	ABC
$sus_{11}\ +\ sus_3sus_{60}$	1	6	12	14	40	73	83.1	B
$sus_{11}sus_{96B}\ +\ sus_{60}$	0	1	1	0	0	2	2.7	AD
$sus_{11}sus_{96B}sus_3sus_{60}$	0	3	4	10	35	52	52.0	A
	52	57	46	34	82	271	350	

*Percentage, among the heterogenotes, of individuals from which the Gal^- segregant examined was pure recipient type.
†Corrected for bias introduced by selecting only those individuals which produced a recipient-type Gal^- segregant.
‡Crossovers on basis of map with prophage order:

$$Gal^A\, sus_{96B}{}^B\, sus_{11}{}^C\, sus_{60}{}^D\, sus_3$$

$\lambda \, dg$ irradiated at the same dose when recipients carrying other mutants were used. This is equivalent to a group of two-factor crosses in each of which the distance between Gal and a given prophage marker is measured. They suggest the order

$$Gal\text{-}sus_{87}\text{-}sus_{12}\text{-}sus_{96B}\text{-}sus_7$$

although these data by themselves would not re-quire us to place all the phage markers on the same side of Gal.

Tables 4–11 show the results obtained with other $\lambda \, dg$ mutants. The linkage between the terminal markers of the vegetative map, suggested in Table 2, becomes undeniable when we use $\lambda \, dg$ mutants whose dg regions extend close to the left end of the lineom. All the data are consistent with the pro-

Table 5 Transduction of W3350 ($\lambda\,sus_{11}sus_{10}sus_{96B}sus_7imm^\lambda$) by $\lambda\,dg_{15}$

UV dose (minutes)	0	1	2	4	8	Total	Corrected† total	Cross- over‡
Number	98	100	100	100	99	497		
% hets	94%	90%	93%	76%	75%			
% hets comp. seg.*	45%	51%	45%	68%	77%			
+ + +	37	28	20	9	6	100	160	Parental
+ + sus_7	2	10	17	18	19	66	77.9	C
sus_{11} + +	0	2	0	0	0	2	3.2	BC
+ $sus_{10}sus_7$	0	0	1	0	3	4	4.7	ABC
sus_{11} + sus_7	0	1	2	2	2	7	8.2	B
$sus_{11}sus_{10}$ +	1	3	1	2	0	7	11.2	AC
$sus_{11}sus_{10}sus_7$	1	1	1	21	27	51	51	A
	41	45	42	52	57	237	316	

*Percentage, among the heterogenotes, of individuals from which the Gal^- segregant examined was pure recipient type.
†Corrected for bias introduced by selecting only those individuals which produced a recipient-type Gal^- segregant.
‡Crossovers on basis of map with prophage order:

$$Gal^A sus_{10}{}^B sus_{11}{}^C sus_7$$

phage order of the genes. Although the numbers are small in some cases, the aggregate result leaves little doubt that it is actually the prophage map which is being followed. The poorest agreement is for the $\lambda\,dg$ mutants of Tables 2 and 4, where single crossovers in region C seem unexpectedly rare. Interestingly, in the study of segregation patterns (Campbell 1963a), it was precisely the corresponding segregant from one of the same $\lambda\,dg$ mutants which was uncommon. Regardless of the precise interpretation, this fact suggests that the segregation experiment and the marker rescue experiment are measuring the same thing, down to a rather detailed level.

UV Sensitivity of Heterogenote Formation

We saw from Fig. 2 that the effect of UV on transducing ability is roughly similar for all $\lambda\,dg$ mutants. However, if instead of looking at the total number of transductants plotted against dose, we consider the fraction that are heterogenotes, we get the curves of Fig. 3. It is clear that the UV sensitivity of heterogenote formation does depend on the $\lambda\,dg$ used. The longer the dg region, the greater the UV sensitivity. A consistent effect on the fraction at zero dose is also seen. Whether this really is a separate effect from that on the slope is uncertain. It is possible that some damage is routinely done to the

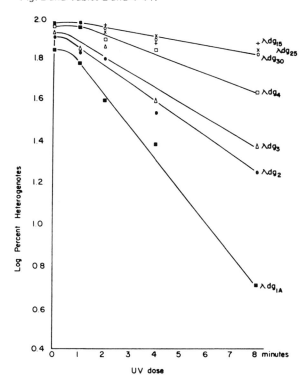

Fig. 3. Percentage of heterogenotes among the transductants. Analysis of colonies from the experiments of Fig. 2 and Tables 2 and 4–11.

Table 6 Transduction of W3350 ($\lambda sus_{11}sus_{96B}imm^{434}sus_3sus_{60}$) by λdg_4

UV dose (minutes)	0	1	2	4	8	Total	Corrected† total	Cross-over‡
Number	100	100	100	98	546	944		
% hets	93%	93%	78%	68%	44%			
% hets comp. seg*	70%	59%	59%	62%	69%			
$+++$	65	31	18	14	18	146	203	Parental
$++sus_{60}$	0	7	3	3	9	22	30.3	BCD
$+sus_3+$	0	5	8	10	32	55	69.3	C
$sus_{11}++$	0	0	0	1	3	4	5.5	ABD
$+sus_3sus_{60}$	1	6	15	10	54	86	103	B
$sus_{11}+sus_{60}$	0	0	0	1	3	4	5.5	ACD
$sus_{11}sus_3+$	0	0	0	0	1	1	1.2	ABC
$sus_{11}sus_3sus_{60}$	0	6	1	3	46	56	56.0	A
	66	55	45	42	166	374	474	

*Percentage, among the heterogenotes, of individuals from which the Gal^- segregant examined was pure recipient type.
†Corrected for bias introduced by selecting only those individuals which produced a recipient-type Gal^- segregant.
‡Crossovers on basis of map with prophage order:

$$Gal^A sus_{11}{}^B sus_{60}{}^C sus_3$$

Table 7 Transduction of W3350 ($\lambda sus_{11}sus_{96B}sus_7imm^{434}sus_3sus_{60}$) by λdg_4

UV dose (minutes)	0	1	2	4	8	Total	Corrected† total	Cross-over‡
Number	100	100	100	100	100	500		
% hets	91%	87%	83%	73%	43%			
% hets comp. seg.*	47%	48%	55%	49%	51%			
$++++$	42	24	23	18	1	108	175	Parental
$++sus_3+$	0	1	1	1	0	3	4.9	CD
$+sus_7++$	1	4	7	1	0	13	18.1	D
$+++sus_{60}$	0	2	0	2	0	4	6.4	BC
$++sus_3sus_{60}$	0	0	1	1	0	2	3.2	BCD
$+sus_7+sus_{60}$	0	0	1	0	1	2	2.8	BCD
$+sus_7sus_3+$	0	7	7	3	7	24	30.1	C
$+sus_7sus_3sus_{60}$	0	4	4	7	6	21	25.2	B
$sus_{11}sus_7sus_3+$	0	0	0	1	0	1	1.2	ABC
$sus_{11}sus_7+sus_{60}$	0	0	0	0	1	1	1.3	ACD
$sus_{11}sus_7sus_3sus_{60}$	0	0	1	2	6	9	9.0	A
	43	42	45	36	22	188	277	

*Percentage, among the heterogenotes, of individuals from which the Gal^- segregant examined was pure recipient type.
†Corrected for bias introduced by selecting only those individuals which produced a recipient-type Gal^- segregant.
‡Crossovers on basis of map with prophage order:

$$Gal^A sus_{11}{}^B sus_{60}{}^C sus_3{}^D sus_7$$

λdg in the preparation of the lysate, and that this damage affects transduction in the same way that UV does, in which case the most sensitive type of λdg would have suffered most before the experiment begins. This possibility is not purely hypo-thetical. We do have evidence that another method of inflicting damage, aging, affects a lysate in a manner similar to UV. The data of column 1 of Table 2 were obtained from three experiments. In two of these, we employed freshly prepared lysates,

Table 8 Transduction of W3350 ($\lambda sus_{11}sus_{96B}sus_7imm^{434}sus_3sus_{60}$) by λdg_3

UV dose (minutes)	0	1	2	4	8	Total	Corrected† total	Cross- over‡
Number	100	99	100	400	100	499		
% hets	86%	71%	73%	40%	24%			
% hets comp. seg.*	—§	55%	50%	—§	—§			
$+ + + +$	—	21	21	—	—	42	68.2	Parental
$+ + + sus_{60}$	—	1	2	—	—	3	4.8	BC
$+ + sus_3 +$	—	1	0	—	—	1	1.6	CD
$+ sus_7 + +$	—	4	4	—	—	8	10.5	D
$+ + sus_3 sus_{60}$	—	0	1	—	—	1	1.6	BD
$+ sus_7 + sus_{60}$	—	1	1	—	—	2	2.8	BCD
$+ sus_7 sus_3 +$	—	6	2	—	—	8	10.1	C
$+ sus_7 sus_3 sus_{60}$	—	2	5	—	—	7	8.4	B
	—	36	36	—	—	72	108	

*Percentage, among the heterogenotes, of individuals from which the Gal^- segregant examined was pure recipient type.
†Corrected for bias introduced by selecting only those individuals which produce a recipient-type Gal^- segregant.
‡Crossovers on basis of map with prophage order:

§Not measured. $Gal^A sus_{11}{}^B sus_{60}{}^C sus_3{}^D sus_7$

Table 9 Transduction of W3350 ($\lambda sus_{11}sus_{96B}sus_7imm^{434}sus_3sus_{60}$) by λdg_2

UV dose (minutes)	0	1	2	4	8	Total	Corrected† total	Cross- over‡
Number	100	100	100	100	100	500		
% hets	81%	68%	63%	35%	18%			
% hets comp. seg.*	—§	58%	52%	—§	—§			
$+ + + +$	—	19	10	—	—	29	47.0	Parental
$+ + + sus_{60}$	—	3	2	—	—	5	8.0	BC
$+ + sus_3 +$	—	2	2	—	—	4	6.5	CD
$+ sus_7 + +$	—	2	7	—	—	9	12.5	D
$+ sus_7 + sus_{60}$	—	1	2	—	—	3	4.1	BCD
$+ sus_7 sus_3 +$	—	6	6	—	—	12	15.1	C
$+ sus_7 sus_3 sus_{60}$	—	5	2	—	—	7	8.4	B
$sus_{11} sus_7 + sus_{60}$	—	0	1	—	—	1	1.3	ACD
	—	38	32	—	—	70	103	

*Percentage, among the heterogenotes, of individuals from which the Gal^- segregant examined was pure recipient type.
†Corrected for bias introduced by selecting only those individuals which produce a recipient-type Gal^- segregant.
‡Crossovers on basis of map with prophage order:

§Not measured. $Gal^A sus_{11}{}^B sus_{60}{}^C sus_3{}^D sus_7$

and in the third, one of the same lysates, which had been stored 3 months in the refrigerator. Of the 80 individuals listed in Table 2, 52 came from the first two experiments, and all were among the 17 cases where all the markers of λdg appeared in the heterogenote. The third experiment produced the other 28 heterogenotes, of which 9 were recombinant types.

The data plotted in Fig. 3 come from Tables 2 and 4–11. Besides the λdg mutants used in these experi-

Table 10 Transduction of W3350 ($\lambda sus_{11}sus_{96B}imm^{434}sus_3sus_{60}$) by λdg_{1A}

UV dose (minutes)	0	1	2	4	8	Total	Corrected† total	Cross-over‡
Number	130	100	100	199	200	729		
% hets	68%	59%	30%	25%	5%			
% hets comp. seg.*	69%	74%	69%	60%	80%			
$++$	51	31	13	15	2	112	155	Parental
$+ sus_{60}$	1	2	2	1	1	7	9.4	AB
$sus_3 +$	2	4	2	6	3	17	20.9	B
sus_3sus_{60}	2	2	3	7	2	16	16.0	A
	56	39	20	29	8	152	201	

*Percentage, among the heterogenotes, of individuals from which the Gal^- segregant examined was pure recipient type.
†Corrected for bias introduced by selecting only those individuals which produce a recipient-type Gal^- segregant.
‡Crossovers on basis of map with prophage order:

$$Gal^A sus_{60}{}^B sus_3$$

Table 11 Transduction of W3350 ($\lambda sus_{11}sus_{96B}sus_7imm^{434}sus_3sus_{60}$) by λdg_{1A}

UV dose (minutes)	0	1	2	4	8	Total	Corrected† total	Cross-over‡
Number	76	124	100	166	34	500		
% hets	71%	60%	49%	24%	6%			
% hets comp. seg.*	59%	63%	54%	55%	100%			
$+++$	25	24	19	6	0	74	120	Parental
$++ sus_{60}$	1	0	1	2	0	4	6.4	AB
$+ sus_3 +$	0	1	1	0	0	2	3.2	BC
$sus_7 ++$	2	10	1	2	0	15	20.7	C
$+ sus_3sus_{60}$	0	1	0	3	0	4	6.4	AC
$sus_7 + sus_{60}$	0	1	1	1	0	3	4.0	ABC
$sus_7sus_3 +$	1	1	3	2	0	7	8.6	B
$sus_7sus_3sus_{60}$	1	8	0	6	2	17	17.0	A
	30	46	26	22	2	126	186	

*Percentage, among the heterogenotes, of individuals from which the Gal^- segregant examined was pure recipient type.
†Corrected for bias introduced by selecting only those individuals which produce a recipient-type Gal^- segregant.
‡Crossovers on basis of map with prophage order:

$$Gal^A sus_{60}{}^B sus_3{}^C sus_7$$

ments, three additional λdg mutants were isolated with end points in region 1, i.e., they are missing all known genes from the left end of the map. All four λdg_1 mutants are shown in Table 12. There is no significant difference between them.

The difference between λdg_1 and λdg_{30} at zero dose is seen not only in transduction of a lysogenic recipient, but equally in transduction of a sensitive recipient, with or without added "helper" phage (Table 13).

DISCUSSION

The experiments reported above can be thought of as crosses between λ and λdg. The λ parent is in the prophage state, while the λdg is introduced into the cell by infection. The recombinant recovered is a λdg in the prophage state. The idea that essentially all such λdg prophages found using an irradiated lysate are indeed recombinants is documented by the evidence presented here.

Table 12 UV sensitivity of heterogenote formation for different λdg_1's

λdg*		UV dose in minutes				
		0	1	2	4	8
1A	% hets	71%	60%	49%	24%	6%
	N	76	124	100	166	34
1A	% hets	68%	59%	30%	25%	5%
	N	130	100	100	199	200
1B	% hets	74%	55%	45%	13%	13%
	N	100	110	213	54	16
1C	% hets	77%	52%	46%	20%	4%
	N	100	88	109	99	99
1D	% hets	81%	55%	42%	18%	3%
	N	100	98	97	98	97

*λdg_{1A}–λdg_{1D} are four different λdg's of independent origin, each missing all known markers at the left end of the λ lineom.

Table 13 Fraction of stable transductants formed by different λdg's

Conditions	Fraction stable transductants*	
	λdg_1	λdg_{30}
Lysogenic recipient	0.26 (132/506)	0.08 (19/249)
Sensitive recipient	0.36 (70/195)	0.17 (33/195)
Sensitive recipient plus helping phage†	0.14 (51/359)	0.03 (8/237)

*For all three conditions, the difference between λdg_1 and λdg_{30} is significant ($P < 0.01$). Each figure is the pooled data from several (2–5) experiments. With sensitive recipients, pairs of lysates (one of λdg_1 and one of λdg_{30}) were made at the same time and in the same way. All individual lysates of λdg_1 gave higher values than the lysates of λdg_{30} for the same conditions.

†An LFT lysate was reconstructed by mixing a high dilution of a HFT lysate with a nontransducing lysate. This mixture was then spread on the surface of an EMB plate together with sensitive Gal^- bacteria. The results indicate that, in picking heterogenotes from an LFT transduction, λdg_{30} mutants are selected preferentially over λdg_1 mutants by a factor of 1.13, a real but extremely small effect.

We think that vegetative λ phage and λ prophage have different genetic maps. We do not know which, if either, of these maps applies to the free phage particle, or to the DNA injected into the cell. We have therefore no expectation as to what might be the genetic map deduced from the above crossing procedure. In fact, we obtain good agreement with the prophage map. This is most readily understood if, at the time it recombines with the λ prophage, the λdg already has the prophage order, or if, at least, the ends of the λdg have already been joined.

If it be true that the ends of the λdg are already joined at the time of the recombinational rescue from UV damage, one may ask next whether they are already joined in the free phage particle, or perhaps become joined automatically following injection. The relevant data are those of Fig. 3, showing the UV sensitivity of heterogenote formation. A heterogenote results from lysogenization by λdg. Hence, the ability to form a heterogenote is the ability for the λdg structure to be rescued by recombination. This ability is very resistant to UV damage, provided that the λdg contains enough λ-specific genetic material on the left end. If the deletion approaches too closely the left end of the λ lineom, heterogenote formation becomes reduced and highly UV sensitive. We do not know the precise extent of the critical region, but it certainly extends into and probably covers the whole A cistron, because λ *dg* 2, 3, and 4 all have deletions terminating within this cistron.

We consider this a strong indication that the lineom of free λ *dg* does indeed have a physical end corresponding to the end of the vegetative map. Regardless of the mechanism of the effect, it is easier to understand how the proximity of the deletion end point to the end of the lineom affects the ability of λ *dg* to be rescued if the lineom really does have an end. The alternative explanation that a region critical to lysogenization just happens to lie at the end of the map cannot be rejected. Its role at any rate is probably structural rather than physiological, since the A cistron seems to be a late genetic block whose function is unnecessary to lysogenization (Campbell 1961, 1962a). Furthermore, if the effect is physiological, the function is unique in being expressed in an immune cell, and the homologous wild-type region of the indigenous prophage must be unable to supply the function for a superinfecting particle.

We would therefore suggest for lysogenization in general, and for rescue of UV'd λ *dg* in particular, the model of Fig. 4. The important feature is that endjoining is prerequisite to stable integration. Therefore, among those individuals which are recovered as heterogenotes, recombination has proceeded according to the prophage order rather than the vegetative order. Whether the multiplying

vegetative phage is itself circular or free-ended need not concern us here.

The probability of endjoining would then be influenced by the presence of λ-specific genetic material at the left end of the lineom. We know very little about how this effect operates. It might be that the endjoining mechanism requires that the ends to be connected have the proper base sequence over a region extending for some distance from the ends. It also seems possible that the effect is indirect. For example, the fact that a λ *dg* with no λ-specific material at the left is completely homologous (under our model) to a linear section of the bacterial lineom might favor the type of pairing which leads to haploidization and thereby prevent or delay endjoining. Irradiation might then either inhibit endjoining or enhance pairing with the bacterial lineom.

In any case, the data allow a large variety of possible interpretations, but there are some which are definitely ruled out. In particular, the difference between λ *dg*$_1$ and λ *dg*$_{30}$ exists even in the absence of another phage in the recipient (Table 13). This rules out the otherwise attractive idea that, as an alternative to the normal endjoining process, a phage can find another phage whose ends are already joined and pair and/or recombine with it, an idea that would explain very well the necessity for some phage-specific genetic material near the end. This possibility is also unlikely from the marker rescue data. The strong linkage *across* the ends implies that the ends of the λ *dg* have been directly

Fig. 4. Model for lysogenization. The frequency of endjoining is affected, directly or indirectly, by the amount of λ-specific genetic material at the left end of the lineom. Marker rescue occurs at or following integration.

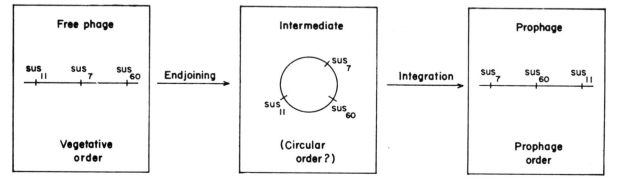

joined with each other, as seen for example, with
λ dg_4 at 8 minutes UV. If there is a recombinational
bypass of the ends, it must involve switches closer
to the ends than the terminal markers sus_{11} and
sus_{60} that we have used.

The data of Table 13 are obtained in the absence
of UV, under which condition the effect is weakest.
With UV'd λ dg in a sensitive recipient, the frequency of heterogenotes is very low; a comparison of the
two λ dg mutants under those circumstances would
seem irrelevant in any case, because, even if endjoining and lysogenization should occur, there is no
possibility of recombinational rescue to form a
heterogenote.

As we consider λ dg mutants with increasingly
larger deletions, their tendency to form heterogenotes becomes progressively less (Fig. 3). If this
reflected a direct need for λ-specific material in the
endjoining process, one might expect that a λ dg in
which the entire left end were missing would be
totally unable to lysogenize and would therefore
never be found. On this basis, we should have expected more variation than was observed among
the four separate λ dg_1 strains studied (Table 12),
as their properties should depend critically on the
exact point between sus_{11} and the end of the lineom
at which their deletion begins. The similarity of
these four strains suggests that they may represent
the most extreme type of λ dg obtainable by the
usual isolation procedures. This could mean that
they are missing the entire left end of the lineom,
or, alternatively, that a certain minimal amount of
λ-specific genetic material at the end is required for
other purposes, e.g., for the "cutting" process by
which the ends become separated in the free phage.
The preferential selection of λ dg mutants with
shorter deletions on the basis of their lysogenization frequencies alone is a small effect and does not
explain the extreme similarity of these four λ dg
mutants (Table 13).

In any case, we regard our inference that the
lineom of free λ dg has physical ends (or, at least,
some structural singularity) at the ends of the normal vegetative map to be more firmly grounded than
any speculations as to how the end effects operate.
The agreement of the genetic results with the prophage map indicates that λ dg can assume the prophage order as well. Beyond that, an interpretation
such as that of Fig. 4 will seem reasonable only if
the spirit of previously expressed views on prophage
integration is accepted (Campbell 1962b).

In Fig. 5 an interpretation of the effect of UV
on transduction by λ dg is presented to show that a
consistent explanation of the facts can be given.
In Fig. 5 we diagram the fate of four λ dg particles
that have been exposed to UV. Particle A has escaped injury. Particles B, C, and D each bear a single
UV lesion.

Particle A transduces by lysogenization in the
normal manner, as previously hypothesized (Campbell 1962b). The resulting heterogenote has acquired
all the donor alleles and will usually segregate them
along with Gal (Campbell 1963).

Particle B has sustained a UV hit close to the end.
If we imagine that it is such hits which inhibit endjoining, then this particle will generally transduce
by pairing with its homolog in the recipient and
substituting for it by double crossing over. As illustrated, this process might occasionally result in a
joint rescue of some phage markers with Gal.

Particle C has been hit in the Gal region, to the
left of the Gal markers employed in these experiments. It circularizes and lysogenizes in the usual
way, but the resulting bacterial lineom cannot multiply as such because it contains a UV lesion. It
therefore crosses over with a second, nonlysogenized
bacterial lineom to give a stable Gal^+ transductant.

Particle D has been hit between the sus_7 and
sus_{60} genes. As with particle C, lysogenization
occurs normally but must be followed by recombination away from the UV hit. This gives rise to a
heterogenote that is a single crossover type according to the prophage map.

With particles C and D, we have introduced a
second bacterial lineom to participate in the second
crossover event. The question arises whether it
might not be simpler to invoke internal recombination of duplicated parts within a single strand, as
was done to explain segregation (Campbell 1963a).
For particle C this is easily done: Internal recombination would give the opportunity for joint rescue
of phage markers among the stable transductants,
as with particle B. But particle D presents more
complications. The transduction involves a net gain
of genetic material. The cell is retaining the prophage it carried already and acquiring a recombinant λ dg. That portion of the λ dg that derives
from the prophage parent is thus appearing twice in
the lineom ultimately produced. There is no way of
doing this by breaking and joining within a single
lineom. It is possible that UV-stimulated recombination involves copy choice rather than breaking and

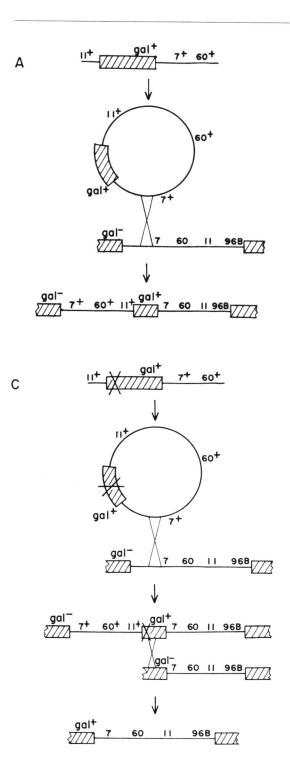

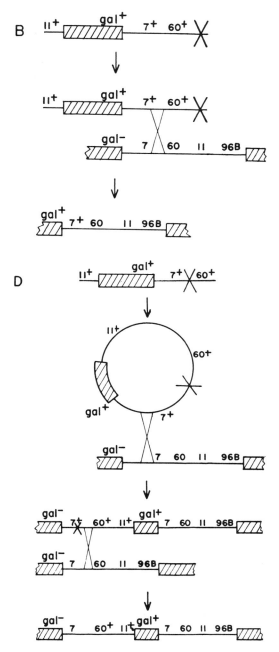

Fig. 5. Interpretative diagram of consequences of UV hits on λ *dg*. UV lesions are represented by heavy crosses. Bacterial genes are shown cross-hatched; phage genes are on thin lines. In each case the UV-damaged λ *dg* is represented as interacting one or more times with the recipient lineom.

joining (Simon 1963). An internal copy-choice mechanism could explain the copying of the same region twice but will produce the observed genetic results only with accessory *ad hoc* assumptions. It is thus simpler in any case to assume that a second bacterial lineom recombines with the lysogenized lineom. This recombination could occur equally well by breakage or by copy choice.

REFERENCES

Arber, W. (1958). Transduction des caractères *gal* par le bactériophage lambda. *Arch. Sci. (Geneva)* 11, 259–338.

Calef, E., and Licciardello, G. (1960). Recombination experiments on prophage host relationships. *Virology* 12, 81–103.

Campbell, A. (1959). Ordering of genetic sites in bacteriophage λ by the use of galactose-transducing defective phages. *Virology* 9, 293–305.

Campbell, A. (1961). Sensitive mutants of bacteriophage λ. *Virology* 14, 23–32.

Campbell, A. (1962a). Effect of 5-fluorouracil on suppressor sensitive mutants of bacteriophage lambda. *Bacteriol. Proc.*, p. 145.

Campbell, A. (1962b). Episomes. *Advan. Genet.* 11, 101–145.

Campbell, A. (1963a). Segregants from lysogenic heterogenotes carrying recombinant lambda prophages. *Virology* 20, 344–356.

Campbell, A. (1963b). Distribution of genetic types of transducing lambda phages. *Genetics* 48, 409–421.

Campbell, A. (1964). Location of the m_6 gene of coliphage lambda. *Bacteriol. Proc.*, in press.

Campbell, A., and Balbinder, E. (1959). Transduction of the galactose region of Escherichia coli K12 by the phages λ and λ-434 hybrid. *Genetics* 44, 309–319.

Simon, E. (1963). Protein synthesis and recombination in bacteriophage T4. *Bacteriol. Proc.*, p. 145.

Part 4. Function of the genetic material

THE RELATIONSHIP BETWEEN GENE AND PHENOTYPE

The one major area we have yet to discuss concerns the function of the genetic material. In some manner, the genetic information, coded within the genes, is translated into the phenotype of the organism. In this part, papers are presented that shed light on the steps involved in this translation — a process called *the heterocatalytic function* of the genetic information. The replication of the genetic material is called *the autocatalytic function.*

The first major step taken in this investigation came only two years after the rediscovery of the Mendelian laws of inheritance. In 1902, Archibald Garrod[20] concluded that the human disease, alkaptonuria, was a metabolic defect inherited as a Mendelian recessive gene. This disease was accompanied by the excretion of large amounts of homogentisic acid, a substance not found in the urine of normal individuals. To Garrod this suggested that the genetic defect was associated with the presence of an abnormal biochemical pathway. Seven years later, Garrod asserted that the alkaptonuria syndrome was caused by the individual's inability to metabolize the amino acid phenylalanine.[21] Since biochemical pathways were known to be controlled by enzymes, he logically concluded that the genetically controlled defect manifested itself in the form of a defective enzyme in the phenylalanine metabolic pathway. This was the first suggestion that genes were responsible for the production of enzymes and that cellular enzyme activity controlled the phenotype of the organism.

At the same time, other human disorders were linked to single gene mutations that displayed Mendelian patterns of transmission. In each case, the disorder could be traced to a defective biochemical pathway and, ultimately, to a defective enzyme. Albinism, for example, an abnormality characterized by the absence of pigmentation, was linked to a defective enzyme necessary for the completion of the normal biochemical pathway in melanin synthesis. Thus, by 1915, the combined knowledge of Mendelian genetics and enzymes presented this view of the gene–phenotype relationship:

Between 1915 and the middle 1930s, barring Garrod's observations, there was little experimentation directed toward revealing the relationship between genes and enzymes. Most of the publications on the topic were too vague and lacked roots in experimentally based observations. The sequence of events leading from genes to their products remained a mystery. Certainly, the major question that had to be answered before progress could be made in establishing their relationship was: What is the quantitative relationship between the gene and the enzyme, i.e., does one gene produce a single enzyme or is one gene capable of directing the production of two or more different enzymes? Without the answer it was impossible to relate the gene product or products to the control of a biochemical reaction or reactions. The answer would come only through the implementation of biochemical studies.

George W. Beadle was one of the men instrumental to the emergence of experimental studies in the field of biochemical genetics. Beadle began his studies as a maize geneticist. However, early in his studies, he took a post-doctoral position at the California Institute of Technology. There he came under the influence of T. H. Morgan and A. H. Sturtevant. Working with Sturtevant, Beadle became interested in eye pigmentation in *Drosophila;* in particular, the large number of separately mapping alleles that controlled the flies' eye color. It seemed reasonable to Beadle that all of these different loci were involved in some way with the control of the specific pheno-type of eye color; and, if one could analyze the manner in which the different gene products interacted biochemically, one might be able to specify the relationship between the gene and the enzyme. The problem confronting Beadle was how to analyze the interaction of mutant gene products in *Drosophila* at the biochemical level.

Boris Ephrussi had developed a technique for transplanting eye tissue from one larva to another. He found that the transplanted tissue would continue to develop under the influence of the host's genotype. This ar-rangement allowed an observer to analyze the biochemical interactions taking place between two different genotypes in the same individual — it was possible to establish whether two mutants functioned autonomously or nonautonomously at the biochemical level. Beadle and Ephrussi applied the latter's technique to discover if defects in pigment metabolism in the transplanted eye tissue could be overcome or bypassed by metabolic prod-ucts in the host organism. They found both autonomous and nonautono-mous relationships between eye color mutants.[1] The pattern of interaction between the various mutants combined within the same host system was such that it seemed to support the idea that separately mapping mutants produced enzymes that affected the same biochemical pathway. However, due to the complexity of biochemical research in *Drosophila,* a diploid, their work could not be extended to establish the exact quantitative rela-tionship between the genes and their enzyme products.

Instead of taking known genetic markers and attempting to match them with subsequently defined biochemical pathways, Beadle wanted to work with known biochemical pathways, analyzing them genetically for mutants that controlled specific steps within the reaction pathway. Therefore, he needed an experimental organism (1) in which biochemical pathways could be easily defined; (2) about which sufficient genetics was known to allow the detection of mutations affecting the biochemical steps within the path-way; and (3) which was not a diploid. In 1941, Beadle and E. L. Tatum, chose to work on the bread mold *Neurospora crassa* — a basically mono-ploid organism that can be cultured easily on defined synthetic medium, enabling biochemical mutants to be detected with ease and analyzed in terms of their effects on the individual steps of biochemical pathways. In this way, Beadle and Tatum were able to experimentally document the validity of the hypothesis that one gene directed the production of one enzyme.[2] Although this hypothesis has undergone subsequent modification — to include the fact that enzymes composed of more than one species of polypeptide are genetically controlled by more than one gene — as originally formulated it was a strong incentive for the creation of models that predicted how genes could control the production of enzymes. However, the mechanism by which a gene could direct the production of a specific polypeptide was still obscure at the end of the 1940s.

Since the chemical nature of the gene was still unknown, few of the speculations made prior to 1950 on the mechanism underlying the heterocatalytic function of the gene relate to our current ideas. Beginning in the early 1950s, proposals began to appear that influenced later models of protein synthesis. Initially, there were several models presented that had protein assembly taking place directly off the DNA molecule.[19] Although these models displayed exceptional creativity, they failed to take into account the experimentally established observations that DNA was restricted to the nucleus and that the majority of protein synthesis took place in the cytoplasm. This spatial discontinuity demanded some form of communication between the genic material in the nucleus and the sites of protein synthesis in the cytoplasm. Attention immediately focused on the second type of nucleic acid, RNA, as a likely candidate for this molecular means of communication. This was not simply a serendipitous choice. Biochemical and histological experiments dating from the 1940s had demonstrated an intimate association between RNA synthesis and protein synthesis. Both Caspersson[12] and Brachet[4] showed a direct correlation between the rates of protein synthesis in cells and the rates of RNA synthesis. No such correlation could be established between DNA synthesis and protein synthesis. Also, these observations were followed by work that suggested RNA was synthesized in the nucleus but migrated almost immediately to the cytoplasm. RNA seemed to be the means of molecular communication between the nucleus and the cytoplasm.

At the same time, RNA was implicated as a necessary component for protein synthesis in work employing acellular experimental protocol. For example, in 1957, Zamecnik and his laboratory used an *in vitro* analysis of a cell-free rat liver system capable of supporting protein synthesis to demonstrate its clear dependency on the presence of a low molecular weight RNA component. This RNA component was purified from the soluble fraction of the cell homogenate,[42] and was directly implicated in the mechanism of protein synthesis when it was found that cellular enzymes first catalyzed the transfer of the amino acids to it.[41] Then these RNA molecules, bound to the amino acids, aggregated at the sites of protein synthesis at which point the amino acids were incorporated into the growing polypeptide chains.

The histological and biochemical observations above and the 1953 model of the structure of DNA provided the platform of information on which, in 1958, Crick constructed his theory of how structural information could flow from nucleic acid to protein molecules.[13] Basically, he proposed: Since both DNA and protein are molecules constructed of sequentially arranged building blocks, transfer of information could take place by certain sequences of nucleotides in DNA directing the sequence of amino acids in proteins. In his model, Crick envisioned two additional components that would be essential for the transfer of information from genes to proteins. He proposed that the nucleotide sequence of the gene, housed within the nucleus, was first transcribed into a complementary sequence of polyribonucleotides. This RNA intermediate would pass from the gene to the site of protein synthesis and there, with the help of the small soluble RNA molecules (now called *transfer RNA* or *tRNA*), the sequence of ribonucleotides would dictate the sequence of amino acids in the nascent polypeptide. Crick's hypothesis stimulated immediate research.

Two predictions had to be considered in evaluating the validity of Crick's hypothesis. First, if Crick's model were valid, one would expect to observe the flow of structural information in a unidirectional manner — passing from nucleic acid molecules to proteins but not in the reverse direction. This prediction has been experimentally supported.[40] Second, since DNA, RNA, and proteins are linear molecules, there should be a colinear relationship between the nucleotide sequence in the nucleic acid molecules and the amino acid sequence in the proteins. For example, if a nucleotide sequence were found in the middle of the gene coding for a specific polypeptide, it should direct the placement of a specific amino acid in the middle of that polypeptide and not at the ends. Experiments were made correlating the map sites of mutants within a specific gene with the amino acid substitutions they specify in the mutant polypeptides. The experiments substantiated that a colinear relationship is maintained between the DNA of the gene and the polypeptide whose production it specifies during protein synthesis.[39] With the acceptance of Crick's hypothesis, the diagram of the relationship between genes and the phenotype appeared as follows:

$$\boxed{\text{Gene (DNA)}} \longrightarrow \boxed{\text{RNA}} \longrightarrow \boxed{\text{Protein}} \longrightarrow \boxed{\text{Phenotype}}$$

It can be seen from this diagram that important questions remained about how genes directed the production of proteins. The questions deal with the mechanics of the transcription of DNA into RNA and the translation of RNA into protein. The transcription of DNA into RNA is fundamentally the same as DNA replication. The reaction is enzyme-catalyzed and the specificity of the nucleotide incorporation is based upon the hydrogen-bonding specificities set forth by Watson and Crick — with the exception that adenine pairs with uracil instead of thymine. During transcription, however, a single-stranded RNA molecule is produced off of one of the strands of the double-stranded DNA molecule.[36]

The first major step in understanding the complexity of translation came

in 1961, with the confirmation of the existence of an unstable RNA intermediate complementary to the DNA strands.[6] This was the first experimental support for the existence of messenger RNA. This RNA fraction,
or synthetically produced RNA molecule, could be used to direct the synthesis of polypeptides *in vitro*,[33] providing not only a means for establishing the coding relationship between RNA and polypeptides but also a
technique for defining the various steps of translation. Within the next ten
years, the intricate steps of translation were worked out at such a rapid
rate that it would be impossible to document each step by a separate paper.
The figure below summarizes some of the details of the transcriptional and
translation scheme; and, Zamecnick's paper[40] presents an overview of the
protein synthesis research.

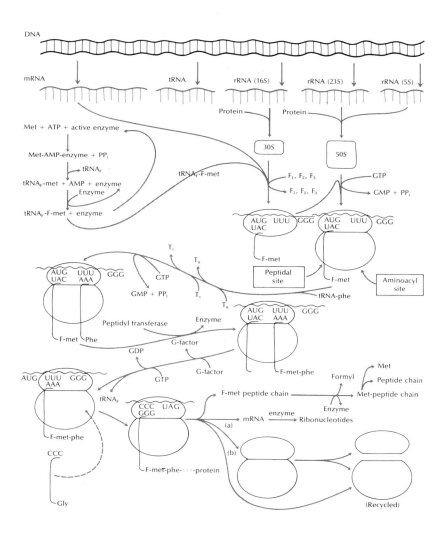

THE GENETIC CODE

With the discovery of the main components in the sequential pathway leading from genes to proteins, we can then ask: How is the sequence of bases in nucleic acids translated into the primary structure of proteins? This is the basic problem of the genetic code. It is like the puzzle of how a language of four nucleic acid bases specifies a language of twenty amino acids.

Because of its cryptic nature, the code has had a unique history. Its appeal has been broadly based and interest in its nature has spawned intriguing, yet often biologically unsound, ideas. We want to look at a few of these ideas because, although some of them did reflect bias or unsupported speculation, they often led to revealing experimentation.

The code has had a brief and fast moving history. Its development can be divided into three phases: the theoretical phase (preceding 1961), the experimental phase (between 1961 and 1966), and the principle phase, i.e., the principles underlying the structure of the code.

THE THEORETICAL PHASE The structure of DNA stimulated much thought on the relationship between DNA and protein. The thought, however, often was not based on hard data and thus had no substantial connection to existing biological systems. This is not to say that the theoretical period (pre-1961) was without merit. During this time many concrete features of the genetic code were formulated and later shown to be correct.

Among the earliest and most innovative of the code theoreticians was the noted astronomer and cosmologist George Gamow. Stimulated by the Watson and Crick studies on the structure of DNA, Gamow suggested a relationship between the linear sequence of bases in DNA and that of amino acids in polypeptides.[19] He suggested a code in which DNA bases formed a three-dimensional diamond into which would fit the various side chains of amino acids. Gamow's so-called diamond code is an overlapping code — any one base pair forms part of more than one coding unit (*codon*). Brenner,[5] studying known sequences of amino acids, showed the unlikelihood of an overlapping code.

Dounce (1952) was the first to suggest that proteins were not synthesized directly off of DNA.[17] He proposed the existence of an intermediary adaptor molecule, later supported and elaborated upon by Crick in 1958.[13] Crick argued that DNA did not have a specific affinity for amino acids, and an adaptor molecule was required to serve as an intermediate between DNA and protein. The discovery of tRNA further supported the adaptor hypothesis.[42]

With the general outline of protein synthesis formulated, i.e., DNA → RNA → Protein, attention was focused on the issue of different coding units in RNA determining specific amino acids. Because there are 4 RNA bases and 20 amino acids, the genetic code took on the dimensions of a cryptogram. Crick postulated a "solution" to the genetic code by addressing the question of which 20 triplet codons of the possible 64 specify amino acids (a sequence of 3 bases specifies an amino acid and there are 64 such triplet sequences). Using an approach unconnected to biological systems, Crick broke the 64 triplet condons into two groups: one group of 20 specifying amino acids and one group of 44 that were nonsense (specified no amino acid). As intriguing as his code was, it was wrong. It failed to prepare the scientific community for the discovery of the first code word: UUU-specifying phenylalanine — UUU was not one of Crick's magic 20 sense codons.

THE EXPERIMENTAL PHASE   The year 1961 was a pivotal year for the genetic code. Crick and his colleagues published a paper[14] on the general nature of the genetic code. They revised Crick's early hypothesis and said that there was a specific starting point and then sequential translation. Almost simultaneous with their paper came one by Nirenberg and Matthaei[33] announcing that mRNA composed only of U (polyuracil), coded for a polypeptide composed exclusively of phenylalanine — so UUU specified phenylalanine. Nirenberg and Matthaei's paper marked the beginning of the experimental period which lasted until 1966.

With the discovery that UUU coded for phenylalanine, there was a flurry of activity focusing on the issue of which amino acids were specified by which codons. Random copolymers of RNA bases were synthesized and used as templates for protein synthesis and the polypeptides synthesized from these templates were analyzed. If the bases are incorporated into the RNA polymer on a random basis, then the proportion of bases in the medium before the synthesis of the synthetic RNA should be the same as that found in the polymeric RNA. Knowing these proportions, the probabilities of certain codons forming is computed. These probabilities are compared with the frequencies of amino acids in the polypeptide synthesized with the least probable codon coding for the amino acid found in lowest frequency, and so on. Unfortunately this technique does not tell us anything about the sequence of bases in a codon — i.e., UAU *or* AUU *or* UUA are sequentially distinct, but proportionally the same two (U's and one A).

Repeating sequence copolymers were also developed (i.e., AUAUAUAU . . . . . .) and they coded for repeating sequences of amino acids. The problem was, however, which of the two possible codons (AUA or UAU) specifies which of the two amino acids.

In 1964, Leder and Nirenberg provided a major breakthrough of the base sequence issue in a codon. They prepared triplet codons of known sequence and determined which amino acids they specified.[32]

THE PRINCIPLE PHASE   A wide variety of techniques contributed to our understanding of what each of the 64 codons specify.[29] Sixty-one of the codons specify amino acids and three of them are termination signals which specify no amino acids. The code is degenerate in the sense that an amino acid can have more than one codon. However, the code is not ambiguous since any one codon specifies only one amino acid with one small exception: AUG is an initiating codon and when found at the beginning of a gene sequence specifies formlymethionine; if AUG is in the middle of a gene sequence, it specifies methionine.

UUU	Phenylalanine	UCU	Serine	UAU	Tyrosine	UGU	Cysteine
UUC		UCC		UAC		UGC	
UUA	Leucine	UCA		UAA	Terminate	UGA	Terminate
UUG		UCG		UAG		UGG	Tryptophan
CUU	Leucine	CCU	Proline	CAU	Histidine	CGU	Arginine
CUC		CCC		CAC		CGC	
CUA		CCA		CAA	Glutamine	CGA	
CUG		CCG		CAG		CGG	
AUU	Isoleucine	ACU	Threonine	AAU	Asparagine	AGU	Serine
AUC		ACC		AAC		AGC	
AUA		ACA		AAA	Lysine	AGA	Arginine
AUG^{init} Methionine		ACG		AAG		AGG	
GUU	Valine	GCU	Alanine	GAU	Aspartic acid	GGU	Glycine
GUC		GCC		GAC		GGC	
GUA		GCA		GAA	Glutamic acid	GGA	
GUG		GCG		GAG		GGG	

To know what all 64 codons specify is not to say that we understand the genetic code. Nor can we realistically assert an understanding of the code by pointing to our knowledge of the protein synthesizing machinery. The code simply cannot be defined exclusively in terms of codon specificity or protein synthesizing machinery. We must know how certain codons came to be associated with certain amino acids. We must know how DNA, three species of RNA, and a host of polypeptides all merged to form the protein synthesizing machinery that we are familiar with today. In order to fully understand the genetic code, we must come to grips with its fundamental nature and with the evolutionary forces that produced it.

It is obvious that the precision and complexity of the genetic code did not appear *de novo;* there was a long period of evolution involved in the formation of this machinery. It is also clear that the mechanism of translation evolved along with the specificity of codon assignments. Thus the codons and translation cannot be considered as separate and distinct areas. Unfortunately, the nature of the genetic code had its beginning in the prebiotic world and gained sophistication in the evolution of prokaryotes and eukaryotes.[38] This is a period of time of which essentially we have no knowledge, only speculation. However, recent advances in the area of prebiotic chemistry have made possible some reasonable arguments concerning the nature of the code, and some of this work is discussed in the Woese paper.

REGULATION OF GENE ACTIVITY

Types and amounts of proteins in cells do not remain constant. In microbial systems different populations of enzymes appear and disappear in response to environmental stimuli. Also, during the transition of a single-celled zygote into a fully differentiated multicellular organism, change is characterized by the continual appearance of new populations of proteins. Since both the microbe and multicellular organism start with a single genome, different genes must function at different times.

Historically, two separate lines of experimentation have supported the existence of differential gene activity. By the turn of the century, it was established that microbes display different patterns of enzyme synthesis depending on the chemical composition of their environment. For example, yeast grown on a galactose-containing medium synthesized enzymes required for galactose metabolism. These enzymes rapidly disappeared when the yeast was transferred to a glucose-containing medium. The cycle could be repeated by returning the yeast to a galactose-supplemented medium. The linking of this environmental control of enzyme production to the regulation of gene activity was not made until after the one gene–one enzyme hypothesis was formulated by Beadle and Tatum in 1945. In the 15 years following their proposal, evidence appeared to support the hypotheses that (1) the substrate-controlled patterns of enzyme production in microbes were based on differential gene activity, and (2) the substrate controlled these patterns at the level of the DNA.

Studies designed to investigate the mechanisms of differentiation in multicellular organisms also supported the existence of differential gene activity. These studies centered on the question of how the different cell types could arise from a single cell, the zygote. Two hypotheses were advanced to explain differentiation in higher organisms. One hypothesis attributed differentiation to a quantitative sorting out of genes during development. Thus, the adult cells possessed only those genes necessary for the production of their terminal structural and functional attributes. The alternative hypothesis stipulated that all adult cells possessed equal complements of genetic information and that differentiation was a result of different combinations of genes functioning in the different adult cells. Support for the differential gene activity hypothesis began to emerge in conjunction with experiments that established equal complements of genetic information in differentiated cells.

Differential gene activity implies that genes can be viewed as separate entities that function independently. As discussed in Part 2, this concept arose in the early 1900s with the experimental support of the unitary nature of the genes. Then, with the establishment of the chromosome theory of inheritance, the chromosomes could be used as markers for determining the presence or absence of sets of genes. Thus, work defining the mechanics of mitosis could be brought to bear on the question of gene constancy in differentiated cell types. By 1900, it was established that mitosis generated new cells having chromosome numbers identical to the parental cell. However, at that time it was not possible to establish whether the new cells qualitatively contained the same chromosomes as the parental cell. A confirmation of chromosome equality was obtained later by tracing the transmission of heteromorphic sex chromosomes and chromosomal rearrangements through mitotic divisions.[7,37] Due to the consistency of distribution during division of these cytologically detectable markers, it was possible to establish that mitosis distributed the same number and type of chromosomes to all cells. On the basis of this work there appeared to be a quantitative equality at the chromosomal level in all somatic tissues. Support for this equality at the level of the gene was initially obtained from analyses of the salivary banding patterns of the polytene chromosomes in *Drosophila.* For example, Bridges demonstrated that the abnormal eye configuration, *Bar eye,* was based upon a chromosomal duplication of the X chromosome.[7] He observed the alteration in the polytene chromosomes in the cells of the salivary gland. Even though these cells were not connected with the development of the eye morphology, they carried the gene duplication responsible for the *Bar* phenotype. Thus, he established that adult cells carried genetic information not involved with their specific pattern of differentiation.

Although the above work supported the hypothesis that fully differentiated cells contained the same genome present in the zygote, it was still possible that small quantitative differences, beyond the level of resolution of the above analyses, might account for differentiation. This doubt was substantially reduced with the initiation of nuclear transplantation experiments.[8,23] These experiments involved the transplantation of nuclei from fully differentiated cells into enucleated eggs. This was done to assess the capacity of the nuclei to control differentiation a second time. If the steps of the original differentiation involved irreversible changes in genomic constitution, i.e., loss of material, then the transplanted nucleus should not have the capacity to again direct complete differentiation. In experiments where the nuclei of fully differentiated *Xenopus* intestinal cells were transplanted into enucleated eggs, the genomic compositions of the transplanted nuclei were capable of directing the emergence of fertile adults. This established that the mechanism of cellular differentiation was not based upon irreversible changes in the genetic composition of the zygote.

By eliminating the possibility that differentiation was due to quantitative irreversible changes in the genetic information, indirect support was obtained for the alternative hypothesis that invoked differential gene activity as a basis for differentiation. However, if one assumes that the presence of messenger RNA is a direct indication of the activity of specific genes, then direct support for differential gene activity is available. DNA–RNA competitive hybridization experiments have confirmed the presence of different populations of messenger RNA molecules in cells from different tissues.[16,18]

So, differential gene activity apparently accounts for both short-term changes within an organism in response to changes in the surrounding environment and more stable changes affiliated with the intricate process of cellular differentiation. With the acceptance of the hypothesis of differential gene activity comes the question of how certain sets of genes can function while neighboring genes remain functionally inactive?

The first significant breakthrough in understanding how gene activity is controlled came in 1961. Jacob and Monod formulated a model to explain how the extracellular presence or absence of galactose could modulate the level of enzymes responsible for its uptake and metabolism in *E. coli*.[27,28] In their model they proposed the existence of a new type of gene. In addition to the structural genes that were transcribed into messenger RNA, they proposed the existence of regulatory genes. These genes had the capacity to recognize the presence or absence of nongenetic signals in the environment and, in response to those signals, to control the activity of the structural genes. Accepting the validity of Crick's 1958 schema (DNA → RNA → Protein), they proposed that their regulators could control the activity of the structural genes during transcription or translation. They concluded that the regulation of the lactose enzymes took place at the level of transcription. A single regulator element (*operator*) determined whether a set of tightly linked structural genes would function. This type of gene regulation was called coordinated gene control because the tightly linked structural genes responded as a single unit. The operator and structural genes, functioning in a coordinated fashion, was called an *operon*. The activities of the different operons depended upon the activity of a second, recombinationally separable, regulatory element — the regulator gene. This regulator gene functioned by coding for a repressor protein. The protein, depending on its functional nature and the nongenetic stimulus present in the environment, regulated the operon by repressing the activity of the operator gene. Jacob and Monod suggested that operons, which functioned only in the presence of an inducer, had regulator genes that produced an active repressor. The inducer, when present, blocked the active repressor and the operon was turned on. Operons that were normally active, unless a corepressor substance appeared in the environment, had regulator genes that coded for inactive repressors. However, the corepressor with the inactive repressor formed an active repressor which turned off the operon.

Although a good deal of experimentation supports the existence of the operon type of gene control in microbial systems, it is obvious now that this is not the only type of function control at the level of transcription. For example, differential gene activity exists during the transcription of the phage T4 genome during vegetative reproduction.[22] Three batteries of genes, i.e., preearly, early and late, each functioning in a coordinated fashion, are transcribed at different times of the life cycle of T4. The pre-early genes are transcribed immediately after infection of the host. The other two sets remain untranscribed until, at a later time, they are transcribed separately. This control appears to be due to the interaction between a polypeptide subunit of the DNA-dependent RNA polymerase (*sigma*) and internal DNA sequences (*promotor regions*) present at the 5' end of the structural genes. Recognition of the promotor region by sigma facilitates polymerase orientation and transcription. Different sigmas recognize and bind to different promotors. The promotor of the T4 preearly genes is recognized by the *E. coli* RNA polymerase sigma factor so that the preearly genes are transcribed immediately following infection. One of the transcriptional products of the preearly genes is a T4 specific sigmalike product, which, when translated, recognizes only the promotor in front of the early genes. This T4 sigma replaces the *E. coli* sigma, turning off the transcription of the preearly genes and turning on the transcription of the early genes. In this way, a temporal sequence of coordinated gene activity occurs without the need for environmental stimuli.

These two types of gene regulation operate at the level of gene transcription. Evidence also exists that, in bacteria and phage, gene activity can be controlled at the time of translation. Due to the complex nature of the translation machinery, many mechanisms may exist for the translational control of gene activity. For example, limiting amounts of the factors required for the initiation, elongation, termination, or release of polypeptide chains could modulate the overall rate of translation. The presence or absence of specific amino acids, transfer RNAs, and synthetases (enzymes which link amino acids to tRNA molecules) could regulate the translation of specific messenger RNAs requiring their presence.[26] There is evidence that ribosome populations are heterogenous and certain ribosomes may bind to particular messenger RNAs more efficiently than others.[24] Evidence also exists that both internal RNA base sequences and secondary and tertiary structural conformations in messenger RNA play a major role in the regulation of translation.[30] Finally, stability control of messenger RNA has been implicated as a means of regulating the activity of genes at the time of translation.[31]

Up to this time, we have considered only data relative to the regulation of gene activity in bacteria and phage. Due to the experimental accessibility of the control systems in these organisms, they were good candidates for use in the initial work defining gene control mechanisms. In addition to this, however, there was a high level of interest in elucidating the mechanisms of gene regulation operating during cellular differentiation in higher organisms. Observations concerning the pattern of gene activity during development revealed the existence of major differences between those patterns in microbial and multicellular organisms. While gene activity in microbial systems could be classified as taking place in a coordinated manner involving linked genes, gene activity in higher systems involved the simultaneous function of noncontiguous genes. In addition, all genes operating at one instant in time, in a multicellular organism, apparently are not turned on or off in unison. Thus, gene activity in higher organisms could be characterized as a noncoordinated type of activity and the prokaryote operon or promotor models of gene control, as they were originally proposed, would not be sufficient to provide such patterns of regulation.

In 1969, Britten and Davidson presented a paper proposing a model that incorporated many of the elements of the operon.[9] It was, however, more complex in that each structural gene was controlled by a long sequence or bank of regulator genes. The important feature of this model was that the bank of regulators allowed for the turning on and off of individual, noncontiguous structural genes by a single environmental stimulus. This stimulus was recognized by a third type of regulatory element, the sensory gene. Although there is still little direct support for this model, there is some indirect evidence that such a system exists in higher organisms. One feature of this model is that it requires repeating sequences of DNA at the sites of the regulatory genes. Britten and Kohne[10] reported that large regions of nucleotide sequence redundancy are known to be present in the genomes of eukaryotes. Another feature of the model is that the redundant regulatory elements, adjacent to the structural genes, are not translated. Chemical studies of the first RNA transcribed off of chromosomal DNA show that it is a high molecular weight RNA.[34] This initial RNA, called *heterogenous RNA* (hetRNA), is rapidly broken down in the nucleus to an RNA molecule possessing a molecular weight characteristic of single structural gene transcripts. This single gene transcript passes to the site of protein synthesis in the cytoplasm.[15] It is tempting to equate this rapidly degraded portion of the heterogenous RNA to a transcript made off of a bank of linked regulator genes responsible for the control of the activity of the linked structural gene. Since these regulator transcripts are not to be translated, they are degraded in the nucleus, leaving only the RNA from the structural gene to pass into the cytoplasm.

Finally, two additional ways of controlling the differential activity of the eucaryote gene must be mentioned. Unlike the previous mode, both mechanisms have substantial experimental support. Because these two mechanisms of control take advantage of structural and functional characteristics unique to eukaryotes, they are not found in prokaryote systems. First, the structural packaging of the genetic information in eukaryotes is more complex than the structural organization of the genetic information of bacteria or phage. Eukaryote chromosomal DNA is complexed with RNA and several classes of proteins while prokaryote DNA is not complexed with these additional classes of molecules. This structural complexity provides a means of regulating the activity of the eukaryote genome. One class of proteins, the histones, has been implicated as repressors of transcription.[3] Regions of DNA, covered by histones are blocked from active transcription. Removal of histones frees the DNA to be used as a template for RNA transcription. Thus, the complexing or decomplexing of histones to DNA provides a means for transcriptional control of gene activity. Due to the poor template activity of intact chromosomes, it appears that the great majority of the DNA in eukaryote chromosomes is complexed with histones — chromosomal DNA then is generally in the repressed condition. Thus, this type of regulation primarily involves derepression. Outside signals must derepress specific regions of the eukaryote chromosome by stimulating the removal of histones, freeing the DNA for transcription. Evidence for such environmentally controlled, localized derepression in eukaryote chromosomes comes from experiments demonstrating the effects of moulting hormones on the puffing patterns of insect chromosomes. These regionalized structural changes have been correlated with increased transcriptional activity. The above puffing phenomenon indicates that it is possible to have a restricted region of the chromosome carrying on active transcription while the remaining portion is repressed.

The second means of control operates at the level of gene replication. In microbial systems, the cellular amplification of specific genes has not been found and the differential gene activity is controlled only at the level of transcription or translation. An exception to this generalization has been found in eukaryotes. It involves the extrachromosomal amplification of genes responsible for the transcription of 18S and 28S ribosomal RNA in amphibian oocytes and lower invertebrates.[11] This "extrachromosomal" amplification of the 450 redundant copies of the ribosomal DNA genes, increases the total number of rDNA genes in the developing oocyte to approximately 6×10^5 copies. This provides more DNA templates for the synthesis of ribosomal RNA. So long as the enzymes and substrates required for transcription are not limiting, this gene amplification provides a means for the regulation of the transcriptional activity of the ribosomal RNA genes (excluding 5S RNA). This type of extrachromosomal gene amplification has not been found in eukaryotes for genes other than the ribosome genes, so that this appears to be a highly specialized means of regulating gene activity.

REFERENCES

1. Beadle, G. W., and G. Ephrussi 1936. The differentiation of eye pigments in *Drosophila* as studied by transplantation. *Genetics* 21:225–47.

2. Beadle, G. W., and E. L. Tatum 1941. Genetic control of biochemical reactions in *Neurospora*. *Proc. Nat. Acad. Sci. U.S.* 27: 499–506.

3. Bonner, J., and J. R. Wu 1973. A proposal for the structure of the *Drosophila* genome. *Proc. Nat. Acad. Sci. U.S.* 70:535–37.

4. Brachet, J. 1941. *La detection histochimique et le microdosage des acides pentosenucléiques.* (*Tissus animaux — developpement embryonnaire des amphibiens*). *Enzymologia* 10:87.

5. Brenner, S. 1957. On the impossibility of all overlapping triplet codes in information transfer from nucleic acids to proteins. *Proc. Nat. Acad. Sci. U.S.* 43:687.

6. Brenner, S., F. Jacob, and M. Meselson 1961. An unstable intermediate carrying information from genes to ribosomes for protein synthesis. *Nature* 190:576-81.

7. Bridges, C. B. 1936. The bar 'gene' a duplication. *Science* 83:210.

8. Briggs, R., and T. J. King 1959. Nucleocytoplasmic interaction in eggs and embryos. In *The cell,* eds. J. Brachet and A. E. Mirsky, vol. 1, pp. 537-617. New York: Academic Press.

9. Britten, R. J., and E. H. Davidson 1969. Gene regulation for higher cells: a theory. *Science* 165:349-57.

10. Britten, R. J., and D. E. Kohne 1968. Repeated sequences in DNA. *Science* 161:529-40.

11. Brown, D. D., and J. B. David 1968. Specific gene amplification in oocytes. *Science* 160:272-80.

12. Caspersson, T. 1941. *Studien uber den Eiweissumsatz der Zelle. Naturwissenschaften.* 29:33.

13. Crick. F. H. C. 1958. On protein synthesis. *Symp. Soc. Exp. Biol.* 12:138.

14. Crick, F. H. C., L. Barnett, S. Brenner, and R. J. Watts-Tobin 1961. General nature of the genetic code for proteins. *Nature* 192:1227-32.

15. Darnell, J. E., L. Philipson, R. Wall and M. Adesnik 1971. Polyadenylic acid sequences: role in conversion of nuclear RNA into messenger RNA. *Science* 174:507-10.

16. Davidson, E. H., M. Crippa, and A. E. Mirsky 1968. Evidence for the appearance of novel gene products during amphibian blastulation. *Proc. Nat. Acad. Sci. U.S.* 60:152.

17. Dounce, A. L. 1952. Duplicating mechanism for peptide chain and nucleic acid synthesis. *Enzymologia* 15:251-58.

18. Flickinger, R. A., R. Greene, D. M. Kohl, and M. Miyagi 1966. Patterns of synthesis of DNA-like RNA in parts of developing frog embryos. *Proc. Nat. Acad. Sci. U.S.* 56:1712.

19. Gamow, G. 1954. Possible relation between deoxyribonucleic acid and protein structure. *Nature* 173:318.

20. Garrod, A. 1902. The incidence of alkaptonuria: a study in chemical individuality. *Lancet* 2:1616-20.

21. Garrod, A. 1909. *Inborn errors of metabolism.* Oxford: Oxford Univ. Press.

22. Guha, A., W. Szybalski, Q. Salser, A. Bolle, E. P. Geiduschek, and J. F. Pulitzer 1971. Controls and polarity of transcription during bacteriophage T4 development. *J. Mol. Biol.* 59:329-49.

23. Gurdon, J. B. 1968. Transplantation nucleic and cell differentiation. *Sci. Am.* 219(6):24-35.

24. Heywood. S. M. 1969. Synthesis of myosin on heterogeneous ribosomes. *Cold Spring Harbor Symp. Quant. Biol.* 34:799-803.

25. Hoagland, M. B., M. L. Stephenson, J. F. Scott, L. I. Hecht, and P. C. Zamecnik 1958. A soluble ribonucleic acid intermediate in protein synthesis. *J. Biol. Chem.* 231:241-57.

26. Ilan, J. 1969. The role of tRNA in translational control of specific mRNA during insect metamorphosis. *Cold Spring Harbor Symp. Quant. Biol.* 34:787-91.

27. Jacob, F., and J. Monod 1961. Genetic regulatory mechanisms in the synthesis of proteins. *J. Mol. Biol.* 3:318-56.

28. Jacob, F., and J. Monod 1965. Genetic mapping of the elements of the lactose region in *Escherichia coli. Biochem. Biophys. Res. Comm.* 18:693-701.

29. Khorana, H. G. 1966-1967. Polynucleotide synthesis and the genetic code. *Harvey Lectures* 62:79-105.

30. Lodish, H., and H. D. Roberston 1969. Regulation of *in vitro* translation of bacteriophage f2 RNA. *Cold Spring Harbor Symp. Quant. Biol.* 33:655-73.

31. Mosteller, R. D., J. K. Rose, and C. Yanofsky 1970. Transcription, initiation and degradation of trp mRNA. *Cold Spring Harbor Symp. Quant. Biol.* 35:461-66.

32. Nirenberg, M., and P. Leder 1964. RNA codewords and protein synthesis. *Science* 145:1399-1407.

33. Nirenberg, M. W. and J. H. Matthaei 1961. The dependence of cell-free protein synthesis in *E. coli* upon naturally occurring or synthetic polyribonucleotides. *Proc. Nat. Acad. Sci. U.S.* 47:1588-1602.

34. Pagoulatos, G. N., and J. E. Darnell, Jr. 1970. Fractionation of heterogenous nuclear RNA: rates of hybridization and chromosomal distribution of reiterated sequences. *J. Mol. Biol.* 54:517-35.

35. Sueoka, N., and T. Kano-Sueoka 1970. Transfer RNA and cell differentiation. *Prog. Nuc. Acid Res.* 10:23-55.

36. Transcription of Genetic Material, 1971. *Cold Spring Harbor Symp. Quant. Biol.* 35. New York: Cold Spring Harbor Laboratory of Quantitative Biology.

37. Wilson, E. B. 1925. *The cell in development and heredity.* New York: Macmillan.

38. Woese, C. R. 1967. *The genetic code: the molecular basis for genetic expression.* New York: Harper and Row.

39. Yanofsky, C., B. C. Carlton, J. R. Guest, D. R. Helinski, and U. Henning 1964. On the colinearity of gene structure and protein structure. *Proc. Nat. Acad. Sci. U.S.* 51:266-72.

40. Zamecnik, P. C. 1969. An historical account of protein synthesis, with current overtones — a personalized view. *Cold Spring Harbor Symp. Quant. Biol.* 34:1-16.

41. Zamecnik, P. C., M. L. Stephenson, and L. I. Hecht 1958. Intermediate reactions in amino acid incorporation. *Proc. Nat. Acad. Sci. U.S.* 44:73.

42. Zamecnik, P. C., M. L. Stephenson, J. F. Scott, and M. B. Hoagland 1957. Incorporation of ^{14}C-ATP into soluble RNA isolated from 105,000 $\times$ g supernatant of rat liver. *Fed. Proc.* 16:275.

Genetic control of biochemical reactions in *Neurospora*

G. W. BEADLE and E. L. TATUM

Reprinted by authors' and publisher's permission from Proceedings of the National Academy of Sciences, U.S., *vol. 27, 1941, pp. 499–506.*

George Beadle was concerned with establishing the quantitative relationship between genes and enzymes, i.e., does a gene control the production of one or more specific enzymes? It was clear that this question had to be resolved before the mechanism of gene function could be understood. As the introduction to this part noted, Beadle began his biochemical genetic studies using *Drosophila*. However, the complexity of that diploid eukaryote organism confounded his work. In the following paper, Beadle and Tatum reported their success in utilizing the common bread mold, *Neurospora crassa*. One should note the special features of *Neurospora* that facilititate its use in biochemical studies.

Beadle and Tatum presented preliminary data that was insufficient to warrant a statement concerning the quantitative relationship of genes and gene products. However, during the next three years, using the above experimental system, they accumulated enough evidence to support the hypothesis that one gene is responsible for the production of one specific enzyme. This paper, placed in context with Beadle's earlier work with *Drosophila*, illustrates the importance of utilizing the organism best suited to answer the specific questions set forth in a research proposal.

From the standpoint of physiological genetics the development and functioning of an organism consist essentially of an integrated system of chemical reactions controlled in some manner by genes. It is entirely tenable to suppose that these genes which are themselves a part of the system, control or regulate specific reactions in the system either by acting directly as enzymes or by determining the specificities of enzymes.[1] Since the components of such a system are likely to be interrelated in complex ways, and since the synthesis of the parts of individual genes is presumably dependent on the functioning of other genes, it would appear that there must exist orders of directness of gene control ranging from simple one-to-one relations to relations of great complexity. In investigating the roles of genes, the physiological geneticist usually attempts to determine the physiological and biochemical bases of already known hereditary traits. This approach, as made in the study of anthocyanin pigments in plants,[2] the fermentation of sugars by yeasts[3] and a number of other instances,[4] has established that many biochemical reactions are in fact controlled in specific ways by specific genes. Furthermore, investigations of this type tend to support the assumption that gene and enzyme specificities are of the same order.[5] There are, however, a number of limitations inherent in this approach. Perhaps the most serious of these is that the investigator must in general confine himself to a study of nonlethal heritable characters. Such characters are likely to involve more or less nonessential so-called terminal reactions.[5] The selection of these for genetic study was perhaps responsible for the now rapidly disappearing belief that genes are concerned only with the control of "superficial" characters. A second difficulty, not unrelated to the first, is that the standard approach to the problem implies the use of the characters with visible manifestations. Many such characters involve morphological variations, and these are likely to be based on systems of biochemical reactions so complex as to make analysis exceedingly difficult.

Considerations such as those just outlined have led us to investigate the general problem of the genetic control of developmental and metabolic reactions by reversing the ordinary procedure and, instead of attempting to work out the chemical bases of known genetic characters, to set out to determine if and how genes control known biochemical reactions. The Ascomycete *Neurospora* offers many advantages for such an approach and is well suited to genetic studies.[6] Accordingly, our program has been built around this organism. The procedure is based on the assumption that x-ray treatment will induce mutations in genes concerned with the control of known specific chemical reactions. If the organism must be able to carry out a certain chemical reaction to survive on a given medium, a mutant unable to do this will obviously be lethal on this medium. Such a mutant can be maintained and studied, however, if it will grow on a medium to which has been added the essential product of the genetically blocked reaction. The experimental procedure based on this reasoning can best be illustrated by considering a hypothetical example. Normal strains of *Neurospora crassa* are able to use sucrose as a carbon source, and are therefore able to carry out the specific and enzymatically controlled reaction involved in the hydrolysis of this sugar. Assuming this reaction to be genetically controlled, it should be possible to induce a gene to mutate to a condition such that the organism could no longer carry out sucrose hydrolysis. A strain carrying this mutant would then be unable to grow on a medium containing sucrose as a sole carbon source but should be able to grow on a medium containing some other normally utilizable carbon source. In other words, it should be possible to establish and maintain such a mutant strain on a medium containing glucose and detect its inability to utilize sucrose by transferring it to a sucrose medium.

Essentially similar procedures can be developed for a great many metabolic processes. For example, ability to synthesize growth factors (vitamins), amino acids and other essential substances should be lost through gene mutation if our assumptions are correct. Theoretically, any such metabolic deficiency can be "bypassed" if the substance lacking can be supplied in the medium and can pass cell walls and protoplasmic membranes.

In terms of specific experimental practice, we have devised a procedure in which x-rayed single-spore cultures are established on a so-called complete medium, i.e., one containing as many of the nor-mally synthesized constituents of the organism as is practicable. Subsequently these are tested by transferring them to a "minimal" medium, i.e., one requiring the organism to carry on all the essential syntheses of which it is capable. In practice the complete medium is made up of agar, inorganic salts, malt extract, yeast extract and glucose. The minimal medium contains agar (optional), inorganic salts and biotin, and a disaccharide, fat or more complex carbon source. Biotin, the one growth factor that wild-type *Neurospora* strains cannot synthesize,[7] is supplied in the form of a commercial concentrate containing 100 micrograms of biotin per cc.* Any loss of ability to synthesize an essential substance present in the complete medium and absent in the minimal medium is indicated by a strain growing on the first and failing to grow on the second medium. Such strains are then tested in a systematic manner to determine what substance or substances they are unable to synthesize. These subsequent tests include attempts to grow mutant strains on the minimal medium with (1) known vitamins added, (2) amino acids added or (3) glucose substituted for the more complex carbon source of the minimal medium.

Single ascospore strains are individually derived from perithecia of *N. crassa* and *N. sitophila* x-rayed prior to meiosis. Among approximately 2000 such strains, three mutants have been found that grow essentially normally on the complete medium and scarcely at all on the minimal medium with sucrose as the carbon source. One of these strains (*N. sitophila*) proved to be unable to synthesize vitamin B_6 (pyridoxine). A second strain (*N. sitophila*) turned out to be unable to synthesize vitamin B_1 (thiamine). Additional tests show that this strain is able to synthesize the pyrimidine half of the B_1 molecule but not the thiazole half. If thiazole alone is added to the minimal medium, the strain grows essentially normally. A third strain (*N. crassa*) has been found to be unable to synthesize para-aminobenzoic acid. This mutant strain appears to be entirely normal when grown on the minimal medium to which *p*-aminobenzoic acid has been added. Only in the case of the "pyridoxinless" strain has an analysis of the inheritance of the induced metabolic defect been investigated. For this reason detailed

---

*The biotin concentrate used was obtained from the S. M. A. Corporation, Chagrin Falls, Ohio.

accounts of the thiamine-deficient and *p*-amino-benzoic acid-deficient strains will be deferred.

Qualitative studies indicate clearly that the pyridoxinless mutant, grown on a medium containing one microgram or more of synthetic vitamin B_6 hydrochloride per 25 cc of medium, closely approaches in rate and characteristics of growth normal strains grown on a similar medium with no B_6. Lower concentrations of B_6 give intermediate growth rates. A preliminary investigation of the quantitative dependence of growth of the mutant on vitamin B_6 in the medium gave the results summarized in Table 1. Additional experiments have given results essentially similar but in only approximate quantitative agreement with those of table 1. It is clear that additional study of the details of culture conditions is necessary before rate of weight increase of this mutant can be used as an accurate assay for vitamin B_6.

It has been found that the progression of the frontier of mycelia of *Neurospora* along a horizontal glass culture tube half-filled with an agar medium provides a convenient method of investigating the quantitative effects of growth factors. Tubes of about 13 mm inside diameter and about 40 cm in length are used. Segments of about 5 cm at the two ends are turned up at an angle of about 45°. Agar medium is poured in so as to fill the tube about half full and is allowed to set with the main segment of the tube in a horizontal position. The turned up ends of the tube are stoppered with cotton plugs. Inoculations are made at one end of the agar surface and the position of the advancing front recorded at convenient intervals. The frontier formed by the advancing mycelia is remarkably well defined, and there is no difficulty in determining its position to within a millimeter or less. Progression along such tubes is strictly linear with time and the rate is independent of tube length (up to 1.5 meters). The rate is not changed by reducing the inside tube diameter to 9 mm, or by sealing one or both ends. It therefore appears that gas diffusion is in no way limiting in such tubes.

The results of growing the pyridoxinless strain in horizontal tubes in which the agar medium contained varying amounts of B_6 are shown graphically in Figures 1 and 2. Rate of progression is clearly a function of vitamin B_6 concentration in the medium.* It is likewise evident that there is no significant difference in rate between the mutant supplied with B_6 and the normal strain growing on a medium without this vitamin. These results are consistent with the assumption that the primary physiological difference between pyridoxinless and normal strains is the inability of the former to carry out the synthesis of vitamin B_6. There is certainly more than one step in this synthesis and accordingly the gene differential involved is presumably concerned with only one specific step in the biosynthesis of vitamin B_6.

In order to ascertain the inheritance of the pyridoxinless character, crosses between normal and mutant strains were made. The techniques for hybridization and ascospore isolation have been worked out and described by Dodge, and by Lindergren.[6] The ascospores from 24 asci of the cross were iso-

Table 1 Growth of pyridoxinless strain of *N. sitophila* on liquid medium containing inorganic salts,* 1% sucrose, and 0.004 microgram biotin per cc. Temperature 25°C. Growth period, 6 days from inoculation with conidia

Micrograms B_6 per 25 cc medium	Strain	Dry weight mycelia, mg
0	Normal	76.7
0	Pyridoxinless	1.0
0.01	Pyridoxinless	4.2
0.03	Pyridoxinless	5.7
0.1	Pyridoxinless	13.7
0.3	Pyridoxinless	25.5
1.0	Pyridoxinless	81.1
3.0	Pyridoxinless	81.1
10.0	Pyridoxinless	65.4
30.0	Pyridoxinless	82.4

*Throughout our work with *Neurospora*, we have used as a salt mixture the one designated number 3 by Fries, N., *Symbolae Bot. Upsalienses*, vol. 3, No. 2, 1–188, 1938. This has the following composition: NH_4 tartrate, 5 g; NH_4NO_3, 1 g; KH_2PO_4, 1 g; $MgSO_4 \cdot 7H_2O$, 0.5 g; NaCl, 0.1 g; $CaCl_2$, 0.1 g; $FeCl_3$, 10 drops 1% solution; H_2O, 1 l. The tartrate cannot be used as a carbon source by *Neurospora*.

*It is planned to investigate further the possibility of using the growth of *Neurospora* strains in the described tube as a basis of vitamin assay, but it should be emphasized that such additional investigation is essential in order to determine the reproducibility and reliability of the method.

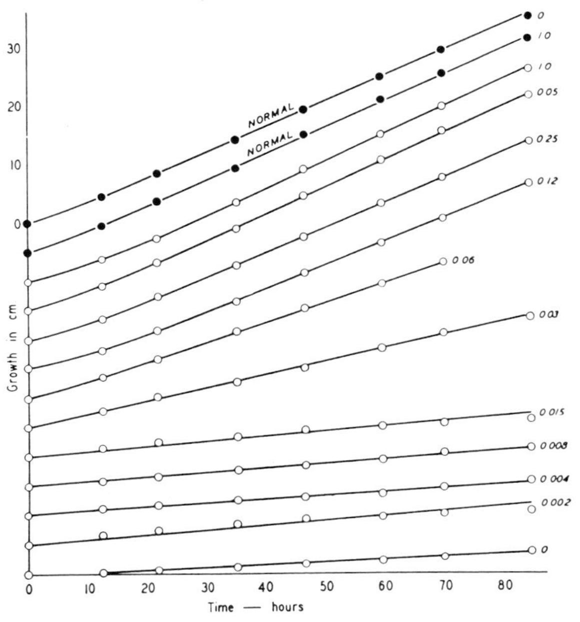

Fig. 1. Growth of normal (top two curves) and pyridoxinless (remaining curves) strains of *Neurospora sitophila* in horizontal tubes. The scale on the ordinate is shifted a fixed amount for each successive curve in the series. The figures at the right of each curve indicate concentration of pyridoxine (B_6) in micrograms per 25 cc medium.

lated and their positions in the asci recorded. For some unknown reason, most of these failed to germinate. From seven asci, however, one or more spores germinated. These were grown on a medium containing glucose, malt extract and yeast extract, and in this they all grew normally. The normal and mutant cultures were differentiated by growing them on a B_6-deficient medium. On this medium the mutant cultures grew very little, while the non-mutant ones grew normally. The results are summarized in Table 2. It is clear from these rather limited data that this inability to synthesize vitamin B_6 is transmitted as it should be if it were differentiated from normal by a single gene.

The preliminary results summarized above appear to us to indicate that the approach outlined may offer considerable promise as a method of learning more about how genes regulate development and function. For example, it should be possible, by finding a number of mutants unable to carry out a particular step in a given synthesis, to determine whether only one gene is ordinarily concerned with the immediate regulation of a given specific chemical reaction.

It is evident, from the standpoints of biochemistry and physiology, that the method outlined is of value as a technique for discovering additional substances of physiological significance. Since the complete medium used can be made up with yeast extract or with an extract of normal *Neurospora*, it is evident that if, through mutation, there is lost the ability to synthesize an essential substance, a test strain is thereby made available for use in isolating the substance. It may, of course, be a substance not previously known to be essential for the growth of any organism. Thus we may expect to discover new vitamins, and in the same way, it should be possible to discover additional essential amino acids if such exist. We have, in fact, found a mutant strain that is

Table 2 Results of classifying single ascospore cultures from the cross of pyridoxinless and normal *N. sitophila*

Ascus number	1	2	3	4	5	6	7	8
17	—	pdx	pdx	pdx	N	N	N	—
18	—	—	N	N	—	—	pdx	pdx
19	—	pdx	—	—	—	—	—	N
20	—	—	N	—	—	—	—	pdx
22	—	—	N	—	—	—	—	—
23	—	*	*	*	N	N	pdx	pdx
24	N	N	N	N	pdx	pdx	pdx	pdx

N, normal growth on B_6-free medium. pdx, slight growth on B_6-free medium. Failure of ascospore germination indicated by dash.
*Spores 2, 3 and 4 isolated but positions confused. Of these, two germinated and both proved to be mutants.

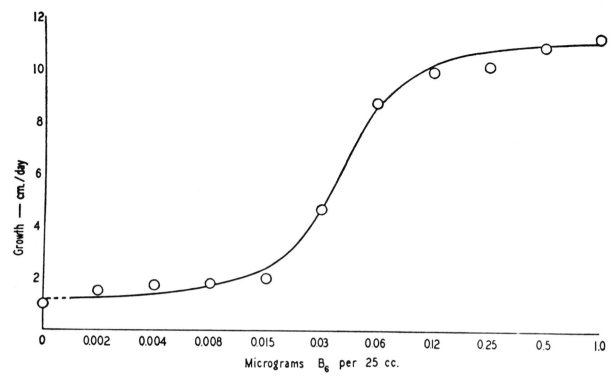

Fig. 2. The relation between growth rate (cm/day) and vitamin B$_6$ concentration.

able to grow on a medium containing Difco yeast extract but unable to grow on any of the synthetic media we have so far tested. Evidently some growth factor present in yeast and as yet unknown to us is essential for *Neurospora.*

SUMMARY

A procedure is outlined by which, using *Neurospora,* one can discover and maintain x-ray induced mutant strains which are characterized by their inability to carry out specific biochemical processes.

Following this method, three mutant strains have been established. In one of these the ability to synthesize vitamin B$_6$ has been wholly or largely lost. In a second the ability to synthesize the thiazole half of the vitamin B$_1$ molecule is absent, and in the third para-aminobenzoic acid is not synthesized. It is therefore clear that all of these substances are essential growth factors for *Neurospora.**

Growth of the pyridoxinless mutant (a mutant unable to synthesize vitamin B$_6$) is a function of the B$_6$ content of the medium on which it is grown. A method is described for measuring the growth by following linear progression of the mycelia along a horizontal tube half-filled with an agar medium.

Inability to synthesize vitamin B$_6$ is apparently differentiated by a single gene from the ability of the organism to elaborate this essential growth substance.

Note Since the manuscript of this paper was sent

*The inference that the three vitamins mentioned are essential for the growth of normal strains is supported by the fact that an extract of the normal strain will serve as a source of vitamin for each of the mutant strains.

to press it has been established that inability to synthesize both thiazole and aminobenzoic acid is also inherited as though differentiated from normal by single genes.

Work supported in part by a grant from the Rockefeller Foundation. The authors are indebted to Doctors B. O. Dodge, C. C. Lindegren and W. S. Malloch for stocks and for advice on techniques, and to Miss Caryl Parker for technical assistance.

REFERENCES

1. The possibility that genes may act through the mediation of enzymes has been suggested by several authors. See Troland, L. T., *Amer. Nat.* 51:321–350, 1917; Wright, S., *Genetics* 12:530–569, 1927; and Haldane, J. B. S., in *Perspectives in Biochemistry,* Cambridge Univ. Press, pp. 1–10, 1937, for discussion and references.
2. Onslow, Scott-Moncrieff and others, see review by Lawrence, W. J. C., and Price, J. R., *Biol. Rev.* 15:35–58, 1940.
3. Winge, O. and Laustsen, O., *Compt. rend. Lab. Carlsberg, Serie physiol.* 22:337–352, 1939.
4. See Goldschmidt, R., *Physiological Genetics,* McGraw-Hill, pp. 1–375, 1939; and Beadle, G. W. and Tatum, E. L., *Amer. Nat.* 75:107–116, 1941, for discussion and references.
5. See Sturtevant, A. H. and Beadle, G. W., *An Introduction to Genetics,* Saunders, pp. 1–391, 1931; and Beadle, G. W. and Tatum, E. L., *loc. cit.,* footnote 4.
6. Dodge, B. O., *Jour. Agric. Res.* 35:289–305, 1927; and Lindegren, C. C., *Bull. Torrey Bot. Club* 59:85–102, 1932.
7. In so far as we have carried them, our investigations on the vitamin requirements of *Neurospora* corroborate those of Butler, E. T., Robbins, W. J., and Dodge, B. O., *Science* 94:262–263, 1941.

Excerpt from: On protein synthesis

F. H. C. CRICK

Reprinted with permission from Symposium of the Society for Experimental Biology, *vol. 12, 1958, pp. 138–163.*

Biological investigation requires testable hypotheses. Specifically, research is designed to evaluate the validity of predictions formulated on hypotheses. These hypotheses may be narrow in scope — limited to one small aspect of life — or they may be of an encompassing nature, stimulating and guiding the development of research in many different disciplines. Crick's paper of 1958 presented a hypothetical model for protein synthesis that, due to the creativity and imaginative power of the author, influenced experimental design and interpretation of data in genetics, biochemistry, cell biology, and other related fields of investigation. In 1902, Garrod was the first to have hypothesized a direct relationship between genes and enzymes; and 56 years later, Crick finally opened the door to an understanding of that relationship. While reading this paper make particular note of Crick's ideas about the genetic code, many of which, though influential, were incorrect. Note also that Crick developed the idea of colinearity between gene and gene product, i.e., the sequence of nucleotides in a gene corresponds to the order of the amino acids in the polypeptide it specifies.

The Nature of Protein Synthesis

The basic dilemma of protein synthesis has been realized by many people, but it has been particularly aptly expressed by Dr. A. L. Dounce (1956):

My interest in templates, and the conviction of their necessity, originated from a question asked me on my Ph.D oral examination by Professor J. B. Sumner. He enquired how I thought proteins might be synthesized. I gave what seemed the obvious answer, namely, that enzymes must be responsible. Professor Sumner then asked me the chemical nature of enzymes, and when I answered that enzymes were proteins or contained proteins as essential components, he asked whether these enzyme proteins were synthesized by other enzymes and so on *ad infinitum.*

The dilemma remained in my mind, causing me to look for possible solutions that would be acceptable, at least from the standpoint of logic. The dilemma, of course, involves the specificity of the protein molecule, which doubtless depends to a considerable degree on the sequence of amino acids in the peptide chains of the protein. The problem is to find a reasonably simple mechanism that could account for specific sequences without demanding the presence of an ever-increasing number of new specific enzymes for the synthesis of each new protein molecule.

It is thus clear that the synthesis of proteins must be radically different from the synthesis of polysaccharides, lipids, coenzymes and other small molecules; that it must be relatively simple, and to a considerable extent uniform throughout Nature; that it must be highly specific, making few mistakes; and that in all probability it must be controlled at not too many removes by the genetic material of the organism.

The Essence of the Problem

A systematic discussion of our present knowledge of protein synthesis could usefully be set out under three headings, each dealing with a flux: the flow of energy, the flow of matter, and the flow of information. I shall not discuss the first of these here. I shall have something to say about the second, but I shall particularly emphasize the third — the flow of information.

By information I mean the specification of the amino acid sequence of the protein. It is conventional at the moment to consider separately the synthesis of the polypeptide chain and its folding. It is of course possible that there is a special mechanism for folding up the chain, but the more likely hypothesis is that the *folding is simply a function of the order of the amino acids*, provided it takes place as the newly formed chain comes off the template. I think myself that this latter idea may well be correct, though I would not be surprised if exceptions existed, especially the globulins and the adaptive enzymes.

Our basic handicap at the moment is that we have no easy and precise technique with which to study how proteins are folded, whereas we can at least make some experimental approach to amino acid sequences. For this reason, if for no other, I shall ignore folding in what follows and concentrate on the determination of sequences. It is as well to realize, however, that the idea that the two processes can be considered separately is in itself an assumption.

The actual chemical step by which any two amino acids (or activated amino acids) are joined together is probably always the same, and may well not differ significantly from any other biological condensation. The unique feature of protein synthesis is that only a single standard set of twenty amino acids can be incorporated, and that for any particular protein the *amino acids must be joined up in the right order*. It is this problem, the problem of "sequentialization," which is the crux of the matter, though it is obviously important to discover the exact chemical steps which lead up to and permit the crucial act of sequentialization.

As in even a small bacterial cell there are probably a thousand different kinds of protein, each containing some hundreds of amino acids in its own rigidly determined sequence, the amount of hereditary information required for sequentialization is quite considerable.

III. RECENT EXPERIMENTAL WORK

The Role of the Nucleic Acids

It is widely believed (though not by everyone) that the nucleic acids are in some way responsible for the control of protein synthesis, either directly or indirectly. The actual evidence for this is rather meager. In the case of deoxyribonucleic acid (DNA)

it rests partly on the T-even bacteriophages, since it has been shown, mainly by Hershey and his colleagues, that whereas the DNA of the infecting phage penetrates into the bacterial cell almost all the protein remains outside (see the review by Hershey 1956); and also on Transforming Factor, which appears to be pure DNA, and which in at least one case, that of the enzyme mannitol phosphate dehydrogenase, controls the synthesis of a protein (Marmur and Hotchkiss 1955). There is also the indirect evidence that DNA is the most constant part of the genetic material, and that genes control proteins. Finally there is the very recent evidence, mainly due to the work of Benzer on the rII locus of bacteriophage, that the functional gene — the "cistron" of Benzer's terminology — consists of many sites arranged strictly *in a linear order* (Benzer 1957) as one might expect if a gene controls the order of the amino acids in some particular protein.

As is well known, the correlation between ribonucleic acid (RNA) and protein synthesis was originally pointed out by Brachet and by Caspersson. Is there any more direct evidence for this connection? In particular is there anything to support the idea that the sequentialization of the amino acids is controlled by the RNA?

The most telling evidence is the recent work on tobacco mosaic virus. A number of strains of the virus are known, and it is not difficult to show (since the protein subunit of the virus is small) that they differ in amino acid composition. Some strains, for example, have histidine in their protein, whereas others have none. Two very significant experiments have been carried out. In one, as first shown by Gierer and Schramm (1956), the RNA of the virus alone, although completely free of protein, appears to be infective, though the infectivity is low. In the other, first done by Fraenkel-Conrat, it has proved possible to separate the RNA from the protein of the virus and then recombine them to produce virus again. In this case the infectivity is comparatively high, though some of it is usually lost. If a recombined virus is made using the RNA of one strain and the protein of another, and then used to infect the plant, the new virus produced in the plant resembles very closely *the strain from which the RNA was taken*. If this strain had a protein which contained no histidine then the offspring will have no histidine either, although the plant had never been in contact with this particular protein before but only with the RNA from that strain. In other words *the viral RNA appears to carry at least part of the information which determines the composition of the viral protein.* Moreover the viral protein which was used to infect the cell was not copied to any appreciable extent (Fraenkel-Conrat 1956).

It has so far not proved possible to carry out this experiment — a model of its kind — in any other system, although very recently it has been claimed that for two animal viruses the RNA alone appears to be infective.

Turnover experiments have shown that while the labeling of DNA is homogeneous that of RNA is not. The RNA of the cell is partly in the nucleus, partly in particles in the cytoplasm and partly as the "soluble" RNA of the cell sap; many workers have shown that all these three fractions turn over differently. It is very important to realize in any discussion of the role of RNA in the cell that it is very inhomogeneous metabolically, and probably of more than one type.

The Site of Protein Synthesis

There is no known case in Nature in which protein synthesis proper (as opposed to protein modification) occurs outside cells, though, as we shall see later, a certain amount of protein can probably be synthesized using broken cells and cell fragments. The first question to ask, therefore, is whether protein synthesis can take place in the nucleus, in the cytoplasm, or in both.

It is almost certain that protein synthesis can take place in the cytoplasm without the presence of the nucleus, and it is probable that it can take place to some extent in the nucleus by itself (see the review by Brachet and Chantrenne 1956). Mirsky and his colleagues (see the review by Mirsky, Osawa, and Allfrey 1956) have produced evidence that some protein synthesis can occur in isolated nuclei, but the subject is technically difficult and in this review I shall quite arbitrarily restrict myself to protein synthesis in the cytoplasm.

In recent years our knowledge of the structure of the cytoplasm has enormously increased, due mainly to the technique of cutting thin sections for the electron microscope. The cytoplasm of many cells contains an "endoplasmic reticulum" of double membranes, consisting mainly of protein and lipid

(see the review of Palade 1956). On one side of each membrane appear small electron-dense particles (Palade 1955). Biochemical studies (Palade and Siekevitz 1956; among others) have shown that these particles, which are about 100–200 A in diameter, consist almost entirely of protein and RNA, in about equal quantities. Moreover the major part of the RNA of the cell is found in these particles.

When such a cell is broken open and the contents fractionated by centrifugation, the particles, together with fragments of the endoplasmic reticulum, are found in the "microsome" fraction, and for this reason I shall refer to them as microsomal particles.

These microsomal particles are found in almost all cells. They are particularly common in cells which are actively synthesizing protein whereas the endoplasmic reticulum is most conspicuously present in (mammalian) cells which are secreting very actively. Thus both the cells of the pancreas and those of an ascites tumour contain large quantities of microsomal particles, but the tumour has little endoplasmic reticulum, whereas the pancreas has a lot. Moreover, there is no endoplasmic reticulum in bacteria.

On the other hand particles of this general description have been found in plant cells (Ts'o, Bonner, and Vinograd 1956), in yeast, and in various bacteria (Schachman, Pardee, and Stanier 1953); in fact in all cells which have been examined for them.

These particles have been isolated from various cells and examined in the ultracentrifuge (Petermann, Mizen, and Hamilton 1952; Schachman *et al.* 1953; among others). The remarkable fact has emerged that they do not have a continuous distribution of sedimentation constants, but usually fall into several well defined groups. Moreover some of the particles are probably simple aggregates of the others (Petermann and Hamilton 1957). This uniformity suggests immediately that the particles, which have "molecular weights" of a few million, have a definite structure. They are, in fact, reminiscent of the small spherical RNA-containing viruses, and Watson and I have suggested that they may have a similar type of substructure (Crick and Watson 1956).

Biologists should contrast the older concept of *microsomes* with the more recent and significant one of *microsomal particles.* Microsomes came in all sizes, and were irregular in composition; microsomal particles occur in a few sizes only, have a more fixed composition and a much higher proportion of RNA. It was hard to identify microsomes in all cells, whereas RNA-rich particles appear to occur in almost every kind of cell. In short, microsomes were rather a mess, whereas microsomal particles appeal immediately to one's imagination. It will be surprising if they do not prove to be of fundamental importance.

It should be noted, however, that Simpson and his colleagues (Simpson and McLean 1955; Simpson, McLean, Cohn, and Brandt 1957) have reported that protein synthesis can take place in mitochondria. It is known that mitochondria contain RNA, and it would be of great interest to know whether this RNA is in some kind of particle. Mitochondria are, of course, very widely distributed but they do not occur in lower forms such as bacteria. Similar remarks about RNA apply to the reported incorporation in chloroplasts (Stephenson, Thimann, and Zamecnik 1956).

Microsomal Particles and Protein Synthesis

It has been shown by the use of radioactive amino acids that during protein synthesis the amino acids appear to flow through the microsomal particles. The most striking experiments are those of Zamecnik and his coworkers on the livers of growing rats (see review by Zamecnik *et al.* 1956).

Two variations of the experiment were made. In the first the rat was given a rather large intravenous dose of a radioactive amino acid. After a predetermined time the animal was sacrificed, the liver extracted, its cells homogenized and the contents fractionated. It was found that the microsomal particle fraction was very rapidly labeled to a constant level.

In the second a very small shot of the radioactive amino acid was given, so that the liver received only a pulse of labeled amino acid, since this small amount was quickly used up. In this case the radioactivity of the microsomal particles rose very quickly *and then fell away*. Making plausible assumptions Zamecnik and his colleagues have shown that this behavior is what one would expect if most of the protein of the microsomal particles were metabolically inert, but 1 or 2% was turning over very rapidly, say within a minute or so.

Very similar results have been obtained by Rabino-

vitz and Olson (1956, 1957) using intact mammalian cells, in this case rabbit reticulocytes. They have also been able to show that the label passed into a well-defined globular protein, namely hemoglobin. Experiments along the same general lines have also been reported for liver by Simkin and Work (1957a).

We thus have direct experimental evidence that the microsomal particles are associated with protein synthesis, though the precise role they play is not clear.

Activating Enzymes

It now seems very likely that the first step in protein synthesis is the activation of each amino acid by means of its special "activating enzyme." The activation requires ATP, and the evidence suggests that the reaction is

amino acid + ATP = AMP – amino acid + pyrophosphate

The activated amino acid, which is probably a mixed anhydride of the form

$$
R-\overset{\overset{\displaystyle NH_2}{|}}{\underset{\underset{\displaystyle H}{|}}{C}}-\overset{\overset{\displaystyle O}{\|}}{C}\diagup O-\overset{\overset{\displaystyle O}{\|}}{\underset{\underset{\displaystyle O}{|}}{P}}-O-Ribose-Adenine
$$

in which the carboxyl group of the amino acid is phosphorylated, appears to be tightly bound to its enzyme and is not found free in solution.

These enzymes were first discovered in the cell-sap fraction of rat liver cells by Hoagland (Hoagland 1955; Hoagland, Keller, and Zamecnik 1956) and in yeast by Berg (1956). They have been shown by DeMoss and Novelli (1956) to be widely distributed in bacteria, and it is surmised that they occur in all cells engaged in protein synthesis. Recently Cole, Coote and Work (1957) have reported their presence in a variety of tissues from a number of animals.

So far good evidence has been found for this reaction for about half the standard twenty amino acids, but it is believed that further research will reveal the full set. Meanwhile Davie, Koningsberger and Lipmann (1956) have purified the tryptophan-activating enzyme. It is specific for trytophan (and certain tryptophan analogues) and will only handle the L-isomer. Isolation of the tyrosine enzyme has also been briefly reported (Koningsberger, van de Ven, and Overbeck 1957; Schweet 1957).

The properties of these enzymes are obviously of the greatest interest, and much work along these lines may be expected in the near future. For example, it has been shown that the tryptophan-activating enzyme contains what is probably a derivative of guanine (perhaps GMP) very tightly bound. It is possible to remove it, however, and to show that its presence is not necessary for the primary activation step. Since the enzyme is probably involved in the next step in protein synthesis it is naturally suspected that the guanine derivative is also required for this reaction, whatever it may be.

In Vitro Incorporation

In order to study the relationship between the activating enzymes and the microsomal particles it has proved necessary to break open the cells and work with certain partly purified fractions. Unfortunately it is rare to obtain substantial net protein synthesis from such systems, and there is a very real danger that the incorporation of the radioactivity does not represent true synthesis but is some kind of partial synthesis or exchange reaction. This distinction has been clearly brought out by Gale (1953). The work to be described, therefore, has to be accepted with reservations. (See the remarks of Simkin and Work, this Symposium.) It has been shown, however, in the work described below, that the amino acid is incorporated into true peptide linkage.

Again the significant results were first obtained by Zamecnik and his coworkers (reviewed in Zamecnik *et al.* 1956). The requirements so far known appear to fall into two parts:

1. The activation of the amino acids for which, in addition to the labelled amino acid, one requires the "pH 5" fraction, containing the activating enzymes, ATP and (usually) an ATP-generating system. There appears to be no requirement for any of the pyrimidine or guanine nucleotides.

2. The transfer to the microsomal particles. For this one requires the previous system plus GTP or GDP (Keller and Zamecnik 1956) and of course the microsomal particles; the endoplasmic reticulum does not appear to be necessary (Littlefield and Keller 1957).

Hultin and Beskow (1956) have reported an experiment which shows clearly that the amino acids become bound in some way. They first incubate the mixture described in (1) above. They then add a great excess of *un*labeled amino acid before adding the microsomal particles. Nevertheless some of the labeled amino acid is incorporated into protein, showing that it was in some place where it could not readily be diluted.

Very recently an intermediate reaction has been suggested by the work of Hoagland, Zamecnik and Stephenson (1957), who have discovered that in the first step the "soluble" RNA contained in the "pH 5" fraction became labeled with the radioactive amino acid. The bond between the amino acid and the RNA appears to be a covalent one. This labeled RNA can be extracted, purified, and then added to the microsomal fraction. In the presence of GTP the labeled amino acid is transferred from the soluble RNA to microsomal protein. This very exciting lead is being actively pursued.

Many other experiments have been carried out on cell-free systems, in particular by Gale and Folkes (1955) and by Spiegelman (see his review, 1957), but I shall not describe them here as their interpretation is difficult. It should be mentioned that Gale (reviewed in Gale 1956) has isolated from hydrolysates of commercial yeast RNA a series of fractions which greatly increase amino acid incorporation. One of them, the so-called glycine incorporation factor has been purified considerably, and an attempt is being made to discover its structure.

RNA Turnover and Protein Synthesis

From many points of view it seems highly likely that the *presence* of RNA is essential for cytoplasmic protein synthesis, or at least for specific protein synthesis. It is by no means clear, however, that the *turnover* of RNA is required.

In discussing this a strong distinction must be made between cells which are growing, and therefore producing new microsomal particles, and cells which are synthesizing without growth, and in which few new microsomal particles are being produced.

This is a difficult aspect of the subject as the evidence is to some extent conflicting. It appears reasonably certain that not *all* the RNA in the cytoplasm is turning over very rapidly — this has been shown, for example, by the Hokins (1954) working on amylase synthesis in slices of pigeon pancreas, though in the light of the recent work of Straub (this Symposium) the choice of amylase was unfortunate. On the other hand, Pardee (1954) has demonstrated that mutants of *Escherichia coli* which require uracil or adenine cannot synthesize β-galactosidase unless the missing base is provided.

Can RNA be synthesized without protein being synthesized? This can be brought about by the use of *chloramphenicol*. In bacterial systems chloramphenicol stops protein synthesis dead, but allows "RNA" synthesis to continue. A very interesting phenomenon has been uncovered in *E. coli* by Pardee and Prestidge (1956), and by Gros and Gros (1956). If a mutant is used which requires, say, leucine, then when the external supply of leucine is exhausted both protein and RNA synthesis cease. If now chloramphenicol is added there is no effect, but if in addition the cells are given a small amount of leucine then rapid RNA synthesis takes place. If the chloramphenicol is removed, so that protein synthesis restarts, then this leucine is built into proteins and then, once again, the synthesis of both protein and RNA is prevented. In other words it appears as if "free" leucine (i.e., not bound into proteins) is required for RNA synthesis. This effect is not peculiar to leucine and has already been found for several amino acids and in several different organisms (Yčas and Brawerman 1957).

As a number of people have pointed out, the most likely interpretation of these results is that protein and RNA require *common intermediates* for their synthesis, consisting in part of amino acids and in part of RNA components such as nucleotides. This is a most valuable idea; it explains a number of otherwise puzzling facts and there is some hope of getting close to it experimentally.

For completeness it should be stated that Anfinsen and his coworkers have some evidence that proteins are not produced from (activated) amino acids in a single step (see the review by Steinberg, Vaughan, and Anfinsen 1956), since they find unequal labeling

between the same amino acid at different points on the polypeptide chain, but this interpretation of their results is not accepted by all workers in the field. This is discussed more fully by Simkin and Work (this Symposium).

Summary of Experimental Work

Both DNA and RNA have been shown to carry some of the specificity for protein synthesis. The RNA of almost all types of cell is found mainly in rather uniform, spherical, viruslike particles in the cytoplasm, known as microsomal particles. Most of their protein and RNA is metabolically rather inert. Amino acids, on their way into protein, have been shown to pass rapidly through these particles.

An enzyme has been isolated which, when supplied with tryptophan and ATP, appears to form an activated tryptophan. There is evidence that there exist similar enzymes for most of the other amino acids. These enzymes are widely distributed in Nature.

Work on cell fractions is difficult to interpret but suggests that the first step in protein synthesis involves these enzymes, and that the subsequent transfer of the activated amino acids to the microsomal particles requires GTP. The soluble RNA also appears to be involved in this process.

Whereas the presence of RNA is probably required for true protein synthesis its rapid turnover does not appear to be necessary, at least not for all the RNA. There is suggestive evidence that common intermediates, containing both amino acids and nucleotides, occur in protein synthesis.

IV. IDEAS ABOUT PROTEIN SYNTHESIS

It is an extremely difficult matter to present current ideas about protein synthesis in a stimulating form. Many of the general ideas on the subject have become rather stale, and an extended discussion of the more detailed theories is not suitable in a paper for nonspecialists. I shall therefore restrict myself to an outline sketch of my own ideas on cytoplasmic protein synthesis, some of which have not been published before. Finally I shall deal briefly with the problem of "coding."

General Principles

My own thinking (and that of many of my colleagues) is based on two general principles, which I shall call the Sequence Hypothesis and the Central Dogma. The direct evidence for both of them is negligible, but I have found them to be of great help in getting to grips with these very complex problems. I present them here in the hope that others can make similar use of them. Their speculative nature is emphasized by their names. It is an instructive exercise to attempt to build a useful theory without using them. One generally ends in the wilderness.

The Sequence Hypothesis This has already been referred to a number of times. In its simplest form it assumes that the specificity of a piece of nucleic acid is expressed solely by the sequence of its bases, and that this sequence is a (simple) code for the amino acid sequence of a particular protein.

This hypothesis appears to be rather widely held. Its virtue is that it unites several remarkable pairs of generalizations: the central biochemical importance of proteins and the dominating biological role of genes, and in particular of their nucleic acid; the linearity of protein molecules (considered covalently) and the genetic linearity within the functional gene, as shown by the work of Benzer (1957) and Pontecorvo (this Symposium); the simplicity of the composition of protein molecules and the simplicity of the nucleic acids. Work is actively proceeding in several laboratories, including our own, in an attempt to provide more direct evidence for this hypothesis.

The Central Dogma This states that once "information" has passed into protein it *cannot get out again*. In more detail, the transfer of information from nucleic acid to nucleic acid, or from nucleic acid to protein may be possible, but transfer from protein to protein, or from protein to nucleic acid is impossible. Information means here the *precise* determination of sequence, either of bases in the nucleic acid or of amino acid residues in the protein.

This is by no means universally held — Sir Macfarlane Burnet, for example, does not subscribe to it — but many workers now think along these lines.

As far as I know it has not been *explicitly* stated before.

Some Ideas on Cytoplasmic Protein Synthesis

From our assumptions it follows that there must be an RNA template in the cytoplasm. The obvious place to locate this is in the microsomal particles, because their uniformity of size suggests that they have a regular structure. It also follows that the synthesis of at least some of the microsomal RNA must be under the control of the DNA of the nucleus. This is because the amino acid sequence of the human hemoglobins, for example, is controlled at least in part by a Mendelian gene, and because spermatozoa contain no RNA. Therefore, granted our hypothesis, the information must be carried by DNA.

What can we guess about the structure of the microsomal particle? On our assumptions the protein component of the particles can have no significant role in determining the amino acid sequence of the proteins which the particles are producing. We therefore assume that their main function is a structural one, though the possibility of some enzyme activity is not excluded. The simplest model then becomes one in which each particle is made of the same protein, or proteins, as every other one in the cell, and has the same basic *arrangement* of the RNA, but that different particles have, in general, different base sequences in their RNA, and therefore produce different proteins. This is exactly the type of structure found in tobacco mosaic virus, where the interaction between RNA and protein does not depend upon the sequence of bases of the RNA (Hart and Smith 1956). In addition Watson and I have suggested (Crick and Watson 1956), by analogy with the spherical viruses, that the protein of microsomal particles is probably made of many identical subunits arranged with cubic symmetry.

On this oversimplified picture, therefore, the microsomal particles in a cell are all the same (except for the base sequence of their RNA) and are metabolically rather inert. The RNA forms the template and the protein supports and protects the RNA.

This idea is in sharp contrast to what one would naturally assume at first glance, namely that the protein of the microsomal particles consists entirely of protein being synthesized. The surmise that most of the protein is structural was derived from considerations about the structure of virus particles and about coding; it was independent of the direct experimental evidence of Zamecnik and his colleagues that only a small fraction of the protein turns over rapidly, so that this agreement between theory and experiment is significant, as far as it goes.

It is obviously of the first importance to know how the RNA of the particles is arranged. It is a natural deduction from the Sequence Hypothesis that the RNA backbone will follow as far as possible a spatially regular path, in this case a helix, essentially because the fundamental operation of making the peptide link is always the same, and we therefore expect any template to be spatially regular.

Although we do not yet know the structure of isolated RNA (which may be an artifact) we do know that a pair of RNA-like molecules can under some circumstances form a double-helical structure, somewhat similar to DNA, because Rich and Davies (1956) have shown that when the two polyribotides, polyadenylic acid and polyuridylic acid (which have the same backbone as RNA) are mixed together they wind round one another to form a double helix, presumably with their bases paired. It would not be surprising, therefore, if the RNA backbone took up a helical configuration similar to that found for DNA.

This suggestion is in contrast to the idea that the RNA and protein interact in a complicated, irregular way to form a "nucleoprotein." As far as I know there is at the moment no direct experimental evidence to decide between these two points of view.

However, even if it turns out that the RNA is (mainly) helical and that the structural protein is made of subunits arranged with cubic symmetry it is not at all obvious how the two could fit together. In abstract terms the problem is how to arrange a long fibrous object inside a regular polyhedron. It is for this reason that the structure of the spherical viruses is of great interest in this context, since we suspect that the same situation occurs there; moreover they are at the moment more amenable to experimental attack. A possible arrangement, for example, is one in which the axes of the RNA helices run radially and clustered in groups of five, though it is always possible that the arrangement of the RNA is irregular.

It would at least be of some help if the approximate location of the RNA in the microsomal particles could be discovered. Is it on the outside or the inside of the particles, for example, or even both? Is the microsomal particle a rather open structure, like a sponge, and if it is what size of molecule can diffuse in and out of it? Some of these points are now ripe for a direct experimental attack.

The Adaptor Hypothesis

Granted that the RNA of the microsomal particles, regularly arranged, is the template, how does it direct the amino acids into the correct order? One's first naive idea is that the RNA will take up a configuration capable of forming twenty different "cavities", one for the side chain of each of the twenty amino acids. If this were so one might expect to be able to play the problem backwards — that is, to find the configuration of RNA by trying to form such cavities. All attempts to do this have failed, and on physical-chemical grounds the idea does not seem in the least plausible (Crick 1957a). Apart from the phosphate-sugar backbone, which we have assumed to be regular and perhaps linked to the structural protein of the particles, RNA presents mainly a sequence of sites where hydrogen bonding could occur. One would expect, therefore, that whatever went on to the template in a *specific* way did so by forming hydrogen bonds. It is therefore a natural hypothesis that the amino acid is carried to the template by an "adaptor" molecule, and that the adaptor is the part which actually fits on to the RNA. In its simplest form one would require twenty adaptors, one for each amino acid.

What sort of molecules such adaptors might be is anybody's guess. They might, for example, be proteins, as suggested by Dounce (1952) and by the Hokins (1954) though personally I think that proteins, being rather large molecules, would take up too much space. They might be quite unsuspected molecules, such as amino sugars. But there is one possibility which seems inherently more likely than any other — that they might contain nucleotides. This would enable them to join on to the RNA template by the same "pairing" of bases as is found in DNA, or in polynucleotides.

If the adaptors were small molecules one would imagine that a separate enzyme would be required to join each adaptor to its own amino acid and that the specificity required to distinguish between, say, leucine, isoleucine and valine would be provided by these enzyme molecules instead of by cavities in the RNA. Enzymes, being made of protein, can probably make such distinctions more easily than can nucleic acid.

An outline picture of the early stages of protein synthesis might be as follows: the template would consist of perhaps a single chain of RNA. (As far as we know a single isolated RNA backbone has no regular configuration [Crick 1957b] and one has to assume that the backbone is supported in a helix of the usual type by the structural protein of the microsomal particles.) Alternatively the template might consist of a pair of chains. Each adaptor molecule containing, say, a di- or trinucleotide would be joined to its own amino acid by a special enzyme. These molecules would then diffuse to the microsomal particles and attach to the proper place on the bases of the RNA by base pairing, so that they would then be in a position for polymerization to take place.

It will be seen that we have arrived at the idea of common intermediates without using the direct experimental evidence in their favour; but there is one important qualification, namely that the nucleotide part of the intermediates must be specific for each amino acid, at least to some extent. It is not sufficient, from this point of view, merely to join adenylic acid to each of the twenty amino acids. Thus one is led to suppose that after the activating step, discovered by Hoagland and described earlier, some other more specific step is needed before the amino acid can reach the template.

The Soluble RNA

If trinucleotides, say, do in fact play the role suggested here their synthesis presents a puzzle, since one would not wish to invoke too many enzymes to do the job. It seems to me plausible, therefore, that the twenty different adaptors may be synthesized by the *breakdown* of RNA, probably the "soluble" RNA. Whether this is in fact the same action which the "activating enzymes" carry out (presumably using GTP in the process) remains to be seen.

From this point of view the RNA with amino acids attached reported recently by Hoagland,

Zamecnik and Stephenson (1957), would be a half-way step in this process of breaking the RNA down to trinucleotides and joining on the amino acids. Of course alternative interpretations are possible. For example, one might surmise that numerous amino acids become attached to this RNA and then proceed to polymerize, perhaps inside the microsomal particles. I do not like these ideas, because the supernatant RNA appears to be too short to code for a complete polypeptide chain, and yet too long to join on to template RNA (in the microsomal particles) by base pairing, since it would take too great a time for a piece of RNA twenty-five nucleotides long, say, to diffuse to the correct place in the correct particles. If it were only a trinucleotide on the other hand, there would be many different "correct" places for it to go to (wherever a valine was required, say), and there would be no undue delay.

Leaving theories on one side, it is obviously of the greatest interest to know what molecules actually pass from the "pH 5 enzymes" to the microsomal particles. Are they small molecules, free in solution, or are they bound to protein? Can they be isolated? This seems at the moment to be one of the most fruitful points at which to attack the problem.

Subsequent Steps

What happens after the common intermediates have entered the microsomal particles is quite obscure. Two views are possible, which might be called the Parallel Path and the Alternative Path theories. In the first an intermediate is used to produce both protein and RNA at about the same time. In the second it is used to produce either protein, or RNA, but not both. If we knew the exact nature of the intermediates we could probably decide which of the two was more likely. At the moment there seems little reason to prefer one theory to the other.

The details of the polymerization step are also quite unknown. One tentative theory, of the Parallel Path type, suggests that the intermediates first polymerize to give an RNA molecule with amino acids attached. This process removes it from the template and it diffuses outside the microsomal particle. There the RNA folds to a new configuration, and the amino acids become polymerized to form a polypeptide chain, which folds up as it is made to produce the finished protein. The RNA, now free of amino acids, is then broken down to produce fresh intermediates. A great variety of theories along these lines can be constructed. I shall not discuss these further here, nor shall I describe the various speculations about the actual details of the chemical steps involved.

Two Types of RNA

It is an essential feature of these ideas that there should be *at least two types of RNA in the cytoplasm*. The first, which we may call "template RNA" is located inside the microsomal particles. It is probably synthesized in the nucleus (Goldstein and Plaut 1955) under the direction of DNA, and carries the information for sequentialization. It is metabolically inert during protein synthesis, though naturally it may show turnover whenever microsomal particles are being synthesized (as in growing cells), or breaking down (as in certain starved cells).

The other postulated type of RNA, which we may call "metabolic RNA," is probably synthesized (from common intermediates) in the microsomal particles, where its sequence is determined by base pairing with the template RNA. Once outside the microsomal particles it becomes "soluble RNA" and is constantly being broken down to form the common intermediates with the amino acids. It is also possible that some of the soluble RNA may be synthesized in a random manner in the cytoplasm; perhaps in bacteria, by the enzyme system of Grunberg-Manago and Ochoa (1955).

One might expect that there would also be metabolic RNA in the nucleus. The existence of these different kinds of RNA may well explain the rather conflicting data on RNA turnover.

The Coding Problem

So much for biochemical ideas. Can anything about protein synthesis be discovered by more abstract arguments? If, as we have assumed, the sequence of bases along the nucleic acid determines the sequence of amino acids of the protein being synthesized, it is not unreasonable to suppose that this interrelationship is a simple one, and to invent abstract descriptions of it. This problem of how, in outline, the

sequence of four bases "codes" the sequence of the twenty amino acids is known as the coding problem. It is regarded as being independent of the biochemical steps involved, and deals only with the transfer of information.

This aspect of protein synthesis appeals mainly to those with a background in the more sophisticated sciences. Most biochemists, in spite of being rather fascinated by the problem, dislike arguments of this kind. It seems to them unfair to construct theories without adequate experimental facts. Cosmologists, on the other hand, appear to lack such inhibitions.

The first scheme of this kind was put forward by Gamow (1954). It was supposedly based on some features of the structure of DNA, but these are irrelevant. The essential features of Gamow's scheme were as follows:

a. Three bases coded one amino acid.

b. Adjacent triplets of bases overlapped. See Fig. 1.

c. More than one triplet of bases stood for a particular amino acid (degeneracy).

In other words it was an overlapping degenerate triplet code. Such a code imposes severe restric-

Fig. 1.   The letters A, B, C and D stand for the four bases of the four common nucleotides. The top row of letters represents an imaginary sequence of them. In the codes illustrated here each set of three letters represents an amino acid. The diagram shows how the first four amino acids of a sequence are coded in the three classes of codes.

```
                        B C A C D D A B A B D C

                       ( B C A
Overlapping            |   C A C
code                   |     A C D
                       (       C D D

                       ( B C A
Partial                |     A C D
overlapping code       |         D D A
                       (             A B A

                       ( B C A
Nonoverlapping         |       C D D
code                   |           A B A
                       (               B D C
```

tions on the amino acid sequences it can produce. It is quite easy to disprove Gamow's code from a study of known sequences — even the sequences of the insulin molecule are sufficient. However, there are a very large number of codes of this general type. It might be thought almost impossible to disprove them all without enumerating them, but this has recently been done by Brenner (1957), using a neat argument. He has shown that the reliable amino acid sequences already known are enough to make *all* codes of this type impossible.

Attempts have been made to discover whether there are any obvious restrictions on the allowed amino acid sequences, although the sequence data available are very meagre (see the review by Gamow, Rich, and Yčas 1955). So far none has been found, and the present feeling is that it may well be that none exists, and that any sequence whatsoever can be produced. This is very far from being established, however, and for all we know there may be quite severe restrictions on the neighbors of the rarer amino acids, such as tryptophan.

If there is indeed a relatively simple code, then one of the most important biological constants is what Watson and I have called "the coding ratio" (Crick and Watson 1956). If B consecutive bases are required to code A consecutive amino acids, the coding ratio is the number B/A, when B and A are large. Thus in Gamow's code its value is unity, since a string of 1000 bases, for example, could code 998 amino acids. (Notice that when the coding ratio is greater than unity stereochemical problems arise, since a polypeptide chain has a distance of only about 3½ A between its residues, which is about the minimum distance between successive bases in nucleic acid. However, it has been pointed out by Brenner [personal communication] that this difficulty may not be serious if the polypeptide chain leaves the template as it is being synthesized.)

If the code were of the nonoverlapping type (see Fig. 1) one would still require a triplet of bases to code for each amino acid, since pairs of bases would only allow 4 X 4 = 16 permutations, though a possible but not very likely way round this has been suggested by Dounce, Morrison and Monty (1955). The use of triplets raises two difficulties. First, why are there not 4 X 4 X 4 = 64 different amino acids? Second, how does one know which of the triplets to read (assuming that one doesn't start at an end)?

For example, if the sequence of bases is . . . , ABA, CDB, BCA, ACC, . . . , where A, B, C and D represent the four bases, and where ABA is supposed to code one amino acid, CDB another one, and so on, how could one read it correctly if the commas were removed?

Very recently Griffith, Orgel and I have suggested an answer to both these difficulties which is of some interest because it *predicts* that there should be only twenty kinds of amino acid in protein (Crick, Griffith, and Orgel 1957). Gamow and Yčas (1955) had previously put forward a code with this property, known as the "combination code" but the physical assumptions underlying their code lack plausibility. We assumed that some of the triplets (like ABA in the example above) correspond to an amino acid — make "sense" as we would say — and some (such as BAC and ACD, etc., above) do not so correspond, or as we would say, make "nonsense."

We asked ourselves how many amino acids we could code if we allowed all possible sequences of amino acids, and yet never accidentally got "sense" when reading the wrong triplets, that is those which included the imaginary commas. We proved that the upper limit is twenty, and moreover we could write down several codes which did in fact code twenty things. One such code of twenty triplets, written compactly, is

$$
\begin{array}{ccccccccc}
 & & A & & A & & A & & A \\
 & & & & & & & & B \\
A & B & & A & & C & B & B & D & C \\
 & & B & & B & & C & & C & D \\
\end{array}
$$

where

$$
\begin{array}{ccc}
 & & A \\
A & B & \\
 & & B \\
\end{array}
$$

means that two of the allowed triplets are ABA and ABB, etc. The example given a little further back has been constructed using this code. You will see that ABA, CDB, BCA and ACC are among the allowed triplets, whereas the false overlapping ones in that example, such as BAC, ACD and DBB, etc., are not. The reader can easily satisfy himself that no sequence of these allowed triplets will ever give one of the allowed triplets in a false position. There are many possible mechanisms of protein synthesis for which this would be an advantage. One of them is described in our paper (Crick *et al.* 1957).

Thus we have deduced the magic number, twenty, in an entirely natural way from the magic number four. Nevertheless, I must confess that I find it impossible to form any considered judgement of this idea. It may be complete nonsense, or it may be the heart of the matter. Only time will show.

V. CONCLUSIONS

I hope I have been able to persuade you that protein synthesis is a central problem for the whole of biology, and that it is in all probability closely related to gene action. What are one's overall impressions of the present state of the subject? Two things strike me particularly. First, the existence of general ideas covering wide aspects of the problem. It is remarkable that one can formulate principles such as the Sequence Hypothesis and the Central Dogma, which explain many striking facts and yet for which proof is completely lacking. This gap between theory and experiment is a great stimulus to the imagination. Second, the extremely active state of the subject experimentally both on the genetical side and the biochemical side. At the moment new and significant results are being reported every few months, and there seems to be no sign of work coming to a standstill because experimental techniques are inadequate. For both these reasons I shall be surprised if the main features of protein synthesis are not discovered within the next ten years.

It is a pleasure to thank Dr. Sydney Brenner, not only for many interesting discussions, but also for much help in redrafting this paper.

REFERENCES

Gamow, G., Rich, A., and Yčas, M. (1955). *Advanc. Biol. Med. Phys.* 4. New York: Academic Press.
Gamow, G., and Yčas, M. (1955). *Proc. Nat. Acad. Sci., Wash.* 41, 1101.
Gierer, A., and Schramm, G. (1956). *Z. Naturf.* 11b, 138; also *Nature, Lond.* 177, 702.
Goldstein, L., and Plaut, W. (1955). *Proc. Nat. Acad. Sci., Wash.* 41, 874.
Gros, F., and Gros, F. (1956). *Biochim. Biophys. Acta,* 22, 200.
Grunberg-Manago, M., and Ochoa, S. (1955). *J. Amer. Chem. Soc.* 77, 3165.

Harris, J. I., Sanger, F., and Naughton, M. A. (1956). *Arch. Biochem. Biophys.* 65, 427.
Hart, R. G., and Smith, J. D. (1956). *Nature, Lond.* 178, 739.
Hershey, A. D. (1956). *Advances in Virus Research.* IV. New York: Academic Press.
Hoagland, M. B. (1955). *Biochim. Biophys. Acta,* 16, 288.
Hoagland, M. B., Keller, E. B., and Zamecnik, P. C. (1956). *J. Biol. Chem.* 218, 345.
Hoagland, M. B., Zamecnik, P. C., and Stephenson, M. L. (1957). *Biochim. Biophys. Acta* 24, 215.
Hokin, L. E., and Hokin, M. R. (1954). *Biochim. Biophys. Acta,* 13, 401.
Hultin, T., and Beskow, G. (1956). *Exp. Cell Res.* 11, 664.
Ingram, V. M. (1956). *Nature, Lond.* 178, 792.
Ingram, V. M. (1957). *Nature, Lond.* 180, 326.
Kamin, H., and Handler, P. (1957). *Annu. Rev. Biochem.* 26, 419.
Keller, E. B., and Zamecnik, P. C. (1956). *J. Biol. Chem.* 221, 45.
Koningsberger, V. V., van de Ven, A. M., and Overbeck, J. Th. G. (1957). *Proc. K. Akad. Wet. Amst.* B, 60, 141.
Littlefield, J. W., and Keller, E. B. (1957). *J. Biol. Chem.* 224, 13.
Marmur, J., and Hotchkiss, R. D. (1955). *J. Biol. Chem.* 214, 383.
Mirsky, A. E., Osawa, S., and Allfrey, V. G. (1956). *Cold Spr. Harb. Symp. Quant. Biol.* 21, 49.
Munier, R., and Cohen, G. N. (1956). *Biochim. Biophys. Acta,* 21, 592.
Palade, G. E. (1955). *J. Biochem. Biophys. Cytol.* 1, 1.
Palade, G. E. (1956). *J. Biochem. Biophys. Cytol.* 2, 85.
Palade, G. E., and Siekevitz, P. (1956). *J. Biochem. Biophys. Cytol.* 2, 171.
Pardee, A. B. (1954). *Proc. Nat. Acad. Sci., Wash.* 40, 263.
Pardee, A. B., and Prestidge, L. S. (1956). *J. Bact.* 71, 677.
Pauling, L., Itano, H. A., Singer, S. J., and Wells, I. C. (1949). *Science,* 110, 543.
Petermann, M. L., and Hamilton, M. G. (1957). *J. Biol. Chem.* 224, 725; also *Fed. Proc.* 16, 232.
Petermann, M. L., Mizen, N. A., and Hamilton, M. G. (1952). *Cancer Res.* 12, 373.
Rabinovitz, M. and Olson, M. E. (1956). *Exp. Cell Res.* 10, 747.

Rabinovitz, M. and Olson, M. E. (1957). *Fed. Proc.* 16, 235.

Rich, A. and Davies, D. R. (1956). *J. Amer. Chem. Soc.* 78, 3548.

Schachman, H. K., Pardee, A. B., and Stanier, R. Y. (1953). *Arch. Biochem. Biophys.* 43, 381.

Schweet, R. (1957). *Fed. Proc.* 16, 244.

Simkin, J. L. and Work, T. S. (1957a). *Biochem. J.* 65, 307.

Simkin, J. L. and Work, T. S. (1957b). *Nature, Lond.* 179, 1214.

Simpson, M. V. and McLean, J. R. (1955). *Biochim. Biophys. Acta,* 18, 573.

Simpson, M. V., McLean, J. R., Cohn, G. I., and Brandt, I. K. (1957). *Fed Proc.* 16, 249.

Spiegelman, S. (1957). In *The Chemical Basis of Heredity*. Ed. McElroy, W. D. and Glass, B. Baltimore: Johns Hopkins Press.

Steinberg, D. and Mihalyi, E. (1957). In *Annu. Rev. Biochem.* 26, 373.

Steinberg, D., Vaughan, M., and Anfinsen, C. B. (1956). *Science,* 124, 389.

Stephenson, M. L., Thimann, K. V., and Zamecnik, P. C. (1956). *Arch. Biochem. Biophys.* 65, 194.

Synge, R. (1957). In *The Origin of Life on the Earth*. U.S.S.R. Acad. of Sciences.

Ts'o, P. O. B., Bonner, J., and Vinograd, J. (1956). *J. Biochem. Biophys. Cytol.* 2, 451.

Wagner, R. P. and Mitchell, H. K. (1955). *Genetics and Metabolism*. New York: John Wiley and Sons.

Yčas, M. and Brawerman, G. (1957). *Arch. Biochem. Biophys.* 68, 118.

Zamecnik, P. C., Keller, E. B., and Littlefield, J. W., Hoagland, M. B. and Loftfield, R. B. (1956). *J. Cell Comp. Physiol.* 47, suppl. 1, 81.

An unstable intermediate carrying information from genes to ribosomes for protein synthesis

DR. S. BRENNER,* DR. F. JACOB,† and DR. M. MESELSON‡

Reprinted by authors' and publisher's permission from Nature, *vol. 190(4776), 1961, pp. 576–581.*

In the preceding paper, Crick structurally tied template RNA (*messenger RNA*) to the ribonucleoprotein particles of the microsomal fraction actively involved in protein synthesis. Because label turnover experiments demonstrated that these ribonucleoprotein particles (*ribosomes*) were stable, it was assumed that template RNA would also be stable. Since the template RNA carried the unique nucleotide sequence of the genes, this suggested that different ribonucleoprotein particles were responsible for the production of different proteins. The data presented in the next paper refutes this interpretation and proposes that the messenger RNA is distinct from the RNA of the ribonucleoprotein particles. Therefore, the authors necessitate the assumption that the stable ribosomes may be used repeatedly, reassociating with newly synthesized unstable RNA. Their work not only predicted our current interpretation of the relationship between messenger RNA and the ribosomes, but it also freed the concept of differential gene activity from the necessity of invoking rapid ribosome turnover — a difficult idea to accept in light of apparent ribosome stability.

A large amount of evidence suggests that genetic information for protein structure is encoded in deoxyribonucleic acid (DNA) while the actual assembling of amino acids into proteins occurs in cytoplasmic ribonucleoprotein particles called ribosomes. The fact that proteins are not synthesized directly on genes demands the existence of an intermediate information carrier. This intermediate template is generally assumed to be a stable ribonucleic acid (RNA) and more specifically the RNA of the ribosomes. According to the present view, each gene controls the synthesis of one kind of specialized ribosome, which in turn, directs the synthesis of the corresponding protein — a scheme which could be epitomized as the one gene–one ribosome–one protein hypothesis. In the past few years, however, this model has encountered some difficulties: (1) The remarkable homogeneity in size[1] and nucleotide composition[2] of the ribosomal RNA reflects neither the range of size of polypeptide chains nor the variation in the nucleotide composition observed in the DNA of different bacterial species.[2,3] (2) The capacity of bacteria to synthesize a given protein does not seem to survive beyond the integrity of the corresponding gene.[4] (3) Regulation of protein synthesis in bacteria seems to operate at the level of the synthesis of the information intermediate by the gene rather than at the level of the synthesis of the protein.[5]

These results are scarcely compatible with the existence of stable RNA intermediates acting as templates for protein synthesis. The paradox, however, can be resolved by the hypothesis, put forward by Jacob and Monod,[5] that the ribosomal RNA is not the intermediate carrier of information from gene to protein, but rather that ribosomes are nonspecialized structures which receive genetic information from the gene in the form of an unstable

*Medical Research Council Unit for Molecular Biology, Cavendish Laboratory, University of Cambridge.
†Institut Pasteur, Paris
‡Gates and Crellin Laboratories of Chemistry, California Institute of Technology, Pasadena, California.

intermediate or "messenger." We present here the results of experiments on phage-infected bacteria which give direct support to this hypothesis.

When growing bacteria are infected with a virulent bacteriophage such as $T2$, synthesis of DNA stops immediately, to resume 7 min later,[6] while protein synthesis continues at a constant rate.[7] After infection many bacterial enzymes are no longer produced;[8] in all likelihood, the new protein is genetically determined by the phage. A large number of new enzymatic activities appears in the infected cell during the first few minutes following infection,[9] and from the tenth minute onwards some 60 percent of the protein synthesized can be accounted for by the proteins of the phage coat.[7] Surprisingly enough, protein synthesis after infection is not accompanied, as in growing cells, by a net synthesis of RNA.[10] Using isotopic labeling, however, Volkin and Astrachan[11] were able to demonstrate high turnover in a minor RNA fraction after phage infection. Most remarkable is the fact that this RNA fraction has an apparent nucleotide composition which corresponds to that of the DNA of the phage and is markedly different from that of the host RNA.[11] Recently, it has been shown that the bulk of this RNA is associated with the ribosomes of the infected cell.[12]

Phage-infected bacteria therefore provide a situation in which the synthesis of protein is suddenly switched from bacterial to phage control and proceeds without the concomitant synthesis of stable RNA. A priori, three types of hypothesis may be considered to account for the known facts of phage protein synthesis (Fig. 1). Model I is the classical model. After infection the bacterial machinery is switched off, and new ribosomes are then synthesized by the phage genes. The ad hoc hypothesis has to be added that these ribosomes are unstable, to account for the turnover of RNA after phage infection. This is, in fact, the model favoured by Nomura et al.[12] Model II assumes that in the particular case of phage the proteins are assembled directly on the DNA; the new RNA is a special molecule which enters old ribosomes and destroys their capacity for protein synthesis. At the same time, synthesis of ribosomes is switched off. Model III implies that a special type of RNA molecule, or "messenger RNA" exists which brings genetic information from genes to nonspecialized ribosomes and that the consequences of phage infection are twofold: (a) to switch off the synthesis of new ribosomes; (b) to

substitute phage messenger RNA for bacterial messenger RNA. This substitution can occur quickly only if messenger RNA is unstable; the RNA made after phage infection does turn over and appears, therefore, as a good candidate for the messenger.

It is possible to distinguish experimentally between these three models in the following way: Bacteria are grown in heavy isotopes so that all cell constituents are uniformly labeled "heavy." They are infected with phage and transferred immediately to a medium containing light isotopes so that all constituents synthesized after infection are "light." The distribution of new RNA and new protein, labeled with radioactive isotopes, is then followed by density-gradient centrifugation[13] of purified ribosomes.

Density-gradient centrifugation was carried out in a preparative centrifuge, and the ribosomes were stabilized by including magnesium acetate (0.01–0.06 M) in the cæsium chloride solution. Ribosomes show two bands, a heavier A band and a lighter B band, the relative proportions of which, for a given preparation, depend on the magnesium concentration used. The lower the magnesium concentration, the smaller the proportion of B band ribosomes and the larger the proportion of A band ribosomes.

In order to show that there is no aggregation of ribosomes during preparation and density gradient centrifugation an experiment was carried out on ribosomes extracted from a mixture of $^{15}N^{13}C$ and $^{14}N^{12}C$ bacteria. The results are shown in Fig. 2, from which it can be seen that ribosomes of different isotopic compositions band independently and that there are no intermediate classes. The same preparation was then dialysed against low magnesium to dissociate the ribosomes into their 50 S and 30 S components and then against high magnesium to reassociate the subunits.[14] This should have resulted in distributing heavy 30 S and 50 S subunits into mixed 70 S and 100 S ribosomes. Surprisingly enough, density-gradient centrifugation of this preparation (Fig. 3) yields the same bands as found in the original ribosomes except for a decrease in the proportion of the B bands. This means that both bands contain units which do not undergo reversible association and dissociation and that the mixed 70 S ribosomes prepared by dialysis separate into their components in the density gradient. Other experiments to be reported elsewhere suggest that the A band is composed of free 50 S and 30 S ribosomes

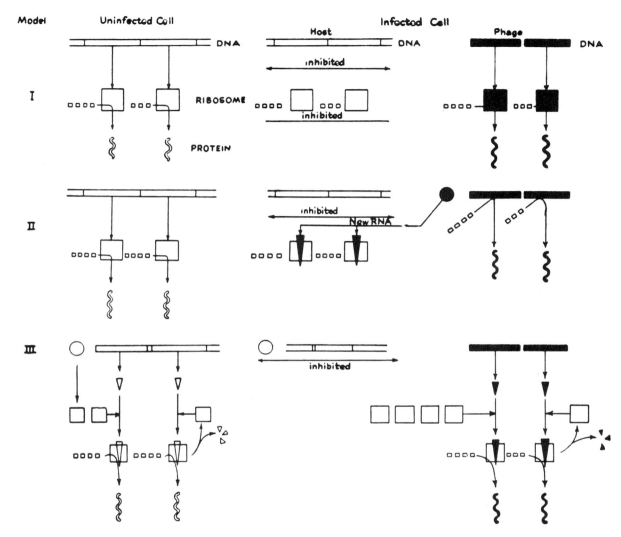

Fig. 1. Three models of information transfer in phage-infected cells

and that the *B* band contains undissociated 70 *S* particles.

The bulk of the RNA synthesized after infection is found in the ribosome fraction, provided that the extraction is carried out in 0.01 *M* magnesium ions.[12] We have confirmed this finding and have studied the distribution of the new RNA among the ribosomal units found in the density gradient. Fig. 4 shows that this RNA, labeled with ^{14}C-uracil, bands in the same position as *B* band ribosomes. There is no peak

corresponding to the *A* band. In addition, there is radioactivity at the bottom of the cell. This is free RNA as its density is greater than 1.8, and, moreover, it must have a reasonably high molecular weight to have sedimented in the gradient. Lowering of the magnesium concentration in the gradient, or dialysing the particles against low magnesium, produces a decrease of the *B* band and an increase of the *A* band. At the same time, the radioactive RNA leaves the *B* band to appear at the bottom of the

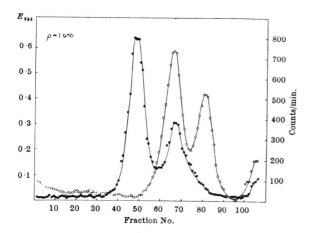

Fig. 2. Distribution of heavy and light ribosomes in a density gradient. *E. coli* B, grown in 5 ml of a medium containing ^{15}N (99 percent) and ^{13}C (60 percent) algal hydrolysate and $^{32}PO_4$, were mixed with a fiftyfold excess of cells grown in nutrient broth, the ribosomes extracted by alumina grinding in the presence of 0.01 M Mg^{++} and purified by centrifugation. 1 mg of ribosomes was centrifuged in 3 ml of cæsium chloride buffered to pH 7.2 with 0.1 M *tris* containing 0.03 M magnesium acetate for 35 hr at 37,000 rpm. in the *SW*39 rotor of the Spinco model *L* ultracentrifuge. After the run, a hole was pierced in the bottom of the tube and drops sequentially collected. Ultraviolet absorption at 254 mµ detects the excess of light ribosomes (○), ^{32}P counts detect the heavy ribosomes (●)

gradient. This shows that the uracil has labeled a species of RNA distinct from that of the bulk of *B* band ribosomes, since the specific activity of the RNA at the bottom of the cell is much higher than that of the *B* band. Fig. 5 shows that this RNA turns over during phage growth. There is a decrease by a factor of four in the specific activity of the *B* band after 16 min of growth in ^{12}C-uridine. Similar results have been obtained using $^{32}PO_4$ as a label.

These results do not distinguish between a messenger fraction and a small proportion of new ribosomes which are fragile in cæsium chloride and which are also metabolically unstable. In order to make the distinction, the experiment was carried out with an isotope transfer, in the following manner: Cells grown in a small volume of $^{15}N^{13}C$ medium were infected with *T* 4, transferred to $^{14}N^{12}C$ medium and fed $^{32}PO_4$ from the second to the seventh minute. They were mixed with a fifty-

fold excess of cells grown and infected in $^{14}N^{12}C^{31}P$ medium. Fig. 6 shows that the RNA formed after infection in the heavy cells has a density greater than that of the *B* band of the carrier. Its peak corresponds exactly with the density of the *B* band of $^{15}N^{13}C$ ribosomes (Fig. 2) although it is skewed to lighter density, and its response to changing the magnesium concentration was that of a *B* band. There is no radioactive peak corresponding to the *B* band of the carrier: this means that no wholly new ribosomes are synthesized after phage infection. As already shown, the new RNA does not represent random labeling of *B* band ribosomes; therefore it constitutes a fraction which is added to pre-existing ribosomes the bulk of the material of which has been assimilated before infection. This result conclusively eliminates model I.

Fig. 3. Distribution of randomized heavy and light ribosomes in a density gradient. The mixture of $^{15}N^{13}C^{32}P$ and $^{14}N^{12}C^{31}P$ ribosomes was dialysed first for 18 hr against 0.0005 M magnesium acetate in 0.01 M phosphate buffer pH 7.0, and then for 24 hr against two changes of 0.01 M magnesium acetate in 0.001 M *tris* buffer pH 7.4. 1 mg of ribosomes was centrifuged for 38 hr at 37,000 rpm. in cæsium chloride containing 0.03 M magnesium acetate. The drops were assayed for ultraviolet absorption (○) and ^{32}P content (●)

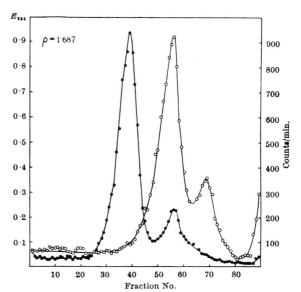

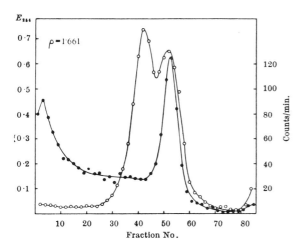

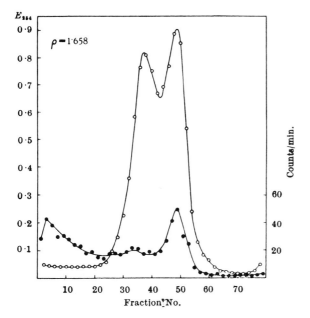

Figs. 4 and 5. Distribution and turnover of RNA formed after phage infection. A 600 ml culture of *E. coli* B6 (mutant requiring arginine and uracil) was infected with *T4D*) (multiplicity 30) and fed ^{14}C-uracil (10 mc/mM) from third to fifth min after infection. One half of the culture was removed and ribosomes prepared (Fig. 4). The other half received a two-hundredfold excess of ^{12}C-uridine for a further 16 min and ribosomes prepared (Fig. 5). In both experiments approximately 3 mg of purified ribosomes were centrifuged for 42 hr at 37,000 rpm. in caesium chloride containing 0.05 *M* magnesium acetate. Alternate drops were collected in *tris*-magnesium buffer for ultra-violet absorption (○) and on to 0.5 ml of frozen 5 percent trichloroacetic acid. These tubes were thawed. 1 mg of serum albumin added, and the precipitates separated and washed by filtration on membrane filters for assay of radioactivity (●)

can be removed by growth in nonradioactive sulphate and methionine (Fig. 8). In this experiment, the incorporation of ^{35}S into the total extract was measured and the amount in the *B* band found to correspond to 10 sec of protein synthesis. This is probably an overestimate since it is unlikely that pool equilibration was attained instantaneously. This value corresponds quite closely with the ribosome passage time of 5–7 sec for the nascent protein of uninfected cells.[15] In addition, electrophoresis of chymotrypsin digests of *B* band ribosomes shows that the radioactivity is already contained in a variety of peptides. It would therefore appear that most, if not all, protein synthesis in the infected cell occurs in ribosomes. The experiment also shows that preexisting ribosomes are used for synthesis a. that no new ribosomes containing stable sulphur-35 are synthesized. This result effectively eliminates model II.

We may summarize our findings as follows: (1) After phage infection no new ribosomes can be detected. (2) A new RNA with a relatively rapid turnover is synthesized after phage infection. This RNA, which has a base composition corresponding to that of the phage DNA, is added to preexisting ribosomes, from which it can be detached in a caesium chloride gradient by lowering the magnesium concentration. (3) Most, and perhaps all, protein synthesis in the infected cell occurs in preexisting ribosomes.

These conclusions are compatible only with model III (Fig. 1), which implies that protein synthesis oc-

To distinguish between models II and III an experiment was carried out to see whether pre-existing ribosomes participate in protein synthesis after phage infection. Cells were grown in ^{15}N medium, infected with phage, transferred to ^{14}N medium and fed [35]SO4 for the first 2 min of phage growth. Fig. 7 shows that only the *B* band of preexisting ribosomes becomes labeled with ^{35}S and there is no peak corresponding to a [14]N *B* band. All this label

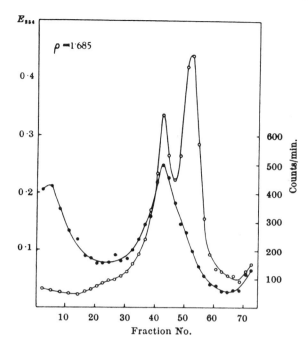

Fig. 6. RNA after isotope transfer in phage-infected cells. *E. coli* B grown in 10 ml of ^{15}N (99 percent) and ^{13}C (60 percent) algal hydrolysate medium were starved in buffer, infected with *T4* and growth initiated by addition of glucose and dephosphorylated broth (^{14}N^{12}C). ^{32}PO$_4$ was fed from the second to seventh min after infection. The culture was mixed with a fiftyfold excess of *E. coli* B grown in nutrient broth, infected in buffer and then grown for 7 min in dephosphorylated broth medium. 1 mg of purified ribosomes was centrifuged for 36 hr at 37,000 rpm. in cæsium chloride containing 0.03 *M* magnesium acetate. Ultraviolet absorption (○) detects the ^{14}N^{12}C^{31}P carrier, radioactivity (●) detects the new RNA in the heavy cells transferred to light medium

Figs. 7 and 8. Distribution and turnover of newly synthesized protein in ribosomes of phage-infected cells after transfer. *E. coli* B grown in 600 ml of a salt glucose medium containing ^{15}NH$_4$Cl (99 percent) were starved in buffer, infected with *T4* and transferred to ^{14}NH$_4$Cl medium. ^{35}SO$_4$ was fed for the first 2 min of infection and one half of the culture removed (Fig. 7). The other half received an excess of ^{32}SO$_4$ and ^{22}S-methionine and growth was continued for a further 8 min (Fig. 8). 1 mg of purified ribosomes was centrifuged for 39 hr at 37,000 rpm in cæsium chloride containing 0.05 *M* magnesium acetate. Drops were assayed for ultraviolet absorption (○) and for radioactivity (●). The arrows mark the expected positions for the peaks of ^{14}N and *A* and *B* bands. The radioactivity at the top of the gradient is contaminating protein

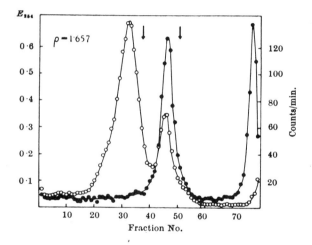

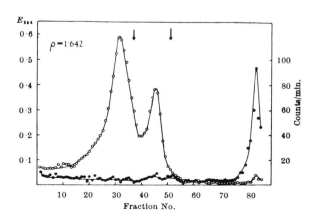

curs by a similar mechanism in uninfected cells. This, indeed, appears to be the case: exposure of uninfected cells to a 10-sec pulse of ^{32}PO$_4$ results in labeling of the RNA in the *B* band and not in the *A* band of ribosomes, and this RNA can be detached from the ribosomes by lowering the concentration of magnesium ions. Similarly the nascent protein can be labelled by a short pulse of ^{35}SO$_4$; it is located in the *B* band and most of the label is removed by growth in nonradioactive sulphate. In contrast to what was observed in infected cells, residual stable radioactivity is found in both bands, reflecting the synthesis of new ribosomes.

In order to act as an intermediate carrier of information from genes to ribosomes, the messenger has to fulfill certain prerequisites of size, turnover and nucleotide composition. In the accompanying article, Gros et al.[16] have analyzed the distribution of pulse-labeled RNA in sucrose gradients. They have shown that in uninfected cells there is an RNA fraction which has a rapid turnover and which can become attached reversibly to ribosomes depending on the magnesium concentration. The T2 phage-specific RNA shows the same behavior and both are physically similar, with sedimentation constants of 14–16 S. We have carried out similar experiments[17] independently and our results confirm their findings. These suggest that, although the messenger RNA is a minor fraction of the total RNA (not more than 4 percent), it is not uniformly distributed over all ribosomes, and may be large enough to code for long polypeptide chains. When ribosomes, from phage-infected cells labeled with $^{32}PO_4$ for five min, are separated by centrifugation in a sucrose density gradient[18] containing 0.01 M Mg^{++}, most of the messenger is found in 70 and 100 S ribosomes, contrary to previous reports.[12] When the magnesium concentration is lowered, the radioactivity is found in three peaks of roughly equal amount: (1) corresponding to a small residual number of 70 S ribosomes; (2) corresponding to 30 S ribosomes; (3) a peak of very high specific activity at 12 S. Separation of the RNA extracted from such ribosomes with detergent shows all the counts to be located in a peak at 12 S, skewed towards the heavier side. These results suggest that the messenger is heterogeneous in size and may have a minimum molecular weight of about ¼ to ½ million. Similar results have been obtained in uninfected cells.[17]

The undissociable 70 S ribosomes are enriched for messenger RNA over the total ribosomes of this type and it has been shown that they are also enriched for the nascent protein.[17] These ribosomes have been called "active 70 S" ribosomes, by Tissières et al.,[19] and they appear to be the only ribosomes which preserve the ability to synthesize protein *in vitro*. This leads one to suspect that there is a series of successive events involved in protein synthesis, and that at any time we investigate a temporal cross section of the process.

The exact determination of the rate of turnover of the messenger RNA should give information about the process of protein synthesis. This might be stoichiometric, in the sense that each messenger molecule functions only once in information transfer before it is destroyed. Its rate of turnover should then be the same as that of the nascent protein; but experiments to test this idea have been limited by difficulties in pool equilibration with nucleotide precursors.

It is a prediction of the hypothesis that the messenger RNA should be a simple copy of the gene, and its nucleotide composition should therefore correspond to that of the DNA. This appears to be the case in phage-infected cells,[11, 20] and recently Yčas and Vincent[21] have found a rapidly labeled RNA fraction with this property in yeast cells. If this turns out to be universally true, interesting implications for coding mechanisms will be raised.

One last point deserves emphasis. Although the details of the mechanism of information transfer by messenger are not clear, the experiments with phage-infected cells show unequivocally that information for protein synthesis cannot be encoded in the chemical sequence of the ribosomal RNA. Ribosomes are nonspecialized structures which synthesize, at a given time, the protein dictated by the messenger they happen to contain. The function of the ribosomal RNA in this process is unknown and there are also no restrictions on its origin in the cell: it may be synthesized by nuclear genes or by enzymes or it may be endowed with self-replicating ability.

This work was initiated while two of us (S. B. and F. J.) were guest investigators in the Division of Biology, California Institute of Technology, Pasadena, during June 1960. We would like to thank Profs. G. W. Beadle and M. Delbrück for their kind hospitality and financial support.

REFERENCES

1. Hall, B. D., and Doty, P., *J. Mol. Biol.*, 1, 111 (1959). Littaner, U. Z., and Eisenberg, H., *Biochim, Biophys. Acta,* 32, 320 (1959). Kurland, C. G., *J. Mol. Biol.*, 2, 83 (1960).
2. Belozersky, A. N., *Intern. Symp. Origin of Life.* 194 (Publishing House of the Academy of Sciences of the U.S.S.R., 1957).
3. Chargaff, E., *The Nucleic Acids,* 1, 307 (Adademic Press, New York, 1955). Lee, K. Y.,

Wahl, R., and Barbu, E., *Ann. Inst. Pasteur*, 91, 212 (1956).

4. Riley, M., Pardee, A. B., Jacob, R., and Monod, J., *J. Mol. Biol.*, 2, 216 (1960).

5. Jacob, F., and Monod, J., *J. Mol. Biol.* (in the press).

6. Cohen, S. S., *J. Biol. Chem.*, 174, 218 (1948). Hershey, A. D., Dixon, J., and Chase, M., *J. Gen. Physiol.* 36, 777 (1953). Vidaver, G. A., and Kozloff, L. M., *J. Biol. Chem.*, 225, 335 (1957).

7. Koch, G., and Hershey, A. D., *J. Mol. Biol.*, 1, 260 (1959).

8. Monod, J., and Wollman, E., *Ann. Inst. Pasteur*, 73, 937 (1947). Cohen, S. S., *Bact. Rev.*, 13, 1 (1949). Pardee, A. B., and Williams, I., *Ann. Inst. Pasteur*, 84, 147 (1953).

9. Kornberg, A., Zimmerman, S. B., Kornberg, S. R., and Josse, J., *Proc. U.S. Nat. Acad. Sci.*, 45, 772 (1959). Flaks, J. G., Lichtenstein, J., and Cohen, S. S., *J. Biol. Chem.* 234, 1507 (1959).

10. Cohen, S. S., *J. Biol. Chem.* 174, 281 (1948). Manson, L. A., *J. Bacteriol.*, 66, 703 (1953).

11. Volkin, E., and Astrachan, L., *Virology*, 2, 149 (1956). Astrachan, L., and Volkin, E. *Biochim. Biophys. Acta.* 29, 544 (1958).

12. Nomura, M., Hall, B. D., and Spiegelman, S., *J. Mol. Biol.* 2, 306 (1960).

13. Meselson, M., Stahl, F. W., and Vinograd, J., *Proc. U.S. Nat. Acad. Sci.*, 43, 581 (1957).

14. Tissières, A., Watson, J. D., Schlessinger, D., and Hollingworth, B. B., *J. Mol. Biol.*, 1, 221 (1959).

15. McQuillen, K., Roberts, R. B., and Britten, R. J., *Proc. U.S. Nat. Acad. Sci.*, 45, 1437 (1959).

16. Gros, F., Hiatt,,H., Gilbert, W., Kurland, C. G., Risebrough, R. W., and Watson, J. D., see following article.

17. Brenner, S., and Eckhart, W. (unpublished results).

18. Britten, R. J., and Roberts, R. B., *Science*, 131, 32 (1960).

19. Tissières, A., Schlessinger, D., and Gros, F., *Proc. U.S. Nat. Acad. Sci.*, 46, 1450 (1960).

20. Volkin, E., Astrachan, L., and Countryman, J. L., *Virology*, 6, 545, (1958).

21. Yčas, M., and Vincent, W. S., *Proc. U.S. Nat. Acad. Sci.*, 46, 804 (1960).

On the colinearity of gene structure and protein structure*

C. YANOFSKY, B. C. CARLTON,[†] J. R. GUEST,[‡] D. R. HELINSKI,[†] and U. HENNING[†, §]

Reprinted by authors' and publisher's permission from the Proceedings of the National Academy of Sciences, U.S., *vol. 51, 1964, pp. 266–272.*

If Crick's model of protein synthesis were valid, then the linear sequence of nucleotides within a gene should correspond to the linear sequence of amino acids in its protein product. This colinear relationship should exist if the subunit sequence in one molecule, DNA, dictates the subunit sequence in another, protein. Thus, a demonstration of colinearity would support Crick's model. However, in 1964, DNA nucleotide sequencing was not possible. Charles Yanofsky by-passed this obstacle by assuming that the recombinational map of different mutants within a gene reflected the relative positioning of the nucleotide alterations in the nucleotide sequence of the gene. Using this assumption, he established recombinationally defined map distances between different mutants and compared these with the positioning of the amino acid substitutions specified by each mutant in the altered protein products. In this way, Yanofsky tested if the relative distances between the mutants in the genetic map paralleled the distances between the amino acid substitutions in the protein. His results and interpretations are presented in the following paper.

The pioneering studies of Beadle and Tatum with *Neurospora crassa*[1] led to the concept that there is a 1:1 relationship between gene and enzyme. Subsequent studies on the structure of proteins[2] and genetic material[3] permitted a restatement of this relationship in molecular terms;[4] the linear sequence of nucleotides in a gene specifies the linear sequence of amino acids in a protein.

Several years ago studies were initiated with the A gene–A protein system of the tryptophan synthetase of *Escherichia coli* with the intention of examining this concept of a colinear relationship between gene structure and protein structure. A large number of mutant strains which produced altered A proteins were isolated, and genetic and protein primary structure studies were performed with these strains to locate the positions of the alterations within the A gene and the A protein.[5-10] It was hoped that with information of this type it would be possible to determine whether a genetic map and the primary structure of the corresponding protein were colinear. Recently, Kaiser[11] has demonstrated the correspondence of the genetic map with the sequence of blocks of nucleotides in DNA. Thus if a colinear relationship could be established between a genetic map and the primary structure of a protein, it would be reasonable to conclude that this relationship extends to the nucleotide sequence corresponding to the genetic map.

In previous reports on studies with the tryptophan synthetase A protein, conclusive evidence was presented for the colinearity of a segment of the A gene and a segment of the A protein.[12, 13] The present communication deals with more extensive data

*This investigation was supported by grants from the National Science Foundation and the U.S. Public Health Service. Communicated by Victor Twitty, December 18, 1963.
†Present address: B. C. Carlton, Department of Biology, Yale University, New Haven, Connecticut; D. R. Helinski, Department of Biology, Princeton University, Princeton, New Jersey; U. Henning, Max-Planck Institut für Zellchemie, Munich, Germany.
‡Guiness Research Fellow, on leave from Department of Biochemistry, Oxford, England.
§Department of Biological Sciences, Stanford University.

with 16 mutants with mutational alterations in one segment of the A gene and the A protein.

MATERIALS AND METHODS

Mutant Strains   Of the A-protein mutants examined in detail in this paper, strains A23, A27, A28, A36, A46, A58, A78, A90, A94, A95, A169, A178, and A187 were isolated following ultraviolet irradiation of the K-12 wild-type strain of *E. coli*, and strain A223 was isolated following treatment of the wild-type strain with ethylmethane-sulfonate. Mutant A446 (previously designated PR8)[14] and mutant A487 were initially isolated as spontaneous second-site reversions and subsequently were separated from the original A mutants with which they had been associated. Strains $anth_1^-$ and $anth_2^-$ are blocked prior to anthranilic acid in the tryptophan pathway and respond to anthranilic acid, indole, or tryptophan. V_1^R and V_1^R $tryp^-$ deletion mutants were isolated by treatment of T1-sensitive populations of the various mutants with phage T1h[+]. All of the V_1^R mutations mentioned are very closely linked to the A gene, and the V_1^R $tryp^-$ deletions include the V_1^R locus and some segment of the A gene.[15] A stock of un-restricted T1 phage (uT1)[16] was kindly supplied by J. R. Christensen.

Protein Studies   The altered A proteins were isolated and examined for primary structure changes as described previously.[6, 7, 9] The ordering of the tryptic peptides mentioned in the paper will be described in detail elsewhere.[17, 18]

Genetic Studies   Recombination experiments were performed with the temperate-transducing phage P1kc.[19] Recombination distances between A mutants were obtained by determining the frequency of appearance of $tryp^+$ transductants. Transduction from $his^- \rightarrow his^+$ was scored in each experiment for internal reference, and the ratio of $tryp^+$/ his^+ transductants calculated.[20] Each value was halved to correct for the difference in relative frequency of transduction in the *his* and *tryp* regions.[20] In transduction experiments with leaky mutants (A169, A223, A446, and A487) the plating medium was supplemented with 0.1 µg/ml DL-5-methyltryptophan to suppress growth of the leaky mutants. This supplement has little or no effect on the growth of wild-type recombinants and does not appear to affect the recombination values obtained with nonleaky mutants. In spite of the presence of 5-methyltryptophan, it was frequently difficult to score recombination in experiments with leaky mutants. Anthranilic acid requirement was scored either by picking and streaking or by replication to appropriate test media. Resistance to phage T1 (V_1^R) was scored by picking, streaking, and spot testing with phage uT1. As shown by Drexler and Christensen,[16] P1 lysogeny does not prevent the multiplication of uT1.

RESULTS AND DISCUSSION

Relative Order of Mutational Alterations in the A Gene   The genetic map based on recombination frequencies, deletion mapping, and three-point crosses is shown in Figure 1.

Recombination Frequency Data   Recombination frequency data alone establishes close linkage of three groups of mutants: (1) A446 and A487; (2) A46, A23, and A187; and (3) A58, A78, and A169. The recombination values obtained also suggest that the order of these groups is as indicated in the figure, but they do not permit ordering within any of the groups.* In most cases recombination experiments were performed in both directions, i.e., each strain served as donor and recipient, and fair agreement was observed between the two values. However, it is evident from the data in Figure 1 that there were some exceptions. The exceptions generally involved leaky mutants, and it is possible that the scoring difficulties encountered with these strains were responsible.

Two of the mutants examined, A23 and A187, gave recombination values considerably higher than those obtained with mutant A46. Other mutants which were independent isolates resembling A23 (A27, A28, A36) and A46 (A95, A178) exhibited the same recombination behavior as the strain they resembled. Extensive mapping experiments with other A mutants and the strains mentioned above indicate that the A23 and A187 recombination values are probably exceptionally high rather than the A46 values exceptionally low. Since the mutational alterations in strains of the A23 and A46 type were probably single nucleotide changes,[21] it would appear that differences of single nucleotide pairs between donor DNA and recipient DNA can influence the frequency of recombinational events. An alternative explanation, that all A23 and A187 double mutants are prototrophic, seems very un-

---

*On the basis of preliminary two-point mapping data with mutant A2, it earlier had tentatively been concluded that the alterations in mutants A58, A78, A90, and A94 were to the left of the alterations in A17 and A46.[5] The more extensive data presented in Figure 1 and the deletion and three-point data indicate that the correct order is as shown in Figure 1.

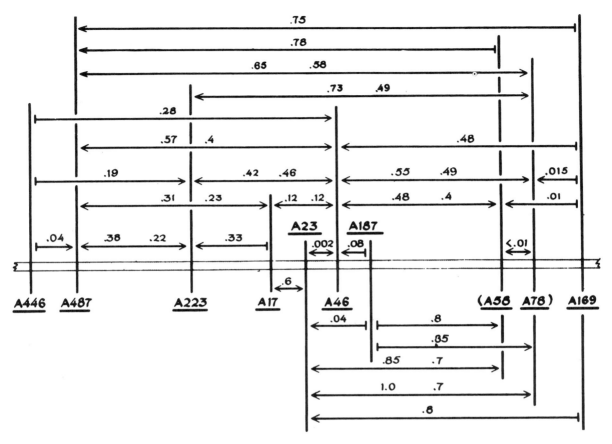

Fig. 1. The order of mutationally altered sites in the A gene based on recombination frequencies, deletion mapping, and three-point genetic tests. The head of each arrow points to the recipient in each transduction cross. If two values are given, the cross was performed in both directions. In such cases each value is placed near the recipient in the cross.

likely on the basis of numerous tests with recombinants from crosses involving A23 and A187.

Recombinants were not obtained in crosses of A23 × A27, A28, or A36; A46 × A95 or A178; or A58 × A90 or A94. On the basis of these tests and the primary structure studies to be described, it was concluded that each of these groups consists of members which arose by repeat identical mutations at the same site. Mutants A58 and A78 are clearly different, but they do not give recombinants at the 0.01 percent level.

Deletion Mapping　Information on the order of the mutational alterations in the relevant mutant strains

was obtained in transduction experiments with a series of V_1^R–*tryp*⁻ deletion mutants (Table 1). These latter strains had regions of the A gene deleted, including, in each case, the V_1 locus which is situated at one end of the A gene.[15] Thus all the deletions extended into the A gene from the same side, and are overlapping, but may have different end points in the A gene. The recovery of tryptophan-independent recombinants from a transduction cross between any A mutant and a deletion mutant would indicate that the mutationally altered site in the A mutant was outside the region of the A gene that was missing in the deletion mutant. Recombination values were determined in transduction

Table 1 Recombination tests with various tryp⁻ deletion mutants

Donor	$his^- V_1^R tryp^-$ deletion mutant recipient					
	T–201	T–5	T–70	T–689	T–211	T–226
A446	0	0	0.08‡	0.08‡		
A487	0	0	0.19*	0.5		
A223	0	0	0.17	0.06	+	+
A23	0	0	0	0	1.3	+
A46	0	0	0	0	0.86	0.77
A187	0	0	0		1.3	+
A58	0	0	0	0	0.09	0.1
A78	0	0	0		0.1	0.1
A169			0		0.1	0.1
A38	+	+	+	+	+	+

+, 0 = recombinants or no recombinants, respectively, in qualitative transduction experiments.
* = each recombination value represents the uncorrected observed ratio of $tryp^+$ to his^+ transductants.
‡A446 gives unusually low recombination values in all experiments.

crosses with some of the deletion mutants to approximate the relative distance from a mutationally altered site in an A mutant to the end point of a deletion.

It is clear from the crosses with T⁻70 and T⁻689 that the altered sites in mutants A23, A46, A187, A58, A78, and A169 — but not those in A446, A487, and A223 — are within the region of the A gene that is missing in these deletion mutants. The relative order of the altered sites in mutants A446, A487, and A223 could not be established by quantitative transduction experiments with the same deletion mutants. The quantitative transduction experiments with deletion mutants T⁻211 and T⁻226 divide the other mutants into two groups: (A46, A23, and A187) and (A58, A78, and A169). The combined results presented suggest the following order of altered sites: A38 — (A446, A487, A223) — (A46, A23, A187) — (A58, A78, A169) — V₁.

Three-point Genetic Tests Genetic tests employing outside markers were performed to determine relative order within each group of closely linked mutants. These tests are summarized in Table 2. The results obtained support the conclusions concerning group order that were arrived at by the previous methods and indicate relative orders within each closely linked group. Crosses 1–7 establish the order anth — A34 — A446 — A487 — A223 — A46,

crosses 8 and 9 the order anth — A46 — A78 — A169, and crosses 10 and 11 the order anth — (A23, A46) — A187. The order anth — A23-A46 had tentatively been assigned on the basis of other data.[21] Crosses 12–15 confirm orders established by other methods. The combined genetic analyses with the mutants examined suggest the sequence of mutational alterations shown in Figure 1 — A446-A487-A223-A23-A46-A187-(A58, A78)-A169.

PRIMARY STRUCTURE STUDIES

Amino acid substitutions have been detected in primary structure studies with each of the 16 A mutants.[6, 7, 14, 17, 18] The substitutions observed and the peptides in which they are present are shown in Figure 2. The conclusions from the genetics studies are also included in the figure (i.e., order of alterations and approximation of recombination distances between alterations). It is apparent that mutants with alterations extremely close to one another in the A gene have amino acid substitutions close to one another in the A protein. Primary structure studies with the A protein[17, 18] have established the linear sequence of peptides TP-11, TP-8, TP-4, TP-18, TP-3, and TP-6, constituting a 75-residue segment of the A protein. This sequence and the sequence of the amino acids of most of these peptides are presented in Figure 3. This seg-

Table 2 Outside-marker ordering of mutationally altered sites

Transduction cross	Nonselective markers	Recombinants detected*		$anth^+$ (%)	Order
(1) 34 $\to$ $anth_2^-$–223	$anth_2$	72 $anth^+$;	332 $anth^-$	18	$anth$–34–223
(2) 46 $\to$ $anth_2^-$–223	$anth_2$	180 $anth^+$;	196 $anth^-$	48	$anth$–223–46
(3) 446 $\to$ $anth_2^-$–223	$anth_2$	14 $anth^+$;	112 $anth^-$	11	$anth$–446–223
(4) 487 $\to$ $anth_2^-$–223	$anth_2$	29 $anth^+$;	118 $anth^-$	20	$anth$–487–223
(5) 34 $\to$ $anth_2^-$–487	$anth_2$	86 $anth^+$;	368 $anth^-$	19	$anth$–34–487
(6) 46 $\to$ $anth_2^-$–487	$anth_2$	27 $anth^+$;	28 $anth^-$	49	$anth$–487–46
(7) 446 $\to$ $anth_2^-$–487	$anth_2$	7 $anth^+$;	31 $anth^-$	23	$anth$–446–487
(8) 46 $\to$ $anth_2^-$–78	$anth_2$	52 $anth^+$;	220 $anth^-$	19	$anth$–46–78
(9) 169 $\to$ $anth_2^-$–78	$anth_2$	22 $anth^+$;	24 $anth^-$	48	$anth$–78–169
(10) 187 $\to$ $anth_1^-$–23	$anth_1$	9 $anth^-$;	42 $anth^+$	83	$anth$–23–187
(11) 187 $\to$ $anth_1^-$–46	$anth_1$	14 $anth^-$;	116 $anth^+$	89	$anth$–46–187
(12) 46 V_1^R $\to$ $anth_1^-$–58 V_1^S	$anth_1$; V_1	22 $anth^-$ V_1^R;	1 $anth^+$ V_1^R	4	$anth$–46–58–V_1
(13) 58 V_1^R $\to$ $anth_1^-$–46 V_1^S	$anth_1$; V_1	10 $anth^+$ V_1^S;	1 $anth^+$ V_1^R	73	$anth$–46–58–V_1
			4 $anth^-$ V_1^S		
(14) 169 V_1^R $\to$ $anth_1^-$–46 V_1^S	$anth_1$; V_1	6 $anth^+$ V_1^S;	1 $anth^-$ V_1^S	90	$anth$–46–169–V_1
(15) 446 $\to$ $anth_1^-$–46	$anth_1$	81 $anth^-$;	22 $anth^+$	21	$anth$–446–46
		20 $anth^-$–	65 $anth^+$		
		446–46;	446–46	76	

Order of markers: $anth$–A34— (A446, etc.) — V_1^R.

*In crosses with the outside marker $anth_2$ the percentage of $anth^+$ recombinants is approximately 20% if the order is

$$\frac{+ \qquad \mathrm{x}}{anth^- \qquad \mathrm{y}}$$

and approximately 50% if the order of x and y is reversed. With the marker $anth_1$ different values are obtained but the order of x and y relative to $anth$ can be clearly established. The explanation for the different values obtained with the two outside markers is not known.

Fig. 2. Amino acid substitutions in the A proteins of various mutants.[6,7,14,17,18] The A58, A78, A90, A94, and A169 substitutions will be described in detail elsewhere.[18]

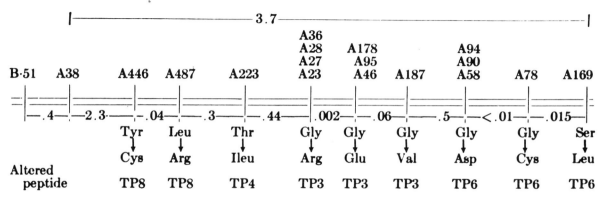

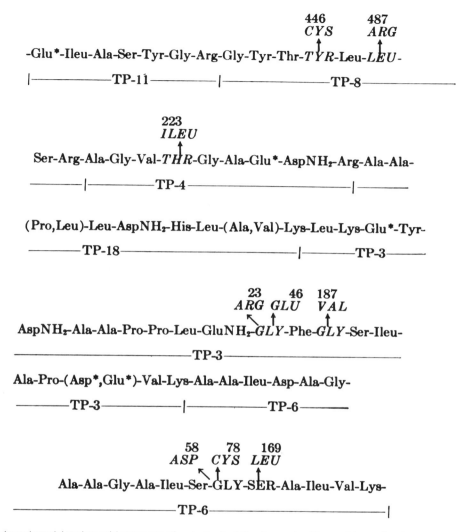

Fig. 3. Peptide and partial amino acid sequence of a segment of the A protein. The positions of amino acid substitutions in the A proteins of the various mutants are indicated. Based on the studies by Carlton and Yanofsky[17] and Guest and Yanofsky.[18] (* = not known whether present as acid or amide.)

ment accounts for approximately one-fourth of the residues in the A protein and is not at the amino or carboxyl end of the protein. The positions of the amino acid replacements in the A proteins of the various mutants are also shown. It is clear that the positions of the amino acid replacements in the segment of the A protein are in the same relative order as the order of the mutationally altered sites of the corresponding mutants in the A gene. These findings convincingly demonstrate a colinear relationship between gene structure and protein structure.

The relationship between the map and residue distances observed is also of interest and can be seen most clearly from the representative values summarized in Table 3. Although the map distance/residue distance ratio varies between 0.01 and 0.05, in most cases this value is approximately 0.02. It would appear, therefore, that distances on the genetic map are representative of distances between amino acid residues in the corresponding protein.

SUMMARY

The concept of colinearity of gene structure and protein structure was examined with 16 mutants with alterations in one segment of the A gene and the A protein of tryptophan synthetase. The results obtained demonstrate a linear correspondence between the two structures and further show that genetic recombination values are representative of the distances between amino acid residues in the corresponding protein.

Table 3 Relationship between map distances and residue distances

Mutant pair	Map distance	Amino acid residue distance	Map distance / Residue distance
A58–A78	< 0.01	0	—
A46–A23	0.002	0	—
A58–A169	0.01	1	0.01
A78–A169	0.015	1	0.015
A46–A187	0.08	2	0.04
A23–A187	0.04	2	0.02
A446–A487	0.04	2	0.02
A487–A223	0.3	6	0.05
A446–A223	0.19	8	0.02
A46–A58	0.44	23	0.02
A46–A78	0.52	23	0.02
A23–A58	0.78	23	0.03
A23–A78	0.85	23	0.04
A46–A169	0.48	24	0.02
A23–A169	0.8	24	0.03
A223–A46	0.44	28	0.02
A487–A46	0.48	34	0.01
A446–A46	0.28	36	0.01
A78–A223	0.61	51	0.01
A169–A487	0.75	58	0.01

Note Added in Proof Dr. Francis Crick has recently forwarded a manuscript which deals with a study of colinearity in another system (Sarabhai *et al. Nature*, in press).

ACKNOWLEDGMENT

The authors are indebted to Virginia Horn for performing the genetic analyses described in this paper. They are also indebted to Deanna Thorpe, Patricia Schroeder, Donald Vinicor, and John Horan for their excellent technical assistance.

REFERENCES

1. Beadle, G. W., and E. L. Tatum, these Proceedings, 27:499 (1941).
2. Sanger, F., and H. Tuppy, *Biochem. J.*, 49:481 (1951).
3. Watson, J. D., and F. H. C. Crick, in *Viruses*, Cold Spring Harbor Symposia on Quantitative Biology, vol. 18 (1953), p. 123.
4. Crick, F. H. C., *Symp. Soc. Exptl. Biol.*, 12:138 (1958).
5. Yanofsky, C., D. R. Helinski, and B. D. Maling, in *Cellular Regulatory Mechanisms*, Cold Spring Harbor Symposia on Quantitative Biology, vol. 26 (1961), p. 11.
6. Helinski, D. R., and C. Yanofsky, these Proceedings, 48:173 (1962).
7. Henning, U., and C. Yanofsky, these Proceedings, 48:183 (1962).
8. *Ibid.*, 1497 (1962).
9. Helinski, D. R., and C. Yanofsky, *Biochim. Biophys. Acta*, 63:10 (1962).
10. Carlton, B. C., and C. Yanofsky, *J. Biol. Chem.*, 238:2390 (1963).
11. Kaiser, A. D., *J. Mol. Biol.*, 4:275 (1962).
12. Yanofsky, C., in *Synthesis and Structure of Macromolecules*, Cold Spring Harbor Symposium on Quantitative Biology, vol. 28 (1963), in press.
13. Yanofsky, C., in *The Bacteria*, ed. I. C. Gunsalus and R. Y. Stanier (Academic Press), vol. 5, in press.
14. Helinski, D. R., and C. Yanofsky, *J. Biol. Chem.*, 238:1043 (1963).

15. Somerville, R., and C. Yanofsky, unpublished observations.
16. Drexler, H., and J. R. Christensen, *Virology*, 13:31 (1961).
17. Carlton, B. C., and C. Yanofsky, in preparation.
18. Guest, J., and C. Yanofsky, in preparation.
19. Lennox, E. S., *Virology*, 1, 190 (1955).
20. Yanofsky, C., and E. S. Lennox, *Virology*, 8:425 (1959).
21. Yanofsky, C., in *Synthesis and Structure of Macromolecules*, Cold Spring Harbor Symposia on Quantitative Biology, vol. 28 (1963), in press.

An historical account of protein synthesis, with current overtones — a personalized view

PAUL C. ZAMECNIK*

Reprinted by author's and publisher's permission from Cold Spring Harbor Symposia on Quantitative Biology, *vol. 33, 1969, pp. 1–19.*

There are four major reasons why we have placed this paper by Paul C. Zamecnik at this point. First, the paper reviews the work that had defined the general outline of how structural information passes from genes to polypeptides. Second, the paper examines the complex events taking place during the translation of the ribonucleotide sequence in messenger RNA into the amino acid sequence of the polypeptide. Third, the paper carefully spells out what is known and what needs to be defined more clearly — providing the reader with insight into the future direction of research in this field of study. Finally, the article is heavily referenced, supplying the interested student with an extensive bibliography relative to the topic of heterocatalysis.

HISTORICAL SECTION

Introduction

A spectacular display of progress in knowledge of the mechanism of protein synthesis has taken place during the past decade, a logarithmic phase of growth which was preceded by an early inoculum of basic facts and an unspectacular lag phase. Just twenty years ago a summary of the extant state of information in this field presented at a Cold Spring Harbor Symposium (Zamecnik and Frantz 1950) ended with the two line poem of Robert Frost —

"We dance round in a ring and suppose,
But the secret sits in the middle and knows."

These were in the pre-cell-free days, and it was still uncertain whether the Bergmann (1942) concept of a reversal of proteolysis or the Lipmann (1941) and Kalckar (1941) suggestion of a phosphorylated intermediate was the key to the direct path from free amino acid to completed protein. As important strands to be woven into the fabric were the observations of Caspersson (1941, 1947) and of Brachet

(1941, 1947) that in rapidly growing cells there was an increase in the content of cytoplasmic ribonucleic acid and a correlation between its presence and protein synthesis. Extensive stable isotopic studies of the Schoenheimer-Rittenberg group (1942) at that time indicated that certain proteins underwent either rapid synthesis and degradation or exchange of free amino acids with protein-bound amino acids. The choice of these alternatives was left in abeyance with reason favoring the first mentioned.

At the same 1950 Cold Spring Harbor Symposium, F. Sanger (1950) presented the first evidence on sequence studies on insulin. The complexity of the amino acid sequence of insulin clearly eliminated the hypothesis that proteins consisted of peptide subunits of simple repeating regular amino acid sequence (Bergmann and Niemann 1937). There were several strong, and some less compelling but appealing, indications that a new set of enzymes might be involved in protein synthesis different from those capable both of proteolysis and of peptide chain-lengthening reactions (Fruton et al. 1951).

In the first place, earlier data (Borsook and Huffman 1938; Borsook and Dubnoff 1940) had indicated that the equilibrium between peptide or peptidelike compound and free amino acid was poised far on the side of hydrolysis of the peptide

*The John Collins Warren Laboratories, Collis P. Huntington Memorial Hospital, Harvard University at the Massachusetts General Hospital, Boston, Massachusetts.

bond. Our colleagues (Frantz et al. 1949) had more currently defined this equilibrium point using ^{14}C-labeled amino acid, unlabeled dipeptide, and a peptidase; they had measured the extent of the back reaction and had calculated the equilibrium constant. Since the equilibrium point was over 99.7% toward hydrolysis, effective reversal of proteolysis seemed unlikely (Zamecnik et al. 1948) even in the presence of coupled reaction sequences (Bergmann 1942). Our colleague, Robert Loftfield, questioned whether the proteolytic enzymes possessed the specificity to discriminate one amino acid side chain from another to the degree requisite for construction of a peptide chain with the detailed precision of insulin. Loftfield and colleagues (1953) found that although an extract of rat liver degraded α-aminobutyrylglycine equally as well as it did valyl-glycine, a rat liver slice built only valine-^{14}C into protein while rigidly excluding α-aminobutyric acid-[14]C. This observation served as an antidote to the interesting finding that methionyl esters could undergo condensation in the presence of a peptidase (Brenner et al. 1950) and shook confidence in the degree of specificity (Bergmann 1939) of the proteolytic enzymes.

At this time several groups of astute biochemists considered that more progress would be made toward an understanding of peptide bond synthesis by study of the requirements for synthesis of oligopeptides such as glutathione (Johnston and Bloch 1949; Hanes et al. 1950) and of peptidelike compounds such as acetylsulfanilamide (Lipmann 1945), glutamine (Speck 1947) and p-aminohippuric acid (Cohen and McGilvery 1947), than by cracking open cells, for a direct confrontation with the entire complex system of protein synthesis. Within a few years a clear-cut requirement for ATP was demonstrated for the synthesis of these simple peptidelike compounds, most of the reactions proceeding, however, by an ATP → ADP route. A remarkable byproduct of the study of the synthesis of acetylsulfanilamide was of course the finding of the new cofactor, coenzyme A (Lipmann 1945).

A shortcoming of this "model peptide" compound approach to protein synthesis was the lack of appreciation at that time that the two steps of (a) activation and (b) determination of future sequence alignment in a peptide chain were linked together in protein synthesis, but not in the case of the synthesis of simple model peptide bonds. Thus, the activation mechanisms of the latter, while highlighting the relevance of a phosphorylated intermediate, did not serve as accurate models for the activation process in the more complex conditions of protein synthesis.

An appealing prediction of those times was that of Chantrenne, who reported from Linderstrøm-Lang's laboratory that it was more likely that an organic phosphate compound (i.e., a di-substituted phosphorous compound) might be an intermediate in protein synthesis as an aminoacyl phosphate anhydride than inorganic phosphate itself, since the former would be a more stable compound (Chantrenne 1948a, 1948b).

EARLY POSTWAR STUDIES

During the early postwar years several groups of biochemists had meanwhile decided to travel the course leading from whole animal to slice to cell-free system in their search for an answer to the mechanism of protein synthesis (Melchior and Tarver 1947a, b; Frantz et al. 1947; Anfinsen et al. 1947; Zamecnik et al. 1948; Greenberg et al. 1948; Borsook et al. 1949; Hultin 1950). The earlier classic work of the Schoenheimer school (1942) had developed techniques and a way of thinking in terms of isotopic tracers as tools for this type of study. The stable isotopes, principally ^{15}N, used earlier with such great profit, were not sufficiently sensitive to permit detailed studies of protein synthesis on the tissue slice level. Carbon 11 had too short a half-life for comfort. It was therefore an enormous boon when ^{14}C was discovered by Ruben and Kamen (1940). As soon as World War II was over, news of its availability stimulated us to form a team to study protein synthesis by means of this new tool. Our colleague Ivan Frantz was clever at arranging counting techniques for Carbon 14, and we drew into our team Robert Loftfield, who was at that time an unusual new hybrid type of organic chemist, unfrightened of employing this untried isotope to solve mechanistic problems in organic chemistry. This he did in what has become known as a "minor classic," the nature of the Faworskii reaction (Loftfield 1951). Using the Strecker cyanide synthesis, Loftfield deftly prepared ^{14}C-amino acids, labeled in the carboxyl group (Loftfield 1947). This location of radioactivity made it possible to hydrolyze the protein with acid, to degrade the amino acids with ninhydrin, and to collect the $^{14}CO_2$ released as a $BaCO_3$ precipitate. The use of ninhydrin before and after acid hydrolysis of a well-washed protein provided us with a degree of assurance that the radioactivity was present in peptide bonding and only releasable after acid hydrolysis (Frantz and Zamecnik 1950; cf. also Greenberg et al. 1948). ^{35}S-labeled cysteine and methionine were already extant, but the hazards of encountering nonpeptide bond linkages (Melchior and Tarver 1947a, b) when employing these amino acids resulted in their general avoidance.

The Artifact Problem   Thus the cardinal difficulty in evaluating early incorporation data obtained in tissue slice experiments and in broken cell preparations was the question of whether the added labeled amino acid was really in α-peptide bonding in a peptide chain (Zamecnik 1950). One says "added," because it was appreciated from the work of the Columbia group (Schoenheimer 1942) that the carbon skeleton of a number of amino acids could be degraded, that transaminations could occur, and that ^{14}C originating in an added amino acid could turn up in a number of unsuspected compounds.

An important tool in following the fate of an added ^{14}C-labeled amino acid was fortunately developed by Moore and Stein (1949). Starch column chromatography made it conveniently possible to say whether the ^{14}C which originated in one added amino acid was still in the same amino acid at the end of the experiment. By 1948 Moore and Stein had carefully worked out the pattern of emergence from potato starch columns of various known mixtures of amino acids; and they permitted Ivan Frantz and me to run the first protein hydrolysate in their laboratory. The evidence provided in this way of conversions of ^{14}C from glycine, alanine, aspartic, and glutamic acids into many metabolic compounds, including other amino acids (Zamecnik and Frantz 1950; Zamecnik et al. 1951), resulted in our choice of the metabolically less active higher aliphatic amino acids — leucine, isoleucine and valine — for labeling (Loftfield and Harris 1956) especially in crucial experiments. In addition, Loftfield resolved the *dl*-amino acids he synthesized, thus avoiding uncertainties due to isotope dilution of the *l*-form and to a variety of other potential artifacts.

A sequence of treatments of a cold trichloroacetic

acid precipitate consisting of lipid extraction, and suspension in hot trichloroacetic acid to degrade nucleic acid, followed by plating of the washed protein, in time replaced the more tedious $BaCO_3$ technique (Zamecnik et al. 1951). As more investigators entered the field and relied on the commercially labeled amino acids beginning to become available (at first glycine, alanine, glutamic, and aspartic acids), they not infrequently were unaware of the above described pitfalls. During the course of several years a number of remarkable artifacts made their way into the literature, including two cell-free systems which did not in reality measure protein synthesis; one measuring formation of amino acid attached to phospholipid, and the other amino acid probably attached to nucleic acid (aa-tRNA!). There were other puzzling observations, such as the incorporation of ^{14}C-*l*-lysine into protein without an added energy source (Borsook et al. 1949). This was in time patiently determined by Schweet (1955) to be due to covalent attachment of the ε-amino group of lysine to existing peptide chains, presumably as a replacement of a phosphorylated group already attached to the peptide chain. A number of pathways by which amino acids become a part of macromolecules other than via the direct line from free amino acid to *de novo* completed protein have emerged over subsequent years and are carefully described in a recent book (Hendler 1968). Fortunately we were not diverted by these side paths.

In answer to the early question of whether a free amino acid could exchange with one linked at both ends in the interior of a peptide chain, no evidence of this possibility was found, and it became established that the incorporation of an amino acid into a peptide chain was an irreversible event; the labeled amino acid resisting efforts to replace it by addition of a large excess of inert amino acid of the same type to the incubation mixture (Littlefield et al. 1955; Zamecnik et al. 1956b).

The Tissue Slice and Dinitrophenol   Employing the above techniques, experimenters had established by 1948 in mammalian tissue slices and cell suspensions that protein synthesis required oxygen (Frantz et al. 1947), was inhibited by azide (Greenberg et al. 1948) and most interestingly, was inhibited by dinitrophenol (Frantz et al. 1948). The latter had been just previously shown to dissociate oxidative phosphorylation from respiration (Loomis and Lipmann 1948), and there was a remarkable parallelism between the concentration curves of dinitrophenol which produced inhibition of phosphorylation and inhibition of protein synthesis. It was, in fact, this point in particular which privately convinced us that a new and undiscovered set of enzymes was responsible for protein synthesis, and that the energy of ATP was in some way fed into amino acid activation.

The Early Homogenates During the next five years, in several laboratories, including our own (Borsook et al. 1950a; Peterson and Greenberg 1952; Gale and Folkes 1953), repeated efforts were made to find a cell-free system in which protein synthesis occurred. In retrospect, it is difficult to evaluate to what extent the early cell-free systems measured *de novo* α-peptide bond synthesis.

In our laboratories we were looking for conditions in which a labeled amino acid became incorporated into internal α-peptide bond linkage by a system which contained no live cells. We spent many months in collaboration with Dr. David Novelli trying to prepare a cell-free extract from *E. coli,* and found it difficult to make it sufficiently free of live cells to be reliable (Zamecnik, Stephenson, and Novelli 1951 unpubl.). We, therefore, returned to the use of rat liver, in which there was greater ease of separation of cell-free extract and whole cells. Furthermore, while the incorporation of labeled amino acid into protein in whole bacterial cells was enormously more rapid than into protein in an apparent cell-free bacterial extract, such was not the case for a damaged liver cell preparation wherein the presence of a small percentage of contaminating intact cells was not crucial. The centrifugal separation of whole cell and homogenate fraction in the case of liver was in addition much easier to make than in the case of live bacteria and bacterial homogenates. Thus there was promise in the cell-free system elaborated with our colleague Siekevitz (Siekevitz and Zamecnik 1951; Siekevitz 1952). The limitations here were (a) that it remained unproven that the radioactivity originating in ^{14}C-alanine was still present as ^{14}C-alanine in protein; (b) that it was really in α-peptide bond linkage in the interior of the labeled protein; and (c) that ATP was the proximal energy donor, since AMP worked as well as ATP when added to the system. It was, however, clearly shown that energy derived from oxidative

phosphorylation carried out by the mitochondrial fraction of the system was being used, and was essential for the incorporation observed, with the microsomal fraction being the site of the most active incorporation. This latter point extended earlier less detailed observations by Borsook et al. (1950b), Hultin (1950), and Keller (1951). The next endeavor was to try to substitute ATP and an ATP-generating system for mitochondria, and to carry out a more rigorous analysis of the incorporation product. At this time our colleague Nancy Bucher had just discovered that gentle homogenization of a liver mince resulted in a cell-free system capable of converting ^{14}C from acetate into cholesterol (Bucher 1953), without an added energy source.

A Useful Cell-free System When we employed the Bucher "loose homogenization" technique and looked for incorporation of ^{14}C-leucine into protein in the cell-free homogenate, the results were negative. Upon addition of either ATP and an ATP-generating system or a glycolytic energy-generating system to this type of cell-free preparation, however, a good and reproducible level of incorporation of amino acid into protein occurred (Zamecnik 1953; Zamecnik and Keller 1954).

This system in time was found to pass the critical criteria for protein synthesis (Littlefield et al. 1955; Keller and Zamecnik 1956), including location in internal peptide bonding (Zamecnik et al. 1956b). What were the special virtues of gentle homogenization? Apparently, soluble enzymes and ribosomes preferentially leaked out of the damaged cells, while ATPases, nucleases and proteases were less readily released. Thus, marginal conditions for measuring protein synthesis and its relationship to energy donors became more reliable ones, responsive to the addition of ATP. In addition, this system was easily dissectable into further subdivisions. It was soon shown that essentially all of the incorporation occurred initially into the 105,000 × *g* sedimentable microsomal (and later, ribosomal) fraction (Keller et al. 1954; Littlefield et al. 1955; Littlefield and Keller 1957).

Our concept of the situation at this time is illustrated in an unpublished figure from the 1953 era (Fig. 1). There were four essential components in the cell-free incorporation system: amino acid(s), ATP, microsomes, and enzymes from the 105,000 × *g* supernatant. The requirement for the last men-

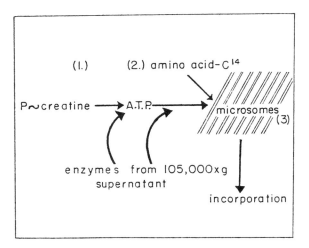

Fig. 1. A 1953 model of protein synthesis.

tioned fourth component was demonstrated by washing the microsomal fraction by resuspension and recentrifugation (from a buffered saline medium), following which the system was dead unless a heat-labile component, largely precipitable at pH 5 ("the pH 5 enzyme fraction") was added. The double set of arrows (Fig. 1) leading from this enzyme denoted the uncertainty of that moment as to where this enzyme fraction interacted. The arrow from the amino acid ^{14}C to the microsomes suggests our initial consideration of the possibility that the amino acid and ATP interacted at the ribosome surface. Otherwise, it appeared, there would have to be a separate supernatant enzyme for each amino acid, a possibility that at the time seemed cumbersome, and out of line with the simpler activating systems of model peptide compounds.

The Activation Step At this time Mahlon Hoagland joined us and undertook to distinguish between these possibilities. He had returned from Lipmann's laboratory, where it had been found that the peptidic bond between β-alanine and pantoic acid was formed via an ATP-participating step in which pyrophosphate was released (Maas and Novelli 1953; Hoagland and Novelli 1954).

The above parallel served as a stimulus to look for amino acid activation by means of ^{32}P-pyrophosphate exchange into ATP in the presence of amino acids and the pH 5 enzyme fraction, and to examine

for hydroxamate formation in the same system. Both of these Dr. Hoagland clearly found, and the first step in protein synthesis was established (Hoagland 1955; Hoagland et al. 1956). At this moment, Paul Berg (1955) was also finding that acetate was activated by an enzyme which interacted with ATP to release pyrophosphate and to form acetyladenylate. Thus the latter compound, rather than the previously favored acetylphosphate, was the activated phosphorylated form of acetate. The formation of an aminoacyl-AMP intermediate was strengthened by the synthesis of *l*-leucyl-AMP (De Moss et al. 1956), and demonstration of its ready conversion to ATP by a purified leucyl amino-acyl synthetase in the presence of PP. The 1955 model of the steps in protein synthesis is shown in Fig. 2.

The Importance of the Ribosomes It had by this time been determined, as stated above, that the ribonucleoprotein part of the microsomes (the ribosome) was the site of synthesis of the new peptide chain. It was also observed that only a small fraction of the ribosomal protein participated in this rapid synthetic process, and the ribosome was thus regarded as the relatively inert marshalling site on which the newly forming chain was built, and from which this nascent protein was subsequently released into the supernatant, soluble protein fraction of the cell (Littlefield et al. 1955).

A Visit with Dr. Watson In the summer of 1954 we inquired of our scientific neighbor, Paul Doty, as to how the RNA of the ribonucleoprotein particle could serve to order the activated amino acids. Dr. Doty mentioned that a Dr. James Watson was visiting him, who with a colleague, Dr. Crick, had recently proposed a model structure for DNA (Watson and Crick 1953). He would send Dr. Watson over to visit me and I could ask him that question. I looked at Dr. Watson's young face above his white Irish turtle neck sweater, then at his wire model of the double-stranded DNA, and inquired how the message of the DNA made its way into the sequence of protein. Unfortunately, it seemed to me privately, the bases were facing in, rather than out, as on the biologically appealing earlier Pauling model. Was protein made directly on a DNA template? Probably not, it appeared, because there was no DNA in the ribonucleoprotein particle. Was RNA made on the DNA template? No answer to this question

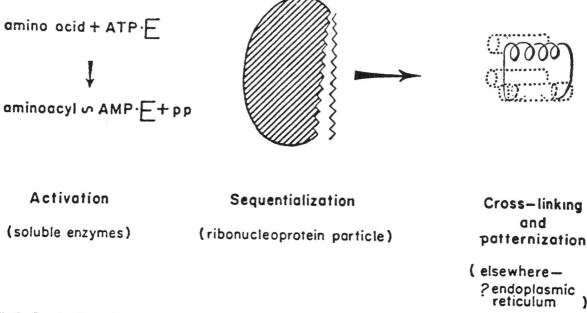

amino acid + ATP·E

$\downarrow$

aminoacyl $\sim$ AMP·E + pp

Activation

(soluble enzymes)

Sequentialization

(ribonucleoprotein particle)

Cross−linking and patternization

(elsewhere—
?endoplasmic
reticulum)

Fig. 2.   Postulated steps in protein synthesis in rat liver cytoplasm, 1955 model (Zamecnik et al. 1956a).

either, although it seemed likely. How did this complicated double helix unwind? There was a gulf between DNA and protein synthesis, Dr. Watson agreed with a diffident smile, as we parted and he took off on vacation to look at birds.

The Finding of GTP as a Cofactor   It was observed around 1955 that when microsomes (or ribosomes) were separated from the supernatant soluble fraction of the cell by two cycles of resuspension and recentrifugation in fresh media, and when the supernatant protein fraction was adjusted to pH 5 and precipitated, then redissolved in fresh medium and reprecipitated, the system was freed of some small molecular weight substance necessary for protein synthesis. This deficiency could be remedied if a crude preparation of ATP was used as an energy source, rather than a purified one. We, therefore, suspected that another mononucleotide or oligonucleotide played a role, and this was pinpointed to GTP, although initially we could not distinguish between GDP and GTP as the active factor (Keller and Zamecnik 1956). Guanine-containing dinucleotides were not able to replace GTP in this function.

It was clear that the site of action of GTP was subsequent to the activation step, for which GTP could be shown not to be requisite. GTP was, however, necessary for the transfer of the aminoacyl group from its association with transfer RNA (tRNA) to its inclusion in a peptide chain (Hoagland et al. 1958). The precise role of GTP in polypeptide polymerization, however, eluded investigators for a decade, and is just now becoming evident, as described abundantly in this symposium.

Amino Acids or Peptides as Intermediates   One of the early questions about the mechanism of protein synthesis had been whether the polypeptide chain grew exclusively by way of one-by-one sequential polymerization of free amino acids or via condensation of small free peptide blocks. This latter possibility, evidence for and against which was reviewed by Borsook (1956), was rendered unlikely by Loftfield and Harris (1956). They induced the synthesis of ferritin *in vivo* in rat liver, while maintaining a constant intracellular specific activity of ^{14}C-leucine, -isoleucine, and -valine. Ferritin isolated after one to three days had these three amino acids with the same specific radioactivity as the intracellular free amino acid pool and much higher radioactivity than other liver proteins. Thus, the ferritin could not have been derived from proteins which existed prior to the experiment. The conclusion was reached that protein synthesis proceeds from free amino acids without significant participation of peptide or protein residues (Loftfield 1957).

The Finding of Transfer RNA   We had for some time wondered whether the same cell-free conditions which served for the study of protein synthesis might also be adequate for the investigation of nucleic acid synthesis. The polynucleotide phosphorylase and formation of poly A by Grünberg-Manago and Ochoa (1955) was a new discovery of this moment, and yet there was the misgiving that the main route of RNA synthesis might follow a different path to account for its presumed precision of sequence. Thus we added ^{14}C-ATP to our cell-free liver system and isolated an RNA fraction, which proved to be labeled with the ^{14}C which had originated in ATP (Zamecnik et al. 1957; Hecht et al. 1958a, b, c). As a control on the possibility of artifact due to failure to wash the RNA fraction adequately, we added ^{14}C-leucine in place of ^{14}C-ATP in one flask, and isolated the RNA. Strangely enough, the RNA fraction was labeled from the amino acid precursor. In spite of careful washing procedures, the amino acid remained tightly bound to the RNA, and we concluded with some wonderment that it must be covalently bound to the RNA. This was a total cell RNA, and we initially presumed that the amino acid was linked to ribosomal RNA. Our colleague Jesse Scott pointed out that there was a small fraction of RNA in the cell, amounting to 10–15% of the total, which did not centrifuge down with the ribosomes, but remained in the soluble enzyme fraction of the cell. This low molecular weight RNA had no known function. When we repeated the experiment, using the 105,000 X *g* supernatant of a cell-free liver homogenate and ^{14}C-ATP, but omitting the ribosomal fraction, the incorporation of ^{14}C- from the ATP into this soluble RNA (sRNA) occurred as well as before, when ribosomal RNA had been present. Furthermore, when ^{14}C-leucine, the 105,000 X *g* supernatant, and ATP were incubated together and the soluble RNA fraction isolated, the ^{14}C-leucine was found to be bound to the sRNA (Zamecnik et al. 1957).

There were already in the air at this time (1955–1956) a few suggestions that something might be missing between the activation and the sequentialization steps of Fig. 2. Hultin and Beskow (1956) had presented evidence for the existence of an unknown intermediate step — a labeled amino acid on the way to completed protein could not be diluted out by a large quantity of unlabeled amino acid. We had also found that a purified tryptophan activating enzyme (Davie et al. 1956), kindly furnished by the Lipmann group, when added to ^{14}C-tryptophan, ATP, and washed ribosomes was inactive in promoting tryptophan incorporation into protein. A rumor had also reached us that at a recent meeting Francis Crick had expressed the opinion that there should be an oligonucleotide intermediate as a language translation piece between activated amino acid and RNA template.

It was at this point that we prepared a ^{14}C-aminoacyl RNA and Mahlon Hoagland plugged it into the cell-free system to determine whether it could substitute for free ^{14}C-amino acid and serve as an intermediate (Hoagland et al. 1957b). As shown in Fig. 3, such was the remarkable finding (Hoagland et al.

Fig. 3.   Transfer of leucine-^{14}C from prelabeled tRNA to microsomal protein (Hoagland et al. 1958).

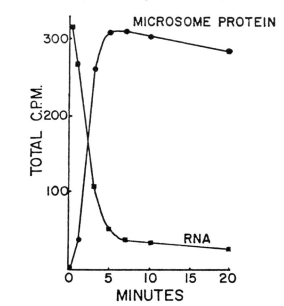

1958). Instead of a triplet oligonucleotide as postulated by Crick, however, there was a rather formidable RNA molecule of sedimentation coefficient around 4 S and a molecular weight of the order of 25,000 — a puzzlingly complex molecule for the job of rapid shuttling and translation. On the other hand, as one later thought of the dual function of this sRNA (or tRNA), the necessity for precision in recognizing the correct aminoacyl synthetase suggested itself, and thereby served to identify the existence of a new problem — the nature of the tRNA-aminoacyl synthetase recognition reaction (Zamecnik 1960, 1966b), one distinct from that of the sequence coding operation. Contemporaneously, Holley, (1957) had reported a ribonuclease-sensitive intermediate step between the activation of alanine (but not other amino acids tested) and the ribosomal polypeptide forming step in protein synthesis. Simultaneously, another group (Ogata and Nohara 1957) had also found evidence for an aminoacyl-RNA intermediate.

In retrospect, it may be asked why we allowed over a year to elapse before reporting on the existence of tRNA, since from the first positive experiment in November 1955 we regarded this molecule as a possible new and dazzling transfer intermediate. We were, however, anxious to remove all doubts that (a) the amino acid was covalently bound, (b) that the aminoacyl nucleotide was a direct line intermediate in the pathway from free amino acid to peptide, and not a side path for storage of activated amino acids, and (c) that the entire complex polynucleotide — and not a much smaller triplet or oligomer which adventitiously aggregated with this larger type of RNA — was the active intermediate.

A Cell-free System for Plants The tobacco mosaic virus (TMV) was known to be an RNA-containing virus, and as another approach to the problem of how RNA might participate in protein synthesis, we undertook an investigation of the synthesis of TMV in tobacco leaves. During the course of this study it was necessary for Dr. Mary Stephenson to work out a cell-free protein synthesizing system in plant material — i.e., in the tobacco leaf. Although this investigation did not find evidence for synthesis of TMV in the broken cell preparations, it did groundbreaking work in providing the first good cell-free system for studying protein synthesis in plants (Stephenson et al. 1956). An interesting, unexpected observation was the ability of chloroplasts to serve as a more self-contained organelle for synthesizing protein than other cell fractions from plants.

The –CCA End Group and the Aminoacyl Esterification Site The finding of an apparent covalent bonding of an aminoacyl group to tRNA immediately raised the question of the site of its attachment. At the maximal labeling level, and with the assumption of an equal percentage of the tRNA pool being associated with each of the 20 amino acids, it appeared that there might be only one aminoacyl group per tRNA molecule. We already knew that both AMP and CMP residues could be added to tRNA as terminal residues by a new pyrophosphorylyzing enzyme present in the soluble protein fraction of the cell. We had in fact initially hoped that we were observing a complete synthesis of tRNA by a pathway different from that of Grunberg-Manago and Ochoa (1955). However, the inability of UTP-^{14}C and GTP-^{14}C to serve as precursors for tRNA, while ATP-^{14}C and CTP-^{14}C were able to do so, indicated that this was an endlabeling of the tRNA (Hecht et al. 1958a) for an unknown purpose.

Isotopic labeling experiments thus showed that there was an end group constellation — CCA — common to all individuals in the tRNA family. In "aged" or preincubated preparations of the pH 5 precipitable portion of the 105,000 $\times$ g supernatant fraction of rat liver or mouse ascites tumor cells, it was found that an aminoacyl group could not be attached to tRNA unless CTP as well as ATP was added to the pH 5 fraction, which contained both aminoacyl synthetases and tRNA (and other enzymes as well). The preincubation served to reverse off the terminal-CCA trinucleotide either partially or completely, by means of the same pyrophosphorolytic enzyme which had attached them. There were definitely two cytidylyl residues interior to the adenylyl residue, and no aminoacyl group could be attached, unless the terminal adenylyl residue was present. The terminal adenylyl residue was thus pinpointed as the site of covalent bonding of the aminoacyl group (Hecht et al. 1958a, b, c; 1959). The question remained as to whether the adenine ring, the ribose group or the internucleotide phosphate was the attachment site.

The last mentioned seemed unlikely because phosphotriesters were notoriously labile, and this aminoacyl-tRNA bond could withstand heating under mildly acidic conditions, and was fairly stable at neutrality. The adenine and ribose rings were left as possibilities. We reported, simultaneously with the Lipmann group at a Gordon Conference in 1958, finding the site of attachment of the aminoacyl group to be in ester linkage on the 3' or 2' hydroxyl group of the terminal adenosyl residue of the tRNA. Our evidence was that periodate oxidation of the terminal 2' and 3' *cis*-hydroxyl groups of the ribosyl group abolished the ability to esterify the aminoacyl group to the tRNA; that the presence of an aminoacyl group on the tRNA prevented periodate oxidation; and that borate, which complexed with the 2', 3'-*cis*-hydroxyl groups on the ribosyl group, inhibited the aminoacyl esterification (Hecht et al. 1959; Zamecnik et al. 1960a). Direct, clear-cut evidence reaching the same conclusion was the finding of aminoacyladenosine by Zachau et al. (1958), as a result of enzymatic hydrolysis of aminoacyl-tRNA by pancreatic ribonuclease. These findings were confirmed by Preiss et al. (1959).

A Convenient Method for Preparing tRNA *in Large Quantity* It had at this moment become desirable to improve the method of preparation of tRNA in order to make it available on a larger scale. We had up to this time disrupted living cells and had obtained tRNA from the 100,000 $\times$ *g* supernatant fraction, using phenol extractions as employed by Gierer and Schramm (1956) for extraction of the RNA of tobacco mosaic virus, and as used by Kirby (1956) for more general work on RNA. Although baker's yeast was a rich source of tRNA convenient to use, the above procedure was long and not readily able to yield large quantities of tRNA. We therefore tried a shortcut on the chance that direct exposure of yeast to phenol would rupture the cells and make available sRNA (tRNA) among a variety of types of RNA for further fractionation. Much less RNA was extracted by this direct phenol method than following disruption of the cells and differential centrifugations. However, we were astonished to find that there was in the extract virtually no high molecular weight RNA and that the yeast cells behaved as if they had become leaky for low molecular weight RNA and yet retained, or precipitated, large molecular weight ribonucleoprotein complexes.

A comparison of the sedimentation constants, chromatographic behavior on DEAE-cellulose columns, base composition, formation of dinitro-

phenylhydrazones, and amino acid acceptor activities of this RNA with those of sRNA prepared in the previous, tedious way revealed very great similarity for both products. Thus, tRNA could be prepared readily from commercial baker's yeast in large quantities (Monier et al. 1960); and was also so prepared, with a minor modification, by Holley et al. (1961). By employing chemical (Zamecnik et al. 1960b), chromatographic (Stephenson and Zamecnik 1961) and countercurrent distribution (Holley et al. 1961; Apgar et al. 1962) techniques for fractionation of the mixed yeast tRNA, highly purified single amino acid-accepting species were then obtained.

The Finding of a Polypeptide Polymerase Step Enzyme When liver microsomes were well washed by centrifugation and resuspension in 0.25 *M* sucrose three times, and the 105,000 $\times$ *g* supernatant from a lysate of ascites cells was adjusted to pH 5, and the precipitate removed, leaving little aminoacyl synthetase, it was found that addition of ^{14}C-aminoacyl-RNA and GTP to these components resulted in incorporation of the amino acid into a polypeptide chain. The enzyme from this supernatant fraction (termed "enzyme 3") appeared to be distinct from aminoacyl synthetase (called "enzyme 1") and provided the first indication for a new catalytic requirement specifically related to polypeptide polymerization (Zamecnik et al. 1958; Zamecnik 1960). It is probable that several enzymes now recognized as being involved in the formation of the polypeptide chain were included in enzyme 3.

To complete this scheme of enzyme numerology it may be mentioned that "enzyme 2" was the –CCA nucleotide-triphosphate pyrophosphorylase, whose finding we have already described (Hecht et al. 1958a). In summary, by 1959 six steps in the pathway from free amino acid to completed protein had been delineated as illustrated in Fig. 4 (Zamecnik 1960).

Development of a Cell-free Bacterial System These early studies on protein synthesis were carried out on mammalian cell-free systems because of the difficulty, already alluded to in this discussion, of obtaining sufficiently cell-free preparations from bacteria. With the background of this earlier failure in mind and the experience of Dr. Stephenson on the plant system (Stephenson et al. 1956) available, Dr.

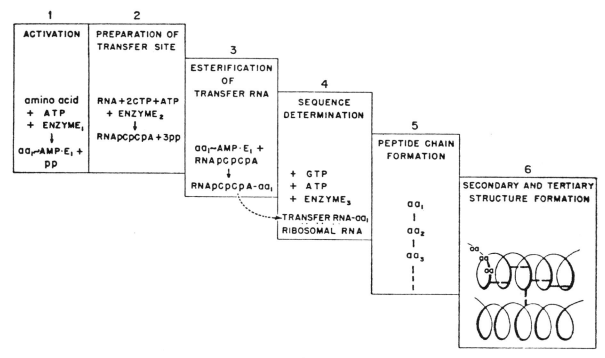

Fig. 4.   Steps between free amino acid and completed protein (Zamecnik 1960).

Lamborg undertook in our laboratories to work out a reliable cell-free system from *E. coli*. Essential to progress in this task was plating out, by serial dilution and counting, the number of live bacteria in these varied preparations of broken cells. It was found that in order for results of incorporation of a labeled amino acid into a broken cell fraction of *E. coli* to be meaningful in cell-free terms even for short incubation times, the concentration of live *E. coli* had to be no greater than 1×10^5 cells per ml. With this figure as a cutoff point for reliability and a monitoring yardstick, the results of experiments became more consistent and the components of the *E. coli* cell-free system were found (Lamborg and Zamecnik 1960) remarkably in agreement with those from rat liver. Other key details for success of the system were (a) the use of mid- or early log phase *E. coli* cells, which were found to contain small concentrations of ATPases and ribonucleases in contrast to the situation in late log and stationary phase; and (b) hand grinding the cells with alumina

in contrast to rupturing of cells by a number of other methods.

Once this cell-free *E. coli* system was established, a new wave of inspired investigators attacked the field and the rate of discovery of new facts took an abrupt turn upward (Tissières et al. 1960; Matthaei and Nirenberg 1961; Nirenberg and Matthaei 1961; Risebrough et al. 1962; Speyer et al. 1963).

Description of a Blind Spot   Let me mention our efforts to examine the polypeptide polymerization site by means of electron microscopy. By good fortune for us, Dr. H. Fernandez-Moran had come to the Massachusetts General Hospital, and in 1960 we began to prepare ribosomes from rabbit reticulocytes which he subjected to high resolution microscopy. We looked hard for our principal objective, the location of tRNA molecules on the ribosomes, but without success. Frequently there were to be seen strands of what appeared to be RNA unraveling from the ribosomes and running from one ribosome to another (Fernandez-Moran and Zamecnik 1961, unpubl.). I felt uneasy about this, considering it to be evidence of roughness in our preparative methods. A year or two later these electron micrographs might have served as textbook pictures of messenger RNA and polysomes. However, at the moment I was fixed on the importance of ribosomal RNA itself as the genetic message and had no eyes for the messenger.

The above historical account will end here, hopefully serving as a description of the "early days" in protein synthesis largely as they touched the activities of a single laboratory. It lacks impersonal judgment of events but, hopefully, may identify for a fresh generation of scientists some of the doors which were consciously or inadvertently opened for others to pass through.

MORE CURRENT INVESTIGATIONS

The story of the discovery of messenger RNA (cf. Watson 1963), of the ability of poly U to serve as an artificial messenger — a breathtaking and likewise fortunate experimental finding in retrospect, in view of the critical importance of initiation factors and proper Mg^{++} concentration — and of the brilliant unfolding of the genetic code (cf. Khorana 1968) are too fresh and well documented for discussion here. As a historical note, it may be recalled that the germinal suggestion of the possibility of a triplet code was made by Dounce (1952). Later, Hoagland et al. (1958, 1959) proposed an adaptor hypothesis involving a nucleotide base-pairing operation between tRNA and template RNA, and Crick (1958) had a similar, more precise idea, involving a triplet code based on theoretical considerations.

The Activation–Esterification Steps Become More Complex Let us, therefore, return to a more current consideration of mechanistic details of the early steps in protein synthesis–activation and aminoacyl-tRNA esterification. After the discovery of transfer RNA there was doubt as to whether one or two enzymes was responsible for these two distinct steps. An aminoacyl synthetase was, however, purified to near homogeneity by Berg and Ofengand (1958), and was clearly shown to carry out combined activation and tRNA esterification. Initially, it was hard

to accept the evidence that a single enzyme could catalyze two such disparate functions. A second enzyme was, however, needed, as mentioned previously, to insure the presence of the –CCA ending common to all tRNAs (Hecht et al. 1958c). In studying the steps from free amino acid to polypeptide in cell fractions, we frequently prepared fractions from more than one biological species and successfully mixed tRNA from yeast with ribosomes from ascites tumor cells or liver (Zamecnik 1960). Thus we were aware of the lack of sharp species specificity in the coding step. We also compared extent of aminoacylation using a single aminoacyl synthetase (from mouse ascites tumor cells); and tRNAs from the same source, from rat liver, calf liver, and yeast (Hecht et al. 1959). Over a range of tRNA concentrations, the extent of labeling with ^{14}C-leucine in 20 min was greatest with ascites tumor tRNA (the homologous interaction), intermediate with rat and calf liver tRNA, and least with yeast tRNA. There was no absolute species specificity. Instead, we observed a homologous species preference, suggesting some unknown species differences in $tRNAs_{leu}$ which impaired their interaction with aminoacyl synthetase in heterologous reactions. We were unaware of isoacceptor tRNAs.

With the finding that ^{18}O initially located in the carboxyl group of an amino acid was transferred to AMP during the activation step (Hoagland et al. 1957a), the formulation of the initial intermediate as an enzyme-bound aminoacyl adenylate had seemed secure. Recently, however, kinetic studies (Loftfield and Eigner 1969) have given substance to the possibility, considered earlier (Boyer 1960), that a general base-stimulated concerted reaction between amino acid, ATP and tRNA in the presence of the enzyme may be a physiological pathway for the aminoacylation reaction. Loftfield's experiments point up the fact that the activation-esterification pathway in protein synthesis is more complex than it initially appeared to be and may not be identical for all aminoacyl tRNA synthetases. Let us mention some of these complexities.

In the case of arginine activation, pyrophosphate exchange can occur only when $tRNA_{arg}$ is added to the reaction mixture [i.e., to synthetase$_{arg}$, arginine, ATP, Mg^{++}, and buffer (Mitra and Mehler 1967; Mehler and Mitra 1967; Marshall and Zamecnik 1969)]. In the analogous situation for glutamic acid the addition of tRNA$_{glu}$ stimulates pyrophosphate exchange but is not required for its occurrence (Ravel et al. 1965). In the case of lysine activation, which we and others have studied in some detail (Stern and Mehler 1965; Kalousek and Rychlik 1965; Haines and Zamecnik 1967; Marshall and Zamecnik 1969), the rate of pyrophosphate exchange is not influenced by the presence or absence of added tRNA over a temperature range from 2°–40°C. When lysyl-tRNA formation is scrutinized in a similar system, both the rate of lysyl-tRNA formation and the final plateau value reached for total lysyl-tRNA formed are influenced markedly by the concentration of lysine employed (i.e., whether in the range 1.5–5 $\times$ 10^{-6} M or in the range 25–200 $\times$ 10^{-6} M). In the latter lysine concentration range, the aminoacylation (at 25°C) is lower and the K_m is higher (27 $\times$ 10^{-6} M); while in the lower concentration range the amino acylation rate is higher and the K_m is lower (2.4 $\times$ 10^{-6} M). These unpublished data suggest, in addition, that the course of lysine activation and esterification may follow either of two alternative kinetic pathways depending on whether the lysyl synthetase combines first with lysine and ATP and subsequently with tRNA$_{lys}$; or, whether the lysyl synthetase combines first with tRNA$_{lys}$ and then with lysine and ATP (Marshall and Zamecnik unpubl.).

At the moment it is difficult to say to what extent our recent kinetic data are complementary and compatible with those mentioned above (Loftfield and Eigner 1969) and with other current kinetic studies of Yarus and Berg (1969). These three investigations do attest, however, to an appreciation that knowledge of the initial two steps in protein synthesis is becoming more detailed. The possibilities of regulation of protein synthesis at this early stage appear correspondingly more appealing and open to experimental evaluation.

The Finding of AP$_4$A. Enzymatic Studies

One such possible regulator is AP$_4$A (and related compounds) which we found several years ago (Zamecnik et al. 1966; Randerath et al. 1966) in studying lysine activation. In essence, we observed that a number of pyrophosphate compounds could compete with inorganic pyrophosphate in the back reaction of the first step (I) in protein synthesis. The back reactions (designated II, III, IV, V and VI) are written in abbreviated form, omitting enzyme (E) and lysine:

$$I - ATP + lysine + E_{lys} \rightleftharpoons$$
$$AMP\text{-}lysine - E + pp$$

$$II - A_{ppp} \rightleftharpoons pp$$

$$III - A_{pppp}A \rightleftharpoons {}_{ppp}A$$

$$IV - A_{pppp}N \rightleftharpoons {}_{ppp}N$$

$$V - A_{ppp}N \rightleftharpoons {}_{pp}N$$

$$VI - A_{ppcp} \leftarrow pcp$$

In the *in vitro* enzyme system above, ATP is initially present in great excess over *pp*, and competes successfully with the latter in the back reaction, with formation of P^1, P^4-di(adenosine-5') tetraphosphate (AP$_4$A, reaction III). N refers to any nucleoside of the ribose or deoxyribose series, and here (reactions IV and V) P^1, P^4-dinucleoside-5', 5'-tetraphosphate (AP$_4$N), or P^1, P^3-dinucleoside-5', 5'-triphosphate (AP$_3$N) are formed. The forward reaction for these compounds is decreased in rate. As discussed below, AP$_4$A has a secondary structure which, once formed, may make this compound (for more than one reason) less amenable to enzymatic attack by aminoacyl synthetase than is ATP. In reaction VI is depicted the formation of the β-γ-methylene analog (APPcP) of ATP when methylene diphosphonate is added to the system in reaction I. APPcP is completely inactive in the forward reaction and the effect of *pcp* is to bring the activation step to a complete halt by converting all of the ATP into APPcP. PcP is, therefore, an interesting irreversible inhibitor of lysine activation (Zamecnik and Stephenson 1969). We have not tested its effect on other activating enzymes and it would be interesting to do so for valine and arginine aminoacyl tRNA ligase, in connection with the question of whether or not a concerted reaction mechanism occurs in these cases.

The compound AP$_4$A has been found (Zamecnik and Stephenson 1968) in *E. coli*, intact rat liver and in ascites tumor cells in a concentration around 10^{-7}–10^{-8} M. Its biological function, if any, is obscure at present.

AP$_4$A Physical Studies

It may be mentioned that we first found evidence of the existence of AP$_4$A while using optical rotatory dispersion in the search for a change in conformation of tRNA when it associated with its cognate aminoacyl synthetase. In the control experiments we omitted the tRNA and found that a new Cotton effect gradually developed when ATP, lysine, and purified lysyl synthetase were incubated together for 15 min at 25°C. This negative Cotton effect occurred in a spectral range with a crossover point close to 260 mμ, suggestive of an alteration related to nucleotide, rather than to protein. Furthermore, it was dialyzable and was identified as AP$_4$A. No other nucleotide then known displayed the negative Cotton effect and the negative circular dichroic pattern of AP$_4$A and AP$_3$A. My colleague, Jesse Scott, constructed a space-filling model of AP$_4$A and related compounds, and in view of their considerable hypochromicity, we considered the possibility of a two ring, stacked configuration (Zamecnik et al. 1967). The presence of the three (in AP$_3$A) or four (in AP$_4$A) phosphate groups provides great flexibility in possible orientations of the adenine rings relative to each other. The unusual sign of the CD pattern suggests that the usual base-stacking arrangement of natural 3′, 5′-dinucleotides and polynucleotides is not present. It appears instead that in the case of AP$_3$A and AP$_4$A in relation to the normal stacking arrangement, one purine ring is rotated 180° so as to appose its opposite face to its neighbor (Scott and Zamecnik, in press). This situation may be likened to rotating one palm of two hands, initially held palm to palm in supplicant position, so as to appose the palm of one hand to the back of the second. Theoretically, this can of course be accomplished in two nonidentical ways, i.e., either by rotating the left or the right hand.

The unusual ORD and CD patterns are concentration independent, and sedimentation equilibrium measurements indicate that one is dealing with a unit of molecular weight of approximately 1000, compatible with the molecular weight of a monomer unit of AP$_4$A. Thus, the ORD and CD patterns represent an intramolecular effect, due to the unusual type of ring stacking mentioned above. Why this should be the preferred stacking orientation when sufficient freedom of rotation of the rings makes a choice possible — it is not easily possible where the normal 3′5′ monophosphate group links two purine ribosides together — is an interesting question in quantum mechanics, which may also prove to be of biological significance. It is notable that AP$_4$A is much more resistant to alkaline degradation than is

ATP. One is tempted to look at AP$_4$A in part as a locked-in, stable form of high energy compound.

Aminoacyl Synthetase-tRNA Recognition: A Central, Unsolved Problem   The question of how an aminoacyl synthetase recognizes its cognate tRNA, and discriminates against all others, to a degree which has not as yet been accurately quantitated, is an important unsolved one, central to molecular biology. Although the existence of this problem has been appreciated for years, it has resisted all efforts due to a lack of appropriate biological tools. An investigator would do well to have at his disposal a purified and sequenced aminoacyl synthetase, and two purified and sequenced cognate tRNAs, one from the species homologous to the synthetase and capable of interacting successfully to form aminoacyl-tRNA; and a second tRNA of the same aminoacyl acceptor type but from a heterologous species, incapable of interacting with the synthetase to form an aminoacyl-tRNA. These two tRNAs would, as part of the ideal situation, both be able in the aminoacylated form to associate with the same ribosomal complex and to donate their aminoacyl groups to a growing peptide chain.

With these biological tools, one possible approach would be to employ a bifunctional reagent and to crosslink the aminoacyl synthetase and the cognate tRNA with which it successfully interacts. Then with partial degradation by enzymatic procedures, one might hope to isolate a smaller, interacting fragment of the enzyme and tRNA. Crystallization of a purified tRNA-aminoacyl synthetase would also fulfill a biologist's dream.

Quite early we attempted to determine whether the anticodon loop might be responsible for aminoacyl synthetase recognition as well as for coding. Using aqueous bromination, we concluded that the anticodon area could not be identical with the recognition site of the tRNA, and expressed the point of view that a more intact secondary (and, we should have added, tertiary) structure of tRNA was necessary for the aminoacylation than for the transfer function (Yu and Zamecnik 1963a, b; 1964). A similar conclusion was reached on the basis of ultraviolet irradiation sensitivity studies of tRNA (Buc and Scott 1966; Aoki et al. 1969). More recent evidence reports the ability of a tRNA to carry out aminoacylation even after an enzymatic excision in the anticodon area (Mirzabekov et al. 1969). Suggestive data implicating the anticodon area in the recognition function came from our laboratory, (cf. also Engelhardt and Kisselev 1966) with the use of osmium tetroxide as a reagent to modify pyrimidine bases, particularly in single-stranded areas of tRNA. A correlation was drawn, in tRNAs exposed to osmium tetroxide, between inhibition of aminoacylation and the presence of a pyrimidine in the middle position of the presumed anticodon triplet (Burton et al. 1966).

No positive evidence relating the GTψCG-containing loop (loop I of Doctor et al. 1969) with the recognition function has been forthcoming. On the contrary, an investigation of Ofengand (1967), in which the ψ residue of this loop (loop I) was chemically altered by acrylonitrile, showed no deleterious effect on the aminoacylation reaction.

Areas in question for the recognition function are the "dihydro U" loop (loop IV) (Shugart and Stulberg 1969), the part of the double-stranded stem region distal to the aminoacyl ester (Imura et al. 1969a, b; Harada et al. 1969), and the vestigial loop (loop II) (cf. Philipps 1969). A more adequate description of the varied experiments which result in the lack of a clear decision of this question is contained in a review by Madison (1968). Both model studies (Cramer et al. 1968; van der Haar and Cramer, pers. commun.), and theoretical considerations (Philipps 1969) based on comparisons of fourteen tRNAs then sequenced, since the initial stunning *tour de force* (Holley et al. 1965), highlight the relationship of the tertiary conformation of tRNA to its recognition function. In this context, the importance of the secondary and tertiary structure of tRNA for its biological activity was prophetically pointed out years ago (Fresco et al. 1960), and served as a precursor for the later clover leaf model.

Mention of the necessity for a correct tertiary structure of a tRNA for success in recognition, points tellingly to a central difficulty in defining its requirements. Chemical modification of a single base in *many* sites in this molecule will alter its tertiary structure, without necessarily being directly involved in the recognition function. The situation is quite analogous to that of identifying an active catalytic site on an enzyme. Since the amino acids making up the active site are not linearly adjacent, modification of a disulfide bridge, some distance away may of course disturb the conformation of the active site.

One is tempted from two points of view to offer what may be by exclusion an obvious suggestion concerning the tRNA side of the aminoacyl synthetase recognition site. The first point of view is that at present no one area of the tRNA molecule has been found to qualify completely as the recognition site, possibly because the latter may not reside in one linear sector of the tRNA. The second point of view is that *there would be more degrees of freedom and of possibility of precision of interaction of* tRNA *with aminoacyl synthetase if the few critical participating bases of the* tRNA *were both nonhydrogen bonded and nonadjacent.* Analogously, nonadjacent aminoacid groups generally constitute the active catalytic site of an enzyme, which is created as an entity by the tertiary structure of the protein. Evolution may very well use the same good trick in two types of macromolecule. *Thus the recognition site may consist of participating bases from single-stranded regions of more than one loop area comprising, by conformational folding, the active recognition area.* The loop IV area, rich in pyrimidines or dihydropyrimidines which do not stack well as a single strand, is appealing as one loop in such a hypothetical multi-loop active site. One may also be a little partial to the evidence which favors the participation of a base from the anticodon since it carries specific amino acid-related information.

It would seem reasonable to imagine that on the side of the aminoacyl synthetase only a few amino acids would constitute its representatives to this mixed macromolecular coding operation. Less work has been done on the recognition site from the point of view of the aminoacyl synthetase than from that of the tRNA. We found (Haines and Zamecnik 1967) that fluorodinitrobenzene, cyanate, and Woodward's reagent *K* had different selective inhibitory activities toward a group of crude *E. coli* aminoacyl synthetases. Very recently, Iaccarino and Berg (1969) reported that n-ethylmaleimide blocks pyrophosphate exchange but has no influence on the association of purified aminoacyl synthetase and purified tRNA; the latter as measured by the convenient membrane filter technique (Yarus and Berg 1967). While little firm ground has been reached in this discussion of the status of knowledge of the aminoacyl synthetase-tRNA recognition reaction, it is an area of enormous potential importance to human genetics and virology, a major investigative battleground for molecular biology in the near future.

PROSPECTS

Our interest in protein synthesis originated from a desire to search for a defect in regulation of growth which might be characteristic of cancer. Edwin J. Cohn, who contributed so much to our knowledge of plasma proteins, told me years ago — if you are interested in a clinical problem, start back about three steps from where you really want to work. Then you will find out something, although it may not be what you were hoping for. Now that so many pieces of the machinery of growth are known, it is perhaps more pertinent to inquire about their relationship to cancer. The RNA oncogenic viruses appear to be a good model for such a study. Here the intruder introduces a messenger RNA whose presence modifies the transfer RNA of the cell, its cell membrane, and other unknown features of the balanced cell economy. One may hope that current knowledge of protein synthesis may contribute to this problem in particular as well as to more general medical mysteries involving hereditary disease and metabolic aberrations.

ACKNOWLEDGMENTS

Over a span of more than twenty years a parade of skillful biochemists has joined the Huntington Laboratories for varying times to contribute to our knowledge of protein synthesis. In describing our common work, I take pleasure in saluting the efforts of M. L. Stephenson, R. B. Loftfield, P. Siekevitz, E. B. Keller, M. B. Hoagland, J. F. Scott, J. W. Littlefield, L. I. Hecht, R. Monier, M. R. Lamborg, C.-T. Yu, D. W. Allen, J. W. Haines, P. Sarin, K. Burton, R. D. Marshall, and P. Schofield.

The work discussed was supported by American Cancer Society grants number E-102 and E-51, United States Public Health Service grant number CA05018 and a grant from the Atomic Energy Commission number AT(30-1)2643.

The author expresses his best thanks to Dr. Mary L. Stephenson for critical review of the manuscript.

This is publication 1355 of the Cancer Commission of Harvard University.

REFERENCES

Anfinson, C. B., A. Beloff, A. B. Hastings, and A. K. Solomon. 1947. The in vitro turnover of dicarboxylic amino acids in liver slice proteins. *J. Biol. Chem.* 168:771.

Aoki, I., T. Idemura, H. Fukutome, and Y. Kawade. 1969. Ultraviolet inactivation of the functions of phenylalanine and lysine transfer RNA's of *Escherichia coli. Biochim. Biophys. Acta* 179:308.

Apgar, J., R. W. Holley, and S. H. Merrill. 1962. Purification of the alanine-, valine-, histidine-, and tyrosine-acceptor ribonucleic acid from yeast. *J. Biol. Chem.* 237:796.

Berg, P. 1955. Participation of adenyl-acetate in the acetate-activating system. *J. Am. Chem. Soc.* 77: 3163.

Berg, P. and E. J. Ofengand. 1958. An enzymatic mechanism for linking amino acids to RNA. *Proc. Nat. Acad. Sci.* 44:78.

Bergmann, M. 1939. Some biological aspects of protein chemistry. *J. Mt. Sinai Hosp.* 6:171.

—. 1942. A classification of proteolytic enzymes. *In* F. F. Nord and C. H. Werkman [ed.] *Advances enzymol.* 2:49. Interscience, New York.

Bergmann, M., and C. Niemann. 1937. Newer biological aspects of protein chemistry. *Science* 86:187.

Borsook, H. 1956. The biosynthesis of peptides and proteins. *J. Cellular Comp. Physiol.* 47:Suppl. 1,35

Borsook, H., C. L. Deasy, A. J. Haagen-Smit, G. Keighly, and P. H. Lowy. 1949. The incorporation of labeled lysine into the proteins of guinea pig liver homogenate. *J. Biol. Chem.* 179:689.

—, —, —, —, —. 1950a. The uptake in vitro of ^{14}C-labeled glycine, 1-leucine, and 1-lysine by different components of guinea pig liver homogenate. *J. Biol. Chem.* 184: 529.

—, —, —, —, —. 1950b. Metabolism of ^{14}C-labeled glycine 1-histidine, 1-leucine, and 1-lysine. *J. Biol. Chem.* 187:839.

Borsook, H., and J. W. Dubnoff. 1940. The biological synthesis of hippuric acid in vitro. *J. Biol. Chem.* 132:307.

Borsook, H., and H. M. Huffman. 1938. *In* C. L. A. Schmidt [ed.] *Chemistry of amino acids and proteins,* p. 865. Thomas, Baltimore.

Boyer, P. D. 1960. Mechanism of enzyme action. *Ann. Rev. Biochem.* 29:15.

Brachet, J. 1941. La detection histochimique et le microdosage des acides pentosenucléiques. (Tissus animaux — developpement embryonnaire des amphibiens). *Enzymologia* 10:87.

—. 1947. *Nucleic acids in the cell and the embryo,* p. 207. Symp. Soc. Exp. Biol. Cambridge University Press.

Brenner, M., H. R. Muller, and R. W. Pfister. 1950. Eine neue enzymatische peptidsynthese. *Helvet. chim. acta.* 33:568.

Buc, M-H., and J. F. Scott. 1966. Effects of ultraviolet light on the biological functions of transfer RNA. *Biochem. Biophys. Res. Commun.* 22:459.

Bucher, N. L. R. 1953. The formation of radioactive cholesterol and fatty acids from ^{14}C-labeled acetate by rat liver homogenates. *J. Amer. Chem. Soc.* 75:498.

Burton, K., N. F. Varney, and P. C. Zamecnik. 1966. Action of osmium tetroxide on amino acid transfer ribonucleic acid. Correlations between the genetic code and the sensitivity of acceptor activity. *Biochem. J.* 99:28C.

Caspersson, T. 1941. Studien uber den Eiweissumsatz der Zelle. *Naturwissenschaften.* 29:33.

——. 1947. *The relations between nucleic acid and protein synthesis,* p. 127. Symp. Soc. Exp. Biol. Cambridge University Press.

Chantrenne, H. 1948a. Mixed anhydrides of benzoic and phosphoric acids. *Comp. Rend. Lab. Carlsberg* 26:297.

——. 1948b. Un modele de synthese peptidique. Proprietes du benzoylphosphate de phenyle. *Biochim. Biophys. Acta* 2:286.

Cohen, P. P., and R. W. McGilvery. 1947. Peptide bond synthesis. III. On the mechanism of *p*-aminohippuric acid synthesis. *J. Biol. Chem.* 171:121.

Cramer, F., H. Doepner, F. van der Haar, E. Schlimme and H. Seidel. 1968. On the conformation of transfer RNA. *Proc. Nat. Acad. Sci.* 61:1384.

Crick, F. H. C. 1958. On protein synthesis. *Soc. Exp. Biol. Symp.* London. 12:138.

Davie, E. W., V. V. Koningsberger, and F. Lipmann. 1956. The isolation of a tryptophan activating enzyme from pancreas. *Arch. Biochem. Biophys.* 65:21.

De Moss, J. A., S. M. Genuth, and G. D. Novelli. 1956. The enzymatic activation of amino acids via their acyl-adenylate derivatives. *Proc. Nat. Acad. Sci.* 42:325.

Doctor, B. P., J. E. Loebel, M. A. Sodd, and D. B. Winter. 1969. Nucleotide sequence of *Escherichia coli* tyrosine transfer ribonucleic acid. *Science* 163:693.

Dounce, A. L. 1952. Duplicating mechanism for peptide chain and nucleic acid synthesis. *Enzymologia* 15:251.

Engelhardt, V. A., and L. L. Kisselev. 1966. Recognition problem: on the specific interaction between coding enzyme and transfer RNA, p. 213. In *Current aspects of biochemical energetics.* Academic Press, N.Y.

Frantz, I. D., Jr., R. B. Loftfield, and W. W. Miller. 1947. Incorporation of ^{14}C from carboxyl-labeled DL-alanine into the proteins of liver slices. *Science* 106:544.

Frantz, I. D. Jr., R. B. Loftfield, and A. S. Werner. 1949. Observations on the equilibrium between glycine and glycyglycine in the presence of liver peptidase. *Fed. Proc.* 1:199.

Frantz, I. D., Jr., and P. C. Zamecnik. 1950. Use of ^{14}C-labeled amino acids in the study of peptide bond synthesis. *Plasma Proteins,* Vol. 2:94. C. C. Thomas, Springfield.

Frantz, I. D., Jr., P. C. Zamecnik, J. W. Reese, and M. L. Stephenson. 1948. The effect of dinitrophenol on the incorporation of ^{14}C-labeled alanine into the proteins of slices of normal and malignant livers. *J. Biol. Chem.* 174:773.

Fresco, J. R., B. M. Alberts, and P. Doty. 1960. Some molecular details of the secondary structure of ribonucleic acid. *Nature* 188:98.

Fruton, J. S., R. B. Johnston, and M. Fried. 1951. Elongation of peptide chains in enzyme-catalyzed transamidation reactions. *J. Biol. Chem.* 190:39.

Gale, E. F., and J. P. Folkes. 1953. The assimilation of amino acids in bacteria. 14. Nucleic acid and protein synthesis in *Staphylococcus aureus. Biochem. J.* 53:483.

Gierer, A. and G. Schramm. 1956. Infectivity of ribonucleic acid from tobacco mosaic virus. *Nature,* 177:702.

Greenberg. D. M., F. Friedberg, M. P. Schulman, and T. Winnick. 1948. Studies on the mechanism of protein synthesis with radioactive carbon-labeled compounds. *Cold Spring Harbor Symp. Quant. Biol.* 13:113.

Grunberg-Manago, M. and S. Ochoa. 1955. Enzymatic synthesis and breakdown of polynucleotides; polynucleotide phosphorylase. *J. Amer. Chem. Soc.* 77:3165.

Haines, J. A. and P. C. Zamecnik. 1967. Chemical modification of aminoacyl ligases and the effect on formation of aminoacyl-tRNA. *Biochim. Biophys. Acta* 146:227.

Hanes, C. S., F. J. R. Hird, and F. A. Isherwood. 1950. Synthesis of peptides in enzymatic reactions involving glutathione. *Nature* 166:288.

Harada, F., F. Kimura, and S. Nishimura. 1969. Nucleotide sequence of oligonucleotides derived from *Escherichia coli* valine transfer RNA by ribonuclease T_1 digestion: Comparison of the sequences of neighboring 3'- and 5'-terminals and anticodon region of *Escherichia coli* valine transfer RNA with those of yeast valine transfer RNA. *Biochim. Biophys. Acta* 182:590.

Hecht, L. I., M. L. Stephenson, and P. C. Zamecnik. 1958b. Formation of nucleotide end groups and incorporation of amino acids into soluble RNA. *Federation Proc.*, 17:239.

—, —, —. 1958c. Dependence of amino acid binding to soluble ribonucleic acid on cytidine triphosphate. *Biochim. Biophys. Acta,* 29:460.

—, —, —. 1959. Binding of amino acids to the end group of a soluble ribonucleic acid. *Proc. Nat. Acad. Sci.* 45:505.

Hecht, L. I., P. C. Zamecnik, M. L. Stephenson, and J. F. Scott. 1958a. Nucleoside triphosphates as precursors of ribonucleic acid end groups. *J. Biol. Chem.* 233:954.

Hendler, R. W. 1968. *Protein biosynthesis and membrane biochemistry.* John Wiley, New York.

Hoagland, M. B. 1955. An enzymatic mechanism for amino acid activation in animal tissues. *Biochim. Biophys. Acta* 16:288.

Hoagland, M. B., E. B. Keller, and P. C. Zamecnik. 1956. Enzymatic carboxyl activation of amino acids. *J. Biol. Chem.* 218:345.

Hoagland, M. B., and G. D. Novelli. 1954. Biosynthesis of coenzyme A from phosphopantetheine and of pantetheine from pantothenate. *J. Biol. Chem.* 207:767.

Hoagland, M. B., M. L. Stephenson, J. F. Scott, L. I. Hecht, and P. C. Zamecnik. 1958. A soluble ribonucleic acid intermediate in protein synthesis. *J. Biol. Chem.* 231:241.

Hoagland, M. B., P. C. Zamecnik, N. Sharon, F. Lipmann, M. P. Stulberg, and P. D. Boyer. 1957a. Oxygen transfer to AMP in the enzymic synthesis of the hydroxamate of tryptophan. *Biochim. Biophys. Acta* 26:215.

Hoagland, M. B., P. C. Zamecnik, and M. L. Stephenson. 1957b. Intermediate reactions in protein biosynthesis. *Biochim. Biophys. Acta* 24:215.

—, —, —. 1959. A hypothesis concerning the roles of particulate and soluble ribonucleic acids in protein synthesis, p. 105. R. E. Zirkle [ed.] *A symposium on molecular biology.* Univ. Chicago Press.

Holley, R. W. 1957. An alanine-dependent, ribonuclease-inhibited conversion of AMP to ATP, and its possible relationship to protein synthesis. *J. Amer. Chem. Soc.* 79:658.

Holley, R. W., J. Apgar, B. P. Doctor, J. Farrow, M. A. Marini, and S. H. Merrill, 1961. A simplified procedure for the preparation of tyrosine- and valine-acceptor fractions of yeast "soluble ribonucleic acid." *J. Biol. Chem.* 236:200.

Holley, R. W., J. Apgar, G. A. Everett, J. T. Madison, M. Marquisee, S. H. Merrill, J. R. Penswick, and A. Zamir. 1965. Structure of a ribonucleic acid. *Science* 147:1462.

Hultin, T. 1950. Incorporation in vivo of ^{15}N-labeled glycine into liver fractions of newly hatched chicks. *Exp. Cell Research* 1:376.

Hultin, T., and G. Beskow. 1956. The incorporation of ^{14}C-1-leucine into rat liver proteins in vitro visualized as a two-step reaction. *Exp. Cell Research* 11:664.

Iaccarino, M., and P. Berg. 1969. Requirement of sulfhydryl groups for the catalytic and tRNA recognition functions of isoleucyl-tRNA synthetase. *J. Mol. Biol.* 42:151.

Imura, N., H. Schwam, and R. W. Chambers. 1969a. Transfer RNA. III. Reconstitution of alanine acceptor activity from fragments produced by specific cleavage of tRNA$_{II}^{Ala}$ at its anticodon. *Proc. Nat. Acad. Sci.* 62:1203.

Imura, N., G. B. Weiss, and R. W. Chambers. 1969b. Reconstitution of alanine acceptor activity from fragments of yeast tRNA$_{II}^{Ala}$. *Nature* 222:1147.

Johnston, R. B., and K. Bloch. 1949. The synthesis of glutathione in cell-free pigeon liver extracts. *J. Biol. Chem.* 179:493.

Kalckar, H. M. 1941. The nature of energetic coupling in biological synthesis. *Chem. Rev.* 28:71.

Kalousek, F., and I. Rychlik. 1965. Purification and properties of lysyl-sRNA synthetase from *Escherichia coli. Collection Czech. Chem. Commun.* 30:3909.

Keller, E. B. 1951. Turnover of proteins of cell fractions of adult rat liver in vivo. *Fed. Proc.* 10: 206.

Keller, E. B., and P. C. Zamecnik. 1956. The effect of guanosine diphosphate and triphosphate on the incorporation of labeled amino acids into proteins. *J. Biol. Chem.* 221:45.

Keller, E. B., P. C. Zamecnik, and R. B. Loftfield. 1954. The role of microsomes in the incorporation of amino acids into proteins. *J. Histochem. Cytochem.* 2:378.

Khorana, H. G. 1968. Synthesis in the study of nucleic acids. *Biochem. J.* 109:709.

Kirby, K. S. 1956. A new method for the isolation of ribonucleic acids from mammalian tissues. *Biochem. J.* 64:405.

Lamborg, M., and P. C. Zamecnik. 1960. Amino acid incorporation by extracts of *E. coli*. *Biochim. Biophys. Acta* 42:206.

Lipmann, F. 1941. Metabolic generation and utilization of phosphate bond energy. *In* F. F. Nord and C. H. Werkman [ed.] *Advances in enzymology and related subjects,* 1:99. Interscience, N. Y.

——. 1945. Acetylation of sulfonilamide by liver homogenates and extracts. *J. Biol. Chem.* 160:173.

Littlefield, J. W., and E. B. Keller, 1957. Incorporation of ^{14}C-amino acids into ribonucleoprotein particles from the Ehrlich mouse ascites tumor. *J. Biol. Chem.* 224:13.

Littlefield, J. W., E. B. Keller, J. Gross, and P. C. Zamecnik. 1955. Studies on cytoplasmic ribonucleoprotein particles from the liver of the rat. *J. Biol. Chem.* 217:111.

Loftfield, R. B. 1947. Preparation of ^{14}C-labeled hydrogen cyanide, alanine, and glycine. *Nucleonics* 1:54.

——. 1951. The alkaline rearrangement of α-haloketones. II. The mechanism of the Faworskii reaction. *J. Amer. Chem. Soc.* 73:4707.

——. 1957. *The biosynthesis of protein. Progress in biophysics and biophysical chemistry,* 8:347. Pergamon Press, New York.

Loftfield, R. B., and E. A. Eigner. 1969. Mechanism of action of amino acid transfer ribonucleic acid ligases. *J. Biol. Chem.* 244:1746.

Loftfield, R. B., J. W. Grover, and M. L. Stephenson. 1953. Possible role of proteolytic enzymes in protein synthesis. *Nature* 171:1024.

Loftfield, R. B., and A. G. Harris. 1956. Participation of free amino acids in protein synthesis. *J. Biol. Chem.* 219:151.

Loomis, W. F., and F. Lipmann. 1948. Reversible inhibition of the coupling between phosphorylation and oxidation. *J. Biol. Chem.* 173:807.

Maas, W. K., and G. D. Novelli. 1953. Synthesis of pantothenic acid by depyrophosphorylation of adenosine triphosphate. *Arch. Biochem.* 43:236.

Madison, J. T. 1968. Primary structure of RNA. *Ann. Rev. Biochem.* 37:131.

Marshall, R. D., and P. C. Zamecnik. 1969. Some physical properties of lysyl and arginyl transfer RNA synthetases of *Escherichia coli B*. *Biochim. Biophys. Acta* 18:452.

Matthaei, J. H., and M. W. Nirenberg. 1961. Charac-

teristics and stabilization of DNAase-sensitive protein synthesis in *E. coli* extracts. *Proc. Nat. Acad. Sci.* 47:1580.

Mehler, A., and S. K. Mitra, 1967. The activation of arginyl transfer ribonucleic acid synthetase by transfer ribonucleic acid. *J. Biol. Chem.* 242:5495.

Melchior, J. B., and H. Tarver. 1947a. Studies in protein synthesis in vitro. I. On the synthesis of labeled cystine (S^{35}) and its attempted use as a tool in the study of protein synthesis. *Arch. Biochem.* 12:301.

——, ——. 1947b. Studies in protein synthesis in vitro. II. Uptake of labeled sulfur by the proteins of liver slices incubated with methionine (S^{35}). *Arch. Biochem.* 12:309.

Mirzabekov, A. D., L. Y. Kazarinova, D. Lastity, and A. A. Bayev. 1969. Enzymatic aminoacylation of dissected molecules of baker's yeast valine tRNA. *Fed. Europ. Biochem. Soc. Letters* 3:268.

Mitra, S. K., and A. H. Mehler. 1967. The arginyl transfer ribonucleic acid synthetase of *Escherichia coli*. *J. Biol. Chem.* 242:5490.

Monier, R., M. L. Stephenson, and P. C. Zamecnik. 1960. The preparation and some properties of a low molecular weight ribonucleic acid from baker's yeast. *Biochim. Biophys. Acta* 43:1.

Moore, S., and W. H. Stein. 1949. Chromatography of amino acids on starch columns. Solvent mixtures for the fractionation of protein hydrolysates. *J. Biol. Chem.* 178:53.

Nirenberg, M. W., and J. H. Matthaei. 1961. The dependence of cell-free protein synthesis in *E. coli* upon naturally occurring or synthetic polyribonucleotides. *Proc. Nat. Acad. Sci.* 47:1588.

Ofengand, J. 1967. The function of pseudouridylic acid in transfer ribonucleic acid. I. The specific cyanoethylation of pseudouridine, inosine, and 4-thiouridine by acrylonitrile. *J. Biol. Chem.* 242:5034.

Ogata, K., and H. Nohara. 1957. The possible role of the ribonucleic acid (RNA) of the pH 5 enzyme in amino acid activation. *Biochim. Biophys. Acta* 25:659.

Peterson, E. A., and Greenberg, D. M. 1952. Characteristics of the amino acid-incorporating system of liver homogenates. *J. Biol. Chem.* 194:359.

Philipps, G. 1969. The primary structure of transfer RNA. *Nature* 223:374.

Preiss, J., P. Berg, E. J. Ofengand, F. H. Bergmann, and M. Dieckmann. 1959. The chemical nature of the RNA-amino acid compound formed by amino acid activating enzymes. *Proc. Nat. Acad. Sci.* 45: 319.

Randerath, K., C. M. Janeway, M. L. Stephenson, and P. C. Zamecnik. 1966. Isolation and characterization of dinucleoside tetra- and triphosphates formed in the presence of lysyl-sRNA synthetase. *Biochem. Biophys. Res. Commun.* 24:98.

Ravel, J. M., S-F. Wang, C. Heinemeyer, and W. Shive. 1965. Glutamyl and glutaminylribonucleic acid synthetases of *Escherichia coli* W. *J. Biol. Chem.* 240:432.

Risebrough, R. W., A. Tissières, and J. D. Watson. 1962. Messenger RNA attachment to active ribosomes. *Proc. Nat. Acad. Sci.* 48:430.

Ruben, S., and M. D. Kamen. 1940. Radioactive carbon of long half-life. *Phys. Rev.* 57:549.

Sanger, F. 1950. Some chemical investigations on the structure of insulin. *Cold Spring Harbor Symp. Quant. Biol.* 14:153.

Schoenheimer, R. 1942. *The dynamic state of body constituents.* Harvard University Press, Cambridge.

Schweet, R. 1955. Incorporation of radioactive lysine into protein. *Fed. Proc.* 14:277.

Scott, J. F., and P. C. Zamecnik. 1969. Some optical properties of diadenosine-5'-phosphates. *Proc. Nat. Acad. Sci.* (in press).

Shugart, L., and M. P. Stulberg. 1969. Borohydride reduction of phenylalanine transfer ribonucleic acid. *J. Biol. Chem.* 244:2806.

Siekevitz, P. and P. C. Zamecnik. 1951. In vitro incorporation of 1-^{14}C-dl-alanine into proteins of rat liver granular fractions. *Fed. Proc.* 10:246.

Siekevitz, P. 1952. Uptake of radioactive alanine in vitro into the proteins of rat liver fractions. *J. Biol. Chem.* 195:549.

Speck, J. F. 1947. The enzymic synthesis of glutamine. *J. Biol. Chem.* 168:403.

Speyer, J. F., P. Lengyel, C. Basilio, A. J. Wahba, R. S. Gardner, and S. Ochoa. 1963. Synthetic polynucleotides and the amino acid code. *Cold Spring Harbor Symp. Quant. Biol.* 28:559.

Stephenson, M. L., K. V. Thimann, and P. C. Zamecnik. 1956. Incorporation of ^{14}C-amino acids into proteins of leaf disks and cell-free fractions of tobacco leaves. *Arch. Biochem. Biophys.* 65:194.

Stephenson, M. L., and P. C. Zamecnik. 1961. Purification of valine transfer ribonucleic acid by combined chromatographic and chemical procedures. *Proc. Nat. Acad. Sci.* 47:1627.

Stern, R., and A. H. Mehler. 1965. Lysyl-sRNA synthetase from *Escherichia coli. Biochem. J.,* 342: 400.

Tissières, A., D. Schlessinger, and F. Gros. 1960. Amino acid incorporation into proteins by *Escherichia coli* ribosomes. *Proc. Nat. Acad. Sci.* 46: 1450.

Watson, J. D. 1963. Involvement of RNA in the synthesis of proteins. *Science* 140:17.

Watson, J. D., and F. H. C. Crick. 1953. Molecular structure of nucleic acids: A structure for deoxyribose nucleic acid. *Nature* 171:737.

Yarus, M., and P. Berg. 1967. Recognition of tRNA by aminoacyl tRNA synthetases. *J. Mol. Biol.* 28: 479.

—, —. 1969. Recognition of tRNA by isoleucyl-tRNA synthetase effect of substrates on the dynamics of tRNA-enzyme interaction. *J. Mol. Biol.* 42:171.

Yu, C-T., and P. C. Zamecnik. 1963a. On the aminoacyl-RNA synthetase recognition sites of yeast and *E. coli* transfer RNA. *Biochem. Biophys. Res. Commun.* 12:457.

—, —. 1963b. Effect of bromination on the amino acid accepting activities of transfer ribonucleic acids. *Biochim. Biophys. Acta* 76:209.

—, —. 1964. Effect of bromination on the biological activities of transfer RNA of *Escherichia coli. Science* 144:856.

Zachau, H. G., G. Acs, and F. Lipmann. 1958. Isolation of adenosine amino acid esters from a ribonuclease digest of soluble, liver ribonucleic acid. *Proc. Nat. Acad. Sci.* 44:885.

Zamecnik, P. C. 1950. The use of labeled amino acids in the study of the protein metabolism of normal and malignant tissues: a review. *Cancer Res.* 10:659.

—. 1953. Incorporation of radioactivity from dl-leucine-1-^{14}C into proteins of rat liver homogenates. *Fed. Proc.* 12:295.

—. 1960. Historical and current aspects of the problem of protein synthesis. *Harvey Lectures* 54:256.

—. 1966a. The mechanism of protein synthesis and its possible alteration in the presence of oncogenic RNA viruses. *Cancer Res.* 26:1.

—. 1966b. Concerning the recognition reaction and transfer RNA in protein synthesis, p.155. *In* E. Mihich [ed.] *Symp. Immun. Chemo.* Academic Press, N. Y.

Zamecnik, P. C., and I. D. Frantz, Jr. 1950. Peptide

bond synthesis in normal and malignant tissues. *Cold Spring Harbor Symp. Quant. Biol.* 14:199.

Zamecnik, P. C., I. D. Frantz, Jr., R. B. Loftfield, and M. L. Stephenson. 1948. Incorporation in vitro of radioactive carbon from carboxyl-labeled d1-alanine and glycine into proteins of normal and malignant rat livers. *J. Biol. Chem.* 175:299.

Zamecnik, P. C., L. I. Hecht, and M. L. Stephenson. 1960a. Intermediate reactions in amino acid incorporation in normal and neoplastic tissues. *Acta Union Internat. Contre Cancer* 16:1121.

Zamecnik, P. C., C. M. Janeway, K. Randerath, and M. L. Stephenson. 1967. A new family of dinucleotides whose formation is catalyzed by 1-lysine, sRNA ligase (AMP), p. 176. *In* V. V. Koningsberger and L. Bosch, [ed.] *Regulation of nucleic acid and protein biosynthesis.* Elsevier, Amsterdam.

Zamecnik, P. C., and E. B. Keller. 1954. Relation between phosphate energy donors and incorporation of labeled amino acids into proteins. *J. Biol. Chem.* 209:337.

Zamecnik, P. C., E. B. Keller, M. B. Hoagland, J. W. Littlefield, and R. B. Loftfield. 1956a. *Ciba Found. Symp. Ionizing Radiat. Cell Metabol.* 161.

Zamecnik, P. C., E. B. Keller, J. W. Littlefield, M. B. Hoagland, and R. B. Loftfield. 1956b. Mechanism of incorporation of labeled amino acids into protein. *J. Cell. Comp. Physiol.* 47:81.

Zamecnik, P. C., R. B. Loftfield, M. L. Stephenson, and J. M. Steele. 1951. Studies on the carbohydrate and protein metabolism of the rat hepatoma. *Cancer Res.* 11:592.

Zamecnik, P. C., and M. L. Stephenson. 1968. A possible regulatory site located at the gateway to protein synthesis, p. 3. *In Regulatory mechanisms for protein synthesis in mammalian cells.* Academic Press, New York.

—, —. 1969. Nucleoside pyrophosphate compounds related to the first step in protein synthesis. *Alfred Benzon Found. Symp.* Elsevier Press. (In press).

Zamecnik, P. C., M. L. Stephenson, and L. I. Hecht. 1958. Intermediate reactions in amino acid incorporation. *Proc. Nat. Acad. Sci.* 44:73.

Zamecnik, P. C., M. L. Stephenson, C. M. Janeway, and K. Randerath. 1966. Enzymatic synthesis of diadenosine tetraphosphate and diadenosine triphosphate with a purified lysyl-sRNA synthetase. *Biochem. Biophys. Res. Commun.* 24:91.

Zamecnik, P. C., M. L. Stephenson, and J. F. Scott. 1960b. Partial purification of soluble RNA. *Proc. Nat. Acad. Sci.* 46:811.

Zamecnik, P. C., M. L. Stephenson, J. F. Scott, and M. B. Hoagland. 1957. Incorporation of ^{14}C-ATP into soluble RNA isolated from 105,000 $\times$ g supernatant of rat liver. *Fed. Proc.* 16:275.

Possible relation between deoxyribonucleic acid and protein structures

*G. GAMOW**

Reprinted with permission from Nature, vol. 173, 1954, p. 318.

The discovery of the structure of DNA stimulated many people, among them the cosmologist, George Gamow, to think about gene function. The next paper presents an abbreviated version of Gamow's ideas about the code. A more extensive version was published in the Proceedings of the Royal Danish Academy.

Gamow suggested that amino acid assembly occurs directly on the DNA molecule itself. He felt that the sequence of DNA bases caused specific types of cavities to appear, into which specific types of amino acids fit. He also suggested an overlapping kind of genetic code wherein any one base pair is part of more than one coding unit.

Gamow's ideas, while in large part incorrect, formed a coherent and abstract theory of coding. He recognized that his overlapping code placed restrictions on the number of possible amino acid sequences. He also realized that his theory could be tested, and encouraged work in that area.

In a communication in *Nature* of May 30, p. 964, J. D. Watson and F. H. C. Crick showed that the molecule of deoxyribonucleic acid, which can be considered as a chromosome fiber, consists of two parallel chains formed by only four different kinds of nucleotides. These are either (1) adenine, or (2) thymine, or (3) guanine, or (4) cytosine with sugar and phosphate molecules attached to them. Thus the hereditary properties of any given organism could be characterized by a long number written in a four-digital system. On the other hand, the enzymes (proteins), the composition of which must be completely determined by the deoxyribonucleic acid molecule, are long peptide chains formed by about twenty different kinds of amino acids, and can be considered as long "words" based on a 20-letter alphabet. Thus the question arises about the way in which four-digital numbers can be translated into such "words."

It seems to me that such translation procedure can be easily established by considering the "key-and-lock" relation between various amino acids, and the rhomb-shaped "holes" formed by various nucleotides in the deoxyribonucleic acid chain.

Fig. 1 shows schematically the structure of the deoxyribonucleic acid molecule as derived by Watson and Crick. We see that each "hole" is defined by only three of the four nucleotides forming it since, indeed, two nucleotides located across the axis of the cylinder are related by $1 \leftarrow \rightarrow 2$ and $3 \leftarrow \rightarrow 4$ binding conditions. It can easily be seen that there are twenty different types of such "holes," as shown in Fig. 2. The first eight of them are simple, while each of the remaining twelve can exist either in right-handed or left-handed modification. It is inviting to associate these "holes" with twenty different amino acids essential for living organisms.

One can speculate that free amino acids from the surrounding medium get caught into the "holes" of deoxyribonucleic acid molecules, and thus unite

Fig. 1.

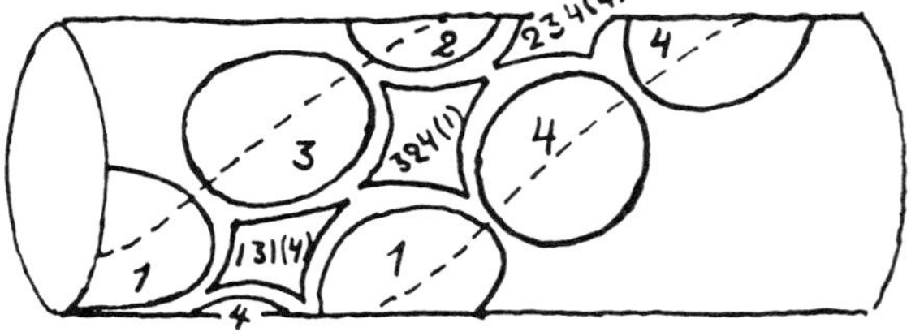

*George Washington University, Washington, D.C. October 22.

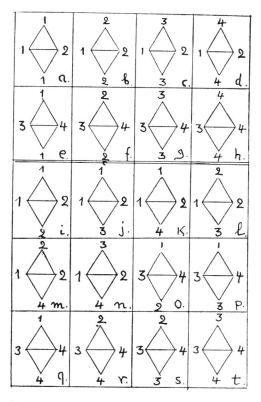

Fig. 2.

into the corresponding peptide chains. If this is true, there must exist a partial correlation between the neighboring amino acids in protein molecules, since the neighboring holes have two common nucleotides. There must also exist a correlation between adenine-to-guanine ratios in different organisms, and the relative amount of various amino acids in the corresponding proteins. The detailed account of the proposed theory will appear in *Kong. Dan. Vid. Selsk.*

General nature of the genetic code for proteins

DR. F. H. C. CRICK, F.R.S., LESLIE BARNETT,
DR. S. BRENNER and DR. R. J. WATTS-TOBIN*

Reprinted by authors' and publisher's permission from Nature, *vol. 192, 1962, pp. 1227–1232.*

This paper lays important intellectual groundwork for the general nature of the genetic code. The research discussed in this paper centered around the use of mutations, and more specifically, frameshift mutations caused by base additions or deletions. Crick and his colleagues concluded and rightly so, that the code is triplet, nonoverlapping, read in fixed sequence, and degenerate.

Especially interesting is this paper's almost simultaneous release with that of Nirenberg and Matthaei. The next to last paragraph of this paper relates Crick's reaction to the discovery that poly U codes for polyphenylalanine.

There is now a mass of indirect evidence which suggests that the amino acid sequence along the polypeptide chain of a protein is determined by the sequence of the bases along some particular part of the nucleic acid of the genetic material. Since there are twenty common amino acids found throughout Nature, but only four common bases, it has often been surmised that the sequence of the four bases is in some way a code for the sequence of the amino acids. In this article we report genetic experiments which, together with the work of others, suggest that the genetic code is of the following general type:

a. A group of three bases (or, less likely, a multiple of three bases) codes one amino acid.

b. The code is not of the overlapping type (see Fig. 1).

c. The sequence of the bases is read from a fixed starting point. This determines how the long sequences of bases are to be correctly read off as triplets. There are no special "commas" to show how to select the right triplets. If the starting point is dis-

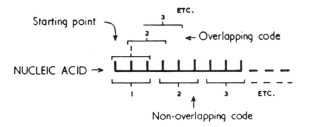

Fig. 1. To show the difference between an overlapping code and a nonoverlapping code. The short vertical lines represent the bases of the nucleic acid. The case illustrated is for a triplet code.

placed by one base, then the reading into triplets is displaced, and thus becomes incorrect.

d. The code is probably "degenerate"; that is, in general, one particular amino acid can be coded by one of several triplets of bases.

The Reading of the Code

The evidence that the genetic code is not overlapping (see Fig. 1) does not come from our work, but from that of Wittmann[1] and of Tsugita and Fraenkel-Conrat[2] on the mutants of tobacco mosaic virus produced by nitrous acid. In an overlapping triplet

*Medical Research Council Unit for Molecular Biology, Cavendish Laboratory, Cambridge.

code, an alteration to one base will in general change three adjacent amino acids in the polypeptide chain. Their work on the alterations produced in the protein of the virus show that usually only one amino acid at a time is changed as a result of treating the ribonucleic acid (RNA) of the virus with nitrous acid. In the rarer cases where two amino acids are altered (owing presumably to two separate deaminations by the nitrous acid on one piece of RNA), the altered amino acids are not in adjacent positions in the polypeptide chain.

Brenner[3] had previously shown that, if the code were universal (that is, the same throughout Nature), then all overlapping triplet codes were impossible. Moreover, all the abnormal human hemoglobins studied in detail[4] show only single amino acid changes. The newer experimental results essentially rule out all simple codes of the overlapping type.

If the code is not overlapping, then there must be some arrangement to show how to select the correct triplets (or quadruplets, or whatever it may be) along the continuous sequence of bases. One obvious suggestion is that, say, every fourth base is a "comma." Another idea is that certain triplets make "sense," whereas others make "nonsense," as in the comma-free codes of Crick, Griffith and Orgel.[5] Alternatively, the correct choice may be made by starting at a fixed point and working along the sequence of bases three (or four, or whatever) at a time. It is this possibility which we now favor.

Experimental Results

Our genetic experiments have been carried out on the B cistron of the r_{II} region of the bacteriophage $T4$, which attacks strains of *Escherichia coli*. This is the system so brilliantly exploited by Benzer.[6, 7]. The r_{II} region consists of two adjacent genes, or "cistrons," called cistron A and cistron B. The wild-type phage will grow on both *E. coli B* (here called B) and on *E. coli K*12 (λ) (here called K), but a phage which has lost the function of either gene will not grow on K. Such a phage produces an r plaque on B. Many point mutations of the genes are known which behave in this way. Deletions of part of the region are also found. Other mutations, known as "leaky," show partial function; that is, they will grow on K but their plaque type on B is not truly wild. We report here our work on the mutant $P\,13$

(now renamed $FC\,0$) in the $B1$ segment of the B cistron. This mutant was originally produced by the action of proflavin.[8]

We[9] have previously argued that acridines such as proflavin act as mutagens because they add or delete a base or bases. The most striking evidence in favor of this is that mutants produced by acridines are seldom "leaky"; they are almost always completely lacking in the function of the gene. Since our note was published, experimental data from two sources have been added to our previous evidence: (1) we have examined a set of 126 r_{II} mutants made with acridine yellow; of these only 6 are leaky (typically about half the mutants made with base analogues are leaky); (2) Streisinger[10] has found that whereas mutants of the lysozyme of phage $T4$ produced by base analogues are usually leaky, all lysozyme mutants produced by proflavin are negative, that is, the function is completely lacking.

If an acridine mutant is produced by, say, adding a base, it should revert to "wild-type" by deleting a base. Our work on revertants of $FC\,0$ shows that it usually reverts not by reversing the original mutation but by producing a second mutation at a nearby point on the genetic map. That is, by a "suppressor" in the same gene. In one case (or possibly two cases) it may have reverted back to true wild, but in at least 18 other cases the "wild type" produced was really a double mutant with a "wild" phenotype. Other workers[11] have found a similar phenomenon with r_{II} mutants, and Jinks[12] has made a detailed analysis of suppressors in the h_{III} gene.

The genetic map of these 18 suppressors of $FC\,0$ is shown in Fig. 2, line a. It will be seen that they all fall in the $B1$ segment of the gene, though not all of them are very close to $FC\,0$. They scatter over a region about, say, one-tenth the size of the B cistron. Not all are at different sites. We have found eight sites in all, but most of them fall into or near two close clusters of sites.

In all cases the suppressor was a nonleaky r. That is, it gave an r plaque on B and would not grow on K. This is the phenotype shown by a complete deletion of the gene, and shows that the function is lacking. The only possible exception was one case where the suppressor appeared to back-mutate so fast that we could not study it.

Each suppressor, as we have said, fails to grow on K. Reversion of each can therefore be studied by the

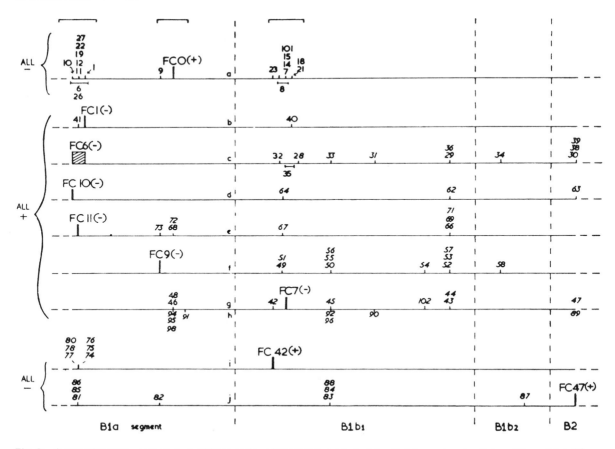

Fig. 2. A tentative map—only very roughly to scale—of the left-hand end of the *B* cistron, showing the position of the *FC* family of mutants. The order of sites within the regions covered by brackets (at the top of the figure) is not known. Mutants in italics have only been located approximately. Each line represents the suppressors picked up from one mutant, namely, that marked on the line in bold figures.

same procedure used for *FC* 0. In a few cases these mutants apparently revert to the original wild-type, but usually they revert by forming a double mutant. Fig. 2, lines *b–g*, shows the mutants produced as suppressors of these suppressors. Again all these new suppressors are nonleaky *r* mutants, and all map within the *B*1 segment except for one site in the *B*2 segment.

Once again we have repeated the process on two of the new suppressors, with the same general results, as shown in Fig. 2. lines *i* and *j*.

All these mutants, except the original *FC* 0, occurred spontaneously. We have, however, produced one set (as suppressors of *FC* 7) using acridine yel-

low as a mutagen. The spectrum of suppressors we get (see Fig. 2, line *h*) is crudely similar to the spontaneous spectrum, and all the mutants are nonleaky *r*'s. We have also tested a (small) selection of all our mutants and shown that their reversion rates are increased by acridine yellow.

Thus in all we have about eighty independent *r* mutants, all suppressors of *FC* 0, or suppressors of suppressors, or suppressors of suppressors of suppressors. They all fall within a limited region of the gene and they are all nonleaky *r* mutants.

The double mutants (which contain a mutation plus its suppressor) which plate on *K* have a variety of plaque types on *B*. Some are indistinguishable from wild, some can be distinguished from wild with difficulty, while others are easily distinguishable and produce plaques rather like *r*.

We have checked in a few cases that the phenomenon is quite distinct from "complementation," since the two mutants which separately are phenotypically *r*, and together are wild or pseudowild, must be put together in the same piece of genetic material. A simultaneous infection of *K* by the two mutants in separate viruses will not do.

The Explanation in Outline

Our explanation of all these facts is based on the theory set out at the beginning of this article. Although we have no direct evidence that the *B* cistron produces a polypeptide chain (probably through an RNA intermediate), in what follows we shall assume this to be so. To fix ideas, we imagine that the string of nucleotide bases is read, triplet by triplet, from a starting point on the left of the *B* cistron. We now suppose that, for example, the mutant *FC* 0 was produced by the insertion of an additional base in the wild-type sequence. Then this addition of a base at the *FC* 0 site will mean that the reading of all the triplets to the right of *FC* 0 will be shifted along one base, and will therefore be incorrect. Thus the amino acid sequence of the protein which the *B* cistron is presumed to produce will be completely altered from that point onwards. This explains why the function of the gene is lacking. To simplify the explanation, we now postulate that a suppressor of *FC* 0 (for example, *FC* 1) is formed by deleting a base. Thus when the *FC* 1 mutation is present by itself, all triplets to the right of *FC* 1 will be read incorrectly and thus the function will be absent. However, when both mutations are present in the same piece of DNA, as in the pseudowild double mutant *FC* (0 + 1), then although the reading of triplets between *FC* 0 and *FC* 1 will be altered, the original reading will be restored to the rest of the gene. This could explain why such double mutants do not always have a true wild phenotype but are often pseudowild, since on our theory a small length of their amino acid sequence is different from that of the wild-type.

For convenience we have designated our original mutant *FC* 0 by the symbol + (this choice is a pure convention at this stage) which we have so far con-

sidered as the addition of a single base. The suppressors of *FC* 0 have therefore been designated –. The suppressors of these suppressors have in the same way been labeled as +, and the suppressors of these last sets have again been labeled – (see Fig. 2).

Double Mutants

We can now ask: What is the character of any double mutant we like to form by putting together in the same gene any pair of mutants from our set of about eighty? Obviously, in some cases we already know the answer, since some combinations of a + with a – were formed in order to isolate the mutants. But, by definition, no pair consisting of one + with another + has been obtained in this way, and there are many combinations of + with – not so far tested.

Now our theory clearly predicts that all combinations of the type + with + (or – with –) should give an *r* phenotype and not plate on *K*. We have put together 14 such pairs of mutants in the cases listed in Table 1 and found this prediction confirmed.

At first sight one would expect that all combinations of the type (+ with –) would be wild or pseudowild, but the situation is a little more intricate than that, and must be considered more closely. This springs from the obvious fact that if the code is made of triplets, any long sequence of bases can be read correctly in one way, but incorrectly (by starting at the wrong point) in two different ways, depending whether the "reading frame" is shifted one place to the right or one place to the left.

If we symbolize a shift, by one place, of the reading frame in one direction by → and in the opposite direction by ←, then we can establish the convention that our + is always at the head of the arrow, and our – at the tail. This is illustrated in Fig. 3.

We must now ask: Why do our suppressors not

Table 1 Double mutants having the *r* phenotype

– With –	+ With +	
FC (1 + 21)	*FC* (0 + 58)	*FC* (40 + 57)
FC (23 + 21)	*FC* (0 + 38)	*FC* (40 + 58)
FC (1 + 23)	*FC* (0 + 40)	*FC* (40 + 55)
FC (1 + 9)	*FC* (0 + 55)	*FC* (40 + 54)
	FC (0 + 54)	*FC* (40 + 38)

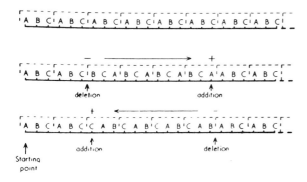

Fig. 3. To show that our convention for arrows is consistent. The letters *A*, *B* and *C* each represent a different base of the nucleic acid. For simplicity a repeating sequence of bases, *ABC*, is shown. (This would code for a polypeptide for which every amino-acid was the same.) A triplet code is assumed. The dotted lines represent the imaginary "reading frame" implying that the sequence is read in sets of three starting on the left.

extend over the whole of the gene? The simplest postulate to make is that the shift of the reading frame produces some triplets the reading of which is "unacceptable"; for example, they may be "nonsense," or stand for "end the chain," or be unacceptable in some other way due to the complications of protein structure. This means that a suppressor of, say, *FC* 0 must be within a region such that no "unacceptable" triplet is produced by the shift in the reading frame between *FC* 0 and its suppressor. But, clearly, since for any sequence there are *two* possible misreadings, we might expect that the "unacceptable" triplets produced by a → shift would occur in different places on the map from those produced by a ← shift.

Examination of the spectra of suppressors (in each case putting in the arrows → or ←) suggests that while the → shift is acceptable anywhere within our region (though not outside it) the shift ←, starting from points near *FC* 0, is acceptable over only a more limited stretch. This is shown in Fig. 4. Somewhere in the left part of our region, between *FC* 0 or *FC* 9 and the *FC* 1 group, there must be one or more unacceptable triplets when a ← shift is made; similarly for the region to the right of the *FC* 21 cluster. Thus we predict that a combination of a + with a − will be wild or pseudowild if it involves a →

shift, but that such pairs involving a ← shift will be phenotypically *r* if the arrow crosses one or more of the forbidden places, since then an unacceptable triplet will be produced.

We have tested this prediction in the 28 cases shown in Table 2. We expected 19 of these to be wild, or pseudowild, and 9 of them to have the *r* phenotype. In all cases our prediction was correct. We regard this as a striking confirmation of our theory. It may be of interest that the theory was constructed before these particular experimental results were obtained.

Rigorous Statement of the Theory

So far we have spoken as if the evidence supported a triplet code, but this was simply for illustration. Exactly the same results would be obtained if the code operated with groups of, say, 5 bases. Moreover, our symbols + and − must not be taken to mean literally the addition or subtraction of a single base.

It is easy to see that our symbolism is more exactly as follows:

$$+ \text{ represents } +m, \text{ modulo } n$$
$$- \text{ represents } -m, \text{ modulo } n$$

where *n* (a positive integer) is the coding ratio (that

Fig. 4. A simplified version of the genetic map of fig. 2. Each line corresponds to the suppressor from one mutant, here underlined. The arrows show the range over which suppressors have so far been found, the extreme mutants being named on the map. Arrows to the right are shown solid, arrows to the left dotted.

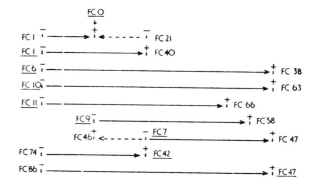

Table 2 Double mutants of the type (+ with −)

−/+	FC 41	FC 0	FC 40	FC 42	FC 58*	FC 63	FC 38
FC 1	W	W	W		W		W
FC 86		W	W	W	W	W	
FC 9	r	W	W	W	W		W
FC 82	r		W	W	W	W	
FC 21	r	W			W		W
FC 88	r	r			W	W	
FC 87	r	r	r	r			W

W, wild or pseudowild phenotype; W, wild or pseudowild combination used to isolate the suppressor; r, r phenotype. *Double mutants formed with FC 58 (or with FC 34) give sharp plaques on K.

is, the number of bases which code one amino acid) and m is any integral number of bases, positive or negative.

It can also be seen that our choice of reading direction is arbitrary, and that the same results (to a first approximation) would be obtained in whichever direction the genetic material was read, that is, whether the starting point is on the right or the left of the gene, as conventionally drawn.

Triple Mutants and the Coding Ratio

The somewhat abstract description given above is necessary for generality, but fortunately we have convincing evidence that the coding ratio is in fact 3 or a multiple of 3.

This we have obtained by constructing triple mutants of the form (+ with + with +) or (− with − with −). One must be careful not to make shifts across the "unacceptable" regions for the ← shifts, but these we can avoid by a proper choice of mutants.

We have so far examined the six cases listed in Table 3 and in all cases the triples are wild or pseudowild.

The rather striking nature of this result can be seen by considering one of them, for example, the triple (FC 0 with FC 40 with FC 38). These three mutants are, by themselves, all of like type (+). We can say this not merely from the way in which they were obtained, but because each of them, when

combined with our mutant FC 9 (−), gives the wild, or pseudowild phenotype. However, either singly or together in pairs they have an r phenotype, and will not grow on K. That is, the function of the gene is absent. Nevertheless, the combination of all three in the same gene partly restores the function and produces a pseudowild phage which grows on K.

This is exactly what one would expect, in favorable cases, if the coding ratio were 3 or a multiple of 3.

Our ability to find the coding ratio thus depends on the fact that, in at least one of our composite mutants which are "wild," at least one amino acid must have been added to or deleted from the polypeptide chain without disturbing the function of the gene product too greatly.

This is a very fortunate situation. The fact that we can make these changes and can study so large a region probably comes about because this part of the protein is not essential for its function. That this is so has already been suggested by Champe and Benzer[13] in their work on complementation in the r_{II} region. By a special test (combined infection on K, followed by plating on B) it is possible to examine the function of the A cistron and the B cistron separately. A particular deletion, 1589 (see Fig. 5) covers the right-hand end of the A cistron and part of the left-hand end of the B cistron. Although 1589 abolishes the A function, they showed that it allows the B function to be expressed to a considerable extent. The region of the B cistron deleted by 1589 is that into which all our FC mutants fall.

Joining Two Genes Together

We have used this deletion to reinforce our idea that the sequence is read in groups from a fixed starting

Table 3 Triple mutants having a wild or pseudowild phenotype

FC (0 + 40 + 88)
FC (0 + 40 + 58)
FC (0 + 40 + 57)
FC (0 + 40 + 54)
FC (0 + 40 + 55)
FC (1 + 21 + 23)

point. Normally, an alteration confined to the *A* cistron (be it a deletion, an acridine mutant, or any other mutant) does not prevent the expression of the *B* cistron. Conversely, no alteration within the *B* cistron prevents the function of the *A* cistron. This implies that there may be a region between the two cistrons which separates them and allows their functions to be expressed individually.

We argued that the deletion 1589 will have lost this separating region and that therefore the two (partly damaged) cistrons should have been joined together. Experiments show this to be the case, for now an alteration to the left-hand end of the *A* cistron, if combined with deletion 1589, can prevent the *B* function from appearing. This is shown in Fig. 5. Either the mutant *P43* or *X142* (both of which revert strongly with acridines) will prevent the *B* function when the two cistrons are joined, although both of these mutants are in the *A* cistron. This is also true of *X142 S1*, a suppressor of *X142* (Fig. 5, case *b*). However, the double mutant (*X142* with *X142 S1*), of the type (+ with −), which by itself is pseudowild, still has the *B* function when combined with 1589 (Fig. 5, case *c*). We have also tested in this way the 10 deletions listed by Benzer,[7] which fall wholly to the left of 1589. Of these, three (386, 168 and 221) prevent the *B* function (Fig. 5, case *f*),

whereas the other seven show it (Fig. 5, case *e*). We surmise that each of these seven has lost a number of bases which is a multiple of 3. There are theoretical reasons for expecting that deletions may not be random in length, but will more often have lost a number of bases equal to an integral multiple of the coding ratio.

It would not surprise us if it were eventually shown that deletion 1589 produces a protein which consists of part of the protein from the *A* cistron and part of that from the *B* cistron, joined together in the same polypeptide chain, and having to some extent the function of the undamaged *B* protein.

Is the Coding Ratio 3 or 6?

It remains to show that the coding ratio is probably 3, rather than a multiple of 3. Previous rather rough estimates[10, 14] of the coding ratio (which are admittedly very unreliable) might suggest that the coding ratio is not far from 6. This would imply, on our theory, that the alteration in *FC* 0 was not to one base, but to two bases (or, more correctly, to an even number of bases).

We have some additional evidence which suggests that this is unlikely. First, in our set of 126 mutants produced by acridine yellow (referred to earlier) we have four independent mutants which fall at or close to the *FC* 9 site. By a suitable choice of partners, we have been able to show that two are + and two are −. Secondly, we have two mutants (*X146* and *X225*), produced by hydrazine,[15] which fall on or near the site *FC* 30. These we have been able to show are both of type −.

Thus unless both acridines and hydrazine usually delete (or add) an even number of bases, this evidence supports a coding ratio of 3. However, as the action of these mutagens is not understood in detail, we cannot be certain that the coding ratio is not 6, although 3 seems more likely.

We have preliminary results which show that other acridine mutants often revert by means of close suppressors, but it is too sketchy to report here. A tentative map of some suppressors of *P 83*, a mutant at the other end of the *B* cistron, in segment *B 9a*, is shown in Fig. 6. They occur within a shorter region than the suppressors of *FC* 0, covering a distance of about one-twentieth of the *B* cistron. The double mutant *WT* (2 + 5) has the *r* phenotype as expected.

Fig. 5. Summary of the results with deletion 1589. The first two lines show that without 1589 a mutation or a deletion in the *A* cistron does not prevent the *B* cistron from functioning. Deletion 1589 (line 3) also allows the *B* cistron to function. The other cases, in some of which an alteration in the *A* cistron prevents the function of the *B* cistron (when 1589 is also present), are discussed in the text. They have been labeled (*a*), (*b*), etc., for convenience of reference, although cases (*a*) and (*d*) are not discussed in this paper. √ implies function; "x" implies no function.

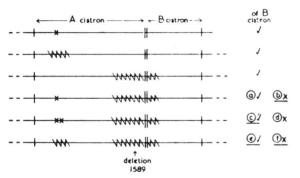

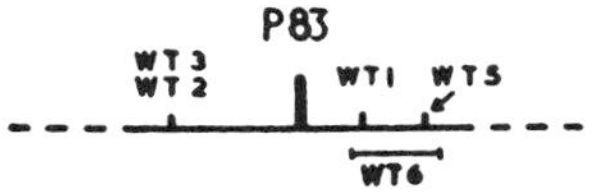

Fig. 6. Genetic map of *P* 83 and its suppressors, *WT* 1, etc. The region falls within segment *B9a* near the right-hand end of the *B* cistron. It is not yet known which way round the map is in relation to the other figures.

Is the Code Degenerate?

If the code is a triplet code, there are 64 (4 × 4 × 4) possible triplets. Our results suggest that it is unlikely that only 20 of these represent the 20 amino acids and that the remaining 44 are nonsense. If this were the case, the region over which suppressors of the *FC* 0 family occur (perhaps a quarter of the *B* cistron) should be very much smaller than we observe, since a shift of frame should then, by chance, produce a nonsense reading at a much closer distance. This argument depends on the size of the protein which we have assumed the *B* cistron to produce. We do not know this, but the length of the cistron suggests that the protein may contain about 200 amino acids. Thus the code is probably "degenerate," that is, in general more than one triplet codes for each amino acid. It is well known that if this were so, one could also account for the major dilemma of the coding problem, namely, that while the base composition of the DNA can be very different in different microorganisms, the amino acid composition of their proteins only changes by a moderate amount.[16] However, exactly how many triplets code amino acids and how many have other functions we are unable to say.

Future Developments

Our theory leads to one very clear prediction. Suppose one could examine the amino acid sequence of the "pseudowild" protein produced by one of our double mutants of the (+ with −) type. Conventional theory suggests that since the gene is only altered in two places, only two amino acids would be changed. Our theory, on the other hand, predicts that a string of amino acids would be altered, covering the region of the polypeptide chain corresponding to the region on the gene between the two mutants. A good pro-

tein on which to test this hypothesis is the lysozyme of the phage, at present being studied chemically by Dreyer[17] and genetically by Streisinger.[10]

At the recent Biochemical Congress at Moscow, the audience of Symposium I was startled by the announcement of Nirenberg that he and Matthaei[18] had produced polyphenylalanine (that is, a polypeptide all the residues of which are phenylalanine) by adding polyuridylic acid (that is, an RNA the bases of which are all uracil) to a cell-free system which can synthesize protein. This implies that a sequence of uracils codes for phenylalanine, and our work suggests that it is probably a triplet of uracils.

It is possible by various devices, either chemical or enzymatic, to synthesize polyribonucleotides with defined or partly defined sequences. If these, too, will produce specific polypeptides, the coding problem is wide open for experimental attack, and in fact many laboratories, including our own, are already working on the problem. If the coding ratio is indeed 3, as our results suggest, and if the code is the same throughout Nature, then the genetic code may well be solved within a year.

We thank Dr. Alice Orgel for certain mutants and for the use of data from her thesis, Dr. Leslie Orgel for many useful discussions, and Dr. Seymour Benzer for supplying us with certain deletions. We are particularly grateful to Prof. C. F. A. Pantin for allowing us to use a room in the Zoological Museum, Cambridge, in which the bulk of this work was done.

REFERENCES

1. Wittman, H. G., Symp. 1, Fifth Intern. Cong. Biochem., 1961, for refs. (in the press).
2. Tsugita, A., and Fraenkel-Conrat, H., *Proc. U.S. Nat. Acad. Sci.*, 46; 636 (1960); *J. Mol. Biol.* (in the press).
3. Brenner, S., *Proc. U.S. Nat. Acad. Sci.*, 43:687 (1957).
4. For refs. see Watson, H. C., and Kendrew, J. C., *Nature*, 190:670 (1961).
5. Crick, F. H. C., Griffith, J. S., and Orgel, L. E., *Proc. U.S. Nat. Acad. Sci.*, 43:416 (1957).
6. Benzer, S., *Proc. U.S. Nat. Acad. Sci.*, 45:1607 (1959), for refs. to earlier papers.
7. Benzer, S., *Proc. U.S. Nat. Acad. Sci.*, 47:403 (1961); see his Fig. 3.

8. Brenner, S., Benzer, S., and Barnett, L., *Nature,* 182, 983 (1958).

9. Brenner, S., Barnett, L., Crick, F. H. C., and Orgel, A., *J. Mol. Biol.,* 3:121 (1961).

10. Streisinger, G. (personal communication and in the press).

11. Feynman, R. P., Benzer, S., Freese, E. (all personal communications).

12. Jinks, J. L., *Heredity,* 16:153, 241 (1961).

13. Champe, S., and Benzer, S. (personal communication and in preparation).

14. Jacob, F., and Wollman, E. L., *Sexuality and the Genetics of Bacteria* (Academic Press, New York, 1961). Levinthal, C. (personal communication).

15. Orgel, A., and Brenner, S. (in preparation).

16. Sueoka, N. *Cold Spring Harb. Symp. Quant. Biol.* (in the press).

17. Dreyer, W. J., Symp. 1, Fifth Intern. Cong. Biochem., 1961 (in the press).

18. Nirenberg, M. W., and Matthaei, J. H., *Proc. U.S. Nat. Acad. Sci.,* 47:1588 (1961).

RNA codewords and protein synthesis, II. Nucleotide sequence of a valine RNA codeword[*]

PHILIP LEDER and MARSHALL NIRENBERG[†]

Reprinted by authors' and publisher's permission from the Proceedings of the National Academy of Sciences, U.S., vol. 52, 1964, pp. 420–427.

The work described in this paper was not only a technical masterpiece; it also settled, once and for all, the triplet nature of the code. Leder and Nirenberg were able to trap, on a filter, the ribosome –mRNA–tRNA complex, with the mRNA being as small as 3 nucleotides but never smaller. Thus UUU, CCC, and AAA cause the absorption of phe-tRNA, pro-tRNA, and lys-tRNA respectively.

This, coupled with data from other sources, supported the idea that codons were triplets and that the code was translated in a nonoverlapping fashion.

Recent studies demonstrate that synthetic polynucleotides, such as poly U, induce with specificity the binding of C^{14}-amino-acyl-sRNA to ribosomes.[1-4] Characteristics of binding in the absence of poly U also have been reported.[5] C^{14}-Phe-sRNA binding induced by poly U is thought to represent an early step in protein synthesis before peptide bond formation; however, the precise nature of the interaction has not been clarified.

To establish the sequence of nucleotides in RNA codewords, we devised a rapid, sensitive method for measuring C^{14}-amino-acyl-sRNA binding to ribosomes, and have investigated both characteristics of binding and the minimum oligonucleotide chain length required to direct such binding.[6] pUpUpU, pApApA, and pCpCpC directed the binding of Phe-, Lys-, and Pro-sRNA, respectively, whereas dinucleotides had no effect. Trinucleotides with 5'-terminal phosphate were more active than those with no terminal phosphate, whereas trinucleotides with 2'-(3')-terminal phosphate were inactive.[6]

In this report, trinucleotides of known sequence were used to direct sRNA binding. GpUpU, but not its sequence isomers, UpGpU and UpUpG, was shown to induce Val-sRNA binding to ribosomes.

These data indicate that the nucleotide sequence of an RNA codeword for valine is GpUpU.

MATERIALS AND METHODS

Analyses of Poly- and Oligonucleotides (a) Paper electrophoresis was performed on Whatman 54 or 3 MM paper in 0.05 M NH$_4$COOH buffer, pH 2.7, at 80 v/cm for 0.5 hr with authentic reference markers. If a marker was not available, the expected mobility was calculated.[7] (b) Descending paper chromatography was performed at room temperature with Whatman 3 MM paper and the following solvents: (A) conc. NH$_4$OH-N-propanol-H$_2$O, 10/55/35, v/v; (B) 40 gm (NH$_4$)$_2$SO$_4$ dissolved in 100 ml 0.1 M sodium phosphate, pH 7.0.[8] Bands were visualized under UV light. (c) Ultraviolet absorption measurements were made in a Zeiss spectrophotometer, and spectra were obtained in a Cary recording spectrophotometer. The absorbency of eluates from paper was read against blank paper eluates. (d) The base ratio of the poly UG preparation (D-132) was 0.74/0.26 (U/G). The base ratio of UpUpG, UpGpU, and GpUpU preparations was determined by incubating 2.0 A^{260} units with 3.5 × 10^{-3} units of T$_2$ RNase[9] in 0.02 ml of 0.5 M ammonium acetate, pH 4.5, for 2.5 hr at 45°. The digestion products were separated by paper electrophoresis, eluted, and the absorbency of each at appropriate λ_{max} was determined.

Preparation and Separation of Oligonucleotides UpU was characterized as described.[6] UpUpG and UpG were obtained by incubating 3 × 10^3 A^{260} units of poly UG

*Communicated by R. B. Roberts, June 30, 1964.
†National Heart Institute, National Institutes of Health, Bethesda, Maryland.

with 1.3 X 10³ units of T_1-RNase (Sankyo Co., Ltd., Tokyo) in 4.7 ml of 0.1 M NH₄HCO₃ at 37° for 6 hr.[10] The reaction mixture was lyophilized, dissolved in 0.1 M (NH₄)₂ CO₃, and terminal phosphates were removed by incubation with *E. coli* alkaline phosphatase, free of diesterase, as described by Heppel *et al.*[11] Oligonucleotide fractions were separated by paper chromatography with solvent A for 18 hr. Ultraviolet-absorbing bands with mobilities of UpG and UpUpG were eluted with H₂O, lyophilyzed, and purified separately by electrophoresis.

GpUpU and GpU were obtained by digesting poly UG with purified pork liver nuclease,[6] removing terminal phosphates, and separating fractions by paper chromatography and electrophoresis are described above. This procedure yielded a fraction containing UpG and GpU and also one containing GpUpU, UpGpU, and UpUpG. Paper chromatography with solvent B for 36 hr resolved each dinucleotide. The trinucleotide fraction was separated into two bands by chromatography with solvent B. The band nearest the origin contained GpUpU of high purity. It was eluted with H₂O, absorbed on acid-washed Norite, and eluted with 45% ethanol 0.5 M NH₄OH. GpUpU was purified again by chromatography with solvent B, eluted, and desalted as described.

A derivative of pancreatic RNase A was used by Dr. Merton Bernfield to catalyze the transfer of uridine-2′, 3′-cyclic phosphate to GpU.[12] UpGpU was purified from the reaction mixture by paper chromatography and electrophoresis as described for UpUpG.

Characterization of Triplets GpUpU, UpGpU, and UpUpG preparations were purified as described. Only one spot was observed after each was subjected to paper chromatography with solvents A or B and to paper electrophoresis. Base ratio and sequence analyses established the purity, chain length, and base sequence of each triplet preparation. Base ratio analyses were as follows: GpUpU, 1.2 U, 1.0 Up, 1.0 Gp; UpGpU, 1.0 U, 1.0 Up, .09 Gp; UpUpG, 2.1 Up, 1.0 G.

Base sequence was determined as follows: 2.0 A²⁶⁰ units of each trinucleotide was digested separately with T_1-RNase (as described earlier) and in other reactions, with 2.0 μg of chromatographically purified pancreatic RNase (Sigma) in 0.02 ml of 0.1 M (NH₄)₂CO₃, at 37° for 6 hr. Reaction products were separated by paper electrophoresis. Dinucleotide products were eluted, lyophilized, and digested with the RNase to which they were susceptible, and the products again were separated by electrophoresis. Nucleosides and nucleotides were identified by their electrophoretic mobility as well as by their spectra.

After digestion of GpUpU, UpGpU, and UpUpG with pancreatic RNase, only the following products were obtained: GpUp + U; Up + GpU; and Up + G, respectively. After digestion of GpUp and GpU with T_1-RNase, Gp + Up and Gp + U were obtained, respectively. T_1-RNase digestion of GpUpU, UpGpU, and UpUpG yielded Gp

+ UpU, UpGp + U, and UpUpG, respectively. Digestion of UpU, UpGp, and UpUpG with pancreatic RNase yielded Up + U, Up + Gp, and Up + G, respectively. No other UV-absorbing material was found. Since a 2% contaminant could have been detected, the purity of each trinucleotide was estimated to be > 98%. GpUpU and UpGpU have not been purified previously.

Ribosomes, sRNA, and E. coli Extracts *E. coli* W 3100 sRNA was prepared from cells grown to late log phase in 0.5% nutrient broth, 1% glucose.[25] *E. coli* B sRNA was obtained from General Biochemicals, Inc. Each C¹⁴-aminoacyl-sRNA was prepared in the presence of 19 C¹²-amino acids as described elsewhere.[6] In additional experiments C¹⁴-Leu-sRNA also was prepared in the absence of other amino acids. C¹⁴-L-Val, C¹⁴-L-Phe, and C¹⁴-L-Leu, uniformly labeled with radioactivities of 293, 229, and 435 cpm/$\mu\mu$mole (Packard Corp. scintillation counter), respectively, were obtained from New England Nuclear Corp. Ribosomes for binding studies and *E. coli* extracts (DNase-treated, preincubated, S-30 fraction) for amino acid incorporation into protein studies were prepared as described elsewhere.[6, 13]

Assays Each 50-μl reaction mixture contained 0.1 M tris-acetate, pH 7.2; 0.02 M magnesium acetate; 0.05 M KCl; and 1.0 A²⁶⁰ units of ribosomes unless otherwise specified. Incubations were at 24° for 20 min unless otherwise indicated. C¹⁴-amino-acyl-sRNA binding to ribosomes was determined, as reported elsewhere,[6] by washing the ribosomes on Millipore filters. Ribosomes with bound sRNA remained on the filter.

C¹⁴-amino acid incorporation into protein was determined as reported previously.[13]

RESULTS

Effect of GpUpU and Poly UG upon the Binding of C¹⁴-valine sRNA to Ribosomes The data of Figure 1 show that the binding of Val-sRNA to ribosomes is stimulated by the addition of GpUpU or poly UG and is dependent upon the concentration of either template. More Val-sRNA is bound to ribosomes in the presence of poly UG than in the presence of an equivalent amount of GpUpU. We have observed similar differences between the activity of pUpUpU and poly U.[6]

In Figure 2, the rates of binding at 0° and 24° are compared, and the specificity of GpUpU and poly UG for amino-acyl-sRNA is shown. At 0°, poly UG markedly stimulated Val- and Phe-sRNA binding to ribosomes, whereas GpUpU stimulated only val-sRNA binding. At 24° the rates of binding directed

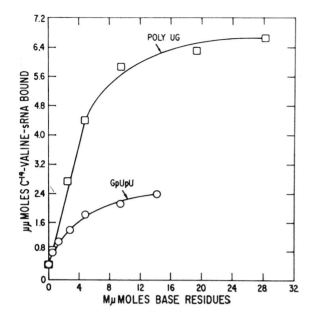

Fig. 1. The relation between GpUpU (○) and poly UG (□) concentration and C^{14}-Val-sRNA binding to ribosomes. Each reaction mixture contained 2.0 A^{260} units of ribosomes and 18.9 $\mu\mu$moles of C^{14}-Val attached to 0.44 A^{260} units of sRNA. Poly UG and GpUpU were added as specified.

by poly UG were higher than at 0°. Slightly higher binding in control reactions was observed (especially of Leu-sRNA). Under the conditions employed, poly UG had little activity in directing Leu-sRNA binding to ribosomes, compared with its ability to direct amino acid incorporation into protein.

Codeword Specificity and the Effect of (Mg^{++}) Concentration: The data of Figure 3 demonstrate that poly UG stimulates binding of Val- and Phe-sRNA to ribosomes optimally at 0.02–0.03 M (Mg^{++}), whereas GpUpU directs binding of Val-sRNA optimally at higher concentrations. At higher (Mg^{++}) concentrations, a small and apparently nonspecific increase in Val-, Leu-, and Phe-sRNA binding was observed in control reaction mixtures (no addition). Tri- and polynucleotides induce specific sRNA binding. GpUpU directed only Val-sRNA binding, whereas UpGpU and UpUpG had no effect upon either Val-, Leu-, or Phe-sRNA. The addition of tri- or polynucleotides did not appreciably in-

crease Leu-sRNA binding over that of control reactions within the range of (Mg^{++}) concentrations tested.

The specificity of di-, tri-, and polynucleotides in directing amino-acyl-sRNA binding is shown in Table 1. Poly UG directed Val- and Phe-sRNA binding, whereas poly U directed the binding only of Phe-sRNA. UpU, GpU, and UpG did not stimulate Val-, Phe-, or Leu-sRNA binding.

The effect of GpUpU was tested upon the binding of 17 other sRNA preparations, each with a different C^{14}-amino accepted (C^{14}-asparagine and C^{14}-glutamine-sRNA not tested). GpUpU was found to direct the binding only of Val-sRNA.

Amino Acid Incorporation into Protein Since poly UG had little effect upon Leu-sRNA binding to ribosomes, we investigated the ability of poly UG to direct amino acid incorporation into protein in *E. coli* extracts at 0° and 24° (also at 37° to make these data comparable to previously reported studies).[14, 15] As shown in Table 2, poly UG directed Phe, Val, and Leu into protein at 37° and 24°, and as reported previously,[14, 15] Leu incorporation almost equaled that of Val. Amino acid incorporation into protein was not detected at 0°.

DISCUSSION

Previously we have shown that UpUpU, ApApA, and pCpCpC specifically induce the binding to ribosomes of Phe-, Lys-, and Pro-sRNA, respectively. The present study demonstrates that the trinucleotide, GpUpU, can induce the binding of Val-sRNA to ribosomes, whereas sequence isomers such as UpGpU and UpUpG, and dinucleotides have no activity. When the specificity of poly UG was compared with that of GpUpU, the polynucleotide was found to direct both Phe- and Val-sRNA binding, but GpUpU induced the binding of only Val-sRNA. We conclude that the nucleotides of a valine codeword in mRNA are arranged in the sequence GpUpU. Since UpUpG does not replace GpUpU, recognition occurs with polarity.

As discussed elsewhere,[6] the ability of tri- but not dinucleotides to induce sRNA binding demonstrates directly that triplets may code for amino acids, and agrees with conclusions derived from both genetic[19] and biochemical studies.[20]

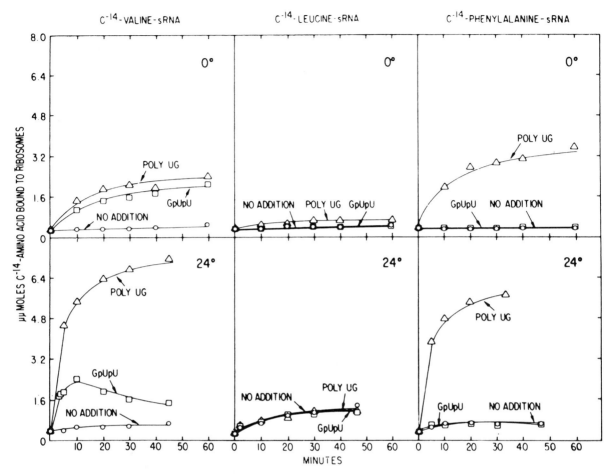

Fig. 2. Effect of GpUpU and poly UG upon the rates of binding of Val-, Leu-, and Phe-sRNA to ribosomes at $0°$ and $24°$. The symbols represent the addition of: (△) poly UG, 4.70 mμmoles of base residues; (□) GpUpU, 2.35 mμmoles of base residues; (○) no addition. Where indicated, 17 mμmoles of C^{14}-Val attached to 0.44 A^{260} units of sRNA, 8.5 $\mu\mu$moles C^{14}-Phe attached to 1.5 A^{260} units sRNA, or 21.0 $\mu\mu$moles of C^{14}-Leu attached to 1.5 A^{260} units of sRNA were added. Samples were incubated at the temperatures and for the times indicated.

Studies of the incorporation of amino acids stimulated by various polynucleotides together with the data on amino acid replacements in mutants provide a basis for assignment of codewords to amino acids.[16-18] Unfortunately, widespread degeneracy and the difficulty of assigning words which contain three different bases prevent unique assignments. Several self-consistent schemes are possible. The additional information provided by the definite knowledge of even a single codeword such as GpUpU

provides a distinction among these different possibilities. The reassignments suggested to make GpUpU the codeword for valine consistent with the other data are listed in Table 3. For example, ApUpU was assigned to isoleucine on the basis of studies employing randomly ordered polynucleotides to direct amino acid incorporation into protein, and the HNO_2-induced replacement of isoleucine by valine in TMV coat protein.

It should be noted that the predicted sequence

Table 1 Codeword specificity

Expt.	Addition, mμmoles base residues	C[14]-Amino-Acyl-sRNA bound to ribosomes, $\mu\mu$moles		
		C[14]-Val-sRNA	C[14]-Phe-sRNA	C[14]-Leu-sRNA
1	None	0.38	0.22	0.37
	4.7 Poly U	0.23	4.73	0.22
	4.7 Poly UG	2.65	1.93	0.24
2	None	0.40	0.22	0.62
	4.7 GpUpU	1.11	0.27	0.56
	4.7 UpGpU	0.40	0.25	0.55
	4.7 UpUpG	0.37	0.25	0.44
3	None	0.18	0.22	0.69
	4.7 GpU	0.20	0.22	0.69
	4.7 UpG	0.19	0.21	0.71
	4.7 UpU	0.18	0.20	0.67

Oligonucleotide specificity in directing sRNA binding to ribosomes. Where indicated, 9.10 $\mu\mu$moles C[14]-Val attached to 0.3 A[260] units of sRNA, 17.0 $\mu\mu$moles C[14]-Phe attached to 0.7 A[260] units of sRNA, or 21.0 $\mu\mu$moles C[14]-Leu attached to 1.5 A[260] units of sRNA were added to reaction mixtures.

Table 2 C[14]-amino acid incorporation into protein

Temperature of incubation	Addition	Amino acid incorporated into protein, mμmoles		
		C[14]-valine	C[14]-phenylalanine	C[14]-leucine
0°	None	< 0.03	< 0.03	< 0.03
	Poly UG, 53 mμmoles	< 0.03	< 0.03	< 0.03
24°	None	< 0.03	0.04	0.04
	Poly UG, 53 mμmoles	0.70	1.57	0.67
37°	None	0.10	0.09	0.08
	Poly UG, 53 mμmoles	1.04	2.05	0.96

Each 0.1-ml reaction mixture contained 0.3 mg *E. coli* protein (DNase-treated, preincubated S-30 fraction),[20] 20 mμmoles of either C[14]-Val, C[14]-Phe, or C[14]-Leu (4.8, 4.0, 4.0 μcuries/μmole, respectively), in addition to components previously described.[20] Reactions were incubated for 10, 20, or 60 min at 37°, 24°, or 0°, respectively.

for isoleucine, ApUpU, and also the GpUpU sequence found for valine do not agree with reported sequence determinations[21] which, on the basis of experiments with uncharacterized polynucleotides, assign the sequence ApUpU to tyrosine, tentatively assign GpUpU to cysteine, and derive the polarity of mRNA for protein synthesis.

Although our results show only marginal stimulation of Leu-sRNA binding by poly UG, we have consistently induced Leu- and also Cys-sRNA binding with poly UG when reactions were incubated at 0° and at a (Mg^{++}) concentration of 0.03 *M*. Leu-

sRNA binding was readily induced by poly UC which confirms results reported by Kaji and Kaji.[3] Further studies with Leu- and Cys-sRNA are in progress.

We have shown that specific binding of sRNA can be induced at 0° and have described preliminary evidence which indicates that intact amino-acyl-sRNA can be eluted from ribosomes after incubation.[6] Although these data suggest that formation of peptide bonds is not required for binding to ribosomes, a requirement for the first transfer enzyme and GTP may exist (cf. refs. 1–4, and 6).

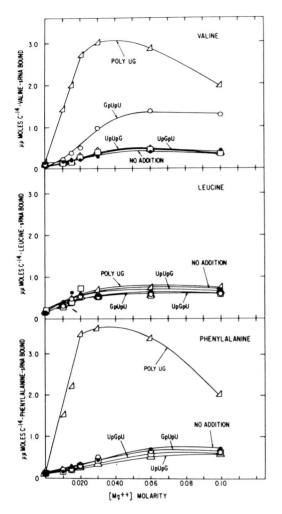

Fig. 3. Relationship between magnesium acetate concentration and C^{14}-amino-acyl-sRNA binding to ribosomes. The symbols represent the addition of: ($\triangle$) poly UG, 4.70 mμmoles of base; ($\bigcirc$) GpUpU, 2.35 mμmoles of base; ($\square$) UpGpU, 2.35 mμmoles of base; ($\triangle$) UpUpG, 2.35 mμmoles of base; ($\bullet$) no addition. 9.10 $\mu\mu$moles C^{14}-Val attached to 0.3 A^{260} units of sRNA. 25.0 $\mu\mu$moles C^{14}-Leu attached to 1.5 A^{260} units of sRNA, or 10.0 $\mu\mu$moles C^{14}-Phe attached to 0.7 A^{260} units of sRNA were added where indicated. The (Mg^{++}) concentration of each reaction mixture is shown on the abscissa. Samples were washed with buffer containing the appropriate concentration of (Mg^{++}).

Table 3 Predicted nucleotide sequences of RNA codewords

Amino acid	Sequence	Reference
Valine	GpUpU	Derived experimentally
Isoleucine	ApUpU Ap (UA)	17, 18
Tyrosine	Up (AU)	
Alanine	GpCpC Gp (CA) Gp (CU)	16
Glutamic acid	GpApA* Gp(AU)	16
Glycine	Gp (GU) Gp (GA) Gp (GC)	16
	Note: UpGpG is ruled out	
Arginine	ApGpA CpGpC (C)Gp(A)	16
Threonine	Ap(CA) (ACC)	17, 18

*The assignment of GpApA to glutamic acid is based upon limited amino acid incorporation into protein data and is, therefore, tentative.[23, 24]

Nucleotides within parentheses have not been arranged in sequence. Amino acid replacements used for these predictions were found in *E. coli* by Yanofsky and coworkers[16] or were induced by HNO$_2$ in TMV by Wittmann and Wittmann-Liebold[17] and also by Tsugita.[18]

Since synonym codewords corresponding to one amino acid often differ in base composition by only one nucleotide,[20] it is possible that bases in common may occupy identical positions within each triplet. A triplet code can be constructed wherein recognition between two out of three nucleotide pairs may, in some cases, suffice for coding; or, alternatively, a base at one position in the triplet may pair optionally with more than one base.[22–24] Ambiguous and synonymous codewords may involve such recognitions. The development of a rapid method for measuring sRNA binding to ribosomes and the use of trinucleotides of known sequence to direct such binding should provide a general method of great simplicity to test these possibilities and also to study the genetic function of other nucleotide sequences and interactions between specific codewords, sRNA, and ribosomes.

SUMMARY

The binding of C^{14}-valine-sRNA to ribosomes was directed both by the trinucleotide, GpUpU, and by poly UG, but not by UpGpU, UpUpG, or dinucleotides. GpUpU had no effect upon the binding to ribosomes of sRNA corresponding to 17 other amino acids. The nucleotide sequence GpUpU was proposed for a valine RNA codeword. The implications of these findings and predictions based on amino acid replacement data are discussed briefly.

ACKNOWLEDGEMENTS

We are grateful to Drs. Leon Heppel and George Rushizky for their advice on oligonucleotide separations. It is also a pleasure to thank Norma Zabriskie and Theresa Caryk for their extremely skillful technical assistance.

The following abbreviations are used: Val, valine; Phe, phenylalanine; Leu, leucine; poly U, polyuridylic acid; poly C, polycytidylic acid; poly A, polyadenylic acid; poly UG, copolymer of uridylic and guanylic acids; TCA, trichloroacetic acid; sRNA, transfer RNA; mRNA, messenger RNA. For mono- and oligonucleotides of specific structure, the "p" to the left of a terminal nucleoside initial indicates a 5'-terminal phosphate; the "p" to the right, a 2'-(3')-terminal phosphate. Internal phosphates of oligonucleotides are (3',5')-linkages.

REFERENCES

1. Arlinghaus, R., G. Favelukes, and R. Schweet, *Biochem. Biophys. Res. Commun.*, 11:92 (1963).
2. Nakamoto, T., T. W. Conway, J. E. Allende, G. J. Spyrides, and F. Lipmann, in *Synthesis and Structure of Macromolecules*, Cold Spring Harbor Symposia on Quantitative Biology, vol. 28 (1963), p. 227.
3. Kaji, A., and H. Kaji, *Biochem. Res. Commun.*, 13:186 (1963); *Federation Proc. (Abstr.)*, 23:478 (1964).
4. Spyrides, G. J., *Federation Proc. (Abstr.)*, 23:318 (1964).
5. Cannon, M., R. Krug, and W. Gilbert, *J. Mol. Biol.*, 7:360 (1963).
6. Nirenberg, M. W., and P. Leder, *Science*, in press.
7. Markham, R., and J. D. Smith, *Biochem. J.*, 52:558 (1952).
8. Rushizky, G. W., and H. A. Sober, *J. Biol. Chem.*, 237: 2883 (1962).
9. *Ibid.*, 238:371 (1963).
10. *Ibid.*, 237:834 (1962).
11. Heppel, L. A., D. R. Harkness, and R. J. Hilmoe, *J. Biol. Chem.*, 237:841 (1962).
12. Bernfield, M. R., and M. W. Nirenberg, *Abstracts*, 148th National Meeting, American Chemical Society, Chicago, Illinois, August 1964.
13. Nirenberg, M. W., in *Methods in Enzymology*, ed. S. P. Colowick and N. O. Kaplan (New York: Academic Press, 1964), vol. 6, p. 17.
14. Nirenberg, M. W., J. H. Matthaei, O. W. Jones, R. G. Martin, and S. H. Barondes, *Federation Proc.*, 22:55 (1963).
15. Singer, M. F., O. W. Jones, and M. W. Nirenberg, these Proceedings, 49:392 (1963).
16. Yanofsky, C., in *Synthesis and Structure of Macromolecules*, Cold Spring Harbor Symposia on Quantitative Biology, vol. 28 (1963), p. 581.
17. Wittmann, H. G., and B. Wittmann-Liebold, in *Synthesis and Structure of Macromolecules*, Cold Spring Harbor Symposia on Quantitative Biology, vol. 28 (1963), p. 589.
18. Tsugita, A., personal communication.
19. Crick, F. H. C., L. Barnett, S. Brenner, and R. J. Watts-Tobin, *Nature*, 192:1227 (1961).
20. Nirenberg, M., O. Jones, P. Leder, B. Clark, W. Sly, and S. Pestka, in *Synthesis and Structure of Macromolecules*, Cold Spring Harbor Symposia on Quantitative Biology, vol. 28 (1963), p. 549.
21. Wahba, A. J., C. Basilio, J. F. Speyer, P. Lengyel, R. S. Miller, and S. Ocho, these Proceedings, 48:1683 (1962).
22. Nirenberg, M. W., and O. W. Jones, in *Informational Macromolecules*, ed., H. Vogel et al., New York: Academic Press.
23. Jones, O. W. and M. W. Nirenberg, *Proc. Natl. Acad. Sci.*, U.S. 48:2115.
24. Gardner, R. S., A. J. Wahba, C. Basilio, R. S. Miller, P. Lengyel, and J. F. Speyer, these Proceedings, 48:2087 (1962).
25. Zubay, G. *J. Mol. Biol.* 4:347 (1962).

On the fundamental nature and evolution of the genetic code

C. R. WOESE, * D. H. DUGRE,[†] S. A. DUGRE,* *
M. KONDO,* and W. C. SAXINGER* *

Reprinted by authors' and publisher's permission from Cold Spring Harbor
Symposia on Quantitative Biology, *vol. 31, 1966, pp. 723–736.*

This paper, which concludes our discussion of the genetic code, is an evaluation of
what that code is and how it might have originated. The paper presents views pertinent
to the third phase of genetic code development — the first two being the theoretical
and the experimental phases. This phase, so far with limited success, seeks the significance of the code's degeneracy and the possibility of a direct relationship between DNA
and amino acids.

The present symposium defines clearly the current state of our knowledge of the genetic code. On one level we possess a high degree of comprehension of the code. And continuation of the present lines of research (based for the most part upon certain *in vitro* systems plus complementing *in vivo* studies) promises soon to complete our understanding of the code on this level. We now know, for example, es-sentially all of the codon assignments; we are mak-ing significant strides in elucidating the genetic sequences concerned with punctuation and modula-tion and with the (possibly) related phenomenon of relative codon usage; and we are beginning to under-stand the complex mechanisms involved in gene transcription and translation.

Yet, on another, more fundamental level, our understanding of the genetic code remains rudimen-tary, if not nonexistent. And it is entirely possible, as we shall see, that this deeper level of understand-ing cannot be attained in terms of the operations and concepts which have heretofore defined our understanding of the genetic code — i.e., in terms of the cell and its components as they exist today. Though we know *what* the codon assignments are, we do not know *why* — why UUU is assigned to phe

(if there be any reason); why certain amino acids are encoded while others (e.g., α-amino-n-butyric acid, norvaline, allothreonine) are not; why certain codons are not normally assigned to amino acids; what processes or mechanisms underlie the codon assignments and the type of translation apparatus we observe; why, in fact, there exists such a phenome-non as translation at all. It should be evident that answers to these sorts of questions are almost cer-tainly closely allied to the answers to questions re-garding the genetic code's evolution.

Unfortunately very little can be said about this latter topic at present. However, in attempting to make some beginning at understanding the funda-mental nature and evolution of the genetic code, it seems worthwhile to start by reviewing what is known about the translating or "decoding" machin-ery of the modern cell. This at least should give us a general feeling for the extent to which the genetic code has evolved, and perhaps also some slight appreciation for the sort of evolution the code has undergone.

THE TRANSLATION APPARATUS OF THE MODERN CELL

The "modern" cell (modern in the sense that its translation apparatus is in all important respects identical to that of cells living today) has undoubt-edly existed for as long as $0.5–2 \times 10^9$ years — i.e.,

*Department of Microbiology, University of Illinois, Ur-
bana, Illinois.
[†]Department of Chemistry and Chemical Engineering,
University of Illinois, Urbana, Illinois.

at least as far back as the evolutionary origin of the metazoans. For, any change in the nature of the translation process would undoubtedly be attended by a drastic change in the pattern of evolution, and this has not occurred during the last 500×10^6 years or more. The "modern" translation apparatus comprises three major classes of macromolecules, the activating enzymes, the sRNAs, and the ribosomes (plus a number of minor but important macromolecular species and, of course, the mRNA "tape" which passes through the translation "tape reader").

The Activating Enzymes

What is so notable about the set of activating enzymes and makes them unique is that they are dual function enzymes, and rather complex ones at that. On the one hand, each enzyme recognizes and activates its specific amino acid:

$$a + ATP + E \rightleftarrows a\text{-}AMP\text{-}E + PP \qquad (1)$$

On the other hand, the enzyme can then recognize certain corresponding sRNA molecules, with which it then charges the amino acid:

$$a\text{-}AMP\text{-}E + sRNA \rightleftarrows a\text{-}sRNA + E + AMP \qquad (2)$$

This overall sRNA charging reaction is carried out with considerable accuracy, the mistake rate (mistaking one amino acid for another naturally occurring one) in charging being, in nearly all cases, at levels below 10^{-3}–10^{-4} mistakes per charging (Berg 1961; Bergmann et al. 1961).

The sRNAs

As our knowledge of this set of macromolecules increases, we become increasingly aware of what truly bizarre and complex molecules the sRNAs really are, both structurally and functionally. They are certainly not the simple "adaptors" they were once thought to be. Like the activating enzymes, the sRNAs are polyfunctional molecules, having sites for interaction with their corresponding activating enzymes, with their corresponding codons, and perhaps with the ribosome and even with one another (during transfer of the growing peptide chain from one sRNA to the next).

Perhaps the most bizarre feature of sRNA is its high content of unusual nucleotides — which can amount to almost 20% of the total nucleotides in the molecule (Madison et al. in this volume). These nucleotides occupy specific and characteristic positions in each species of sRNA. They are introduced as posttranscriptional modifications (Mandel and Borek 1961), and so their introduction demands that each sRNA species have a particular entourage of rather sophisticated enzymes capable of performing the various posttranscriptional modifications. It is reasonable to assume that nearly all of the various odd nucleotides present in sRNA are there because they function in some way better than their normal counterparts, A, C, G, or U. (Their introduction as posttranscriptional modifications is a considerable "expense" to the cell, not only energetically, but also in terms of the extent of evolution required to develop the enzymes performing these tasks.) And, indeed, we see elsewhere in this volume evidence demonstrating the importance of the unusual nucleotides. It has been shown that the methylated bases play a definite (but as yet undefined) role in both the sRNA charging and the decoding processes. The former is suggested by the failure of a yeast activating enzyme to charge demethylated sRNA of *E. coli*, whose normal counterpart it will charge (Peterkofsky et al., in this volume). The latter is shown by the increased frequency with which demethylated sRNAs "recognize" incorrect codons (Littauer et al., in this volume).

It is particularly interesting to note the presence of inosine and pseudouridine in the anticodon portions of sRNAs. The role of the former appears to be associated with the creation of a particular pattern of ambiguous codon recognition. Both the position of I in the anticodon and the patterns of degeneracy observed for codon-anticodon recognition are, for the most part, in accord with a mechanism for such a degeneracy suggested by Crick (1966) based upon alternate types of base pairing.

The role of pseudouridine (found in the II position of the yeast sRNA_{try} anticodon) seems to be a rather different one, however. Pseudouridine, ψ, shows one or two properties not possessed by U. The pronounced secondary structure of poly ψ suggests that ψ may have base-stacking properties which U does not have (Pochon et al. 1964). More importantly, perhaps, ψ is capable of base pairing with A in either of two ways, as shown in Fig. 1, — a

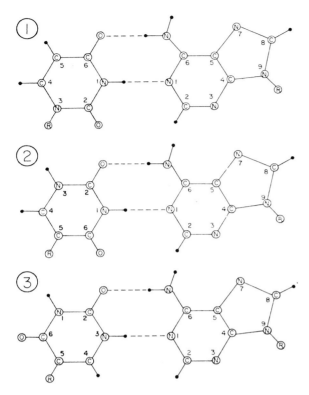

Fig. 1. Alternate types of base pairing between pseudo-uridylic acid and adenylic acid: (1) typical A —. U pair; (2) analogous A —. ψ pair – "normal" configuration; (3) alternate A —: ψ pair ("flipped" configuration). The symbols used are obvious, with the possible exception of R which denotes the sugar-phosphate portion of the nucleotides.

"normal" configuration, analogous to the Watson-Crick U—A pairing, and a "flipped" configuration in which the same groups on A are involved in pairing, but the non-H-bonded keto-group on ψ is no longer adjacent to A. If this latter configuration were the one adopted in codon-anticodon interactions, it would probably serve to reduce greatly the probability of codon-anticodon mispairings, as the "flipped" configuration of ψ is incapable of a number of alternative (incorrect) base pairings which are possible with the "normal" configuration. As the reader can visualize, a G—"U" pairing involving two H-bonds is possible only in the "normal" configuration. This is true, also, for one of the two

possible U—"U" pairings. And the removal of the non-H-bonded keto-group of ψ from the vicinity of A would more readily permit the evolution of an anticodon which sterically blocks the remaining incorrect base pairing C — "U" (e.g., by insertion of a sufficiently large nonpolar grouping where the keto-group on ψ would otherwise have been, in the "normal" configuration).

The tertiary structure of sRNA must be complex. Although no sRNA tertiary structures are known at present, it is nevertheless evident from the known primary structures that no simple overall geometry of the molecule – e.g., the double helical "hairpin" model that was last year's vogue – is possible (Madison et al. and Zachau et al., this volume). It is easy, and perhaps even useful, to envisage sRNA as a nucleic acid that is trying to be a protein. Aliphatic substitution on many of its bases and its dihydrouridine residues confer upon sRNA nonpolar characteristics more reminiscent of proteins than nucleic acids. And in some cases even the possibility for –S—S– bridges in sRNA exists (Carbon et al. 1965; Lipsett, this volume), as well as for functional configurational changes (Fresco et al., this volume), ostensibly analogous to the allosteric transitions seen in proteins. All in all, sRNAs are seen to be most sophisticated molecules.

The Ribosome

Too little is known about the ribosome's function to characterize it to any great extent in terms relevant to the present context. By its size, and the number of different kinds of proteins it contains — which could exceed 30 — the ribosome is certainly a complex entity. Furthermore, the ribosome undoubtedly has a fairly intricate role to play in the translation process. The fact that certain ribosomal mutations can alter the accuracy of translation is one indication of this (Gorini and Kataja 1964). Moreover, some factors important to peptide chain initiation also appear to reside with the ribosome (Wahba et al., this volume).

Conclusions

From the above survey of the components of the modern translation apparatus, two very simple and obvious conclusions can be drawn regarding the evolution of the genetic code: The translation apparatus in all its parts is a most intricate mechanism, subtly and finely adjusted in order to achieve a high accuracy of translation. Therefore, (1) the translation apparatus must have had a long and involved evolutionary history, in the early stages the mechanism being far less complex and (2) the simpler evolutionary precursors of the modern translation mechanism must have been far more prone to translation errors than is the present day apparatus. (To give some idea of the error levels involved here, consider that, in order to achieve a 70–95% level of perfect translations of a gene of typical size, the translational mistake rate has to be in the range of 10^{-3}–10^{-4} errors/codon/translation; merely raising the mistake rate 10-fold to 10^{-2}, would make perfect translations of most genes almost impossible.)

These two conclusions, in turn, lead to some interesting consequences. For one, they predict that up to a certain point in evolution (i.e., until translation achieved a sufficiently high degree of accuracy) the entire course of evolution was defined, and severely limited by the accuracy of translation; and in one sense, this early evolution was concerned for the most part with the development of a more accurate translation process. This meant, of course, cells were at one time far less complex than now — proteins were probably smaller, enzyme functions simpler and far less precise, and the subtleties of cellular regulation processes — which are the hallmark of life today — were beyond the capabilities of primitive cells (Woese 1965). This, in turn, demands that the evolutionary pattern to which we are accustomed — i.e., the creation of species, which requires that the states of a cell or organism be defined within very narrow limits, was not possible for primitive cells.

Of particular interest in the context of the present discussion are two further specific points which follow from the above. First, since accurate translation nessitates activating enzymes and the synthesis of (accurate) activating enzymes seemingly necessitates accurate translation, it is reasonable to assume that activating enzymes played no part in the earliest efforts of cells to translate, and that these enzymes only arose at some later stage in evolution. Second, in view of the extensive evolution the translation mechanism must have undergone, it is entirely possible that translation mechanisms have changed

so much during the course of evolution that interactions important to the early translation machinery no longer exist, no longer play a role in the "modern" apparatus, (This point is particularly pertinent to consideration of amino-oligonucleotide "recognition" interactions, discussed below.)

THE CENTRAL ISSUE

The problems surrounding the fundamental nature and evolution of the genetic code appear to center about one main issue. This can be stated in various ways, of which the following is one: The basic point at issue is whether the genetic code is in essence monolithic — all (or nearly all) of its many facets reflecting some basic underlying mechanism or principle — or whether the genetic code is basically a *pot pourri*, being derived in the main by stochastic processes operating within the framework of the biological selection principle and therefore reflecting *no* basic underlying mechanism. More particularly, the central issue seems to boil down to whether or not the genetic code and its evolution rest upon any sort of "recognition" phenomena involving amino acids and oligonucleotides. Are particular amino acids and particular oligonucleotides predestined to become associated with one another during evolution because they interact "optimally" in some way?

The issue of amino acid-oligonucleotide "recognition" has been in controversy right from the inception of the genetic coding field. Initially, Gamow (1954) unquestioningly assumed that nucleic acid "templates" for the various amino acids existed. This was in keeping with the prevailing thinking on "complementarity" of biological structures (Pauling and Delbrück 1940), although it was a completely unproven assertion. The only person who seemed disquieted about "templates" was Crick who, in formulating the adaptor hypothesis, brought the dogma sharply into question. Crick could see no reasons for, and so flatly denied the existence of, any sort of amino acid-oligonucleotide "recognition." And from this basic premise he deduced the necessity for there being "adaptors" in the process of translating a nucleic acid sequence into an amino acid sequence. ["... If one considers the physico-chemical nature of the amino acid side chains we do not find complementary features on the nucleic acid. Where are the knobby hydrophobic surfaces to distinguish valine from leucine or isoleucine? Where are the charged groups, in specific positions, to go with the acidic and basic amino acids? . . . . I don't think that anybody looking at DNA or RNA would think of them as templates for amino acids. . . . What the DNA structure *does* show — and probably RNA will do the same — is a specific pattern of hydrogen bonds, and very little else. . . . (This suggests that) each amino acid would combine chemically, at a special enzyme, with a small molecule which, having a specific hydrogen-bonding surface, would combine specifically with the nucleic acid template. . . . (In other words) each amino acid is fitted with an *adaptor* to go on the template." (Crick, as quoted in Hoagland 1960.)]

When evidence consistent with the adaptor hypothesis was adduced (see Hoagland 1960), the adaptor hypothesis and its postulate of the impossibility of amino acid-nucleic recognition themselves became unquestioned dogma, and have remained so more or less to this day. [e.g., "There is no specific affinity between the side groups of many amino acids and the purine and pyrimidine bases found in RNA. For example, the hydrocarbon side groups of the amino acids alanine, valine, leucine, and isoleucine do not form hydrogen bonds and would be actively repelled by the amino and keto groups of the various nucleotide bases. . . ." (Watson 1965).]

However, although there exists evidence consistent with the adaptor hypothesis, there is no evidence even remotely supporting the basic postulate that "template" interactions are impossible. [Attempts to detect interactions between amino acids and nucleic acids by equilibrium dialysis (Zubay and Doty 1958) or by chromatography on polynucleotide columns (R. J. Britten and Woese unpubl.) have given negative results — but such results are not uninterpretable.] Therefore, when considering the fundamental nature of the genetic code, one must eliminate from one's mind any proscriptions regarding the "impossibility" of amino acid-oligonucleotide "recognition" phenomena, and approach this possibility *tabula rasa*.

THE STRUCTURE OF THE CATALOGUE OF CODON ASSIGNMENTS

At the present moment it appears that the best of the few leads we possess regarding the fundamental

nature of the genetic code is the structure of the set of codon assignments, the "codon catalogue." Thus it is worthwhile to detail the constraints characterizing this catalogue, shown in Table 1. The constraints can be put into two classes — constraints governing the assignment of codons to any given amino acid, and constraints governing the assignment of codons to "related" amino acids. The former are blatant: (1) all codons assigned to the same amino acid differ from one another whenever possible (i.e., when an amino acid has assigned four or less codons) in the III position of the codon only. (2) In general, an amino acid has either two or four codons. (The exceptions to this rule are met, trp, ileu, ser, arg, and perhaps cys. Met and trp have one codon apiece, ending in G. Ileu and perhaps cys have

Table 1 Catalogue of codon assignments

UUU	phe	UCU		UAU	tyr	UGU	cys
UUC		UCC		UAC		UGC	
			ser				
UUA	leu	UCA		UAA	CT*	UGA	†
UUG		UCG		UAG		UGG	trp
CUU		CCU		CAU	his	CGU	
CUC		CCC		CAC		CGC	
	leu		pro				arg
CUA		CCA		CAA	gln	CGA	
CUG		CCG		CAG		CGG	
AUU	ileu	ACU		AAU	asn	AGU	ser
AUC		ACC		AAC		AGC	
			thr				
AUA	ileu	ACA		AAA	lys	AGA	arg
AUG	met	ACG		AAG		AGG	
GUU		GCU		GAU	asp	GGU	
GUC		GCC		GAC		GGC	
	val		ala				gly
GUA		GCA		GAA	glu	GGA	
GUG		GCG		GAG		GGG	

*Unassigned codons which result in termination of peptide chain synthesis.

†Assignment uncertain — probably assigned but not to tryptophan.

Codon assignments as determined from the triplet binding method and from peptide synthesis directed by polyribonucleotides of simple known sequence (Nirenberg et al. 1965; Söll et al. 1965; Khorana et al.; Wahba et al.; and Matthaei et al., this volume.)

three codons apiece, ending in U, C, and A. Ser and arg each have six codons.) In the former case the bases in the III codon position are always either U and C or A and G.

Regarding those constraints on codons assigned to "related" amino acids we can make less definite statements. The reason is precisely because we have no adequate definition of what "related amino acids" are — a definition which is relative, is context dependent, in any case. Nevertheless, it is clear that "related" amino acids tend to possess related codon assignments. Looking crudely at what amino acids are grouped by codon assignment, it is apparent that those amino acids having U_{II} codons are among the most hydrophobic, while those having A_I, or G_{II} codons for the most part have relatively "reactive" side chains (i.e., groups which tend to play a major role in protein tertiary structure and/or function). The only question here concerns the exact nature of this type of constraint, and how strictly it orders the amino acids. We shall have the answer only when we define "related" amino acids in the proper quantitative terms.

Possible Correlates of the Structure of the Codon Catalogue in the Structure and/or Function of the Translation Apparatus

The constraints defining the structure of the codon catalogue seem to have certain counterparts in structural and/or functional aspects of the translation mechanism, particularly with regard to the sRNA molecule. The degeneracy in codon assignments, the constraint seen in the III position, has its counterpart in a codon-anticodon recognition degeneracy. More precisely, the same sRNA molecule can in many cases recognize many or all of the codons assigned to an amino acid when (and only when) these differ from one another in the III codon position (Nirenberg et al., this volume; Söll et al. 1966). What we do not yet know in this case is the causal relationship between the constraint in codon assignments and the degeneracy in codon recognition.

The "related amino acid" codon assignment constraint seems to have some sort of counterpart in a detectable, but as yet poorly characterized property of sRNAs. So far, this property manifests itself most clearly in the ordering of sRNAs by various separation procedures, particularly by the order of their elution from a MAK (methylated albumin on kiesel-

guhr) column and their ordering in certain CCD (countercurrent distribution). Figure 2 presents the MAK ordering of sRNAs from four unrelated organisms, as well as the CCD ordering for one. (Experimental details are outlined in the figure caption.) It is clear that the order of elution of sRNAs from MAK in each case approximately follows this rule: namely, for those sRNAs responding to codons having a common base in the II codon position, the sRNA responding to the G_I codons elutes first, then those responding to the A_I codons, then those for the C_I codons next, and finally those for the U_I codons. There are exceptions to this rule for each organism, but the rule is obeyed almost exactly when derived by "averaging" over all four organisms. (Some uncertainty exists for the U_{II} codons, however.) It seems amazing that any such pattern emerges at all, considering the fact that these separation procedures probably do not measure solely the biological function of the sRNA molecule. Although the above pattern is seen for both MAK and CCD, no discernible pattern holds for DEAE (Cherayil and Bock 1965) or hydroxyapatite (Pearson and Kelmers 1966) separations; Nathenson et al. (1965) have noted some correlation between codon assignment and the column elution profile for a Sephadex system employing a Zachau solvent system. It is useless to speculate on the meaning of this sort of ordering until more is known about the properties of sRNA which give rise to the pattern.

Hypotheses Accounting for the Structure of the Codon Catalogue

Although the degree of order exhibited by the codon catalogue is certainly extreme enough to rule out a good many of the conceivable explanations for the nature and evolution of the genetic code, we shall see that these constraints still do not permit us to decide whether the foundation of the code is basically mechanistic (particularly with regard to amino acid-oligonucleotide "recognition") or stochastic.

To one who has a naive orientation regarding biology, it is likely that the first model suggested by the structure of the codon catalogue would be one which has a mechanistic basis: one in which some sort of specific amino acid-"codon" interaction determines the structure of the genetic code. This general sort of model we shall refer to as a "codon-amino acid pairing" model. The more sophisticated

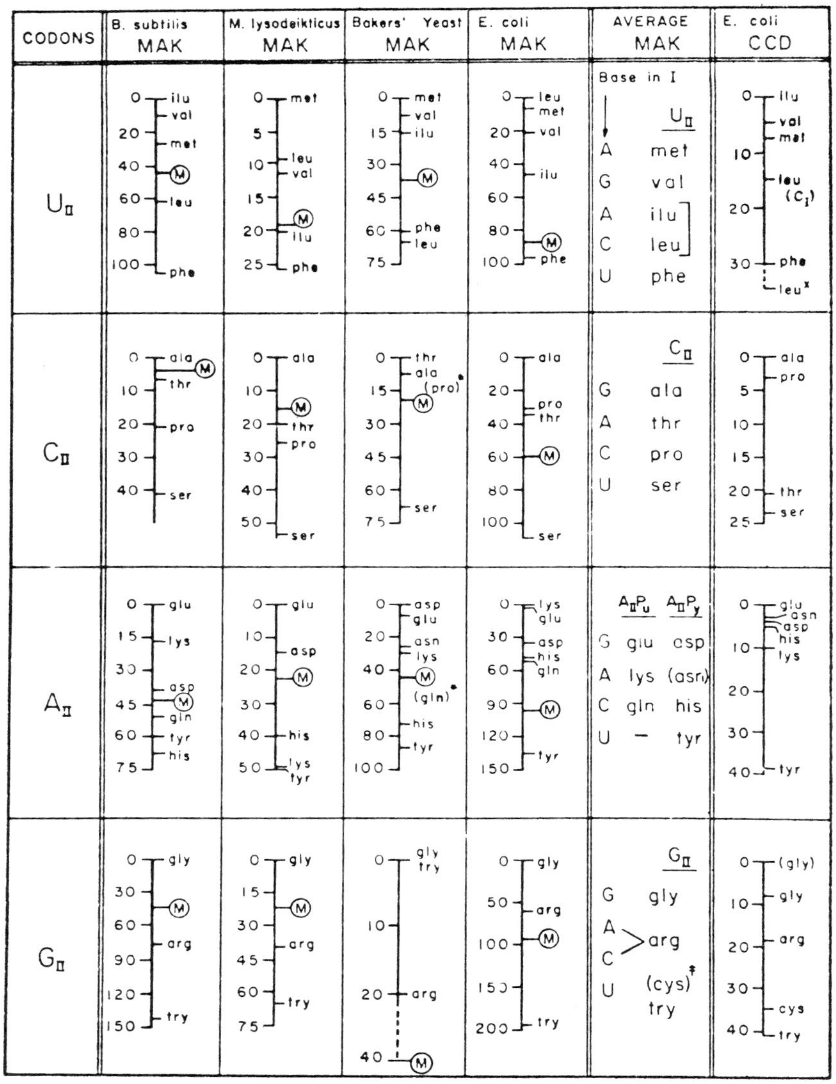

Fig. 2. Ordering of sRNAs by elution from MAK column. The NaCl gradient method is used to elute deacylated total sRNA from MAK, into 200–300 tubes. Portions of each tube are then charged separately with each of 19 C^{14} amino acids (cys omitted), thus determining the elution profile for the sRNAs corresponding to each amino acid. For each amino acid, the peak tube of its first major sRNA to elute is taken as a measure of the position of that amino acid's sRNA in the elution pattern. For each organism those sRNAs responding to codons with a common base in the II position are plotted together. The scale in each case is the difference, in number of tubes, between the first sRNA to elute in any one group and the remaining ones. For each organism, M is an arbitrary "normalizing point," a common tube, permitting a comparison from one group of sRNA to any other.

*Relative order correct, position approximate (data in these cases from Yamane and Sueoka, unpubl.).

^{x}sRNA$_{leu}$ responding to UUG codon.

‡Relative order inferred from CCD ranking (CCD ordering taken from data of Goldstein et al. 1964, and Söll et al. 1965).

biologist, conversant with the selection process, can devise a class of basically stochastic explanations for the structure of the codon catalogue, however. These we term "amino acid replacement" models, for reasons which will become obvious.

The mechanistic, codon-amino acid pairing type of model is self-evident and needs no further discussion. It supposes that some sort of interaction between amino acids and "codons" exists, that certain amino acids interact "optimally" with certain "codons," and that this interaction determines, somehow, the codon assignments. ("Codon" is not used here in the usual sense, merely as a trinucleotide sequence found in mRNA. By "codon" we mean the conventional codon or any oligonucleotide which is a simple transformation of it, e.g., the "codon" for *phe* need not be UUU, it could be AAA, the Watson-Crick base-pairing transform of UUU.)

The most easily explained of the stochastic models is the amino acid replacement model proposed by Sonneborn (1965) and briefly summarized as follows: It is clear that mutations are deleterious or, at best, neutral events to the cell (beneficial mutations are so rare that they can be neglected, in this context). Thus there would be some selective advantage to the cell were it to minimize its burden of deleterious mutations. One way this could conceivably be done would be to arrange codon assignments so that (1) the probability is *maximized* that (base substitution) mutations change codons in such a way that both the original and the mutated codon are assigned to the same amino acid, and (2) when a mutation does happen to result in an amino acid replacement, the original and the new amino acid are, on the average, as closely related (functionally) to one another as possible. These two constraints will produce a codon catalogue in which all codons assigned to the same amino acid are "maximally connected," each codon in such a group differing from all others by a single base substitution whenever possible. This means that all codons assigned to the same amino acid will, whenever possible, differ from one another in only *one* of the three codon positions, and that the position in which the codons differ from one another for any one amino acid will be the same for all or many other amino acids. It is conceivable for nature to create such a codon catalogue through the workings of the selection process, starting even from a completely unordered set of codon assignments.

THE CASE FOR THE VARIOUS HYPOTHESES

A Codon-Amino Acid Pairing Basis for Genetic Code

In view of the existence of a reasonable explanation for the structure of the genetic code in stochastic terms, in view of the apparent validity of the adaptor hypothesis, and in view of the fact that no direct evidence for the existence of codon-amino acid pairing has ever been found, it is necessary to produce some evidence in keeping with a codon-amino acid pairing basis for the genetic code in order to make the hypothesis credible.

If specific "recognitions" between amino acids and the nucleic acid bases are possible, then it must follow that many other heterocyclic bases can interact with amino acids in a similar fashion. While these latter interactions may not generally be of biological significance, such interactions could be very useful because one can work with bases like pyridine under conditions, and in quantities that solubility, cost, etc., would preclude for the biologically important bases. For this reason it seems sensible, in attempting to demonstrate the existence of the elusive codon-amino acid pairing interactions, to begin by attempting to demonstrate the existence of, and to characterize interactions between amino acids and those heterocyclic bases most easily worked with.

Amino acid-pyridine interactions can be shown to exist, and can be demonstrated readily by simple paper chromatographic procedures. Such interactions, as might be expected, involve both the pyridine ring nitrogens (a polar interaction) and the ring carbons (a nonpolar interaction) (Woese et al. 1966). The crucial point now becomes whether such interactions have any relevance to the genetic code and its evolution. If such were the case, we should expect that amino acids defined as "related" by their characteristics in pyridine chromatography should also be "related" in having related codon assignments.

The definition of "related" (in the context of amino acid chromatography in pyridine solvents) we have used is based upon the slope of the straight line which results when log amino acid R_M [R_M is defined here as $(1 - R_F/R_F)$] is plotted against log mole fraction of water in the pyridine solvent (see Fig. 3). Amino acids showing similar values of this slope, which we shall henceforth call their "polar

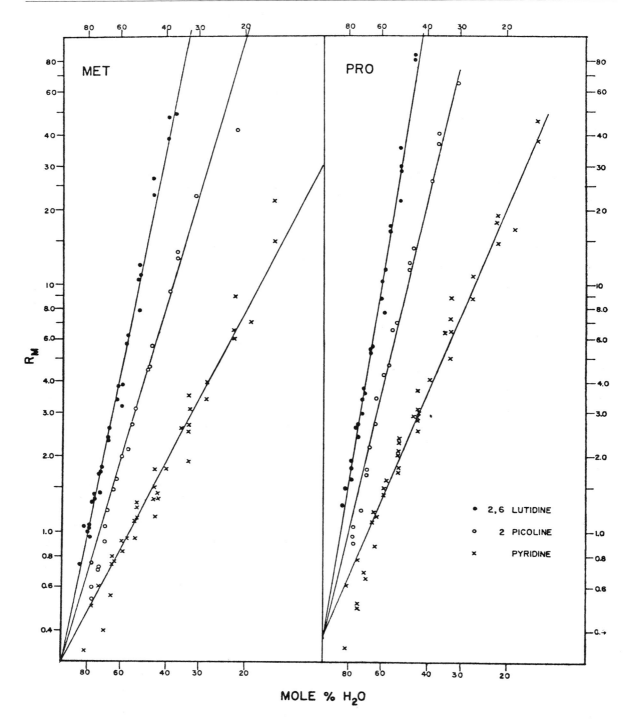

Fig. 3. Log amino acid R_M versus log mole fraction water in chromatographic solvent.

requirement," are considered to be "related." This definition of "related" has the advantage of permitting a beginning to quantitation of amino acid relatedness.

Table 2 shows the amino acids arranged, in the now familiar fashion, according to codon assignment, and gives for each amino acid the value of its polar requirement (in this case, determined in 2,6 dimethylpyridine:water solvent systems). [Essentially the same relative ordering of amino acids by polar requirement is obtained with pyridine:water and a variety of aliphatic substituted pyridine:water systems (Woese et al. 1966, and unpubl.)] It is clear that the amino acid ordering by codon assignment bears a striking relation to ordering by polar requirement. Perhaps the most impressive feature of this correlation is that all amino acids, whose codons differ from one another in the III position only, have nearly identical polar requirements. The amino acid pairs in question are his-gln, asn-lys, asp-glu, and cys-trp. (As the reader will recall, it was previously apparent that changing a base in the III codon posi-

tion often did not change the amino acid assignment at all. Now we see that when such a base alteration does change the amino acid assignment the two amino acids in question are always "closely related" to one another.) A second striking feature of the table is that all amino acids with codons containing U in the II position constitute a closely related group, as do those amino acids with C_{II} codons. These results are entirely compatible with and certainly suggestive of some sort of codon-amino acid pairing basis to the genetic code.

The possibility of a codon-amino acid pairing basis to the genetic code makes one wonder whether it is actually possible for a trinucleotide to assume a configuration where all (or at least two) of its bases are in intimate contact with an amino acid; for it is certainly conceivable that the constraints imposed by the covalent linkages in a trinucleotide structure could preclude the formation of such a structure. Having obviously no experimental approach to this question, one turns to molecular model building to test whether the possibility is sterically (if not

Table 2 Amino acid polar requirements

UUU	phe 5.0	UCU		UAU	tyr 5.4	UGU	cys 4.8
UUC		UCC	ser 7.5	UAC		UGC	
UUA	(leu)	UCA		UAA		UGA	trp 5.2
UUG		UCG		UAG		UGG	
CUU	leu 4.9	CCU	pro 6.6	CAU	his 8.4	CGU	arg 9.1
CUC		CCC		CAC		CGC	
CUA		CCA		CAA	gln 8.6	CGA	
CUG		CCG		CAG		CGG	
AUU	ileu 4.9	ACU	thr 6.6	AAU	asn 10.0	AGU	(ser)
AUC		ACC		AAC		AGC	
AUA	ileu	ACA		AAA	lys 10.1	AGA	(arg)
AUG	met 5.3	ACG		AAG		AGG	
GUU	val 5.6	GCU	ala 7.0	GAU	asp 13.0	GGU	gly 7.9
GUC		GCC		GAC		GGC	
GUA		GCA		GAA	glu 12.5	GGA	
GUG		GCG		GAG		GGG	

Polar requirements for the codon-assigned amino acids, determined in 2,6 dimethylpyridine: water solvents, as described in the text, and in Woese et al. (1966).

energetically) feasible. Granting that one should be wary of model building in the present state of ignorance, we feel it is justified in this instance because the question asked is a simple one.

Several attempts at this approach have been reported. A number of years ago (actually, when only the phe-UUU assignment had been determined) one of us (C. W.), in collaboration with J. Brown of the General Electric Research Laboratory, attempted to construct from a trinucleotide model a clathrate-like structure that would house an amino acid (see Woese 1963). A trinucleotide model can indeed assume a cuplike configuration. The phosphate-sugar backbone of the trinucleotide makes the rim of the "cup," and the bases make the bowl; in general, a number of H-bonds (indicated by lines in the figure) are available for holding the bases together. Many of the amino acids will fit snugly into such "cups," R group down, the carboxyl-amino axis residing in the vicinity of the phosphate-sugar "rim." No serious attempt to show specificity of particular trinucleotides for particular amino acids was, or could be made at that time, however.

Very recently, Pelc and Welton (1966) have reported more extensive and ambitious attempts at this sort of model building. In building a model, they require that the amino acid COO^-–NH_3^+ axis bind to the base in the III position, and then score how many bonds of either the H-bond type or nonpolar type an amino acid can make with any given trinucleotide. Using the number of such bonds as an index of "goodness of fit," they report that many amino acids fit their assigned codons better than the remaining (unassigned) ones. It is very doubtful whether these models are correct, if for no other reason than that these authors have ignored all interactions save one particular set of amino acid–base interactions. They have failed to consider interactions among the bases themselves, and interactions involving the OH groups on the pentoses, and those involving the phosphate moieties as well. Nevertheless, these models further strengthen the idea that an interaction between an amino acid and three bases of a trinucleotide is at least sterically feasible. And even if their models are not correct, the correlation they report, if valid, is a strong suggestion that there may indeed be properties of certain codons which allow them to interact "optimally" with particular amino acids. (See note added in proof.)

An Amino Acid Replacement Explanation for the Genetic Code

Unfortunately, it is impossible to give an adequate definition of "related amino acids" in terms of amino acid function in proteins, so we cannot make a definite assessment of any amino acid replacement explanation for the genetic code. However, it is not likely that all the same amino acids defined as related in terms of their interactions with heterocyclic bases would also be related when the definition is in the context of protein structure and/or function. Let us then ask specifically whether an amino acid replacement model can account for the relatedness of such amino acids, for example, as asn and lys, gln and his, arg and ser, cys and trp, or ala and pro — amino acids defined as related by their codon assignments. The little evidence we have, from amino acid replacement data, suggests that at least in some cases the amino acids within each pair are not similar to one another in their roles in protein. A clear instance of this occurs in the particular locus in the tryptophan synthetase A protein where so many amino acid replacements have been studied. It is known in this case that the replacement of gly (the wild-type amino acid) by any of the following amino acids still permits a functional or partially functional protein to exist: ala, ser, asn, thr, val, and ileu; but replacement of gly by glu, asp, arg, or presumably lys (since this last amino acid has never occurred among the more than 20 examples of functional revertants obtained from the functionless arg-containing protein) leads to nonfunctional proteins (Yanofsky, this volume). Other less extensive amino acid replacement data also lead to a similar conclusion. While an amino acid replacement explanation for the genetic code cannot be positively ruled out on the basis of such data, it does become rather suspect. In a negative sense, then, the above contradictions also lend further support to the codon-amino acid pairing explanation for the genetic code (which is in accord with all but one of the above pairs of amino acids defined as "related" by codon assignment).

GENERAL DISCUSSION

The case for a codon-amino acid pairing basis for the genetic code is now sufficiently strong that it is worthwhile considering it and its ramifications seriously. One prediction of such a model, as mentioned, is that a "codon" (in the broad sense) must have some special properties making it interact optimally with the particular amino acid to which it is assigned. Thus by defining characteristics of an amino acid, one automatically defines corresponding characteristics for its codons. This in turn means that we should also be able to detect certain relationships between the properties of the amino acids and the properties of the bases in their "codons." (In this respect the codon-amino acid pairing model for the genetic code differs sharply from an amino acid-replacement sort of explanation, the latter's constraints governing merely *relative* orderings of codons and amino acids, not *absolute* codon assignments.) This prediction of codon-amino acid pairing models is impossible to check in very much detail at the moment, mainly because the nucleic acid bases have not been characterized in polar-nonpolar terms analogous to the amino acid characterization by polar requirement. However, as we shall now see, the small amount of data available are certainly consistent with the prediction of the codon-amino acid pairing models.

The polar requirement criterion we have used to characterize the amino acids appears to measure (loosely speaking) some difference between the maximum number of water molecules an amino acid can bind firmly and the amino acid's capacity to interact in a nonpolar fashion with the chromatographic solvent. Thus, for example, gly, ala, and phe all presumably bind the same number of water molecules maximally, but phe's capacity to interact in a nonpolar fashion with the solvent gives it the lowest polar requirement of all (gly-8, ala-7, phe-5, see above). However, for such an amino acid as his, both its capacity to interact in a nonpolar fashion with the solvent and the maximum number of water molecules it can bind should be greater than is the case for gly, but the one interaction tends to "cancel" the other; thus it is not surprising to find gly and his having about the same polar requirement (8 versus 8½ respectively). For such an amino acid as asp, however, capable of binding many water molecules by virtue of its two carboxyl groups, and interacting but slightly with the solvent and in a nonpolar fashion, one expects and does find a relatively high polar requirement, i.e., 13.

The correlations of Pelc and Welton, discussed above, strongly suggest that the "codon" with which

an amino acid interacts is the same as the conventional codon — not some transform of it. Over and above this the polar requirement data suggest the following interpretation: For simple amino acids, i.e., those whose codons have a pyrimidine in the II position, only the base in the II codon position plays a major role in determining the codon assignment, for only when this base is changed is a marked effect on polar requirement of the amino acids noted. Further, for those amino acids generally possessing polarity in their side chains, *both* the II and the I position bases appear important in determining the codon assignment, for changing either base generally has a pronounced effect on polar requirement. Thus there should be properties of the II position base in some instance or the II plus I position bases in others that somehow correspond to properties of the amino acids to which the various codons are assigned.

The amino acids which have the most nonpolar side chains are phe, leu, ileu, met and val; as expected, these have the lowest polar requirements. All possess U_{II} codons. Turning to the amino acids with purine$_{II}$ codons, we find that those having the most nonpolar, least polar side chains are those having U_I codons (i.e., tyr, trp, and cys). In fact, for the amino acids with A_{II} codons, there appears to be a general gradient of decreasing nonpolarity and/or increasing polarity of the amino acid side chains as one proceeds in the codon series from U_I (tyr) to C_I (his, gln) to A_I (asn, lys) to G_I (asp, glu) codons. In any case, there is a strong suggestion in all this that U is the base associated with the strongly nonpolar amino acid side chains. While U may not be the most nonpolar of all four bases, it certainly is the least polar of all — as evidenced by Huckel calculations regarding its various electron donor properties (Pullman 1965) (keto, ring N, etc.), and by its chromatographic behavior (Saxinger and Woese, unpubl.). (It would not be surprising if some uracil derivative capable of stronger nonpolar interactions than U (e.g., thymine) were eventually found to be the base originally used in the genetic code in the place of U.)

Another question one would ask of a codon-amino acid pairing model is to what extent it orders the codon assignments. We have already seen that the most nonpolar, least polar amino acids all possess U_{II} codons. However, is it significant (not a chance occurrence) that phe possesses UU codons while val possesses GU ones, or could a phe GU codon assignment just as easily have evolved? In other words, is the codon catalogue strictly, or only approximately ordered by codon-amino acid pairing? In one instance, the polar requirement data suggest the former possibility to be the case. As we have just seen, the amino acids with A_{II} codons clearly rank the bases (in the I codon position) according to some ill-defined scale in the order U < C < A < G. The polar requirements for amino acids with U_{II} codons, given in Table 2, are so similar to one another that it is not possible to tell how these rank the I position bases. However, by resorting to the device of determining polar requirements in a substituted pyridine system (e.g., 4-ethylpyridine) in which the pyridine seems to interact strongly with amino acids in a nonpolar fashion, one clearly can rank the U_{II} amino acids by relative polar requirement: thus phe (U_I) < leu (C_I) < ileu (A_I) < met (A_I) < val (G_I), which is the same ranking of the I position bases as obtained with the A_{II} amino acids. (Similar, but nonidentical rankings are seen for the C_{II} and G_{II} amino acids.) The regularity in the various rankings of the I position base suggests that assignment of this base is not a chance matter. It is further of interest to note that the probable ranking of the bases suggested by our data (i.e., U < C < A < G) is also the same ranking of bases as is observed by either decreasing ionization potential or increasing polarizability (B. Pullman, pers. commun.).

A codon-amino acid pairing model for the genetic code offers obvious and simple explanations why various amino acids are not encoded at all, and perhaps why certain codons have remained unassigned. In the first case, some aspect of the geometry and/or energetics of the hypothetical codon-amino acid complex is not "optimal" for the unassigned amino acids. For example, it is clear that the U_{II} amino acids all have polar requirements closely clustered (i.e., 5.2 ± 0.4) and the C_{II} amino acids likewise have polar requirements clustered in the range 7.0 ± 0.5. The unassigned amino acid, α-amino-n-butyric acid, happens to have a polar requirement, 5.9, higher than the highest one of the U_{II} amino acids, but lower than the lowest of the C_{II} amino acids, and so should not "pair" with codons as well as the assigned amino acids. Unassigned codons could then be those which do not interact "optimally" with any amino acid, or perhaps do so equally well with more than one amino acid, which latter condition could create ambiguity.

Finally, attention should be directed briefly to

the problem of how codon-amino acid pairing could fit into the biological context. Does it play a coding role in cells today, or was its function confined to the recesses of evolution? And how did the translation apparatus evolve — through what states, what sorts of mechanisms, did it pass? There exist no clear-cut answers to any of these questions, so we shall have to restrict discussion to emphasizing the pertinent issues.

We have seen above that the translation apparatus must have had a most extensive evolution, and that in its earlier stages it must have been a very different process from the one we know now. In particular, primitive translation had to be far more error-prone than its modern counterpart, and the actual mechanisms employed would have been far simpler than those now in use. Therefore, our knowledge of the modern translation apparatus may not only be of little use in understanding the evolution of that process, but it may even be a positive detriment in so far as we are imbued with the dogma of ribosome-mRNA-sRNA-activating-enzyme. Thus it is wise to rid one's mind of the dogmata of modern translation when looking at the evolutionary aspects of the genetic code.

One of the main issues is whether codon-amino acid pairing types of interactions exist in the modern translation process. There is no evidence showing that it does. The work of Chapeville et al. (1962), demonstrating that the codon recognition properties of an sRNA (sRNA cys) are independent of the amino acid it happens to be carrying, practically rule out codon-amino acid pairing as playing any role in the actual decoding step. However, with regard to sRNA charging, no such clear-cut answer exists. The evidence here, which we shall not go into (see, e.g., Baldwin and Berg 1966 and previous work), can be interpreted to suggest that codon-amino acid pairing is involved in the sRNA charging step, but by no means proves the point. Thus one of the key points regarding the nature of the genetic code remains unsettled.

Concerning the nature of primitive translation processes, nothing positive can be said, but some reasonable negative conclusions can be drawn. Modern translation involves the participation of no less than four major macromolecular species — the activating enzymes, the sRNAs, the ribosome (itself comprising many types of macromolecules), and the mRNAs. In lieu of evidence to the contrary, the principle of Occam's razor demands that translation began in association with no more than one (perhaps two) macromolecular species, which probably was an evolutionary precursor of one (or more) of the components of the modern translation apparatus. This of course, says that primitive translation could not be accomplished via an mRNA "tape reading" type of mechanism involving "adaptors" for the amino acids.

Since primitive translation almost certainly utilized very different mechanisms from its modern counterpart, one should question whether translation began as "translation," (i.e., as the mapping of a primary structure of a nucleic acid into that of a polypeptide). It is certainly conceivable that translation evolved out of some different sort of nucleic acid–amino acid or peptide relationship, which initially had nothing to do with translation. The studies of Beljanski and Beljanski (1963) are of interest in this respect, as they deal with peptide synthesis which could involve RNA in a "nontranslational" capacity. There seems little point in attempting to develop models for primitive translation at this time. Plausible mechanisms involving either an evolutionary precursor of sRNA or one of rRNA can be constructed. (For such a discussion see Woese 1967.)

To summarize briefly; at the present state of our knowledge (or better, ignorance) the following very general features concerning the fundamental nature and evolution of the genetic code seem to be emerging: (1) The genetic code is likely to be based upon some rudimentary sort of "recognition" interactions between specific amino acids and specific oligonucleotides, or "codon-amino acid pairing." (2) It is likely that the primitive attempts at translation were unlike the tape-reading system employing amino acid "adaptors" in use in "modern" cells. It is reasonable to suppose that such primitive mechanisms utilized no more than one (or at most two) class of macromolecules; which perhaps was the evolutionary precursor of some component(s) in the modern translation apparatus. (3) Primitive attempts at translation must have been highly inaccurate, placing severe limitations upon the nature of primitive cells, and their immediate potential for evolving.

Note Added in Proof

We have recently attempted to repeat some of the work of Pelc and Welton. Using their published constraints, we still find that a greater number of amino acids make a "good fit" with a given trinucleotide model in general, than they report. Also in our hands trinucleotide models do not tend to show a better fit with l-amino acids than with their d-counterparts. We do not wish to imply that the correlation Pelc and Welton report is invalid. Our own work is not yet extensive enough to determine this. However, we do feel that the lack of specific detail regarding manipulation of models, structures which yield "good fit," and exact criteria for fit in their papers prevents checking their work, a checking which is essential in view of the import of their conclusions.

REFERENCES

Baldwin, A. N., and P. Berg. 1966 tRNA induced, hydrolysis of valyladenylate bound to isoleucyl RNA synthetase. *J. Biol. Chem.* 241:839–845.

Beljanski, M., and M. Beljanski, 1963. Acide aminé-acide ribonucléique, intermédiare dans la synthèse des laisons peptiduque. VI. *Biochim. et Biophys. Acta* 72:585–597.

Berg, P. 1961. Specificity in protein synthesis. *Ann. Rev. Biochem.* 30:293–324.

Bergmann, F. H., P. Berg, and M. Dieckmann. 1961. The enzymatic synthesis of amino acyl derivatives of ribonucleic acid. III. *J. Biol. Chem.* 236:1735–1740.

Carbon, J. A., L. Hung, and D. S. Jones. 1965. A reversible oxidative inactivation of specific transfer RNA species. *Proc. Natl. Acad. Sci.* 53:979–986.

Chapville, F., F. Lipmann, G. von Ehrenstein, B. Weisblum, W. Ray, and S. Benzer, 1962. On the role of soluble ribonucleic acid in coding for amino acids. *Proc. Natl. Acad. Sci.* 48:1086–1092.

Cherayil, J. D., and R. M. Bock, 1965. A column chromatographic procedure for the fractionation of sRNA. *Biochemistry* 4:1174–1182.

Crick, F. H. C. 1966. Codon-anticodon pairing; the wobble hypothesis. *J. Mol. Biol.* 19:548–555.

Gamow, G. 1954. Possible relation between deoxyribonucleic acid and protein structure. *Nature* 173:318.

Goldstein, J., T. P. Bennett, and L. C. Craig, 1964. Countercurrent distribution studies of *E. coli* B sRNA. *Proc. Natl. Acad. Sci.* 51:119–125.

Gorini, L., and E. Kataja. 1964. Streptomycin-induced oversuppression in *E. coli. Proc. Natl. Acad. Sci.* 51:995–1001.

Hoagland, M. B. 1960. The relationship of nucleic acid and protein synthesis as revealed by studies in cell-free systems, p. 249–408. *In* E. Chargoff and J. N. Davidson [ed.] *The nucleic acids,* v. 3. Academic Press, New York.

Mandel, L. R., and E. Borek, 1961. The source of the methyl group for thymine of RNA. *Biochem. Biophys. Res. Commun.* 6:138–140.

Nathenson, S. G., F. C. Dohan, Jr., H. H. Richards, and G. L. Cantoni. 1965. Partition chromatography of yeast and *Escherichia coli* soluble ribonucleic acid. Relation of coding properties to fractionation. *Biochemistry* 4:2412–2418.

Nirenberg, M., P. Leder, M. Bernfield, R. Brimacombe, J. Trupin, F. Rottman, and C. O'Neal, 1965. RNA codewords and protein synthesis. VII. On the general nature of the RNA code. *Proc. Natl. Acad. Sci.* 53:1161–1168.

Pauling, L., and M. Delbrück. 1950. The nature of the intermolecular forces operative in biological processes. *Science* 92:77–81.

Pearson, R. L., and A. D. Kelmers. 1966. Separation of transfer ribonucleic acids by hydroxyapatite columns. *J. Biol. Chem.* 241:767–769.

Pelc, S. R., and M. G. E. Welton. 1966. Stereochemical relationships between coding triplets and amino acids. *Nature* 209:868–870.

Pochon, F., A. M. Michelson, M. Grunberg-Manago, W. E. Cohn, and L. Dondon. 1964. Polynucleotide analogues. III. Polypseudouridylic acid: synthesis and some physicochemical and biological properties. *Biochim., Biophys. Acta* 80:441–447.

Pullman, B. 1965. Some recent developments in the quantum mechanical studies on the electronic structure of the nucleic acids. *J. Chem. Phys.* 43:5233–5243.

Söll, D., D. S. Jones, E. Ohtsuka, R. D. Faulkner, R. Lohrmann, H. Hayatsu, H. G. Khorana, J. D. Cherayil, A. Hempel, and R. M. Bock. 1966. Specificity of sRNA for recognition of codons as studied by ribosomal binding technique. *J. Mol. Biol.,* in press.

Söll, D., E. Ohtsuka, D. S. Jones, R. Lohrmann, H. Hayatsu, S. Nishimura, and H. G. Khorana. 1965. Studies on polynucleotides. XLIX. Stimulation of the binding of aminoacyl-sRNS's to ribosomes by ribotrinucleotides and a survey of codon assignments for 20 amino acids. *Proc. Natl. Acad. Sci.* 54:1378–1385.

Sonneborn, T. M. 1965. Degeneracy of the genetic code: extent, nature and genetic implications, p. 377–397. *In* V. Bryson, and H. J. Vogel [ed.] *Evolving genes and proteins.* Academic Press, New York.

Watson, J. D. 1965. *Molecular biology of the gene.* W. A. Benjamin, New York. 494 p.

Woese, C. 1963. The genetic code-1963. *Intern. Counc. Sci. Unions Rev.* 5:210–252.

——. 1965. Evolution of the genetic code. *Proc. Natl. Acad. Sci.* 54:1546–1553.

——. 1967. *The genetic code,* Harper and Row, in press.

Woese, C. R., D. H. Dugre, W. C. Saxinger, and S. A. Dugre. 1966. The molecular basis to the genetic code. *Proc. Natl. Acad. Sci.* 55: in press.

Zubay, G., and P. Doty. 1958. Nucleic acid interactions with metal ions and amino acids. *Biochim. Biophys. Acta* 29:47–58.

DISCUSSION

I. B. Weinstein: I would like to comment on the question of whether a chemical "fit" between nucleotides and amino acids existed only during early evolution of the code or whether this fit continues to operate during protein synthesis today. At this symposium we have heard a good deal of evidence that to a first approximation the code is universal and at the same time we have heard evidence for the prevalence of suppressor genes and their ability to alter markedly codon assignments. If suppressor genes have been functioning during a considerable portion of evolution then one might expect the code to have drifted to such an extent that it would no longer be universal. The fact that this has not happened to an appreciable extent favors, I believe, the possibility that chemical restraints continue to operate, at the present time, perhaps during the attachment of amino acids to sRNA (I. B. Weinstein 1963, Cold Spring Harbor Symp. Quant. Biol. 28:579).

C. R. Woese: The older argument for codon-amino acid pairing (CAP) based upon universality of the genetic code (referred to by Dr. Weinstein in his reference), and the newer one based upon the characteristics of suppressor tRNAs, referred to by Dr. Weinstein in his comment, are appealing, but unfortunately not compelling. I will not go into all the questions which surround this issue at this time.

However, all that one need say in general terms to circumvent a CAP explanation for a universal genetic code existing in the face of suppressor tRNAs which change codon assignments, is that for some reason these altered tRNAs put the cell at a selective disadvantage, which prevents its surviving in nature. Since this general explanation is not too satisfying, a specific model might help: Suppose that the site on tRNA (AERS) which recognizes the activating enzyme includes the site on tRNA (CRS) which recognizes the codon. A mutation that changes CRS would also then change AERS. This could give rise to a reduced affinity of the tRNA in question for its (normal) activating enzyme, which in turn could put the cell at some selective disadvantage. (A CRS alteration which also affects the AERS may exist in the arg-gly suppressor discussed by Carbon et al. this volume.) In this way, this class of suppressor mutations might be uniformly lethal (in the long run) and so have a negligible tendency to randomize the codon assignments. No CAP need be invoked for this sort of explanation.

These next two papers should be read together. They illustrate the formulation and sophistication of the "operon" type control of gene activity in microbial systems. The Beckwith paper answered most of the questions raised, but left unanswered, by Jacob and Monod e.g., what was the chemical nature of the repressor and the exact mode of function of the operator. It is also valuable for updating the operon model for the control of inducible enzyme systems. In this respect, appreciation can be gained for the insight and creative thinking of Jacob and Monod in light of the minor theoretical changes arising between their 1961 paper and Beckwith's 1967 comprehensive review.

The Goldberger–Kovach paper discusses in depth the mechanics of a repressible enzyme system. Although the paper ends with a question, no separate regulator gene has been found that produces an inactive repressor for the control of the histidine biosynthetic pathway. On the contrary, it has been established by Vogel et al. that the inactive repressor substance, activated by the corepressor histidyl tRNA, is in fact the first enzyme of the biochemical pathway.

Regulation of the *lac* operon

Recent studies on the regulation of lactose metabolism in *Escherichia coli* support the operon model

JONATHAN R. BECKWITH*

Reprinted by author's and publisher's permission from Science, *vol. 156, 1967, pp. 597–604. Copyright © 1967 by the American Association for the Advancement of Science.*

The mechanism of regulation of gene expression is today one of the most actively studied problems in molecular biology, in good part as a result of the pioneering work of Jacob and Monod on the control of the genes involved in lactose metabolism in the bacterium *Escherichia coli*. Since 1961, when Jacob and Monod first proposed the operon model for gene regulation,[1] a number of alternative suggestions have been published for the ways in which genes are controlled,[2-5] some of them radically different from the Jacob-Monod model.

The lactose (*lac*) system is still the best-studied example of gene regulation. In the years since 1961, there has been a considerable amount of new information, both genetic and biochemical, on the *lac* operon. On the basis of the most recent information, which will be discussed in this article, it appears that the original formulation of the operon, in most of its aspects, is still the simplest model fitting all the known facts about the *lac* system.[6]

LACTOSE METABOLISM IN ESCHERICHIA COLI

The initial steps in the metabolism of lactose by *E. coli* involve two protein components: (i) a membrane-bound protein [M-protein[7] or permease][8] which is probably responsible for both the transport of lactose into the bacterial cell and for its concentration therein; and (ii) the enzyme β-galactosidase which catalyzes the hydrolysis of lactose within the

cell to glucose and galactose. The structure of these two proteins is determined by two chromosomal genes, *y* for the permease and *z* for β-galactosidase. In wild-type strains of *E. coli* grown on almost any carbon source but lactose, the activities of these genes are repressed, their products being found in only very small amounts. However, growth on lactose as sole carbon source, or addition to the growth medium of various compounds structurally related to lactose, results in the induction of gene expression, with an increase in the amounts of these proteins of as much as 1000 times. Under these optimum conditions, β-galactosidase represents approximately 3 percent of the total protein of the cell. Induction also results in the appearance of another enzymic activity, thiogalactoside transacetylase,[9] corresponding to a third structural gene, *a*. Although the *a* gene is regulated in parallel with the *z* and *y* genes, the enzyme plays no essential role in lactose metabolism.[10] These three structural genes of the lactose system lie next to one another on the chromosome, mapping in the order *zya* (Fig. 1).

THE MODEL

The Jacob-Monod operon model of control, with some additions and modifications resulting from recent work on the *lac* system (Fig. 1) may be described as follows. In the absence of any regulation, the expression of the three *lac* structural genes involves two steps. First, the information from these genes is transcribed into a single RNA (messenger RNA) molecule. The information from each gene-

*Department of Bacteriology and Immunology, Harvard Medical School, Boston, Massachusetts.

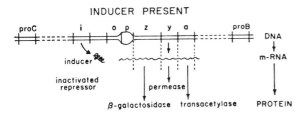

Fig. 1. The control of expression of the *lac* operon. The *lac* region and two of the markers surrounding it (genes involved in proline biosynthesis) on the chromosome are represented as regions of double-stranded DNA. The picture of the operator and promoter in this figure is only one of many ways of visualizing their interaction. The promoter is shown serving as an initiation site for mRNA synthesis only when its two DNA strands are not held together by hydrogen bonding in the "closed" configuration. The repressor binds to and closes the operator, resulting in a closing of the promoter. An inducer alters the repressor so that it can no longer interact with the operator in this way. Pro C and pro B indicate two genes involved in the biosynthesis of proline.

copy within the RNA is then translated by the protein-synthesizing machinery into the structure of the three protein products. The synthesis of this mRNA molecule is initiated at the promoter site, p, which is adjacent to, or part of the z gene. The structure of the promoter, or of a site in the same region, determines the rate at which the message is transcribed from these genes under conditions of maximal expression. The control of the expression of the *lac* structural genes is effected by a repressor molecule, the protein product of the closely linked i gene. The repressor acts on the DNA to inhibit the transcription of these genes. The combination of the repressor with the operator site, o, which is adjacent to p, in some way inhibits the initiation process. Compounds that cause induction of the expression of the *lac* genes act by either destroying or altering the repressor-operator complex, thus

allowing initiation of mRNA synthesis at the promoter.

A group of genes whose activity is coordinated by an operator is termed an operon. According to this model, the operator is not essential for operon activity, but rather serves as a controlling site superimposed on a functioning unit.

MAPPING OF THE ELEMENTS OF THE *LAC* OPERON

Jacob and Monod have presented evidence from three-factor crosses demonstrating the gene order *iozy*.[11] In addition, strains which carry deletions extending into the *lac* genes from outside either end of the *lac* operon have been isolated, providing further confirmation for the order *iozya* (see Fig. 4 and note p. 417).[10] Class-I deletions remove the i gene and leave the operator and structural genes intact, while class IVa deletions remove the a gene but leave all other sites in the operon intact.

TRANSLATION OF THE *LAC* OPERON

The Translation Process To discuss certain aspects of *lac* operon control, I must first describe what is known about the mode of translation of mRNA information into protein. The mRNA copies of genes are thought to be translated by the following process.[13] Ribosomes attach to one end of the message (corresponding to the amino-terminal end of the protein), and, in conjunction with the other components of the protein-synthesizing machinery, begin to move along the message as the peptide bonds are formed. New ribosomes continually attach to this end of the message and proceed in this way, so that at any one time the mRNA will carry many ribosomes along its length. This complex of ribosomes and mRNA is called a polysome. At the end of the gene-copy in the mRNA, the translation machinery meets a codon that signals termination and release of the completed polypeptide chain.

As a result of mutation, chain-terminating codons may be introduced into the gene at various points preceding the normal site of chain termination. The two well-studied chain-terminating codons arising by mutation are the amber and ochre codons, UAG and UAA (for uracil, adenine, and guanine) respectively.[15] Chain termination by an amber mutation (and probably by an ochre mutation also) results in a quantitative release of an NH_2-terminal fragment of the protein coded for by a particular gene. The length of the protein fragment depends upon the distance of the mutation from the beginning of the gene (corresponding to the NH_2-terminal end of the protein).

Another type of mutation that interferes with translation is the frameshift mutation.[16] These mutations, through the addition or removal of one or more base pairs from a gene, cause the reading of an incorrect sequence of codons in the mRNA gene-copy.

The Operon Messenger RNA There is still no direct evidence that all the information from the structural genes of the *lac* operon is contained in a single piece of mRNA. However, studies on polarity (described later) do suggest that this is so. Also, Kiho and Rich[17] have presented evidence that an amber mutation prematurely terminating translation in the *y* gene affects the size of the polysome on which β-galactosidase is made. If the *y* and *z* gene-copies were on different mRNA molecules, no effect should have been observed. We shall assume that there is a single operon mRNA.

The Amino-terminal End of the z *Gene* A knowledge of the direction of translation of the structural genes of the operon is important in interpreting the various experiments on operon expression. The direction of translation indicates the direction of transcription of the operon, since it has been shown by biochemical and genetic experiments that both these processes begin at the 5' end of the RNA molecules.[18] Studies by Fowler and Zabin[19] on three chain-terminating mutants of the *z* gene[20] indicate that translation is initiated at the proximal* end of the *z* gene-copy in the mRNA. These mutants, one of which lies very close to the *y* gene, map at the distal end of the *z* gene.

Each mutant makes a large amount of a protein that is immunologically similar to β-galactosidase. If this distal end of the *z* gene corresponded to the

NH$_2$-terminal end of the protein, translation should have begun at this end, and, in these mutants, terminated within a short distance, releasing a small polypeptide fragment. The finding of large amounts of a large protein molecule is thus very strong evidence that translation begins at the proximal end of the *z* gene, and that in these mutants, most of the *z* polypeptide chain is made before the chain-terminating triplet is read and the protein released. As mentioned earlier, this conclusion also leads to the further conclusion that operon mRNA synthesis is initiated at this end of the operon. Even stronger evidence for the operator-proximal end of a gene corresponding to the NH_2-terminal end of the protein comes from studies on the tryptophan operon by Yanofsky and co-workers.[21]

POLAR MUTANTS

Two different models for the translation of the gene-copies in an operon mRNA have been considered (Fig. 2). (i) A ribosome can enter onto the mRNA at only one end of the molecule;[2] in the *lac* operon, this end would correspond to the NH_2-terminal end of the *z* gene. Then, during translation, the ribosome proceeds down the message and must complete passage through the *z* gene in order to start translation of *y* and *a*. (ii) Ribosomes can enter independently at the starting points for all three gene-copies in the mRNA, *z*, *y*, and *a*, without

Fig. 2. Two possible models for translation of the *lac* operon mRNA. The circles represent ribosomes, and the squiggly lines attached to them represent the growing polypeptide chain.

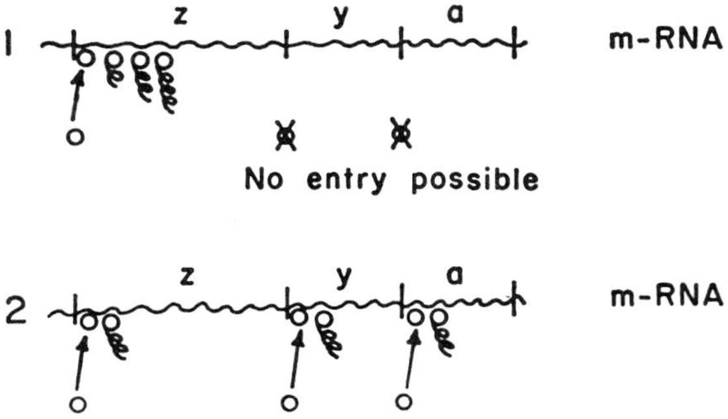

---

*In this paper, the proximal end of a gene or of the operon refers to that end which is closest to the operator. A distal end is that farthest from the operator.

any requirement for having translated a previous gene-copy.[22]

There is no conclusive evidence to distinguish between the two models. However, any model for the translation of the *lac* operon mRNA must take into account the class of mutants known as polar mutants. Polar mutations in the *z* gene of the *lac* operon are usually point mutations that not only abolish *z* gene activity, but, in addition, reduce or abolish the expression of the *y* and *a* genes.[23, 24] Such mutations in the *y* gene affect the activity of the *a* gene but not of the *z* gene. Although many polar mutants are not well characterized, a large percentage are amber and ochre mutations.[22] Since these chain-terminating mutations exert their effects on translation, the polarity must be due primarily to an interference with the translation of the operon mRNA.

The polarity of a chain-terminating mutation in the *z* gene depends upon its position within the gene (Fig. 3).[22] Amber and ochre mutants mapping toward the distal end of the *z* gene are not very strongly polar, whereas those chain-terminating mutants mapping in a proximal segment of the *z* gene do not make any detectable amounts of the *y* and *a* gene products. The extent of polarity of a chain-terminating mutation in *z* is determined by the distance of the mutation from the boundary between *z* and *y*, and not by its proximity to the operator. Thus, when an extremely polar mutation, mapping at the proximal end of the *z* gene, is combined with a succeeding deletion in the gene, effectively moving the mutation closer to the *y* gene, the polarity effects are markedly reduced.[22, 25, 26] The distance from the operator end of the *z* gene to this mutation has not changed. This finding shows that, even if translation of the operon is terminated by a chain-terminating mutation very shortly after it begins, the rest of the operon can still be expressed if this mutation is very close to the *y* gene.

Polar mutations in the *z* gene not only reduce the amounts of the proteins, permease and transacetylase, but also reduce the amount of *lac* mRNA present in the cell.[27] Again, the reduction of the amount of *lac* mRNA is correlated with the position of the mutation in the *z* gene. Extracts of induced cells carrying one of the extremely polar mutations at the proximal end of the *z* gene contain no detectable amounts of *lac* mRNA; cells harboring *z*⁻

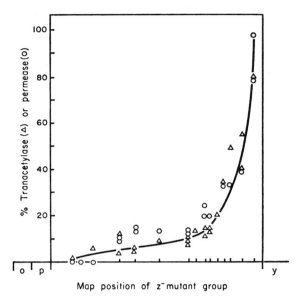

Fig. 3. The gradient of polarity of chain-terminating mutants in the *z* gene. This figure is copied from Newton *et al.*[22] The permease activity (O) and transacetylase activity (△) of many *z*⁻ amber and ochre mutants are plotted against the position within the *z* gene. These positions are only rough approximations of true location on the genetic map location. However, the order of the mutants is unambiguous.

polar mutations closer to the *y* gene contain detectable, but still reduced, levels of *lac* mRNA. There is some indication that in these mutants there are normal amounts of the fragment of *lac* mRNA corresponding to the proximal segment of the *z* gene preceding the mutation, but that complete operon mRNA molecules are present only in reduced amounts. Thus, as a result of the interference by polar mutations with the translation process, a polarity effect on *lac* mRNA levels also results.

The results of studies on polar mutants fit both of the models for operon mRNA translation considered above. If the ribosomes can only enter at the beginning of *z* (model 1), then, after encountering an amber or ochre mutation in the *z* gene-copy, ribosomes would continue to pass down the message. However, since these ribosomes are no longer engaged in translation, there is an increased probability of their falling off the mRNA. The longer the distance to be traversed after the site of chain-termination, the lower the probability that the ribosome will reach the y gene-copy and there initiate translation. Recent studies of Malamy[28] on the strong polarity effects of what appear to be frameshift mutations in the distal segment of the z gene suggest that ribosomes after encountering chain-terminating mutations do continue moving along the mRNA. He finds that double mutants, in which the "frameshift" is preceded by an ochre or amber mutation in z, are still just as polar as the original frameshift mutant. If ribosomes fell off the mRNA at chain-terminating codons, they should never encounter the frameshift, and the polarity effects should be reversed. However, it should be noted that apparent frameshift mutations in one gene of the histidine operon in *Salmonella typhimurium* do not have the strong polarity effects seen in similar mutants in the *lac* operon.[29] Confirmation of the proposal that the extreme polarity effects of these mutants in the *lac* operon are due to their frameshift character would be strong supporting evidence that ribosomes continue their movement after chain-terminating codons.

If ribosomes cannot continue along an mRNA after a chain-terminating codon, then, in order to explain why most amber and ochre mutants allow some expression of y and a, we must admit that ribosomes can enter the y and a gene-copies independently of the completion of passage of the z gene-copy (model 2). The polarity effects of chain-terminating mutants can be accounted for by one of the following additional hypotheses.

1. Initially, the operon mRNA sticks to the DNA. As ribosomes move down the mRNA during translation and get closer to the beginning of the y gene-copy, there is an increasing chance that they will cause release of this part of the mRNA, thus freeing it for ribosome entry.[22] Premature chain-termination in the z gene and falling off of ribosomes would make it more likely that the mRNA will stick to the DNA.

2. The introduction of a chain-terminating mutation in the z gene, leaving an untranslated portion of the mRNA, results in a more rapid destruction of the distal segment of the mRNA.[15, 30] The amount of destruction could depend upon the segment of mRNA which is not engaged in translation.

3. The secondary structure of the RNA surrounding the ribosome entry points for y and a is such that ribosomes cannot enter unless there has been enough progress of ribosomes on the z gene-copy to disrupt this structure.[31] Premature chain-termination in the z gene and falling off of ribosomes would reduce the probability of exposing the entry site on y.

Although all of these hypotheses must include some explanation for the decreased amounts of *lac* mRNA in polar mutants, it is not a very important consideration in formulating the models. If this decrease in mRNA is not directly predicted by the model for polarity, then the suggestion can always be added that, under the particular conditions, the mRNA is destroyed very rapidly. If, of course, it is proved that the mRNA is not rapidly destroyed in these mutants, then certain models of polarity must be discarded.

Certain important findings do come out of the studies on polarity. First, the conclusion that the amount of *lac* mRNA in a cell somehow depends upon the translation of the operon has led to new speculations on mechanisms of repression of operon activity involving effects on translation. Second, it can be concluded that translation of an early part of the z message is not necessary for the release of the operon mRNA from the DNA, except in so far as this region is distant from the beginning of the y part of the mRNA. Stent,[3] Yanofsky,[32] and others have suggested that there is a critical region at the beginning of the first gene-copy in an operon mRNA, and this region must be translated in order to release the entire mRNA from the DNA. In the *lac* operon, this model cannot account for the extreme polar mutants.

SEQUENTIAL EXPRESSION OF THE *LAC* GENES

When the kinetics of expression of the *lac* operon are followed after the addition of inducer to a culture of *E. coli*, β-galactosidase activity begins to increase a minute or so before transacetylase activity rises.[33, 34] Since the appearance of an enzymic activity requires transcription, translation, and assembly of a protein into the correct configuration, any or all of these processes could be responsible for this lag. Experiments designed to determine the basis of this lag do show that there is a measurable time lapse between the transcription of the z gene and of the a gene.[34, 35] However, it is not clear yet whether this delay is responsible for the lag in appearance of transacetylase activity. In any case, the results showing the sequence of transcription of the *lac* genes are another indication that transcription begins at the operator end of the operon.

UNBALANCED TRANSLATION OF *LAC* GENES

Zabin has shown that the molar amounts of transacetylase synthesized are five or more times lower than the molar amounts of β-galactosidase.[36] One explanation for this finding derives from the studies on polar mutants which show that the distance between a chain-termination event in one gene and the initiation of translation in the next determines how much of the next protein is made. It is possible that, after the chain-terminating triplet in either z or y, there is a "dead space" of untranslated nucleotide sequence, the length of which determines the amount of transacetylase made.

THE OPERATOR

I have described evidence that the transcription of the *lac* operon starts at the operator end. But is transcription actually initiated at the operator itself? In the original operon hypothesis, it was proposed that the operator has three functions. (i) It forms the first part of the z gene; (ii) it is the site of repressor action; and (iii) it is the site of initiation of transcription. The second function is confirmed by the existence of operator-constitutive mutants (O^c),[1] which have partly lost sensitivity to repressor. As a result, these mutants make large amounts of the *lac* enzymes in the absence of inducer, but can still be induced to make more. The nature of the operator is indicated by the finding that O^c mutations cause constitutivity only for the *lac* genes which are linked to that operator. Thus, in a partial diploid strain carrying two copies of the *lac* region, one of which is O^c, the genes attached to the O^+ operator are still normally repressible, while those linked to the O^c operator are still partially constitutive.

There is reason to believe that all O^c's are dele-

tions.[37] First, certain of the O^c's can be shown to be deletions, since they remove the i gene also (as discussed later).[37] Second, the frequency of O^c's is not increased by mutagens which only cause base substitutions, but it is increased by treatment with x-rays, which does cause deletions.[37] Third, no suppressible O^c's have yet been found.[38] With another operon there is also mutagenic evidence that O^c's are always deletions.[39]

The third function for the operator was based on the existence of a second class of mutants ("O^0") mapping at the beginning of the z gene, which permanently shut off the z, y, and a genes.[1] However, it was subsequently shown that complete activity is restored to the y and a genes by deletion of the "O^0" mutant site.[40] Thus, this region of the operon is not necessary for operon expression and cannot be the site of initiation of mRNA transcription for the operon. In addition, it was shown that "O^0" mutations do not, in fact, lie in the operator region, but are only extreme examples of the polar mutants.[15, 22, 30, 40] Therefore, the term "O^0" is a misnomer, since these mutants have nothing to do with operator control. As a result of these studies and further work indicating that O^c mutations do not affect the properties of β-galactosidase,[37, 40, 41] it has been concluded that the operator lies outside the first structural gene of the operon.

THE PROMOTER

The following discussion summarizes evidence which suggests that transcription begins not at the operator, but at the adjacent promoter site. It is an elaboration of the argument for the promoter presented by Jacob, Ullman, and Monod.[37] This concept of operon function is based on the properties of mutants in which various components of the *lac* operon are deleted. Starting with strains in which the *lac* operon has been inactivated by certain mutations, it is possible to select O^c mutations that render the operon partially or fully constitutive.[37] In addition, by selecting for O^c mutations under conditions where only y gene function is required, O^c's are found which are $z^- y^+$.[37, 40] In all cases examined, O^c mutations of this latter type (Fig. 4, class III) are a result of deletions which cover a proximal segment of the z gene, the operator, and the i gene, and presumably fuse the intact y and a genes to another operon or gene. In most cases, the gene or

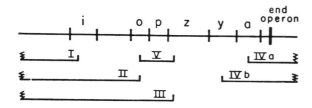

Fig. 4. Deletions of the *lac* operon.

genes to which *lac* has become fused have not been identified. However, two sets of deletions of this type have been isolated in which *lac* has been fused to known operons. In one case, the *y* and *a* genes have come under the control of an operator for two genes involved in purine biosynthesis,[42] and, in the other, the *lac* genes have come under the control of the operon determining tryptophan biosynthesis (Fig. 5).[12] In these cases, derepression of the tryptophan (*try*) or purine (*pur*) operons leads to a corresponding derepression of the remaining *lac* genes. In addition, the presence of purines or of tryptophan in the growth medium of the two types of strains results in strong repression of the synthesis of permease and transacetylase.

These results show that when a deletion removes the *lac* operator and a proximal segment of the *z* gene and fuses the *y* and *a* genes to an operon in the repressed state, the *y* and *a* genes are not expressed. Thus, it is not possible to restore *y* and *a* function to an inactivated *lac* operon merely by deleting the proximal end of the operon without regard to where the other end of the deletion lies. On the contrary, it seems likely that, to allow expression of the *y* and *a* genes, the deletions must connect these genes to another functioning gene or operon. Further evidence that this is the case comes from the failure to find, among a large number of O^c mutations, deletions covering part of the *z* gene which do not extend past the *i* gene. If it were possible to connect the *y* and *a* genes to any region

of the DNA and allow function, one would have expected to find among the O^c-z^- deletions some in which the deletion either ended in the operator or in whatever space might exist between *i* and *o* (Fig. 4, class V). These results suggest that the *lac* operon has an associated site which is essential for expression and that removal of this site, without the substitution of a new one, inactivates the operon. Apparently, deletions of class V inactivate the operon, since they are not found. Therefore, this essential site, in the case of the *lac* operon, must lie between *o* and *z*. On the basis of this argument, Jacob, Ullman, and Monod[37] have proposed this location for the promoter site.

SECOND FUNCTION FOR THE PROMOTER

The studies on deletions of the *lac* operon also provide information on the site or sites which determine the rate of expression of the operon under conditions of full induction. All deletions of class III (Fig. 4) result in a marked reduction in the amount of expression of the remaining *y* and *a* genes. In most cases, the activities of permease and transacetylase in these strains are only 10 percent or less of the activities seen in fully induced wild-type strains.[37,40,42] Several explanations can account for these findings.

1. The *lac* operon may ordinarily be transcribed at a very fast rate compared to that of most other genes and operons, either because of repression of the other genes acting on transcription, or because of lower intrinsic potentials for transcription of the other genes. According to this proposal, the *y* and *a* genes are always likely to function at a lower rate when connected to a different gene or operon.

2. The *lac* operon mRNA is translated at a very fast rate compared to that of other genes or operons. This argument concerning the rate of translation is analogous to the first argument.

3. The deletions fusing *z* to some other gene create a sequence of nucleotides which results in polarity effects, thus reducing the expression of *y* and *a*. For example, the deletion may create a frameshift in the operon or a nonsense mutation at the site of fusion. This explanation could account for low *y*

Fig. 5. Fused *lac* operons. The *site* of fusion of the *lac* and *try* operons was determined by E. R. Signer.[51]

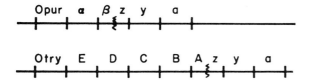

and *a* in some cases, but it seems very unlikely that every deletion creates just the correct conditions for such an effect.

Another set of deletions was isolated which removed only part of the operator but again deleted the *i* gene (class II),[37] thus connecting the *lac* operon, through its operator, to some nearby segment of the chromosome. Deletions of class II are far more frequent than those of class III. None of these class II deletions completely abolishes operator function, as measured by its sensitivity to repressor. In the presence of the wild-type allele of the *i* gene, these O^c mutants are still partly repressible. But, the striking property of these deletions, in contrast to those of class III, is that they all produce maximally exactly the same amounts (100 percent) of *lac* gene products as the wild-type strain. Thus, since extensive deletion of the operator does not reduce the maximal rate of operon expression, the operator* cannot be the site which sets the maximal potential for expression of the *lac* operon.

How can we explain the difference between deletions of classes II and III? Since we have concluded that a site *p* lying between *o* and *z* is essential for expression of the operon, we may now suggest that this site (or a site in the same region) is also involved in determining the maximum rate of this expression.[43] Deletions of class III remove this site and connect the *lac* operon to some other gene or operon with a promoter which functions at a lower rate. Deletions of class II, in contrast, do not delete the promoter, and so the site determining the maximum rate of expression of the operon is left fully functional. In addition, there are no sites at the distal end of the operon determining the maximum level of operon expression, since class IVa and IVb deletions have no effect on the expression of the intact proximal genes.[12]

REGULATION OF OPERON TRANSCRIPTION

The information obtained from study of these deletions is critical in analyzing the various mechanisms proposed for regulation and expression of the operon. These arguments can be reduced to the question of whether or not the operator (of some site previous to it) is the starting point of mRNA synthesis. I now consider these two possibilities for operon regulation and their implications.

1. The operator, in addition to carrying the information determining the sensitivity to repressor, is also the site of initiation of mRNA synthesis for the operon. Since we have concluded that the operator does not set the maximum level of operon expression, it follows from this model that the rate of initiation of mRNA synthesis is not the rate-limiting factor in the setting of this maximum. In other words, all genes or segments of the DNA to which the operon has become attached by deletions of class III are transcribed at the same rate as the *lac* operon is normally (when fully induced). The variation in maximal levels of gene expression is always a result of different rates of translation. An implication of this model is that *E. coli* has not evolved a mechanism for setting different maximal rates of RNA synthesis for different genes and operons. Although there is no strong evidence against this possibility, it is rather unattractive.

2. The only function of the operator is that of being the repressor-sensitive site. The promoter site is the initiation point for mRNA synthesis, and its structure determines how much transcription takes place in the absence of repression. According to this model, in contrast to the first possibility, different rates of mRNA synthesis can be set by different sequences of nucleotides (promoters) on the DNA.

In view of the implications of model 1, I favor this latter picture of transcription of the *lac* operon. Since the promoter would be the site of initiation of mRNA synthesis, according to this model, the operator cannot be transcribed as part of the operon mRNA. Therefore, the operon mRNA does not contain a repressor-recognition site. Then, in order to affect the expression of the *lac* structural genes, the repressor must recognize the operator on the chromosome. If repression does take place on the chromosome, the repressor could either directly repress the initiation of synthesis of operon mRNA or interact with the mRNA-DNA complex (or even a ribosome-RNA-DNA complex) to prevent further synthesis of operon mRNA.

---

*One might consider that since no 100 percent O^c's have been found, a right-hand extremity of the operator is involved in determining the rate of operon expression. This argument is not very different from the one discussed in that rather than one site with two distinguishable regions there are two sites.

It is very likely that a promoter region plays some part in determining the activity of the *lac* operon. It should be possible, therefore, to isolate mutants of the promoter which alter this maximal level. Mutants have been found which reduce the maximum expression of the *lac* operon and which have other properties expected of promoter mutants.[43] These are mutants which, when fully induced, make only 5 to 10 percent of the normal fully induced levels of all three gene products. The mutants are not connected with the *i-o* control system, do not appear to be polar mutants, and map in the same region as the promoter.

THE REPRESSOR

The existence of two classes of mutations affecting *lac* regulation and defining the *i* locus, led to the concept of the *i* gene product as a repressor molecule which interacts both with the operator and the inducer.[1] Strains carrying *i*⁻ mutations which abolish activity of the *i* gene result in a maximal synthesis of the *lac* proteins, even in the absence of inducer. Moreover, in partial diploid strains carrying both an *i*⁻ mutation and the wild-type *i*⁺ allele of the *i* gene, the operon becomes repressible.[1,44] Thus, the *i* gene appears to be the structural gene for a diffusible product which is responsible for the repression of the operon. The *i*ˢ (super-repressed) mutations, in contrast, result in noninducibility of the *lac* operon,[23,45] in spite of the presence of unaltered operator, promoter, and structural genes. These mutants are thought to produce a repressor which has no affinity for inducer, so that the operon is permanently shut off. However, as pointed out by Brenner,[5] on the basis of only these two classes of mutations one can devise somewhat more complex pictures of *i* gene action; for example, the *i* gene product is not itself the repressor but is an enzyme catalyzing the synthesis of a small molecule which is part of the repressor. However, if there were another molecule involved in repression, one would have expected to find mutations in another locus resulting in constitutivity. No such mutants have been found. Thus, the original Jacob-Monod idea of *i* gene action is still the simplest.

There is now both genetic and chemical evidence that the *i* gene ultimately directs the synthesis of a protein molecule. The genetic evidence is that there exist suppressible amber mutants (*i*⁻) of the *i*

gene.[38,46] Since it is known that these mutants affect translation, the *i* gene messenger must be translated into protein. Recently, Gilbert and Muller-Hill[47] have isolated the *i* gene product and have shown that it is at least partially composed of protein. The isolation was achieved by purification of a fraction of *E. coli* protein which binds the inducer IPTG (isopropyl-β-*d*-thiogalactoside) with the expected affinity constant. The identification of this protein as the *i* gene product was established first by the demonstration that this protein could not be detected by this technique in *i*⁻, *i*ˢ and *i*-deletion strains; more importantly, the affinity constant of this protein for IPTG was altered in a strain carrying an *i* mutation leading to an increased efficiency of IPTG as an inducer.

OLD THEORIES AND FUTURE EXPERIMENTS

I have discussed the genetic work on the *lac* operon and its implications in terms of the Jacob-Monod operon hypothesis. As mentioned earlier, several alternative models of control have been proposed. In addition to the Jacob-Monod model, in which the operator is not transcribed into the operon mRNA, the following possibilities have been suggested. The first possibility is a model in which the operator is transcribed into the operon mRNA, but is not translated. For example, (i) the repressor binds to the mRNA copy of the operator, inhibiting the initiation of protein synthesis (at the promoter?), which, in turn, is necessary for continuing synthesis and release of the mRNA from the DNA. In this model, repression during translation inhibits the transcription process.[33] (ii) The operon mRNA is synthesized in constant amounts, but the repressor inhibits initiation of protein synthesis and thus causes a very rapid destruction of the mRNA.[15,30] (iii) The operon mRNA is made in constant amounts, but the repressor is a ribonuclease which destroys the mRNA.[5,30]

The second possibility includes models in which the operator is transcribed into the operon mRNA and then must be translated in order to allow translation of the structural genes of the operon.[2-4] Again, the failure to translate may result either in the mRNA's sticking to the DNA or in rapid destruction of the mRNA.

All of these models take into account the experiments of Attardi *et al.*[27] which show that repression results in a disappearance of *lac* mRNA from the cell. This finding, like the similar finding with the strong polar mutants, must be considered in formulating a model for *lac* operon regulation, but it does not severely limit the number of possibilities.

The thesis of this article has been that the current knowledge of the *lac* operon suggests that the operator is not transcribed into the operon mRNA and that, therefore, the Jacob-Monod suggestion for operon control is most likely to be correct for this system. Some of the other models listed are difficult to reconcile with some of the evidence discussed. One of the most striking of the recent findings is that most and probably all O^c's are deletions but still retain some sensitivity to repressor. Although this finding makes any picture of the repressor-operator interaction somewhat difficult to visualize, it makes particularly unattractive models in which the operator is translated into protein or is the initiation site for operon protein synthesis. One would expect deletions of the operator to have drastic effects on operon functioning in such models, and this does not appear to be so.

Final proof of one model or another will probably have to come from biochemical experiments on operon functioning. It is possible to set up a system *in vitro* in which *lac* mRNA is made, with RNA polymerase and DNA preparations that contain a high proportion of *lac* genes.[48] In such experiments, other mRNA species are made also, but these can be eliminated by annealing with DNA preparations from appropriate strains. Then, the amount of *lac* mRNA made can be estimated. Using this system, Gilbert and Muller-Hill[49] are attempting to ascertain whether the *lac* repressor protein will inhibit synthesis of *lac* mRNA. In the same system, we are attempting to see whether the potential promoter mutants affect *lac* mRNA synthesis *in vitro*.

OTHER REGULATORY SYSTEMS

Of the regulatory systems that have been studied in detail, the *lac* operon appears to be one of the simplest. Although the operon model can account for all the information concerning *lac*, in some other systems the control is clearly more complex, and even the basic control mechanism may be entirely different.[50] However, in none of these cases is there yet any strong evidence against the operon model.

REFERENCES

1. F. Jacob and J. Monod, J. Mol. Biol. 3:318 (1961).
2. B. N. Ames and P. E. Hartman, Cold Spring Harbor Symp. Quant. Biol. 28:349 (1963).
3. G. Stent, Science 144:816 (1964).
4. W. Maas and E. McFall, Ann. Rev. Microbiol. 18:95 (1964).
5. S. Brenner, Brit. Med. Bull. 21:244 (1965).
6. Both Jacob [Science 152:1470 (1966)] and Monod [*ibid.* 154:475 (1966)] have in their Nobel laureate lectures discussed many historical and recent aspects of the *lac* operon and its regulation.
7. C. F. Fox and E. P. Kennedy, Proc. Nat. Acad. Sci. U.S. 54:891 (1965).
8. H. V. Rickenberg, G. N. Cohen, G. Buttin, J. Monod, Ann. Inst. Pasteur 91:829 (1956).
9. I. Zabin, A. Kepes, J. Monod, J. Biol. Chem. 237:253 (1962).
10. C. F. Fox, J. R. Beckwith, W. Epstein, E. R. Signer, J. Mol. Biol. 19:576 (1966).
11. F. Jacob and J. Monod, Biochem. Biophys. Res. Commun. 18:693 (1965).
12. J. R. Beckwith, E. R. Signer, W. Epstein, Cold Spring Harbor Symp. Quant. Biol. 31:393 (1967).
13. W. Gilbert, J. Mol. Biol. 6:374 (1963); A. Gierer, *ibid.*, p. 148; J. R. Warner, P. Knopf, A. Rich, Proc. Nat. Acad. Sci. U.S. 49:122 (1963); F. O. Wettstein, T. Staehelin, H. Noll, Nature 197:430 (1963); H. M. Dintzis, Proc. Nat. Acad. Sci. U.S. 48:247 (1961).
14. S. Benzer and S. P. Champe, Proc. Nat. Acad. Sci. U.S. 48:1114 (1962); A. Garen and O. Siddiqi, *ibid.*, p. 1121; A. S. Sarabhai, A. O. W. Stretton, S. Brenner, A. Bolle, Nature 201:13 (1964); S. Brenner, A. O. W. Stretton, S. Kaplan, *ibid.* 206:994 (1965); M. G. Weigert and A. Garen, *ibid.*, p. 992.
15. S. Brenner and J. R. Beckwith, J. Mol. Biol. 13:629 (1965).
16. F. H. C. Crick, L. Barnett, S. Brenner, R. J. Watts-Tobin, Nature 192:1227 (1961).
17. Y. Kiho and A. Rich, Proc. Nat. Acad. Sci. U.S. 54:1751 (1965).
18. R. E. Thach, M. A. Cecere, T. A. Sundarajan, P. Doty, *ibid.*, p. 1167; M. Salas, M. A. Smith, W. M. Stanley, A. J. Wahba, S. Ochoa, J. Biol. Chem. 240:3988 (1965); G. Streisinger, Y.

Okada, E. Terzaghi, J. Emrich, A. Tsugita, M. Inouye, Cold Spring Harbor Symp. Quant. Biol., in press; A. Goldstein, J. Kirschbaum, A. Roman, Proc. Nat. Acad. Sci. U.S. 54:1669 (1965); V. Maitra and J. Hurwitz, *ibid.*, p. 815; H. Bremer, M. W. Konrad, K. Gaines, G. S. Stent, J. Mol. Biol. 13:540 (1965); J. R. Guest and C. Yanofsky, Nature 210:799 (1966).

19. A. V. Fowler and I. Zabin, Science 154:1027 (1966).

20. Studies on complementation with z^- mutants suggest the possibility that the z gene is composed of three or more genes specifying distinct polypeptide chains [A. Ullman, D. Perrin, F. Jacob, J. Monod, J. Mol. Biol. 12:918 (1965)]. Although this question has not been conclusively resolved, most of the evidence indicates that the z gene directs the synthesis of only *one* polypeptide chain [J. L. Brown, S. Koorajian, J. Katze, I. Zabin, J. Biol. Chem. 241:2826 (1966); G. Craven, personal communication; D. Fan, personal communication].

21. C. Yanofsky, B. C. Carlton, J. R. Guest, D. R. Helinski, U. Henning, Proc. Nat. Acad. Sci. U.S. 51:266 (1964).

22. W. A. Newton, J. R. Beckwith, D. Zipser, S. Brenner, J. Mol. Biol. 14:290 (1965).

23. F. Jacob and J. Monod, Cold Spring Harbor Symp. Quant. Biol. 26:193 (1961).

24. N. C. Franklin and S. E. Luria, Virology 15:299 (1961).

25. W. A. Newton, Cold Spring Harbor Symp. Quant. Biol. 31:181 (1967).

26. D. Zipser and W. A. Newton, J. Mol. Biol., in press.

27. G. S. Attardi, S. Naono, J. Rouviere, F. Jacob, F. Gros, Cold Spring Harbor Symp. Quant. Biol. 28:363 (1963); G. Contesse, S. Naono, F. Gros, Compt. Rend. 263:1007 (1966).

28. M. Malamy, Cold Spring Harbor Symp. Quant. Biol. 31:89 (1967).

29. R. G. Martin, D. F. Silbert, D. W. E. Smith, H. J. Whitfield, J. Mol. Biol. 21:357 (1966).

30. J. R. Beckwith, Abhandl. Deutsch. Akad. Wiss. Berlin Kl. Med. 4:119 (1964).

31. M. R. Capecchi, J. Mol. Biol. 21:173 (1966); G. N. Gussin, *ibid.*, p. 435.

32. C. Yanofsky and J. Ito, *ibid.*, p. 313.

33. D. H. Alpers and G. M. Tomkins, Proc. Nat. Acad. Sci. U.S. 53:797 (1965).

34. A. Kepes, in preparation.

35. D. H. Alpers and G. M. Tomkins, J. Biol. Chem. 241:4434 (1966).

36. I. Zabin, Cold Spring Harbor Symp. Quant. Biol. 28:431 (1963); W. Epstein, personal communication; D. Perrin, personal communication.

37. F. Jacob, A. Ullman, J. Monod, Compt. Rend. 258:3125 (1964).

38. S. Bourgeois, M. Cohn, L. E. Orgel, J. Mol. Biol. 14:300 (1965); S. Bourgeois, personal communication.

39. T. Ramakrishnan and E. A. Adelberg, J. Bacteriol. 87:566 (1964).

40. J. R. Beckwith, J. Mol. Biol. 8:427 (1964).

41. E. Steers, G. R. Craven, C. B. Anfinsen, Proc. Nat. Acad. Sci. U.S. 54:1174 (1965).

42. F. Jacob, A. Ullman, J. Monod, J. Mol. Biol. 13:704 (1965).

43. J. Scaife and J. R. Beckwith, Cold Spring Harbor Symp. Quant. Biol. 31:403 (1967).

44. A. B. Pardee, F. Jacob, J. Monod, J. Mol. Biol. 1:165 (1959).

45. C. Willson, D. Perrin, M. Cohn, F. Jacob, J. Monod, *ibid.*, 8:582 (1964).

46. B. Muller-Hill, *ibid.* 15:374 (1966).

47. W. Gilbert and B. Muller-Hill, Proc. Nat. Acad. Sci. U.S. 56:1891 (1966).

48. High-frequency transducing lysates for the *lac* genes have been isolated with the restricted transducing phage φ80 [J. R. Beckwith and E. R. Signer, J. Mol. Biol. 19:254 (1966)]. Isolation of DNA from phage lysates results in a DNA preparation which contains genetic material, the *lac* region, and a small portion of the chromosome surrounding *lac*.

49. W. Gilbert and B. Muller-Hill, personal communication.

50. E. Engelsberg, J. Irr, J. Power, N. Lee, J. Bacteriol. 90:946 (1965) (arabinose metabolism); H. Echols, A. Garen, S. Garen, A. Torriani, J. Mol. Biol. 3:425 (1961); A. Garen and N. Otsuji, *ibid.* 8:841 (1964) (alkaline phosphatase); M. Schwartz, personal communication (maltose metabolism).

51. E. R. Signer, unpublished data.

Regulation of histidine biosynthesis in *Salmonella typhimurium*

ROBERT F. GOLDBERGER and JOHN S. KOVACH*

Reprinted by permission of the authors and Academic Press Inc. from Current
Topics in Cellular Regulation, *vol. 5, 1972, pp. 285–308.*

I. INTRODUCTION

Regulation of histidine biosynthesis in *Salmonella
typhimurium* has been the subject of intensive investigation in a number of laboratories for more
than a decade. These studies have brought to light
some of the basic phenomena of molecular biology.
Recently Brenner and Ames[23] presented an excellent and comprehensive review on histidine biosynthesis. In the present review we give a description
of the system to provide background for a more
lengthy treatment of the subject of repression,
stressing work done since the review of Brenner
and Ames.[23]

The pathway for histidine biosynthesis in *Salmonella typhimurium* consists of a series of ten reactions, each catalyzed by a specific enzyme.[3,7,67,89,90]
The rate at which these enzymes convert ATP and 5-
phosphoribosyl 1-pyrophosphate (PRPP) into the
amino acid, histidine, is regulated by feedback inhibition: the activity of the enzyme which catalyzes
the first step of the pathway is inhibited by the end
product, histidine.[7,57,66,85] The structural genes for
the ten enzymes of histidine biosynthesis, together
with an operator gene, are clustered on the *Salmonella* chromosome in the region known as the histidine operon.[39,41] The rate at which the information
encoded in the operon is utilized for synthesis of the
enzymes is under repression control.[2–6] As originally shown by Ames and Garry,[2] the operon functions
as a unit in response to the level of histidine available to the organism. Thus, under conditions in
which ample histidine is available, as in the wild-type
organism, the operon is utilized at a relatively low
rate, maintaining a basal intracellular level of the
histidine enzymes; when the availability of histidine

limits the growth rate of the organism, the operon is
utilized at a higher rate, causing an increase in the
intracellular concentrations of the enzymes, up to a
level approximately 20-fold higher than the basal
level.

II. THE ENZYMES OF HISTIDINE BIOSYNTHESIS

The structures of the intermediates in the pathway
for histidine biosynthesis are shown in Fig. 1. This
pathway, elucidated largely through the efforts of
B. N. Ames and his colleagues,[3,7,67,90,91] is required
only for the biosynthesis of histidine; it is not essential for the production of any other metabolite.
There is, however, one byproduct of the pathway:
phosphoribosyl aminoimidazole carboxamide, which
is also an intermediate in the *de novo* pathway for
purine biosynthesis.[62] Although there are ten steps
in the histidine pathway, two of them (the seventh
and ninth steps) are catalyzed by a single complex
enzyme, the structure of which is specified by a
single gene of the operon, the *hisB* gene.[51,94] The
molecular weights of all the enzymes have been
estimated from sucrose density-gradient centrifugation.[101]

By using these molecular weights for the enzymes
(discounting identical subunits), and applying the
coding ratio (nucleotides per amino acid) of 3, the
length of histidine operon can be calculated to be
approximately 10,000 nucleotides, a figure which
agrees quite closely with the length of the operon
calculated on the basis of other data.[13,16,23,56] Isolation of four of the enzymes into homogeneous form
has been reported. They are the enzymes specified
by the *hisG* gene,[15,57,97,99] *hisA* gene,[55] *hisC* gene,[59]
and *hisD* gene,[52,53,104,108] which catalyze the first,
fourth, eighth, and tenth steps of the pathway,

*Laboratory of Chemical Biology, National Institute of
Arthritis and Metabolic Diseases, National Institutes of
Health, Bethesda, Maryland.

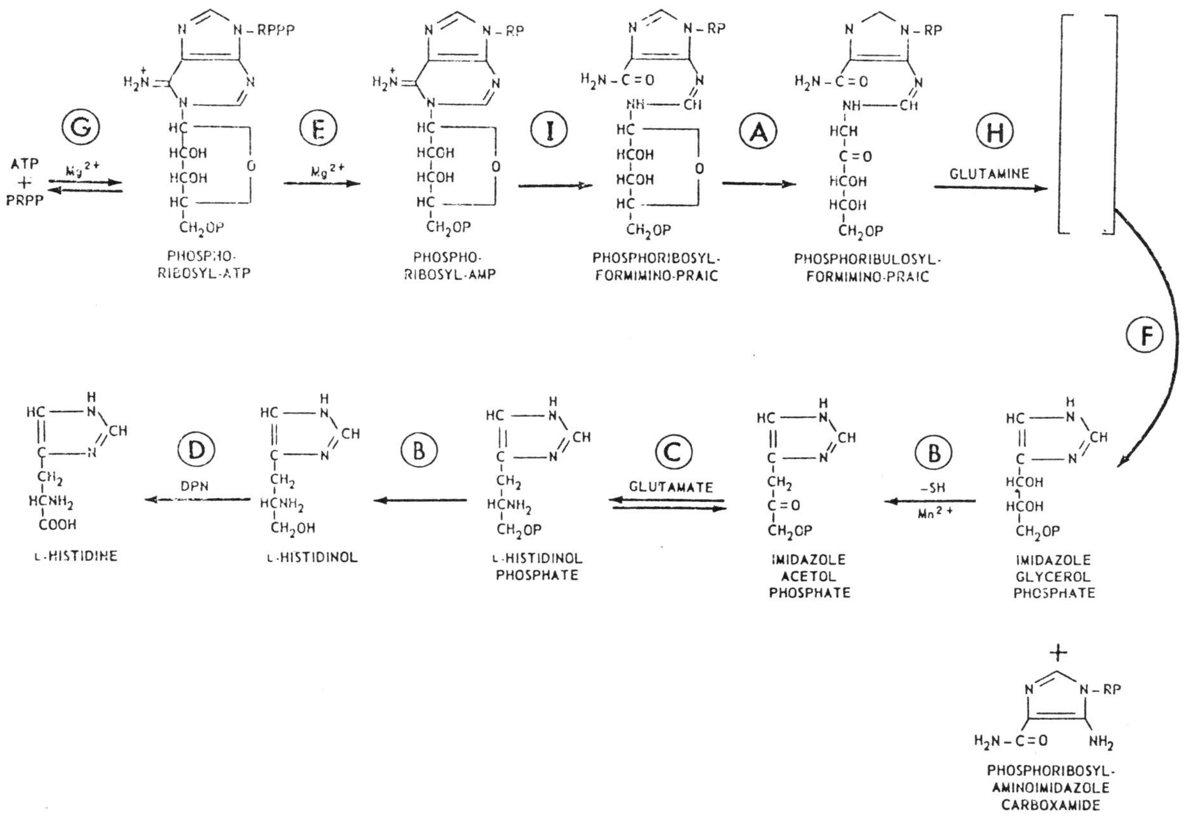

Fig. 1. Pathway for histidine biosynthesis in *S. typhimurium*. The capital letters within circles represent genes of the histidine operon which specify structures of enzymes catalyzing these reactions. Brackets indicate an intermediate of unknown structure.

respectively. Recently, a protein, comprising most of the D and C enzymes covalently bound together, has been isolated[71,109] from a mutant which is discussed more fully below (see Section III, C). This fused protein displays the catalytic activities characteristic of both enzymes. Details of the purification and assay of the enzymes for histidine biosynthesis and isolation of the intermediates of the pathway are given in a recent review.[58]

III. THE OPERON

A. General

Isolation of several thousand mutants requiring histidine has allowed P. E. Hartman and his colleagues to construct a detailed map of the histidine operon, the outlines of which are shown in Fig. 2.[40,54] The order of the nine structural genes in the operon does not correspond with the sequence in which the enzymes function in the metabolic pathway.[5] The single exception is the first gene, the *hisG* gene, which does specify the structure of the enzyme catalyzing the first step of the pathway. Because a correlation between first gene and first enzyme is also found in other systems for amino acid biosynthesis,[82] it is tempting to speculate that it reflects an important feature of the control of biosynthetic pathways (see Section V, D, 3).

The histidine operon is transcribed into a single molecule of messenger RNA. Evidence for the existence of this polycistronic message was first presented by Martin[56] and was later confirmed by Vene-

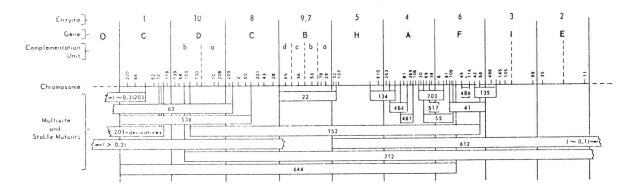

Fig. 2. Map of the histidine operon of *Salmonella typhimurium*. The numbers at the top of the diagram indicate the steps in the pathway for histidine biosynthesis catalyzed by the various enzymes. The structural genes for these enzymes and the operator gene, designated by capital letters, are shown in the second line of the diagram. Below this are shown, in descending order, complementation units (designated by small letters), a few of the point mutants (designated by arabic numerals), and a few of the multisite (deletion) mutants (designated by arabic numerals enclosed within boxes, the lengths of which indicate the sizes of the deletions). For a detailed map of the histidine operon, see Hartman *et al.*[40]

tianer *et al.*,[96] who demonstrated that a species of RNA with a length corresponding to that of the histidine operon is present in an organism containing the histidine operon but absent from an organism missing the histidine operon.

The polar effect of nonsense mutations of the histidine operon provides additional evidence that the histidine message is polycistronic.[5] A nonsense mutant lacks the enzyme activity corresponding to the gene in which the mutation occurs because the mutation causes premature polypeptide chain termination during translation of the message.[5,36,43,83] In addition, a nonsense mutation results in a diminished frequency of translation of that part of the message distal to the site of mutation (considering the operator-promoter region to be the most proximal portion of the operon). The mechanism responsible for the latter effect is not known; it may involve premature degradation of that portion of the message which is distal to the site of mutation.[65]

B. Intercistronic Region

Studies of frameshift mutants of the histidine operon have recently led to several interesting findings. Rechler and Martin,[74] Rechler *et al.*,[72,73] and Bruni *et al.*[25] studied a frameshift mutant in which the mutation was located very close to the end of the

hisD gene (the second structural gene of the operon). By examining the amino acid sequence of the carboxyl terminal portion of the *D* enzyme in the wild-type organism and in the mutant, they were able to show that the shift in the reading frame allowed translation to continue through the normal termination signal at the end of the *hisD* gene. Their data allowed them to deduce that there is an intercistronic region between the *hisD* and *hisC* genes, at least one nucleotide in length. Thus, intercistronic regions, previously shown by Steitz[93] to exist in RNA phage, also exist in bacterial cells.

C. Fused Enzymes

Recently, Yourno *et al.*[108] and Rechler and Bruni[71] have reported the isolation of a single protein with the catalytic activities normally associated with two distinct enzymes, those specified by the *hisD* and *hisC* genes. This covalently fused protein was found in a revertant of a frameshift mutant in which the mutation is located near the end of the *hisD* gene. Although the mechanism of this reversion has not been defined, it is presumably either due to a multisite mutation excising the original frameshift mutation and connecting the *hisD* and *hisC* genes, or due to a second mutation located even closer to the end of the *hisD* gene than is the first mutation, presumably in the intercistronic region. Thus, both cistrons are translated correctly, except for a small portion at the end of the first cistron. The product synthesized by translation of the two cistrons is a single polypeptide chain comprised of the two polypeptides specified by the two cistrons, joined together. In the region of fusion, the carboxyl terminal portion of the first polypeptide is replaced by a short "bridge" of abnormal amino acid sequence resulting from translation through the termination signal. Perhaps the most remarkable feature of the findings with this mutant is that the fused polypeptide chain is able to fold in such a way as to arrive at a three-dimensional conformation in which one part resembles the native structure of one enzyme and another part resembles the native structure of the other closely enough to allow the fused protein to display the catalytic activities of both enzymes.[71,108]

D. Frameshift Suppression

Another recent finding of interest involves phenotypic reversion of frameshift mutants by an unlinked suppressor. Yourno et al.[105] have shown that this suppressor does not suppress nonsense or missense mutations; it appears to be a suppressor of a new class.[105] By analysis of the amino acid sequences of the products of the hisD gene in a frameshift mutant and in the same mutant containing the suppressor mutation in addition, Yourno and coworkers[106,107,109] have shown that the unlinked suppressor mutation corrects the reading frame in a +1 frameshift mutant. Yourno and Tanemura[109] have suggested that this suppression may be due to a transfer RNA which has been altered in such a way that it binds four bases[104] in the messenger RNA, taking up the slack in the message and allowing subsequent translation to proceed in the correct frame. Riddle and Roth[75] have recently shown that most frameshift mutations in the histidine operon which appear to be of the +1 type are externally suppressible. They have presented evidence that suppressed strains of this type contain any of several alterations in transfer RNA.[76]

E. Operator and Promoter

The operator gene of the histidine operon, first identified by Roth et al.,[76] is located at one end of the operon, as shown in Fig. 2. Organisms with mutations in this region have elevated intracellular concentrations of the enzymes for histidine biosynthesis even when grown in the presence of excess histidine — that is, they are constitutively derepressed.[78] Recently, Fankhauser et al.[33] have mapped the operator in greater detail, and have estimated that its size is between 75 and a few hundred base pairs. It is noteworthy that some of the operator mutants are point mutants, a finding which indicates that the change of only a single base can drastically affect the properties of the operator region.[33]

Atkins and Loper[11] and Fankhauser et al.[33] have isolated mutants with characteristics which have been attributed to mutations in the promoter region — the region to which RNA polymerase binds.[32] These mutations, like those in the operator, are *cis* dominant; but unlike operator mutations, they result in very low intracellular concentrations of the enzymes for histidine biosynthesis, both in the presence and absence of exogenous histidine. An unexpected feature of the data of Fankhauser et al.[33] is that the promoter maps within the operator region. This brings up the interesting possibility that a repressor, bound to the operator region of the histidine operon, may not only prevent polymerase from transcribing the operon, but may also prevent polymerase from binding to the promoter. Atkins and Loper[11] have reported the presence of two other sites within the histidine operon of the wild-type organism where transcription may be initiated. The efficiency of these "promoters" (located in, or at the terminal region of, the hisC and hisF genes), is so low that their physiological significance is uncertain.

F. Specialized Transducing Phages

Recently, Voll[98] has succeeded in introducing the histidine operon of *Salmonella typhimurium* into bacteriophage $\phi 80i^{\lambda C}$ 1857. The *Escherichia coli* lysogen presumably contains a $\phi 80$-λ hybrid in tandem with a defective $\phi 80$ carrying the histidine operon. Voll[98] has obtained lysates in which one in 10^3 phage particles transduces the histidine operon. Bruni[24a] and Kasai[43a] have now succeeded in purifying this defective *his* transducing phage by banding in cesium chloride. Specialized transducing phages for the histidine operon of *E. coli* have recently been isolated by Blasi[19] and by Wolf.[102a] A preparation of DNA highly enriched for the histidine operon provides an important new way to study regulation of the histidine system.

IV. FEEDBACK INHIBITION

Like many biosynthetic pathways in bacterial cells, the histidine pathway is subject to feedback inhibition.[7,57,66] This regulatory mechanism provides the cell with an efficient means of preventing overproduction of histidine, thereby conserving substrates and other energy sources. The interaction between the first enzyme of the pathway with the end product of the pathway, histidine, is strong enough [K_i approximately 7×10^{-5} M][99] to permit feedback inhibition to play a role in regulating histidine biosynthesis in the wild-type organism, since the internal concentration of histidine in the wild-type organism is approximately 1.5×10^{-5} M.[8] When an exogenous supply of histidine is available, the pathway stops functioning. This occurs even at very low concentrations of exogenous histidine because of the highly efficient permease system. Ferro-Luzzi Ames[8] has shown that there are several systems by which histidine is transported into the cell. One of these permeases is shared with the aromatic amino acids, whereas another is specific for histidine. The K_m of the histidine-specific permease is 8×10^{-8} M.[8,9] Thus, the organism utilizes exogenous histidine when the concentration in the medium is higher than approximately 10^{-7} M.

Martin[57] showed that histidine interacts with the first enzyme at a site distinct from the sites for binding the substrates, ATP and PRPP. This finding was expected on the basis of the lack of structural homology between the substrates and histidine. In keeping with the idea that inhibition by histidine is due to an allosteric transition in the enzyme, kinetic analysis of the enzyme reaction by Whitfield[99] showed that the mechanism of the inhibition is "uncompetitive" with respect to the substrate, PRPP. Modified Hill plots[12,63] of the inhibition data gave slopes which indicate cooperativity in the binding of histidine.[99] Blasi et al.[20] showed that, as expected for an allosteric interaction, binding of histidine was accompanied by a conformational change in the enzyme, and they observed that this effect of histidine also showed cooperativity.

Sheppard[86] isolated several missense mutants in which a defect in the first enzyme rendered it less sensitive to inhibition by histidine than is the wild-type enzyme. These mutations, which were mapped in two separate regions of the hisG gene,[40,86] cause the organism to excrete histidine. This finding provides evidence that feedback inhibition normally regulates the rate of histidine production in the wild-type organism. The first enzyme isolated from feedback resistant mutants has an altered K_i for histidine.[99] Although the mutant enzyme binds histidine,[20,99] it fails to display the inhibition characteristic of the wild-type enzyme apparently because the alteration in its structure prevents the phenomenon of histidine binding from being translated into conformational change.[20]

V. REPRESSION

A. General

Ames and Gary[2] showed that when growth rate is limited by the availability of histidine, the intracellular concentrations of the enzymes for histidine biosynthesis increase (derepress) in a coordinate manner. When an adequate supply of histidine is made available to the derepressed organism, the intracellular concentrations of the enzymes fall (repress). This responsiveness of an operon to the availability of the end product of a metabolic pathway is characteristic of repression control. The changes in enzyme concentration which accompany repression and derepression appear to be due to changes in the rate at which the operon is transcribed into messenger RNA (mRNA). Venetianer[95] has shown that the concentration of mRNA transcribed from the histidine operon rises as the growth rate is diminished by increasingly severe histidine limitation. The polycistronic character of this message[56,96] is responsible for the coordinated enzyme levels; when the rate of histidine message synthesis is increased (or decreased) by a given factor, the intracellular concentrations of all the enzymes specified by the message increase (or decrease) to the same degree.[2]

B. Kinetics of Repression and Derepression

Kinetic studies of the changes in enzyme levels which occur when the organism is grown on histidine at a limiting concentration have been helpful in understanding the mechanism by which translation of polycistronic messages is initiated.[17],[18],[38] Berberich et al.[17] showed that derepression may proceed by two different modes, depending upon the availability of formyl groups. Figure 3, Panel A, shows that when the formylating capacity of the cell is low, the enzymes become derepressed in a temporal sequence which corresponds with the positional sequence of genes in the histidine operon. This sequential derepression is evidently due to the fact that translation of the polycistronic message is initiated at the operator end by ribosomes which translate the message sequentially from one end to the other. Judging from the time interval between derepression of the enzymes corresponding to the first and last genes of the operon, it appears to take about 15 minutes for ribosomes to traverse the entire message, a translation rate which is in close agreement with translation rates calculated on the basis of other data.[35],[64] When the formylating capacity of the cell is adequate — that is, when the one carbon pool of the cell has been augmented in some way, such as by adding formyl donors to the culture medium — then the kinetic pattern of derepression is quite different, as shown in Fig. 3, Panel B. In this case all the enzymes begin to derepress at essentially the same time. Berberich et al.[17] have concluded that the simultaneous derepression of the histidine enzymes reflects a change in the mechanism by which translation of the polycistronic histidine message is initiated. With an adequate capacity to formylate methionyl-tRNA for polypeptide chain initiation, translation of the message may be initiated at several sites. Studies on the arabinose operon of E. coli have shown that the kinetic pattern of induction of the enzymes which degrade arabinose also changes from sequential to simultaneous when the one carbon pool of the cell has been augmented.[48a] Similar observations have been made on the enzymes of the lactose operon.[14]

Studies on the kinetics of repression of the histidine operon have been helpful in understanding the mechanism by which translation of polycistronic messages ceases. When histidine is added to a derepressed culture, the intracellular concentrations of the enzymes begin to fall in a temporal sequence which corresponds with the positional sequence of genes in the histidine operon,[44] as shown in Fig. 4. Sequential patterns have also been found in repres-

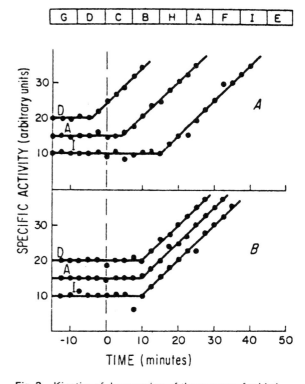

Fig. 3. Kinetics of derepression of the enzymes for histidine biosynthesis. At the top is shown the order of the structural genes of the histidine operon. Below this are plotted the specific activities of the enzymes specified by the *hisD*, *hisA*, and *hisI* genes during growth of the organism in excess histidine (when the operon is repressed and the enzymes are synthesized at a basal rate) and during growth of the organism under conditions in which the rate of growth is limited by the availability of histidine (when the operon becomes derepressed and the enzymes are synthesized at an increased rate). The vertical dashed line denotes the time of change in growth rate. Panel A shows the results of an experiment with an organism in which there is a reduced formylating capacity.[17] Although the results for only three enzymes are shown, all seven of those examined (the enzymes specified by the *hisG*, *hisD*, *hisC*, *hisB*, *hisA*, *hisF*, and *hisI* genes) followed the same pattern, in which the temporal sequence of derepression corresponded with the positional sequence of genes in the operon. Panel B shows the results of an experiment with an organism in which the formylating capacity has been augmented by addition of one carbon donors to the medium. Although the results for only three of the enzymes are shown, all seven of those examined followed the same pattern, in which all the enzymes became derepressed at essentially the same time.[17]

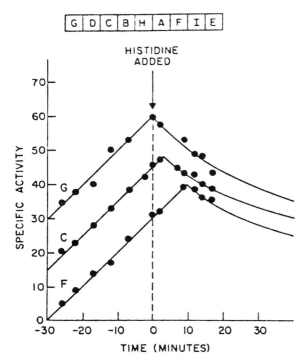

Fig. 4. Kinetics of repression of the enzymes for histidine biosynthesis. At the top is shown the order of the structural genes of the histidine operon. Below are plotted the specific activities of the enzymes specified by the *hisG*, *hisC*, and *hisF* genes during growth of the organism under conditions in which the rate of growth is limited by the availability of histidine (when the operon is derepressed and the intracellular concentrations of the enzymes are increasing) and following the addition of excess histidine to the medium (when the operon becomes repressed and the concentrations of the enzymes begin to fall). Although the results for only three enzymes are shown, all seven of those examined (the enzymes specified by the *hisC*, *hisD*, *hisC*, *hisB*, *hisA*, *hisF*, and *hisI* genes) followed the same pattern, in which the temporal sequence of repression corresponded with the positional sequence of the genes in the operon.[44] *(Reprinted by permission of the American Society for Microbiology, Washington, D. C.)*

sion[64] and deinduction[1] of other polycistronic operons. The sequential pattern of repression of the histidine operon is independent of the formylating capacity of the cell. It results from cessation of translation of preexisting and nascent mRNA at the operator end, with continuing translation by those ribosomes already engaged in translating the message.[44]

C. Role of Transfer RNA in Repression

Although much of the early work on repression of the histidine system was understood in terms of the availability of histidine, it is now known that histidine exerts its effect in the form of aminoacylated tRNA[His]. Schlesinger and Magasanik[84] showed that the histidine analog, α-methylhistidine, causes derepression of the histidine operon in the wild-type organism. This analog inhibits aminoacylation of tRNA[His] by competing with histidine for the synthetase. Thus, it causes derepression even though the intracellular concentration of histidine is normal, indicating that the effector in the repression process is not histidine itself, but aminoacylated tRNA[His].

More direct evidence for the participation of tRNA in the repression process was provided by the isolation of mutants defective in the synthesis or aminoacylation of tRNA[His].[10,77-80,87] The mutations in such strains have been mapped in five loci, none of which is linked to the histidine operon. They cause derepression of the histidine operon and, in the cases tested (*hisR*, *hisW*, and *hisT*) are recessive to the corresponding wild-type alleles.[34]

The *hisR* gene is the structural gene for tRNA[His]. Each *hisR* mutant contains about half as much tRNA[His] as there is in the wild-type organism.[24,50,87] Furthermore, when an episome bearing a normal *hisR* gene is introduced into a *hisR* mutant the concentration of tRNA[His] rises to one and a half times as much as that in the wild-type organism.[24] An explanation which has been offered for these phenomena is that there are two chromosomal copies (tandem duplication) of the *hisR* gene in the wild-type organism; mutation of one copy halves the production of tRNA[His], whereas mutation of both is lethal.[24] The fact that a structural gene for tRNA[His] has not been found at any site other than the *hisR* locus is consistent with the finding that there is only one species of tRNA[His] in *Salmonella*.[24,87] Recently, Singer and Smith[88] have purified and determined the nucleotide sequence of this tRNA. The structure is shown in Fig. 5. It should be noted that there are eight base pairs in the CCA end of the molecule and that there are two residues of pseudouridylic acid near the anticodon.

The *hisS* gene is the structural gene for histidyl tRNA synthetase.[30,77] Various mutations in *hisS* have been correlated with enzyme alterations of several types.[30] Roth and Ames[77] originally showed that certain *hisS* mutations cause an alteration in

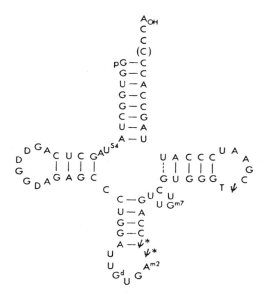

Fig. 5. Nucleotide sequence of the histidyl species of transfer ribonucleic acid of *Salmonella typhimurium*, strain LT-2, determined by Singer *et al.*[89] A: adenylic acid; A^{m2}: 2-methyladenylic acid; C: cytidylic acid; D: 5,6-dihydrouridylic acid; G: guanylic acid; G^d: a deriative of guanylic acid; G^{m7}: 7-methylguanylic acid; T: 5-methyluridylic acid; U: uridylic acid; U^{S4}: 4-thiouridylic acid; and ψ: pseudouridylic acid. The presence of the cytidylic acid residue shown in parentheses (C) is uncertain. The two starred pseudouridylic acid residues (ψ^*) are absent from the histidyl species of transfer ribonucleic acid of the mutant, *hisT1504*; in their places are two uridylic acid residues.[89]

the K_m of the enzyme for histidine. Because of this defect the amount of tRNA^{His} in the aminoacylated form is lower than that in the wild-type organism. Thus, such mutants become derepressed despite a high endogenous concentration of histidine. However, if the endogenous concentration of histidine is raised enough by the addition of large amounts of histidine to the growth medium, the histidine operon in some of these mutants does become repressed. Under such conditions, the synthetase, despite its defect, can aminoacylate a sufficient amount of tRNA^{His} to repress the histidine operon. DeLorenzo and Ames[29] and DeLorenzo *et al.*[30] have purified the synthetase from the wild-type organism and from certain mutants, and have shown that in addition to those mutations which cause an alteration in the K_m for histidine, certain mutations cause an al-

teration in the K_m for ATP or for tRNA^{His}. As expected, mutants of the latter two types cannot be repressed by raising the intracellular concentration of histidine.[30] Williams and Neidhardt[102] have reported that in *E. coli,* under conditions in which the histidine operon becomes derepressed, the rate of synthesis of histidyl tRNA synthetase increases. However, under such conditions an increase in the intracellular concentration of the synthetase does not occur because of an increased rate of destruction of the enzyme.[102] These findings, which have been shown to apply to other aminoacyl tRNA synthetases as well,[102] must have some significance for regulation of these systems, although the significance remains to be defined.

In many cases, the primary product of transcription of a gene for transfer RNA is not capable of fulfilling its physiological function without further modification. A large number of such modifications has been identified.[92] Presumably, each modification is catalyzed by a specific enzyme. The *hisT* gene appears to be the structural gene for an enzyme of this type, the function of which is to modify tRNA^{His} and perhaps other species of tRNA as well.[26,28]

Mutants of the *hisT* gene were isolated as organisms constitutive for the histidine operon, but without any demonstrable defect in the synthesis or aminoacylation of tRNA^{His}.[50,78] Chang *et al.*[27] isolated temperature-sensitive and nonsense mutants of the *hisT* gene, indicating that the gene product is a protein that is not essential for survival. Brenner and Ames[24] showed that the tRNA^{His} in *hisT* mutants behaves differently on chromatography from that in the wild-type organism. These findings led Ames and his colleagues to suggest that the *hisT* gene product is an enzyme which modifies tRNA so that it can function in repression.[23,26] Recently, Singer *et al.*[89] have proved that the structure of tRNA^{His} in *hisT* mutants is different from that in the wild-type organism. They demonstrated that the tRNA^{His} in the wild-type organism contains two pseudouridylic acid residues near the anticodon, whereas the tRNA^{His} in a *hisT* mutant contains, instead, two uridylic acid residues, as shown in Fig. 5. Thus, the enzyme which is altered in *hisT* mutants is evidently required for conversion of certain uridylic acid residues of tRNA to pseudouridylic acid. The defective tRNA^{His} produced in *hisT* mutants is present in normal amounts, is aminoacylated normally, and supports normal growth of the organism.[50] Thus, the only gross functional abnormality of this defective tRNA^{His} is its inability to act in repression.

Like the *hisT* gene, the *hisU* and *hisW* genes appear to be structural genes for enzymes required for maturation of tRNA^{His}.[10,22,50] Lewis and Ames[50] have shown that *hisU* and *hisW* mutants contain less tRNA which can be aminoacylated by histidine than does the wild-type organism. However, at least in the case of *hisU*, the structure of tRNA^{His} is identical to that of the wild-type organism.[89] Lewis and Ames[50] have suggested that owing to the defective enzymes in these mutants, only a portion of the tRNA^{His} reaches the form in which it can be aminoacylated by histidine. Although this amount of aminoacylated tRNA^{His} is sufficient to support almost normal growth rates, it is not sufficient to repress the histidine operon.[50] An interesting feature of *hisU* and *hisW* mutants is that the amount of certain other species of tRNA which can be aminoacylated is also reduced, indicating that the enzymes specified by the *hisU* and *hisW* genes are involved in modification of several species of tRNA.[24,50]

In summary, isolation and characterization of mutants defective in the synthesis or aminoacylation of tRNA^{His} have provided ample evidence that repression of the histidine operon requires the participation of aminoacylated tRNA^{His}, not histidine. Furthermore, the repression mechanism is sensitive only to the intracellular concentration of aminoacylated tRNA^{His}, not to the concentration of total tRNA^{His} or to the proportion of tRNA^{His} which is aminoacylated.[50] Finally, the intracellular concentration of aminoacylated tRNA^{His} required to maintain the normal growth rate of the organism is less than that required to maintain repression of the histidine operon, indicating that the repression mechanism is more senstive to the concentration of aminoacylated tRNA^{His} than is the protein-synthesizing system.

D. Mechanism of Repression

1. General In 1961, Jacob and Monod[42] proposed a model for induction and repression of enzyme synthesis in bacterial cells. According to this model, in its more recent form, an inducer is a small molecule, usually related to the substrate of a degradative pathway, which combines with the repressor protein, rendering it unable to bind to the operator gene

of an inducible operon; a corepressor is a small molecule, usually related to the end product of a biosynthetic pathway, which combines with the aporepressor protein, endowing it with the capacity to bind to the operator gene of a repressible operon. Isolation of the repressors of several inducible systems,[37,68,70] together with other studies on the mechanism of induction, has demonstrated that the model of Jacob and Monod[42] is correct for inducible systems. However, the idea that every repressible system is regulated by a repressor which is composed of an aporepressor (protein) and a corepressor (small molecule) has yet to be substantiated. In the histidine system, for example, although the role of aminoacylated tRNA^{His} as "corepressor" has been well documented, no aporepressor has been identified.

An aporepressor is ordinarily identified by mutations in its structural gene which result in loss of repression control. Although many organisms with such mutations have been isolated, they have all been found to have defects in the operator gene or in one of the genes controlling the synthesis of the "corepressor," aminoacylated tRNA^{His}.[10,77-80,87] Indeed, there is no evidence which rules out the possibility that aminoacylated tRNA^{His} itself is the repressor of the histidine operon. However, this possibility has not been accepted as an explanation for the mechanism of repression of the histidine operon largely because an aporepressor gene is thought to exist in other systems for amino acid biosynthesis and because of the widely held view that the regulatory function required of a repressor necessitates the participation of a large and complex molecule capable of undergoing reversible modifications in structure — a molecule with characteristics ordinarily possessed only by proteins. Thus, efforts to identify an aporepressor for the histidine operon continue.

If there is an aporepressor for the histidine operon, one must ask why no such molecule has yet been identified. One possible explanation is that there are two (duplicate) genes for this protein. Mutation in both genes would be an exceedingly rare combination of events, likely to elude detection. Another possibility is that the method of selection could not reveal aporepressor mutants. All constitutive mutants of the histidine operon reported in the literature have been isolated with the use of the histidine analogue, 1,2,4-triazole-3-alanine.[23,49,50,78,85,87] This method selects mutants constitutive for the histidine operon, but selects against organisms unable to grow in minimal medium. If the aporepressor of the histidine operon has a second function — a function required for growth on minimal medium — no aporepressor mutant could be detected by the standard triazolalanine method.[78] This consideration leads us to suggest that if there is an aporepressor for the histidine operon it has two distinct functions, one involved in repression control, the other required for growth on minimal medium. The histidyl tRNA synthetase and one of the enzymes for histidine biosynthesis are obvious candidates.

2. Possibility of Synthetase as Aporepressor Brenner and Ames[24] and DeLorenzo and Ames[29] have proposed the idea that the repressor of the histidine operon may be the complex formed between histidyl tRNA synthetase and aminoacylated tRNA^{His}. In support of this possibility, they cite the fact that from the high affinity of the synthetase for aminoacylated tRNA^{His} and from the concentrations of these two cellular components, one can calculate that a large portion of the synthetase is complexed with aminoacylated tRNA^{His} in the cell.[29] This calculation suggests that the synthetase, if it is not itself the aporepressor, must at least compete with the aporepressor for the corepressor, aminoacylated tRNA^{His}. Mutations in the gene for the synthetase (*hisS* gene) result in the phenotype expected for aporepressor mutants: derepression of the histidine operon.[77] However, the derepression seen in all *hisS* mutants can be explained entirely on the basis of defective synthesis of the corepressor — that is, the synthetase of all *hisS* mutants has been found to be impaired in its ability to catalyze aminoacylation of tRNA^{His}.[29,30] The only way to identify the synthetase as the aporepressor by genetic means alone would be to isolate a *hisS* mutant which is either derepressed or hyperrepressed for the histidine operon but in which the intracellular concentration of aminoacylated tRNA^{His} is normal. No such mutant has yet been isolated.

3. Possibility of First Enzyme as Aporepressor Over the past several years it has become clear that the first enzyme for histidine biosynthesis plays a role in regulation of the histidine operon.[22,44–46] However, it is not certain that the first enzyme functions as an aporepressor. The first indication that this enzyme plays a role in regulation of the histidine operon was obtained by Kovach *et al.,*[44] who studied the kinetic pattern of repression in various histidine auxotrophs (see Section V, B). They observed that the normal sequential pattern of repression was altered when the organism was grown in the presence of thiazolalanine. Thiazolalanine is an analogue of histidine which inhibits the first enzyme but does not aminoacylate tRNA^{His}.[7,57,66] Therefore, Kovach *et al.*[44] concluded that the first enzyme has some effect on the repression process, and that the feedback-sensitive site of the enzyme is involved in this effect. Indeed, they found that the pattern of repression was also altered in feedback-resistant mutants — that is, mutants in which the first enzyme is catalytically active but has a damaged feedback site.

More striking evidence that the first enzyme plays a role in regulation of the histidine operon was the finding that the histidine analogue, triazolalanine, cannot cause repression in feedback-resistant mutants.[46] Triazolalanine has been known for some time to mimic the effect of histidine in repressing the histidine operon.[49] This effect is based on the ability of triazolalanine to aminoacylate tRNA^{His},[50,87] forming a corepressor similar to that formed by histidine itself (see Section V, C). However, the different effects of triazolalanine and histidine on repression of the histidine operon in feedback-resistant mutants demonstrates that triazolalanyl tRNA^{His} is a less effective corepressor than is histidyl tRNA^{His}. Apparently, it is this reduced effectiveness of triazolalanyl tRNA^{His} as a corepressor which allowed Kovach *et al.*[46] to detect the altered ability of feedback-resistant mutants to be repressed. They suggested that the physiological corepressor, histidyl tRNA^{His}, must interact with the first enzyme in order to express its regulatory function. This suggestion was offered to explain the data from the experiments with triazolalanine and feedback-resistant mutants: defects in both parts of the regulatory complex (feedback-resistant first enzyme and triazolalanyl tRNA^{His}) would be expected to function less well than would a complex in which there is a defect in only one of the two parts.[46]

The idea that the first enzyme must interact with the corepressor to express its regulatory function is supported by studies on the interaction between the enzyme and tRNA^{His} *n vitro*.[21,45] These studies show that the first enzyme has a high affinity for aminoacylated tRNA^{His}.[45] Furthermore, the enzyme shows a preference for histidyl tRNA over tRNA aminoacylated with other amino acids[45] and shows a preference for aminoacylated tRNA^{His} over tRNA^{His}.[96a] Blasi et al.[21] have reported that aminoacylated tRNA^{His} binds to the enzyme at a site distinct from both the feedback-sensitive site and the catalytic site. An interesting correlation between the physiological experiments of Kovach et al.[44,46] on the altered repressibility of feedback-resistant mutants and the in vitro binding studies[21,45] is that the affinity of purified first enzyme from a feedback-resistant mutant for aminoacylated tRNA^{His} is much lower than that of the enzyme from the wild-type organism.[21] Thus, the altered repression system in feedback-resistant mutants may be due to the fact that the interaction of the first enzyme with the corepressor is impaired in such mutants.

The effect of the wild-type hisG gene in allowing repression by triazolalanine is trans to that of a hisG gene containing a mutation to feedback resistance.[44a] This finding demonstrates that the first enzyme exerts its effect as a freely diffusible gene product, as expected for a regulatory molecule such as an aporepressor.

A regulatory molecule which acts at the genetic level would be expected to interact with the DNA of the operon in vitro. With the recent availability of a phage carrying the histidine operon of Salmonella typhimurium, it has been possible to study the interaction between the first enzyme and histidine operon DNA. Preliminary experiments show that the enzyme binds ϕ80dhis DNA in preference to ϕ80 DNA.[46a] This finding demonstrates that the enzyme binds specifically to some bacterial gene(s) carried in the defective phage genome, presumably some portion of the histidine operon. Thus, it appears that the regulatory function of the first enzyme is carried out at the genetic level.

If the first enzyme fulfills its role in regulation as the aporepressor, one would expect that certain mutations in the hisG gene of the histidine operon would result in constitutive derepression. Among the many constitutive histidine mutants which have been isolated, none is of this type. However, any mutation of the hisG gene resulting in loss of aporepressor function might be expected to result in loss of catalytic function as well. Because the triazolalanine method selects against histidine auxo-

trophs, such mutants could not have been detected. Nonetheless, procedures designed to select constitutive histidine auxotrophs have also failed to reveal constitutive mutants of the hisG gene.[43b] Recently, Rothman-Denes and Martin[81] have reported the isolation of a mutant which, though not of the type predicted above, is pertinent to this discussion. Due to a single mutation in the hisG gene, the concentrations of the histidine enzymes in this organism are more than twice as high as in the wild-type organism, and aspartic acid (as well as histidine) represses the histidine operon. Since aspartic acid has no effect in isogenic strains with a wild-type hisG gene, Rothman-Denes and Martin[81] have concluded that the first enzyme plays a role in regulation of the histidine operon.

Patthy and Dénes[69] have reported the isolation of a mutant of E. coli which produces a feedback-resistant first enzyme and has a faulty repression mechanism. They attribute both these effects to a single mutation in the hisG gene.

A serious objection to the possibility that the regulatory role of the first enzyme is that of aporepressor of the histidine operon is the finding that two organisms containing nonsense mutations in the hisG gene and two containing frameshift mutations in the hisG gene are repressible by histidine.[44a] One would not ordinarily expect the incomplete protein produced in such mutants to retain any of its functions, although it is conceivable that a relatively short portion of the first enzyme suffices as the aporepressor.

It is important to consider the possibility that the first enzyme may play some regulatory role other than that of aporepressor for the histidine operon.[21,44-46] This possibility is discussed in Section VI.

4. Summary Failure to identify a classical aporepressor for the histidine operon has led some workers in this field to consider the idea that the aporepressor, which forms active repressor upon combination with aminoacylated tRNA^{His}, is a protein with more than one function. The "other" function of this protein is thought to be one required for growth on minimal medium. Among the candidates most seriously considered are the histidyl tRNA synthetase and the first enzyme for histidine biosynthesis. It is noteworthy that both of these proteins are present in the cell at concentrations very much higher (as much as several hundredfold) than those of the repressors of inducible systems.[29,37,100] For this reason, a repression system involving the synthetase or the first enzyme, both of which have a high affinity for the corepressor,[29,45] would have to be one in which the repressor has a relatively low affinity for the histidine operator. The possibilities remain, however, that some other protein, as yet undetected, is the aporepressor of the histidine operon, that the aporepressor is comprised of more than one protein (for example, the first enzyme and the synthetase together), or that there is no aporepressor, aminoacylated tRNA^{His} acting as repressor by itself.

VI. OTHER POSSIBLE MECHANISMS OF REGULATION

We have discussed the idea that the synthetase or the first enzyme functions as aporepressor for the histidine system. However, one must consider the possibility that either or both of these molecules participate in a regulatory mechanism other than repression or influence the repression system without themselves being a part of it. In keeping with this idea, Rothman-Denes and Martin[81] have suggested, on theoretical grounds, that the first enzyme, in combination with histidine or aminoacylated tRNA^{His}, fulfills its regulatory function by interacting with the promoter of the histidine operon.

We have discussed regulation of histidine biosynthesis as a negatively controlled system. However, it is possible that the histidine system is positively controlled, in the manner first described by Englesberg and his colleagues[31] in relation to the arabinose operon of E. coli. This possibility does not solve the mystery of the missing aporepressor: the missing molecule simply becomes the positive activator protein. The reasons discussed above for considering the first enzyme and the synthetase as aporepressor are equally applicable for considering them as positive activator protein. In order to accommodate the data on the role of aminoacylated tRNA^{His} in regulation of the histidine system, one would guess that if there is a positive activator protein it loses its activator function upon interaction with aminoacylated tRNA^{His}, and may even take on the new activity of repressor. Wyche[102b] has introduced an

episome of Salmonella abony carrying a hisS$^+$ gene into Salmonella typhimurium. Thus, the merodiploid contains at least two copies of the hisS$^+$ gene, the structural gene for the histidyl tRNA synthetase. Assay of extracts of this strain have revealed that the strain contains two to three times as much synthetase as does the wild-type organism and that its histidine operon is 2- to 3-fold derepressed. A possible explanation is that the synthetase is a positive activator of the histidine system. Among the alternative explanations for the derepression of the merodiploid is that the extra quantity of synthetase in this strain may bind an amount of aminoacylated tRNA^{His} sufficient to reduce the intracellular concentration of free aminoacylated tRNA^{His} to the level at which derepression occurs.

Another possibility which remains open is that the histidine system is regulated at the level of translation, either completely or in combination with regulation at the genetic level. Although this possibility has not been thoroughly explored in the case of the histidine system, translational control has been suggested for regulation of arginine[47,61] and tryptophan[48] biosynthesis.

REFERENCES

1. Alpers, D. H., and Tomkins, G. M., *Proc. Nat. Acad. Sci. U.S.* 53, 797 (1965).
2. Ames, B. N., and Garry, B. J., *Proc. Nat. Acad. Sci. U.S.* 45, 1453 (1959).
3. Ames, B. N., Garry, B., and Herzenberg, L. A., *J. Gen. Microbiol.* 22, 369 (1960).
4. Ames, B. N., and Hartman, P. E., *in* "The Molecular Basis of Neoplasia," p. 322. Univ. of Texas Press, Austin, 1962.
5. Ames, B. N., and Hartman, P. E., *Cold Spring Harbor Symp. Quant. Biol.* 28, 349 (1963).
6. Ames, B. N., Hartman, P. E., and Jacob, F., *J. Mol. Biol.* 7, 23 (1963).
7. Ames, B. N., Martin, R. G., and Garry, B., *J. Biol. Chem.* 236, 2019 (1961).
8. Ames, G. F.-L., *Arch. Biochem. Biophys.* 104, 1 (1964).
9. Ames, G. F.-L., and Lever, J., *Proc. Nat. Acad. Sci. U.S.* 66, 1096 (1970).
10. Anton, D. N., *J. Mol. Biol.* 33, 533 (1968).

11. Atkins, J. F., and Loper, J. C., *Proc. Nat. Acad. Sci. U.S.* 65, 925 (1970).
12. Atkinson, D. E., Hathaway, J. A., and Smith, E. C., *J. Biol. Chem.* 240, 2682 (1965).
13. Bagdasarian, M., Ciesla, Z., and Sendecki, W., *J. Mol. Biol.* 48, 53 (1970).
14. Ballesteros-Olmo, A., Kovach, J. S., Van Knippenberg, P., and Goldberger, R. F., *J. Bacteriol.* 98, 1232 (1969).
15. Bell, R. M., and Koshland, D. E., Jr., *Biochem. Biophys. Res. Commun.* 38, 539 (1970).
16. Benzinger, R., and Hartman, P. E., *Virology* 18, 614 (1962).
17. Berberich, M. A., Kovach, J. S., and Goldberger, R. F., *Proc. Nat. Acad. Sci. U.S.* 57, 1857 (1967).
18. Berberich, M. A., Venetianer, P., and Goldberger, R. F., *J. Biol. Chem.* 241, 4426 (1966).
19. Blasi, F., personal communication.
20. Blasi, F., Aloj, S. M., and Goldberger, R. F., *Biochemistry* 10, 1409 (1971).
21. Blasi, F., Barton, R. W., Kovach, J. S., and Goldberger, R. F., *J. Bacteriol.* 106, 508 (1971).
22. Brenchley, J. E., Ph.D. thesis, University of California, Davis (1970).
23. Brenner, M., and Ames, B. N., *Metab. Pathways* 5, 349 (1971).
24. Brenner, M., and Ames, B. N., *J. Biol. Chem.* (1972) (in press).
24a. Bruni, C. B., personal communication.
25. Bruni, C. B., Martin, R. G., and Rechler, M. M., in preparation.
26. Chang, G. W., and Ames, B. N., personal communication.
27. Chang, G. W., Roth, J. R., and Ames, B. N., *J. Bacteriol.* 108, 410 (1971).
28. Cortese, R., and Ames, B. N., personal communication.
29. DeLorenzo, F., and Ames, B. N., *J. Biol. Chem.* 245, 1710 (1970).
30. DeLorenzo, F., Straus, D. S., and Ames, B. N., *J. Biol. Chem.* (1972) (in press).
31. Englesberg, E., Irr, J., Power, J., and Lee, N., *J. Bacteriol.* 90, 946 (1965).
32. Epstein, W., and Beckwith, J. R., *Annu. Rev. Biochem.* 37, 411 (1968).
33. Fankhauser, D. B., Ely, B., and Hartman, P. E., *Genetics* 68, 518 (1971).
34. Fink, G. P., and Roth, J. R., *J. Mol. Biol.* 33, 547 (1968).
35. Forchhammer, J., and Lindahl, L., *J. Mol. Biol.* 55, 563 (1971).
36. Franklin, N. C., and Luria, S. E., *Virology* 15, 299 (1961).
37. Gilbert, W., and Müller-Hill, B., *Proc. Nat. Acad. Sci. U.S.* 56, 1891 (1966).
38. Goldberger, R. F., and Berberich, M. A., *Proc. Nat. Acad. Sci. U.S.* 54, 279 (1965).
39. Hartman, P. E., Hartman, Z., and Šerman, D., *J. Gen. Microbiol.* 22, 354 (1960).
40. Hartman, P. E., Hartman, Z., Stahl, R. C., and Ames, B. N., *Advan. Genet.* 16, 1 (1971).
41. Hartman, P. E., Loper, J. C., and Šerman, D., *J. Gen. Microbiol.* 22, 323 (1960).
42. Jacob, F., and Monod, J., *J. Mol. Biol.* 3, 318 (1961).
43. Jacob, F., and Monod, J., *Cold Spring Harbor Symp. Quant. Biol.* 26, 193 (1961).
43a. Kasai, T., personal communication.
43b. Kovach, J. S., unpublished data.
44. Kovach, J. S., Berberich, M. A., Venetianer, P., and Goldberger, R. F., *J. Bacteriol.* 97, 1283 (1969).
44a. Kovach, J. S., and Goldberger, R., unpublished data.
45. Kovach, J. S., Phang, J. M., Blasi, F., Barton, R. W., Ballesteros-Olmo, A., and Goldberger, R. F., *J. Bacteriol.* 104, 787 (1970).
46. Kovach, J. S., Phang, J. M., Ference, M., and Goldberger, R. F., *Proc. Nat. Acad. Sci. U.S.* 63, 481 (1969).
46a. Kovach, J. S., Vogel, T., Meyers, M., Levinthal, M., Blasi, F., Bruni, C., and Goldberger, R. F., unpublished data.
47. Lavalle, R., *J. Mol. Biol.* 51, 449 (1970).
48. Lavalle, R., and De Hauwer, G., *J. Mol. Biol.* 51, 435 (1970).
48a. Lee, N., and Patrick, J., personal communication.
49. Levin, A. P., and Hartman, P. E., *J. Bacteriol.* 86, 820 (1963).
50. Lewis, J. A., and Ames, B. N., *J. Mol. Biol.* (1972) (in press).
51. Loper, J. C., *Proc. Nat. Acad. Sci. U.S.* 47, 1440 (1961).
52. Loper, J. C., *J. Biol. Chem.* 243, 3264 (1968).
53. Loper, J. C., and Adams, E., *J. Biol. Chem.* 240, 788 (1965).
54. Loper, J. C., Grabnar, M., Stahl, R. C., Hart-

man, Z., and Hartman, P. E., *Brookhaven Symp. Biol.* 17, 15 (1964).

55. Margolies, M. N., and Goldberger, R. F., *J. Biol. Chem.* 241, 3262 (1966).
56. Martin, R. G., *Cold Spring Harbor Symp. Quant. Biol.* 28, 357 (1963).
57. Martin, R. G., *J. Biol. Chem.* 238, 257 (1963).
58. Martin, R. G., Berberich, M. A., Ames, B. N., Davis, W. W., Goldberger, R. F., and Yourno, J. D., *Methods Enzymol.* 17, 3–44 (1971).
59. Martin, R. G., and Goldberger, R. F., *J. Biol. Chem.* 242, 1168 (1967).
60. Martin, R. G., Silbert, D. F., Smith, D. W. E., and Whitfield, H. J., Jr., *J. Mol. Biol.* 21, 357 (1966).
61. McLellan, W. L., and Vogel, H. J., *Proc. Nat. Acad. Sci. U.S.* 67, 1703 (1970).
62. Moat, A. G., and Friedman, H., *Bacteriol. Rev.* 24, 309 (1960).
63. Monod, J., Wyman, J., and Changeux, J.-P., *J. Mol. Biol.* 12, 88 (1965).
64. Morse, D. E., Mosteller, R. D., and Yanofsky, C., *Cold Spring Harbor Symp. Quant. Biol.* 34, 725 (1969).
65. Morse, D. E., and Yanofsky, C., *Nature* (London) 224, 329 (1969).
66. Moyed, H. S., *J. Biol. Chem.* 236, 2261 (1961).
67. Moyed, H. S., and Magasanik, B., *J. Biol. Chem.* 235, 149 (1960).
68. Parks, J. S., Gottesman, M., Shimada, K., Weisberg, R. A., Perlman, R. L., and Pastan, I., *Proc. Nat. Acad. Sci. U.S.* 68, 1891 (1971).
69. Patthy, L., and Denes, G., *Acta Biochim. Biophys.* 5, 147 (1970).
70. Ptashne, M., *Proc. Nat. Acad. Sci. U.S.* 57, 306 (1967).
71. Rechler, M. M., and Bruni, C. B., *J. Biol. Chem.* 246, 1806 (1971).
72. Rechler, M. M., Bruni, C. B., and Martin, R. G., in preparation.
73. Rechler, M. M., Bruni, C. B., Martin, R. G., Gayer, R., Poy, G., Rogersen, D., and Terry, W., in preparation.
74. Rechler, M. M., and Martin, R. G., *Nature* (London) 226, 908 (1970).
75. Riddle, D., and Roth, J. R., *J. Mol. Biol.* 54, 131 (1970).
76. Riddle, D., and Roth, J. R., *Genetics* 68, Suppl. 1, Part 2, S54 (1971).
77. Roth, J. R., and Ames, B. N., *J. Mol. Biol.* 22, 325 (1966).
78. Roth, J. R., Anton, D. N., and Hartman, P. E., *J. Mol. Biol.* 22, 305 (1966).
79. Roth, J. R., and Hartman, P. E., *Virology* 27, 297 (1965).
80. Roth, J. R., and Sanderson, K. E., *Genetics* 53, 971 (1966).
81. Rothman-Denes, L., and Martin, R. G., *J. Bacteriol.* 106, 227 (1971).
82. Sanderson, K. E., *Bacteriol. Rev.* 31, 354 (1967).
83. Sarabhai, A. S., Stretton, A. O. W., Brenner, S., and Bolle, A., *Nature* (London) 201, 13 (1964).
84. Schlesinger, S., and Magasanik, B., *J. Mol. Biol.* 9, 670 (1964).
85. Schlesinger, S., and Schlesinger, M. J., *J. Biol. Chem.* 242, 3369 (1967).
86. Sheppard, D. E., *Genetics* 50, 611 (1964).
87. Silbert, D. F., Fink, G. R., and Ames, B. N., *J. Mol. Biol.* 22, 335 (1966).
88. Singer, C. E., and Smith, G. R., *J. Biol. Chem.* (1972) (in press).
89. Singer, C. E., Smith, G. R., Cortese, R., and Ames, B. N., *Nature* (London) *(New Biol.)* (1972) (in press).
90. Smith, D. W. E., and Ames, B. N., *J. Biol. Chem.* 239, 1848 (1964).
91. Smith, D. W. E., and Ames, B. N., *J. Biol. Chem.* 240, 3056 (1965).
92. Söll, D., *Science* 173, 293 (1971).
93. Steitz, J. A., *Nature* (London) 224, 957 (1969).
94. Vasington, F. D., and LeBeau, P., *Biochem. Biophys. Res. Commun.* 26, 153 (1967).
95. Venetianer, P., *J. Mol. Biol.* 45, 375 (1969).
96. Venetianer, P., Berberich, M. A., and Goldberger, R. F., *Biochim. Biophys. Acta* 166, 124 (1968).
96a. Vogel, T., unpublished data.
97. Voll, M. J., Appella, E., and Martin, R. G., *J. Biol. Chem.* 242, 1760 (1967).
98. Voll, M. J., in press.
99. Whitfield, H. J., Jr., *J. Biol. Chem.* 246, 899 (1971).
100. Whitfield, H. J., Jr., Gutnick, D. L., Margolies, M. N., Martin, R. G., Rechler, M. M., and Voll, M. J., *J. Mol. Biol.* 49, 245 (1970).
101. Whitfield, H. J., Jr., Smith, D. W. E., and Martin, R. G., *J. Biol. Chem.* 239, 3288 (1964).

102. Williams, L. S., and Neidhardt, F. C., *J. Mol. Biol.* 43, 529 (1969).

102a. Wolf, R. E., personal communication.

102b. Wyche, J., personal communication.

103. Yourno, J., *J. Biol. Chem.* 243, 3277 (1968).

104. Yourno, J., *J. Mol. Biol.* (1972) (in press).

105. Yourno, J., Barr, D., and Tanemura, S., *J. Bacteriol.* 100, 453 (1969).

106. Yourno, J., and Heath, S., *J. Bacteriol.* 100, 460 (1969).

107. Yourno, J., and Ino, I., *J. Biol. Chem.* 243, 3273 (1968).

108. Yourno, J., Kohno, T., and Roth, J. R., *Nature* (London) 228, 820 (1970).

109. Yourno, J., and Tanemura, S., *Nature* (London) 225, 422 (1970).

Positive control of transcription by a bacteriophage *sigma* factor

ANDREW A. TRAVERS

Reprinted by author's and publisher's permission from Nature, vol. 225, 1970, pp. 1009–1012.

Travers' paper describes a second type of transcriptional control. This control differs fundamentally from the "operon" control in that sigma, a protein, must be present for transcription to occur. In the absence of sigma, structural genes are not transcribed. This is called *positive control*. In operon control, described in the previous two papers, if an active repressor is not present, the structural genes will be transcribed continuously. So, the regulator gene turns off transcription. This is *negative control*. A second function of sigma involves the control of asymmetrical transcription. For the one gene-one polypeptide generalization to be valid, assuming that there is no selective strand degradation of messenger RNAs, only one of the polydeoxyribonucleotide strands of a gene can be transcribed. As Travers points out, sigma orients the initiation of transcription so that one strand of a gene is used consistently as the template for the RNA polymerase.

A sigma specificity factor induced by infection with bacteriophage T4 directs the *in vitro* synthesis of RNA species corresponding to those switched on *in vivo* about 2 min after infection.

During the course of bacteriophage T4 development, different classes of viral RNA are synthesized in a defined temporal sequence.[1-5] The earliest studies[1-3] distinguished two principal classes of T4-specific RNA: early and late, transcribed predominantly before and during DNA replication, respectively. More recent work[4,5] has shown, however, that early RNA can itself be separated into at least two distinct classes. The first of these, class I or preearly, appears immediately after phage infection while the second, class II or early, is switched on about 2 min after infection at 30°C. The synthesis of all classes of phage RNA, except preearly RNA, is dependent on phage specific protein synthesis,[4,5] indicating that the switching on of RNA species characteristic of later stages of the infection cycle requires at least one component of the transcription machinery to be phage-coded.

Recently a molecular mechanism for such positive control of transcription was suggested by the elucidation of the role of the RNA polymerase σ factor.[6,7] This factor directs the accurate initiation of selective transcription by the core RNA polymerase, the enzymatic machinery for synthesizing RNA (refs. 8, 9). For example, with T4 DNA as template, σ factor directs the initiation of preearly RNA species *in vitro*.[8,10,11] Because the host core polymerase is conserved, at least in part, throughout the infection cycle (ref. 12 and personal communication from C. G. Goff), it seemed probable that phage coded factors analogous to σ might direct the synthesis of the later species of RNA. Such a factor was recently identified in T4 infected cells.[13] This factor, termed σ^{T4}, stimulates transcription only on T-even phage DNA and even depresses the nonspecific synthesis by the core polymerase on other DNA templates. An additional alteration of the host transcription machinery after infection with phage T4 is the modification of the α subunit of the core polymerase.[14-16] We show here that σ^{T4} directs the synthesis of early RNA and further that the specificity of initiation is determined primarily by the type of factor and is not affected by the modification of the core polymerase.

*The Biological Laboratories, Harvard University.

437

Isolation of σ^{T4}

Because the original procedure described for the isolation of a T4-coded sigma factor is unsuitable for processing large quantities of cells, we have developed an alternative method. The cells are grown, infected and harvested as previously described and are then lysed by grinding briefly with alumina instead of by a lysozyme-freeze/thaw treatment. After removal of alumina and cell debris by centrifugation, the crude cell extract is fractionated into a supernatant fraction and a ribosomal pellet. This pellet is then washed with a buffer of high ionic strength to elute absorbed protein. With both methods of lysis the σ^{T4} activity is normally found preferentially in the ribosomal wash, although in about 30 percent of all preparations the factor was uniformly distributed between the ribosomal pellet and the supernatant fraction. The σ^{T4} factor can be separated from the core polymerase by fractionation of the ribosomal wash by precipitation with ammonium sulphate (Table 1). Most of the RNA polymerase activity is recovered in the fraction precipitating between 25 and 40 percent saturation

with ammonium sulphate. This polymerase preparation transcribes calf thymus DNA well and exhibits a low but significant activity on T4 DNA. It thus probably contains mainly core polymerase with a small amount of σ^{T4}. Addition of the fraction precipitating between 40 and 55 percent saturation to the polymerase fraction greatly enhances the activity on T4 DNA, however, while RNA synthesis on calf thymus DNA is depressed. Because the 40–55 percent cut has negligible RNA polymerase activity it must contain σ^{T4}, and was used as a source of this factor in these experiments.

Asymmetric Transcription by Reconstituted System

Previously we have shown that the crude mixture of T4 factor and RNA polymerase in the ribosomal wash transcribes mature T4 DNA in a highly asymmetric manner, only the l-strand of the T4 DNA being copied.[13] Bautz et al.[8] have demonstrated, however, that the core polymerase, whether modified or not, transcribes T4 DNA as though RNA

Table 1 Template specificity of transcription during purification of σ^{T4}

Fraction DNA template:	^{14}C-AMP incorporated (pmoles/0.1 ml)		Ratio of activity, T4 DNA/calf thymus DNA
	T4 (37 μg/ml)	Calf thymus (37 μg/ml)	
Ribosomal wash (42 μg)	132	36	3.7
Ammonium sulphate precipitates:			
25–40 percent (16 μg)	64	352	0.18
40–55 percent (13 μg)	23	15	—
25–40 percent cut (16 μg) + 40–55 percent cut (13 μg)	248	69	3.6

To separate core polymerase and σ^{T4} by fractional precipitation with ammonium sulphate, the ribosomal wash from T4 infected cells was prepared as previously described[13] except that the cells were lysed by grinding them for 2 min with twice the cell wet weight of alumina. The paste of alumina and lysed cells was resuspended in 1 volume of standard buffer (0.01 M Tris-HCl, pH 7.9 at $25°C$, 0.005 M $MgCl_2$, 0.005 M 2-mercaptoethanol, 0.0001 M EDTA, 0.05 M KCl) per wet weight of cells. 10 ml of ribosomal wash was brought up 25 percent saturation with ammonium sulphate by the addition of a saturated solution, pH 8, and allowed to stand 15 min at $4°C$. The solution was centrifuged and the supernatant was then brought to 40 percent saturation with ammonium sulphate, allowed to equilibrate for 15 min, and finally centrifuged. The pellet was redissolved in 1.5 ml standard buffer while the second supernatant was brought to 55 percent saturation with ammonium sulphate. Once again the precipitate was collected by centrifugation and redissolved in 1.5 ml standard buffer. The redissolved pellets were assayed for RNA polymerase activity on calf thymus and T4 DNA as previously described.[13] The background incorporation in the absence of added DNA has been subtracted from all data; in no case did this background exceed 9 percent of the maximum observed incorporation. The amounts of protein added to the assay mixture are indicated in parentheses.

Table 2 Asymmetry of *in vitro* transcription by reconstituted core polymerase-factor complexes

Core polymerase	σ factor	Input cpm	Hybridization		Hybridization (ratio r : l)
			r-strand (cpm.)	l-strand (cpm.)	
E. coli	E. coli	1,340	45	1,019	4.2 : 95.8
E. coli	T4	1,810	5	1,532	0.3 : 99.7
T4	E. coli	1,260	16	1,109	1.4 : 98.6
T4	T4	1,090	18	775	2.3 : 97.7

T4 SH-RNA was prepared as previously described,[13] the labeled precursor being ^{3}H-ATP (specific activity 24 Ci/mmole). The reaction mixtures contained per ml, as appropriate, *Escherichia coli* core polymerase, 25 μg; *E. coli* σ, 50 μg, T4 core polymerase, 36 μg; σ^{T4} (as the 40–55 percent ammonium sulphate fraction), 156 μg. The core polymerases were prepared by the method of Burgess.[24] Incubation of the reaction mixtures was for 4 min at 37°C. Aliquots of the ^{3}H-RNA were annealed in 0.03 M sodium citrate − 0.3 M NaCl for 4 h at 66°C with 5 μg/ml of either the l- or r-strand of T4 DNA in a reaction volume of 200 μl. The DNA–RNA hybrids were measured as previously described.[13] Data have been corrected for radioactivity retained by the nitrocellulose filter in the absence of added DNA. These backgrounds were in the range of 10–20 cpm.

chains were initiated almost at random along the entire DNA molecule. By measuring the asymmetry of RNA synthesis in such reconstituted systems, we have asked whether the addition of σ^{T4} to core polymerase restores the specificity of transcription of the T4 genome. The results (Table 2) show that when σ^{T4} directs RNA synthesis either by host core polymerase or by T4-modified core polymerase more than 90 percent of the RNA product hybridizes to the l-strand of T4 DNA. Similarly, RNA synthesized by either host or modified core polymerase under the direction of the host σ factor also hybridizes almost exclusively to the l-strand.

σ^{T4} directs the Synthesis of Early RNA

During normal phage development, RNA, complementary to the l-strand of T4 DNA is synthesized at all times after infection whereas transcription from the r-strand does not become significant until 10 min after infection.[17] To determine whether σ^{T4} directed the synthesis *in vitro* of RNA species containing nucleotide sequences homologous to those in *in vivo* T4 RNA species, we competed labeled σ^{T4} directed RNA and also host σ directed RNA with unlabeled *in vivo* T4 RNA isolated at 3 min, 8 min and 16 min after infection. In addition, competition was carried out with T4 RNA isolated from cells treated with chloramphenicol

(CM) before infection. CM RNA is identical to preearly RNA, and presumably only the host transcription machinery is required for its synthesis *in vivo*. The results (Fig. 1) show that whereas CM RNA is a relatively poor competitor of σ^{T4}-directed RNA, only competing for about 40 percent of the radioactive label, the RNAs at 3 min, 8 min and 16 min are good competitors, virtually 100 percent competition being observed. Because the efficiency of hybridization in conditions of DNA excess is high (about 80–90 percent) this complete competition demonstrates that all the nucleotide sequences in σ^{T4} directed RNA are homologous to *in vivo* messages. The corollary of this conclusion is that little or no antimessenger is synthesized in the presence of σ^{T4}. By marked contrast, when core polymerase transcribes T4 DNA in the absence of any σ factor, about 30 percent of the resulting RNA is self-complementary.[8] Unlike σ^{T4}-directed RNA, host σ-directed RNA was well competed by CM RNA. Similarly 3-min and 8-min RNA were good competitors but 16-min RNA had slightly diminished competing capacity. Maximum competition was achieved by 3-min RNA where more than 97 percent of the labeled RNA could be diluted out.

The difference in the ability of host σ-directed RNA and σ^{T4}-directed RNA to be competed by CM RNA confirms the previous observation[13] that the two σ factors direct the synthesis of qualitatively different sets of RNA. The good competition of

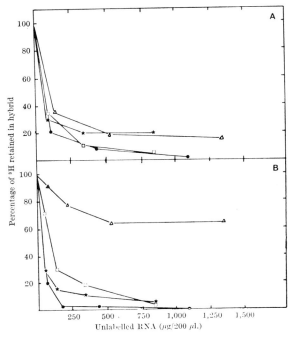

Fig. 1. RNA–DNA hybridization competition experiments comparing σ^{T4} directed T4 RNA and *E. coli* σ directed RNA with *in vivo* T4 RNA. The *in vivo* RNA was extracted from T4 infected cells at 3 min, 8 min and 16 min after infection by the method of Bolle *et al.*[25] T4 CM RNA was extracted 3 min after infection from cells to which chloramphenicol (100 μg) had been added 1 min before infection. T4 ^{3}H-RNA was prepared as described in the legend to Table 2. Hybridization of *in vitro* T4 ^{3}H-RNA to 10 μg/ml T4 l-strand in the presence of increasing amounts of RNA was performed as described in the legend to Table 2. *A*, Competition of RNA synthesized *in vitro* by complex of *E. coli* σ and *E. coli* core polymerase. Input, 2,680 cpm.; hybridization efficiency, 24 percent. *B*, Competition of RNA synthesized *in vitro* by σ^{T4} and T4 core polymerase. Input, 2,180 cpm.; hybridization efficiency, 31 percent. Competitor RNA species: CM RNA ($\triangle$); 3-min RNA ($\square$); 8-min RNA ($\bullet$); 16-min RNA ($\star$).

host-directed RNA by CM RNA strongly supports the demonstration by Bautz *et al.*[8] that σ directs the synthesis of preearly RNA. Because most of the σ^{T4}-directed RNA species are competed by 3-min RNA but not by CM RNA, we conclude that σ^{T4} must direct the synthesis of early RNA which is switched on between 1.75 min and 2.5 min after infection.[4,5]

Modification of Core Polymerase does not affect Specificity

We have shown that the host core + host σ and T4-modified core + σ^{T4} complexes synthesize different classes of T4 RNA. The recognition unit for the correct initiation of transcription is probably the complex of core polymerase and factor, so we asked whether it was necessary to change both components in order to change the specificity of RNA synthesis. Accordingly we tested RNA synthesized *in vitro* by host core + σ^{T4} or T4-modified core + host σ reconstituted complexes by competing these RNA samples with *in vivo* T4 RNA. The results (Fig. 2) demonstrate that σ^{T4} directs host core polymerase to synthesize early RNA and host σ directs T4-modified core polymerase to synthesize preearly RNA. The amount of competition by the *in vivo* species is almost identical in this case to that

Fig. 2. *A*, Competition of RNA synthesized *in vitro* by *E. coli* core polymerase with σ^{T4}. Input, 3,620 cpm.; hybridization efficiency, 28 percent. *B*, Competition of RNA synthesized *in vitro* by T4 core polymerase with *E. coli* σ. Input, 2,520 cpm.; hybridization efficiency, 24 percent. $\triangle$ and $\bullet$ as in Fig. 1.

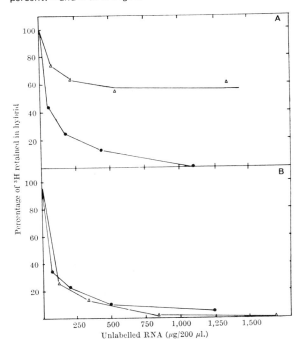

observed during competition of RNA synthesized by homologous core-factor complexes. Thus it is clear that in this system the specificity of transcription is determined primarily by the type of factor and not by the modification of the core.

The Role of σ^{T4}

Two models have been proposed to explain why different classes of phage messenger RNA appear sequentially early in the infection cycle.[4, 5, 10] This effect could arise either from sequential nonsynchronous initiation of different transcription units or from the propagation of transcription along synchronously initiated transcription units of appreciable length. The experiments reported here clearly strongly support the first hypothesis. σ^{T4} must act at the level of initiation, for in addition to stimulating RNA synthesis on T4 DNA it depresses random initiation by core polymerase on such templates as T7 and calf thymus DNA (ref. 13). Milanesi et al.[10] have shown, however, that the sequential appearance of preearly and early RNA in vivo can be exactly paralleled in vitro using only the host complex of factor and core together with T4 DNA. This observation can be explained by the lack of natural RNA chain termination in this highly purified in vitro system, because in the presence of the RNA-synthesis termination factor, ρ (ref. 18), host σ directs the synthesis of preearly RNA but without it both preearly and early RNA sequences are transcribed (Table 3). That in the absence of ρ, host σ initiated core polymerase does read through into early sequences is confirmed by the observation that host σ-directed RNA synthesized for 10 min at 37°C is a good competitor of σ^{T4}-directed RNA whereas host σ-directed RNA synthesized for 5 min competes similarly to CM RNA (Fig. 3). All these results strongly suggest that preearly and early RNA sequences occur on adjacent regions of the T4 chromosome and, further, that their arrangement must be such that the promoters for early RNA species are close to the termination signals for preearly RNA sequences. Kasai and Bautz[19] have shown that promoters for preearly species occur at widely separated points on the T4 genome and also that early genes are similarly scattered. Both host σ and σ^{T4} must therefore each direct the synthesis of several different T4 RNA species.

Table 3 Effect of ρ on host σ directed RNA synthesis in vitro

In vitro RNA	Competition by in vivo RNA	
	CM RNA	8-min RNA
E. coli σ and core polymerase + ρ	89	94
E. coli σ and core polymerase alone	51	93

T4 ^{14}C-RNA was synthesized in vitro in the presence and absence of purified ρ factor for 20 min at 37°C. The reaction mixtures contained per ml as appropriate: E. coli complex of core polymerase and σ, 30 μg; ρ factor, 20 μg; T4 DNA, 15 μg. Other conditions were as specified in the legend to Table 2 except that [KC] = 0.1 M. The reactions were terminated by the addition of phenol and the ^{14}C-RNA was annealed to 10 μg/ml T4 l-strand in the presence of increasing amounts of unlabeled in vivo T4 RNA. Hybridization was performed as described in the legend to Fig. 2. The figures for extent of competition by CM RNA (1,350 μg/200 μl.), and 8-min RNA (1,100 μg/200 μl) are plateau values. The experiment was performed in collaboration with Dr J. Roberts.

Do host σ and σ^{T4} direct the synthesis of independent or overlapping sets of RNA species? Because host σ-directed RNA is not merely a subset of σ^{T4}-directed RNA (ref. 13), the 40 percent competition of σ^{T4} directed in vitro T4 RNA by in vivo CM RNA suggests that the set are overlapping. Support for this idea is provided by the results of Hosoda and Levinthal,[20] who have shown that whereas some of the T4 proteins that are synthesized immediately after phage infection are switched off after 8 min at 25°C, others continue to be synthesized until 20 min. These latter proteins behave similarly in this respect to another class of proteins which is switched on about 3 min after infection and which is probably coded by early RNA species. It is perhaps significant that when cells are infected with a T4 phage mutant in gene 55 (such a phage is unable to synthesize late RNA (ref. 21)) the RNA synthesized between 8 min and 11 min after infection is quantitatively similar in its ability to be competed by CM RNA (ref. 5) to σ^{T4} directed in vitro RNA. Thus the characteristics of in vivo transcription in this phage can be explained simply by assuming that σ^{T4} is responsible for directing all

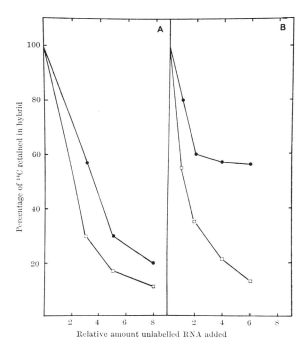

Fig. 3. Readthrough by *E. coli* σ-initiated RNA polymerase into sequences transcribed by σ^{T4}-initiated polymerase. ^{14}C-labeled precursor was ^{14}C-ATP (specific activity 35 Ci/mole) and unlabeled RNA samples were synthesized in parallel tubes as described in the legend to Table 2. In each case ^{14}C-RNA was annealed to 0.2 μg/ml T4 l-strand in the presence of increasing amounts of unlabeled RNA. *A*, □, Competition of 10-min ^{14}C-labeled *E. coli* σ-directed RNA by unlabeled 10-min *E. coli* σ-directed RNA; input; 3,410 cpm.; hybridization efficiency, 33 percent. ●, Competition of 5-min ^{14}C-labeled T4 σ-directed RNA by unlabeled 10-min *E. coli*-directed RNA; input, 2,880 cpm.; hybridization efficiency, 27 percent. *B*, □, competition of 5-min ^{14}C-labeled *E. coli* σ-directed RNA by unlabeled 5-min *E coli* σ-directed RNA; input, 3,200 cpm.; hybridization efficiency 31 percent. ●, Competition of 5-min ^{14}C-labeled T4 σ-directed RNA by unlabeled 5-min *E. coli* σ-directed RNA; input, 2,880 cpm.; hybridization efficiency, 27 percent. The times given are the length of synthesis at 37°C.

transcription after the host σ factor ceases to function.

We can tentatively formulate a model for the events which take place immediately after infection by phage T4. On injection of the phage DNA into the bacterium the host core–factor complex rapidly binds to this DNA (ref. 22) and initiates the synthesis of preearly RNA. Among the proteins coded by this RNA would be the σ^{T4} and possibly the enzyme responsible for the modification of the core polymerase. σ^{T4} then directs the synthesis of delayed early RNA together with certain immediate early RNA species. Meanwhile, the synthesis of those preearly species not directed by the T4 factor would be switched off.[19] This switch-off would result from the functional replacement of the host factor by the T4 factor. The mechanism of such a replacement is obscure, but it is noteworthy that host σ factor activity is only detectable in T4 infected cells immediately after infection and is not found at late times.[13-15] One possibility is that the modification of the core polymerase reduces the affinity for the host factor, thereby allowing the T4 factor to bind preferentially to the core polymerase.

The only experimental finding which is not compatible with this model is the timing of the appearance of the σ^{T4} factor in infected cells. *In vivo*, early RNA is switched on about 2 min after infection at 30°C (refs. 4 and 5), while the T4 sigma factor cannot be detected until after 5 min from infection.[13] It remains possible, however, that the assay for the factor is not sufficiently sensitive to detect it earlier. Indeed, Schmidt, Mazaitis and Bautz (personal communication) have tentatively identified an early promoter in the rIIB region which is active *in vivo* about 2 min after infection and which is not read by the *E. coli* σ factor.[8] The obvious conclusion is that this early promoter site is recognized by a phage-specific factor.

The experiments reported here confirm the model of positive gene control[13, 23] in which transcription can be regulated by altering the initiation specificity of RNA polymerase. This specificity is determined primarily by the σ factors and is altered during the development of phage T4 and also T7 (ref. 23) by the substitution of a phage-induced factor for the host factor.

I thank Mrs. Christine Roberts for technical assistance and Professors J. D. Watson, W. Gilbert and K. Weber and Dr J. Roberts and C. Goff for discussions. This work was carried out during the tenure of a Damon Runyon Memorial fellowship and was supported in part by a grant from the US Public Health Service.*

*Received February 10, 1970.

REFERENCES

1. Kano-Sueoka, T., and Spiegelman, S., *Proc. US Nat. Acad. Sci.*, 48, 1942 (1962).
2. Khesin, R. B., Gorlenko, Zh. M., Shemyakin, M. F., Bass, I. A., and Prozorov, A. A., *Biokhimiya*, 28, 1070 (1963).
3. Hall, B. D., Nygaard, A. P., and Green, M. H., *J. Mol. Biol.*, 9, 143 (1964).
4. Grasso, R. J., and Buchanan, J. M., *Nature*, 224, 882 (1969).
5. Salser, W., Boller, A., and Epstein, R., *J. Mol. Biol.* (in the press).
6. Travers, A. A., and Burgess, R. R., *Nature*, 222, 537 (1969).
7. Burgess, R. R., Travers, A. A., Dunn, J. J., and Bautz, E. K. F., *Nature*, 221, 43 (1969).
8. Bautz, E. K. F., Bautz, F. A., and Dunn, J. J., *Nature*, 223, 1022 (1969).
9. Goff, C. G., and Minkley, E. G., in *First Lepetit Colloquium on RNA Polymerase and Transcription* (North-Holland, Amsterdam, 1970).
10. Milanesi, G., Brody, E. N., and Geiduschek, E. P., *Nature*, 221, 1014 (1969).
11. Crouch, R., Hall, B. D., and Hager, G., *Nature*, 223, 476 (1969).
12. Haselkorn, R., Vogel, M., and Brown, R. D., *Nature*, 221, 836 (1969).
13. Travers, A. A., *Nature*, 223, 1107 (1969).
14. Walter, G., Seifert, W., and Zillig, W., *Biochem. Biophys. Res. Commun.*, 30, 240 (1968).
15. Bautz, E. K. F., and Dunn, J. J., *Biochem. Biophys. Res. Commun.*, 34, 230 (1969).
16. Seifert, W., Qasba, P., Walter, G., Palm, P., Schachner, M., and Zillig, W., *Europ. J. Biochem.*, 9, 319 (1969).
17. Guha, A., and Szybalski, W., *Virology*, 34, 608 (1968).
18. Roberts, J. W., *Nature*, 224, 1168 (1969).
19. Kasai, T., and Bautz, E. K. F., *J. Mol. Biol.*, 41, 401 (1969).
20. Hosoda, J., and Levinthal, C., *Virology*, 34, 709 (1968).
21. Bolle, A., Epstein, R. H., Salser, W., and Geiduschek, E. P., *J. Mol. Biol.*, 33, 339 (1968).
22. Oleson, A. E., Pipsa, J. P., and Buchanan, J. M., *Proc. US Nat. Acad. Sci.*, 63, 473 (1969).
23. Summers, W. C., and Siegel, R. B., *Nature*, 223, 1111 (1969).
24. Burgess, R. R., *J. Biol. Chem.*, 244, 6160 (1969).
25. Bolle, A., Epstein, R. H., Salser, W., and Geiduschek, E. P., *J. Mol. Biol.*, 31, 325 (1968).

Gene regulation for higher cells: a theory

New facts regarding the organization of the genome provide clues to the nature of gene regulation.

ROY J. BRITTEN and ERIC H. DAVIDSON*

Reprinted by authors' and publisher's permission from Science, *vol. 165, 1969, pp. 349–357. Copyright © 1969 by the American Association for the Advancement of Science.*

We are using Britten and Davidson's paper to introduce the topic of gene control in higher organisms because it provides the reader with the following basic information: (1) the major differences in the structural organization and the patterns of gene activity between prokaryote and eukaryote organisms; (2) how these structural and functional differences make the microbial regulatory systems inadequate for the control of gene function in higher organisms; (3) the speculative attributes necessary to a regulatory system in order for it to control the experimentally established patterns of gene activity in higher organisms; and (4) the presentation of a model that reflects the unique structural complexities of eukaryote genetic information and that possesses the control attributes necessary to regulate gene activity in a noncoordinated manner. Regardless of whether future research supports or refutes the Britten and Davidson model, it is apparent that any system of eukaryote gene regulation will have to incorporate many of the basic considerations presented here.

Cell differentiation is based almost certainly on the regulation of gene activity, so that for each state of differentiation a certain set of genes is active in transcription and other genes are inactive. The establishment of this concept[1] has depended on evidence indicating that the cells of an organism generally contain identical genomes.[2] Direct support for the idea that regulation of gene activity underlies cell differentiation comes from evidence that much of the genome in higher cell types is inactive[3] and that different ribonucleic acids (RNA) are synthesized in different cell types.[4]

Little is known, however, of the molecular mechanisms by which gene expression is controlled in differentiated cells. As far as we are aware no theoretical concepts have been advanced which provide an interpretation of certain of the salient features of genomic structure and function in higher organisms. We consider here experimental evidence relating to

these features. (i) Change in state of differentiation in higher cell types is often mediated by simple external signals, as, for example, in the action of hormones or embryonic inductive agents. (ii) A given state of differentiation tends to require the integrated activation of a very large number of noncontiguous genes. (iii) There exists a significant class of genomic sequences which are transcribed in the nuclei of higher cell types but appear to be absent from cytoplasmic RNA's. (iv) The genome present in higher cell types is extremely large, compared to that in bacteria. (v) This genome differs strikingly from the bacterial genome due to the presence of large fractions of repetitive nucleotide sequences which are scattered throughout the genome. (vi) Furthermore, these repetitive sequences are transcribed in differentiated cells according to cell type-specific patterns.

In this article we propose a new set of regulatory mechanisms for the cells of higher organisms such that multiple changes in gene activity can result from a single initiatory event. These proposals are presented in the form of a specific, relatively de-

*Dr. Britten is on the staff of the Department of Terrestrial Magnetism, Carnegie Institution of Washington. Washington, D.C. 20015, and Dr. Davidson is on the staff of Rockefeller University, New York 10021.

tailed model at the level of complexity which appears to us to be required for the genomic regulatory machinery of higher cells. We make no attempt to arrive at definitive statements regarding these proposed mechanisms; obviously evidence is not now available to support any model in detail. Our purpose in presenting an explicit theory is to describe the regulatory system proposed in terms of elements and processes which are capable of facing direct experimental test. It is hoped that our relatively detailed commitment will induce discussion and experiment, and it is expected that major modifications in concept will result.

Undoubtedly important regulatory processes occur at all levels of biological organization. We emphasize that this theory is restricted to processes of cell regulation at the level of genomic transcription.

We begin by describing our usage of certain terms and their role in the model, and then present the model itself. We then consider relevant experimental observations and certain testable implications of the model. Finally, some general implications of the model for evolutionary theory are mentioned.

Elements of the Model

The following definitions are intended only to clarify the usage of certain terms in our discussion of this model.

Gene A region of the genome with a narrowly definable or elementary function. It need not contain information for specifying the primary structure of a protein.

Producer Gene A region of the genome transcribed to yield a template RNA molecule or other species of RNA molecules, except those engaged directly in genomic regulation. We are using this term in a manner analogous to that in which the term "structural gene" has been used in the context of certain bacterial regulation systems.[5] Products of the producer gene include all RNA's other than those exclusively performing genomic regulation by recognition of a specific sequence. Among producer genes, for example, are the genes on which the messenger RNA template for a hemoglobin subunit is synthesized, and also the genes on which transfer RNA molecules are synthesized.

Receptor Gene A DNA sequence linked to a producer gene which causes transcription of the producer gene to occur when a sequence-specific complex is formed between the receptor sequence and an RNA molecule called an activator RNA. We do not, in this model, wish to specify a mode of action for the receptor gene — that is, the nature of the molecular events occurring between the DNA, histones, polymerases, and so forth, present in the receptor complex. This model is concerned primarily with interrelations among the DNA sequences present in the genome.

Activator RNA The RNA molecules which form a sequence-specific complex with receptor genes linked to producer genes. The complex suggested here is between native (double-stranded) DNA and a single-stranded RNA molecule.[6] The role proposed for activator RNA could well be carried out by protein molecules coded by these RNA's without changing the formal structure of the model.* Decisive evidence is lacking in higher cells, and we have chosen the simpler alternative.[7] As the discussion of the evolutionary implications of this model will indicate, however, the probability of formation of new batteries of genes in evolution appears to differ greatly between these two alternatives.

Integrator Gene A gene whose function is the synthesis of an activator-RNA. The term integrator is intended to emphasize the role of these genes in leading, by way of their activator RNA's, to the coordinated activity of a number of producer genes. A set of linked integrator genes is activated together in response to a specific initiating event, resulting in the concerted activity of a number of producer genes not sharing a given receptor gene sequence.

Sensor Gene A sequence serving as a binding site for agents which induce the occurrence of specific patterns of activity in the genome. Binding of these inducing agents is a sequence-specific phenomenon dependent on the sensor gene sequence, and it

---

*In this case it would be desirable to use a distinctive term other than producer gene for the DNA sequences coding for such regulative protein molecules. The producer genes would then be defined as those sequences coding for RNA's other than those translated to produce proteins which recognize the receptor sequences.

results in the activation of the integrator gene or genes linked to the sensor gene. Such agents include, for example, hormones and other molecules active in intercellular relations as well as in intracellular control. Most will not bind to sensor gene DNA, and an intermediary structure such as a specific protein molecule will be required. This structure must complex with the inducing agent and must bind to the sensor gene DNA in a sequence-specific way.

Battery of Genes　The set of producer genes which is activated when a particular sensor gene activates its set of integrator genes. A particular cell state will usually require the operation of many batteries.

Integrative Function of the Model

The concerted activation of one or more batteries of producer genes is considered to underlie the existence of diverse states of differentiation. Examples of two basic aspects of the proposed integrative function appear in Fig. 1. In each case, the producer genes shown are integrated into three different, very small batteries. Sensor gene S_1 and its integrator specify the activation of producer genes P_A, P_B, and P_C; S_2 that of P_A and P_B; and S_3 that of P_A and P_C.

In Fig. 1A, the control pattern depends on the existence of redundant receptor sequences in the receptor gene sets of the three producer genes. Inclusion of a particular producer gene in each of the batteries calling on it depends on the presence of the appropriate receptor gene adjacent to the producer gene. Thus, in the case where there is only one integrator gene per sensor as in Fig. 1A, there will be as many copies of a given receptor gene sequence as there are producer genes in a battery.

In the case shown in Fig. 1B, however, redundancy is present between the integrator genes of different integrator sets. A particular producer gene, in this example, is included in each of several batteries calling on it by virtue of the inclusion of the same integrator gene adjacent to each of the appropriate sensor genes. Here there will be as many copies of a given integrator gene as there are batteries that call on its producer gene. For certain commonly required genes, for example those used in the fundamental biochemistry of each cell, this could be a very large number indeed.

Systems of the type portrayed in Fig. 1A might be most useful in the case where the producer genes to be integrated direct the synthesis of enzymes whose function is tightly coordinated physiologically, for example, the ten enzymes of the urea synthesis system. Where the system is needed, all the genes would be needed. The system portrayed in Fig. 1B is a more powerful integrative system since it can govern a larger diversity of producer genes. The number of receptor sequences governing each producer sequence is probably small since transcription of a producer gene sequence is not likely to be activated from a great distance along the DNA strand. There is no reason *a priori*, on the other hand, to restrict the number of integrator genes per integrator set, except for the requirement that the integrator genes not be so distant that there is a high probability of their being separated by translocation.

In this model, regulation is accomplished by sequence-specific binding of an activator RNA and not by sequence recognition on the part of histones. The latter seem clearly to be the general inhibitors of transcription in the genome, but evidently these general inhibitors do not possess sufficient diversity to be considered as sequence-specific regulatory elements themselves.[8,9] We have assumed that, unless otherwise specified, the state of the higher cell genome is histone-mediated repression and that regulation is accomplished by specific activation of otherwise repressed sites, rather than by repression of otherwise active sites.

Figure 2 combines the elements and systems we have thus far described. In the remainder of our discussion we consider various properties and consequences of the minimal model, as portrayed in this figure. The magnitude of the producer gene batteries is only suggested by the diagram in Fig. 2, and of course no attempt has been made to portray the actual complexity of the system, that is, to illustrate the number of elements whose function is likely to be integrated in a living cell. Obviously, the coordination of many batteries of genes is required in order to account for massive changes in differentiated state, such as the neogenesis of a tissue during development. We visualize such phenomena as being mediated by sensor genes sensitive to the products of integrator genes in other integrative sets. In other words, a single inducing agent could lead to the activation of a number of sensor-

A. Example using redundancy in receptor genes

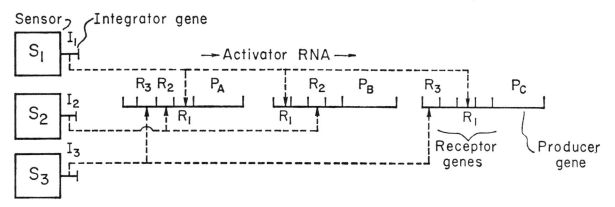

B. Example using redundancy in integrator genes

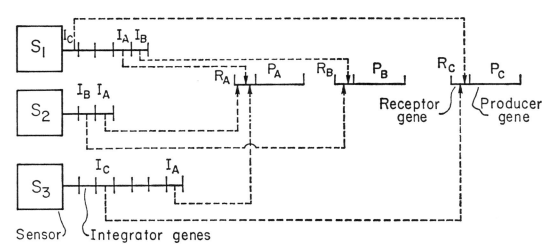

Fig. 1. Types of integrative system within the model. (A) Integrative system depending on redundancy among the regulator genes. (B) Integrative system depending on redundancy among the integrator genes. These diagrams schematize the events that occur after the three sensor genes have initiated transcription of their integrator genes. Activator RNA's diffuse (symbolized by dotted line) from their sites of synthesis — the integrator genes — to receptor genes. The formation of a complex between them leads to active transcription of the producer genes P_A, P_B, and P_C.

integrator sets, activating a vast number of producer genes.

Sequential patterns of gene activation, as in development, could result if certain sensors respond to the products of producer genes. In addition, the protein of a newly effective sensor assembly is, in the model, a product of a previously activated producer gene. Stabilization of a cell type in a given state of differentiation might also be explained in this way. Living systems continuously adjust their activities in accordance with their internal state, and it is evident that a requirement for sensors sensitive to feedback control by certain producer-gene products exists as well.

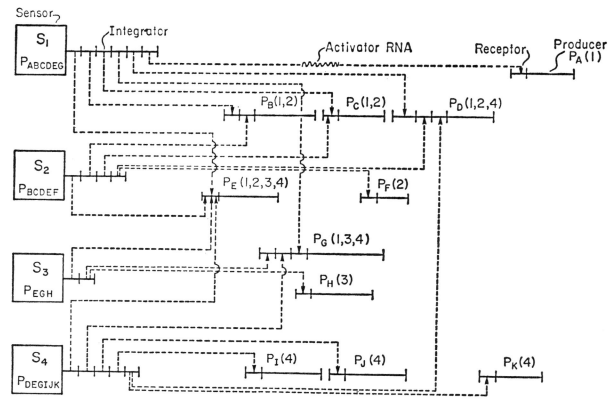

Fig. 2. This diagram is intended to suggest the existence of overlapping batteries of genes and to show how, according to the model, control of their transcription might occur. The dotted lines symbolize the diffusion of activator RNA from its sites of synthesis, the integrator genes, to the receptor genes. The numbers in parentheses show which sensor genes control the transcription of the producer genes. At each sensor the battery of producer genes activated by that sensor is listed. In reality many batteries will be much larger than those shown and some genes will be part of hundreds of batteries.

Broadly speaking, genome size increases with the grade of organization of eukaryotes, as first pointed out by Mirsky and Ris in 1951.[10, 11] The wide range of genome sizes often observed among closely related creatures obscures the correlation. Organisms with large genomes presumably have a requirement for genomic information similar to that of their relatives with smaller genomes. This implies the evolutionary multiplication of the genome of ancestors possessing the minimum amount of DNA required to effect each grade of organization. It is thus useful to consider the minimum amount of DNA observable at each grade of organization. Figure 3 shows the minimum genome size[12] for

some major steps in evolution between viruses and the higher chordates.

A reasonable explanation has not been advanced for the large genome sizes occurring at the higher organizational levels. Most of the known biosynthetic pathways are already represented in unicellular organisms. It is not possible to estimate the increase in number of producer genes required to specify structure and chemistry at the higher levels of organization. Nonetheless, it seems unlikely that the 30-fold increase from poriferan to mammal can be attributed to a 30-fold increase in the number of producer genes. This problem cannot be escaped by attributing the large genome size to redundancy.

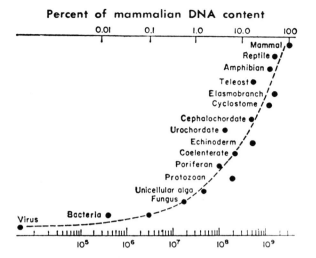

Fig. 3. The minimum amount of DNA that has been observed for species[12] at various grades of organization. Each point represents the measured DNA content per cell for a haploid set of chromosomes. In the cases of mammals, amphibians, teleosts, bacteria, and viruses enough measurements exist to give the minimum value meaning. However for the intermediate grades few measurements are available, and the values shown may not be truly minimal. No measurements were unearthed for acoela, pseudocoela and mesozoa. The ordinate is not a numerical scale, and the exact shape of the curve has little significance. The figure shows that a great increase in DNA content is a necessary concomitant to increased complexity of organization.

Fifty-five percent of the DNA of the calf, for example, occurs in nonrepetitive sequences.[13] This is enough DNA to provide almost 10^7 diverse producer-gene sequences the size of the gene coding for the beta chain of hemoglobin. A few other measurements have been made which indicate that such diversity in DNA sequence is general.[14]

Quite possibly, the principal difference between a poriferan and a mammal could lie in the degree of integrated cellular activity, and thus in a vastly increased complexity of regulation rather than a vastly increased number of producer genes.* Much

---

*The choice of a mammal in this comparison could be objected to on the grounds that an enormous number of producer genes could be required for the construction of the supercomplex central nervous system and immuno-response systems of higher chordates. However, it seems unlikely to us that the enormous increase in genome size

of the DNA accumulating in the genomes toward the upper end of the curve in Fig. 3 might then have a regulative function. The model also suggests that a large amount of DNA could be devoted to regulatory function: consider integrator and receptor sequences which are not redundant. In this case a battery of producer genes would require a distinct integrator gene for each producer gene. Producer genes occurring in several batteries would require receptor genes corresponding to each battery. The resulting multiplicity of integrator and receptor genes might result in a much larger quantity of DNA in regulatory sequences than in producer sequences. It is likely that an ever growing library of different combinations of groups of producer genes is needed as more complex organisms evolve. An effective way of storing the information specifying these combinations in the genome is to make use of sensors responsive to the activator RNA's of other integrative sets. Thus we propose that a higher level of integrator gene sets is accumulated. Each of these, when activated, could specify a very large program of producer gene activations by specifying the activity of a network of other sensor-integrator sets. Thereby many batteries of genes of the sort shown in Fig. 2 could be activated.

Experimental Justification of the Elements of the Model

There are five important classes of elements in this model: sensor genes, integrator genes, activator RNA, receptor genes, and producer genes. Is this degree of complexity really necessary? The particular set of elements we have postulated may of course not be the required ones. Five, however, is the minimum number of classes of elements which can carry out the following formally described process: (i) response to an external signal; (ii) production of a second signal; (iii) transmission of the second signal to a number of receptors unresponsive to the original signal; (iv) reception of the second signal; and (v) response to this event by activation of a producer gene and its transcription to provide the cell with the producer gene product. In the follow-

---

which has occurred in evolution could be explained in this way.

ing sections we examine evidence that such a description is applicable to gene regulation in higher organisms, and explore evidence that suggests the existence of the elements of the model.

Integration of Physically Unlinked Producer-Gene Activity

We have assumed that a given state of differentiation depends on the coordinated activity of a number of biochemical systems. Each of these systems will probably contain a number of components. As an example, Table 1 lists some of the enzyme systems operating in one cell type, mammalian liver.

An underlying principle of this model is that producer genes active in any given tissue need not be physically linked in the genome. For physically adjacent producer genes, integration of activity could be based on the operation of gigantic polycistronic tissue-specific operons. There are good reasons for believing that this is not the case in eukaryotes. Some producer genes are called into activity in a number of different tissues, as illustrated in Table 2.

Table 2 shows the overlapping pattern of activity

Table 1 Several of the functionally linked enzyme systems present in liver (ref. 15, chapter 12; *36*). Uridine monophosphate, UMP; adenosine monophosphate, AMP.

System	Number of enzymes
Glycogen synthesis	5
Galactose synthesis	6
Phosphogluconate oxidation	11
Glycolysis	12
Citric acid cycle	17
Lecithin synthesis	8
Fatty acid breakdown	5
Lanosterol synthesis	10
Phenylalanine oxidation	8
Methionine to cysteine	10
Methionine to aspartic acid	10
Urea formation	10
Coenzyme A synthesis	6
Heme synthesis	9
Pyrimidine synthesis (to UMP)	6
Purine synthesis (to AMP)	14

for 17 enzymes in 8 tissues. Direct contiguity of active producer genes could not produce this set of patterns if a single copy of each gene were present in the genome. Genetic evidence does not at present indicate the presence of multiple producer genes yielding identical products, except for ribosomal RNA and transfer RNA. An equally strong point can be made that control of the producer gene sets for the systems listed in Table 1 cannot be based on physical linkage of one set to the next in the liver genome. In other tissues, some but not all of these systems are functional.[15] In other words, even where the producer genes within a physiologically coordinated enzyme system (Table 1) are linked, the same formal problem remains: a mechanism is required for coordinating the activity of the noncontiguous systems of producer genes characteristic of each state of differentiation. In at least some instances the integrated producer genes within each physiologically coordinated set are known to be noncontiguous in higher organisms. As an example, in the human the producer genes coding for the alpha and beta subunits of hemoglobin are unlinked.[16] Another case concerns two of the enzymes of the phosphogluconic acid oxidation pathway (system No. 3 of Table 1) in *Drosophila melanogaster*. These are glucose-6-phosphate dehydrogenase (E.C.1.1.1.49) and 6-phosphogluconate dehydrogenase (E.C.1.1.1.43) whose genes are located on separate linkage groups.[17] Evidently producer genes whose activity must be functionally integrated in the most intimate way can be located far apart in the genome. We conclude that within at least some functionally integrated producer-gene systems as well as among these systems, specification of particular patterns of activity requires a method of control other than one depending on contiguity of the producer loci.

The data considered so far provide instances of the type of pattern which our model is designed to interpret, but they do not indicate the extensiveness of the producer-gene batteries called forth in given conditions of differentiation. Table 3 lists some of the effects of estrogen on the uterus, an estrogen target tissue. Although we are ignorant of the diverse proteins involved in effecting these changes it is obvious that there must be many. Though in Table 3, of course, we present only a partial list, and the number of diverse producer genes required

Table 2 Distribution of various enzymes in tissues of one organism, rat (ref. 15, Table XIII). + Means enzyme is present in amount 40 to 100 percent of that in the tissue where it is most plentiful; 0 means enzyme is essentially absent, that is, less than 8 percent the level of the tissue with the highest activity. If the level falls between 8 and 40 percent, or if data are lacking, space is left blank.

E.C. No.	Enzyme	Liver	Kidney	Spleen	Heart	Skeletal muscle	Small intestine	Pancreas	Brain
1.1.1.30	3-Hydroxybutyrate dehydrogenase	+		0	0	0		0	0
1.1.1.37	Malate dehydrogenase		0		+				
1.5.1.1	Pyrroline-2-carboxylate reductase		+	0	0				+
1.11.1.6	Catalase	+	+			0		0	0
1.11.1.7	Peroxidase	0	0	+	0	0	+		0
1.13.1.5	Homogentisate oxidase	+	+	0	0	0	0		0
2.1.1.6	Catechol methyltransferase	+		0	0	0	0		0
2.1.1.3	Dimethythetin-homocysteine methyltransferase	+		0		0			0
2.7.7.16	Ribonuclease	0	0		0	0		+	
3.1.1.1	Carboxylesterase		0	0		0		+	
3.1.1.5	Phospholipase	+		+	0	0	+		0
3.1.1.7	Acetylcholinesterase	0	0			0			+
3.1.1.8⎫ 3.1.1.9⎭	Cholinesterases		0		+	0	+		
3.1.3.1	Alkaline phosphatase	0	+	0		0	+	0	0
3.1.3.2	Acid phosphatase			+	0	0			
3.1.3.9	Glucose-6-phosphatase	+		0	0	0			0
3.2.1.25	β-Mannosidase	+	+			0	0	+	0
3.2.1.30	β-Acetylamino deoxyglucosidase		+		0	0		0	0
3.2.1.31	β-Glucuronidase	+		+		0		0	
3.5.3.1	Arginase	+		0	0	0	0	0	0
3.5.4.3	Guanine deaminase	+	+	+		0		+	0
4.1.2.7	Aldolase	0	0	0	0	+			0
4.1.3.7	Citrate synthase	0	0		+	0			
4.2.1.3	Aconitate hydratase		+		+				0
6.3.1.2	Glutamine synthetase	+	0	0		0		0	+

for each item on the list can only be guessed at present, this table provides a more realistic description of the magnitude of the problem of producer-gene integration than Tables 1 or 2 do. Analogous problems exist in explaining the integration of a multitude of producer genes in every cell lineage during development. As diverse cell lineages differentiate, a huge variety of qualitatively novel properties appear together. As in the case of the hormones, such processes of differentiation appear to require a mechanism for the simultaneous activation of many systems, such as proposed in this model.

Evidence for the Existence of Sensor Elements

There are many chemically defined agents that have the evident property of inducing large-scale changes in the producer-gene activity of specific target tissues. These agents now include steroid hormones, polypeptide hormones, several plant hormones, several vitamins, and several embryonic inductive agents.[18] Frequently, the responsible agents also produce an alteration in the spectrum of RNA's being transcribed in the target tissues, as indicated

Table 3 Some effects of estrogen on uterine cells.

Effect	Ref.
Increase in total cell protein	(33)
Increase in transport of amino acids into cell	(33, 34)
Increase in protein synthesis activity per unit amount of polyribosomes	(33)
Increased synthesis of new ribosomes	(33)
Alteration of amounts of nuclear protein to nucleus	(33)
Increased amount of polyribosomes per cell	(33)
Increase in nucleolar mass and number	(35)
Increase in activity of two RNA polymerases	(37)
Increase in synthesis of contractile proteins	(36)
Imbibition of water	(37)
Increased synthesis of many phospholipids	(38)
Increased *de novo* synthesis of purines (dependent on new enzyme synthesis)	(39)
Alteration in membrane excitability	(36)
Alteration in glucose metabolism	(40)
Increase in synthesis of various mucopolysaccharides	(41)

by data obtained with RNA–DNA hybridization and studies *in vitro* of chromatin template activity;[18] and these agents have been identified in the nuclear apparatus of the target cells.[18] The most intensively studied system is perhaps estrogen response (Table 3). All of the above-mentioned forms of evidence exist for this hormone.[18]

In addition, Maurer and Chalkey[19] have isolated from calf endometrial chromatin a protein that binds 17 β-estradiol. The binding is stereospecific, noncovalent, and strong (the Michaelis constant, K_m, for binding is $2 \times 10^{-8} M$); and the responsible protein appears not to be a histone. It does not bind steroids as closely related as 17 α-estradiol or diethylstilbestrol. Such a protein, in combination with the specific external agent for which it is the receptor, must interact with the genome in a sequence-specific way, since this interaction results in the activation of only a certain group of genes. Consider a system in which the genomic binding sites are simply adjacent to all the producer genes activated by the external chemical agent. Such a system would appear to possess a limited integrative function which might be utilized for certain small gene

batteries. However, the binding of an external agent to a sequence-specific site on the genome could lead to the activation of a large number of distant producer genes. This is exactly the role the sensor elements of this model carry out. Implicit in the available data on hormone action are genomic elements performing some of the functions of the producer and integrator genes of the model.

Evidence Suggesting the Existence of Activator RNA's

Many of the properties attributed to the RNA in our model are actually those of a certain class of RNA molecules already described extensively; yet no known function has so far been attributed to this class of RNA's. The activator RNA molecules of the model have the following properties that can be tested. (i) They will, in the main, be confined to the nucleus, that is, they are not precursors of cytoplasmic polysomes. (ii) When observed in their functional role, they would be found in chromatin, bound to DNA in a sequence-specific manner. (iii) They are often the product of the redundant fraction of the genome. (iv) They include sequences not present in the polysomes carrying producer-gene templates, that is, most or all cytoplasmic polysomes. Table 4 summarizes some recent studies of RNA's which seem to fulfill condition (i). These RNA's have the suggestive properties of nuclear location, heterogeneity, and probable lack of "precursor" relation to cytoplasmic polysomal templates.

The hybridization experiments of McCarthy and Shearer[20] (Table 4) were performed at relatively low concentrations of nucleic acid and at short incubation time. Therefore, the RNA's they describe are the products of the redundant fraction of the genome. The presence of sequences specific to the nucleus and their absence from the cytoplasm is indicated by competition experiments. Furthermore, the nuclear RNA's contain sequences binding as much as five times more DNA than the cytoplasmic RNA at empirical saturation of the DNA with RNA. These and the other data of Table 4 show that RNA's are already known which might fulfill the functions we have assigned to activator RNA's.

At the heart of the model regulation system lies nucleus-confined RNA which determines the pat-

Table 4 Nucleus-confined, apparently heterogeneous RNA's of unknown function.

Source	Size of RNA	Compositional peculiarity if any	Turn-over	Evidence against precursor relation with cytoplasmic RNA	Reference
HeLa cell nuclei	Heterodisperse 10S to 65S	29 to 32% uridine	Rapid	Composition; size; absence of association with nascent proteins	(42)
L-cell nuclei			Rapid	Sequences present in nuclear RNA absent in cytoplasmic RNA	(20)
Kidney and liver cell nuclei				Sequences present in nuclear RNA absent in cytoplasmic RNA	(43)
Reticulocyte nuclei	Heterodisperse 30S to 80S	29 to 31% uridine	Rapid	Kinetics (cytoplasmic mRNA does not turn over at all); size; base composition	(44)
HeLa cell nuclei	100 to 180 nucleotides	Extensive methylation	Extremely low	Composition; small size	(45)
Pea seedling nuclei	40 to 60 nucleotides	Presence of dihydro-pyrimidines		Sequences present in nuclear RNA absent from cytoplasmic RNA; composition	(46)

tern of cellular gene activity, and this remains a key area of uncertainty. Bekhor, Kung, and Bonner,[9] and also Huang and Huang,[9] have presented evidence suggesting that sequence-specific binding between chromosomal RNA and genomic DNA determines the sites at which the transcription-inhibiting chromatin proteins bind to the DNA.* Thus, according to these experiments, sequence recognition between the special chromosomal RNA's and the DNA specifies the pattern of gene activity.[9] Furthermore, the chromosomal RNA's have the property of binding, in what is apparently a sequence-specific way, to double-stranded native DNA.† The significance of this line of investigation

*These experiments are carried out by dissociating the chromatin in high salt and allowing the DNA and the proteins to reassociate by gradually lowering salt concentration in the presence of 5M urea. If the chromosomal RNA (RNA found initially bound to chromatin) is destroyed by ribonuclease or zinc hydrolysis, or if conditions are such as to prevent sequence-specific binding of the RNA to the DNA during the reconstitution procedure, the proteins reassociate in random positions and the reconstituted chromatin makes all the species of gene products made by totally deproteinized DNA. If the chromosomal RNA is allowed to associate with the DNA during the reconstitution process, on the other hand, the proteins of the chromatin evidently return to positions on the DNA close to those in the starting material, so that the reconstituted chromatin now synthesizes a spectrum of RNA's homologous to that of the original chromatin.

†See note, p. 445.

for experimental test of the idea of activator RNA's is obvious.

Large Changes in Transcription of Redundant Sequences

It is a striking fact that very large changes in the spectrum of RNA's deriving from repetitive sequences are observed when the state of differentiation alters. This knowledge is derived from RNA–DNA hybridization experiments carried out at relatively low concentrations of nucleic acid and short annealing times, so that reaction of RNA with any but the repetitive sequences in the genome is precluded. The spectrum of RNA's present or in the process of being synthesized in different tissues,[4, 21] both in hormone response[22] and in embryonic development and differentiation,[23, 24] has been investigated with competition procedures. In these experiments RNA from a cell type in one state of differentiation is used to compete with RNA from a cell type in another state of differentiation for binding sites in the repetitive fraction of the DNA. This type of analysis has shown that different families of repetitive genomic sequences are represented in the RNA of cells in diverse states of differentiation. Changes as large as 100 percent (apparent complete lack of homology) in the measured RNA's have been observed — for example, in successive stages of the embryogenesis of *Xenopus*.[24] It is not particularly obvious why such changes should be detected, since the populations of producer genes active in each state of differentiation might be expected in general to be strongly overlapping. One possible explanation would be that much of the pulse-labeled RNA monitored in these studies is the rapidly turning over product of different regulatory genes such as the integrator genes of this model.

Regulatory Genes Known in Higher Organisms

The model suggests that a sizable portion of the functional genes in differentiated cell types may be regulatory genes (integrator and receptor genes). If this is so, it might be expected that, despite the difficulty of detecting such genes with classical genetic procedures, a certain number of apparent regulatory mutations would be known in higher organisms. The distinguishing characteristic of such regulatory loci would be pleiotropic effects on the activity of a number of producer genes, particularly with reference to a pattern of integration on the part of the latter. A number of good cases of this genre actually exist, particularly for drosophila and maize. A notable example is the Notch series of X chromosome deficiencies,[25] some of which are sharply localized. Notch mutants display a very large variety of developmental abnormalities — all affecting early embryonic organization — for example, failure to form a complete gut, failure of mesodermal differentiations to occur, overly large neural structure, and subnormal ectodermal skin production. Their effects are clearly pleiotropic. The multiplicity of the actual primary failures of these mutants is unknown. That is, no comparison can be made of the number of diverse producer genes affected simultaneously, as opposed to the array of sequential effects that follow the initial primary effects. Nonetheless, the effect of the Notch genes on the organization of the embryo is consistent with what would be expected of mutations in integrator gene sets. Many similar cases are known in which specific organizational lesions result from simple mutations affecting a small region of the genome.[26] Studies with drosophila imaginal disk cell determination and transdetermination carried out by Hadorn and his associates[27] also demonstrate the existence of an apparatus in the genome for specifying integrated patterns of activity in the various cell types deriving from the disk cells. In experimental imaginal disk systems, highly exact specification of the patterns of producer gene activity is heritable through many cell divisions and is separated in time from producer gene function *per se* (that is, manifest differentiation).

Genes are known in maize which display control over producer genes and are located in the genome at sites distant from the producer genes that they control.[28] In addition, McClintock and others have demonstrated the presence of other control sites adjacent in the genome to the same producer genes as those controlled by the distant regulatory elements.[28] Control of the expression of the producer gene is accomplished through the interaction of the distant regulatory gene with the contiguous regulatory gene. This point has been demonstrated by insertion of the contiguous regulatory genes at different sites in the genome, near known genes, which then respond to the same control system governed by the distant regulatory unit. An example is the system termed Ac-Ds. Here the distant regulatory element Ac (which behaves as an integrator gene of this model) can be made to govern producer genes in other chromosomes such as the gene series for synthesis of anthocyanin pigment. Establishment of Ac control over the pigment synthesis system is accomplished by transposing the contiguous regulatory element responsive to Ac (Ds) to the loci of the anthocyanin producer genes (Ds thus behaves like a receptor gene of this model). In several ways, these and other data presented by McClintock[29] would seem easily to fit a model such as that presented here.

DNA Sequence Repetition

The existence of repeated sequences in higher organisms led us independently to consider models of gene regulation of the type we describe here. This model depends in part on the general presence of repeated DNA sequences. The model suggests a present-day function for these repeated DNA sequences in addition to their possible evolutionary role as the raw material for creation of novel producer gene sequences. The apparently universal occurrence of large quantities of sequence repetition in the genomes of higher organisms[13] suggests strongly that they have an important current function.

The quantity of DNA in repeated sequences, the frequency of repetition (that is, number of times a given sequence is present per genome), and the precision of the repetition show great variation among species. Frequencies from 100 to 1,000,000 have been observed, and the quantities of DNA involved range from 15 to 80 percent of the total DNA. The usual relation between repeated sequences is not that of a perfect copy,* but the sharing of most of the nucleotides in a sequence extending for at least a few hundred nucleotides. Repeated sequence families in the DNA are observed, with degrees of similarity varying from perfect matching to matching of perhaps only two-thirds of the nucleotides. Expression of families of repeated sequences by transcription into RNA shows tissue specificity (as mentioned above) in spite of the fact that the individual families contain these widely divergent sequences.

In the cases studied there is good evidence that the repeated sequences are scattered throughout the DNA. For example in bovine DNA, 75 percent of all fragments about 5000 nucleotides long contain a segment of repeated DNA.[30] When the fragment size is reduced to about 500 nucleotides, only 45 percent contain repeated sequences. Therefore, the typical bovine DNA fragment of 5000 nucleotides is a composite of lengths of repeated sequence and nonrepeated sequence. For longer fragments (20,000 or so nucleotides), there is suggestive evidence[13] that more than 95 percent contain repetitive sequences. Therefore, for bovine DNA (and probably that of other organisms) repeated sequences are intimately interspersed with nonrepeated sequences, throughout the length of the genome. This is precisely the pattern required in our model if repeated sequences are usually or often regulatory in function.

Evolutionary Implications of the Model

Any evolutionary changes in the phenotype of an organism require, in addition to changes in the producer genes, consistent changes in the regulatory system. Not only must the changes be compatible with the interplay of regulatory processes in the adult, but also during the events of development and differentiation. At higher grades of organization, evolution might indeed be considered principally in terms of changes in the regulatory systems. It is therefore a requirement of a theory of genetic regulation that it supply a means of visualizing the process of evolution.

Inactivity of New Genetic Material

A characteristic of this model is that DNA sequences are inactive in transcription, unless specifically activated. Thus the genome of an organism can accommodate new and even useless or dangerous segments of DNA sequence such as might result from a saltatory replication.[31] Initially these sequences would not be transcribed, and thus would not be subject to adverse selection. Only by inclusion in integrated producer-gene batteries (through translocation of receptor genes) would their usefulness as producer genes be tested.

Formation of New Integrative Relations

A peculiar combination of conservatism and flexibility is supplied by the model system. Preexisting useful batteries of genes will tend to remain integrated in function. At the same time, there is the potentiality of formation of new integrative combinations of preexisting producer genes. These combinations would be the result of translocations, principally among the integrator gene sets. Less often, new producer-gene batteries would result from events in which receptor genes are translocated into positions contiguous to other producer genes.

We visualize many of the integrator genes and receptor genes as being members of families of repeated DNA sequences. It is known that new repeated sequence families have originated periodically in the course of evolution.[31] The new families of repeated sequences might well be utilized to form integrator and receptor gene sets specifying novel batteries of producer genes. Thus saltatory replications can be considered the source of new regulatory DNA. All that is required for regulatory function in this model is sequence complementarity

---

*Since the relationship is commonly imperfect and nothing is yet known of the functional importance of sequence multiplicity, the phrase "redundant sequences" cannot be applied in its strict sense. Nevertheless redundant and repeated are used interchangeably since no properly applicable word exists.

(translocation of members of the same repetitive sequence family to integrator and receptor positions). Almost any set of nucleotide sequences would suffice. *The likelihood of utilization of new DNA for regulation is thus far greater than the likelihood of invention of a new and useful amino acid sequence*, since for the latter case great restrictions on the nucleotide sequence exist.

Changes in the integrator systems make possible the origin of new functions and possibly even of new tissues and organs. In other words, the model supplies an avenue for the appearance of novelty in evolution by combining into new systems the already functioning parts of preexisting systems.

Divergence within Repeated Sequence Families

Individual sequences may differ from others in a family as a result of many base changes. We presume that binding of activator RNA to the receptor genes will occur for a degree of sequence homology far short of perfect complementarity. However, at some degree of divergence, binding would be lost, and a producer gene would fail to be activated as a part of its previous battery. Eventually, the process of divergence might yield regulatory DNA in which the original patterns of repetition are no longer observable. In this way, nonrepeated (unique) regulatory DNA could arise, leading to the situation discussed earlier with respect to the fraction of the genome utilized for regulation.

The possibility of increasing sequence divergence among integrator and receptor genes suggests a novel evolutionary mechanism. The divergence of regulatory sequences can be expected to be reversible. If the degree of complementarity required for binding between activator RNA and receptor sequence is fairly low then a reasonably good probability would exist for a subsequent base change to restore the complementarity lost by an earlier change. Intermediate degrees of transcription of certain producer genes will probably result since sequences with a degree of complementarity near some critical value will bind only part of the time. Natural selection could then reversibly affect the integration of individual producer genes into batteries. The potentiality for smoothly changing patterns of integration among many sets of producer genes supplies a

mechanism for direct adjustment by natural selection of the organization of systems of cellular activity. In other words, the model implies that selective factors can influence the integrative configurations in which an organism uses its genes.

The families of repeated sequences that appear and remain in the genome of a species affect the rate at which newly integrated systems of producer genes will arise. Thereby, the rate of evolution is affected. It follows that the rate of evolution will be acted on by natural selection.

The issues raised in considering the evolution of the regulatory systems themselves are of a magnitude which is really out of reach in this brief discussion. However, the model offers interesting and surprising predictions. The properties of the model regulatory system suggest that *both the rate and the direction of evolution (for example, toward greater or lesser complexity) may be subject to control by natural selection.*

Summary

A theory for the genomic regulation systems of higher organisms is described. Batteries of producer genes are regulated by activator RNA molecules synthesized on integrator genes. The effect of the integrator genes is to induce transcription of many producer genes in response to a single molecular event. Current evidence suggesting the existence of elements of this model is summarized. Some evolutionary implications are indicated.

REFERENCES AND NOTES

1. The variable gene activity theory of cell differentiation was explicitly proposed in the early 1950s, by A. E. Mirsky [in *Genetics in the Twentieth Century*, L. C. Dunn, Ed. (Macmillan, New York, 1951), p. 127] and by E. Stedman and E. Stedman [*Nature* 166, 780 (1950)]. T. H. Morgan, among several others, had earlier considered this idea [*Embryology and Genetics* (Columbia Univ. Press, New York, 1934)].
2. Evidence for the equivalence of differentiated cell genomes comes from a variety of sources, including regeneration experiments, nuclear

transplantation experiments, early embryological studies in which nuclei were positioned in cells other than those normally receiving them, measurements of DNA content per cell, and so forth. [See E. H. Davidson, *Gene Activity in Early Development* (Academic Press, New York, 1968), pp. 3–9.] Critical recent evidence has been provided by Gurdon's demonstration that differentiated intestinal cell nuclei from *Xenopus* tadpoles can direct the development of whole frogs when reimplanted into enucleate eggs. [J. B. Gurdon, *Develop. Biol.* 4, 256 (1962)], and by DNA reassociation studies in which the DNA of different tissues was shown to be indistinguishable in sequence content [B. J. McCarthy and B. H. Hoyer, *Proc. Nat. Acad. Sci. U.S.* 52, 915 (1964)].

3. V. G. Allfrey and A. E. Mirsky, *Proc. Nat. Acad. Sci. U.S.* 44, 981 (1958); *ibid.* 48, 1590 (1960); V. C. Littau, V. G. Allfrey, A. E. Mirsky, *ibid.* 52, 93 (1964); J. Bonner, M. E. Dahmus, D. Fambrough, R. C. Huang, K. Marushige, Y. H. Yuan, *Science* 159, 47 (1968).

4. J. Paul and R. S. Gilmour, *Nature* 210, 992 (1966); B. J. McCarthy and B. H. Hoyer, *Proc. Nat. Acad. Sci. U.S.* 52, 915 (1964).

5. F. Jacob and J. Monod, *J. Mol. Biol.* 3, 318 (1961).

6. I. Bekhor, J. Bonner, G. K. Dahmus, *Proc. Nat. Acad. Sci. U.S.* 62, 271 (1969).

7. J. H. Frenster [*Nature* 206, 1269 (1965)] has proposed that RNA could act as a "derepressor." There is only a formal similarity to one part of this model — the activator RNA. An RNA molecule, fully complementary to the "unread" strand was supposed to maintain the DNA of a gene in a strand-separated condition and thus permit its transcription.

8. E. W. Johns and J. A. V. Butler, *Nature* 204, 853 (1964); K. Murray, *Ann. Rev. Biochem.* 34, 209 (1965); L. S. Hnilica, *Biochem. Biophys. Acta* 117, 163 (1966); and H. A. Kappler, *Fed. Proc.* 24, Pt. I (2), 601 (1965); J. L. Beeson and E. L. Triplett, *Exp. Cell. Res.* 48, 61 (1967); D. M. Fambrough and J. Bonner, *J. Biol. Chem.* 243, 4434 (1968); R. J. deLange, D. M. Fambrough, E. L. Smith, J. Bonner, *J. Biol. Chem.* 243, 5906 (1968). In addition to lack of diversity of tissue specificity of histones shown by the above references, information

also exists which indicates directly that histone-DNA recognition does not control the sequence-specific pattern of transcription in chromatin in vitro. [See J. Paul and R. S. Gilmour, *J. Mol. Biol.* 34, 305 (1968)].

9. I. Bekhor, G. M. Kung, J. Bonner, *J. Mol. Biol.* 39, 351 (1969); R. C. Huang and P. C. Huang, *ibid.*, p. 365.

10. A. E. Mirsky and H. Ris, *J. Gen. Physiol.* 34, 451 (1951).

11. A. E. Mirsky, *Harvey Lec.* Ser. 46, 98 (1950–51).

12. Species whose genome sizes appear in Fig. 3 are as follows (from top to bottom): *Bos bos* [C. Leuchtenberger, R. Leuchtenberger, C. Vendrely, R. Vendrely, *Exp. Cell Res.* 3, 240 (1952)]; *Chelonia mydas* [H. Ris and A. E. Mirsky, *J. Gen. Physiol.* 33, 125 (1949)]; *Scaphiopus couchi* [E. Sexsmith, thesis, Univ. of Toronto (1968)]; *Tetraodon fluviatilis* [R. Hinegardner, *Amer. Natur.* 102, 517 (1968)]; *Carcharias obscurus* (*11*); *Lampetra planeri* and *Amphioxus lanceolatus* [N. B. Atkin and S. Ohno, *Chromosoma* 23, 10 (1967)]; *Acidea atra* (*11*); *Paracentrotus lividus* [R. Vendrely, *Compt. Rend. Soc. Biol.* 143, 1386 (1949)]; *Cassiopea* and *Dysidea crawshagi* (*11*); *Amoeba hystolytica* (A. Gelderman, unpublished data); *Sacharomyces* [M. Ogur, S. Minckler, G. Lindegren, C. G. Lindergren, *Arch. Biochem. Biophys.* 40, 175 (1952)]; *Escherichia coli* [J. Cairns, *J. Mol. Biol.* 4, 407 (1962)]; *Mycoplasma* [H. Bode and H. J. Morowitz, *ibid.* 23, 191 (1967)]; *Simian Virus* 40 [T. Ben-Porat and A. S. Kaplan, *Virology* 16, 261 (1962)].

13. R. J. Britten and D. E. Kohne, *Science* 161, 529 (1968).

14. In addition to calf and mouse (*14*), the following species have been shown to contain relatively large quantities of nonrepetitive DNA: *Xenopus laevis* (E. H. Davidson and B. R. Hough, *Proc. Nat. Acad. Sci. U.S.*, in press); Sea urchin and chicken (D. E. Kohne, unpublished data); *Drosophila*, 3 species (C. D. Laird and B. J. McCarthy, in press); Human (R. J. Britten, unpublished data). Even *Amphiuma* with a genome 30 times the size of the mammalian appears to have about 20 percent nonrepeated DNA [R. J. Britten, *Carnegie Inst. Wash. Year B.* 66, 75 (1968)].

15. M. Dixon and E. C. Webb, *Enzymes* (Academic Press, New York, 1964).

16. C. J. Epstein and A. G. Motulsky, in *Progr. Med. Genet.* 4, 97 (1965).

17. W. J. Young, *J. Hered.* 57, 58 (1966).

18. Action of the steroid hormones aldosterone, cortisone, testosterone, and estrogen are known to depend on gene activity for their effects in target tissues: for testosterone. S. Liao, R. W. Barton, A. H. Lin [*Proc. Nat. Acad. Sci. U.S.* 55, 1593 (1966)] and S. Liao [*J. Biol. Chem.* 240, 1236 (1965)] demonstrate increase in prostate template-active RNA, and actinomycin-sensitive increase in RNA synthesis, with change in base composition. Template activity of muscle chromatin is also increased [C. B. Brewer and J. R. Florini, *Biochemistry* 5, 3857 (1966)] and testosterone binds to a protein or proteins of prostate chromatin in vivo [N. Bruchovsky and J. D. Wilson, *J. Biol. Chem.* 243, 5953 (1968)]. A similar range of evidence exists for the effect of estrogen on uterus [see T. H. Hamilton, *Science* 161, 649 (1968)]. Striking increase in RNA synthesis in uterus occurs almost instantly with estrogen [A. R. Means and T. H. Hamilton, *Proc. Nat. Acad. Sci. U.S.* 56, 1594 (1966)] and the spectrum of gene products made is different [B. O'Malley, W. L. McGuire, P. A. Middleton, *Nature* 218, 1249 (1968)]. Corticosteroids cause many liver enzymes to be synthesized [G. Weber, S. K. Srivastava, L. Radhey, *J. Biol. Chem.* 240, 750 (1965); D. Greengard and G. Acs, *Biochim. Biophys. Acta* 61, 652 (1962)] and cause sharp increases in liver nuclear RNA synthesis [G. T. Kenny and F. J. Kull, *Proc. Nat. Acad. Sci. U.S.* 50, 493 (1963)], resulting from both increased polymerase [O. Barnabei, B. Romano, G. Bitono. V. Tomasi, F. Sereni, *Biochim. Biophys. Acta* 113, 478 (1966); W. Schmid, D. Gallwitz, C. E. Sekeris, *ibid.* 134, 85 (1967)] and from increased template activity in isolated chromatin [M. E. Dahmus and J. Bonner, *Proc. Nat. Acad. Sci. U.S.* 54, 1370 (1965)]. The specific effect of aldosterone on isolated bladder is abolished by actinomycin, and this hormone localizes in the bladder cell nuclei [G. A. Porter, R. Bogorach, I. S. Edelman, *Proc. Nat. Acad. Sci. U.S.* 52, 1326 (1964); I. S. Edelman, R. Bogorach, G. A. Porter, *ibid.* 50, 1169 (1963)]. The polypeptide gonadotropins fol-licular stimulating hormone and lutenizing hormone stimulate protein synthesis in ovary in an actinomycin-sensitive way [M. Civen, C. B. Brown, J. Hilliard, *Biochim. Biophys. Acta* 114, 127 (1966)], and adrenocorticotropin (ACTH) promotes actinomycin-sensitive increase in adrenal RNA synthesis and steroidogenesis [R. B. Farese, *Fed. Proc.* 24 Pt. I (2), 306 (1965)]. Mobilization of Ca^{++} from bone, promoted by the polypeptide parathyroid hormone, is prevented by actinomycin [H. Rasmussen, D. Arnaud, C. Hawker, *Science* 144, 1019 (1969)]. Several specific plant hormones, including giberellins, auxin, and cytokinins [reviewed by A. Trewas, in *Progr. Phytochem.* 1, 114 (1968)] as well as antheridogens (B. R. Voeller, personal communication) also appear to act at the genomic level in tissues affected by them. Transport of Ca^{++} mediated by vitamin D in intestinal mucosa is blocked by actinomycin [J. E. Zull, E. Czarnowska-Misztall, H. F. deLuca, *Science* 149, 183 (1965)]. Various embryonic processes known to occur as a result of specific inducing agents [see review by T. Yamada, in *Compr. Biochem.* 28, 113 (1967)] are known to be sensitive to actinomycin treatment of the competent tissues, for example, in mouse pancreas induction [N. K. Wessels and F. H. Wilt, *J. Mol. Biol.* 13, 767 (1965)] and neurulation in the frog [J. Brachet, H. Denis, F. deVitry, *Develop. Biol.* 9, 398 (1964)].

19. H. R. Maurer and G. R. Chalkey, *J. Mol. Biol.* 27, 431 (1967).

20. R. W. Shearer and B. J. McCarthy, *Biochemistry* 6, 283 (1967).

21. N. Miyagi, D. Kohl, R. A. Flickinger, *J. Exp. Zool.* 165, 147 (1967); H. Ursprung, K. D. Smith, W. H. Sofer, D. T. Sullivan, *Science* 160, 1075 (1968).

22. B. O'Malley, W. L. McGuire, P. A. Middleton, *Nature* 218, 1249 (1968).

23. R. B. Church and B. J. McCarthy, *J. Mol. Biol.* 23, 459 (1967); V. R. Glisin, M. V. Glisin, P. Doty, *Proc. Nat. Acad. Sci. U.S.* 56, 285 (1966); A. H. Whitely, B. J. McCarthy, H. R. Whitely, *ibid.* 55, 519 (1966).

24. H. Denis, *J. Mol. Biol.* 22, 285 (1966); E. H. Davidson, M. Crippa, A. E. Mirsky, *Proc. Nat. Acad. Sci. U.S.* 60, 152 (1968).

25. D. F. Poulson, *Amer. Natur.* 79, 340 (1945).

26. Some of the most striking examples are the *t* mutant series in the mouse [L. C. Dunn and S. Gluecksohn-Waelsh, *Genetics* 38, 261 (1953); P. Chesley and L. C. Dunn, *ibid*. 21, 525 (1936)]; the creeper gene in the chicken [W. Landauer, *J. Genet.* 25, 367 (1932)]; the control of a variety of morphogenetic patterns in insects such as bristle formation and wing color pattern [see review by D. H. Waddington, *New Patterns in Genetics and Development* (Columbia Univ. Press, New York, 1962), chap. 6]. A spectacular case of organizational control loci is provided by the "bithorax" series in Drosophila [E. B. Lewis, in *The Role of Chromosomes in Development*, M. Locke, Ed. (Academic Press, New York, 1964), p. 231].

27. S. Hadorn, *Brookhaven Symp. Biol.* 18, 148 (1965).

28. B. McClintock, *ibid.*, p. 162.

29. —, *Develop. Biol. (Suppl.)* 1, 84 (1967).

30. R. J. Britten, in preparation.

31. — and D. E. Kohne, *Carnegie Inst. Wash. Year B.* 66, 83 (1968).

32. C. Rouiller, Ed., *The Liver: Morphology, Biochemistry, Physiology* (Academic Press, New York, 1963), vol. 1, chaps 9–11.

33. T. H. Hamilton, *Science* 161, 649 (1968).

34. M. W. Noall and W. M. Allen, *J. Biol. Chem.* 236, 2987 (1961).

35. R. J. Laquens, *Ultrastructures Res.* 10, 578 (1964).

36. A. Csapo, *Ann. N.Y. Acad. Sci.* 75, 740 (1959).

37. G. Pincus, Ed., *The Hormones: Physiology, Chemistry, and Applications* (Academic Press, New York, 1964), vol. 5, pp. 787–793.

38. Y. Aizawa and G. C. Mueller, *J. Biol. Chem.* 236, 381 (1961).

39. G. C. Mueller, J. Gorski, Y. Aizawa, in *Mechanism of Action of Steroid Hormones*, C. A. Ville and L. L. Engel, Eds. (Pergamon, Oxford, 1961), pp. 181–184.

40. J. A. Nicolette and J. Gorski, *Arch. Biochem. Biophys.* 107, 279 (1964).

41. H. Sinohara and H. H. Sky-peck, *ibid.* 106, 138 (1964).

42. R. Soeiro, H. C. Birnboim, J. E. Darnell, *J. Mol. Biol.* 19, 362 (1966).

43. R. B. Church and B. J. McCarthy, *Proc. Nat. Acad. Sci. U.S.* 58, 1548 (1967).

44. G. Attardi, H. Parnas, M-I. H. Hwang, B. Attardi, *J. Mol. Biol.* 20, 145 (1966).

45. R. A. Weinberg and S. Penman, *ibid.* 38, 289 (1968).

46. J. Bonner and J. Widholm, *Proc. Nat. Acad. Sci. U.S.* 57, 1379 (1967); I. Bekhor, J. Bonner, G. K. Dahmus, *ibid.*, in press.

Biogenesis of mRNA: genetic regulation in mammalian cells

In mammalian cells, unlike bacteria, messenger RNA arises from modified nuclear RNA after transcription.

JAMES E. DARNELL, WARREN R. JELINEK, GEORGE R. MOLLOY*

Reprinted by authors' and publisher's permission from Science, vol. 181, *173, pp. 1215–1221. Copyright © 1973 by the American Association for the Advancement of Science.*

This paper, which concludes the part on gene function regulation, is a good example of the creative efforts being made in the area of eukaryote genetic control systems. The paper shows clearly that there are fundamental differences between prokaryote and eukaryote genetic control. The formation in eukaryotes of heterogenous RNA (*HnRNA*) that is tailored down to shorter length mRNA is one such major difference. This tailoring of HnRNA is posttranscriptional regulation, and should be contrasted with bacterial operon models. The authors of the paper have also presented an updated interpretation of the biological significance of polyadenine (*polyA*), a monopolymer whose function in living systems has remained obscure.

The experimental description of the mechanics as well as the regulation of bacterial protein synthesis appears to have entered a final stage. Virtually every step involved in the assembly of amino acids into protein — chain initiation, elongation, termination — can be performed in the test tube with components purified from bacterial cells.[1] The steps in bacterial messenger RNA (mRNA) biogenesis are almost as well understood. The functional RNA polymerase and several regulatory proteins have been isolated and shown to participate in the control of synthesis of specific mRNA, both under conditions where only RNA is produced and in coupled systems where all the steps occur between the initiation of mRNA production and completion of functional enzyme protein.[1] These awesome biochemical achievements with bacterial preparations probably would not have been possible, certainly not at this time, if a decade of successful bacterial genetics had not provided evidence concerning the mode of regulation that exists — namely, control of transcription of mRNA by regulatory proteins.

Presented with the challenge of understanding protein synthesis in mammalian cells in detail comparable to the present understanding in bacteria, the cell biologist might, after a first appraisal, despair. The genetic methods for isolating and characterizing mutants which might affect the availability of mRNA do not exist for mammalian cells and, if developed, will certainly present greater practical difficulties in their characterization and experimental use compared to bacteria. Even in yeasts, perhaps the most easily manipulated genetic system among eukaryotic cells, regulatory mutants known to affect the availability of mRNA have not yet been reported.

The purpose of this article is to claim reason for optimism in the attempts to understand regulation of gene expression in cultured eukaryotic cells through biochemical studies in spite of the lack of genetic details. An apparent first order understanding of gene expression in cultured eukaryotic cells seems possible through biochemical studies with cultured cells without the necessity for regulatory mutants.

As will be discussed, such studies have already

*Dr. Darnell is professor and Dr. Jelinek and Dr. Molloy are research associates in the Department of Biological Sciences at Columbia University, New York 10027.

contributed to our knowledge about mRNA bio-genesis. When experiments concerning mRNA pro-duction can be coupled with experiments concerning the changing rate of specific protein synthesis in cells which respond to a known extracellular influ-ence (such as hormones) at least some elementary information about modes of regulation should result.

The Biogenesis of mRNA in Mammalian Cells

It now appears that mRNA in mammalian cells,[2–4] like the predominant types of RNA in these cells [that is, ribosomal RNA (rRNA) and transfer RNA (tRNA)], is derived by posttranscriptional modifi-cation of larger RNA precursor molecules.[5] The mRNA precursors are part of a class of high-molecu-lar-weight nuclear RNA molecules (between 5000 and 50,000 nucleotides long) whose base composi-tion resembles that of DNA (U* substituted for T).[6–9] Most of the labeled RNA that can be isolated from cells after brief exposures to radioactive RNA precursors is contained in this high-molecular-weight nuclear RNA, yet this material does not accumulate in large amounts, that is, it "turns over" rapidly. We have used a descriptive term for this class of RNA, heterogeneous nuclear RNA, abbre-viated HnRNA, rather than call it mRNA precursor because, as will be described, only a small portion of it becomes mRNA.

The question was raised some time ago whether mRNA might be derived from HnRNA[6, 10] in a manner similar to that in which rRNA and tRNA are derived from precursor molecules.[5, 6, 11, 12] Two properties of rRNA and precursor rRNA were cru-cial to the proof of their relation to each other. (i) The 18S and 28S rRNA constitutes most of the cellular RNA and both rRNA and the 45S precursor rRNA are individual molecular species, so that it was relatively easy to isolate these RNA molecules in pure form.[6, 9, 12, 13] (ii) It was found that rRNA and its precursor, in addition to being similar in base composition,[6, 9, 13] share chemical markers, for example, methyl groups,[11, 14] which have been utilized to prove detailed sequence similarities.[2, 11, 13]

Even if HnRNA were a precursor to mRNA, both classes of RNA would be expected to be composed of many chemically different molecules, increasing the difficulty of testing the potential precursor-product relationship. However, two recent findings concerning sequence similarities between HnRNA and mRNA have provided evidence that mRNA is derived from HnRNA.

First, it was shown that cells transformed by the small DNA tumor virus SV40 contain the virus DNA as part of the cell genome.[15] In such trans-formed cells, HnRNA molecules considerably larger than virus-specific polysomal mRNA were found to contain regions complementary to virus DNA.[2] Whereas the HnRNA molecules containing virus-specific sequences were heterogeneous in size and contained host cell sequences,[16] the virus-specific mRNA was discrete in size and lacked host cell sequences.[16] Evidence has also been obtained in cells lytically infected with DNA viruses that nuclear virus-specific sequences exist which are much larger than cytoplasmic sequences.[2] Thus, it appears that a processing step occurs which cleaves the HnRNA to yield discrete mRNA fragments containing only a portion of an HnRNA molecule.

The second sequence that has been shown to be shared by HnRNA and mRNA was discovered many years ago,[17–20] but was not recognized to be part of HnRNA and mRNA molecules until recently. This sequence is about 200 nucleotides long and is un-usual because it contains only adenylic acid residues (see below); this polyadenylic acid region is termed poly(A). That such a special sequence is present in both HnRNA and mRNA (but not in other cellular RNA molecules — for example, rRNA or tRNA) fortifies the idea that mRNA is derived from Hn-RNA.

Presence of Poly(A) in Both HnRNA and mRNA

Kates[21] refocused attention on poly(A) when he found that vaccinia mRNA contained at its 3'OH end a relatively uniform, ribonuclease resistant segment that consisted largely of adenylic acid residues. This finding recalled earlier reports[17–20] of ribonuclease resistant adenylate-rich fragments

---

*Abbreviations used: C, cytidylate; A, adenylate; G, guanyl-ate; U, uridylate; T, thymidylate; Py, pyrimidine; and, X, nucleotide.

in thymus and liver cells. More recent work has established that both HnRNA and mRNA contain a fairly discrete poly(A) segment as part of the polynucleotide chain (that is, not dissociable by treatment of the RNA molecules with dimethyl-sulfoxide).[22–25] When cells are briefly exposed to labeled adenosine, the labeled poly(A) obtained from HnRNA and mRNA is the same size as determined by migration during gel electrophoresis[26] (Fig. 1).

3'OH Location of Poly(A)

Evidence from several types of experiments indicate that the poly(A) of both mRNA and HnRNA molecules exists only at the 3'OH end of these molecules. Poly(A) obtained after ribonuclease digestion from either mRNA or HnRNA contains 1 adenosine residue per 200 adenylic acid residues, indicating that poly(A) is 200 nucleotides long and is located at the 3'OH terminus.[27] Digestion of mRNA and HnRNA with an exonuclease that requires a free 3'OH quickly removes the poly(A) from both types

of molecules, giving further support to a 3'OH localization.[28, 29] These findings agree with recent demonstrations that a considerable fraction of the 3'OH adenosine termini of poly(A) becomes labeled when whole HnRNA and mRNA molecules containing poly(A) are oxidized with periodate and are reduced with [^{3}H]borohydride.[30]

Further, recent experiments with poly(A) derived by T1 or pancreatic ribonuclease (or both) give conclusive evidence for the location of the poly(A) and the 3' terminus of both HnRNA and mRNA as well as information about the composition of the nucleotides adjacent to poly(A) from the two sources.[31] T1 ribonuclease cleaves RNA on the 3' side of guanylic acid (that is, 5' — GpXp — 3') and pancreatic ribonuclease on the 3' side of pyrimidines (5'— PypXp — 3'). T1 ribonuclease derived poly(A) segments contained no guanylic acid; pancreatic ribonuclease derived poly(A) segments contained no pyrimidines, indicating that poly(A) consists entirely of adenylic acid and that poly(A) is present only at the 3' terminus of mRNA and HnRNA. Poly(A) segments derived by T1 ribonuclease showed that the structure of the 3' end of

Fig. 1. Poly(A) from HnRNA and mRNA of HeLa cells. The left and center portions show the isolation by sucrose-gradient sedimentation of labeled HnRNA from the nucleus and mRNA from the polyribosomes of HeLa cells. The majority species of RNA, 45*S* and 32*S* ribosomal precursor RNA in the nucleus, and 28*S* and 18*S* rRNA in the cytoplasm serve as absorbancy markers in the course of fractionation.[8, 11] The right portion of the graph demonstrates that ribonuclease digests of HnRNA or mRNA labeled with adenosine for very brief times (20 minutes or less) contain poly(A) that migrates identically during gel electrophoresis [–●–, nuclear poly(A); –○–, cytoplasmic poly(A)] .[26]

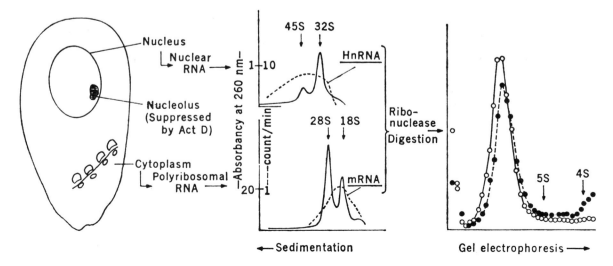

poly(A)-terminated molecules (both HnRNA and mRNA) was $G(C_2U)A_{200}$. This finding is consistent with the idea that mRNA is derived from HnRNA. The pyrimidine nucleotides adjacent to poly(A) may be a defined sequence or a limited set of sequences in both HnRNA and mRNA molecules, a question that can be answered by sequence analysis of the 5′ end of T1 derived poly(A).

How Is Poly(A) Synthesized?

Poly(A) was discovered during the study of an enzyme isolated from thymus nuclei that specifically incorporates adenylic acid into a polyribonucleotide primer without a DNA template.[17] This is in contrast to DNA directed RNA polymerases (bacterial or mammalian), which require a DNA template and which can initiate RNA chains with a 5′ terminal nucleotide and propagate them toward the 3′OH terminus.[5] Thus, if poly(A) is localized at the 3′OH terminus of HnRNA and mRNA molecules, it might be synthesized by a non-DNA dependent enzyme (such as the Edmonds-Abrams enzyme).[17] Three findings suggest that such a posttranscriptional addition of poly(A) is likely. (i) In cultured cells, actinomycin D, which prevents DNA-dependent RNA synthesis by more than 80 to 90 percent within 1 to 2 minutes after its addition to a cell culture, has almost no effect on poly(A) synthesis during the ensuing 1 to 2 minutes (Fig. 2).[3,4] Thus, it seems that the normal movement (some 5 to 10,000 nucleotides in 1 to 2 minutes)[5] of the RNA polymerase along the DNA template is not required for continued poly(A) synthesis. (ii) The DNA of adenovirus type 2, from which large, nuclear, virus-specific RNA sequences are transcribed during virus replication, contains no region to which poly(A) will hybridize; that is, it contains no long stretches of poly dT.[32] Nevertheless, virus-specific nuclear RNA, as well as the smaller virus-specific mRNA molecules in the polyribosomes, contain poly(A) of the same size as the cellular poly(A). (iii) Deoxypyrimidine nucleotide stretches can be isolated from DNA because they are resistant to acid hydrolysis.[33] The DNA of CELO virus, an avian adenovirus,[34] as well as HeLa cell DNA[35] do not contain pyrimidine-rich regions large enough to be transcribed into the 200 nucleotide segments of poly(A). It appears, therefore, that poly(A) is

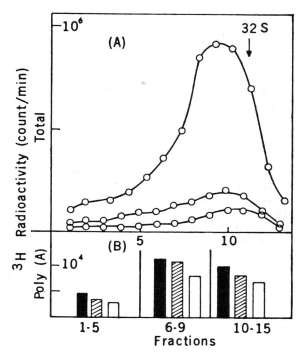

Fig. 2. Effect of actinomycin on total HnRNA synthesis compared to poly(A) synthesis. (A) Growing cells, and cells previously treated with a high dose of actinomycin D (7.5 µg/ml) for 1 or 2 minutes, were exposed to [³H] adenosine for 1.5 minutes, and labeled nuclear RNA was separated on a sucrose gradient. The direction of sedimentation is from right to left. The effect on total incorporation (–○–) is shown in the top graph (top curve, no actinomycin; middle curve, 1-minute treatment with actinomycin; bottom curve, 2-minute treatment with actinomycin). (B) Molecules of various sizes (fractions 1 to 5, 6 to 9, and 10 to 13) were collected from each of the three RNA preparations. The poly(A) content of each size class from all three preparations was then determined. Solid bar, control; hatched bar, 1-minute treatment with actinomycin; open bar, 2-minute treatment with actinomycin. (The details of such experiments are described in refs. 3 and 4.)

probably added to HnRNA molecules by a DNA-independent process after transcription.

Because of the rapid rate of RNA chain synthesis, it is difficult to distinguish experimentally the stepwise addition of individual nucleotides from the union of two polynucleotides. Nevertheless, because after very short label times poly(A) is found entirely associated with HnRNA, and not as a separate entity, it is suggested that poly(A) is synthesized by the stepwise addition of single adenylate residues to preexisting HnRNA molecules.[4]

Poly(A) Is Added in the Nucleus and Has a Nuclear Role

When HeLa cells are labeled with [³H] adenosine for 1.5 minutes or less almost all (more than 95 percent) of the poly(A) is in the nucleus as part of HnRNA molecules.[4] This result implies that the nucleus is by far the most active, if not the only, site of synthesis of the 200-nucleotide poly(A) segment. After 20 minutes of labeling there are equal amounts of labeled poly(A) in the nucleus and cytoplasm. In each succeeding 20-minute interval, the radioactivity in cytoplasmic poly(A) increases by an amount equal to the total labeled nuclear poly(A), while the nuclear poly(A) increased only about 25 percent. This leads to a fourfold greater amount of labeled cytoplasmic poly(A) compared to that in the nucleus by 120 to 160 minutes of labeling. The greater accumulation of labeled poly(A) in the cytoplasm was obtained both with growing cells and with cells in which ribosomal RNA synthesis had been stopped by a low dose of actinomycin D. In contrast, the total radioactivity in HnRNA remains much greater than that in mRNA for many hours, a circumstance that originally led to the conclusion that most of the HnRNA turned over,[9, 36] (see Fig.1). Therefore, during the processing of mRNA from HnRNA, the conservation of poly(A) in mRNA is far greater than the conservation of the HnRNA molecule. These results are consistent with the transfer of most of the poly(A) to the cytoplasm, but the possibility of some poly(A) turnover in the nucleus cannot be excluded.

The exploration of the physiological role of poly(A) in the nuclear biosynthesis and transport to the cytoplasm of mRNA has been facilitated by use of the drug cordycepin, which is 3′deoxyadenosine (3′dA). Penman, Rosbash, and Penman[37] found that in HeLa cells, 3′dA, which acts to terminate RNA chains prematurely[38] stopped the synthesis of some RNA molecules but not of others. For example, in cells treated with 3′dA, rRNA synthesis was quickly and completely halted, but incorporation of labeled precursor into HnRNA was not affected; mRNA, however, failed to appear in cyto-

plasmic polyribosomes. An explanation for these results and a suggestion of the importance of poly (A) in mRNA biogenesis came from finding that the synthesis of the 200-nucleotide poly(A) segment of HnRNA was stopped by the drug.[3, 4, 27] In agreement with earlier results, labeled mRNA appearance was reduced in the presence of 3′dA by about 80 percent (Fig. 3A) and virtually none of the 200-nucleotide poly(A) segment became labeled in mRNA.[39]

An additional experiment indicated that 3′dA affected mRNA biogenesis at a step after transcription. If cells were briefly labeled with [³H] uridine and then treated with actinomycin to stop further transcription, a measurable amount of previously synthesized RNA (presumably HnRNA) subsequently appeared in the polyribosomes. The cytoplasmic appearance of this "preformed mRNA" was blocked by the simultaneous addition of actinomycin and 3′dA (Fig. 3B).[39] Thus the posttranscriptional addition of poly(A) must be allowed to proceed or some step in the derivation of mRNA from HnRNA does not occur. These results suggest a nuclear role for poly(A) either in the proper cleavage of HnRNA or the transport of mRNA (or both) to the cytoplasm.

The fact that viruses which replicate in the cytoplasm also contain poly(A) does not necessarily conflict with the prediction of a nuclear role of poly(A). The mRNA molecules of such viruses may encounter a similar problem to that of cell mRNA in getting from the site of manufacture to the site of translation. For example, the replication of poliovirus RNA (which contains a 3′ terminal segment of poly(A) about 90 nucleotides long)[40] occurs on smooth membranes, whereas virus protein synthesis occurs on "rough" membranes, that is, those membranes bearing ribosomes.[41]

While the precise role or roles of the poly(A) remains unknown, it is possible that it has both a nuclear and a cytoplasmic function. For example, recent experiments have shown that after arrival in the cytoplasm, the poly(A) segment gradually becomes shorter.[26, 27] It has also been suggested that poly(A) may have specific proteins bound to it while in the polyribosomes.[42] Obviously further work with isolated components of protein synthesizing systems is needed to learn what, if any, cytoplasmic functions may be served by poly(A).

It might be asked whether the pathway of mRNA

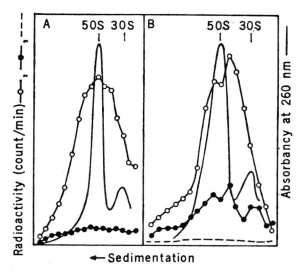

Fig. 3. Effect of 3'-deoxyadenosine on mRNA synthesis. (A) Cells in which rRNA formation was suppressed (fig. 1) were labeled for 25 minutes with [^{3}H] uridine either after no further treatment (control, -●-) or after a 10-minute exposure to 3'-deoxyadenosine (-○-). Messenger RNA from polyribosomes (see fig. 1) was assayed by sucrose-gradient sedimentation as radioactive polyribosomal RNA released by ethylenediaminetetraacetate (EDTA) into material sedimenting from 20*S* to 60*S*. The profile of absorbancy at 260 nm comes from ribosomal subunits released from polyribosomes. (B) Growing HeLa cells were labeled for 5 minutes with [^{3}H] uridine, and one-third of the culture was removed onto frozen medium. Actinomycin D (7.5 µg/ml) was added to the remaining two-thirds, which was then equally divided; one portion received in addition to the actinomycin, 3'-deoxyadenosine (100 µg/ml). After 25 minutes at 370°C, the drug-treated samples were rapidly chilled. The polyribosomes from each sample were isolated and their content of labeled mRNA was assayed by EDTA release as in (A). —, 5-minute labeling; -○-, 5-minute labeling followed by treatment with actinomycin for 25 minutes; -●-, 5-minute labeling plus treatment for 25 minutes with actinomycin plus 3'-deoxyadenosine (for details of experiments see refs. 3 and 39).

formation involving poly(A) synthesis is an abnormal aberration of cultured mammalian cells. This is certainly not the case. Edmonds and Abrams originally demonstrated the existence of adenylate-rich fragments in thymus nyclei,[17] Hadjivasilou and Brawerman demonstrated its existence in rat liver,[18]

and Lim and Canellakis[22] reported such fragments in reticulocytes. Messenger RNA's responsible for the synthesis of hemoglobin,[43] immunoglobulins,[44] and ovalbumin[45] have been isolated from cells not grown in culture and shown to contain poly(A). These findings as well as the fact that 75 to 90 percent of the mRNA from polyribosomes of cultured cells contain poly(A)[23, 24, 46, 47] indicate that a major pathway of mRNA biogenesis in mammalian cells involves the addition of poly(A). Recently, invertebrates[48] and slime molds[49] have been shown to possess poly(A) both in nuclear and mRNA fractions; thus, all eukaryotic cells may employ a mechanism of mRNA biosynthesis that involves the addition of poly(A) to nuclear molecules.

One interesting exception to the finding that most all mRNA from eukaryotic cells may contain poly (A) is the group of mRNA's which direct histone synthesis.[46, 47] These small mRNA molecules, which are formed only during the DNA synthesis phase of the cell cycle, exit more rapidly from the nucleus as compared to the majority of mRNA's and appear to lack poly(A).[46, 50] Whether histone mRNA is derived from a larger precursor molecule is not known. The lack of poly(A) in histone mRNA suggests that at least two mechanisms for the manufacture of mRNA exist, a nonpoly(A) and a poly(A) pathway, with the latter being much more common.

Uncertainties about the HnRNA-mRNA Relationship

Many important points about the proposed HnRNA-mRNA conversion are uncertain and deserve mention here. Some time ago, kinetic analysis indicated that the majority of the nucleotides incorporated into HnRNA never reached the cytoplasm, and thus a large part of each HnRNA or all of some HnRNA molecules must turn over in the nucleus.[6, 7, 9, 36] Every HnRNA molecule as isolated from cell nuclei does not contain a poly(A) sequence,[4, 47] but it remains unknown which of the following explanations accounts for this finding. (i) Some or most HnRNA's are not precursors to mRNA and, therefore, "turn over" without ever containing poly(A); (ii) some HnRNA molecules are nascent — that is, not yet complete to the 3'OH end — but might eventually contain poly(A); and (iii) some HnRNA molecules are cleavage products from the 5' portion of larger molecules and don't ever contain poly(A). If explanations (ii) or (iii), or both, are correct, every HnRNA might be transcribed for the purpose of yielding one mRNA. While it is possible to conclude that the 3'OH end of HnRNA molecules is the location at which poly(A) is added and from which mRNA is then derived, it is not certain that transcription actually ceases at the point where poly(A) is added. It is possible that an internal cleavage in the HnRNA exposes a poly(A) addition site. Some of these uncertainties will be discussed later in the section on possible models of regulation.

Regulation of Gene Expression in Mammalian Cells

The preceding sections have summarized some of the details of mRNA manufacture in eukaryotic cells. From the number of steps involved between initiation of transcription and subsequent participation of mRNA in protein synthesis, it seems clear that many sites for the control of protein synthesis may exist. Unfortunately even almost complete knowledge of the steps in mRNA synthesis and transport will not guarantee understanding of the control of protein synthesis. Before going further into the discussion of regulation of gene expression, it is necessary to recognize two major levels at which control is possible: (i) the provision of mRNA to the protein synthesizing apparatus, and (ii) the extent of use of the mRNA molecules for protein synthesis.

Events within the cell which allow the accumulation of mRNA molecules will be called regulation: control of the production of mRNA during RNA polymerase action will be called *transcriptional regulation*; control of the number of mRNA molecules made available after the completion of transcription but before translation will be called *posttranscriptional regulation*. Changes in the output of protein directed by a given amount of mRNA will be called *translational modulation*.

An early and lucid description of these possible levels of control was provided by Scherrer and Marcaud[51] who used the term "cascade regulation" to embrace all these possibilities. They argued that, because mammalian cells and their genomes were so much more complex than bacteria, control at all levels should be expected.

Because of the inability to measure a specific mRNA, most attempts to gain information about fluctuating levels of mRNA in mammalian cells have been indirect. Various workers have studied enzyme activities that fluctuate after a cell is stimulated with a specific molecule, or have studied differentiating cells for the appearance of specific differentiated protein products. Studies of this sort have uncovered substantial evidence of translational modulation, but have not yet clarified the details of transcriptional or posttranscriptional regulation of mRNA formation.

One of the best cells for the study of protein synthesis regulation is a derivative of a liver hepatoma in which an increased synthesis of tyrosine aminotransferase (TAT) occurs after the cells are treated with corticosteroids.[52] From work on this system, several conclusions can be made about control of enzyme synthesis which may be generally applicable to protein synthesis in mammalian cells.

1. Induction (increased rate of de novo enzyme synthesis) is accomplished by the accumulation of additional mRNA molecules that direct the increased synthesis of enzyme protein.[52] Induction of ovalbumin synthesis by estrogen treatment and hemoglobin synthesis by erythropoietin also show evidence of increased mRNA molecules as the basis for increased specific protein synthesis.[45, 53, 54]

2. The interruption of enzyme synthesis on cytoplasmic polyribosomes (deinduction) requires a positive action involving further synthetic events, or at least additional RNA synthesis and perhaps additional protein synthesis.[52] Simply stopping the synthesis of RNA does not result in a prompt cessation of protein synthesis as is the case in bacteria, undoubtedly because the half-life of mRNA in mammalian cells is longer than in bacteria.[10, 55] These findings recall earlier experiments showing that the normal interruption of virus-specific synthesis of thymidine kinase in cells infected with vaccinia-virus required the independent synthesis of additional RNA and probably protein molecules.[56]

3. Induction and deinduction of ongoing TAT synthesis are possible only during a portion (from midway in the G1 phase through the S phase) of the cell division cycle.[57]

The complexities of the control of TAT synthesis, involving both a mechanism of mRNA accumulation and a mechanism for interruption of mRNA function, suggest caution in the interpretation of results in differentiating cells. It has been reported recently that erythroid precursor cells, which can be stimulated to enter hemoglobin production, contain no detectable hemoglobin mRNA before stimulation, while hemoglobin mRNA could be detected after stimulation.[53] These results might indicate a control of transcription, but other means of control may also be involved, as suggested by analogy with TAT induction. For example, the unstimulated erythroid precursor cell might transcribe an HnRNA precursor to hemoglobin mRNA at only a brief period during the cell cycle. Such a precursor might escape detection since, without the stimulating agent, the precursor might be destroyed and not processed into hemoglobin mRNA. Thus mRNA would not accumulate, and regulation would have involved both a transcriptional and a posttranscriptional event.

Models for mRNA Formation and Regulation in Mammalian Cells

The two areas of experimentation discussed so far — that is, the pathway of mRNA biogenesis and changing rate of synthesis of particular proteins in the cytoplasm have not yet been connected by experimental results. Nevertheless, since it is now possible to describe with some confidence the physical form of the transcription product from which mRNA arises and to measure some specific mRNA molecules,[45, 53, 58] it is appropriate to reconsider the points at which regulation of formation of a specific mRNA might occur. Furthermore, in view of recent advances we can specify necessary, realizable experiments in order to choose between the possible modes of regulation. We should point out that the models we describe below are concerned with production of mRNA molecules and not with events involved in the modulation of protein synthesis.

In Fig. 4 we show two models in which mRNA production is regulated by transcription only and two models where mRNA production is regulated by both transcriptional and posttranscriptional events. In all of these models we assume that controlled initiation of transcription is necessary if

proper chain synthesis is to proceed. Although it is theoretically possible that chain initiation is not controlled, this possibility seems *a priori* so unlikely that regulated initiation of transcription is included in all models. In those eukaryotic cells where transcription of HnRNA at particular chromosomal loci can be studied (the specific puffing patterns on insect chromosomes), transcription occurs only at certain times in the life cycle of the organism.[59] Thus it seems reasonable to assume that regulation of the initiation of transcription plays a role in determining the portions of DNA to be transcribed.

Models in Fig. 4, A and B, depict two situations where transcriptional regulation of HnRNA synthesis is automatically followed by mRNA biogenesis involving poly(A) addition at the completed 3' end of the HnRNA. Two cases are outlined. First, where only one mRNA per HnRNA exists, termination of RNA synthesis would be automatic when the RNA polymerase "reached the end" (designated t for termination). Biogenesis of mRNA would also follow automatically through poly(A) addition and cleavage (at site c). Each mRNA region or structural gene region is thus bounded by a cleavage and termination site.

The second possibility is that several mRNA regions per HnRNA might exist either scattered through the HnRNA or as a polycistronic region at the 3' end (Fig. 4B). In such cases, both a proper termination signal would be needed at the 3' end of any mRNA region and a regulated ability (a "readthrough" signal) would be needed so that the RNA polymerase would only stop after the proper mRNA region had been synthesized. Again the 3' terminal mRNA region would automatically be processed into a usable mRNA molecule. It is also conceivable that, if more than one mRNA existed in an HnRNA molecule, the processing of the 3' terminal mRNA might expose a second mRNA which would then be processed.

In these models the regulation of the initiation of transcription would not necessarily be governed in a manner identical to that in bacteria. For example, it is known that genes which code for enzymes on the same biosynthetic pathway form one operon in bacteria, but are scattered throughout the genome in eukaryotes. This correlates with the finding of polycistronic bacterial mRNA's from entire operons, while individual mRNA's are probably the rule in eukaryotic cells.[5, 60] Also the induction of bacterial

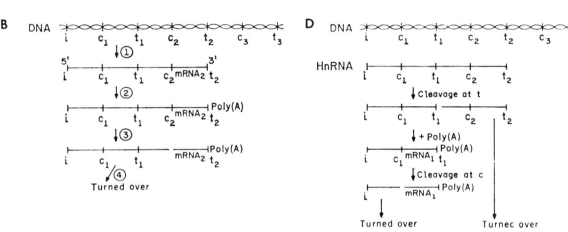

Fig. 4. Four models of regulation of mRNA formation in eukaryotic cells. The symbols are: i, initiation of transcription; c, cleavage points at $5'$ end of mRNA; t, for termination at $3'$ end of DNA-encoded region of mRNA; c_1, t_1, c_2, t_2, and so on indicate multiple sites on same molecule. The four steps in mRNA biosynthesis from HnRNA are: ① DNA dependent transcription by RNA polymerase; ② posttranscriptional addition of poly(A); ③ enzymatic cleavage at $5'$ end of mRNA; ④ turnover of unused region(s) of HnRNA. (A) Transcriptional regulation at initiation site only. (B) Transcriptional regulation at initiation site and termination site (t_1 is passed by in favor of t_2). (C) Posttranscriptional regulation where one-half of the HnRNA molecules yield an mRNA and one-half are destroyed. (D) Posttranscriptional regulation where specific cleavage at t_1 reveals the $3'$-terminus of mRNA, for processing and the $mRNA_2$ region is discarded.

enzymes is possible throughout the cell cycle although this may not be true in eukaryotes. These and other potential differences in eukaryotic cells have provoked considerable speculation about the details of transcriptional regulation in eukaryotic cells.[61]

Our present purpose, however, is not to concentrate on possible differences between prokaryotes and eukaryotes in the details of transcription, but

to call attention to the fact that the point of regulation of mRNA production could be the same, that is, transcription. This is true even though the mechanical steps in the manufacture of mRNA in a mammalian cell are different from those in a bacterium.

The second class of models describing how cells generate mRNA involve posttranscriptional regulatory steps as well as regulated transcription. In these models (Fig. 4, C and D) the correct processing of any particular HnRNA molecules containing a potential mRNA would require a successful encounter with a posttranscriptional regulator. HnRNA molecules might contain only one possible mRNA (structural gene), in which case regulation could be accomplished by either destroying or not destroying a completed HnRNA (Fig. 4C). If an HnRNA contained more than one "structural gene" region, the regulation might involve an initial cleavage at the proper internal site (for example, at the site labeled "t") followed by poly(A) addition and cleavage at a second site "c."

In considering how posttranscriptional regulation of mRNA production might occur, one could suggest as possible regulatory events poly(A) addition, HnRNA cleavage, or transport of mRNA to the cytoplasm. Past that point, the fate of the mRNA falls into the category of translational modulation.

If posttranscriptional regulation does, in fact, occur, an important point needs specific emphasis. Posttranscriptional regulation necessarily implies transcriptional overproduction of potential mRNA. If potential mRNA's in excess of what eventually serve in the cytoplasm are not transcribed, then posttranscriptional regulation cannot occur. A corollary to this point is that even if excess transcription occurs but there is always a fixed probability of a potential mRNA getting to and functioning in the cytoplasm, then again posttranscriptional events might take place but posttranscriptional regulation in the sense we are using the term would not occur. In connection with this emphasis on the necessity for excess transcription, one proposed category of translation modulation requires special comment. In oogenesis it has been proposed that mRNA storage occurs in the egg cytoplasm, and only after fertilization is the mRNA utilized.[62] It is true that such a series of events might seem to qualify as regulation of the availability of mRNA even though overproduction had not occurred but, in fact, if the egg becomes fertilized, the fate of the mRNA is to be used. Such a situation may simply be viewed as an exaggerated case of transcriptional modulation.

How would any of these models function in a situation where increased mRNA production occurred, for example, in TAT-mRNA induction? If the synthesis of TAT-mRNA were regulated entirely by transcription, the assumption would be that, during late G1 and the S phases, TAT-HnRNA would be normally formed at a low rate in uninduced cells, followed by the automatic processing of the HnRNA to yield the uninduced amount of TAT-mRNA and enzymes. Induction by steroid hormone would effect the proposed cytoplasmic stabilization of TAT-mRNA[52] and the promotion of an increased transcription rate. (This could involve either a small number of polymerases moving faster or more polymerase molecules synthesizing TAT-HnRNA.) Implicit in this scheme is the synthesis of Tat-HnRNA only during periods of the cell cycle when TAT-mRNA molecules can be accumulated to increased levels.

If, alternatively, the manufacture of additional TAT-mRNA molecules involved posttranscriptional regulation, then the major regulating event would be the preservation of more TAT-mRNA from the TAT-HnRNA during the time in the cell cycle when TAT-HnRNA production normally occurs. It is also possible that TAT-HnRNA might be synthesized at all times but might only be in a form where processing could occur during late G1 and S phase. Increased processing of TAT-HnRNA into TAT-mRNA during the G1-S phase would again be the regulatory device responsible for induction.

Proposed Experiments and Conclusion

The models described raise experimental questions in two areas where possible answers may be at hand. The first has to do with the structure of the HnRNA. Do HnRNA molecules contain only one mRNA region and is the potential mRNA sequence always at the 3' end? If so, then models involving correct RNA polymerase read through or correct internal cleavage (models in Fig. 4, B and D) become unlikely. More difficult but perhaps also answerable is whether, even though all mRNA sequences may be at the 3' end of HnRNA, an HnRNA can contain several potential different mRNA's at the 3' end. These questions are being or can be studied in two systems where specific mRNA and HnRNA can be recognized and isolated: (i) cells transformed by DNA viruses,[16] and (ii) cells producing large amounts of two or more proteins the genes for which are linked (for example, beta and delta chains of human hemoglobin).[63]

The second area of experimentation which could help one choose among the models of Fig. 4 involves systems like TAT induction or stimulated erythropoiesis. Several experimental questions could be clearly phrased if the mRNA and the HnRNA for the protein in question could be accurately measured. (i) Is the RNA for an inducible mRNA synthesized during a phase in the cell cycle when induction is not possible (for example, hepatoma cells during the G2 or early G1 phase of the cell cycle)? (ii) When the number of mRNA molecules of a given type is being increased is there a larger amount of corresponding HnRNA?

Answers to these questions should go a long way toward ascertaining whether or not posttranscriptional regulation is just an often suggested possibility or a reality in eukaryotic cells. What seems clear at the moment is that the biochemical mechanisms of mRNA formation in eukaryotes differs radically from bacteria. It would surprise a great many people if the types of regulation didn't differ also. The challenge, however, is not to settle for this latter possibility as likely but to prove or disprove it.

REFERENCES AND NOTES

We thank G. M. Tompkins for suggestions about this manuscript.

1. P. Lengyel and D. Soll, *Bacteriol. Rev.* 33, 264 (1969); J. Lucas-Lenard and F. Lipman, *Annu. Rev. Biochem.* 40, 409 (1971); M. Nomura, *Science* 179, 864 (1972); *Cold Spring Harbor Symp. Quant. Biol.* 35, 47–497 (1970).
2. U. Lindberg and J. E. Darnell, *Proc. Nat. Acad. Sci. U.S.A.* 65, 1089 (1970).
3. J. E. Darnell, L. Philipson, R. Wall, M. Adesnik, *Science* 174, 507 (1971).
4. W. Jelinek, M. Adesnik, M. Salditt, D. Sheiness, R. Wall, G. Molloy, L. Philipson, J. E. Darnell, *J. Mol. Biol.*, in press.
5. Papers in *Cold Spring Harbor Symp. Quant. Biol.* 35, 505–737 (1970); J. E. Darnell, *Bacteriol. Rev.* 32, 262 (1968); D. Bernhardt and J. E. Darnell, *J. Mol. Biol.* 42, 43 (1969); R. H. Burdon and A. E. Clason, *ibid.* 39, 113 (1969).
6. K. Scherrer, H. Latham, J. E. Darnell, *Proc. Nat. Acad. Sci. U.S.A.* 49, 240 (1963); K. Scherrer and L. Marcaud, *Bull. Soc. Chim. Biol.* 47, 1967 (1965).
7. G. P. Georgiev, O. P. Samarina, M. I. Lerman, M. N. Smirnov, A. N. Severitzov, *Nature* 200, 192 (1963).
8. J. F. Houssais and G. Attardi, *Proc. Nat. Acad. Sci. U.S.A.* 56, 616 (1966).
9. R. Sociro, H. C. Birnboim, J. E. Darnell, *J. Mol. Biol.* 19, 362 (1966); R. Sociro, M. H. Vaughan, J. R. Warner, J. E. Darnell, *J. Cell Biol.* 39, 112 (1968).
10. S. Penman, K. Scherrer, Y. Becker, J. E. Darnell, *Proc. Nat. Acad. Sci. U.S.A.* 49, 654 (1963).
11. B. E. H. Maden, *Progr. Biophys. Mol. Biol.* 22, 129 (1971); M. Salim, B. E. H. Maden, R. Williamson, *Fed. Eur. Biochem. Soc. Lett.* 12, 109 (1970).
12. S. Penman, *J. Mol. Biol.* 17, 177 (1966).
13. P. Jeanteur, F. Amaldi, G. Attardi, *J. Mol. Biol.* 33, 757 (1968).
14. H. Greenberg and S. Penman, *ibid.* 21, 527 (1966); E. F. Zimmerman and R. W. Holler, *ibid.* 23, 149 (1967).
15. J. Sambrook, H. Westphal, P. R. Srinivasan, R. Dulbecco, *Proc. Nat. Acad. Sci. U.S.A.* 60, 1288 (1968).
16. R. Wall, J. E. Darnell, *Nature New Biol.* 232, 73 (1971); R. Wall, J. Weber, Z. Gage, J. E. Darnell, *J. Virol.*, in press; S. Tonegawa, G. Walter, A. Bernardini, R. Dulbecco, *Cold Spring Harbor Symp. Quant. Biol.* 35, 823 (1970).
17. M. Edmonds and R. Abrams, *J. Biol. Chem.* 235, 1142 (1960); *ibid.* 238, 1186 (1963).
18. A. Hadjivassiliou and G. Brawerman, *J. Mol. Biol.* 20, 1 (1966).
19. M. Edmonds and M. G. Caramela, *J. Biol. Chem.* 244, 1314 (1969).
20. N. Canellakis, E. S. Canellakis, L. Lim, *Biochim. Biophys. Acta* 209, 128 (1970).
21. J. Kates and J. Beeson, *J. Mol. Biol.* 50, 19 (1970); J. Kates, *Cold Spring Harbor Symp. Quant. Biol.* 35, 743 (1970).
22. L. Lim and E. S. Canellakis, *Nature* 227, 710 (1970).
23. M. P. Edmonds, M. H. Vaughan, H. Nakazato, *Proc. Nat. Acad. Sci. U.S.A.* 68, 1136 (1971); H. Nakazato and M. Edmonds, *J. Biol. Chem.* 247, 3365 (1972).
24. Y. Lee, J. Mendecki, G. Brawerman, *Proc. Nat. Acad. Sci. U.S.A.* 68, 1331 (1971).
25. J. E. Darnell, R. Wall, R. J. Tushinski, *ibid.* 68, 1321 (1971).

26. D. Sheiness and J. E. Darnell, *Nature New Biol.* 241, 265 (1973).

27. J. Mendecki, Y. Lee, G. Brawerman, *Biochemistry* 11, 792 (1972).

28. G. R. Molloy, M. Sporn, D. E. Kelley, R. P. Perry, *ibid.*, p. 3256.

29. R. Sheldon, J. Kates, D. Kelley, R. P. Perry, *ibid.*, p. 3829.

30. H. Nakazato, D. Kopp, M. Edmonds, *J. Biol. Chem.*, in press.

31. G. R. Molloy and J. E. Darnell, *Biochemistry*, in press.

32. L. Philipson, R. Wall, G. Glickman, J. E. Darnell, *Proc. Nat. Acad. Sci. U.S.A.* 68, 2806 (1971).

33. K. Burton and G. B. Peterson, *Biochem. Biophys. Acta* 26, 667 (1957).

34. A. J. D. Beliett, H. B. Younghusband, S. B. Primrose, *Virology* 50, 35 (1972).

35. H. C. Birnboim. *Proc. Nat. Acad. Sci. U.S.A.*, in press.

36. H. Harris, *Biochem. J.* 84, 60 (1962).

37. S. Penman, M. Rosbach, M. Penman, *Proc. Nat. Acad. Sci. U.S.A.* 67, 1878 (1970).

38. H. T. Shiqeura and G. E. Boxer, *Biochem. Biophys. Res. Commun.* 17, 758 (1964); H. Klenow and S. Frederiksen, *Biochim. Biophys. Acta* 87, 495 (1964).

39. M. Adesnik, M. Salditt, W. Thomas, J. E. Darnell, *J. Mol. Biol.* 71, 21 (1972).

40. Y. Yogo and E. Wimmer, *Proc. Nat. Acad. Sci. U.S.A.* 69, 1877 (1972).

41. L. A. Caliguiri and I. Tamm, *Virology* 42, 112 (1970).

42. S. Kwan and G. Brawerman, *Proc. Nat. Acad. Sci. U.S.A.* 69, 3247 (1972); G. Blobel, *ibid.* 70, 924 (1973).

43. R. Pemberton and C. Baglioni, *J. Mol. Biol.* 65, 531 (1972); G. Brawerman, J. Mendecki, S. Y. Lee, *Biochemistry* 11, 637 (1972).

44. D. Swan, H. Aviv, P. Leder, *Proc. Nat. Acad. Sci. U.S.A.* 69, 1967 (1972).

45. A. R. Means, J. P. Comstock, G. C. Rosenfield, B. W. O'Malley, *ibid.*, p. 1146.

46. M. Adesnik and J. E. Darnell, *J. Mol. Biol.* 67, 397 (1972).

47. J. R. Greenberg and R. P. Perry, *ibid.* 72, Bank, P. A. Marks, *ibid.*, p. 3575.

48. S. Penman, in preparation.

49. R. A. Firtel, A. Jacobson, H. F. Lodish, *Nature New Biol.* 239, 225 (1972).

50. G. Schochetman and R. P. Perry, *J. Mol. Biol.* 63, 591 (1972).

51. R. Scherrer and L. Marcaud, *J. Cell Physiol.* 72 (suppl. 1), 181 (1968).

52. G. M. Tompkins, T. Gelehrter, D. Granner, D. W. Martin, H. Samuels, E. B. Thompson, *Science* 166, 1474 (1969).

53. J. Ross, Y. Ikawa, P. Leder, *Proc. Nat. Acad. Sci. U.S.A.* 69, 3260 (1972); M. Terada, L. Cantor, S. Metafora, R. A. Rifkind, A. Bank, P. A. Marks, *ibid.*, p. 3575.

54. G. M. Tompkins, *Harvey Lect.*, in press.

55. R. B. Singer and S. Penman, *J. Mol. Biol.*, in press; J. Greenberg, *Nature* 240, 102 (1972); W. Murphy and G. Attardi, *Proc. Nat. Acad. Sci. U.S.A.* 70, 115 (1973).

56. B. R. McAuslan, *Virology* 21, 383 (1963).

57. D. W. Martin, Jr., G. M. Tompkins, D. Graner, *Proc. Nat. Acad. Sci. U.S.A.* 62, 248 (1969).

58. R. Palacios, D. Sullivan, N. M. Summers, M. L. Kiely, R. T. Schimke, *J. Biol. Chem.* 248, 540 (1973).

59. C. Pelling, *Cold Spring Harbor Symp. Quant. Biol.* 35, 521 (1970); B. Daneholt, J.-E. Edstrom, E. Egyhazi, B. Lambert, V. Ringborg, *ibid.*, p. 513.

60. E. L. Kuff and N. E. Roberts, *J. Mol. Biol.* 26, 211 (1967).

61. R. J. Britten and E. H. Davidson, *Science* 165, 349 (1969).

62. P. R. Gross, *Annu. Rev. Biochem.* 37, 631 (1968).

63. C. Baglioni, *Proc. Nat. Acad. Sci. U.S.A.* 48, 1880 (1962); S. H. Boyer, D. L. Rucknagel, D. J. Weatherall, E. J. Watson-Williams, *Amer. J. Hum. Genet.* 15, 438 (1963).